The History of Modern Physics

1800 – 1950

Volume 6

TITLES IN SERIES

VOLUME 1
Alsos
by Samuel A. Goudsmit

VOLUME 2
Project Y: The Los Alamos Story
Part I: Toward Trinity
by David Hawkins
Part II: Beyond Trinity
by Edith C. Truslow and Ralph Carlisle Smith

VOLUME 3
American Physics in Transition: A History of Conceptual Change in the Late Nineteenth Century
by Albert E. Moyer

VOLUME 4
The Question of the Atom: From the Karlsruhe Congress to the First Solvay Conference, 1860–1911
by Mary Jo Nye

VOLUME 5
Physics For a New Century: Papers Presented at the 1904 St. Louis Congress

VOLUME 6
Basic Bethe: Seminal Articles on Nuclear Physics, 1936–1937
by Hans A. Bethe, Robert F. Bacher, and M. Stanley Livingston

INTRODUCTORY NOTE

The Tomash/American Institute of Physics series in the History of Modern Physics offers the opportunity to follow the evolution of physics from its classical period in the nineteenth century when it emerged as a distinct discipline, through the early decades of the twentieth century when its modern roots were established, into the middle years of this century when physicists continued to develop extraordinary theories and techniques. The one hundred and fifty years covered by the series, 1800 to 1950, were crucial to all mankind not only because profound evolutionary advances occurred but also because some of these led to such applications as the release of nuclear energy. Our primary intent has been to choose a collection of historically important literature which would make this most significant period readily accessible.

We believe that the history of physics is more than just the narrative of the development of theoretical concepts and experimental results: it is also about the physicists individually and as a group—how they pursued their separate tasks, their means of support and avenues of communication, and how they interacted with other elements of their contemporary society. To express these interwoven themes we have identified and selected four types of works: reprints of "classics" no longer readily available; original monographs and works of primary scholarship, some previously only privately circulated, which warrant wider distribution; anthologies of important articles here collected in one place; and dissertations, recently written, revised, and enhanced. Each book is prefaced by an introductory essay written by an acknowledged scholar, which, by placing the material in its historical context, makes the volume more valuable as a reference work.

The books in the series are all noteworthy additions to the literature of the history of physics. They have been selected for their merit, distinction, and uniqueness. We believe that they will be of interest not only to the advanced scholar in the history of physics, but to a much broader, less specialized group of readers who may wish to understand a science that has become a central force in society and an integral part of our twentieth-century culture. Taken in its entirety, the series will bring to the reader a comprehensive picture of this major discipline not readily achieved in any one work. Taken individually, the works selected will surely be enjoyed and valued in themselves.

The History of Modern Physics
1800 - 1950

Volume 6

Basic Bethe

Seminal Articles on Nuclear Physics, 1936–1937

HANS A. BETHE

ROBERT F. BACHER

M. STANLEY LIVINGSTON

Preface by
HANS A. BETHE

Introduction by
ROGER H. STUEWER

Tomash Publishers

American Institute of Physics

PUBLISHERS' NOTE

The papers appearing in this volume were originally published in the *Reviews of Modern Physics*, © The American Physical Society, and are reprinted with permission.

Printed in the United States of America.

Library of Congress Cataloging in Publication Data

Bethe, Hans Albrecht, 1906–
Basic Bethe.

(History of modern physics, 1800–1950;)
Includes bibliographies.
1. Nuclear physics. I. Bacher, Robert F. (Robert Fox), 1905– . II. Livingston, M. Stanley (Milton Stanley) III. Title. IV. Series.
QC776.B47 1986 539.7 86-10712
ISBN 0-88318-495-8

CONTENTS

Hans A. Bethe

PREFACE

When I arrived in the United States in early 1935, I found great enthusiasm and interest in nuclear physics. In many universities, including Cornell, new laboratories were being started for experimental research in nuclear physics, with very competent experimenters in charge. However, the physicists concerned knew relatively little about nuclear theory and about the advances in nuclear physics since the discovery of the neutron.

After talking individually to many nuclear physicists in this country, I decided that it would be more effective if I would write a comprehensive article on the status of nuclear physics. The *Reviews of Modern Physics* kindly offered to publish this article, which soon developed into three installments.

It was of great help to me that Stanley Livingston, the experimental physicist in charge of the baby Cornell cyclotron, had a nearly complete card file of all the articles in nuclear physics that had appeared since 1932. I also asked him to join me in writing a large part of the third installment of my article, "Nuclear Dynamics, Experimental." Soon after I started on this task, Robert F. Bacher joined the faculty at Cornell, having previously worked most intensively on atomic spectra, including their hyperfine structure. I asked him to write the chapter on hyperfine structure.

Of enormous help also were my two research associates who came to Cornell during the writing of the articles, Emil Konopinski and Morris E. Rose. We did much of the work together, and several original papers in *Physical Review*, by them and by myself, many of them jointly, were byproducts of the articles in *Reviews of Modern Physics*. Konopinski remained a great expert in nuclear beta disintegrations and wrote a comprehensive book on this subject. M. E. Rose, before his untimely death, made significant contributions to the theory of internal conversion and its application to nuclear spectroscopy, his work also being incorporated in several important books.

Another piece of good fortune was that, during the writing of the articles, Niels Bohr conceived the theory of the compound nucleus and Breit and Wigner published their theory of the shape of nuclear resonances. The exploitation of these ideas formed an important part of the second article, "Nuclear Dynamics, Theoretical." Unfortunately, in one chapter of that article, on alpha disintegration, I went overboard and greatly exaggerated the influence of the compound nucleus on this phenomenon. That chapter should be disregarded, especially my claim of extremely large nuclear radii.

Some sections of the three-part article have been superseded by more recent books that take into account the wealth of information, experimental and theoretical, gathered since 1937. The book that covers most of the subjects in my articles is that by Blatt and Weisskopf, which appeared in 1952. The most comprehensive and now most useful texts are the two books by Bohr and Mottelson, *Nuclear Structure* (Volume I published in 1969 and Volume II in 1975). They contain many ideas and subjects that were unknown in 1937, and contain a lot of the answers to questions in our minds in 1936 and 1937. There are still a few sections in the now reissued *Review* articles that have not been replaced and may still be useful. While the entire articles may not be of intense usefulness to nuclear physicists, they are of historical value as a real view of the state of the art then.

By now, nuclear physics has had so many ramifications that it would be almost impossible to write a comprehensive book about it, as was possible in 1935–37. The beginning of a systematic explanation of the atomic nucleus in those years was a wonderful time, full of enthusiasm and expectation. It was a heady and most satisfactory task to write a summary of the then existing knowledge.

Hans A. Bethe
Cornell University

December 1985

INTRODUCTION

Hans Bethe's trilogy of articles, published fifty years ago in 1936 and 1937, occupy a unique position in the history of nuclear physics. They were the articles to which nuclear physicists turned to learn the subject. M. Stanley Livingston remarked to an interviewer that at Cornell Bethe "gave me a feeling for the fundamentals of physics, and what was going on in nuclear physics. With him for the first time I sensed how deep the field was, how involved it was, and how much we needed in the way of new information."[1] When Bethe put pen to paper, the orientation he provided personally to Livingston was provided to everyone in the field. Nuclear physicists immediately termed the result the Bethe Bible.

In the early 1930s scientific and political currents crossed with unparalleled force. The discovery of deuterium, the neutron, and the positron, and the invention of the Cockcroft–Walton accelerator and the cyclotron burst the floodgates in nuclear physics by the end of 1932. A few months later, the Nazi Civil Service Law exiled numerous scholars, including many physicists, from Germany. By the mid-1930s, in England and America, a profound symbiosis was occurring: native-born and émigré nuclear physicists were coming together at universities and conferences, discussing current problems and writing joint papers, and exposing graduate students to their combined talents.

It was in this atmosphere that Bethe undertook to write his *Reviews of Modern Physics* articles. However, as R. F. Bacher and V. F. Weisskopf noted in Bethe's Festschrift,[2] these articles were not "reviews in the ordinary sense of the word"; they were "systematic recreations of the knowledge of the time, supplemented by original work wherever there were gaps and omissions in the current knowledge." They encompassed virtually all of nuclear physics, both theoretical and experimental. They established a common knowledge base, and in so doing contributed fundamentally to the intellectual symbiosis then taking place in nuclear physics.

In the half century following its publication, the Bethe Bible has been read by one generation after the other of nuclear physicists and their graduate students, not to mention historians of nuclear physics and their graduate students. In every physics library the original issues have been underlined, annotated, inadvertently torn, and repaired by transparent tape. It is wonderful to have a replacement so that this process can begin anew.

Roger H. Stuewer
University of Minnesota

April 1986

1. Interview with M. Stanley Livingston by C. Weiner and Niel Goldman, Aug. 21, 1967, American Institute of Physics, Center for History of Physics.
2. *Perspectives in Modern Physics: Essays in Honor of Hans A. Bethe,* edited by R. F. Bacher and V. F. Weisskopf (Interscience, New York, 1966).

NUCLEAR PHYSICS

Part A

Stationary States of Nuclei

Nuclear Physics

A. Stationary States of Nuclei[1]

H. A. BETHE AND R. F. BACHER, *Cornell University*

TABLE OF CONTENTS

[1] The present issue contains Part A of this report. Part B, treating the dynamics of nuclei, particularly transmutations, will appear in a later issue of these *Reviews*.

I. Fundamental Properties of Nuclei

§1. Charge, Weight (A4)[2]

OF all properties of the nucleus the charge is by far the most important for *atomic* physics. It determines the number of electrons of an atom in its neutral state, the energy levels of the atom, its chemical actions; in short, all the properties of the atom except for very small corrections such as hyperfine structure, isotope shift of spectral lines, etc.

It is well known that the charge of any nucleus is an integral multiple of the charge e of the proton, let us say Ze. Every integer Z corresponds to a certain chemical element. All elements corresponding to values of Z from 1 to 92 have been actually found in nature, except the elements 85 and 87.[3] Recently, the elements 93 and probably 94 have been produced by disintegration experiments (F6, F9).

With our present knowledge about atomic physics, the nuclear charge of an element can be easily inferred from its chemical or spectroscopic properties (periodic system). The most direct measurement of the nuclear charge of an atom consists, however, in the determination of its x-ray spectrum, in other words, of the binding energy of its inner electrons. Another method is based on the large-angle scattering of α-particles or protons by the nuclei of the element. It is less accurate but of great historical importance because it was this experiment which led Rutherford to the concept of the nuclear atom (R15, C4).

The second nuclear quantity, the atomic weight, was known long before the existence of nuclei was discovered. Moreover, it was suggested as early as 1813 by Prout that all atomic weights are integral multiples of the weight of the hydrogen atom. Later, it was seen that this rule held more accurately when one-sixteenth of the atomic weight of oxygen was taken as the unit. However, some bad exceptions from the integral—weight—rule remained, e.g., chlorine.

We know now that all the elements not conforming to this rule, and, indeed, many of those apparently conforming, consist of several isotopes. The nuclei of two isotopic atoms possess the same charge, but different weight. Since the charge alone determines the chemical and spectroscopic properties of the atom, two isotopes have the same chemical behavior and (practically) the same spectrum. The atomic weight of every isotope is very nearly an integer, while the mixed element which is found in nature will, of course, in general[4] have a non-integral atomic weight.

The analysis of the isotopic constitution of an element as well as the determination of the atomic weights of the single isotopes requires the use of a mass spectrograph (A4). Mainly through the work of Aston, we know at present about 280 different isotopes which occur in nature, corresponding to about 3 isotopes per element. The highest number of isotopes for any single element is found for Sn (10 isotopes).

The atomic weights of the known isotopes run from 1 (light hydrogen) up to 238 (uranium). From 1 to 212, there exists at least one isotope for every integral atomic weight,[5] with the only exception of the atomic weights 5 and 8. In many cases, the same atomic weight is found for two isotopes of two different elements; e.g., one of the isotopes of argon as well as one of the calcium isotopes has the atomic weight 40. Such nuclei which have equal atomic weight but different nuclear charge are called isobars. There are 44 pairs of such isobars known, excluding about 10 pairs for which one of the two isobars is doubtful. In at least two cases, the existence of three isobars seems to be definitely established (atomic weight 96 (Zr, Mo, Ru), and atomic weight 124: tin, tellurium and xenon). The actual number of existing pairs of isobars is certainly larger than the number found thus far, because the isobars in the region of the rare earths are practically unknown. This is because it is very difficult to separate the various rare

[2] A letter and a number, e.g., A4, are used for references to original papers. A list of references is found at the end of the paper.

[3] The discovery of illinium (61) seems to be still disputed.

[4] In some cases all the isotopes except one are very rare, e.g., the isotopes 17 and 18 of oxygen compared to O^{16}. Then the atomic weight of the natural mixed element is approximately equal to that of the abundant isotope and therefore is nearly an integer.

[5] Cf. J. Mattauch (M11).

earths and thus to tell whether a given mass found in the mass spectrograph is an isotope of the element actually investigated or of another rare earth which occurs as an impurity.

The notation accepted for denoting a given isotope, consists in putting the atomic weight of the isotope as a superscript on the chemical symbol of the element the isotope belongs to, e.g., H^2, A^{40}, Te^{124}. Some writers put in addition the nuclear charge in the lower left-hand corner, thus: ${}_{18}A^{40}$, or ${}^{40}_{18}A$. We shall not do so because the charge is uniquely determined by the chemical symbol. It need hardly be noted that the upper index is not the *exact* atomic weight of the isotope which deviates slightly from an integer but the "mass number," i.e., the integer nearest to the atomic weight.

The atomic weights of the natural elements are remarkably independent of the source where the element is found. This means that the various

TABLE I. *Known stable isotopes.*

A = Atomic weight, Z = Nuclear charge, Ch = Chemical symbol.

A	*Z*	*Ch*
1	1	H
2	1	H
3	1	H
4	2	He
5	—	—
6	3	Li
7	3	Li
8	—	—
9	4	Be
10	5	B
11	5	B
12	6	C
13	6	C
14	7	N
15	7	N
16	8	O
17	8	O
18	8	O
19	9	F
20	10	Ne
21	10	Ne
22	10	Ne
23	11	Na
24	12	Mg
25	12	Mg
26	12	Mg
27	13	Al
28	14	Si
29	14	Si
30	14	Si
31	15	P
32	16	S
33	16	S
34	16	S
35	17	Cl
36	18	A
37	17	Cl
38	18	A
39	19	K
40	18	A
	20	Ca
41	19	K
42	20	Ca
43	20	Ca
44	20	Ca
45	21	Sc
46	22	Ti
47	22	Ti
48	22	Ti

A	*Z*	*Ch*
49	22	Ti
50	22	Ti
	24	Mo
51	23	V
52	24	Cr
53	24	Cr
54	24	Cr
	26	Fe
55	25	Mn
56	26	Fe
57	26	Fe
58	28	Ni
59	27	Co
60	28	Ni
61	28	Ni
62	28	Ni
63	29	Cu
64	30	Zn
65	29	Cu
66	30	Zn
67	30	Zn
68	30	Zn
69	31	Ga
70	30	Zn
	32	Ge
71	31	Ga
72	32	Ge
73	32	Ge
74	32	Ge
	34	Se
75	33	As
76	32	Ge
	34	Se
77	34	Se
78	34	Se
	36	Kr
79	35	Br
80	34	Se
	36	Kr
81	35	Br
82	34	Se
	36	Kr
83	36	Kr
84	36	Kr
	38	Sr
85	37	Rb
86	36	Kr
	38	Sr
87	37	Rb*

A	*Z*	*Ch*
	38	Sr
88	38	Sr
89	39	Y
90	40	Zr
91	40	Zr
92	40	Zr
	42	Mo
93	41	Cb
94	40	Zr
	42	Mo
95	42	Mo
96	40	Zr
	42	Mo
	44	Ru
97	42	Mo
98	42	Mo
	44	Ru
99	44	Ru
	(43	Ms?)
100	42	Mo
	44	Ru
101	44	Ru
102	44	Ru
	46	Pd
103	45	Rh
104	44	Ru
	46	Pd
105	46	Pd
106	46	Pd
	48	Cd
107	47	Ag
108	46	Pd
	48	Cd
109	47	Ag
110	48	Cd
111	48	Cd
112	48	Cd
	50	Sn
113	48	Cd
	49	In
114	48	Cd
	50	Sn
115	49	In
	50	Sn
116	48	Cd
	50	Sn
117	50	Sn
118	48	Cd
	50	Sn

A	*Z*	*Ch*
119	50	Sn
120	50	Sn
121	51	Sb
122	50	Sn
	52	Te
123	51	Sb
	52	Te
124	50	Sn
	52	Te
	54	Xe
125	52	Te
126	52	Te
	54	Xe
127	53	I
128	52	Te
	54	Xe
129	54	Xe
130	52	Te
	54	Xe
131	54	Xe
132	54	Xe
133	55	Cs
134	54	Xe
	56	Ba
135	56	Ba
136	54	Xe
	56	Ba
137	56	Ba
138	56	Ba
139	57	La
140	58	Ce
141	59	Pr
142	58	Ce
	60	Nd
143	60	Nd
144	60	Nd
	62	Sm
145	60	Nd
146	60	Nd
	(62	Sm)
147	62	Sm
148	62	Sm
149	62	Sm
150	62	Sm
151	63	Eu
152	62	Sm
153	63	Eu
154	62	Sm
155	64	Gd

A	*Z*	*Ch*
156	64	Gd
157	64	Gd
158	64	Gd
159	65	Tb
160	64	Gd
161	66	Dy
162	66	Dy
163	66	Dy
164	66	Dy
165	67	Ho
166	68	Er
167	68	Er
168	68	Er
169	69	Tu
170	68	Er
171	70	Yb
172	70	Yb
173	70	Yb
174	70	Yb
175	71	Cp
176	70	Yb
	72	Hf
177	72	Hf
178	72	Hf
179	72	Hf
180	72	Hf
181	73	Ta
182	74	W
183	74	W
184	74	W
185	75	Re
186	74	W
	76	Os
187	75	Re
188	76	Os
189	76	Os
190	76	Os
191	77	Ir
192	76	Os
	78	Pt
193	77	Ir
194	78	Pt
195	78	Pt
196	78	Pt
	80	Hg
197	79	Au
198	78	Pt
	80	Hg
199	80	Hg

A	*Z*	*Ch*
200	80	Hg
201	80	Hg
202	80	Hg
203	81	Tl
204	80	Hg
	82	Pb
205	81	Tl
206	82	Pb
		(U Pb)
207	82	Pb
		(AcPb)
208	82	Pb
		(ThPb)
209	83	Bi
210	84	Po
		(Ra F)
211	—	
212	84	Th C′
213	83	Ac C
214	84	Ra C′
215	—	
216	84	Th A
217	84	Ac A
218	84	Ra A
219	—	
220	86	ThEm
221	86	AcEm
222	86	RaEm
223	—	
224	88	Th X
225	88	Ac X
226	88	Ra
227	—	
228	90	RdTh
229	90	RdAc
230	90	Io
231	—	
232	90	Th
233	91	Pa
234	92	U II
235	92	U?
236	—	
237	—	
238	92	U I

Remarks: β-radioactive isotopes have not been included in the table, except Rb^{87}, which is possibly β-active, and Ac C which has a very weak β-activity besides a strong α-activity. Radioactive α-emitters have generally been included.

isotopes occur always in about the same ratio in the natural element. Exceptions from this rule are hydrogen, lead and, to a very small extent, boron. The content of heavy hydrogen H^2 (deuterium) in H from different sources, varies from about 1 part in 3500 to 1 in 5000. This great variation is due to the very large relative difference in atomic weights. For boron, this relative difference is, of course, much smaller than for hydrogen but larger than for most other elements. The ratio $B^{10} : B^{11}$ has been found (B29) to vary from about 1 : 4 to 1 : $3\frac{3}{4}$. On the other hand, lead in uranium ores is produced by the radioactive decay of uranium and is thus, in an ideal case, the pure isotope Pb^{206}.

Since every isotope contains its own characteristic nucleus, the separation of isotopes is very important for nuclear physics. The most complete separation is achieved in the mass spectrograph; the main disadvantage being that extremely high currents or long times are required to separate an appreciable amount of an element into its isotopes. Light and heavy hydrogen can be separated comparatively easily by repeated electrolysis of water. Repeated diffusion has been successful for neon. Repeated chemical actions, distillations, etc., are all possible methods but it is hard to obtain more than a partial separation of the isotopes with their help, except in the case of hydrogen.

It remains to interpret the rules of (nearly) integral atomic weights and (exactly) integral nuclear charge where the latter is always smaller than the former. The interpretation which suggests itself is that any nucleus consists of particles of unit atomic weight some of which are positively charged while others are neutral. In fact, we know two nuclei of atomic weight unity, *viz.*, the proton and the neutron (C5), the first bearing unit positive charge while the second is neutral. We are thus led to the hypothesis that every nucleus consists of protons and neutrons;[6] (H7, also H2, I1). The total number of elementary particles, protons and neutrons together, is then equal to the atomic weight of the nucleus or more exactly to its mass number A, i.e., the integer nearest to the atomic weight. The number of protons must be given by the nuclear charge Z, wherefrom the number of neutrons follows as being

$$N = A - Z. \tag{1}$$

Since for many of the lighter nuclei the number of neutrons is approximately equal to the number of protons, it is sometimes useful to introduce the "isotopic number" I (cf. H2), i.e., the excess of the number of neutrons over the number of protons, *viz.*,

$$I = N - Z = A - 2Z. \tag{2}$$

§2. Energy (A4, A5, B5, B13)

The mass spectrograph has shown that the atomic weight of every nucleus is approximately an integer, thus giving support to the hypothesis that any nucleus is constituted of neutrons and protons. At the same time, however, the mass spectrograph revealed that the atomic weights of separated isotopes are not exactly integers, e.g., $H^1 = 1.00807$, $Li^7 = 7.0164$, $Kr^{80} = 79.926$ and $Tl^{205} = 205.037$. It is seen that the lightest atoms have atomic weights M slightly higher, those of medium atomic weight (from about 20 up to 200) such somewhat lower and the very heaviest atoms again weights slightly higher than the next integer A.

The difference between exact atomic weight M and "mass number" A, the so-called "mass defect"

$$-\Delta = A - M \tag{3}$$

is far outside the experimental error, being for the lightest atoms about 100 times, for the heavier ones about 10 times the probable error. On the other hand, the mass defect is much too small and depends much too regularly on the mass number A to allow the abandoning of the rule of integral atomic weights and thus of our hypothesis about the constitution of the nuclei. It must therefore be concluded that protons and neutrons bound together in a nucleus have a weight different, and more precisely smaller, than the same number of free protons and neutrons. This can be interpreted by Einstein's law of equivalence of mass and energy, as show-

[6] This hypothesis has been used for the first time as the basis of a thorough nuclear theory by Heisenberg (H7); however, it had been suggested earlier as a convenient manner of describing the existing isotopes by Harkins (H2) and Rutherford.

ing that the binding of neutrons and protons in a nucleus decreases the total energy. Thus the mass defect gives us direct information about the binding energy of the particles in a nucleus.

This information is extremely useful.[7] It serves to determine the total binding energy of the elementary particles of a nucleus by comparing its weight to that of an equal number of neutrons and protons. It also serves to determine the binding energy of the last neutron, proton or α-particle in a nucleus by comparing the weight of the nucleus to that of another nucleus containing one neutron, proton or α-particle, respectively, less than the given nucleus. In this way, it can be decided whether a given nucleus is stable or not. Furthermore, it can be deduced whether a given nuclear reaction will be endothermic or exothermic, thus providing a great help to the experimental investigator of nuclear reactions, etc.

As an example, let us take the nucleus Li^6. Its atomic weight (B13) is 6.01614. The nucleus consists of three neutrons, of atomic weight 1.00845 each, and three protons (atomic weight of the hydrogen atom 1.00807). These six particles together in the free state would have a weight of 6.04956 which is 0.03342 units more than the weight of the Li^6 atom. The binding energy of Li^6 is thus 0.03342 "mass units," one mass unit being the energy corresponding, according to Einstein's law of equivalence, to one-sixteenth of the mass of the oxygen atom. To convert this energy into more familiar units, we note that the energy corresponding to the mass m, is mc^2 according to Einstein's law, c being the velocity of light.

Thus
$$1 \text{ mass unit} = 8.99\cdot 10^{20}\cdot(1/16)\cdot M_O \text{ ergs} = 1.49\cdot 10^{-3} \text{ erg},$$

where $M_O = 2.64\cdot 10^{-22}$ gram is the mass of the oxygen atom O^{16}.

More useful than ergs are electron volts as units of nuclear energies, because the kinetic energies of projectiles used for nuclear disintegrations are measured directly in these units. Let $-10e$ denote the charge of the electron in *international* coulombs, so that $-e/q$ is the electronic charge in absolute electromagnetic units, and q the conversion factor of international into absolute coulombs.[8] Let furthermore pq be the conversion factor of international into absolute volts; then the energy in international electron volts corresponding to one mass unit is

$$M = \frac{c^2 M_O}{16eq}\frac{10^{-8}}{pq} = \frac{10^{-9}c^2}{Fpq^2},$$

where $F = 10e/(M_O/16)$ is the "Faraday." With Birge's values for the constants, we find

$$M = \frac{10^{-9}\cdot(2.99796\pm 0.00004)^2\cdot 10^{20}}{(96494\pm 1)(1.00051\pm 0.00002)} \times (0.99995\pm 0.00005)^2 \text{ev},$$

$$= 931.05 \pm 0.15 \text{ MV} \quad (4)$$

where MV denotes million electron volts.[9] (The greatest uncertainty arises from the conversion factor q.)

With this conversion factor, we find that the binding energy of the Li^6 nucleus, compared to the free elementary particles, is

$$0.03342\cdot 931.0 = 31.11 \text{ MV}.$$

To assure the stability of Li^6, it must also be lighter than, e.g., $H^3 + He^3$ or $H^2 + He^3$, or generally, lighter than the sum of any two nuclei which between them contain as many neutrons and protons as Li^6. We find for the atomic weights:

[7] It is, in fact, the only point in which our information about nuclear properties is superior to the information about atomic properties: The total binding energy of all the electrons in an atom, which knowledge is quite worth while, can only be inferred by measuring the successive ionization potentials of the atom, a procedure extremely hard to carry out with heavy atoms; there is no way of determining this total binding energy directly. The reason why the nuclear binding energies show up in the atomic weights, while the binding energies of the atomic electrons practically do not, is of course the very large magnitude of nuclear binding energies, *viz.*, several million electron volts per nuclear particle compared to about 300,000 electron volts binding energy of all the electrons in the uranium atom together.

[8] Cf. Birge, Rev. Mod. Phys. **1**, 1 (1929).

[9] It seems to us that a newly introduced abbreviation should be as short as possible without giving rise to confusion with other abbreviations. It seems unnecessary to show explicitly in the abbreviation that *electron* volts are meant especially since this follows clearly from the text in any given case. On the other hand, it has always been customary to denote volts by a capital V, also mega- (or million) by a capital M, in contrast to "milli"- (which is usually denoted by m).

$He^4 = 4.00336$	$He^4 = 3.01699$
$H^2 = 2.01423$	$H^3 = 3.01610$
6.01759	6.03309
$Li^6 = 6.01614$	$Li^6 = 6.01614$
bind. en. = 0.00145 mass unit	0.01685 m.u.
= 1.35 MV	= 15.69 MV

Thus we see that the Li^6 nucleus is very stable against spontaneous disintegration into a He^3 and a H^3 nucleus, but much less stable against disintegration into an α-particle and a deuteron.

It has probably been noted that we have given the *atomic* weights of all the particles concerned rather than the nuclear weights. The atomic weight of an atom of nuclear charge Z contains, besides the weight of the nucleus, that of Z orbital electrons (atomic weight of one electron 0.000548). In ordinary nuclear reactions, the nuclear charge must always balance up, so that the atoms produced in the reaction contain just as many orbital electrons as the atoms originally present. Thus no error is introduced in the energy balance of a nuclear reaction if atomic masses are used instead of nuclear masses, since the number of electrons contained in the atomic masses does not change in the reaction. The same is true in the stability considerations carried out above. A given nucleus must always be compared to a number of other nuclei whose total charge is equal to its own charge.

Even for the energy evolved in radioactive β-decay, the use of atomic rather than nuclear masses is legitimate: If a nucleus of charge Z transforms into one of charge $Z+1$ plus a negative electron, the energy set free is obviously equal[10] to the mass of the nucleus Z, minus the sum of the masses of nucleus $Z+1$ and one electron. Now the atomic weight of atom Z contains the weight of Z electrons, while the weight of atom $Z+1$ contains the weight of one more electron; thus the difference of the two atomic weights is equal to the difference of the nuclear weights minus the mass of one electron, which is exactly the energy[10] set free in the β-decay. The mass of the neutrino (cf. §39) has been assumed to be zero.

Only in the case of positron radioactivity the difference of the weights of original and final atom does not give immediately the energy set free. Let us assume that an atom of nuclear charge Z and *atomic* weight M_Z emits a positron and thus transforms into an atom $Z-1$ with atomic weight M_{Z-1}. Calling the electron mass m, the masses of the two *nuclei* are $M_Z - Zm$ and $M_{Z-1} - (Z-1)m$. The energy set free[10] is equal to the difference of these nuclear masses, minus the mass of the positron emitted, *viz.*,

$$E = (M_Z - Zm) - (M_{Z-1} - (Z-1)m) - m$$
$$= M_Z - M_{Z-1} - 2m. \qquad (5)$$

The energy evolved is thus equivalent to the difference of the atomic weights of the two atoms, minus *twice* the mass of an electron, or 0.00110 mass unit. The factor 2 comes in, because firstly a positron is created, and secondly one more electron is contained in the atomic weight M_Z than in M_{Z-1}.

It is sometimes useful to define the mass defect per elementary particle contained in the nucleus, the so-called packing fraction

$$P = \Delta/A = (M-A)/A. \qquad (6)$$

Fig. 1 gives the packing fraction as a function of the mass number. The packing fractions of proton and neutron are $+0.00807$ and $+0.00845$, respectively. The packing fraction then decreases with increasing mass number, indicating stronger binding of the nuclear particles. P reaches, by definition, zero for O^{16}; then it becomes negative and almost constant, equal to -0.001 over a large region. This indicates that the mass defect and therefore the binding energy of all the nuclei from O up to Hg is very nearly proportional to the number of particles in the nucleus, which is a very important theorem (cf. §7). The total binding energy of any given nucleus in this region is

$$Z\Delta_H + (A-Z)\Delta_n - \Delta_A \approx A\left(\frac{P_H + P_n}{2} - P_A\right)$$
$$= 0.009A \text{ mass units} = 8\tfrac{1}{2}A \text{ MV}, \qquad (7)$$

where Δ_H, Δ_n, Δ_A are the "mass excesses" of proton, neutron and atom A, as defined in (6); $P_H = \Delta_H$, $P_n = \Delta_n$ and P_A are the respective packing fractions; and it has been taken into account

[10] The factor c^2 is omitted.

that the number of protons in atom A is approximately equal to the number of neutrons, thus: $Z \approx A-Z \approx \frac{1}{2}A$. The value of the binding energy (7) is small compared to the energy corresponding to the nuclear mass (cf. (4)). Therefrom we may conclude that the velocities of nuclear particles are small compared to the velocity of light, so that nonrelativistic quantum mechanics can be applied to the motion of neutrons and protons in the nucleus.

§3. Size (G2)

The radii of nuclei range from about 2 or $3\cdot 10^{-13}$ cm for the α-particle, up to about $9\cdot 10^{-13}$ cm for the uranium nucleus. It seems that the volume of a nucleus is approximately proportional to its mass number, so that the volume per elementary particle is about the same in every nucleus. Only the very lightest nuclei seem to be exceptions from this rule.

The size of nuclei has been determined by the interaction of nuclei with small nuclear entities, such as proton, neutron, deuteron, α-particle. If a nucleus and a positively charged particle are a considerable distance apart, there is an electrostatic repulsion between them, calculable from the well-known Coulomb law. However, when the two particles get closer together, it is found that deviations from Coulomb's law set in, and finally an attractive force is found to exist between nucleus and external particle. The existence of this attractive force can be inferred from the fact that nuclear particles can become attached to an existing nucleus, forming a new stable nucleus. The point where the repulsion changes into an attraction gives a possible definition of the nuclear radius. It seems that the specifically nuclear forces between nucleus and external particle fall off very rapidly when the distance between the two becomes larger than the nuclear radius, or, more exactly, the sum of the radii of nucleus and particle, so that the Coulomb law holds if the distance is only just larger than this sum of the radii. Thus the boundary of a nucleus is quite well defined; better, at any rate, than the boundary of an atom.

The interaction between a neutron and nucleus is, of course, zero at large distances, the neutron being not subject to electric forces. In this case, there is only the "specifically nuclear" interaction which sets in when the distance of the two particles becomes of the order of the nuclear radius or smaller.

The most exact determination of nuclear radii is afforded by the lifetime of radioactive nuclei, emitting α-rays (cf. Chapter IX). The information about the sizes of light nuclei has been obtained from the scattering of α-particles and protons in the nuclear fields and from the probability of nuclear disintegrations as a function of the energy of the bombarding particles (Chapter X). Similar results can be obtained from the scattering of fast neutrons by nuclei (Chapter XII).

The size of the nuclei is of the order of magnitude which one should expect from the binding energy of the nuclear particles, assuming that the nucleus consists of protons and neutrons. The wave-length of a proton or neutron (mass M) is

$$\lambda = 2\pi\hbar/Mv = 2\pi\hbar(2ME_{\text{kin}})^{-\frac{1}{2}}, \tag{8}$$

where $2\pi\hbar$ is Planck's constant, v the velocity and E_{kin} the kinetic energy of the particle. It is safe to assume that E_{kin} is of the same order of magnitude as the binding energy ϵ of the particle which is, according to the end of §2, equal to about $8\frac{1}{2}$ MV for atoms of medium atomic weight. We obtain therefore[11]

$$\lambda\!\!\!^{-} = \lambda/2\pi = \hbar(2M\epsilon)^{-\frac{1}{2}}$$
$$= 1.04\cdot 10^{-27}(2\cdot 1.66\cdot 10^{-24}\cdot 8.5\cdot 10^{6}\cdot 1.59\cdot 10^{-12})^{-\frac{1}{2}}$$
$$= 1.55\cdot 10^{-13} \text{ cm}.$$

This is of the same order but somewhat smaller than nuclear radii. $\lambda\!\!\!^{-}$ should, indeed, be somewhat smaller than the nuclear radius because heavier nuclei contain many protons and neutrons, most of which must be in excited quantum states. Their wave functions must then have several nodes inside the nucleus. Thus the observed size of nuclei again lends support to the hypothesis that protons and neutrons are the elementary particles constituting a nucleus.

If we want to apply the considerations corresponding to Eq. (8) to electrons, we have to use the relativistic relation between wave-length

[11] For obtaining estimates of the order of magnitude, $\lambda\!\!\!^{-}$ is a much more suitable quantity than the wave-length λ.

and kinetic energy; *viz.*,

$$\lambda\!\!\!^{-} = \hbar/p = \hbar c[(2mc^2 + E_{\mathrm{kin}})E_{\mathrm{kin}}]^{-\frac{1}{2}} \tag{9}$$

(m = electron mass, p = momentum). The radius of middle-sized nuclei, *viz.*, $5\cdot 10^{-13}$, is certainly larger than $\lambda\!\!\!^{-}$ according to the foregoing. Therefore, if we put $\lambda\!\!\!^{-} = 5\cdot 10^{-13}$, we certainly obtain too small a value for E_{kin}. We find then

$$E_{\mathrm{kin}} > 2\pi\hbar c/\lambda\!\!\!^{-} = 1.05\cdot 10^{-27}\cdot 3\cdot 10^{10}/5\cdot 10^{-13} = 6.3\cdot 10^{-5}\ \mathrm{erg} = 40\ \mathrm{MV}.$$

This value is much larger than nuclear binding energies, and the actual value of E_{kin} should be even bigger. This does not seem plausible and is thus a strong point against the assumption of the existence of electrons in the nucleus. A more rigorous disproof of this assumption will be given in §38.

§4. Statistics (E1)

Any kind of particles in nature obeys either the Fermi-Dirac or the Bose-Einstein statistics. In the first case, the Pauli principle holds for the particular sort of particles under consideration, i.e., there can never be two particles of this kind in the same quantum state. Notable examples are electrons (positive and negative), protons, and, as we shall see, neutrons and neutrinos (§39). It seems, in fact, to turn out that every "elementary" particle obeys Fermi statistics.

Particles obeying Bose statistics are allowed to be in the same quantum state; indeed, they have even what may be called a preference for being in the same state. Photons are the most well-known example; deuterons, α-particles and a great many other nuclei also belong to this category.

The most rigorous and most fruitful definition of the statistics is based upon properties of the wave function of a system of particles of a given kind. Let us suppose we have n identical particles $1, 2, \cdots, i, \cdots n$; then the wave function describing their motion will be a certain function $\psi(x_1\cdots x_i\cdots x_n)$ depending on the coordinates[12] of the particles. If the coordinates of two particles are interchanged, e.g., the coordinate x_j of particle j inserted in place of x_i, and x_i in place of x_j, another function of the coordinates is obtained. This new function is (a) identical with the original function, if the particles obey Bose statistics, (b) identical except for a change of sign, if the particles obey Fermi statistics. Functions of kind (a) are called symmetrical in the particles, such of kind (b) antisymmetrical. In formulae, we have

$$\begin{aligned}
&\psi(x_1\cdots x_j\cdots x_i\cdots x_n)\\
&= +\psi(x_1\cdots x_i\cdots x_j\cdots x_n)\ \text{(Bose statistics)},\\
&\psi(x_1\cdots x_j\cdots x_i\cdots x_n)\\
&= -\psi(x_1\cdots x_i\cdots x_j\cdots x_n)\ \text{(Fermi statistics)}.
\end{aligned} \tag{10}$$

The experimental determination of the statistics of nuclei is based on alternating intensities in rotational band spectra, most conveniently of diatomic molecules. The theory of this determination will be described in §47. It has been found that the proton obeys Fermi statistics, the deuteron, the α-particle, the nuclei of N^{14}, O^{16}, etc., Bose statistics, the Li^7 nucleus again Fermi statistics. From the observations the general rule *can be inferred that all nuclei with even atomic weight follow Bose statistics, those of odd atomic weight Fermi statistics.* This is in accord with the assumption that all nuclei are composed of protons and neutrons, and that the *neutron has Fermi statistics.*

To prove this statement, we have to show that a system composed of elementary particles each obeying Fermi statistics, obeys Bose or Fermi statistics according to whether the number of elementary particles in the system is even or odd. We consider two systems (nuclei), α and β, each containing m elementary particles of one sort (protons) and n particles of another sort (neutrons). We assume that the first m protons and the first n neutrons constitute the nucleus α which is situated near the point r_α, the second set of m protons and n neutrons are bound up in the second nucleus β near r_β. This state of affairs will be described by a certain wave function depending on the coordinates of all particles. We now exchange one proton of nucleus α and one of nucleus β; when doing this, the wave function of the whole system is multiplied by -1 because the protons obey Fermi statistics. We then exchange another pair of particles and continue this process until all

[12] x_i is meant to symbolize all three coordinates of the particle i, and, if the particles have spin, also the spin coordinate.

the m protons and n neutrons originally constituting nucleus α have been brought to the point r_β and *vice versa*. Every exchange multiplies the wave function by -1, thus after the exchange of all $m+n$ particles of nucleus α against the $m+n$ particles constituting β the wave function has been multiplied by $(-1)^{m+n}$. On the other hand, our process corresponds to an exchange of the entire nuclei α and β. Thus the wave function is multiplied by $(-1)^{m+n}$ when the coordinates of our two nuclei are interchanged, which means that the wave function is symmetrical (antisymmetrical) in the coordinates of the two nuclei if the total number of particles $m+n$ in each nucleus is even (odd). This is the theorem which we wanted to prove.

The proof given above is however not quite rigorous, because it is not possible to construct a wave function which is antisymmetrical in all protons and neutrons and which at the same time assigns definite particles to a definite nucleus. A rigorous proof has been given by Ehrenfest and Oppenheimer (E1). It should be noted that at the time when Ehrenfest and Oppenheimer's paper was written, the neutron had not yet been discovered and it was therefore believed that nuclei consist of protons and *electrons*. Thus the word "electrons" in their paper should be replaced by "neutrons" throughout. (If this is done, the contradiction between theoretical and experimental results concerning the statistics of N^{14} disappears.)

The first use we make of our theorem is to deduce the statistics of neutrons from the experimental fact that the proton obeys Fermi and the deuteron Bose statistics. Assuming the deuteron to consist of one neutron and one proton, we must conclude from our theorem that the neutron follows the Fermi statistics. Knowing now the statistics of the neutron, we can predict that any nucleus of even atomic weight will obey Bose statistics because the total number of particles, protons and neutrons together, is equal to the atomic weight, and it is this total number which determines the statistics. Similarly, all nuclei with odd atomic weight must obey Fermi statistics. No exception has been found to this rule. The nuclei investigated for statistics are H^1, H^2, He^4, Li^7, N^{14}, O^{16}, Na^{23}, P^{31}, S^{32}, Cl^{35} and K^{39}.

The old nuclear theory assumed the nuclei to consist of protons and electrons. The number of protons then had to be assumed equal to the atomic weight A, because only the protons contributed to the weight, while $A-Z$ electrons had to be assumed in order to neutralize the charges of $A-Z$ protons and leave a resultant charge of only Z. The total number of particles was thus $2A-Z$ and was thus even or odd according to whether the nuclear charge was even or odd. Accordingly, all elements with odd Z, e.g., H and N, should have obeyed Fermi statistics, irrespective of their atomic weight. This was in direct contradiction to the experimental result for H^2 [13] and N^{14} (R7) which contradiction constituted another strong argument against the "electron theory" of nuclear constitution.

§5. Spin and Magnetic Moment (cf. chapter VIII)

The intrinsic angular momentum (spin) of nuclei can be determined from hyperfine structure, molecular ray analysis, depolarization of resonance radiation, and alternating intensities in rotational band spectra of molecules. These methods except for the last also determine the magnetic moment associated with the spin. A detailed description of methods and results will be given in chapter VIII.

The most important result for a general theory of the nuclei is that the spins of all nuclei of odd atomic weight seem to be half-integer multiples of $\hbar$, while all nuclei of even atomic weight have integer spin, most of them probably having spin zero. The total spin of a nucleus is the resultant of all the angular momenta of the orbital motions of all particles inside the nucleus, and of all the spins of the nuclear particles. The resultant has to be taken according to the rules of the vector model of quantum theory. Now the *orbital* angular momenta are always integers (in units $\hbar$). Thus the appearance of half-integer values for the total spin of some nuclei, must be attributed to half-integer values for the *spins* of the individual nuclear particles.

[13] Actually the statistics of deuterons were only determined after the discovery of the neutron.

The empirical rule connecting atomic weight and nuclear spin must thus be interpreted as showing that both the proton and the neutron have half-integer spins. In the case of the protons it can be shown experimentally that the spin is exactly $\frac{1}{2}$. The neutron spin might, from experimental evidence, be just as well $\frac{3}{2}$ as $\frac{1}{2}$. However, simplicity is a strong argument in favor of the value $\frac{1}{2}$ which we shall, therefore, assume throughout this article. It seems to be a general rule that this value of the spin is true for all elementary particles known, *viz.*, proton, neutron, electron (positive and negative) and neutrino (§39).

If both proton and neutron have spin $\frac{1}{2}$, then the resultant of the spins of A elementary particles, neutrons and protons, will be integer or half-integer according to whether the atomic weight A is even or odd. This conclusion from the vector model of quantum theory is in accord with all existing observations. It is analogous to the statement about the statistics of nuclei in the preceding paragraph, such that nuclei obeying Bose statistics have integer spins and such obeying Fermi statistics have half-integer spins. The two statements are, however, independent of each other—at least as long as we do not understand the connection between statistics and spin of a particle which seems to exist but has thus far not been explained.

It need hardly be pointed out that the old nuclear theory which assumed the nuclei to consist of protons and electrons, faced with respect to spin a difficulty analogous to that regarding the statistics. The experimental situation was even worse in the case of the nuclear spin, because a great number of spins of nuclei with even nuclear charge and odd atomic weight had been measured and were found to be half-integer in contradiction to the "electron theory" and conforming to the "neutron theory" of nuclei, whereas the statistics had actually been determined for only one nucleus with even A and odd Z.

The magnetic moment of the proton has been determined by Stern and Estermann (E4) and by Rabi, Kellogg, and Zacharias (R2). It does not have the value of one nuclear "magneton,"

$$\mu_0 = \hbar e/2Mc = 5.02 \cdot 10^{-24} \text{ gauss cm}^3 \quad (11)$$

(M = mass of the proton). μ_0 would be the magnetic moment which would be expected if Dirac's theory would hold for protons. The value actually observed is about $2.9\mu_0$,[14] i.e., much greater than the "theoretical" value μ_0.[15] Attempts to explain this discrepancy will be explained in §45. It has been proved recently by Kellogg, Rabi and Zacharias (K3) that the magnetic moment points in the *same* direction as the angular momentum of the proton spin, as would be expected for a positive charge (for electrons, the directions of magnetic moment and spin are opposite).

The magnetic moment of the neutron is hardly accessible to a direct measurement. It has to be inferred from the moments of other, more complex, nuclei. The simplest of these is the deuteron which consists of one proton and one neutron. Its spin has been measured and turns out to be unity. It can safely be assumed that the deuteron has no orbital angular momentum in its ground state (§12); thus its observed angular momentum 1 must be attributed to the *spins* of the proton and the neutron in the deuteron. This means that the spins of proton and neutron are parallel in the deuteron. Thus the magnetic moment of the deuteron is the sum of the magnetic moments of proton and neutron.

The most exact experimental determination of the magnetic moment of the deuteron is that of Kalckar and Teller (theory of the method, K1) and of Farkas, Farkas and Harteck (experiment, F1). Its basis is the measurement of the velocity of the conversion of orthohydrogen into parahydrogen by the action of paramagnetic gases. This velocity is, among other things, proportional to the square of the magnetic moment of the hydrogen nuclei. By measuring the velocity of conversion for light hydrogen (H_2) and heavy hydrogen (D_2), the ratio of the magnetic moments of H and D can be determined. The result is

$$\mu_D : \mu_H = 1 : 4 \quad (12)$$

[14] New measurements of Rabi, Kellogg and Zacharias (private communication).

[15] In spite of this fact, the magnetic moment of the proton is, of course, much smaller than the Bohr magneton of an electron, $\mu_1 = \hbar e/2mc = 1838\mu_0$ (m = electron mass).

with an accuracy of about 5 percent.* The magnetic moment of the deuteron is also directed in the same direction as the spin (R4), therefore, we find

$$\mu_{\text{neutron}} = \mu_D - \mu_H = -\tfrac{3}{4}\mu_H = -2.2\mu_0. \quad (13)$$

Thus the magnetic moment of the neutron has the direction which would be expected if the neutron had a negative charge. Its magnitude is of the same order as the magnetic moment of the proton.

II. Qualitative Arguments about Nuclear Forces

In the theory of *atomic* structure, we deal with electric particles, *viz.*, nucleus and electrons, and we therefore know the forces acting between them. The problem of atomic physics has therefore not been to determine the forces between atomic particles but to find out how electrons move if subjected to a known force. This problem has been solved by quantum theory.

In nuclear theory we can have confidence that the quantum theory holds for the motion of the neutrons and protons in the nucleus. This assumption is strongly supported by the relation between size and binding energy of nuclei (cf. §3), which is just what should be expected from quantum theory, and also by the success of actual calculations (chapters III, IV, etc.). Furthermore, we can safely assume that relativity corrections are small because the binding energies are small compared to the energy corresponding to the rest mass of proton and neutron (end of §2).

On the other hand, we do not know the forces between the nuclear particles, with the exception of the Coulomb repulsion between the protons in the nucleus, which, however, plays only the role of a correction (§8). The principal attractive forces are certainly not electric in nature because they act upon neutrons which bear no charge. What the nature of these forces is, how they depend on the distance of nuclear particles, on their spin and possibly other quantities, has to be inferred from experimental data. We shall do that in this chapter in a qualitative way, and in later chapters apply the knowledge thus obtained to special problems which will furnish more quantitative data on the nuclear forces.

§6. The Ratio of Atomic Weight to Nuclear Charge (H7)

When the periodic system was first discovered, nothing was known about nuclear charge. The atoms were ordered according to their atomic weight and this led to the discovery of the periodic system. This fact alone shows that the atomic weight is closely connected with the nuclear charge. Indeed, the known stable isotopes of any given element do not vary greatly in atomic weight, the variation being only 10 percent even for an element with so many isotopes as tin. Thus, to a first approximation, we may speak of a definite relation between atomic weight and nuclear charge.

For the light elements, up to about argon, this relation is very simple indeed. The atomic weights are very nearly twice as large as the nuclear charge, e.g., C^{12}, N^{14}, O^{16}, etc. Therefore the number of neutrons $N=A-Z$ in any of these light nuclei is approximately equal to the number of protons Z.

This experimental rule must be interpreted as showing that the largest attractive forces in the nucleus are forces between neutrons and protons. If this were not the case, e.g., if two neutrons would attract each other more strongly than a neutron and a proton, the most stable nuclei would obviously be composed exclusively of neutrons. We can, of course, not deduce from our empirical rule that there are no forces between a pair of neutrons or a pair of protons[16] at all, but if there are such forces, they must be smaller than the force between a proton and a neutron.

Our rule $N=Z$ tells us even more about the forces between like particles (two protons, or two neutrons). Provided such forces exist at all,

* *Note added in proof:* Recent experiments of Rabi, Kellogg and Zacharias seem to show that the ratio of the moments is smaller, about 1 : 3.5. This corresponds to a deuteron moment of $0.85\mu_0$ and a neutron moment of about $-2.0\mu_0$.

[16] We refer here to forces between two protons *beside* the Coulomb repulsion.

they must be very nearly equal, i.e., the force between two neutrons must be nearly equal to that between two protons, leaving out the electrostatic repulsion between the latter. For instance, if the attraction between two protons were larger than that between two neutrons, the maximum of stability would not occur for equal numbers of neutrons and protons but would be shifted towards a relatively larger number of protons in the nucleus. Explicitly, if (nn), (pp), (np) denote the binding energies between two neutrons, two protons, and one neutron and one proton, respectively, the energy of a nucleus containing N neutrons and Z protons would be, neglecting 1 compared to N and Z:

$$\tfrac{1}{2}N^2(nn)+\tfrac{1}{2}Z^2(pp)+NZ(pp), \qquad (14a)$$

which reaches its maximum, for a given atomic weight $A=N+Z$, when

$$N-Z(=A-2Z)=\tfrac{1}{2}A\frac{(nn)-(pp)}{(np)-\frac{1}{2}(nn)-\frac{1}{2}(pp)}. \qquad (14)$$

Experimentally, the difference $N-Z$ does, for the most stable light nuclei, certainly not exceed 10 percent of the nuclear charge $Z\approx\frac{1}{2}A$. According to (14), the difference $(nn)-(pp)$ must then certainly be smaller than one-tenth of $(np)-\frac{1}{2}(nn)-\frac{1}{2}(pp)$. The most satisfactory assumption from the standpoint of symmetry as well as from the experimental evidence, is that the force between two neutrons and that between two protons are *exactly* equal—disregarding, of course, the Coulomb energy between the protons, and provided there are any forces between like particles at all.

§7. Saturation of Nuclear Forces (H7, M6)

If every particle in the nucleus is supposed to interact with every other particle, the interaction energy, and therefore the binding energy holding the nucleus together, would be roughly proportional to the number of interacting pairs, i.e., to the *square* of the number of particles in the nucleus. If any deviation could be expected from this law, it would be in the direction of a more rapid increase of the binding energy with the number of particles, because with increasing interaction between them, the particles will draw closer together, and this will lead to an increase of the interaction even between a single pair of particles, making the total binding energy increase faster than A^2.[17]

Actually, it is found experimentally that the mass defects of nuclei, and therefore the binding energies, increase only linearly with increasing number of particles (§2). This fact may be compared to the behavior of a liquid or solid containing many atoms where the total chemical binding energy is sensibly proportional to the number of atoms present.

We therefore try to be guided by the chemical analogy. How does the proportionality of the chemical binding energy to the number of atoms arise? There are essentially three possible reasons for this, corresponding to the three possible types of chemical binding: polar binding, homopolar binding and van der Waals (polarization) binding.

The van der Waals type is most clearly realized in the case of rare gases in the liquid or solid state. Between any two rare gas atoms, we have an attractive force (van der Waals force) when the atoms are more than a certain distance r_0 apart. The attractive force falls rapidly, approximately as r^{-7}, when the distance r between the atoms increases. For distances smaller than r_0, a strong repulsive force begins to act which prevents any appreciable interpenetration of the two atoms. In the liquid, only near neighbors have any appreciable interaction, because of the rapid decrease of the attractive force with increasing atomic distance. Any given atom thus interacts only with a small number of neighbors, however large the total number of atoms may be. On the other hand, the repulsive forces prevent any increase in density which would allow more atoms to interact with any given atom. Thus, the repulsive forces which prevent the interpenetration of atoms are, in this case, primarily responsible for the binding energy being proportional to the number of atoms. However, it would seem very unsatisfactory to transfer such a mechanism to nuclei: it would involve the assumption of a force between *elementary* particles, *viz.*, protons and neutrons, which would be attractive at large distances and repulsive at small distances, an assumption which one would make only very reluctantly. (For particles with internal structure, such as atoms or the α-particle, the assumption of such a force is, of course, not objectionable but results directly from *simple* assumptions about the forces between elementary particles.)

The polar binding is realized in salts, e.g., NaCl. Two unlike atoms (Na and Cl) attract each other with a force

[17] The conditions outlined in this paragraph are actually found for the electrons in atoms. The total binding energy of all electrons in an atom increases roughly as the 7/3 power of the number of electrons, i.e., faster than Z^2 (cf. e.g., reference S23).

which decreases very slowly with increasing distance, two like particles repel each other with a similar force. The counteraction of these two forces keeps the binding energy from increasing quadratically with the number of particles, in spite of the fact that there is appreciable interaction between far distant ions. The assumption of repulsive forces between like particles in the nucleus is, however, impossible for other reasons (§10); this is *a fortiori* true for the assumption that these repulsive forces are equal in magnitude to the attractive forces between neutron and proton. Thus the analogy to the polar binding must also be rejected.

The homopolar binding is most clearly represented by elements like hydrogen. There is strong binding between two hydrogen atoms. A third atom, however, would not be strongly attached to the H_2 molecule. We say, the H_2 molecule is saturated. An assembly of many hydrogen atoms, e.g., in a drop of liquid hydrogen, therefore has an energy approximately equal to that of the corresponding number of hydrogen molecules, and therefore proportional to the number of atoms present. It is true that the binding energy of a hydrogen droplet will be slightly greater than that of separated molecules because of van der Waals forces between the H_2 molecules, but they again give an energy proportional to the first power of the number of atoms.

We thus see that we shall obtain the correct dependence of nuclear binding energies on the number of particles in the nucleus, if we assume forces between the nuclear particles which show saturation, in much the same way as the forces of homopolar chemical binding. It is at once clear that the association of nuclear particles which will correspond most nearly to a saturated molecule, is the α-particle. In fact, the binding energy of the α-particle, as deduced from its mass defect, is 28 MV, or 7 MV per elementary particle. The binding energy of the nuclei which have the highest packing fractions, *viz.*, those with Z round 30, is $8\frac{1}{2}$ MV per elementary particle. In our analogy, this means that 7 MV of these $8\frac{1}{2}$ are due to the chemical "binding energy of the molecule" He^4, while the remaining $1\frac{1}{2}$ MV are to be attributed to "van der Waals" forces between the α-particles.

The "van der Waals" forces between the α-particles can, without difficulty, be assumed analogous to those between rare gas atoms or H_2 molecules; i.e., an attraction at larger distances, falling off very rapidly with increasing distance, and a repulsion at small distances, giving something like mutual impenetrability of two α-particles. A force of this type has actually been deduced by Heisenberg (§31). The assumptions will, of course, lead to a binding energy between α-particles which is approximately proportional to the number of such particles.

The α-particle contains 2 neutrons and 2 protons. The forces between neutrons and protons must thus be such that they are saturated when 2 neutrons and 2 protons are brought together and are practically nil when a third proton or neutron is brought into the neighborhood of the first 4 particles. Now 2 neutrons and 2 protons can, on account of Pauli's principle, just be placed into the same quantum state as regards their motion in space, due to the possibility of two different states of spin. We shall therefore assume that *protons and neutrons exert strong forces upon each other only if they are in the same*, or approximately the same, quantum state *with regard to their motion in space*, i.e., if their wave functions depend in approximately the same way on the spacial coordinates.

However, we must not assume that the forces depend critically upon the relative spin of the two particles. If we did so, e.g., if we would assume that a proton and a neutron would only interact strongly when their spins are parallel, then the nuclear forces would already be saturated in the deuteron (proton plus neutron with parallel spins). In reality, the binding energy of the deuteron is only slightly over 2 MV, compared to 28 MV for the α-particle. This shows that the deuteron can certainly not be regarded as saturated. *The forces between proton and neutron can therefore depend only slightly, if at all, upon the relative spin directions of the two particles.*[18]

Turning to the mathematical representation of the interaction, we may again be guided by the chemical analogy. The forces of homopolar

[18] Heisenberg (H7) had originally assumed an interaction which was attractive for particles with parallel spins, repulsive for opposite spins. According to the foregoing consideration, such an interaction would make the deuteron a saturated structure. This was pointed out by Majorana (M6) and the assumed interaction changed accordingly.

binding are "*exchange*" forces, connected with the changing places of the electrons from one atom in the molecules to the other. We know from the chemical analogy that such exchange forces show saturation. *We therefore assume that the nuclear forces* also *have the character of exchange forces between neutron and proton.* Just as in the case of molecules, but of course not due to the same mechanism, an electron[19] passes from neutron to proton so that the former neutron is transformed into a proton and the former proton into a neutron. This can be considered as an exchange of the coordinates of neutron and proton: Thus our "exchange forces" mean that neutron and proton interchange their positions whenever they interact. The mathematical formulation of these ideas will be deferred to §11.

Since this paragraph contains the clue to all nuclear theory, we want to sum up. The proportionality of nuclear binding energy and number of particles in the nucleus requires the assumption of exchange forces between the nuclear particles which show saturation. The high binding energy of the α-particle, compared to the deuteron, requires these forces *not* to show saturation for the deuteron, and therefore not to depend to any considerable extent upon the relative spin directions of the interacting particles.

From our analogy between nuclear and chemical forces, we can draw a conclusion about the size of nuclei, as a function of the number of particles contained in them. We know that the volume of a droplet of a liquid, or of a solid, is proportional to the number of atoms contained in it, each atom occupying about the same volume. Since the nuclei are held together by forces similar to chemical forces, we may expect them also to have a volume proportional to the number of particles in the nucleus. This is in agreement with the experimental evidence referred to in §3.

It remains to be said that no exact proportionality between binding energy and number of particles is to be expected. Not only will there be irregular variations depending on the special structure of any particular nucleus, but also a regular trend towards slightly increasing binding energy per particle, with increasing size of the nucleus. This effect is analogous to the surface tension of a droplet of a liquid. The atoms on the surface of the droplet do not receive the full attraction which atoms in the interior would receive, and do therefore not contribute their full share to the binding energy. The same is true for the particles at the surface of a nucleus. Since the number of the surface particles, as a fraction of the total number of particles, decreases when the total number increases, we expect a slight increase of the binding energy per particle with increasing atomic weight. This is actually shown by Fig. 1 which represents the packing fractions as function of A.

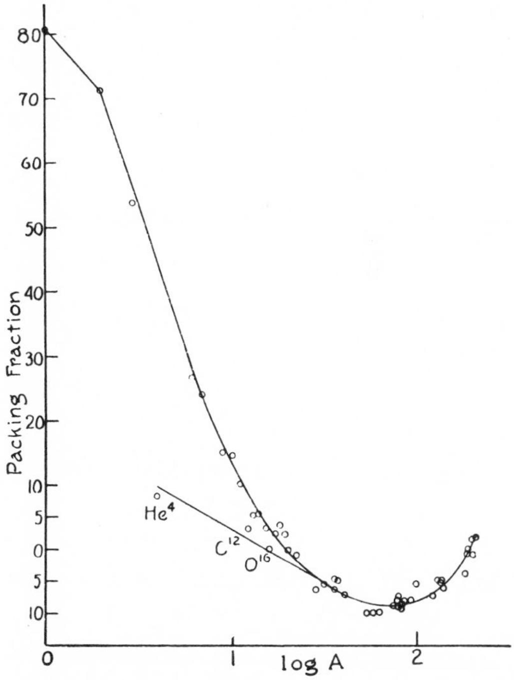

FIG. 1. Packing fraction as a function of atomic weight. In order to represent light and heavy atoms in the same diagram, a logarithmic scale has been chosen for the absicissa. It is seen that among the light nuclei the α-particle and its multiples (C^{12}, O^{16}) have much smaller packing fractions than all the other nuclei. (The packing fractions are expressed in ten-thousandths of a mass unit.)

[19] Together with electron, a neutrino must pass over. The "passage" is, of course, not meant literally in the sense that the electron and neutrino are bound in the neutron and then pass bodily to the proton. We might assume that electron and neutrino are just "created" for a very short time, of the order $e^2/mc^3 = 10^{-24}$ sec., and are then reabsorbed. For details see §44.

§8. The Electrostatic Repulsion of the Protons. Stability Against α-Decay (H7, W2)

The preceding two paragraphs have led us to assume (1) that the binding energy of nuclei is approximately proportional to the number A of the particles in the nucleus and (2) that the binding energy, for given total number of particles, is a maximum when neutrons and protons are present in equal numbers. This can be expressed by the rough formula

$$E_n=(Z+N)\epsilon[1-\alpha((N-Z)/(N+Z))^2], \quad (15)$$

where ϵ is the binding energy per particle for a nucleus containing equally many protons and neutrons, and α a constant which measures the dependence of the binding energy on the "isotopic number" $I=N-Z$.

α might, in principle, still depend on the total number of particles, A. It would not do so, however, if we assume the binding energy to be proportional to (14a) *for a given value of* A. It is true that (14a) has been derived by assuming every particle in the nucleus to interact with every other particle which is not possible when the forces show saturation. On the other hand, the saturation will affect primarily the dependence of the binding energy on the total number of particles, rather than that on the ratio of the numbers of neutrons and protons. We therefore accept (15) as a preliminary expression, deferring a more accurate treatment to chapter V.

In our discussion thus far, we have neglected the electrostatic forces between the protons. Taking for the mutual distance of two protons r approximately the nuclear radius (cf. §3), i.e., about $5\cdot10^{-13}$ cm $\approx 2e^2/mc^2$ (m = electron mass), we find for their electrostatic potential energy $e^2/r\approx\frac{1}{2}mc^2=\frac{1}{4}$ MV. This is indeed negligible compared to the average binding energy per particle, *viz.*, $8\frac{1}{2}$ MV (cf. §2).

However, because of the saturation character of the specifically nuclear forces the Coulomb repulsion between the protons becomes important for heavy nuclei in spite of its smallness for a single pair of protons. For the Coulomb force shows, of course, no saturation. Therefore the total energy of the Coulomb interaction is actually equal to the number of pairs of protons in the nucleus, i.e., $\frac{1}{2}Z(Z-1)$, times the potential energy of a single proton pair. The latter is in the average $(6/5)e^2/R$ with R the radius of the nucleus, if the protons are considered as distributed uniformly over the nucleus.[20] Thus the total electrostatic energy of the protons is

$$E_{el}=\tfrac{3}{5}Z^2e^2/R \quad (16)$$

if 1 is neglected in comparison to Z. Since R is proportional to the cube root of the atomic weight (cf. §3, and end of §7), and Z proportional to the atomic weight itself, the electrostatic energy is proportional to $A^{5/3}$. On the other hand, the binding energy E_n due to the specifically nuclear forces is only proportional to the first power of the atomic weight A. Thus the relative importance of the electrostatic forces increases with increasing atomic weight, roughly as $A^{2/3}$.

The consequences of this are twofold: Firstly, we shall obtain for heavier nuclei a deviation from the rule $N=Z$, in the sense that stable nuclei contain fewer protons than neutrons, because the replacement of a proton by a neutron decreases the electrostatic repulsion and thus the total energy of the nucleus. This effect is well-known experimentally: The ratio N/Z, i.e., number of neutrons to number of protons, increases from 1 for light nuclei gradually to 1.6 for uranium.

Secondly, the binding energy per particle will decrease, on account of the electrostatic forces, with increasing atomic weight. This effect works in the opposite direction from the "surface tension" discussed at the end of §7. The surface tension will be more important as long as the nucleus is still small, therefore we get a decrease of the packing fraction with increasing atomic

[20] If $v=4\pi R^3/3$ is the volume of the nucleus, $1/v$ will be the charge density due to one proton distributed uniformly. The electrostatic potential due to this charge density, at a distance r from the center of the nucleus, is according to ordinary electrostatics,

$$V(r)=e\frac{4\pi}{v}\left(\int_0^r\frac{\rho^2d\rho}{r}+\int_r^R\frac{\rho^2d\rho}{\rho}\right)=\frac{4\pi e}{v}\left(\frac{1}{2}R^2-\frac{1}{6}r^2\right).$$

The energy of a second proton, uniformly distributed, in this potential is

$$w=\frac{e}{v}\int_0^R 4\pi r^2dr\,V(r)=\left(\frac{4\pi e}{v}\right)^2\left(\frac{1}{2}R^2\frac{1}{3}R^3-\frac{1}{6}\frac{1}{5}R^5\right)$$
$$=\left(\frac{3e}{R^3}\right)^2\cdot\frac{1}{6}\frac{4}{5}R^5=\frac{6}{5}\frac{e^2}{R}.$$

weight in that region. On the other hand, for heavy nuclei the electrostatic repulsion between the protons will be more important so that the packing fraction rises again towards the end of the periodic table. This has actually been observed (cf. Fig. 1). For very heavy nuclei, this rise becomes so pronounced that the nuclei become unstable against α-disintegration.

The total binding energy of a nucleus is the difference of the binding energy given in (15) and the Coulomb repulsion (16). To write the latter in a suitable form, we put

$$R = r_0 A^{\frac{1}{3}}, \tag{17}$$

where r_0 can be determined from the known radii of radioactive nuclei: The average of these radii as determined from the lifetimes of α-decaying nuclei, is $9 \cdot 10^{-13}$ cm. The average atomic weight of the nuclei concerned is 222, therefore

$$r_0 = 9 \cdot 10^{-13} \cdot 222^{-\frac{1}{3}} = 1.48 \cdot 10^{-13} \text{ cm} \tag{17a}$$

and

$$\frac{3}{5}\frac{e^2}{r_0} = 0.60 mc^2 \frac{e^2/mc^2}{r_0}$$

$$= 0.307 \cdot \frac{2.80 \cdot 10^{-13}}{1.48 \cdot 10^{-13}} \text{MV} = 0.58 \text{ MV}. \tag{17b}$$

We shall abbreviate this expression by the letter γ.

Inserting (17), (17b) into (15) and (16), we obtain for the binding energy

$$E = A\epsilon - \alpha\epsilon(A - 2Z)^2/A - \gamma A^{-\frac{1}{3}} Z^2. \tag{18}$$

The maximum of this expression for given A is obtained when

$$2\alpha\epsilon(A - 2Z)/A - \gamma A^{-\frac{1}{3}} Z = 0$$

or

$$\frac{I}{Z} = \frac{N-Z}{Z} = \frac{A-2Z}{Z} = \frac{\gamma}{2\alpha\epsilon} A^{\frac{2}{3}}, \tag{18a}$$

$$I/A = \gamma A^{\frac{2}{3}} / (4\alpha\epsilon + \gamma A^{\frac{2}{3}}). \tag{18b}$$

The ratio of the isotopic number $I = N - Z$ to the nuclear charge Z, is thus proportional to the two-third power of the atomic weight. This relation is illustrated by Fig. 2, in which $(A - 2Z)$ is plotted against A, for known stable nuclei. It can be seen that the observed points fall near, and on both sides of the solid line.[21] That line represents the relation (18b), with

$$\gamma/2\alpha\epsilon = 0.0146. \tag{18c}$$

This value is so chosen that the line passes

[21] For deviations from that line (periodicities) cf. §34.

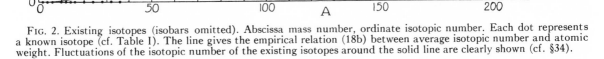

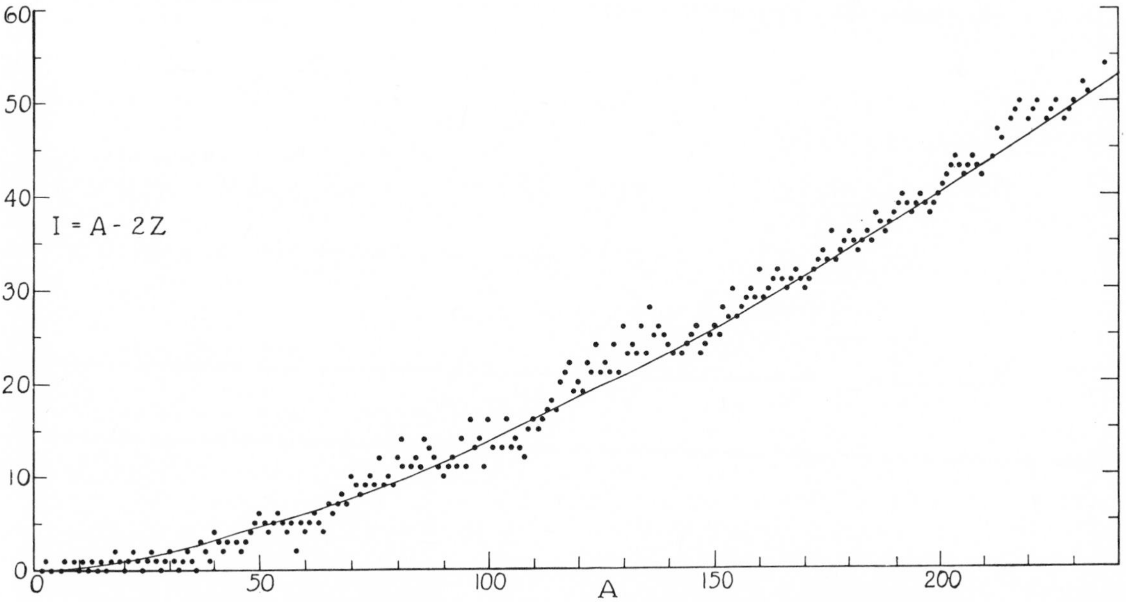

FIG. 2. Existing isotopes (isobars omitted). Abscissa mass number, ordinate isotopic number. Each dot represents a known isotope (cf. Table I). The line gives the empirical relation (18b) between average isotopic number and atomic weight. Fluctuations of the isotopic number of the existing isotopes around the solid line are clearly shown (cf. §34).

through Hg^{200}, for which $A=200$, $Z=80$, therefore $(A-2Z)/Z=0.50$ and

$$\gamma/2\alpha\epsilon=0.50/(200)^{\frac{2}{3}}=0.0146. \qquad (18d)$$

We may now calculate the total binding energy of a nucleus whose charge has the "most favorable" value for the given mass of nucleus. To do this, we insert the value (18b) for $(A-2Z)/A$ into (18) and find

$$E_{max}(A)=A\epsilon-A\alpha\epsilon[\gamma A^{\frac{2}{3}}/(4\alpha\epsilon+\gamma A^{\frac{2}{3}})]^2 -\gamma A^{5/3}[2\alpha\epsilon/(4\alpha\epsilon+\gamma A^{\frac{2}{3}})]^2, \qquad (19)$$

$$E_{max}(A)=A\epsilon[1-\alpha\gamma A^{\frac{2}{3}}/(4\alpha\epsilon+\gamma A^{\frac{2}{3}})],$$

or, denoting by Z_A the "most favorable charge" for the atomic weight A, we get from (18b)

$$E_{max}(A)=A\epsilon-(A-2Z_A)\alpha\epsilon. \qquad (19a)$$

Now the value of $\alpha\epsilon$ can be deduced from (17b) and (18c):

$$\alpha\epsilon=0.58/2\cdot 0.0146=20 \text{ MV}. \qquad (19b)$$

Inserting this and the observed value of the binding energy for Hg^{200} into (19) we can deduce ϵ. The binding energy can be calculated from the atomic weight of Hg^{200}, 200.016, and the combined weight of 80 protons and $200-80=120$ neutrons, which gives (cf. 75a)

$$\begin{aligned} E(Hg^{200}) &= 120\cdot 1.00846+80\cdot 1.00807 \\ &\quad -200.016=1.645 \text{ mass units} \qquad (19c) \\ &= 931\cdot 1.645 \text{ MV}=1530 \text{ MV}. \end{aligned}$$

Therefore, from (19a) and (19b)

$$\epsilon=(1530+40\cdot 20)/200=11.6_5 \text{ MV}. \qquad (19d)$$

This is rather higher than the binding energy per particle for medium sized nuclei (8.5 MV). The reason is that the actual binding energy is reduced due to surface tension as well as due to the electrostatic repulsion of the protons, and even for medium sized nuclei both these effects are quite appreciable and reduce the "naive" binding energy, which would be $11.6A$, by over 25 percent. For mercury, the observed binding energy per particle is only $1530/200=7.6_5$ MV, which is 35 percent less than the binding energy would be if there were no electrostatic forces. *Thus the electrostatic forces amount to 35 percent of the "specifically nuclear" forces for mercury.*

From (19b) and (19c) we find furthermore

$$\alpha=20/11.6_5=1.72,$$

which is quite reasonable (cf. §30, Eq. (185)).

The most interesting question to be answered approximately by our rough formulae is the probable energy of α-particles which might be emitted by radioactive atoms. Of course, we can only give an "average" value for this energy which depends smoothly on the atomic weight while the actual α-energies vary irregularly from one radioactive atom to the other, which variation could only be deduced from a more refined theory. Let us suppose the α-emitter has a nuclear charge "most favorable" for its atomic weight; its energy is thus given by (19a). The nucleus produced in the α-disintegration will have a charge slightly different (too small) from the most favorable charge for its weight. But since the binding energy has, for given A, a maximum at $Z=Z_A$, it varies only quadratically with the difference $Z-Z_A$. Thus we may assume that (19a) is very nearly true even for the product nucleus. Therefore we have:

(1) binding energy of the α-emitter (atomic weight A, nuclear charge Z_A)

$$E_1=A\epsilon-(A-2Z_A)\alpha\epsilon;$$

(2) binding energy of the product nucleus (atomic weight $A-4$, nuclear charge approximately Z_{A-4})

$$\begin{aligned} E_2 &= (A-4)\epsilon-(A-4-2Z_{A-4})\alpha\epsilon \\ &= E_1-4\epsilon+4\alpha\epsilon d(A-2Z_A)/dA \\ &= E_1-4\epsilon+4\alpha\epsilon\gamma A^{\frac{2}{3}}\frac{(20/3)\alpha\epsilon+\gamma A^{\frac{2}{3}}}{(4\alpha\epsilon+\gamma A^{\frac{2}{3}})^2}; \qquad (20a) \end{aligned}$$

(3) binding energy of the α-particle

$$E_3=4\epsilon', \quad \text{where} \quad \epsilon'=6.9 \text{ MV};$$

therefore energy of a would-be-emitted α-particle

$$E_\alpha=E_2+E_3-E_1=-4(\epsilon-\epsilon') +4\alpha\epsilon\gamma A^{\frac{2}{3}}\frac{(20/3)\alpha\epsilon+\gamma A^{\frac{2}{3}}}{(4\alpha\epsilon+\gamma A^{\frac{2}{3}})^2}. \qquad (20)$$

For $A=222$ (radon; this atomic weight corresponds approximately to the average for all radioactive elements) we find, with the values (17b), (19b), (19d), (20a) for γ, α, ϵ, ϵ':

$$E_\alpha = -4\cdot 4.7_5 + 80\cdot 0.58\cdot 222^{2/3} \cdot \frac{133+0.58\cdot 222^{2/3}}{(80+0.58\cdot 222^{2/3})^2} = 6.6 \text{ MV}. \quad (20b)$$

This is indeed fairly close to, but slightly larger than, the average kinetic energy of the α-particles emitted by radioactive substances. The surface effect, which will be discussed in §29, 30, will decrease the theoretical value (20b) to 3.8 MV.

From (20) it is obvious that α-radioactivity will, in general, only be possible if the atomic weight exceeds a certain critical value A_0 which is determined by $E_\alpha(A_0)=0$, or

$$\frac{x(5/3+x)}{(1+x)^2} = \frac{\epsilon-\epsilon'}{\alpha\epsilon} = \frac{4.7_5}{20} = 0.275$$

with $x=\gamma A_0^{2/3}/4\alpha\epsilon$. The solution is $x=0.176$, or

$$A_0 = 119. \quad (20c)$$

Thus nuclei of higher atomic weight than 120 should, in the average, be unstable against α-decay. "In the average" means that the binding energy of α-particles in nuclei of atomic weight around 120 should be positive in about as many causes as it is negative. The stability limit will be shifted to slightly higher atomic weights if the "surface tension" is taken into account (§30), but only to 14.7 Why, then, has actual α-radioactivity only been found for much higher atomic weights (lowest observed: polonium, $A=210$)? The answer is that the lifetime of an α-radioactive nucleus becomes extremely long when the kinetic energy of the α-particle when emitted is small (chapter IX). Thus a nucleus is practically stable against α-decay, although not perfectly stable, if the decay energy is not very large. Indeed, no α-particles of kinetic energies less than 2 MV have actually been observed.

This explains why actually only the nuclei heavier than about 200 have an observable α-radioactivity. A notable exception is one samarium isotope, of atomic weight near 140: In this case, we obviously have a fairly large deviation of an individual binding energy from the "average" binding energy prevailing in that region of atomic weights.

§9. Deuteron and α-Particle: The Form of the Potential Function (W12)

It is known experimentally that the mass defect of the α-particle is about 13 times as large as that of the deuteron, *viz.*, 27.7 MV compared to 2.14 MV. On the other hand, we have proved in §6 that forces between like particles, if they exist at all, must be smaller than the forces between proton and neutron. (In §21 we shall show that the ratio of these two kinds of forces is about 2 : 3.) Thus we would, from a naive consideration, expect that the α-particle has only slightly more than 4 times, and certainly less than 6 times, the binding energy of the deuteron: For we have in the deuteron one pair of interacting particles, in the α-particle each of the two neutrons interacts with each of the two protons, which gives 4 times the deuteron interaction.

The solution of this problem has been given by Wigner (W12). We have to assume that the forces between neutron and proton are very strong when the two particles are close together, but fall off very rapidly when the distance between them becomes larger than a certain, small distance a. We thus assume a strong, short-range force between the two particles.

To make a *complete* use of such a short range force, the nuclear particles have to get very close, more accurately, inside the reach of the force. If their wave function is to be confined in that small region, its wave-length must be of the order of the range of the forces, i.e., very small. Accordingly, the momentum, and the kinetic energy of the particles must become very large; the larger, the smaller the range of the force. The kinetic energy of the particles in the deuteron may in this way become even larger than the potential energy at close distances; if so, the two particles *cannot* be confined within the range of the forces between them; the particles will actually travel over a larger region in space, in which way their kinetic energy may be kept down. But when this is the case, also the time during which the particles actually are near enough to exert a strong attraction is reduced, and thus the binding energy will come out quite small compared to the potential energy between the particles.

If we now take the α-particle, four to five times the attractive forces are available, while the number of moving particles is only twice as large as for the deuteron. Thus it is well conceivable that now the attractive potential will suffice to overcome the kinetic energy, and to actually draw the particles into the range of their mutual forces. Then full use can be made of the large interaction potential, and the binding energy will be of the same order of magnitude as the interaction potential.

This shows that with a deep and narrow hole representing the potential energy between proton and neutron, the binding energy (mass defect) of the α-particle can be made very much larger than that of the deuteron. Thomas (T2) has actually shown that the ratio of the mass defects of H^3 and H^2 becomes infinitely large if the range of the forces is reduced to zero, and at the same time the magnitude of the potential energy increased in such a way that it yields the observed binding energy of the deuteron (§19). If the binding energy of H^3 tends to infinity, this is, of course, *a fortiori* true for that of the α-particle. Thus Thomas' calculation shows that any desired value may be obtained for the ratio of the mass defects of α-particle and deuteron, by a suitable choice of the range of the forces.

The actual determination of the range of the forces from the given mass defects requires, of course, the solution of the Schrödinger equation for the α-particle and the deuteron. A suitable form must be assumed for the potential energy between neutron and proton as a function of the distance, leaving two parameters free which determine the width and the depth of the potential hole (range and magnitude of the force). Then the Schrödinger equations for H^2 and He^4 have to be solved with this potential. The first equation, for the deuteron, can easily be solved rigorously (chapter III). That for the α-particle has to be treated by approximate methods, e.g., by the Ritz method based on the variation principle. Unfortunately, this method converges very slowly, so that the results thus far obtained are not very certain although much work has been put into the attempt of solving the problem, especially by Feenberg (F2, F3). The range of the forces resulting from his calculations is about $2\cdot 10^{-13}$ cm, i.e., approximately the radius of a sphere whose volume is equal to the volume per particle of heavier nuclei, and considerably less than the radius of the deuteron ($4.36\cdot 10^{-13}$ cm, cf. §12). Details of the calculation will be given in chapter IV.

The determination of the analytical *form* of the dependence of the nuclear forces upon the distance between the nuclear particles, is at present quite hopeless. Any rapidly decreasing function, whether $e^{-\alpha r^2}$, $e^{-\beta r}$, a rectangular potential hole or a more complicated function having the same characteristic behavior, will fit the experimental data equally well as long as no *very* accurate calculations of the binding energies expected for a given force, are available. It has been suggested (cf. §44) that the potential should be proportional to some high negative power of the distance for large r, and become more or less constant at small r. For the present, however, the potential can be represented by a function which is most convenient for the integration of the Schrödinger equation, without introducing any error comparable to that due to the insufficiency of our present methods for integrating that equation.

§10. Forces Between Like Particles. Odd and Even Isotopes (Y1)

The considerations of §§6–8 have given us an idea about the general dependence of the binding energy of nuclei upon the atomic weight and the nuclear charge. Experimental evidence about this general dependence was used to fix some constants in the assumed expression for the binding energy. However, no reference has been made to any details in the distribution of known isotopes.

There is one such detail which is very outstanding and which strikes one immediately if one glances at the table of known isotopes (Table I): While there are 154 isotopes known[22]

[22] β-emitting substances have been excluded. The figure 148 includes 16 radioactive α-emitters.

with even nuclear charge and even atomic weight, there are only 4 with *odd* nuclear charge and even weight, and all of these latter have atomic weights smaller than 14. In the remainder of the periodic system, from $A = 14$ up to 238, there is not a single stable isotope with odd charge and even weight.—The isotopes of odd atomic weight occupy, as far as their number is concerned, an intermediate position: There are 106 stable isotopes of this kind well established, of which 55 have even nuclear charge and 52 odd charge (in addition, 7 α-emitters of odd atomic weight are known).

What is the reason for the striking difference between nuclei with even weight and even charge, and such with even weight and odd charge? To account for this difference, in fact, to make any theory of nuclear stability, it is necessary to know the condition for stability. If we would accept the, obviously necessary, condition that the removal of any neutron or proton from the nucleus must require energy, then practically any pair of values A, Z would lead to a stable nucleus. More stringent is the condition of stability against α-emission (cf. §9) but even this would allow a wide variety of nuclear charges for any given atomic weight. Actually, the most important condition is stability against β-transformation, i.e., against emission or absorption of electrons.

The emission of an electron by a nucleus leads to a new nucleus whose mass number is identical with that of the original nucleus while its charge is one unit higher. The β-emission can take place energetically, if the energy of the original nucleus is higher than that of the produced nucleus plus mc^2, where m is the mass of the electron; in other words, if the exact atomic weight of the original nucleus is higher than that of the nucleus produced in the β-decay (cf. §2). In stating this condition, the mass of the neutrino has been assumed equal to zero (cf. §39). Now the experimental evidence about β-decay seems to show that, whenever β-decay is energetically possible, the decay occurs almost always in a reasonably short time, ranging from fractions of a second up to a few years. Some notable exceptions, primarily radioactive potassium and rubidium, which have lifetimes of the order of 10^8 years, can be accounted for without serious difficulty (§43). We shall thus assume that in general any substance which is energetically unstable against β-decay, will disintegrate in a time very short compared to the life of the earth, and will thus not be found among the existing isotopes in nature. (For the explanation of exceptions, and the conditions therefore, see §43.)

From the standpoint of nuclear theory, the nucleus produced in β-disintegraton differs from the original nucleus by containing one proton more and one neutron less. So we get the rule: *A nucleus is unstable* against β-disintegration, *if the replacement of a neutron in the nucleus by a proton would make the energy of the corresponding atom smaller.*

A similar rule holds for the replacement of a proton by a neutron. This replacement is brought about when the nucleus absorbs an electron, e.g., one of the orbital electrons of its own atom. That such an absorption of external electrons by a nucleus is possible, can be inferred with practical certainty from the fact that positron-emitting radioactive nuclei are known in great number: The emission of a positron can, according to Dirac's theory, be considered as the absorption of an electron which has been in a state of negative energy. If this process is possible, there is no conceivable reason why electrons in states of positive energy could not be absorbed by nuclei. The energetical condition for such an absorption is obviously that the energy of the absorbing nucleus, plus the intrinsic energy mc^2 of the absorbed electron, is larger than the energy of the nucleus produced by the absorption. Thus *a nucleus is unstable against absorption of electrons if the atomic weight is decreased when a proton in the nucleus is replaced by a neutron.*

For complete stability, a nucleus must therefore *be lighter than both the two neighboring isobars*, i.e., the two nuclei whose mass number is the same and whose charge is by one unit less or greater than that of the given nucleus.

We thus conclude that *the energy of any nucleus with even atomic weight and odd nuclear charge is larger than* that of at least one of its neighboring isobars, which would have *even mass and even charge.* If this theorem can be proved, it follows at once that all stable nuclei with even mass *must* have even charge, in agreement with experiment. (Of course, it must also be proved why the four light nuclei H^2, Li^6, B^{10} and N^{14} are exceptions from the rule.)

Nuclei having even mass and even charge, obviously contain even numbers of neutrons and protons. For even mass and odd charge, we would have an odd number of neutrons as well as protons. Thus we can express our empirical rule by saying that even numbers of neutrons and protons lead to a lower total energy of the nucleus than odd numbers, in other words that a pair of neutrons or protons has much the same function in nuclear physics as closed shells in atomic physics, insofar as it leads to a specially low energy of the system. That two neutrons (or protons) may form a "closed shell," is plausible because two particles with opposite spin may just be placed in exactly the same quantum state with respect to orbital motion. A third neutron would have to go into the next higher quantum state, and would therefore be less strongly bound. It is true that this rule differs appreciably from the rule valid in atomic physics where we find groups of $2(2l+1)$ electron states with the same n and l (principal and azimuthal quantum number) all having sensibly the same energy. The difference seems to be due to the fact that in an atom we have practically a central field which we have not in nuclei. This problem will be discussed in more detail in chapter VI, where we shall also show that there is evidence for other periodicities in the structure of nuclei, with longer periods than 2, which are more similar to the electron shells in atoms.

For our present discussion we simply accept that every state of orbital motion of a proton or neutron has its own energy, differing from the energy of all other states, so that two neutrons or two protons form a "closed shell." With this assumption, it is quite simple to prove that no nucleus containing an odd number of protons and neutrons can be stable, except in the very beginning of the periodic system.

Let us take any nucleus of even atomic weight A and even nuclear charge Z and call it the standard nucleus; e.g., we might choose Ni^{60} for this purpose. In the field of this standard nucleus, there will be certain energy levels for neutrons and other levels for protons. By "level" or "state" we refer, in this discussion, to the state of orbital motion only, so that each state can take two particles. The $\frac{1}{2}Z$ lowest proton states, and the $\frac{1}{2}N$ lowest neutron states are occupied in our standard nucleus. Of the empty states, either the lowest proton or the lowest neutron state will be lower. If we construct the nucleus of atomic weight $A+1$, this nucleus will have $Z+1$ protons and N neutrons if the proton state is the lower of the two, or Z protons and $N+1$ neutrons if the neutron state is the lower. Both cases will occur with about equal probability, therefore we expect that *for odd atomic weight nuclei with even charge are about as numerous as such with odd charge.* This is actually true, the statistics we have mentioned before showed 52 known nuclei with odd weight and odd charge and 55 stable nuclei with odd weight and even charge. In the case of our "standard" nucleus Ni^{60}, the addition of a neutron leads to the stable nucleus, Ni^{61}. Now let us add a second particle to our standard nucleus. If the first unoccupied neutron level in the standard nucleus lies lower than the first unoccupied proton level, the most stable nucleus of weight $A+2$ will be obtained by adding two neutrons to the standard nucleus so that it has nuclear charge Z. The addition of one neutron and one proton will lead to a less stable nucleus (charge $Z+1$), because the proton level lies higher. The addition of two protons will give us a nucleus (charge $Z+2$) which is even less stable. Conversely, if the proton level is the lower of the two, the most stable nucleus will be obtained by adding two protons, a less stable one by adding one proton and one neutron, and the least stable if we add two neutrons. Thus *in both cases the most stable resulting nucleus of atomic weight $A+2$*, has even nuclear charge, in one case the same charge Z as the "standard" nucleus, in the other case, the charge $Z+2$. In no case will the nucleus of charge $Z+1$ be the most stable. (In our case of Ni^{60}, we know from the first step of adding one particle that the neutron level lies lower. Thus we expect the stable nucleus of atomic weight 62 to be nickel again which actually is the case.)

Thus it seems that we have proved our theorem from a simple consideration about neutron and proton levels. However, we know that in many cases *isobars* exist, i.e., several stable nuclei having the same atomic weight, and nuclear charges usually differing by two units. This means that, starting from our "standard nucleus" of weight A and charge Z it often happens that both the addition of two neutrons and the addition of two protons lead to stable nuclei, while the addition of one proton and one neutron never does. This is quite inexplicable from our previous considerations, which led us to expect that the energy of the nucleus of atomic weight $A+2$ and charge $Z+1$ always lies in between the energies of its isobars with charges Z and $Z+2$.

The fact that both the nuclei Z and $Z+2$ may have energies lower than the intermediate nucleus $Z+1$, thus requires a special explanation. We must obviously assume that there is some *attraction between* the *two neutrons or* the *two protons* which we added to the "standard" nucleus. With such an attraction, the energies of the nuclei Z and $Z+2$ become depressed below the value expected from our previous considerations which assumed the additional two particles to move independently from each other. The attraction between particles of equal kind therefore allows the existence of two stable isobars, both with even charges, differing by two units, while the intermediate nucleus of odd charge is unstable.

The objection might be raised that the neutron and the proton, which we have added to the standard nucleus in order to obtain the nucleus $(A+2, Z+1)$ should also show an interaction. In fact, we have even proved (§6) that the interaction between a neutron and a proton must be larger than that between a pair of neutrons or a pair of protons. This would mean the energy of the nucleus $(A+2, Z+1)$ would be more decreased by the interaction of the two additional particles than that of the nuclei $(A+2, Z)$ or $(A+2, Z+2)$. From such a reasoning, we would therefore conclude that the nucleus of *odd* charge and even weight $(A+2, Z+1)$ is stable and the nuclei of even charge and weight, $(A+2, Z)$ and $(A+2, Z+2)$ are not, in contradiction to experience.

The fallacy in this argument is due to the fact that a neutron and a proton interact strongly only if they are in approximately the same quantum state, because the forces are "saturation" forces (§7). We know, however, that the first empty neutron state in the field of our standard nucleus is quite different from the first empty proton state. If we add a neutron and a proton, they will therefore have practically no interaction. On the other hand, two added neutrons (or two protons) will move in the same orbit and therefore have full interaction, irrespective of what we assume about the forces between like particles—i.e., whether they also act only between particles in the same quantum state, or between any pair of light particles (cf. §24). Thus we have accounted for the rule that all stable isotopes with even weight have even charge, and also for the existence of isobars of this type.

Our argument shows us, however, also the reason for the exceptions from that rule: If the added proton and neutron would move in the same quantum state, they would have a strong interaction, and therefore the atom of even weight and *odd charge* would be stable. The condition for this is obviously that equally many proton states are occupied in the "standard" nucleus as neutron states, in other words, that the standard nucleus contains exactly equal numbers of neutrons and protons. This is then also true for the nucleus which we obtain by adding a neutron and a proton to the standard nucleus. Therefore *nuclei of even weight and odd charge will be stable if they contain exactly as many protons as neutrons.* This is true for all the four stable nuclei of the type, *viz.*, H^2, Li^6, B^{10}, N^{14}. All these nuclei are very light, in fact, they are the lightest possible nuclei of their type. Only for light nuclei, can the number of neutrons be exactly equal to that of the protons. As soon as the nuclei get heavier, the Coulomb repulsion between the protons begins to become appreciable and to make the number of neutrons greater than that of protons in any given nucleus. Then the existence of nuclei with even weight and odd charge becomes impossible. This is true already for the atomic weight 18, for which the nuclear charge 8 rather than 9 leads to a stable nucleus (O^{18}, not F^{18}).

A convenient way of visualizing the results of this paragraph is to plot the energy of isobars (stable and unstable) as a function of their nuclear charge. If we do this for odd atomic

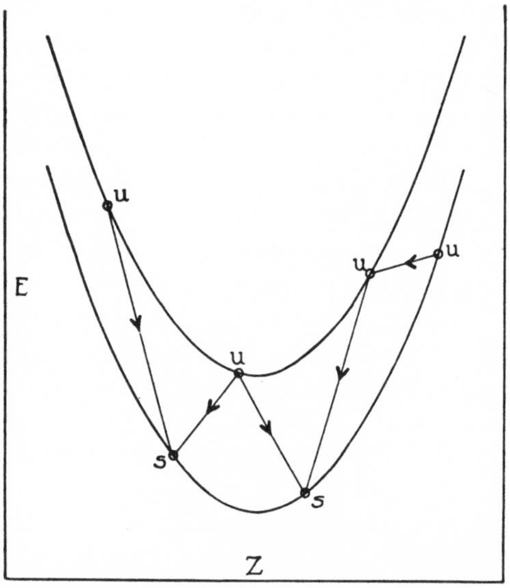

FIG. 3. Schematic graph representing the energy of isobars of even atomic weight as a function of nuclear charge Z. The upper parabola contains the nuclei of odd charge, the lower one those of even charge. The arrows represent possible β-disintegrations. The two "even" nuclei marked S are stable, all the others unstable.

weight, we shall get a smooth curve. The nucleus nearest to the minimum of that curve will be the stable nucleus for the given atomic weight. Chances are equal that this stable nucleus has even or odd charge.

For even atomic weight, on the other hand, we obtain two separate curves for nuclei with odd and such with even charge. We may assume that each of these curves is smooth. On the lower curve (even charge) we find always one, but in many cases several, stable nuclei, i.e., nuclei which have less energy than both their neighbors (Fig. 3).

We may try to determine the magnitude of the forces between like particles from the statistics of isobars. It is reasonable to assume that the two curves representing the energies of even charge and odd charge isobars, are parabolas just shifted vertically by an amount δ. The minimum of the parabolas will in general lie at a fractional value of the nuclear charge, Z_A. Let us denote by β the difference between Z_A and the nearest *odd* number; β obviously may be any number between -1 and $+1$. The curvature of the parabola may be determined from our general formula (18) for nuclear energies:

$$\kappa = -\tfrac{1}{2}\partial^2 E/\partial Z^2 = (4\alpha\epsilon + \gamma A^{2/3})/A. \quad (21)$$

Then we have for the energies of a nucleus of weight A and charge Z:

$$\begin{aligned} E &= E(Z_A) + \kappa(Z - Z_A)^2 && \text{if } Z \text{ is even,} \\ E &= E(Z_A) + \delta + \kappa(Z - Z_A)^2 && \text{if } Z \text{ is odd;} \end{aligned} \quad (21a)$$

e.g., for the odd nucleus whose charge is nearest to Z_A:

$$E_{\text{odd}} = E(Z_A) + \delta + \kappa\beta^2 \quad (21b)$$

for the two even nuclei nearest to Z_A:

$$\begin{aligned} E^{+}_{\text{even}} &= E(Z_A) + \kappa(1+\beta)^2, \\ E^{-}_{\text{even}} &= E(Z_A) + \kappa(1-\beta)^2. \end{aligned} \quad (21c)$$

The conditions that both the even nuclei are stable, read therefore

$$\delta > \kappa(1+2\beta) \quad \text{and} \quad \delta > \kappa(1-2\beta),$$

or, both conditions in one:

$$\delta > \kappa(1+2|\beta|). \quad (21d)$$

To find out the meaning of δ, we remember that we have proved that the energy of the "odd" nucleus must be midway between the two even ones if there are no forces between like particles. In this case, therefore, we would have

$$E'_{\text{odd}} = E(Z_A) + \kappa(1+\beta^2),$$

so that, in the absence of forces between like particles, $\delta' = \kappa$. Therefore the effect of the forces between like particles is given by

$$\delta'' = \delta - \delta' = \delta - \kappa \quad (21e)$$

and, according to (21d), we have two stable isobars if

$$\delta'' > 2\kappa|\beta| \quad (21f)$$

and only one stable element of atomic weight A if

$$\delta'' < 2\kappa|\beta|. \quad (21g)$$

To determine the critical value of $|\beta|$ above which the existence of isobars becomes impossible, we use the statistics of the known isotopes of atomic weight between 110 and 140. The reason for choosing this region is that the elements of higher atomic weight are not well explored as regards their isotopic structure, because they are rare earths, while among the elements of lower weight too few isobars are found. Of the 15 even mass numbers from 112 to

140, 10 are occupied by two isobars,[23] 5 by one nucleus only. We thus conclude that $\beta_0=10/15=2/3$ is the critical value for $|\beta|$ such that for $|\beta|>\beta_0$ only one nucleus of weight A is stable while for $|\beta|<\beta_0$ two stable isobars exist. Inserting $A=125$ into (21), we have (cf. 19b)

$$\kappa=\frac{4\cdot 20+0.58\cdot 25}{125}=\frac{80+15}{125}=0.76\text{ MV}$$

and
$$\delta''=2\beta_0\kappa=1.01\text{ MV}. \tag{22}$$

From formula (21f) we can conclude that the existence of isobars is the more probable the smaller κ. According to (21), κ decreases with increasing atomic weight. Thus we expect the more isobars the heavier the nuclei. This is actually true, at least up to $A\approx 150$. In the region of the rare earths, there are probably a great number of isobaric pairs yet unknown. For still higher atomic weight, stability against α-decay probably also plays an important role so that our rule cannot be expected to hold.

There is another indication of forces between like particles from the scattering of protons by protons (§18). That scattering is not in agreement with the expectation from a purely electrostatic interaction between the protons. Strong evidence for the forces between like particles comes also from the quantitative calculations of the binding energies of H^3 and He^4 (§21).

III. Theory of the Deuteron

§11. The Wave Equations of Heisenberg, Wigner and Majorana (H7, W12, M6, B16)

We have shown in §7 that the force between neutrons and protons must be exchange forces and must not depend on the relative spin directions of the two particles. This type of force was first suggested by Majorana (M6). Earlier, Heisenberg had suggested saturation forces which did depend on the relative spins (H7), and Wigner ordinary forces which did not show saturation (W12).[24]

Although we have given good evidence for the Majorana type of force, we want to write down the wave equations for the two other suggested types as well. The reasons are on one side to facilitate comparisons, on the other hand (and this seems even more important) it seems probable that a small force of the Heisenberg type is superposed upon the main Majorana force (cf. §14).

To lead up to the wave equation, we consider again the chemical analogy. The system most nearly comparable to a system of a neutron and a proton, is the hydrogen molecular ion, H_2^+.[25] Its wave function is the product of an electronic wave function φ and a wave function ψ describing the motions of the two nuclei. φ depends on the distances ρ_α and ρ_β of the (single) electron of the H_2^+ from the two protons α and β. φ may be either symmetrical or antisymmetrical with respect to ρ_α and ρ_β, and therefore with respect to the coordinates of the two protons. Since the protons obey Fermi statistics, ψ must be antisymmetrical in the two protons if φ is symmetrical, and *vice versa*. We consider in particular those electron states of H_2^+ which go over into a hydrogen atom in the ground state, plus a proton. There are just two states of this type, one whose eigenfunction φ is symmetrical in the two proton coordinates and one antisymmetrical state. Only the former leads to binding. The electronic energy of the system may be written:

$$V(r)=-C(r)\mp A(r), \tag{23}$$

where $C(r)$ is the Coulomb interaction between a hydrogen atom in the ground state and a proton at a distance r from the nucleus of the hydrogen atom, while $A(r)$ is the "exchange integral" which measures how often the electron changes its place, going over from one proton to the other. The upper sign in (23) is related to the symmetrical electron wave function φ, the lower sign to the antisymmetrical φ.

The electron energy for fixed nuclear distance r must, as is well known, be regarded as a potential energy for the motion of the nuclei in the molecule. That motion is described by the wave function ψ. For antisymmetrical ψ the upper sign in (23) has to be taken, for symmetrical ψ the lower. We have thus two different Schrödinger equations for ψ according to the symmetry of ψ. We can, however, formally write them into one, by using the fact that

$$\begin{aligned}\psi(r_\beta r_\alpha)&=+\psi(r_\alpha r_\beta)\text{ for symmetrical }\psi,\\ \psi(r_\beta r_\alpha)&=-\psi(r_\alpha r_\beta)\text{ for antisymmetrical }\psi,\end{aligned} \tag{23a}$$

according to the definition of symmetry and antisymmetry

[23] One mass number (124) is actually occupied by 3 isobars.

[24] Recently, Bartlett (B10) has pointed out that there is still another type of force, *viz.*, "ordinary" forces depending on the relative spin. These forces would lead to both the difficulties of the Heisenberg and the Wigner theory, *viz.*, saturation at H^2 and too large binding energies for heavy nuclei.

[25] Cf. *Handbuch der Physik*, Vol. 24/1, p. 524.

($r_\alpha r_\beta$ are the coordinates of the two nuclei, including spin). We can thus write the Schrödinger equation:

$$(\hbar^2/2M)(\Delta_\alpha+\Delta_\beta)\psi(r_\alpha r_\beta)+E\psi(r_\alpha r_\beta)$$
$$=-C(r)\psi(r_\alpha r_\beta)+A(r)\psi(r_\beta r_\alpha). \quad (23b)$$

In the second term on the right-hand side the coordinates of the two nuclei have been interchanged. The term is equivalent to $+A(r)\psi(r_\alpha r_\beta)$ if ψ is symmetrical, and $-A(r)\psi(r_\alpha r_\beta)$ if ψ is antisymmetrical. Any reference to the electron has disappeared from (23b), therefore we can take it over directly into nuclear theory.

It must be emphasized at this point that the analogy to the hydrogen molecular ion must in no way be regarded as a deduction or justification of the wave equation for the deuteron. The forces between neutron and proton are an entirely new phenomenon, not connected in any way with forces familiar in atomic physics. A particular form chosen for the interaction between neutron and proton can therefore only be justified by comparison of the results deduced from this interaction with experimental data about nuclei. The theory of the hydrogen molecular ion serves only to suggest a possible form of the interaction which leads to saturation of the forces. The analogy to the H_2^+ has not been introduced because we think the neutron is in any way comparable to a small hydrogen atom but only because we know from qualitative considerations that the nuclear forces show saturation.

We now write down the wave equation for a neutron and a proton, interacting with each other, in analogy to (23b):

$$(\hbar^2/2M)(\Delta_x+\Delta_\xi)\psi(xs,\,\xi\sigma)$$
$$+E\psi(xs,\,\xi\sigma)=J(r)\psi(\xi\sigma,\,xs). \quad (24)$$

M is the mass of the proton which is sensibly the same as that of the neutron. The first two arguments in the wave function denote the position and the spin coordinate[26] of the proton, the last two refer to position and spin of the neutron. $|\psi(xs,\,\xi\sigma)|^2$ thus means the probability that the proton is found at the point x (x is, of course, a vector) with spin s, while the neutron is at the point ξ and has spin σ. Δ_x and Δ_ξ are the Laplacian operators with respect to the coordinates of neutron and proton, E the total energy and $J(r)$ the potential energy as a function of the distance $r=x-\xi$ between proton and neutron. (In comparing (24) to (23b), it may be noted that we have assumed no "ordinary" interaction $C(r)$ between proton and neutron but only "exchange" interaction $A(r)=J(r)$.)

Eq. (24) is that originally proposed by Heisenberg. It shows saturation effects (cf. §7) but it corresponds to an interaction which depends upon the relative spin directions of proton and neutron. To see this, we write the wave function ψ as a product of a function depending on the positions of the two particles, and a function depending on spin only:

$$\psi(xs,\,\xi\sigma)=\varphi(x\xi)\chi(s\sigma). \quad (24a)$$

If the spins of proton and neutron are parallel, the spin wave function χ is symmetrical[27] in the two spin coordinates s and σ, *viz.*,

$$\chi(\sigma s)=\chi(s\sigma) \qquad \text{(parallel spins)} \quad (24b)$$

if the spins are antiparallel, s is antisymmetrical, *viz.*,

$$\chi(\sigma s)=-\chi(s\sigma) \qquad \text{(antiparallel spins).} \quad (24c)$$

Therefore (24) goes over into an equation involving the "spatial" wave function φ only:

$$(\hbar^2/2M)(\Delta_x+\Delta_\xi)\varphi(x\xi)+E\varphi(x\xi)$$
$$=\pm J(r)\varphi(\xi x), \quad (25)$$

the upper sign holding for parallel spin of neutron and proton, the lower for antiparallel spin. Thus, if $J(r)$ is negative, we get an attraction between a neutron and a proton with parallel spins, but a repulsion for antiparallel spins. If $J(r)$ is positive, the reverse is true. In any case, the force between the two particles depends on the relative spin directions, and saturation is obtained when a single neutron is bound to a proton, the neutron spin being parallel or antiparallel to the proton spin according to the sign of J. A second neutron could not be bound to the proton but would even be repelled. The deuteron would be the "saturated" nucleus instead of the α-particle, in contradiction to experience.

This is avoided in the wave equation of Majorana (M6)[28]

$$(\hbar^2/2M)(\Delta_x+\Delta_\xi)\psi(xs,\,\xi\sigma)+E\psi(xs,\,\xi\sigma)$$
$$=J(r)\psi(\xi s,\,x\sigma). \quad (26)$$

It differs from Heisenberg's equation in the interaction term on the right-hand side: While in Heisenberg's equation spatial coordinates as well

---

[26] As such we may choose the spin component in a given direction z, which may have either of the two values $+\frac{1}{2}$ or $-\frac{1}{2}$.

[27] Cf., e.g., *Handbuch der Physik*, Vol. 24/1, p. 325.

[28] For some difficulty connected with the difference in mass between proton and neutron, cf. references B27, P6.

as spin coordinates of neutron and proton are interchanged in the interaction term, only the spatial coordinates are interchanged in Majorana's equation. The spin of the proton is s in the interaction term of (26) as well as on the left-hand side of that equation, the spin of the neutron is σ on both sides of the equation. Therefore, if we again write the wave function in the form (24a), we have

$$\psi(\xi s, x\sigma)=\varphi(\xi x)\chi(s\sigma);\quad \psi(xs, \xi\sigma)=\varphi(x\xi)\chi(s\sigma) \tag{26a}$$

in other words, the *same* spin function occurs on both sides of Eq. (26). Irrespective of the relative spin directions, Eq. (26) therefore reduces to

$$(\hbar^2/2M)(\Delta_x+\Delta_\xi)\varphi(x\xi)+E\varphi(x\xi)=J(r)\varphi(\xi x). \tag{27}$$

A negative interaction potential J leads to binding, for any spin directions.

Both Majorana's and Heisenberg's equation assume "exchange" forces which is necessary to explain the observed "saturation" of nuclear forces (§7). Forces of the ordinary type (the term "ordinary forces" will be used in distinction from exchange forces) have been assumed by Wigner (W12). Wigner's wave equation for the deuteron thus reads

$$(\hbar^2/2M)(\Delta_x+\Delta_\xi)\psi(xs, \xi\sigma)+E\psi(xs, \xi\sigma)=J(r)\psi(xs, \xi\sigma), \tag{28}$$

or, after separation of spin

$$(\hbar^2/2M)(\Delta_x+\Delta_\xi)\varphi(x\xi)+E\varphi(x\xi)=J(r)\varphi(x\xi). \tag{28a}$$

The generalizations of the Eqs. (24), (26), (28) to more than two particles are obvious. On the right-hand side there appear interaction terms for each pair of particles, similar to those in (24), (26), (28). Explicitly, we have for a nucleus containing N neutrons and Z protons:

$$\frac{\hbar^2}{2M}\left(\sum_{i=1}^{Z}\Delta_{xi}+\sum_{k=1}^{N}\Delta_{\xi k}\right)\psi(x_1s_1\cdots x_is_i\cdots\xi_k\sigma_k\cdots\xi_n\sigma_n)+E\psi(\cdots x_is_i\cdots\xi_k\sigma_k\cdots)$$

$$=\sum_{i=1}^{Z}\sum_{k=1}^{N}J(r_{ik})\psi(x_1s_1\cdots\xi_k\sigma_k\cdots x_is_i\cdots\xi_n\sigma_n)\ \text{(Heisenberg)}, \tag{29a}$$

$$=\sum_{i=1}^{Z}\sum_{k=1}^{N}J(r_{ik})\psi(x_1s_1\cdots\xi_ks_i\cdots x_i\sigma_k\cdots\xi_n\sigma_n)\ \text{(Majorana)}, \tag{29b}$$

$$=\sum_{i=1}^{Z}\sum_{k=1}^{N}J(r_{ik})\psi(x_1s_1\cdots x_is_i\cdots\xi_k\sigma_k\cdots\xi_n\sigma_n)\ \text{(Wigner)}. \tag{29c}$$

For the application to the deuteron, it is convenient to separate the motion of the center of gravity of the deuteron from the relative motion of the two particles in the deuteron. We are only interested in the latter. It will be described by a wave function u which depends upon the relative coordinate $\mathbf{r}=\mathbf{x}-\boldsymbol{\xi}$ of the proton with respect to the neutron. Interchanging the coordinates of the two particles means replacing $\mathbf{r}$ by $\boldsymbol{\xi}-\mathbf{x}=-\mathbf{r}$, in other words, changing the sign of the relative coordinates. Then we have for u the wave equation

$$\frac{\hbar^2}{M}\Delta u(\mathbf{r})+Eu(\mathbf{r})=$$

$$\pm J(r)u(-\mathbf{r})\ \text{Heisenberg}, \tag{30a}$$

$$J(r)u(-\mathbf{r})\ \text{Majorana}, \tag{30b}$$

$$J(r)u(\mathbf{r})\ \text{Wigner}. \tag{30c}$$

u depends only on the three relative coordinates $\mathbf{r}$, and Δ is therefore the ordinary Laplacian operator in three dimensions. The proton mass M which appeared in (24) to (29) has been replaced by the reduced mass $\frac{1}{2}M$ since we are dealing with relative motion. The signs in (30a) refer to parallel (+) and antiparallel (−) spins of the proton and neutron.

§12. Ground State of the Deuteron (B16)

The deuteron plays in nuclear physics the same role as the hydrogen atom in atomic physics. It consists of two elementary particles, one proton and one neutron. It is well known that any two-body problem can be integrated explicitly if the force between the two particles is a known function of the distance of the particles. Thus the theoretical results about the deuteron are free from approximations made to simplify the mathematical treatment. They are, as we shall see, also to a very large extent independent of the assumptions we make about details of the force between neutron and proton, i.e., of the function $J(r)$ in (30). The theory of the deuteron is thus more suitable for quantitative comparisons with experiment, and therefore for a check of the underlying ideas about nuclear structure, than any other part of nuclear theory.

Using the Majorana hypothesis about the forces between neutron and proton, we have obtained the wave equation

$$(\hbar^2/M)\Delta U(\mathbf{r})+EU(\mathbf{r})=J(r)U(-\mathbf{r}), \quad (30b)$$

where r is the relative coordinate of the proton with respect to the neutron, M the mass of proton or neutron and E the energy of the system. If E is negative, $\epsilon=-E$ is the binding energy of the deuteron.

The potential energy $J(r)$ is spherically symmetrical. Therefore (30b) can be separated in polar coordinates r, θ, φ by putting

$$U(\mathbf{r})=(u_l(r)/r)P_{lm}(\theta)e^{im\varphi}, \quad (31)$$

where P_{lm} is a spherical harmonic which we assume to be normalized unless otherwise stated. In (30b) the function $U(-\mathbf{r})$ enters as well as $U(\mathbf{r})$. If the polar coordinates of the point $\mathbf{r}=(x, y, z)$ are r, θ, φ, then those of the point $-\mathbf{r}=(-x, -y, -z)$ are r, $\pi-\theta$, $\pi+\varphi$. Now it is easy to show[29] that

$$P_{lm}(\pi-\theta)e^{im(\pi+\varphi)}=(-1)^lP_{lm}(\theta)e^{im\varphi}. \quad (31a)$$

Therefore the wave equation for u becomes[30]

$$\frac{\hbar^2}{M}\left(\frac{d^2u_l}{dr^2}-\frac{l(l+1)}{r^2}u_l\right)+Eu_l=(-1)^lJ(r)u_l. \quad (32)$$

If $J(r)$ is assumed to be negative, it is seen that the right-hand side of (32) corresponds to an attractive potential energy, if l is even, and to a repulsive potential, if l is odd. This alternation of the sign of the force for even and odd l, is a characteristic feature of exchange forces and could, in principle, be used to decide whether the forces acting between neutron and proton are exchange forces or ordinary ones (§14, reference W9).

The lowest quantum state will be obtained for $l=0$, if we assume $J(r)$ to be negative. Its eigenfunction obeys the wave equation

$$d^2u_0/dr^2=(M/\hbar^2)(J(r)-E)u_0. \quad (33)$$

u_0 is subject to the condition that it vanishes for small r as r itself, because otherwise $U(r)$ would, according to (31), become infinite at small distances r. Furthermore, u_0 must not become infinite for large r.

First we shall discuss the behavior of u_0 qualitatively, making very general assumptions about the interaction potential $J(r)$. We know from the ratio of the binding energies of α-particle and deuteron (§9) that $J(r)$ must be very large for small r, and must fall off very steeply at larger distances. We may therefore define a range a of the force such that $|J(r)|$ is negligible compared to E if $r>a$. If r is, by a sufficient amount, smaller than a, $|J(r)|$ will be large compared to $|E|$. The behavior of u_0 up to $r=a$ will therefore be determined almost exclusively by J and will not depend to a large extent upon E. On the other hand, beyond a the energy E alone will determine u_0.

Since $J(r)$ is negative and, for the most part of the region $r<a$, absolutely greater than E, the right-hand side of (33) is negative in this region and therefore u_0 is concave towards the r axis (cf. Fig. 4). By integrating (33) up to $r=a$, we

[29] $P_{lm}(\theta)$ is an even or odd function of cos θ, according to whether $l-m$ is even or odd (see, e.g., Jahnke-Emde, table of functions, p. 173). Since $\cos(\pi-\theta)=-\cos\theta$, we have $P_{lm}(\pi-\theta)=(-1)^{l-m}P_{lm}(\theta)$. Furthermore, $e^{im(\pi+\varphi)}=(-1)^me^{im\varphi}$.

[30] For the algebra involved in the separation of the Schrödinger equation in polar coordinates, cf. e.g., *Handbuch der Physik*, Vol. 24, 1, p. 275.

can determine u_0 and its first derivative u_0' at the point $r=a$. We put

$$(u_0'/u_0)_{r=a} = -\alpha. \tag{34}$$

α is a reciprocal length characteristic for the potential $J(r)$. According to the foregoing, α does not depend sensitively on the energy E. Of course, the sign of α as well as its absolute magnitude will depend on the strength of the forces. If, e.g., $J(r)$ is very small, the curvature of u_0 will be very small, and u_0 will still be increasing when we arrive at $r=a$ for any negative value of E. In that case, u_0'/u_0 would be positive, therefore α negative, and, as we shall see, no stable state ($E<0$) of the deuteron would be possible.

For $r>a$, the potential energy $J(r)$ is supposedly negligible. Therefore Eq. (33) can be solved immediately. Assuming E to be negative, *viz.*,

$$E=-\epsilon \tag{35a}$$

we have $$d^2u_0/dr^2=(M\epsilon/\hbar^2)u_0; \tag{35}$$

therefore $$u_0=ce^{-(M\epsilon)^{\frac{1}{2}}r/\hbar} \tag{36}$$

because the alternative solution, $e^{+(M\epsilon)^{\frac{1}{2}}r/\hbar}$, is to be excluded. c is a coefficient to be determined by normalization.

At $r=a$, the two solutions obtained by intetrating the Schrödinger equation "inside" ($r<a$) and "outside" ($r>a$) must join smoothly, so that

$$(u_0'/u_0)_{\text{outside}}=(u_0'/u_0)_{\text{inside}}. \tag{36a}$$

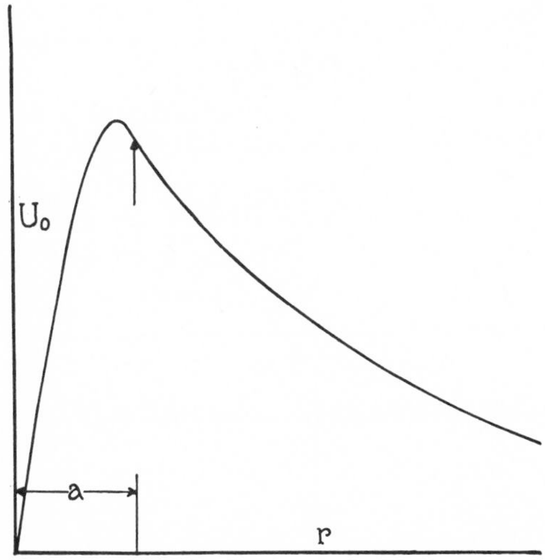

FIG. 4. Eigenfunction of the ground state of the deuteron. Ordinate: r times eigenfunction; abscissa: r. Simple potential hole of width a.

Now (34) gives us the "inside" value of u_0'/u_0 while the outside value follows by differentiating (36). We have therefore

$$(M\epsilon)^{\frac{1}{2}}/\hbar=\alpha. \tag{37}$$

The constant α which is directly connected to the given force field, thus determines the binding energy ϵ. Or, conversely, we may use *the observed value of ϵ to deduce α and thus to get some information about the forces between neutron and proton.*

If α were negative, i.e., if the force is too weak (see above), the inside wave function could not be joined smoothly to an exponentially decreasing wave function outside, but only to an exponentially increasing function. Such a function being excluded because of becoming infinite for $r=\infty$, we find that for negative α no solution with a positive binding energy ϵ can be found. (For positive energy E, there is, of course, always a solution which behaves, for $r>a$, like sin $((ME)^{\frac{1}{2}}r/\hbar)$ or cos $((ME)^{\frac{1}{2}}r/\hbar)$ instead of exponentially. From the mere existence of the deuteron we can therefore conclude that the forces between neutron and proton must be strong enough to make α positive, i.e., to make the curvature of $u_0(r)$ large enough so that u_0 *decreases* with increasing r, at $r=a$.

We can even deduce from this condition a quantitative estimate of $J(r)$ no matter what the particular dependence of J on r. We have

$$u_0''=(M/\hbar^2)(J(r)u_0+\epsilon u_0). \tag{37a}$$

Integrating over r from 0 to a, we find

$$(u_0')_{r=a}-(u_0')_{r=0}=M\hbar^{-2}\int_0^a J(r)u_0dr + M\hbar^{-2}\epsilon\int_0^a u_0dr. \tag{37b}$$

Now $(u_0')_{r=0}=1$, if u_0 is suitably normalized. u_0 is, for small r, proportional to r (cf. after Eq. (33)). This rule will no longer hold exactly for $r=a$, but it might serve as an approximation. Then $(u_0')_{r=a}\approx-\alpha a$, which is negligible compared to unity since the range of the forces, a, is supposed to be small compared to the (known) quantity $1/\alpha$ (cf. 44a). Furthermore, the last term in (37b) can be neglected compared to unity, because it is approximately $\frac{1}{2}M\epsilon a^2/\hbar^2\approx\frac{1}{2}a^2\alpha^2\ll 1$ (cf. (37)). Therefore, we obtain approximately

$$\int_0^a J(r)rdr=-\hbar^2/M. \tag{37c}$$

Actually, we should expect a somewhat larger absolute value for the integral on the left of (37c), because $u_0<r$. [In the special case of the rectangular potential hole, we have, according to (40), $\int_0^a J(r)rdr=-\frac{1}{2}V_0a^2=-\pi^2\hbar^2/8M$, which is rather close to (37c).]

We now discuss the solutions for a few simple forms of the potential function $J(r)$:

(a) *Rectangular hole:*

$$J(r)=-V_0 \quad \text{for} \quad r<a, \qquad J(r)=0 \quad \text{for} \quad r>a. \tag{38}$$

Solution

$$u_0=b\sin\left[M^{\frac{1}{2}}(V_0-\epsilon)^{\frac{1}{2}}r/\hbar\right] \quad \text{for } r<a,$$
$$u_0=c\exp\left[-(M\epsilon)^{\frac{1}{2}}(r-a)/\hbar\right] \quad \text{for } r>a. \tag{38a}$$

Joining solutions:

$$(u_0'/u_0)_{\text{inside}}=M^{\frac{1}{2}}(V_0-\epsilon)^{\frac{1}{2}}\hbar^{-1} \times \cot\left[M^{\frac{1}{2}}(V_0-\epsilon)^{\frac{1}{2}}a/\hbar\right],$$
$$(u_0'/u_0)_{\text{outside}}=-(M\epsilon)^{\frac{1}{2}}/\hbar.$$

Therefore

$$\cot\left[M^{\frac{1}{2}}(V_0-\epsilon)^{\frac{1}{2}}a/\hbar\right]=-(\epsilon/V_0-\epsilon)^{\frac{1}{2}}. \tag{39}$$

Since $V_0\gg\epsilon$ (cf. §9), the right-hand side of (39) is small. Therefore in first approximation

$$M^{\frac{1}{2}}(V_0-\epsilon)^{\frac{1}{2}}a/\hbar=\tfrac{1}{2}\pi+(\epsilon/V_0)^{\frac{1}{2}}+0(\epsilon/V_0)^{\frac{3}{2}},$$
$$V_0=\frac{\hbar^2}{Ma^2}\left(\frac{\pi}{2}+\left(\frac{\epsilon}{V_0}\right)^{\frac{1}{2}}\right)^2+\epsilon+0\left(\frac{\epsilon^{\frac{3}{2}}}{V_0^{\frac{1}{2}}}\right), \tag{39a}$$

$$V_0=\frac{\pi^2}{4}\frac{\hbar^2}{Ma^2}+2\left(\frac{\hbar}{Ma^2}\epsilon\right)^{\frac{1}{2}}+\left(1-\frac{4}{\pi^2}\right)\epsilon+0\left(\frac{\epsilon^{\frac{3}{2}}}{(\hbar/Ma^2)^{\frac{1}{2}}}\right). \tag{39b}$$

In very rough approximation,

$$V_0a^2=\pi^2\hbar^2/4M. \tag{40}$$

This means that the product of the depth and the square of the breadth of the potential hole can be determined from the mere existence of the deuteron. For a separate determination of breadth and depth, it is necessary to consider the binding energies of other nuclei, e.g., the α-particle, as well. But whatever the value of a and V_0, the product V_0a^2 will not differ very much from the value (40) which is a universal constant. The smaller the range a of the forces, the more exactly will (40) be true.

Normalization of the wave function: From (38a) (39) we have

$$b=c/\sin\left(M^{\frac{1}{2}}(V_0-\epsilon)^{\frac{1}{2}}a/\hbar\right)=c(V_0/V_0-\epsilon)^{\frac{1}{2}}, \tag{41}$$

$$\int_0^\infty u_0^2dr=b^2\int_0^a\left[\tfrac{1}{2}-\tfrac{1}{2}\cos\left(2M^{\frac{1}{2}}(V_0-\epsilon)^{\frac{1}{2}}r/\hbar\right)\right]dr+c^2\int_a^\infty dr\exp\left[-2M^{\frac{1}{2}}\epsilon^{\frac{1}{2}}(r-a)/\hbar\right]$$
$$=b^2\left(\tfrac{1}{2}a-\frac{\hbar}{4M^{\frac{1}{2}}(V_0-\epsilon)^{\frac{1}{2}}}\sin 2\frac{M^{\frac{1}{2}}(V_0-\epsilon)^{\frac{1}{2}}a}{\hbar}\right)+c^2\hbar/2(M\epsilon)^{\frac{1}{2}}$$
$$=\tfrac{1}{2}c^2\frac{V_0}{V_0-\epsilon}\left[\frac{\hbar}{(M\epsilon)^{\frac{1}{2}}}+a\right]. \tag{41a}$$

Normalizing to unity, we have (cf. 37)

$$c=\left(\frac{2(V_0-\epsilon)}{V_0}\right)^{\frac{1}{2}}\left(\frac{\hbar}{(M\epsilon)^{\frac{1}{2}}}+a\right)^{-\frac{1}{2}}=\left(\frac{2(V_0-\epsilon)\alpha}{V_0(1+\alpha a)}\right)^{\frac{1}{2}} \tag{41b}$$

and, according to (31), (38a), (41)

$$U(\mathbf{r})=\frac{U_0}{(4\pi)^{\frac{1}{2}}r}$$
$$=\begin{cases}\left(\dfrac{\alpha}{2\pi(1+\alpha a)}\right)^{\frac{1}{2}}\dfrac{\sin\left(M^{\frac{1}{2}}(V_0-\epsilon)^{\frac{1}{2}}r/\hbar\right)}{r},\\[2ex] \left(\dfrac{\alpha}{2\pi(1+\alpha a)}\right)^{\frac{1}{2}}\left(\dfrac{V_0-\epsilon}{V_0}\right)^{\frac{1}{2}}e^{-\alpha(r-a)}.\end{cases} \tag{41c}$$

(b) *Exponential potential*

$$J(r)=-V_0e^{-r/a}. \tag{42}$$

We introduce the independent variable

$$x=e^{-r/a} \tag{42a}$$

so that $x=1$ for $r=0$, $x=1/e$ for $r=a$, $x=0$ for $r=\infty$

and $\qquad d/dr=-(1/a)x(d/dx).$

Then the Schrödinger Eq. (33) becomes

$$x\frac{d}{dx}\left(x\frac{du_0}{dx}\right)-\frac{Ma^2}{\hbar^2}(-V_0x+\epsilon)u_0=0,$$
$$\frac{d^2u_0}{dx^2}+\frac{1}{x}\frac{du_0}{dx}+\left(\frac{Ma^2V_0}{\hbar^2}\frac{1}{x}-\frac{Ma^2\epsilon}{\hbar^2}\frac{1}{x^2}\right)u_0=0. \tag{42b}$$

This is the differential equation for a Bessel function (cf. Jahnke-Emde, p. 214), its solution is

$$u_0=cJ_p(2M^{\frac{1}{2}}V_0^{\frac{1}{2}}a\hbar^{-1}x^{\frac{1}{2}})$$

$$=cJ_p(2M^{\frac{1}{2}}V_0^{\frac{1}{2}}a\hbar^{-1}e^{-r/2a}), \quad (42c)$$

where $$p=2a(M\epsilon)^{\frac{1}{2}}/\hbar \quad (42d)$$

is the order of the Bessel function. At large distances r, the argument of the Bessel function is small so that the first term of the ordinary expansion of J in a power series is sufficient. We have then:

$$u_0=\frac{c}{p!}\left(\frac{2(MV_0)^{\frac{1}{2}}a}{\hbar}\right)^p \exp\left[-(M\epsilon)^{\frac{1}{2}}r/\hbar\right] \quad (r\gg a), \quad (42e)$$

which is, apart from the constant factor, identical with (38a). The eigenvalue ϵ is determined by the condition that $u_0(r=0)$ must vanish. For a given V_0, we have therefore to find p from the condition

$$J_p(2(MV_0)^{\frac{1}{2}}a/\hbar)=0 \quad (42f)$$

and then to calculate ϵ from p with the help of (42d). In order that (42f) has a solution p at all, it is necessary that V_0 be greater than a certain limit. This limit follows readily from the fact that the first zero x_p of $J_p(x)$ moves towards smaller values of x when p decreases. Therefore certainly $x_p>x_0$. Now the first zero of the Bessel function of order zero is $x_0=2.4048$ (Jahnke-Emde, p. 237). Therefore (42f) has a solution p only if

$$2(MV_0)^{\frac{1}{2}}a/\hbar>2.4048;$$

$$V_0>(\hbar^2/Ma^2)\cdot 1.4457. \quad (42g)$$

If $V_0=1.4457\ \hbar^2/Ma^2$, the solution of (42f) will be $p=0$ and therefore (cf. 42d) $\epsilon=0$, i.e., just no binding energy. If the binding energy remains small compared to V_0, which we have to assume (cf. §9), then V_0 must be only slightly larger than the value (42g).

Table II gives the values of $MV_0a^2\hbar^{-2}$ for different ranges a of the force.

(c) *"Error function" potential*

$$J(r)=-Be^{-r^2/a^2}. \quad (43)$$

The solution u_0 has to be obtained by numerical integration of the differential equation. As in case (a) and (b), and in the qualitative discussion, Ba^2 must be larger than a certain universal constant, in our case 2.65 $\hbar^2/M$, to give any binding for the deuteron at all. Table III gives the relation between $MBa^2/\hbar^2$ and the range a according to Feenberg (F2, F3). This table allows us to determine Ba^2 if a is known. Ba^2 changes only slightly with changing range of the forces a.

TABLE II. *Relation between width a and depth V_0 of exponential force $V_0e^{-r/a}$* (deuteron energy 2.15 MV).

$a=$ 0	0.5	1.0	1.5	2.0	2.5	3.0	$\cdot 10^{-13}$ cm
$p=$ 0	0.228	0.456	0.684	0.912	1.140	1.368	index of Bessel function
$MV_0a^2\hbar^{-2}=$ 1.446	1.888	2.370	2.890	3.466	4.039	4.664	
$V_0=59.5a^{-2}$	310	97	53	35.4	26.5	21.3	MV

TABLE III. *Relation between width a and depth B of "Gaussian" potential Be^{-r^2/a^2}* (deuteron energy 2.15 MV).

a (in 10^{-13} cm)	0	0.8	1.0	1.2	1.4	1.6	1.8	2.0	2.25
$T=\hbar^2/Ma^2$ (MV)	∞	64.3	41	28.5	20.9	16.1	12.7	10.25	8.10
B/T (pure number)	2.70	3.22	3.37	3.53	3.68	3.84	4.01	4.18	4.40
B (MV)	∞	207	138	100.5	74.8	61.8	51.0	42.8	35.6
$\bar{B}/T$*	2.70	3.09	3.20	3.32	3.43	3.55	3.68	3.81	3.97
a	2.5	2.75	3	3.5	4	4.5	5	5.5	6
T	6.56	5.43	4.55	3.35	2.56	2.02	1.64	1.35	1.14
B/T	4.62	4.85	5.08	5.55	6.04	6.56	7.13	7.70	8.28
B	30.3	26.3	23.1	18.6	15.45	13.2	11.7	10.4	9.4
$\bar{B}/T$*	4.14	4.31	4.48	4.84	5.20	5.59	6.02	6.45	6.88

* $\bar{B}$ represents the Majorana force between proton and neutron, plus *half* the Heisenberg force (cf. §14, end). This combination enters the theory of H^3, He^4 and heavier nuclei (§20).

All our results about the ground state of the deuteron remain unchanged if we assume either the Wigner or the Heisenberg interaction between neutron and proton instead of the Majorana force. For the ground state, and indeed for any S state, $U(-\mathbf{r})=U(\mathbf{r})$ (cf. (31), (31a)) so that the Wigner equation (30c) and the Majorana equation (30b) become identical. It is only for odd azimuthal quantum number l that there exists any difference between the Wigner and the Majorana theory (cf. 32). The Heisenberg theory also becomes identical with the Majorana theory, if we restrict ourselves to states in which the spins of neutron and proton are parallel, as is the case for the ground state of the deuteron (experimental value of the deuteron spin = one unit).

In conclusion, we like to emphasize that the eigenfunction of the ground state outside of the range of the forces (i.e., for $r>a$) is completely determined by the binding energy ϵ of the deuteron, as is shown by Eq. (36). With the observed value of that binding energy, *viz:*

$$\epsilon=2.15 \text{ MV} \tag{44}$$

we find

$$\frac{1}{\alpha}=\frac{\hbar}{(M\epsilon)^{\frac{1}{2}}}=\frac{1.042\cdot 10^{-27}}{(1.66_5\cdot 10^{-24}\cdot 2.15\cdot 1.59\cdot 10^{-6})^{\frac{1}{2}}}=4.36\cdot 10^{-13}\text{ cm.} \tag{44a}$$

$1/\alpha$ may be regarded as the "radius" of the deuteron (cf. the wave function (36)). The range of the forces a is probably about $2\cdot 10^{-13}$ cm, which is considerably smaller than the radius of the deuteron. We can therefore say that the solution (36) represents the eigenfunction of the ground state of the deuteron over the greater part of space. For this reason, matrix elements, normalization integrals, etc., may be calculated to a good approximation by assuming (36) to be valid throughout. Then the normalization integral becomes

$$\int u_0^2 dr=c^2\int_0^\infty e^{-\alpha r}dr=c^2/2\alpha=1$$

or

$$u_0=(2\alpha)^{\frac{1}{2}}e^{-\alpha r}, \tag{44b}$$

$$U(r)=\frac{u_0}{(4\pi)^{\frac{1}{2}}r}=\left(\frac{\alpha}{2\pi}\right)^{\frac{1}{2}}\frac{e^{-\alpha r}}{r} \tag{44c}$$

independent of the form of the potential $J(r)$.

In (41c) we have derived the exact normalization factor for the special case of a rectangular hole potential. It differs from that in (44c) by a factor

$$\left(\frac{V_0-\epsilon}{V_0}\right)^{\frac{1}{2}}\frac{e^{\alpha a}}{(1+\alpha a)^{\frac{1}{2}}}=1+\tfrac{1}{2}\alpha a-O(\alpha a)^2. \tag{44d}$$

(ϵ/V_0 would be of the order $(\alpha a)^2$.) Thus (44c) corresponds to putting the range of the forces equal to zero.

§13. Excited States of the Deuteron

It can easily be shown that no stable excited states of the deuteron exist, if we disregard the spin and make the same assumptions about the forces as in the preceding section, *viz.*, strong forces acting only over a limited range a. For simplicity, we choose the rectangular hole as representing the potential.

The following excited states might *a priori* be expected: p states ($l=1$), d states ($l=2$), etc., or higher s states ($l=0$).

(a) *p states:* In the Majorana theory there would be repulsion between proton and neutron, if the angular momentum of their relative motion is $l=1$ (cf. 32). For Wigner forces, and for Heisenberg forces in the case of antiparallel spins of the two particles, the forces are attractive. Thus in the Majorana theory which we have accepted, stable p states of the deuteron are entirely impossible. But even in the Wigner theory their impossibility can be concluded from the known binding energy of the ground state. We have the wave equation for p states

$$\frac{d^2u_1}{dr^2}-\frac{2}{r^2}u_1=\frac{M}{\hbar^2}\cdot\begin{cases}(-V_0+\epsilon)u_1 & \text{for } r<a,\\ \epsilon u_1 & \text{for } r>a,\end{cases} \tag{45}$$

with the solution

$$u_1=b\left(\frac{\sin kr}{kr}-\cos kr\right) \quad \text{with } k=M^{\frac{1}{2}}(V_0-\epsilon)^{\frac{1}{2}}/\hbar, \quad r<a,$$
$$u_1=ce^{-\alpha(r-a)}(1+1/\alpha r) \quad \text{with } \alpha=(M\epsilon)^{\frac{1}{2}}/\hbar, \quad r>a. \tag{45a}$$

We equate the expressions for u_1'/u_1, obtained from these two expressions, for $r=a$:

$$k\frac{(\cos ka)/ka+\sin ka[1-(ka)^{-2}]}{(\sin ka)/ka-\cos ka}=-\alpha\frac{1+(\alpha a)^{-1}+(\alpha a)^{-2}}{1+(\alpha a)^{-1}}. \tag{45b}$$

Now we certainly find the minimum V_0 necessary to give a bound p state, by putting the binding energy $\epsilon=0$ and therefore $\alpha=0$. When doing so, the right-hand side of

(45b) becomes $-1/a$. Therefore putting $ka=x$, we must have:

$$\cos x+\sin x(x-1/x)=-(\sin x/x-\cos x); \quad (45c)$$

$$ka=x=\pi; \quad V_0a^2=\pi^2\hbar^2/M. \quad (45d)$$

This value is irreconcilable with our previous conclusion (cf. (40)) that V_0a^2 is only slightly larger than $(\pi^2/4)\hbar^2/M$, which followed from the fact that V_0 is large compared to the binding energy of the deuteron. (From the theory of the α-particle, a value V_0a^2 of about $4\hbar^2/M$ can be deduced, §21.) We therefore have to conclude that there is no stable p state of the deuteron even if we assume Wigner forces.

(b) *d states*. For Wigner forces, s, p and d states form a monotonous sequence with increasing "centrifugal force" $l(l+1)/r^2$. Therefore, if p states do not exist for Wigner forces, this is *a fortiori* true for d states. Since the Majorana and Wigner equations are identical for d states ($l=2$), there is no stable d state in the case of Majorana forces either.

(c) *Higher s states*. From (39) it follows that one stable s state may be found for $M^{\frac{1}{2}}(V_0-\epsilon)^{\frac{1}{2}}a/\hbar$ between $\pi/2$ and π, another for $M^{\frac{1}{2}}(V_0-\epsilon)^{\frac{1}{2}}a/\hbar$ between $3\pi/2$ and 2π, etc. If the second bound s state is to exist, we must certainly have

$$(MV_0)^{\frac{1}{2}}a/\hbar>3\pi/2; \quad V_0a^2>(9\pi^2/4)\hbar^2/M, \quad (45e)$$

which is again impossible for the same reason as excited p states.

Therefore *no excited states of the deuteron exist which differ from the ground state with respect to orbital motion.*

However, we should expect to find a second state of the deuteron which differs from the ground state with respect to the total spin. In the ground state the spin of the deuteron is 1 unit, i.e., the spins of proton and neutron are parallel. We should expect another state with antiparallel spins of the two particles and therefore total spin $s=0$. This second state would be a singlet state in spectroscopic nomenclature (nondegenerate state, statistical weight 1), while the ground state is a triplet state (triply degenerate because of three possible orientations of the deuteron spin in an external magnetic field).

If we assume the pure Majorana interaction, the energies of singlet and triplet state are, in first approximation, equal. Their difference arises only from the magnetic interaction between the two spins. We assume the classical interaction between the magnetic moments

$$W=\mu_n\mu_p[(\boldsymbol{\sigma}_n\boldsymbol{\sigma}_p)r^2-3(\boldsymbol{\sigma}_n\mathbf{r})(\boldsymbol{\sigma}_p\mathbf{r})]r^{-5}, \quad (46)$$

where μ_n and μ_p are the magnetic moments of neutron and proton, $\boldsymbol{\sigma}_n$ and $\boldsymbol{\sigma}_p$ the respective spin operators and $\mathbf{r}$ the relative coordinate of the proton with respect to the neutron. We may write explicitly

$$W=\mu_n\mu_p[\sigma_{nz}\sigma_{pz}r^{-3}(1-3\cos^2\theta)+\cdots]. \quad (46a)$$

The diagonal matrix element of W with respect to an s state vanishes because of the dependence of W upon the angular coordinates θ, φ of the point $\mathbf{r}$. The splitting δE of an s state due to the magnetic spin interaction is therefore a second-order effect. By the ordinary Schrödinger perturbation theory it can be shown that the order of magnitude of the splitting is about

$$\delta E\approx W^2(a)/V_0 \quad (46c)$$

where $W(a)$ is some average value of the magnetic interaction if the particles are a distance a apart, *viz.*,

$$W(a)\approx\mu_n\mu_pa^{-3}=2.9\cdot 2.0(e\hbar/2Mc)^2a^{-3}=1.5(\hbar/Mca)^2e^2/a.$$

With $\hbar/Mc=0.21\cdot 10^{-13}$, $a=10^{-13}$ and $e^2/mc^2=2.80\cdot 10^{-13}$ (m = electron mass), we have

$$W(a)=1.5mc^2\cdot(0.21)^2\cdot 2.80=0.10 \text{ MV}, \quad (46d)$$

$$V_0=\pi^2\hbar^2/4Ma^2=100 \text{ MV},$$

$$\delta E=100 \text{ volts}. \quad (46e)$$

Actually, the range of the forces is rather larger than $1\cdot 10^{-13}$ cm which would make δE even smaller.

We should therefore expect an energy difference of the order of 100 volts between the singlet and triplet state of a deuteron, if we assume it to be due only to magnetic interaction between the spins, and if we use the classical formula for this interaction. Experimentally, there is strong reason to believe that the singlet state lies about 2 million volts higher than the triplet state (§14). This can apparently not be explained by magnetic interaction. We may, however, assume that the nuclear forces themselves depend to a certain extent upon the relative spin directions of the interacting proton and neutron. In other words, we assume that *small Heisenberg forces exist*, after all, *besides the principal forces of the Majorana type*.

If $J(r)$ is the "Majorana" potential and $K(r)$ the "Heisenberg" potential, we have then, according to (30)

$$(\hbar^2/M)\Delta U(\mathbf{r})+EU(\mathbf{r})=(J(r)\pm K(r))U(-\mathbf{r}), \quad (47)$$

where the upper sign holds for a triplet, the lower for a singlet state. To explain that the triplet state of the deuteron lies lower than the singlet state, we have to assume that K is negative as well as J. The magnitude of K may be determined by assuming that K is represented by a rectangular hole of the same width as J, such that

$$K(r) = -V_1 \quad \text{for} \quad r<a,$$
$$K(r) = 0 \quad \text{for} \quad r>a. \quad (47a)$$

Then we have, analogous to (39)

$$\cot\left[M^{\frac{1}{2}}(V_0+V_1-\epsilon_t)^{\frac{1}{2}}a/\hbar\right] = -(\epsilon_t/V_0-V_1-\epsilon_t)^{\frac{1}{2}} \quad \text{(triplet state)}, \quad (47b)$$
$$\cot\left[M^{\frac{1}{2}}(V_0-V_1-\epsilon_s)^{\frac{1}{2}}a/\hbar\right] = -(\epsilon_s/V_0+V_1-\epsilon_s)^{\frac{1}{2}} \quad \text{(singlet state)}.$$

To deduce the numerical value of V_1, we use the result of the scattering experiments (§14) that ϵ_s is very nearly zero, *viz.*, $\epsilon_s \approx 40{,}000$ volts. Therefore we may neglect ϵ_s entirely compared to ϵ_t (2.15 MV). Then we have, in analogy to (39b)

$$V_0-V_1 = (\pi^2/4)(\hbar^2/Ma^2),$$
$$V_0+V_1 = (\pi^2/4)\hbar^2/Ma^2 + 2\epsilon_t^{\frac{1}{2}}(\hbar/Ma^2)^{\frac{1}{2}} \quad (47c)$$

or
$$V_1 = (\epsilon_t\hbar/Ma^2)^{\frac{1}{2}} = (2/\pi)(V_0\epsilon)^{\frac{1}{2}}. \quad (48)$$

Since the most probable value for V_0 is about 30 MV (§21) and $\epsilon = 2.15$ MV, we have

$$V_1 \approx 5 \text{ MV}. \quad (48a)$$

The magnitude of the Heisenberg force is of the order of the geometric mean between the Majorana potential and the deuteron binding energy.

§14. Scattering of Neutrons by Protons. I: Cross Section (W13, B18, M8, T1)

Closely related to the deuteron problem is the scattering of neutrons by protons. Here again we have just two interacting particles, one proton and one neutron, the only difference being that the system has *positive* energy E. Since all our calculations refer to the relative motion of the two particles, E is the kinetic energy of the two particles in a coordinate system in which the center of gravity of the particles is at rest (C system). If v is the relative velocity of the particles, $\frac{1}{2}v$ will be the velocity of each particle in the C system, and therefore

$$E = \tfrac{1}{4}Mv^2. \quad (49)$$

In general, the experimental arrangement will be such that neutrons of a given velocity v are shot against protons at rest. The kinetic energy of the neutrons in a coordinate system which is at rest (R system) is then

$$E_0 = \tfrac{1}{2}Mv^2 = 2E. \quad (49a)$$

The velocity of the center of gravity in the R system is $\frac{1}{2}v$.

In the scattering process, the neutron may be deflected by an angle θ in the C system. Then its velocity component in its original direction of motion will be $\frac{1}{2}v\cos\theta$ in the C system, or $\frac{1}{2}v(1+\cos\theta)$ in the R system. The velocity of the proton in the C system must be opposite and equal to that of the neutron, it has therefore the components $-\frac{1}{2}v\cos\theta$, $\frac{1}{2}v\sin\theta$ parallel and perpendicular to the direction of motion of the incident neutron, respectively. The velocity of the proton in the R system has therefore the components

$$v_{||}' = \tfrac{1}{2}v(1-\cos\theta) = v\sin^2\tfrac{1}{2}\theta,$$
$$v_{\perp}' = \tfrac{1}{2}v\sin\theta = v\sin\tfrac{1}{2}\theta\cos\tfrac{1}{2}\theta. \quad (49b)$$

The angle between the motion of the proton after collision and that of the neutron before collision is therefore given by

$$\varphi = \tfrac{1}{2}\pi - \tfrac{1}{2}\theta. \quad (49c)$$

If the neutron suffers a small deflection θ, the proton goes off at right angles. A proton emitted in the direction of the incident neutron corresponds to a reversal of the motion of the neutron in the C system ($\theta = 180°$). The energy of the recoil proton is

$$E' = \tfrac{1}{2}Mv'^2 = \tfrac{1}{2}Mv^2\sin^2\tfrac{1}{2}\theta = E_0\cos^2\varphi; \quad (49d)$$

therefore the energy of the neutron after collision

$$E'' = E_0\sin^2\varphi. \quad (49e)$$

In a head-on collision ($\varphi = 0°$, $\theta = 180°$) all the energy is transferred to the proton. In a soft collision ($\varphi \approx 90°$, $\theta \approx 0°$) practically no energy is lost by the neutron.

The wave function of two particles interacting with a central force can always be expanded in a series of spherical harmonics, *viz*:

$$U(\mathbf{r}) = \sum_l c_l[u_l(r)/r]P_l(\cos\theta). \quad (50)$$

We choose the direction of motion of the incident neutron as axis of our polar coordinate system. The wave function representing the scattering of the two particles will obviously have axial symmetry round that direction, therefore (50) contains only the ordinary Legendre polynomials $P_l(\theta)$, not any associated functions $P_{lm}(\theta)e^{im\varphi}$. The radial functions u_l satisfy the equation

$$\frac{\hbar^2}{M}\left(\frac{d^2u_l}{dr^2} - \frac{l(l+1)}{r^2}u_l\right) + (E-(-1)^lJ(r))u_l = 0 \quad (32)$$

if the Majorana theory is accepted.

Asymptotically for large r, we may neglect the terms $l(l+1)/r^2$ and $J(r)$ in (32) so that the solution of (32) is

$$u_l = c \sin (kr - \tfrac{1}{2}l\pi + \delta_l) \quad (51)$$

with $k = (ME)^{\frac{1}{2}}/\hbar = Mv/2\hbar = (\frac{1}{2}ME_0)^{\frac{1}{2}}/\hbar. \quad (52)$

The "phase" δ_l is a constant which has to be determined by integration of (32). If $J(r)=0$, i.e., if no force acts between the two particles, all δ_l's turn out to be zero. The knowledge of the δ_l's is sufficient to determine the scattering cross section for a given angle θ (cf. Mott and Massey, *Atomic Collisions*, p. 24), *viz.*:

$$d\sigma = (\pi/2k^2)\,|\sum_l (2l+1)P_l(\theta)(e^{2i\delta_l}-1)|^2 \times \sin\theta d\theta. \quad (53)$$

The cross section $d\sigma$ is defined as the number of neutrons scattered per unit time through an angle between θ and $\theta+d\theta$, if there is one neutron crossing unit area per unit time in the incident beam.

We know (§9, §12) that the forces between proton and neutron are restricted to a very small range a of the order of $2\cdot 10^{-13}$ cm. This is considerably less than the wave-length of all neutrons which have been used for scattering experiments: The fastest neutrons thus far used for such experiments have an energy of about $E_0=4$ MV. The neutron wave-length λ is given by

$$\frac{\lambda}{2\pi} = \frac{1}{k} = \frac{\hbar}{(\frac{1}{2}ME_0)^{\frac{1}{2}}} = \frac{1.04\cdot 10^{-27}}{(\frac{1}{2}\cdot 1.66\cdot 10^{-24}\cdot 1.59\cdot 10^{-6}E_0^{\mathrm{MV}})^{\frac{1}{2}}} = \frac{9.0_5\cdot 10^{-13}}{(E_0^{\mathrm{MV}})^{\frac{1}{2}}}\,\mathrm{cm} \quad (54)$$

in the system where the center of gravity of neutron and proton is at rest, E_0^{MV} being the kinetic energy of the neutron, expressed in MV. For $E_0=4$ MV, we have

$$\bar{\lambda} = \lambda/2\pi = 4.5\cdot 10^{13}\ \mathrm{cm}, \quad (54a)$$

which is more than twice the range of the forces, q.e.d. (We have purposely calculated $\bar{\lambda}=\lambda/2\pi$ rather than λ itself, because $\bar{\lambda}$ is the quantity which enters directly into the following considerations about the magnitude of the δ_l's.)

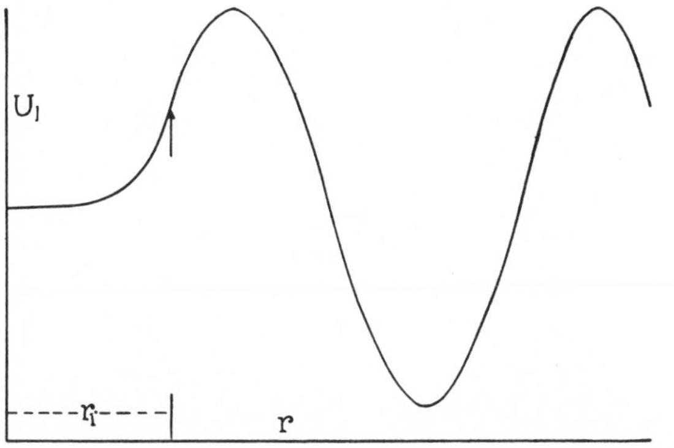

FIG. 5. The wave function of a particle of high angular momentum ($l=4$) in a potential field. The wave function is concave towards the line $U_l=0$ for $r>r_l$, convex and very small for $r<r_l$. (The axis $U_l=0$ has been omitted and should be the continuation of the beginning of the curve.)

From the fact that $a \ll \bar{\lambda}$, one can easily deduce that *all phases δ_l will be small except δ_0.* To see this, we first prove that any wave function u_l becomes small if r is small compared to the "classical impact parameter" $r_l = \bar{\lambda}l = l\hbar/(\frac{1}{2}Mv)$ which is the distance at which particles of the angular momentum $l\hbar$ and the linear momentum $\frac{1}{2}Mv$ would pass each other. ($\frac{1}{2}M$=reduced mass.) In quantum mechanics, r_l marks a change in the behavior of the wave function u_l. For $r>r_l$, u_l has the character of a wave, i.e., it is concave towards the r axis, while u_l is convex (has exponential character) for $r<r_l$ (see Fig. 5). This follows immediately from (32). We neglect $J(r)$ which we can do for $r>a$, and therefore certainly for $r=l\bar{\lambda}$ if $l\neq 0$. Of the remaining terms, the term $-\hbar^2 l(l+1)u/Mr^2$ in (32) will be larger than Eu_l if

$$r^2 < l(l+1)\hbar^2/ME = l(l+1)\bar{\lambda}^2. \quad (54b)$$

This is certainly fulfilled if $r<r_l=l\bar{\lambda}$. Neglecting accordingly Eu_l and Ju_l, we have the differential equation

$$d^2u_l/dr^2 - [l(l+1)/r^2]u_l = 0 \quad \text{for} \quad r<r_l=l\bar{\lambda}, \quad (54c)$$

whose solution is $\quad u_l = cr^{l+1} \quad$ for $\quad r<r_l \quad (54d)$

(c a constant). Thus u_l decreases rapidly with decreasing r when r becomes smaller than $l\bar{\lambda}$. Since the range a of the potential energy $J(r)$ is supposed to be small compared to $\bar{\lambda}$, these considerations apply certainly for $r=a$, if only l is different from zero; i.e., $u_l(l\neq 0)$ will be very small in the region where the force $J(r)$ is acting, which minimizes the effect of the potential $J(r)$ upon the wave function. Since δ_l is a measure of that effect, δ_l must be small. The case $l=0$ is an exception because here the "critical distance" r_l would be zero so that our argument does not apply. (For the quantitative proof of our argument, see §15.)

In the scattering cross section (53), only the term $l=0$ will therefore be important. From this result, which is based solely on the fact that the forces between proton and neutron have a very small range a, smaller than the neutron wave-length λ, two important conclusions can be drawn.

(1) The scattering cross section will be independent of the angle θ, in other words, *the scattering will be spherically symmetrical* in the C system, i.e., *in the system in which the center of gravity of neutron and proton is at rest.*

(2) The total scattering cross section can be obtained by calculating only the phase δ_0. We have, putting all δ_l's in (53) except δ_0 equal to zero,

$$d\sigma = 2\pi k^{-2} \sin^2 \delta_0 \sin \theta d\theta \quad (55)$$

or for the total cross section

$$\sigma = \int d\sigma = 4\pi k^{-2} \sin^2 \delta_0. \quad (55a)$$

We defer the discussion of the angular distribution to §15, and begin with the calculation of δ_0, starting from our treatment of the ground state of the deuteron. We know that for the ground state, i.e., for the energy $-\epsilon$, the slope of the eigenfunction $l=0$ is given by (34), *viz.*:

$$[(1/u_0)(du_0/dr)]_{r=a} = -\alpha \quad \text{for} \quad E = -\epsilon. \quad (34)$$

We show first that this relation still holds approximately for not too large *positive* values of E, of the order ϵ, whatever the forces between neutron and proton may be. To show this, we write down the wave equation for the wave functions u_0^E and $u_0^{-\epsilon}$:

$$d^2u_0^E/dr^2 + (M/\hbar^2)(E - J(r))u_0^E = 0, \quad (55b)$$
$$d^2u_0^{-\epsilon}/dr^2 + (M/\hbar^2)(-\epsilon - J(r))u_0^{-\epsilon} = 0.$$

Multiplying the first equation by $u_0^{-\epsilon}$, the second by u_0^E, and subtracting, we have

$$u_0^{-\epsilon}\frac{d^2u_0^E}{dr^2} - u_0^E\frac{d^2u_0^{-\epsilon}}{dr^2} = -\frac{M}{\hbar^2}(E+\epsilon)u_0^E u_0^{-\epsilon} \quad (55c)$$

or integrated from 0 to a:

$$\left| u_0^{-\epsilon}\frac{du_0^E}{dr} - u_0^E\frac{du_0^{-\epsilon}}{dr} \right|_0^{r=a} = -\frac{M}{\hbar^2}(E+\epsilon)\int_0^a u_0^E u_0^{-\epsilon} dr. \quad (55d)$$

The expression on the left-hand side vanishes for $r=0$, because $u_0^{-\epsilon}$ as well as u_0^E vanish as r itself for small r [cf. the remark after Eq. (33)]. Dividing by $u_0^{-\epsilon}(a)u_0^E(a)$, we obtain therefore

$$\left(\frac{1}{u_0^E}\frac{du_0^E}{dr}\right)_{r=a} - \left(\frac{1}{u_0^{-\epsilon}}\frac{du_0^{-\epsilon}}{dr}\right)_{r=a} = -\frac{M}{\hbar^2}\frac{E+\epsilon}{u_0^E(a)u_0^{-\epsilon}(a)}\int_0^a u_0^E u_0^{-\epsilon} dr. \quad (56)$$

The integral on the right-hand side is of the order $a \cdot u_0^E(a)u_0^{-\epsilon}(a)$. It actually is somewhat smaller because $u_0^{-\epsilon}$ and u_0^E vanish at $r=0$ and their value at $r=a$ is practically equal to their maximum value (cf. the explicit wave function 38a, for $r<a$). Therefore we put

$$\int_0^a u_0^E u_0^{-\epsilon} dr = \gamma a u_0^E(a) u_0^{-\epsilon}(a). \quad (56a)$$

For the special case of a rectangular potential hole, $\gamma = \frac{1}{2}$, see below. Inserting the value (34) for $du_0^{-\epsilon}/u_0^{-\epsilon}dr$, we have then

$$\left(\frac{1}{u_0^E}\frac{du_0^E}{dr}\right)_{r=a} = -\alpha - \frac{\gamma Ma}{\hbar^2}(E+\epsilon). \quad (57)$$

The right-hand side of (57) reaches the value -2α when E has the value

$$E' \approx \frac{\hbar^2\alpha}{Ma\gamma} = \frac{1}{\gamma}\left(\frac{\hbar^2}{Ma^2}\epsilon\right)^{\frac{1}{2}} \quad (58)$$

remembering (37). In the case of a rectangular potential hole we obtain, using (40) and $\gamma = \frac{1}{2}$:

$$E' = (4/\pi)(V_0\epsilon)^{\frac{1}{2}}. \quad (58a)$$

If we assume $V_0 = 30$ MV (§21) and take the observed value $\epsilon = 2.15$ MV, we find $E' = 10$ MV. For energies $E \ll E'$ the value of $(1/u_0)(du_0/dr)$ will be approximately equal to $-\alpha$.

Since E is only one-half of the kinetic energy E_0 of the neutron, we may put

$$(1/u_0)(du_0/dr) = -\alpha \quad (57a)$$

for all neutron energies E_0 small compared to $2E' = 20$ MV, i.e., for all neutron energies thus far available.

Assuming now (57a) to hold, we can easily calculate δ_0. For $r>a$, $J(r)$ is zero so that

$$d^2u_0/dr^2 = -(ME/\hbar^2)u_0 = -k^2u_0$$

(cf. (52)). Therefore

$$u_0 = c \sin (kr + \delta_0), \quad (51_0)$$

where c is a constant. Joining this solution to the solution for $r<a$, we have

$$[(1/u_0)(du_0/dr)]_{r=a} = k \cot (ka + \delta_0) = -\alpha, \quad (59)$$

which yields

$$\delta_0 = \tfrac{1}{2}\pi + \arctan (\alpha/k) - ka. \quad (59a)$$

Now we have assumed throughout this section that the range of the forces a is small compared to $\lambda = 1/k$. Therefore we neglect ka in (59a) and have

$$\cot \delta_0 = -\alpha/k. \quad (59b)$$

Inserting this into (55a), we obtain for the total cross section

$$\sigma = 4\pi/(\alpha^2 + k^2) \quad (60)$$

or, inserting the values of α and k from (37) and (52):

$$\sigma = 4\pi\hbar^2/M(\epsilon+\tfrac{1}{2}E_0), \qquad (61)$$

or, with $\epsilon = 2.15$ MV:

$$\sigma = 2.39\cdot 10^{-24}\epsilon/(\epsilon+\tfrac{1}{2}E_0)\ \text{cm}^2. \qquad (61a)$$

This value for the cross section is in fair agreement with experimental determinations for fast neutrons, considering the difficulty of the experiment, which is chiefly that of obtaining a beam of neutrons of well-defined energy. For $E_0 = 4.3$ MV Chadwick (C6) found a cross section between 0.5 and $0.8\cdot 10^{-24}$ cm^2 compared to a theoretical value of $1.2\cdot 10^{-24}$; for $E_0 = 2.1$ MV the agreement is better, *viz.*, 1.1 to $1.5\cdot 10^{-24}$ experimental and $1.6\cdot 10^{-24}$ theoretical.[31]

However, for slow neutrons our formula fails completely. The experimental cross section, observed by Dunning, Pegram, Fink and Mitchell* (D4) is about $35\cdot 10^{-24}$ cm^2, i.e., more than 14 times as large as our theoretical cross section (61a) for $E_0 \ll \epsilon$. Since the theoretical value depends only upon the binding energy ϵ and not upon any details about the force between neutron and proton, this discrepancy looks at first sight very serious. This is the more true because our assumption that the range of the forces is small compared to the wave-length is much better fulfilled for slow neutrons than for fast ones.

To solve this difficulty, it has been pointed out by Wigner (private communication) that the observed binding energy of the deuteron refers only to the binding of a proton and a neutron with parallel spins while nothing can be deduced from it about the interaction of protons and neutrons with opposite spins. The binding energy ϵ' of a deuteron in a singlet state, i.e., with the spins of the two constituent particles antiparallel, may well be assumed to be very small. This assumption is sufficient to make the probability that a slow neutron is scattered by a proton with opposite spin extremely large, according to (61).

Accepting this explanation, we may use the experimental value of the cross section to determine the binding energy ϵ' of the deuteron in the singlet state. The cross section for the scattering of a neutron by a proton with spin parallel to that of the neutron, will be given by (61); the cross section for antiparallel spins will have the same form, only with ϵ' instead of ϵ. Now it is just 3 times as probable that the spins of a given neutron and a given proton are parallel, as that they are antiparallel.[32] Therefore the average cross section for the scattering of neutrons by protons, averaged over the possible directions of spin, is

$$\sigma^a = \frac{4\pi\hbar^2}{M}\left(\frac{1}{4}\frac{1}{\epsilon'+\frac{1}{2}E_0} + \frac{3}{4}\frac{1}{\epsilon+\frac{1}{2}E_0}\right). \qquad (62)$$

From the observed cross section for $E_0 = 0$, *viz.*, $35\cdot 10^{-24}$ cm^2, and the known value of $\epsilon = 2.15$ MV, we can deduce

$$\epsilon' = 40{,}000 \text{ ev}. \qquad (62a)$$

The binding energy of the singlet state of the deuteron must therefore be supposed to be very small, compared to that of the ground (triplet) state.

Fig. 6 shows the cross section (62) as a function of the neutron energy E_0. At high energies, the difference between (62) and (61) is hardly noticeable because the binding energies ϵ and ϵ' can be neglected compared to E_0. This explains why formula (61) was found to agree with

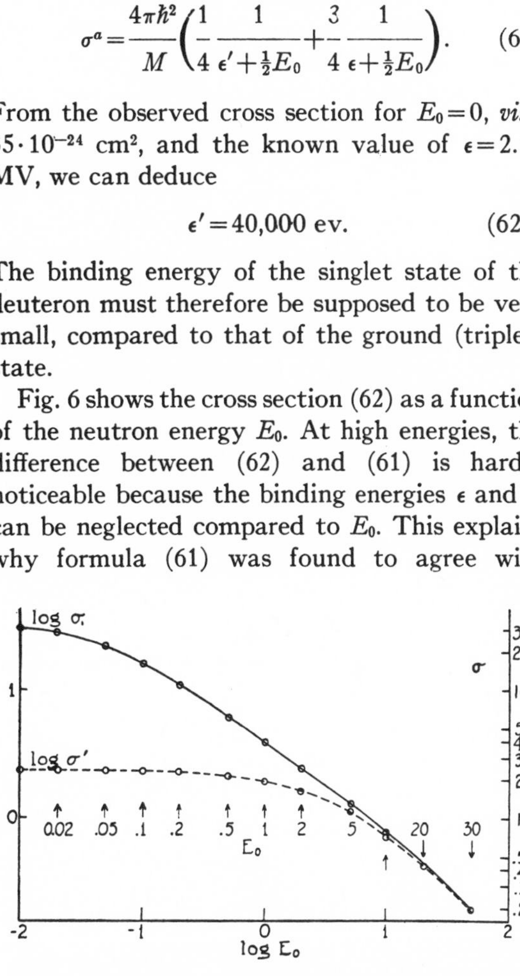

FIG. 6. Scattering of neutrons by protons. Abscissa: energy in MV, ordinate: cross section; both on a logarithmic scale. Solid curve: actual theoretical cross section. Dotted curve: cross section, if interaction is independent of the relative spin direction of the particles.

* *Note added in proof:* Recent experiments of Amaldi and Fermi give an even larger cross section, *viz.* about $80\cdot 10^{-24}$ cm^2. The reason for this discrepancy is not clear.

[31] Dunning, Pegram, Fink and Mitchell give $1.68\cdot 10^{-24}$ cm^2 as the cross section for "fast" neutrons from Be$+\alpha$ $= C^{12}+n$ (mixed velocities).

[32] If the spins are parallel, the total spin of the system (proton$+$neutron) is 1 unit. This total spin can orient itself in three different ways with respect to a given direction, e.g., the direction of an external field, the three orientations corresponding to a spin component $+1$, 0, or -1 in the given direction. There are therefore three possible spin states if the two spins are parallel. In the case of antiparallel spins, the total spin is 0, and only one quantum state exists.

experiments about fast neutrons. At about $E_0=2$ MV, the actual cross section (62) will begin to become considerably larger than the cross section (61) which we would expect if singlet and triplet state had the same binding energy. The cross section then increases rapidly with decreasing neutron energy, reaching half the "slow neutron value" for E_0 about 100,000 volts.

The bearing of the large difference between the binding energies of singlet and triplet state of the deuteron, upon the problem of nuclear forces, has been discussed in §13. At this place, we want only to point out that it cannot be inferred from the scattering cross section whether there actually *is* a *stable* singlet state of the deuteron. For the formula (60) for the cross section contains only α^2 or rather the corresponding quantity for the singlet state which we may call β^2. We can therefore only infer the magnitude of β from the scattering but not its sign. Only for positive β will there be a stable singlet state, for negative β the singlet state would be just unstable. Which of the two alternatives is true, does not make much difference for the conclusions about the Heisenberg forces drawn in §13, because there we needed only to use the fact that $|\beta|\ll\alpha$. There will, however, be some difference in the probability for the capture of slow neutrons by protons (§17) and that effect may lead to a decision whether the singlet state is just stable or just unstable.

Another means to decide the sign of the energy of the singlet level, and at the same time to test the whole theory, has been pointed out by Teller (T1). We have assumed that the intensity of the scattering of neutrons by protons depends strongly on their relative spin orientations. This makes the scattering of slow neutrons by parahydrogen molecules quite different from that by orthohydrogen. In the latter case, both protons in the molecule have their spins *parallel* to each other: Therefore a neutron will either be scattered strongly by both the protons, or weakly by both of them; in either case, the scattered intensity will be the same for both protons. Consequently, there will be strong interference effects, if the wave-length of the neutrons is of the same order as the distance of the protons in the H_2 molecule, which is actually the case for neutrons of about thermal energy. The scattered neutrons will have an angular distribution identical to that of x-rays of the same wave-length scattered by a diatomic molecule.

The scattering by parahydrogen will be quite different: There we have in each molecule one proton whose spin is parallel to that of the incident neutron while the spin of the other proton is opposite. In first approximation, only the *second* proton will scatter so that we get no interference effects. The presence of interference in the scattering of neutrons by orthohydrogen and the absence in the scattering by parahydrogen would be a direct test of our assumption about the dependence of the scattering on the spin orientation.

Now we consider the case of parahydrogen in second approximation. There will be *some* scattering from the proton having its spin *parallel* to the neutron. The ratio of the amplitudes of the waves scattered by the "parallel" and by the "antiparallel" proton is for small energy of the neutron:

$$\alpha/\beta=(\epsilon'/\epsilon)^{\frac{1}{2}}=1/7, \quad \text{approximately.} \qquad (62\text{b})$$

Now if the singlet state is a real bound state, the phases of the waves scattered by the two protons will be the same, whereas they will be opposite if the singlet state is a virtual state. In the first case, the amplitude of the neutron wave scattered by a parahydrogen molecule in the *forward* direction (i.e., through a very small angle) will be 8/7 times that scattered by an isolated proton having its spin opposite to the neutron; in the second case, the amplitude will be only 6/7. The ratio of the scattered intensities in the two cases will therefore be

$$\frac{\sigma_{\text{real singlet state}}}{\sigma_{\text{virtual state}}}=\left(\frac{8}{6}\right)^2\approx 1.8. \qquad (62\text{c})$$

On the other hand, if we could observe at such an angle that the difference in path between the neutrons scattered by the two protons of the molecule, is equal to half a wave-length, the ratio of the scattering intensities would be reversed. Actually, such an angle cannot be found because the molecules rotate in space. The result is that the scattering through large angles should be almost equal in the two cases (real and virtual singlet state).

We must now discuss the accuracy of our formulae. We have made two neglections which both amount to assuming the range a of the forces to be zero. On one hand, we have neglected ka in (59a), on the other hand, $\gamma Ma(E+\epsilon)/\hbar^2\alpha$ in (57a). It seems in order, to calculate the necessary corrections to our formulae by using the simple rectangular hole potential. We have, then, the following expressions for the wave function

$$\psi = b \sin \kappa r \quad \text{for } r<a, \qquad (63)$$
$$\psi = c \sin (kr+\delta_0) \quad \text{for } r>a$$

with $\kappa = M^{\frac{1}{2}}(V_0+E)^{\frac{1}{2}}/\hbar, \quad k=(ME)^{\frac{1}{2}}/\hbar.$ (63a)

Joining at $r=a$ yields

$$\cot (ka+\delta_0) = (\kappa/k) \cot \kappa a. \qquad (63b)$$

We now use the fact that for the ground state of the deuteron (cf. (39) (37))

$$\kappa_0 \cot \kappa_0 a = -\alpha \qquad (63c)$$

with $\kappa_0 = M^{\frac{1}{2}}(V_0-\epsilon)^{\frac{1}{2}}/\hbar, \quad \alpha=(M\epsilon)^{\frac{1}{2}}/\hbar.$ (63d)

Expanding in powers of k, we have

$$\kappa = \kappa_0 + (\alpha^2+k^2)/2\kappa_0,$$
$$\kappa \cot \kappa a = \kappa_0 \cot \kappa_0 a + \tfrac{1}{2}(\alpha^2+k^2)\kappa_0^{-1} \times (\cot \kappa_0 a - \kappa_0 a \sin^{-2} \kappa_0 a). \qquad (63e)$$

Now $\cot \kappa_0 a$ is very small compared to unity, and can therefore be neglected in the second term of (63e), while $\sin^2 \kappa_0 a$ is practically 1. Thus:

$$\kappa \cot \kappa a = -\alpha - \tfrac{1}{2}(\alpha^2+k^2)a. \qquad (63f)$$

By using the definition of α and k, this is identical with (57), if γ in that equation is put equal to $\frac{1}{2}$. Inserting (63f) into (63b), and expanding the left-hand side, we have

$$\cot \delta_0 - ka(\sin \delta_0)^{-2} = -\alpha/k - (\alpha^2+k^2)a/2k. \qquad (63g)$$

In the second term on the left, we may insert for $\cot \delta_0$ the approximate value $-\alpha/k$. Then we find

$$\cot \delta_0 = -\frac{\alpha}{k} + ka\left(1+\frac{\alpha^2}{k^2}\right) - \frac{(\alpha^2+k^2)a}{2k}$$
$$= -\frac{\alpha}{k} + \frac{(\alpha^2+k^2)a}{2k} \qquad (63h)$$

and for the cross section

$$\sigma = \frac{4\pi}{k^2}\sin^2 \delta_0 = \frac{4\pi}{k^2(1+\cot^2 \delta_0)}$$
$$= \frac{4\pi}{k^2[1+(\alpha/k)^2 - \alpha a(\alpha^2+k^2)k^{-2}]} \approx \frac{4\pi(1+\alpha a)}{\alpha^2+k^2}, \qquad (63i)$$

neglecting higher powers of αa than the first. Thus we see that the cross section is simply multiplied by a constant factor $1+\alpha a$. In accord with the theory of H^3 and He^4 (§21) we assume the range a to be about $2\cdot 10^{-13}$ cm. With the experimental value $1/\alpha = 4.36\cdot 10^{-13}$ cm, we obtain then $1+\alpha a = 3/2$ for the scattering of neutrons by protons with parallel spin. For opposite spin, the correction factor would be only $1+\beta a$, where $\beta^2 = M\epsilon'/\hbar^2$. With $\epsilon' = 40{,}000$ volts, this gives

$$1/\beta = 32\cdot 10^{-13} \text{ cm} \qquad (63k)$$

and $1+\beta a = 1.06$. The correction is thus small for the larger part of the cross section.

For extremely high energies, the expansion in powers of E/V_0 is, of course, no longer legitimate. If $E \gg V_0$, we can expand in powers of V_0/E instead. With

$$\gamma = (MV_0)^{\frac{1}{2}}/\hbar \qquad (64a)$$

we have then

$$\kappa = k + \gamma^2/2k;$$
$$(\kappa/k) \cot \kappa a = \cot ka + \tfrac{1}{2}(\gamma/k)^2(\cot ka - ka \sin^{-2} ka). \qquad (64b)$$

Therefore, from (63b):

$$\delta_0 = -\tfrac{1}{2}\sin^2 ka(\gamma/k)^2(\cot ka - ka \sin^{-2} ka)$$
$$= (\gamma/2k)\gamma a(1-\sin 2ka/2ka). \qquad (64c)$$

The last term may be neglected for large ka, and γa be put equal to $\pi/2$, according to (40). Thus we find

$$\delta_0 = (\pi/4)(\gamma/k) = (\pi/4)(V_0/E)^{\frac{1}{2}}; \qquad (64d)$$

therefore $\sigma_0 = \frac{\pi^3}{4}\frac{\gamma^2}{k^4} = \frac{\pi^3}{4}\frac{\hbar^2}{M}\frac{V_0}{E^2}.$ (64)

However, in this case also the contribution of the partial waves with nonvanishing angular momentum contribute appreciably to the scattering (§15, end).

§15. Scattering of Neutrons by Protons. II: Angular Distribution (W12, W9, B18)

We have already shown in the preceding section that the angular distribution of the neutrons scattered by protons should be practically spherically symmetrical in a coordinate system in which the center of gravity of the two particles is at rest. We shall now discuss the deviations from spherical symmetry. To do this, we have to calculate the phases δ_l in the scattering formula (53), for $l \neq 0$.

Since we know already these phases to be small, we may calculate them by a perturbation method. We do this for the particular case $l=1$. Let v_1 denote the wave function for $l=1$ in the case of vanishing potential energy $J(r)$, i.e., the solution of the equation

$$\frac{\hbar^2}{M}\left(\frac{d^2v_1}{dr^2} - \frac{2}{r^2}v_1\right) + Ev_1 = 0. \qquad (65)$$

Multiplying (65) by u_1, i.e., the wave function for nonvanishing J, and the Eq. (32) for $l=1$ by v_1 and subtracting, we have

$$v_1\frac{d^2u_1}{dr^2} - u_1\frac{d^2v_1}{dr^2} = -\frac{M}{\hbar^2}J(r)u_1v_1. \qquad (65a)$$

We integrate from zero to a very large radius R, remembering that at the lower limit $r=0$ both the functions u_1 and v_1 vanish (cf. (55d)), and divide by $u_1(R)v_1(R)$:

$$\left(\frac{1}{u_1}\frac{du_1}{dr}\right)_{r=R} - \left(\frac{1}{v_1}\frac{dv_1}{dr}\right)_{r=R} = -\frac{M}{\hbar^2}\frac{\int_0^R J(r)u_1v_1 dr}{u_1(R)v_1(R)}. \qquad (65b)$$

Now at large distances u_1 as well as v_1 behave like sine waves; we put (cf. 51)

$$v_1=\sin\,(kr-\tfrac{1}{2}\pi)=-\cos kr,\quad u_1=-\cos\,(kr+\delta). \tag{65c}$$

The amplitudes have been assumed unity in both cases, which can always be achieved by suitable normalization of u_1 and v_1. We apply (65b) to a point where v_1 has one of its maxima, i.e., kR is a multiple of 2π. Then

$$[(1/u_1)(du_1/dr)]_{r=R}=-k\tan\,(kR+\delta)=-k\tan\delta;\quad [(1/v_1)(dv_1/dr)]_{r=R}=0;\quad u_1v_1\approx 1. \tag{65d}$$

Then from (65b)

$$\tan\delta=M\hbar^{-2}k^{-1}\int_0^R J(r)u_1v_1dr. \tag{66}$$

The potential J is only large for $r<a$. In that region we know, however, the wave functions u_1 and v_1 to be small [cf. the discussion following Eq. (54)]. This shows that δ must be small. A quantitative estimate may be obtained by putting u_1 in the integral equal to v_1. The latter function, i.e., the solution of the wave equation (65) for a free particle with unit angular momentum, is well known. It is:

$$v_1=-\cos kr+\sin kr/kr. \tag{66a}$$

For $kr\ll 1$ we find by expansion:

$$v_1=\tfrac{1}{3}(kr)^2. \tag{66b}$$

Inserting this into (66), we have

$$\delta_1=(1/9)M\hbar^{-2}k^3\int_0^a J(r)r^4dr. \tag{66e}$$

The integral can be estimated with the help of Eq. (37c) which is based upon the fact that a stable state of the deuteron with comparatively low binding energy exists. We may estimate

$$\int_0^a J(r)r^4dr=\tfrac{1}{2}\mu a^3\int_0^a J(r)rdr=-\tfrac{1}{2}\mu a^3\hbar^2/M, \tag{66d}$$

where μ is a constant of the order of magnitude unity. For the particular case of a rectangular hole potential, μ may be determined by explicit solution of the wave equation, it turns out to be

$$\mu=6\left(1+\frac{12}{\pi^2}\right)-\frac{36}{\pi}\frac{1+e^{-\pi}}{1-e^{-\pi}}=0.89. \tag{66c}$$

Inserting (66d) into (66c) we find

$$\delta_1=-(1/18)\mu(ka)^3. \tag{67}$$

δ_1 is therefore very small compared to unity, as long as the wave-length $1/k$ of the neutron is large compared to the range of the forces a, i.e., as long as the neutron energy is small compared to the depth V_0 of the potential hole (cf. (40), (52)). This is true for any neutrons thus far available since V_0 is of the order 30 MV. The higher δ_l's (for $l>1$) are, of course, even smaller than δ_1. A calculation similar to the above yields

$$\delta_l\approx\frac{(-1)^l}{1^2\cdot 3^2\cdot\;\cdots(2l+1)^2}\mu_l(ka)^{2l+1} \tag{67a}$$

with μ_l a constant rather smaller than unity.

In calculating the angular distribution of scattered neutrons for available velocities, we may therefore neglect the contributions of all l's larger than one, and also neglect δ_1^2. Then we find from (53)

$$\begin{aligned}d\sigma&=(\pi/2k^2)\,|\cos 2\delta_0-1+i\sin 2\delta_0+6i\delta_1\cos\theta|^2\sin\theta d\theta\\&=(\pi/2k^2)(4\sin^2\delta_0+12\delta_1\sin 2\delta_0\cos\theta)\sin\theta d\theta,\end{aligned} \tag{68}$$

$$d\sigma=(2\pi/k^2)\sin^2\delta_0(1+6\delta_1\cot\delta_0\cos\theta)\sin\theta d\theta.$$

The parenthesis determines the deviation of the scattering from spherical symmetry. According to (67), δ_1 is negative. Thus we shall find the scattering backwards ($\theta>90°$) greater than the scattering forwards ($\theta<90°$) provided $\cot\delta_0$ is positive. If $\cot\delta_0$ is negative, the reverse will be the case. Now $\cot\delta_0$ has been calculated in (63h). It is obviously negative for small neutron energies, and positive for high energies. Cot δ_0 vanishes when

$$\alpha^2+k^2=2\alpha/a, \tag{69}$$

or, introducing energies according to (37), (40), (52)[33]

$$\tfrac{1}{2}E_s{}^0+\epsilon=2(\epsilon\hbar^2/Ma^2)^{\frac{1}{2}}=(4/\pi)(\epsilon V_0)^{\frac{1}{2}}. \tag{69a}$$

If neutron and proton have parallel spin, we have to insert $\epsilon=2.15$ MV, and V_0 about 30 MV, so that

$$E_s{}^0=20\text{ MV}. \tag{69b}$$

For opposite spins, the asymmetry is negligible. Neutrons of energy less than about 20 MV have, therefore, to be considered as "slow" as regards the sign of the deviation from spherically symmetrical scattering.

To calculate the deviation from spherical symmetry explicitly, we have to add the cross sections for parallel and antiparallel spin. From (60), (62), (68) and (63h) we have

[33] We introduce the actual kinetic energy E_0 of the neutron, rather than the kinetic energy in the system where the center of gravity of neutron and proton is at rest.

$$d\sigma = \frac{3}{4}\frac{2\pi}{k^2+\alpha^2}\left(1-\frac{6\delta_1\cos\theta}{k}\times[\alpha-\tfrac{1}{2}a(\alpha^2+k^2)]\right)\sin\theta d\theta + \frac{1}{4}\frac{2\pi}{k^2+\beta^2}\left(1-\frac{6\delta_1\cos\theta}{k}\times[\beta-\tfrac{1}{2}a(\beta^2+k^2)]\right)\sin\theta d\theta. \quad (70)$$

Inserting δ_1 from (67), we find that the asymmetry can only be appreciable for comparatively high energies. We therefore neglect in (70) terms of the relative order $(\alpha/k)^4$. Neglecting also β compared to α_1, we find that the relative asymmetry

$$A=[\sigma(\theta=0)-\sigma(\theta=\pi)]/\sigma(\theta=\tfrac{1}{2}\pi) \quad (70a)$$

has a maximum for

$$k_0^2=(3\alpha/4a)(1-\tfrac{1}{2}a\alpha) \quad (70b)$$

corresponding to an energy of about 3 MV. For this energy, the asymmetry becomes

$$A_{\max}=\frac{9}{48}\mu(a\alpha)^2\left(1-\frac{5}{3}a\alpha\right)=0.85 \text{ percent} \quad (71)$$

if we put $a=2\cdot10^{-13}$ cm and $1/\alpha=4.4\cdot10^{-13}$ cm. The asymmetry of scattering should therefore, even at the maximum, not exceed the minute amount of one percent. Such a small asymmetry is, with the present methods, quite unobservable.

An appreciable asymmetry in the scattering should only be found for neutron energies higher than 20 MV (cf. (69b)), which are at present unavailable. These high energy neutrons should preferentially be scattered backwards by protons (cf. Eq. (68)), quite in contrast to other scattering processes. This unusual behavior is due to the exchange type of the forces between neutron and proton. The scattering process can be interpreted by saying that the incident particle actually is only deflected by a small angle but has, in the process of scattering, changed roles with the scattering particle so that it goes off as a proton if it was a neutron before the collision.

It was first pointed out by Wick that by observing the asymmetry of the scattering of neutrons by protons, one could decide whether the forces between the particles are of the "ordinary" or the "exchange" type; e.g., at high energies one would expect to find the neutrons to be scattered preferentially in the forward direction, if ordinary forces act, while we have seen that they are scattered preferentially backwards by exchange forces. At low energies, the reverse would be true, because of the negative sign of cot δ_0 in (68). Unfortunately, the asymmetry of the scattering should be much too small for the available neutron energies to allow any decision about the forces.

Our main result is thus that the scattering of neutrons by protons should, theoretically, be spherically symmetrical in the coordinate system, in which the center of gravity of neutron and proton is at rest. In the "ordinary" coordinate system, in which the proton is initially at rest, the distribution should be

$$d\sigma=\text{const}\cdot\sin\theta d\theta=2\ \text{const}\cdot\sin\varphi\cos\varphi d\varphi, \quad (72)$$

i.e., the number of scattered protons per unit solid angle $\sin\varphi d\varphi$ should have a flat maximum in the forward direction $\varphi=0$. The experimental results are highly contradictory. While Chadwick (C6), Monod-Herzen (M15), Kurie (K15) and Barton, Mueller and Lampson (unpublished[34] preliminary results in photographic emulsion) found spherical symmetry within the limits of their experimental error in accord with theory, Harkins, Gans, Kamen and Newson (H3) found many more protons at small angles φ with respect to the incident neutron, than at large angles. This would mean that large deflections θ of the neutrons are much more probable than small ones. This is quite irreconcilable with our considerations. Thus far, the possible experimental error is still very large. Should, however, the deviation from spherical symmetry as observed by Harkins and others, be confirmed by more extensive experiments, it would make our assumption about the short range of the forces between neutron and proton quite untenable. There would then arise a very grave difficulty in how to explain the large mass defect of the α-particle as compared to the deuteron.

[34] We are indebted to Mr. Mueller for communicating to us these preliminary results, based on the observation of 105 tracks.

Before concluding this section, we want to calculate the cross section for very high neutron energies, $E \gg V_0$. In this case, we may apply Born's approximate method for calculating the scattering cross section which gives

$$d\sigma = (\tfrac{1}{2}M)^2/2\pi\hbar^4 \cdot |\int U'^*(\mathbf{r})J(r)U(-\mathbf{r})d\tau|^2 \sin\theta d\theta, \quad (73)$$

where $U(\mathbf{r})$ and $U'(\mathbf{r})$ are the wave functions of incident and scattered neutron, in zero approximation. It should be noted that $U(-\mathbf{r})$ enters rather than $U(\mathbf{r})$ because of the exchange nature of the forces. Denoting the wave vectors of incident and scattered neutron by $\mathbf{k}$ and $\mathbf{k}'$, we have

$$U(\mathbf{r}) = e^{i(\mathbf{k}\cdot\mathbf{r})}, \quad U'(\mathbf{r}) = e^{i(\mathbf{k}'\cdot\mathbf{r})},$$

$$d\sigma = (M^2/8\pi\hbar^4)|\int J(r)e^{-i(\mathbf{k}+\mathbf{k}')\cdot\mathbf{r}}d\tau|^2 \sin\theta d\theta. \quad (73a)$$

We assume $J(r)$ to depend exponentially[35] upon r, *viz.*:

$$J(r) = -V_0 e^{-r/a} \quad (73b)$$

and obtain by an elementary calculation

$$d\sigma = \frac{2\pi M^2 V_0^2 \sin\theta d\theta}{\hbar^4 a^2[(\mathbf{k}+\mathbf{k}')^2 + a^{-2}]^4}. \quad (73c)$$

We consider now that (cf. Table II)

$$(\mathbf{k}+\mathbf{k}')^2 = 2k^2(1+\cos\theta), \quad 2\hbar^2k^2/M = E_0, \quad (73d)$$

$$V_0 = A\hbar^2/Ma^2, \quad \text{where} \quad A \approx \sqrt{8}.$$

We have, except for very small values of $1+\cos\theta$ (of the order V_0/E_0):

$$d\sigma \approx (\pi/4)a^2(V_0/E_0)^4(1+\cos\theta)^{-4}\sin\theta d\theta, \quad (74)$$

or, using the angle φ between the recoil proton and the incident neutron ($\varphi = \tfrac{1}{2}(\pi-\theta)$)

$$d\sigma = (\pi/64)a^2(V_0/E_0)^4 \sin^{-8}\varphi \cos\varphi \sin\varphi d\varphi. \quad (74a)$$

The total cross section is, neglecting higher powers of V_0/E_0:

$$\sigma = \frac{\pi}{3k^2}\left(\frac{MV_0a^2}{\hbar^2}\right)^2 \approx \frac{\pi A^2}{3k^2} = \frac{16\pi}{3}\frac{\hbar^2}{ME_0}. \quad (74b)$$

For neutrons of 10^9 volts energy, this gives a cross section of $0.68 \cdot 10^{-26}$ cm² which is rather large.

[35] The rectangular potential hole gives for $d\sigma$ an expression which is a rapidly oscillating function of $(\mathbf{k}+\mathbf{k}')$.

It was first pointed out by Bhabha (B19) that the collisions between protons and neutrons of very high energy have an important application for cosmic rays. Suppose there are protons of about 10^9 volt energy in cosmic rays. Such protons have a fair chance of knocking a neutron out of a nucleus they encounter. This neutron would, due to the exchange character of the forces, carry away almost all the energy of the proton. The cross section for the occurrence of this process in a collision between a proton and, say, a nitrogen nucleus, may safely be assumed to be 7 times the cross section for a proton and a free neutron, because the nuclear binding energies are small compared to the kinetic energy of the proton. The cross section becomes thus $\tfrac{1}{2} \cdot 10^{-25}$ cm² for N which means that the process should happen about once per 5 meter water equivalent of the atmosphere. Cosmic-ray protons would in this way "become" neutrons after traversing a comparatively small amount of air, and become protons again after going through more air. An incident proton radiation would therefore be about half protons half neutrons when reaching sea level. Moreover, the thickness of matter required to stop protons of a given energy, would be doubled by this process, since neutrons suffer no appreciable energy losses.

§16. Photoelectric Disintegration of the Deuteron (C7, C8, B16, F8, H1)

Chadwick and Goldhaber (C7, C8) have observed that the deuteron can be disintegrated into a neutron and a proton by the γ-rays from Th C′, of energy $h\nu = 2.62$ MV. This experiment, besides being of high interest in itself, gives at the same time the most exact determination of the binding energy of the deuteron. For this purpose, Chadwick and Goldhaber have determined the number of ions formed by the proton which is produced in the disintegration. This number turns out to be about 7200, with an accuracy of, say, ±20 percent. The average energy spent by the proton in producing one ion in air is certainly close to 33 volts, which figure holds for the ionization by α-particles as well as electrons. The kinetic energy of the proton is therefore $33 \cdot 7200 = 240{,}000$ volts ±20 percent. The neutron formed in the disintegration must, because of momentum conservation, receive the same energy and go off in the opposite direction from the proton, since the momentum of the γ-ray quantum is negligible compared to that of neutron and proton.[36] The kinetic energy of neutron plus proton is thus $2 \cdot 0.24$ MV$=0.48$ MV, with an accuracy of ± 20 percent$=\pm 0.10$ MV.* Therefore the binding energy of the deuteron is

$$\epsilon = 2.62 - 0.48 \pm 0.10 = 2.14 \pm 0.10 \text{ MV} \quad (75)$$

a value which we have made use of repeatedly.

From this binding energy, we can also determine the mass of the neutron, since the masses of both deuteron and proton are known from Bainbridge's mass-spectroscopic determinations. Choosing the values suggested by Bethe (B13), we have

$$H^2 = 2.01423 \pm 0.0002,$$

$$H^1 = 1.00807,$$

$$\epsilon = (2.14 \pm 0.10)/931 \quad (75a)$$
$$= 0.00230 \pm 0.00010 \text{ mass unit},$$

$$n^1 = H^2 + \epsilon - H^1 = 1.00846 \pm 0.0002.$$

The mass of the neutron thus turns out to be 0.00039 mass units$=0.36\pm 0.20$ MV larger than that of the hydrogen atom. This result is only based upon the ratio of the atomic weights of heavy and light hydrogen, as determined by Bainbridge (B5); it is independent of the absolute value of the atomic weight of, say, deuterium compared to oxygen. An error in ϵ as large as 0.36 MV can almost certainly be excluded; thus the only error which could materially influence the difference between the atomic weights of neutron and proton would be an error in Bainbridge mass determinations which would have to be much larger than the probable error deduced from the internal consistency of his results.[37] At present, it seems more likely that the result (75a) is correct i.e., that the neutron is really considerably heavier than the hydrogen atom. A free neutron must then disintegrate spontaneously into a proton plus an electron (§43), the lifetime being about a month.

Returning to the disintegration of the deuteron by γ-rays, it is obvious that this effect is closely analogous to the photoelectric effect in atoms. The electric field of the γ-ray produces an optical transition of the deuteron from the ground state to a state of positive energy

$$E = h\nu - \epsilon, \quad (76)$$

E being the sum of the kinetic energies of proton and neutron produced in the process. The cross section for the photoelectric effect is given by the well-known formula

$$\sigma = 8\pi^3 \nu |M_{0E}^{el}|^2/c, \quad (77)$$

where M_{0E} is the matrix element of the electric moment of the deuteron relative to its center of gravity and in the direction of polarization of the γ-ray, the matrix element referring to the transition from the ground state to the state of energy E. (The transition can also be produced by a *magnetic* moment, this "magnetic dipole" photoelectric effect is, however, small compared to the "electric dipole" effect discussed here, except for very low energies E. Cf. the end of this §, and §17.) Since only the proton has a charge e, and since its coordinate relative to the center of gravity of the deuteron is $\frac{1}{2}\mathbf{r}$, we have

$$M_{0E}^{el} = \tfrac{1}{2} e \int U_0 z U_E d\tau \quad (77a)$$

if the γ-ray is polarized in the z direction. Here U_0 is the wave function of the ground state, as given in (44c), and U_E is the wave function of the final state, normalized per unit energy.

From the familiar selection rule for the angular

[36] A simple momentum consideration shows that the energy of the proton should vary from 0.21 MV, if it is emitted in the direction opposite to the incident γ-ray, to 0.27 MV in the forward direction, 0.24 MV being assumed as the average proton energy.

* *Note added in proof:* Feather (Nature **136**, 467 (1935)) measured the range of the protons and deduced from it a kinetic energy of 0.18 MV for the protons. The range-energy relation in present use is likely to give too low values for the energy (Chapter XV) so that the correct energy may be about 0.20 MV. This would raise ϵ to 2.22 MV, raise the weight of the neutron to 1.00855, lower $1/\alpha$ (cf. 44a) to $4.29 \cdot 10^{-13}$ cm, lower the numerical factor in (61a) to $2.31 \cdot 10^{-24}$ and the factor in (80) to $1.12 \cdot 10^{-26}$. Furthermore, in (80b) $\gamma - 1$ should be 0.178 and $\sigma = 5.3 \cdot 10^{-28}$ cm^2, in perfect agreement with the observed value $5 \cdot 10^{-28}$ cm^2. However, the finite range of the forces (paragraph below (80b)) and the addition of the photomagnetic effect (81) would again raise the theoretical cross section for the disintegration of the deuteron to nearly twice the experimental value, which is just within experimental error.

[37] With Aston's recent "preliminary" determination of the mass ratio $D : H$ (A5), the neutron mass would come out as high as 1.0090, i.e., 0.9 MV larger than that of the hydrogen atom.

momentum we infer that the final state U_E must have the angular momentum $l=1$, i.e., in spectroscopic notation, it must be a P state.[38] We know, however, from the discussion in §13 and §15 that P states are practically uninfluenced by the force between neutron and proton, provided the range of the force a is small compared to the wavelength $\lambda\!\!\!^{-}=\hbar/(ME)^{\frac{1}{2}}$ corresponding to the state E. This condition is well fulfilled for our case ($E=0.5$ MV, $\lambda\!\!\!^{-}=9\cdot 10^{-13}$ cm$=4a$). Therefore the wave function for our P state will have the same form as if the neutron and proton were free, *viz.*:

$$U_E=(3/4\pi)^{\frac{1}{2}}\cos\theta(2/\pi)^{\frac{1}{2}}(dk/dE)^{\frac{1}{2}} \times r^{-1}(-\cos kr+\sin kr/kr). \tag{77b}$$

Here $(3/4\pi)^{\frac{1}{2}}\cos\theta$ represents the (normalized) first spherical harmonic. The bracket is the radial wave function as given in (66a), while the factor $1/r$ has been introduced in (50). The factor $(2/\pi)^{\frac{1}{2}}$ normalizes the radial wave function "per unit wave number dk"[39] while the factor

$$(dk/dE)^{\frac{1}{2}}=(M/2\hbar^2k)^{\frac{1}{2}} \tag{77c}$$

transforms to normalization per unit energy. We rewrite (77b):

$$U_E=\frac{3^{\frac{1}{2}}}{2\pi\hbar}\cos\theta\left(\frac{M}{k}\right)^{\frac{1}{2}}\frac{1}{kr^2}\,\mathrm{Re}\,[e^{ikr}(-i-kr)], \tag{77d}$$

"Re" denoting the real part. We assume the axis of our polar coordinate system to be parallel z, so that

$$z=r\cos\theta. \tag{77e}$$

Furthermore, we use the fact the wave function of the ground state is, for the larger part of the space, represented by (44c). Then we obtain from (77a)

$$M_{0E}{}^{\mathrm{el}}=\tfrac{1}{2}e\frac{3^{\frac{1}{2}}}{(2\pi)^{\frac{3}{2}}\hbar}\left(\frac{\alpha M}{k}\right)^{\frac{1}{2}}\int 4\pi r^3dr\,\overline{\cos^2\theta}\frac{1}{kr^3} \times\mathrm{Re}\,[(-i-kr)e^{-\alpha r+ikr}]. \tag{78}$$

$\overline{\cos^2\theta}$ is the average of $\cos^2\theta$ over all directions in space, *viz.*, $\frac{1}{3}$. We carry out the integration:

$$M_{0E}{}^{\mathrm{el}}=\frac{e}{(6\pi)^{\frac{1}{2}}\hbar}\left(\frac{\alpha M}{k}\right)^{\frac{1}{2}} \times\mathrm{Re}\left(-\frac{i}{k(\alpha-ik)}-\frac{1}{(\alpha-ik)^2}\right)$$

$$=\frac{e}{(6\pi)^{\frac{1}{2}}\hbar}\left(\frac{\alpha M}{k}\right)^{\frac{1}{2}}\left(\frac{1}{\alpha^2+k^2}-\frac{\alpha^2-k^2}{(\alpha^2+k^2)^2}\right)$$

$$=\left(\frac{2}{3\pi}\right)^{\frac{1}{2}}\frac{e}{\hbar}(\alpha M)^{\frac{1}{2}}\frac{k^{\frac{3}{2}}}{(\alpha^2+k^2)^2}. \tag{78a}$$

Inserting this into (77), we find for the cross section

$$\sigma=\frac{8\pi}{3}\frac{2\pi\nu}{c}\frac{e^2M}{\hbar^2}\frac{\alpha k^3}{(\alpha^2+k^2)^4}. \tag{78b}$$

Here we use (76):

$$\hbar\omega=2\pi\hbar\nu=E+\epsilon=\hbar^2(\alpha^2+k^2)/M \tag{78c}$$

and express α and k in terms of ϵ and E (cf. (37), (52)):

$$\sigma=\frac{8\pi}{3}\frac{e^2}{\hbar c}\frac{\hbar^2}{M}\frac{\epsilon^{\frac{1}{2}}E^{\frac{3}{2}}}{(E+\epsilon)^3}. \tag{79}$$

Introducing the ratio of the energy of the γ-ray to the binding energy of the deuteron, *viz.*,

$$\gamma=h\nu/\epsilon=(E+\epsilon)/\epsilon \tag{79a}$$

we have, using (44a)

$$\sigma=\frac{8\pi}{3}\frac{1}{137}\frac{1}{\alpha^2}\frac{(\gamma-1)^{\frac{3}{2}}}{\gamma^3}$$

$$=1.16\cdot 10^{-26}(\gamma-1)^{\frac{3}{2}}\gamma^{-3}\ \mathrm{cm}^2. \tag{80}$$

This cross section vanishes for $\gamma=1$, i.e., if the energy of the γ-ray is just sufficient to produce disintegration. σ then increases slowly with increasing γ-ray energy, and reaches a maximum for $\gamma=2$, i.e., when the kinetic energy of the disintegration products E is just as large as the binding energy ϵ of the deuteron. This maximum is, according to (80),

$$\sigma_{\max}=14.5\cdot 10^{-28}\ \mathrm{cm}^2. \tag{80a}$$

For still more energetic γ-rays, the cross section decreases again.

[38] More accurately, we may say it to be a ^3P state, since the spin remains unchanged in the photoelectric transition, and since the ground state is a triplet state.

[39] Cf., e.g., *Handbuch der Physik*, Vol. 24, p. 292, Eq. (4.18).

For the γ-rays of Th C′, we have (see above) $E=0.48$ MV, therefore

$$\gamma-1=E/\epsilon=0.224$$

and
$$\sigma=6.7\cdot 10^{-28}\ \text{cm}^2. \tag{80b}$$

The experiments of Chadwick and Goldhaber give a cross section of $5\cdot 10^{-28}$ cm² with an uncertainty of a factor of 2 in either direction. The agreement between theory and experiment is therefore satisfactory.

It must again be emphasized that the theoretical formula contains no assumptions except our usual one that the range of the forces between neutron and proton is small compared to the radius of the deuteron. The error introduced by this assumption is mainly due to the normalizing factor of the eigenfunction of the ground state, which should, according to (44d), be multiplied by the factor $1+\frac{1}{2}\alpha a$, giving a factor $1+\alpha a\approx 1.4$ to the cross section (cf. H1). (The correction necessary in the calculation of the matrix element M_{0E} itself, apart from the change in normalization, is very small because the contribution of the region $r<a$, for which the expressions (44c) and (77d) fail to hold, to the matrix element is only of the relative order of magnitude $(ka)^3$, i.e., about $\frac{1}{2}$ percent.) The theoretical cross section should thus be somewhat greater than (80).

Furthermore, we have neglected the possible "photomagnetic" effect, i.e., the transition due to the magnetic dipole moment (F8). In the following §, we shall calculate the corresponding dipole moment M_{0E}^{magn} (cf. 34). The ratio is (cf. 78a, 37, 52, 78c, 94)

$$\tau=\frac{1}{3}\left(\frac{M_{0E}^{\text{magn}}}{M_{0E}^{\text{el}}}\right)^2 =\frac{(\mu_p-\mu_n)^2}{4}\frac{(E+\epsilon)^2}{E+\epsilon'}\frac{(\epsilon^{\frac{1}{2}}\mp\epsilon'^{\frac{1}{2}})^2}{EMc^2}. \tag{81}$$

The factor $\frac{1}{3}$ arises from the fact that each of the three magnetic substates of the ground state of the deuteron may be disintegrated by a γ-ray of given polarization in the photo*electric* effect while only *one* of the three substates may be disintegrated photo*magnetic*ally (cf. the remarks after (92b)). For $\mu_p-\mu_n=4.9$ (cf. §5); $E=0.48$, $\epsilon=2.14$, $\epsilon'=0.040$ and $Mc^2=931$ MV, (81) gives 0.31 or 0.56, according to whether the negative or the positive sign holds. (The negative sign stands if there is a stable singlet state of the deuteron, cf. §13, §15, §17; the positive sign if there is no such state.) For Th C′ γ-rays, we should therefore add about 31 percent (or 56 percent) to the cross section (80b) for the photoelectric disintegration. For γ-rays of smaller energy which are just capable of producing disintegration, the ratio (81) would be higher, becoming $\tau=1$ for $E_c=210{,}000$ or 300,000 volts respectively, according to the sign standing in (81). If the disintegration products get less energy than E_c, the magnetic effect will predominate and will cause the cross section to tend less rapidly to zero with decreasing energy E. The complete formula for the photoelectric cross section, including electric and magnetic effect, is

$$\sigma=\frac{8\pi}{3}\frac{e^2}{\hbar c}\frac{\hbar^2}{M}\left(\frac{\epsilon^{\frac{1}{2}}E^{\frac{3}{2}}}{(E+\epsilon)^3}+\frac{(\mu_p-\mu_n)^2}{4}\frac{E^{\frac{1}{2}}\epsilon^{\frac{1}{2}}(\epsilon^{\frac{1}{2}}\pm\epsilon'^{\frac{1}{2}})^2}{(E+\epsilon)(E+\epsilon')Mc^2}\right), \tag{81a}$$

the minus and plus sign standing according to whether a stable singlet state of the deuteron exists or not.

The most important influence of the "photomagnetic" effect is that upon the *angular distribution* of the protons and neutrons produced in the disintegration of the deuteron. As we have seen, the photo*electric* effect leads to a P state in the continuous spectrum, more exactly to *that* P state which has no angular momentum around the direction of polarization of the incident γ-ray. In this case, the number of protons (or neutrons) emitted per unit solid angle should be proportional to $\cos^2\theta$, θ being the angle between the direction of the proton and the polarization of the γ-ray. Averaging over the directions of polarization of the γ-ray, we obtain a distribution proportional to $\sin^2\Theta$, where Θ is the angle between the direction of *propagation* of the γ-ray and the motion of the proton.

On the other hand, the magnetic effect leads (cf. §17) to a ^1S state in the continuous spectrum of the deuteron, i.e., to a uniform angular distribution of the protons. If τ is the ratio of probabilities of magnetic to electric effect, we shall therefore expect the number of protons emitted into the solid angle $\sin \Theta d\Theta$ to be proportional to

$$\sigma(\Theta) \sin \Theta d\Theta = (\sin^2 \Theta + \tfrac{2}{3}\tau) \sin \Theta d\Theta. \quad (82)$$

The magnetic effect can therefore be verified by observing the number of protons (or neutrons) projected in the direction of propagation of the γ-ray. This number should vanish if the magnetic effect were absent. By actually measuring the number in the forward direction ($\Theta = 0$) as well as perpendicularly to the γ-ray ($\Theta = 90°$) one might determine the coefficient τ quantitatively. Inserting τ into (81), one might then decide which sign in (81) corresponds to reality, in other words, whether a stable singlet state of the deuteron exists or not.

The disintegration of the deuteron can also be brought about by electron bombardment of deuterium. It can be shown (B16) that the electric field of an electron is approximately as effective as 1/137 of a light quantum. If the electron energy W is large compared to the binding energy ϵ of the deuteron, the cross section is (B16, Eq. (28))

$$\sigma_E = \frac{2\pi}{3\alpha^2}\left(\frac{e^2}{\hbar c}\right)^2 \left(\log \frac{W^2}{\epsilon mc^2} - 1.432\right) \quad (83)$$

$$= 2.1 \cdot 10^{-29}\left(\log \frac{W^2}{\epsilon mc^2} - 1.432\right) \text{cm}^2, \quad (83a)$$

where m is the electron mass and log denotes the natural logarithm. It seems not impossible to observe this effect. In fact, the disintegration of the beryllium nucleus was first carried out by electron bombardment (B21). However, it must be borne in mind that fast electrons produce a great number of x-rays (continuous x-ray spectrum). These secondary x-rays may be more effective in producing nuclear disintegrations than the primary electrons.

The scattering of light by a deuteron has been calculated (B16). It is found to be smaller than that by a free proton and is thus hardly observable.

§17. Capture of Neutrons by Protons (F8, W3, D4)

Neutrons may be captured by protons with the emission of a γ-ray which carries away the surplus energy

$$h\nu = E + \epsilon = \tfrac{1}{2}E_0 + \epsilon. \quad (76a)$$

[E_0 = kinetic energy of the neutron in the "ordinary" coordinate system (R system of §14), E = kinetic energy in the system in which the center of gravity of neutron and proton is at rest (C system of §14).] This process is the inverse of the photoelectric disintegration discussed in the preceding section, and the probabilities of the two processes are therefore connected by thermodynamic relations.

In each of the two processes we have in the initial state an incident particle which is in one case the neutron in the other case the light quantum. The cross section of any process is, quite generally, given by the number of the processes occurring per second, divided by the incident current. The number of processes per second is proportional to the square of the matrix element, and to the number of states of the final system per unit energy. Now we have in the final state an outgoing particle, either a light quantum or a neutron. The number of possible states of a particle of momentum p and energy E, per dE and per unit volume, is

$$4\pi p^2 dp/h^3 dE. \quad (84)$$

By using the relation

$$E^2 = c^2p^2 + m^2c^4, \quad (84a)$$

where E is the energy including the "rest" energy mc^2, (84) becomes

$$4\pi pE/h^3c^2. \quad (84b)$$

If therefore the indices 1 refer to the incident particle, 2 to the particle produced, the cross section becomes

$$\sigma_{12} \propto (p_2E_2/c^2v_1)\,|M|^2, \quad (85)$$

M being the matrix element for unit density of both particles and v_1 the velocity of the first particle. The cross section for the reverse process is

$$\sigma_{21} \propto (p_1E_1/c^2v_2)\,|M|^2. \quad (85a)$$

The matrix elements being the same in both cases, we have *cet. par.*

$$\sigma_{21}/\sigma_{12} = p_1E_1v_1/p_2E_2v_2. \quad (86)$$

Using the relativistic formula

$$Ev = pc^2 \qquad (86a)$$

we have $$\sigma_{21}/\sigma_{12} = p_1^2/p_2^2. \qquad (87)$$

The cross sections are proportional to the square of the momenta of the particles produced in the respective reactions.

However, formula (84) gives only the number of possible states of *motion* of the particle produced. In the case of light quanta, we must multiply this by a factor 2 because of the two possible directions of polarization. Furthermore, we must multiply the cross section of the capture process by the probability that the spin of the incident neutron is suitable for capture. Denoting this probability by g_σ we have

$$\frac{\sigma_{\text{capture}}}{\sigma_{\text{disintegr.}}} = 2g_\sigma\left(\frac{p_{\text{quantum}}}{p_{\text{neutron}}}\right)^2 = 2g_\sigma\left(\frac{h\nu/c}{\hbar k}\right)^2 = 2g_\sigma\left(\frac{\omega}{kc}\right)^2, \qquad (88)$$

where $\omega = 2\pi\nu$. Using expression (77) for $\sigma_{\text{disintegr.}}$, we have

$$\sigma_{\text{capt.}} = 8\pi^2 g_\sigma \omega^3 k^{-2} c^{-3} |M_{0E}|^2. \qquad (89)$$

(It should be noted that in (89) M_{0E} is the matrix element of the electric moment in *one* direction, in accord with (77).) Now the reverse of the photoelectric effect of the deuteron can occur if the spin of incident neutron is parallel to the spin of the capturing proton, because this is the case for the ground state. Therefore $g_\sigma = \frac{3}{4}$ and, using (78b)

$$\sigma_c = \frac{3}{2}\left(\frac{\omega}{kc}\right)^2 \sigma = 4\pi \frac{\omega^3}{c^3}\frac{Me^2}{\hbar^2}\frac{\alpha k}{(\alpha^2+k^2)^4}. \qquad (89a)$$

With (78c) this reduces to

$$\sigma_c = 4\pi \frac{e^2}{Mc^2}\frac{\hbar}{Mc}\frac{(E\epsilon)^{\frac{1}{2}}}{E+\epsilon}. \qquad (90)$$

The formula (90) for the capture cross section contains the Compton wave-length of the proton,

$$\hbar/Mc = 2.09 \cdot 10^{-14}\text{ cm} \qquad (90a)$$

and the "classical proton radius"

$$e^2/Mc^2 = (1/137)(\hbar/Mc) = 1.52_5 \cdot 10^{-16}\text{ cm}. \qquad (90b)$$

As a function of the energy E, the maximum of the cross section occurs for $E=\epsilon$ and has the value

$$\sigma_c^{\max} = 2\pi(e^2/Mc^2)(\hbar/Mc) = 2.00 \cdot 10^{-29}\text{ cm}^2, \qquad (90c)$$

which is extremely small. For slow neutrons ($E \ll \epsilon$) the cross section would be even smaller.

On the other hand, it has been observed experimentally by various authors (W3, D4) that slow neutrons are absorbed by hydrogen-containing substances such as water and paraffin. Therefore there must be another mechanism of capture which is more efficient than that due to the electric dipole transition, and particularly so at low velocity.

The latter fact gives us a valuable clue. The decrease of the capture cross section (90) with decreasing neutron energy E is due to the small amplitude of the p wave function of a slow neutron at small distances r between neutron and proton. In other words, a slow neutron with angular momentum $l=1$ has little chance to get sufficiently near the proton to be captured. This would be different if the incident neutron had zero angular momentum (cf. §§14, 15), i.e., for "s neutrons." In this case, particularly if the spins of neutron and proton are opposite, we know that the chance of the neutron coming sufficiently near to the proton is very large, resulting in a very large scattering cross section (§14, end). We shall therefore expect a very large capture cross section as well, if there is any process leading to the capture of a neutron with no angular momentum by a proton having opposite spin. Such a capture would correspond to a transition of the system consisting of neutron and proton, from a ^1S state to a ^3S state (ground state of the deuteron), while the capture process considered in (90) corresponds to a transition from a ^3P to a ^3S state.

The required mechanism giving rise to the $^1S \rightarrow {}^3S$ transition is found in the *magnetic dipole radiation*. In the transition from a singlet to a triplet state, the spin of either the proton or the neutron must change its direction; the magnetic moments of neutron and proton will therefore have matrix elements referring to that transition, and this magnetic dipole moment causes the emission of radiation.

To calculate the transition probability, we have simply to replace the matrix element of the electric dipole moment, M_{0E}^{el} (cf. 78a) by the matrix element of the magnetic dipole moment. We denote by μ_n and μ_p the magnetic moments of neutron and proton, in units of the "nuclear magneton"

$$\mu_0 = e\hbar/2Mc; \tag{91}$$

then, according to §5, $\mu_p=2.9$ and $\mu_n=-2.0$. Furthermore, we introduce the spin operators $\boldsymbol{\sigma}_p$ and $\boldsymbol{\sigma}_n$ of proton and neutron so that $\mu_0\mu_p\sigma_p$ would be, as to magnitude as well as direction, the magnetic moment of the proton. Finally, we introduce the spin wave functions χ_0 and χ_E for ground state and excited state, respectively, then we have

$$M_{0E}^{\text{magn}} = \mu_0 \int \sum_\sigma U_0 \chi_0 (\mu_p \boldsymbol{\sigma}_p + \mu_n \boldsymbol{\sigma}_n) U_E \chi_E d\tau, \tag{92}$$

the $\sum_\sigma$ denoting summation over the spin coordinates of proton and neutron.

The excited state being a singlet state, we have for its spin function[40]

$$\chi_E = 2^{-\frac{1}{2}}[\alpha(p)\beta(n) - \alpha(n)\beta(p)], \tag{92a}$$

p and n denoting the spin coordinates of neutron and proton, respectively. (α means spin parallel to a given direction z, β = spin opposite to z.) The ground state has three substates (triplet) corresponding to spin components in the z direction of $m=+1$, 0 and -1. The z components of the spin operators $\boldsymbol{\sigma}_p$ and $\boldsymbol{\sigma}_n$ in (92) will cause transitions from the state U_E to the substate $M=0$ of the ground state, the x and y components of the magnetic moment transitions to the substates $M=\pm 1$. We calculate the transition produced by σ_p^z and σ_n^z and have therefore for the spin wave function of the ground state

$$\chi_0 = 2^{-\frac{1}{2}}[\alpha(p)\beta(n) + \alpha(n)\beta(p)]. \tag{92b}$$

According to the remark after Eq. (89), we must only calculate the transition probability due to the magnetic moment in *one* direction.

From the definition of the spin wave functions it follows that

$$\sigma_p^z\alpha(p) = \alpha(p) \quad \sigma_n^z\beta(n) = -\beta(n), \quad \text{etc.,} \tag{92c}$$

therefore

$$\chi_0(\mu_p\boldsymbol{\sigma}_p + \mu_n\boldsymbol{\sigma}_n)\chi_E = \tfrac{1}{2}[\alpha(p)\beta(n) + \alpha(n)\beta(p)] \cdot (\mu_p - \mu_n)[\alpha(p)\beta(n) + \alpha(n)\beta(p)]. \tag{92d}$$

[40] Cf., e.g., *Handbuch der Physik*, Vol. 24, 1, p. 372.

For the summation over the spin coordinates we have to remember that

$$\sum_\sigma \alpha^2(p)\beta^2(n) = \sum \beta^2(p)\alpha^2(n) = 1; \quad \sum_\sigma \alpha(p)\beta(p)\alpha(n)\beta(n) = 0 \tag{92e}$$

and obtain

$$\chi_0(\mu_p\boldsymbol{\sigma}_p + \mu_n\boldsymbol{\sigma}_n)\chi_E = \mu_p - \mu_n \tag{92f}$$

$$M_{0E}^{\text{magn}} = \frac{e\hbar}{2Mc}(\mu_p - \mu_n)\int U_0 U_E d\tau. \tag{93}$$

For the eigenfunction of the ground state we take, as in §16, the expression (44c) which is valid for $r>a$. For U_E we have to insert the wave function corresponding to energy E, angular momentum $l=0$ and opposite spin of proton and neutron, normalized to unit energy. This function is, according to (50), (51)

$$U_E = u_0/r = (M/k)^{\frac{1}{2}} \sin(kr + \delta_0')/2\pi\hbar r \tag{93a}$$

where δ_0' is to be calculated from a formula analogous to (59a), only with α replaced by β since we are dealing with a singlet state (cf. end of §14). The normalization is identical to (77d). We have now

$$\int U_0 U_E d\tau = \frac{(\alpha M)^{\frac{1}{2}}}{(2\pi)^{\frac{1}{2}}\hbar k^{\frac{1}{2}}} \int 4\pi r^2 dr \frac{e^{-\alpha r}}{r}\frac{1}{r}\sin(kr+\delta_0')$$
$$= \frac{(2\alpha M)^{\frac{1}{2}}}{\pi^{\frac{1}{2}}\hbar k^{\frac{1}{2}}} \operatorname{Im}\left(\frac{e^{i\delta_0'}}{\alpha - ik}\right)$$
$$= \frac{(2\alpha M)^{\frac{1}{2}}}{\pi^{\frac{1}{2}}\hbar k^{\frac{1}{2}}} \frac{k\cos\delta_0' + \alpha\sin\delta_0'}{\alpha^2 + k^2}, \tag{93b}$$

where Im denotes the imaginary part. δ_0' is given by (59b) where, however, α has to be replaced by β (singlet state!). Therefore we have

$$\int U_0 U_E d\tau = \frac{(2\alpha M k)^{\frac{1}{2}}}{\pi^{\frac{1}{2}}\hbar} \frac{\alpha - \beta}{(\alpha^2+k^2)(\beta^2+k^2)^{\frac{1}{2}}}. \tag{93c}$$

It should be noted that this expression is only different from zero because $\alpha \neq \beta$, i.e., because the forces between proton and neutron are considerably different according to whether the spins of the particles are parallel or opposite (cf. §13, §14). This fact makes the Schrödinger equations for U_0 and U_E different; were this not the case, U_0 and U_E would be orthogonal and the integral $\int U_0 U_E d\tau$ would vanish. The capture of neutrons by protons by our mechanism is therefore only due to the dependence of the neutron-proton force upon spin.

Inserting (93c) into (93), we have

$$M_{0E}^{\text{magn}} = \frac{e}{c}\left(\frac{\alpha k}{2\pi M}\right)^{\frac{1}{2}} \frac{\alpha - \beta}{(\alpha^2+k^2)(\beta^2+k^2)^{\frac{1}{2}}}(\mu_p - \mu_n). \tag{94}$$

We insert this into (89) and consider that only one in every four neutrons has its spin opposite to a given proton so that $g_\sigma = \frac{1}{4}$. Then we find

$$\sigma_c = \pi \frac{e^2\omega^3}{Mc^5} \frac{\alpha}{k} \frac{(\alpha-\beta)^2(\mu_p-\mu_n)^2}{(\alpha^2+k^2)^2(\beta^2+k^2)}. \tag{94a}$$

By using again (78c) and expressing the α, β, k in terms of the energies ϵ, ϵ', $E_0=\frac{1}{2}E$, this reduces to

$$\sigma_c=\pi\frac{e^2}{Mc^2}\frac{\hbar}{Mc}\left(\frac{2\epsilon}{E_0}\right)^{\frac{1}{2}}\times\frac{(\epsilon^{\frac{1}{2}}\mp\epsilon'^{\frac{1}{2}})^2(\epsilon+\frac{1}{2}E_0)}{(\epsilon'+\frac{1}{2}E_0)Mc^2}(\mu_p-\mu_n)^2. \quad (95)$$

The $\mp$ sign stands according to whether the singlet state of the deuteron is stable or virtual.

It will be noticed that the capture cross section *increases* with decreasing kinetic energy E_0 of the neutron. We are therefore particularly interested in the cross section for very small E_0, *viz.*, $E_0\ll\epsilon'$. We compare, for this case, the capture cross section (95) to the cross section for elastic scattering given in (62). We have

$$\kappa=\left(\frac{\sigma_c}{\sigma_{el}}\right)_{E_0\ll\epsilon'}=\left(\frac{\epsilon}{8E_0}\right)^{\frac{1}{2}}\frac{e^2}{\hbar c}\frac{\epsilon^2}{\frac{1}{4}\epsilon+\frac{3}{4}\epsilon'}\times\frac{(\epsilon^{\frac{1}{2}}\mp\epsilon'^{\frac{1}{2}})^2}{(Mc^2)^2}(\mu_p-\mu_n)^2. \quad (95a)$$

We insert the numerical values $\mu_p=2.9$, $\mu_n=2.0$, $e^2/\hbar c=1/137$, $Mc^2=931$ MV, $\epsilon=2.15$ MV, $\epsilon'=0.040$ MV. Then, if we express E_0 in volts, we find

$$\kappa=\frac{\sigma_c}{\sigma_a}=\left(\frac{2.15\cdot 10^6}{8E_0}\right)^{\frac{1}{2}}\frac{4.9^2}{137}\times\frac{2.15^2(\sqrt{2.15}\mp\sqrt{0.040})^2}{(\frac{1}{4}\cdot 2.15+\frac{3}{4}\cdot 0.040)\cdot 931^2}=\begin{cases}0.00138E_0^{-\frac{1}{2}},\\ 0.00227E_0^{-\frac{1}{2}},\end{cases} \quad (96)$$

the upper value holding if there is a stable singlet state of the deuteron, the lower, if there is none. In particular, we may calculate the ratio κ for the case that the neutrons are in thermal equilibrium with the protons which can be achieved by multiple scattering (chapter XII, reference C12). Then in the average, $E_0=kT$ which is equal to 1/40 volt for room temperature ($T=290°$). In this case

$$\kappa=\begin{cases}1/118 \text{ if there is a stable singlet state of the deuteron}\\ 1/71 \text{ if there is no stable singlet state of the deuteron.}\end{cases} \quad (96a)$$

This means that neutrons of thermal velocity will make, in the average, 71 or 118 elastic collisions with protons before they are captured. The probability of capture is thus fairly large. The cross section is 0.5 or $0.3\cdot 10^{-24}$ cm² in the two cases, respectively.

The lifetime τ of a neutron in a substance containing hydrogen, such as water, can easily be calculated. The "number of captures per second" is, by definition

$$1/\tau=N\sigma_c v, \quad (97)$$

$v=(2E_0/M)^{\frac{1}{2}}$ being the velocity of the neutrons and N the number of hydrogen atoms per cubic centimeter, which is

$$N=6.73\cdot 10^{22} \quad \text{for } H_2O.$$

We have from (95), (97)

$$\tau=\begin{cases}2.84\cdot 10^{-4} \text{ sec.,}\\ \qquad \text{if stable singlet state exists,}\\ 1.63\cdot 10^{-4} \text{ sec.,}\\ \qquad \text{if stable singlet state does not exist.}\end{cases} \quad (97a)$$

The theory seems to be in sufficient agreement with experiments of Amaldi and Fermi (F10). These authors investigated the diffusion of slow neutrons (absorbable by Cd, energy probably of the order kT) in paraffin. They inserted Cd absorbers into a paraffin block, at various distances from a Rh detector, and observed the decrease in activity of the detector as a function of this distance. In this way, Fermi and Amaldi could determine the average distance which the neutrons would have traveled had they not been absorbed in the Cd. The distance was found to be 2 cm. Theoretically, this distance is (F10) equal to $\lambda(3\kappa)^{-\frac{1}{2}}$ where λ is the mean free path for elastic collisions. With the value of the collision cross section as determined by Dunning and others (D4), *viz.* $\sigma=35\cdot 10^{-24}$ cm², we find $\lambda=0.35$ cm and therefore $\kappa=100$.

The correction to our results for the finite range of the forces between neutron and proton amounts to an increase in the capture cross section proportional to the increase in the elastic scattering (end of §14) and to a very slight change of the ratio of the capture cross section to the elastic cross section, changing the values in (96a) to 1/113 and 1/75, respectively.

In conclusion, it might be noted that there would also be capture of neutrons by protons with *parallel* spin, if the singlet state of the deuteron is stable. The cross section for this capture which is a $^3S \rightarrow {}^1S$ transition, is however very small: It is obtained by interchanging ϵ and ϵ' in (95), and is therefore, at low energies E_0, smaller by a factor $(\epsilon'/\epsilon)^{5/2} = 1/15,000$ than the capture into the ground state discussed here.

§18. Scattering of Protons by Protons (W8, T11, P7)

If there is no other interaction between a pair of protons but the Coulomb repulsion, the scattering cross section is given by the Rutherford formula as modified by Mott to take account of the possibility of exchange of the two protons. The cross section for a deflection by an angle between θ and $\theta + d\theta$ is[41]

$$2\pi\sigma_0(\theta) \sin\theta d\theta = \frac{e^4}{M^2v^4}\left(\frac{1}{\sin^4\theta} + \frac{1}{\cos^4\theta} - \frac{\cos((e^2/\hbar v)\log\tan^2\theta)}{\sin^2\theta\cos^2\theta}\right) 2\pi \sin 2\theta d(2\theta). \quad (99)$$

2θ is the deflection of the incident proton in the coordinate system in which the center of gravity of the two protons is at rest (C system, cf. the beginning of §14), therefore the element of solid angle in that system is $2\pi \sin 2\theta d(2\theta) = 8\pi \cos\theta \sin\theta d\theta$. The first term in the bracket in (99) gives the number of incident protons deflected by an angle θ (i.e., $\frac{1}{2}\theta$ in the C system) according to the Rutherford formula. (Cf. Mott and Massey, p. 36. Note that in that formula the reduced mass $\frac{1}{2}M$ has to be inserted!) The second term gives the number of incident protons deflected by an angle $\pi/2 - \theta$; each of these is accompanied by a recoil proton at right angles to its own motion (cf. §14), i.e., making an angle θ with the incident beam; these recoil protons are also counted among the protons scattered through θ. The last term in the bracket is the effect of exchange between incident and scattering proton. We shall be particularly interested in fast protons, having velocities of the order one-twentieth of the velocity of light (energy ≈ 1 MV); for these, $e^2/\hbar v \ll 1$ and

$$\cos((e^2/\hbar v)\log\tan^2\theta) \approx 1 \quad (99a)$$

unless θ is *very* near or $\pi/2$. We therefore have, with $E_0 = \frac{1}{2}Mv^2$

$$\sigma_0(\theta) = \pi\frac{e^4}{E_0^2}\left(\frac{1}{\sin^4\theta} + \frac{1}{\cos^4\theta} - \frac{1}{\sin^2\theta\cos^2\theta}\right)\cos\theta. \quad (99b)$$

Actually, experiments of White (W8) and of Tuve, Heydenburg, and Hafstad do not agree at all with this formula, but indicate that there are actually considerably more protons at 45° than according to (99b). This proves that there is another force acting besides the Coulomb force. We have already deduced the existence of a force between two protons, and more exactly of an attractive force, from the existence of isobars of even atomic weight and charge (§10). It seems reasonable to assume that this force has the same characteristics as that between neutron and proton, since they no doubt originate from the same reason, i.e., we assume the force between two protons to be also restricted to a small range a and to be large inside that range. Then we can, similarly to the neutron scattering, conclude that only the partial wave function $l=0$ will be materially influenced by the "nuclear forces" between the two protons.

To deduce the scattering including such a force, we have to calculate the wave function of the relative motion of the two protons. There are two cases to distinguish. Either the spins of the protons are parallel or antiparallel. In the first case, the wave function of the system will be symmetrical in the two spins; it must then necessarily be antisymmetrical in the spatial coordinates of the proton, since protons obey the Fermi statistics, i.e., the wave function of a system of two protons changes sign when both the spatial and the spin coordinates of the protons are interchanged. If the spins of the two protons are opposite, the spatial wave function will be symmetrical.

Now we have seen in §14, Eq. (32) that a wave function describing the relative motion of two

[41] Cf. Mott and Massey, *Atomic Collisions*, p. 75, 76, Eqs. (25), (26).

particles, of the form

$$\psi = u_l(r)P_{lm}(\cos\theta)e^{im\varphi}$$

is symmetrical in the coordinates of the two particles, if l is even, and antisymmetrical if l is odd. Therefore, for parallel spins of the protons, the wave function contains only odd azimuthal quantum numbers l, and will therefore not be materially influenced by short range forces. Only for opposite spins of the two protons, i.e., only in 1/4 of all cases, will there be an appreciable influence of the "nuclear" forces between the protons, since in this case the wave function contains the term $l=0$, which is the only term strongly influenced by the nuclear forces. We can calculate the total wave function including the nuclear forces by simply calculating the total wave function without these forces, subtracting from it the part corresponding to $l=0$ without forces and adding the wave function for $l=0$ including the nuclear forces.

We first have to write down, for two protons with Coulomb interaction between them, a wave function symmetrical in the spatial coordinates of the protons. According to Mott and Massey, p. 35, Eq. (16) and p. 72, an *un*symmetrical wave function would be

$$\psi = e^{ikz+i\alpha\log k(r-z)} + e^2/(Mv^2 r\sin^2\tfrac{1}{2}\Theta) \cdot e^{ikr-i\alpha\log 2kr - i\alpha\log\sin^2\frac{1}{2}\Theta + i\pi + 2i\eta_0}, \tag{100}$$

where $\qquad \alpha = e^2/\hbar v, \quad k = Mv/2\hbar, \qquad$ (100a)

r = distance of the two protons, z = difference of their z coordinates, $\cos\Theta = z/r$, $e^{i\eta_0} = \Gamma(1+i\alpha)/|\Gamma(1+i\alpha)|$. The first term in (100) represents the incident wave, a plane wave moving in the z direction corrected by a small phase shift due to the Coulomb field. The second term is the scattered spherical wave. A *symmetrical* wave function is

$$\Psi(x, y, z) = \psi(x, y, z) + \psi(-x, -y, -z), \tag{100b}$$

because changing the sign of all the relative coordinates $x = x_1 - x_2$, $y = y_1 - y_2$, $z = z_1 - z_2$ of the first proton relative to the second, corresponds to an interchange of the coordinates of the two protons. Now in (100) only the z coordinate occurs explicitly, therefore we have to replace z by $-z$, $\cos\Theta$ by $-\cos\Theta$, and therefore $\sin^2\frac{1}{2}\Theta = \frac{1}{2}(1-\cos\Theta)$ by $\frac{1}{2}(1+\cos\Theta) = \cos^2\frac{1}{2}\Theta$, and then to add the function thus obtained to (100). The resulting function will represent two beams of protons proceeding in the positive and the negative z direction, each containing one proton per cm^3.

Next we consider the part of the wave function (100b) corresponding to $l=0$. If we put, as usual, $\psi_0 = v_0/r$, the function v_0 will be a solution of the radial Schrödinger equation

$$d^2v_0/dr^2 + (M/\hbar^2)(\tfrac{1}{2}E_0 - e^2/r)v_0 = 0, \tag{101}$$

which has, for large distances r, the asymptotic form (cf. Mott and Massey, p. 33, Eq. (7) for $n=0$).

$$v_0 = 2e^{i\eta_0}\sin(kr+\eta_0-\alpha\log 2kr)/kr$$
$$= (i/kr)(e^{-i(kr-\alpha\log 2kr)} + e^{i(kr+\pi+2\eta_0-\alpha\log 2kr)}). \tag{101a}$$

v_0 is, according to (86), normalized correctly, i.e., in such a way that the first term in (101a) is the first term (the term not depending on the angle) of the expansion of the function (100b) in spherical harmonics.

If a force acts between the two protons at close distances, the phase of the wave function will be shifted by a certain amount δ_0, analogous to the case of scattering of neutrons by protons. Thus the wave function v_0 has to be replaced by

$$u_0 = (2c/kr)e^{i\eta_0}\sin(kr+\eta_0-\alpha\log 2kr+\delta_0). \tag{101b}$$

The constant c has to be fixed in such a way that the term containing e^{-ikr} is identical with (101a), so that $u_0 - v_0$ contains only an *outgoing* spherical wave, i.e., one proportional to e^{+ikr}, which represents the scattering. This is achieved by putting $c = e^{i\delta_0}$ so that

$$u_0 = (i/kr)[e^{-i(kr-\alpha\log 2kr)} + e^{i(kr+\pi+2\eta_0+2\delta_0-\alpha\log 2kr)}]. \tag{101e}$$

The effect of the force between the two protons is now simply to replace the function v_0 (valid for no force) by u_0 (which takes the force into account), i.e., to add $u_0 - v_0$ to the "unperturbed" wave function (100b):

$$\Psi' = \Psi + u_0 - v_0 = e^{ikz+i\alpha\log k(r-z)} + e^{-ikz-i\alpha\log k(r-z)}$$
$$+ \frac{1}{r}e^{ikr-i\alpha\log 2kr+i\pi+2i\eta_0}\left[\frac{e^2}{Mv^2}\left(\frac{e^{-i\alpha\log\sin^2\frac{1}{2}\Theta}}{\sin^2\frac{1}{2}\Theta} + \frac{e^{-i\alpha\log\cos^2\frac{1}{2}\Theta}}{\cos^2\frac{1}{2}\Theta}\right) + \frac{i}{k}(e^{2i\delta_0}-1)\right]. \tag{102}$$

The scattering cross section per unit solid angle (in the C system) for a given angle Θ is equal to the absolute square of the square bracket. Inserting $k = Mv/2\hbar$ and neglecting $\alpha\log\sin^2\frac{1}{2}\Theta$ because $\alpha = e^2/\hbar v \ll 1$, we have

$$\sigma(\Theta) = \left(\frac{e^2}{Mv^2}\right)^2\left[\left(\frac{1}{\sin^2\frac{1}{2}\Theta} + \frac{1}{\cos^2\frac{1}{2}\Theta} - \frac{2\hbar v}{e^2}\sin 2\delta_0\right)^2 + \left(\frac{2\hbar v}{e^2}\right)^2(1-\cos 2\delta_0)^2\right]. \tag{103a}$$

If there were only the Coulomb field, the cross section $\sigma_0(\Theta)$ would be obtained by putting $\delta_0 = 0$. We have therefore, with $\frac{1}{2}\Theta = \theta$,

$$\sigma(\Theta) - \sigma_0(\Theta) = \left(\frac{e^2}{Mv^2}\right)^2\left[\left(\frac{4\hbar v}{e^2}\right)^2\sin^2\delta_0 - \frac{8\hbar v}{e^2}\sin\delta_0\cos\delta_0\left(\frac{1}{\sin^2\theta} + \frac{1}{\cos^2\theta}\right)\right]. \tag{103b}$$

We now consider the fact that only one-quarter of all proton pairs have spin opposite, therefore the additional cross section (103b) will only be present in 1/4 of all cases. In the average, we have to add 1/4 of (103b). Furthermore, we calculate the cross section per unit solid angle in the *ordinary* coordinate system, by multiplying (103b) with $4\cos\theta$ (see after (99)).

Thus we find for the complete cross section

$$\sigma(\theta)=\sigma_0(\theta)+\cos\theta(\sigma(\Theta)-\sigma_0(\Theta)),$$

$$\sigma(\theta)=\frac{e^4}{E^2}\cos\theta\left[\frac{1}{\sin^4\theta}+\frac{1}{\cos^4\theta}-\frac{1}{\sin^2\theta\cos^2\theta}-\frac{2\hbar v}{e^2}\frac{\sin\delta_0\cos\delta_0}{\sin^2\theta\cos^2\theta}+\left(\frac{2\hbar v}{e^2}\right)^2\sin^2\delta_0\right].\quad(103)$$

The actual magnitude of this expression depends, of course, upon δ_0. This phase shift δ_0 might, of course, be quite small. But even then it will contribute appreciably to (103), since $\sin^2\delta_0$ is multiplied by the very large factor $(2\hbar v/e^2)^2$ which is 160 for protons of 1 MV energy. Therefore the scattering of protons by protons is extremely sensitive to the existence of a force between the protons other than the Coulomb force. This is particularly true for angles θ near 45° where $\cos\theta$ as well as $\sin\theta$ are comparatively large. A good measure of the effect of the "nuclear" proton-proton forces is therefore the ratio of the actual scattering observed at 45° and the scattering following from the Mott formula (99). This ratio is, for $\sin^2\theta=\cos^2\theta=\frac{1}{2}$:

$$\frac{\sigma(\theta)}{\sigma_0(\theta)}=\left(\frac{\hbar v}{e^2}\sin\delta_0\right)^2-2\frac{\hbar v}{e^2}\sin\delta_0\cos\delta_0+1.\quad(104)$$

The first experiments on proton-proton scattering were carried out by White (W8). The protons were accelerated in a cyclotron and their tracks observed in a hydrogen-filled cloud chamber. Since the scattering through large angles is a rare event, and the experimental arrangement does not yield high intensities of the incident proton beam, the number of large deflections observed by White is very small; e.g., only 5 protons were observed which had been deflected by more than 40°. Even so, the experiments seem to show that the scattering through 45° is many times as probable than would be expected from (99b). The ratio of observed scattering to theoretical scattering in a Coulomb field at 45° is 9 in White's experiments, for an average proton energy of, probably, about 750 kv.[42] The angular distribution does not agree well with (104) (cf. 97, and calculations of Serber mentioned by White) which is probably due to the very small number of data taken by White.

Much more extensive experiments have been performed by Tuve, Heydenburg and Hafstad.[43] At 920 kv, they find a ratio of observed scattering to Coulomb scattering of 4.65 at 40°. An energy of 920 kv corresponds to a velocity $v=0.0445c$ (c = velocity of light) so that $\hbar v/e^2=137\cdot 0.0445=6.09$. Inserting this and the observed ratio 4.65 into (103), we find

$$\delta_0=30.8°\quad(104a)\quad\text{or}\quad\delta_0=-12.2°.\quad(104b)$$

Inserting these values into (103), the angular distributions may be found. They are given in curves A and R in Fig. 7. Curve A corresponds to the positive value of δ_0 and therefore (see below) to an *attractive* potential between the protons, R corresponds to the negative δ_0

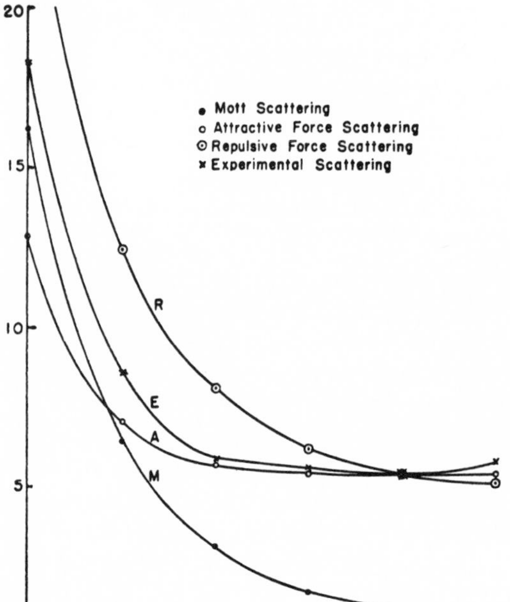

FIG. 7. Scattering of protons by protons. Abscissa: angle of deflection, ordinate: Number of particles scattered through an angle between θ and $\theta+d\theta$, divided by $\sin 2\theta d\theta$. The scale of the ordinate has been chosen so that the Mott scattering in the Coulomb field alone has the value unity at 45°. The theoretical curves are made to agree with the experiments at 40°. Proton energy = 920 kv.

[42] White gives 675 kv, this figure being apparently deduced from the range of protons with the help of the range-energy relation of the Cavendish laboratory. This relation gives too low values for the proton energy.

[43] We are indebted to Messrs. Tuve, Heydenburg and Hafstad for communicating their results to us before publication.

(*repulsive* potential). Curve A is seen to be very flat between 30 and 40°, while R decreases continuously with increasing angle θ. Curve R lies higher than A throughout, except at 45° where the two curves are made to give the experimental value, *viz.*, 4.65 times the Coulomb scattering. The scattering in the pure Coulomb field, according to the Mott formula (99b), is represented in curve M; it lies between the scattering for attractive and repulsive fields for small θ, but falls, of course, below both of them for larger deflections. The experimental points of Tuve, Heydenburg and Hafstad (marked by crosses) follow between 30° and 45° rather closely the curve for the attractive potential. At small angles, the agreement between experimental points and curve A is not perfect; the experimental scattering being too large. However, it is unmistakable that curve A agrees with the experimental data very much better than R; and that the Mott scattering curve M is entirely out of question.

Thus the scattering of protons by protons shows conclusively:

(1) There must be a force between two protons besides the Coulomb force.
(2) This force must be attractive.
(3) Reasonable agreement with experiments is obtained by assuming the force to have a short range.

We have already come to the conclusions (1) and (2) in §10 when discussing the even-odd rule of isotopes. Moreover, the existence of an attractive force between two protons is shown by the calculations of the binding energies of H^3, He^3 and He^4 (§21). We shall even find that the quantitative results for the magnitude of this force as calculated from the binding energies of H^3, He^3, He^4, and from the proton-proton scattering, agree quite well [see (107), and (128)].

The existence of a force between two protons necessitates the assumption of a force between two *neutrons* of practically the same strength. This is shown by the fact that the numbers of neutrons and protons in light nuclei are equal (§6) and, even more accurately, by the difference of the binding energies of H^3 and He^3 (§22).

From (104a), we may calculate a quantity γ which is the analog of the quantities α and β for the neutron-proton scattering (cf. 34). In other words, γ is defined by

$$\gamma = -((1/u_0)(du_0/dr))_{r=a}, \qquad (105a)$$

u_0 being the partial wave function $l=0$, and a the range of the forces. We have (cf. 59b)

$$k \cot (kr_0+\delta_0) = -\gamma. \qquad (105b)$$

Assuming the range r_0 to be very small, we may neglect kr_0 compared to δ_0. For $E_0=920$ kv, we have (cf. 54) $k=1.06\cdot 10^{12}$ cm^{-1}. With δ_0 given by (104a), we have therefore

$$\gamma = -1.06\cdot 10^{12} \cot 30.8°$$

$$= -1.78\cdot 10^{12}\ \text{cm}^{-1} \qquad (105)$$

or

$$1/\gamma = 5.6\cdot 10^{-13}\ \text{cm}, \qquad (105c)$$

$$T = \hbar^2\gamma^2/M = 1.30\ \text{MV}. \qquad (105d)$$

γ turns out to be negative which shows that there is no stable energy level of a "di-proton." (From symmetry arguments, we can then conclude that there exists no stable "di-neutron" either.) In contrast to the scattering of *neutrons* by protons, the sign of γ can be determined from the proton-proton scattering. The reason is that there is interference between the scattering due to the Coulomb field and that due to the specifically nuclear forces. The resultant angular distribution depends therefore on the sign of δ_0 (cf. the curves A and R in Fig. 7).

The "virtual energy level" of the system of two protons lies, according to (105d), at 1.15 MV kinetic energy, i.e., appreciably higher than the singlet level of the deuteron (near zero, cf. 62a) but in the same general region. The force between two protons is therefore smaller, but not very much smaller than the force between neutron and proton (cf. §44).

From (105), (105b), we may deduce the value of δ_0 for other values of the energy. We find, neglecting kr_0:

$$\cot \delta_0 = \gamma/k. \qquad (105e)$$

Inserting this value into (104), we find

$$R = \frac{\sigma(45°) - \sigma_0(45°)}{\sigma_0(45°)} = \frac{E_0}{C(T+\frac{1}{2}E_0)} \times (E_0 - 2(TC)^{\frac{1}{2}}) \qquad (106)$$

with

$$C = Me^4/\hbar^2 = 49{,}800\ \text{volts}. \qquad (106a)$$

It can be seen from (106) that the ratio R of "excess scattering" to Coulomb scattering at 45° increases rapidly with increasing energy E_0, being about proportional to E_0 for high energies ($E_0 \gg T$) and to E_0^2 for $E_0 \ll T$. This increase of R is due to the fact that the Coulomb field gives, especially for high energies, a particularly low scattering at 45°, which will be increased by almost any perturbation of the Coulomb field. At very *low* energies, (106) will be negative, corresponding to a *smaller* scattering at 45° than the Coulomb scattering: The reversal of sign occurs for

$$E_0^r = 2(TC)^{\frac{1}{2}} = 510{,}000 \text{ volts} \qquad (106b)$$

with our values (105d), (106a) for T and C. The smallest value of R is obtained for 240,000 volts, in this case,

$$(\sigma(45°)/\sigma_0(45°))_{E_0=0.23\text{ MV}} = 0.09, \qquad (106c)$$

i.e., the scattering at 45° should be only one-eleventh of the Coulomb scattering. For $E_0 = 0.74$ MV, we find

$$R = 2.0, \qquad \sigma(45°)/\sigma_0(45°) = 3.0, \qquad (106d)$$

which is considerably larger than the results of Tuve, Heydenburg and Hafstad ($R = 0.65$).

The calculations above should be corrected for two reasons: Firstly, the wave function in the Coulomb field is not a plane wave. Therefore (105b), (105e) hold only approximately. This is particularly true for small energy of the proton. The effect of this correction is to make the ratio R decrease even faster with decreasing energy E_0.

Secondly, the range of the nuclear forces is not zero. This fact has the opposite effect on the change of R with decreasing energy. If we take for the range of the forces the value derived in (128), *viz.*,

$$a = 2.3 \cdot 10^{-13} \text{ cm} \qquad (107a)$$

we have $kr_0 = 0.244 = 14.0°$, and therefore instead of (105):

$$\gamma = -1.06 \cdot 10^{12} \cot 44.8° = -1.09 \cdot 10^{12} \text{ cm}^{-1} \qquad (107b)$$

$$T = 0.49 \text{ MV}, \qquad (107c)$$

i.e., the virtual level lies much nearer to zero. If we now assume a simple potential hole of width a and depth

$$V_0 = \hbar^2\kappa^2/M \qquad (107d)$$

we may determine κ from the condition (cf. 39)

$$(\kappa^2 + k^2)^{\frac{1}{2}} \cot([\kappa^2 + k^2]^{\frac{1}{2}} r_0) = -\gamma \qquad (107e)$$

which gives

$$\kappa = 5.94 \cdot 10^{12} \text{ cm}^{-1}, \qquad (107f)$$

$$V_0 = 14.5 \text{ MV}. \qquad (107)$$

This depth may be compared to the depth of the hole for a proton and a neutron with opposite spin, *viz.*, $V_0 = 18.7$ MV. (This depth is necessary to give a quantum state at zero energy.) The potential energy between two protons is therefore about 80 percent of that between neutron and proton of opposite spin, or about 55 percent of the interaction between neutron and proton of parallel spin. This agrees satisfactorily with the result of the theory of H^3 and He^4 (Eq. 128).

IV. Theory of H^3, He^3 and He^4

§19. Thomas' Proof of the Finite Range of Nuclear Forces (T2)

We have shown in our discussion of the deuteron that the binding energy and the physical properties of that nucleus depend essentially only on a certain combination of the depth V_0 and the width a of the potential hole representing the interaction between neutron and proton. The combination involved is approximately V_0a^2, for details cf. §12. The same properties of the deuteron can be obtained with a deep and narrow hole, or with a shallow and wide hole. To fix depth and width separately, we have therefore to use the properties of other nuclei. The most suited for the purpose appear to be the nuclei immediately following upon the deuteron in complication, i.e., the nuclei of mass three H^3 and He^3, and the α-particle He^4.

We have already pointed out in §9 that the mass defects of these nuclei are very much larger than that of the deuteron, and that this fact may be explained by assuming the potential hole to be deep and narrow (Wigner, W12). However, no indication could be given at that time as to *how* deep and narrow the hole must be chosen in order to explain the observed mass defects. In fact, we cannot even say whether *any* finite depth and width of the hole will suffice to explain the very large mass defect of the α-particle which is as much as 14 times the mass defect of the deuteron. It might be necessary to assume an infinitely deep and narrow hole for that purpose. This point has been cleared up by Thomas (T1) in favor of the *finite* depth and width of the hole. Thomas has shown that an infinite binding energy would be obtained for H^3 if the hole were assumed to be infinitely deep and narrow, and if at the same time the product V_0a^2 were kept constant so that the binding energy of the deuteron would retain its observed value. From his calculations, Thomas estimates that the range a of the forces cannot be less than $1\cdot 10^{-13}$ cm. Thomas' proof is the more gratifying since we have up to the present no method for the explicit calculation of the binding energies of H^3 and He^4 which is at the same time rigorous and practical for obtaining numerical results.

The assumptions of Thomas are very general. No forces are assumed to act between two neutrons. This makes the conclusion hold *a fortiori* if there are attractive forces between like particles, as suggested by the odd-even-rule of isotopes (§10) and the scattering of protons by protons (§18) because such forces will lower the energy of H^3 even further. Repulsive forces between like particles can be considered as ruled out with certainty by the considerations of §10; but even small forces of this kind would not alter Thomas' conclusions. Finally, the force between protons and neutrons may be either an ordinary (Wigner) or an exchange (Majorana) force.

The method adopted by Thomas is based on the Schrödinger variation principle. The eigenfunctions inserted into the variational integral in order to calculate the energy are chosen symmetrical in proton and neutron whenever these particles are close enough to interact. Under these circumstances it does not make any difference whether the Wigner or Majorana type of forces is chosen. We shall therefore speak in the following in terms of ordinary forces, for simplicity.

The interaction between proton and neutron is supposed to vanish if the distance between them is larger than a certain value a. For $r<a$, we assume a potential energy

$$V(r) = -a^{-2}f(r/a)+O(a^{-1}) \qquad r<a. \quad (108)$$

$f(r/a)$ is an arbitrary function of its argument except that it shall give the correct binding energy of the deuteron. (108) has been chosen so that Va^2 stays approximately constant when a changes, which is necessary to make the binding energy of the deuteron independent of a (cf. theory of the deuteron, §12).

The solution of the wave equation for the deuteron in the potential (108) will be (cf. §12)

$$\psi = \varphi(r), \quad (108a)$$

where $\varphi(r) = Ae^{-\alpha r}/r$ for $r>a$ (108b)

with $\alpha^2 = M\epsilon\hbar^{-2}$, (108c)

ϵ being the binding energy of the deuteron. The eigenfunction φ is of course supposed to be normalized.

We shall later on need the contribution of the inner region ($r<a$) to the "kinetic energy" of the deuteron, *viz.*,

$$T = 4\pi\int_0^a (d\varphi/dr)^2 r^2 dr. \quad (108d)$$

When the range of the forces decreases, the normalizing factor A in (108b) stays almost constant because the normalization is determined chiefly by the "outside" part of the wave function (cf. §12). Therefore $\varphi(a)$ increases about as $1/a$, and so will the wave function for $r<a$. From our assumption that V retains its shape (cf. 108) becoming only proportionately larger with decreasing range a, the same follows for φ so that we may put

$$\varphi(r) = a^{-1}\chi(r/a) \qquad (r<a), \quad (108e)$$

which makes $T = K/a$, (108f)

where $K = 4\pi\int_0^1 (d\chi/dx)x^2 dx$ (108g)

is a function depending on the shape of the potential hole. For the rectangular hole,

$$K = \alpha\pi^2/4. \quad (108h)$$

The H^3 nucleus consists of one proton (coordinate $\mathbf{r}_3$) and two neutrons (coordinates $\mathbf{r}_1\mathbf{r}_2$). The wave equation of H^3 is

$$(\hbar^2/2M)(\Delta_1+\Delta_2+\Delta_3)\psi + (E-V(r_{13})-V(r_{23}))\psi=0. \quad (109)$$

We introduce the relative coordinates

$$\mathbf{s}_1=\mathbf{r}_{13}=\mathbf{r}_1-\mathbf{r}_3; \qquad \mathbf{s}_2=\mathbf{r}_2-\mathbf{r}_3 \quad (109a)$$

and suppose that ψ depends only on these relative coordinates; i.e., we leave out the motion of the center of gravity. Then (109) transforms into

$$(\hbar^2/M)(\Delta_1\psi+(\text{div}_1\cdot\text{grad}_2)\psi+\Delta_2\psi) + (E-V(s_1)-V(s_2))\psi=0, \quad (109b)$$

where Δ_1 and Δ_2 now refer to differentiations with respect to $\mathbf{s}_1$ and $\mathbf{s}_2$ rather than $\mathbf{r}_1$ and $\mathbf{r}_2$. The Schrödinger variation principle equivalent to (109b) is

$$E<[\int d\tau_1 d\tau_2\psi^2]^{-1}\int d\tau_1 d\tau_2\{(\hbar^2/M)[(\text{grad}_1\,\psi)^2 + (\text{grad}_1\,\psi\cdot\text{grad}_2\,\psi)+(\text{grad}_2\,\psi)^2] + [V(s_1)+V(s_2)]\psi^2\}.$$

This unequality is true if ψ is *any* continuous function of $\mathbf{s}_1$, $\mathbf{s}_2$. If ψ happens to be the correct eigenfunction, the "equal" sign stands instead of the "<" sign.

The success of the variational method depends on a good choice of the approximate eigenfunction ψ in (110.) First of all, we shall find the *exact* solution of the wave equation (109b) for the regions in which the potential energy is zero, i.e., for large distances between the particles ("outer wave function," region I). Then we shall set up a wave function for *small* distances s_1 between first neutron and proton and large distances s_2 of the second neutron (region II), which resembles closely the wave function of the deuteron, and join it on to the "outer" wave function. A corresponding function will hold for small s_2 and large s_1 (region III). In region IV both neutrons are near the proton.

Region I: We assume both s_1 and s_2 to be larger than the range a of the forces. Then the potential energies $V(s_1)$ and $V(s_2)$ vanish and (109b) reduces to

$$\Delta_1\psi+\text{div}_1\,\text{grad}_2\,\psi+\Delta_2\psi=\mu^2\psi \quad (111)$$

with

$$\mu^2=-ME/\hbar^2. \quad (111a)$$

The solution of (111) is

$$\psi=3^{\frac{1}{2}}K_0(\mu s)\left[\frac{\arccos(s_1/s)}{s_1(s^2-s_1^2)^{\frac{1}{2}}}+\frac{\arccos(s_2/s)}{s_2(s^2-s_2^2)^{\frac{1}{2}}}\right], \quad (111b)$$

where

$$s^2=\tfrac{2}{3}(r_{12}^2+r_{23}^2+r_{31}^2)=\tfrac{4}{3}(s_1^2-(\mathbf{s}_1\cdot\mathbf{s}_2)+s_2^2) \quad (111c)$$

is, except for the factor 2/3, the sum of the squares of the three sides of the triangle formed by the three particles, and $K_0(x)$ is the Hankel function of imaginary argument and zero order which vanishes exponentially for large arguments, *viz.*[44]:

$$K_0(x)=\tfrac{1}{2}\pi i H_0^{(1)}(ix). \quad (111d)$$

The proof that (111b) is the solution of (111) is found in Thomas' paper. We might use (111b) everywhere provided only $s_1>a$ and $s_2>a$. The region thus defined would, however, not be convenient for the integration of the variational integral (110). We shall therefore use (111b) only if, besides the conditions $s_1>a$ and $s_2>a$, the further condition $s>l$ is fulfilled where l is an auxiliary length large compared to the range a of the forces but small compared to $1/\mu$, the "radius of H^3" (definition of region I).

Region II: We assume $s_1<a$, but $s>l$. In this region we put

$$\psi=K_0(\mu s)\left[c(s)\varphi(s_1)+3^{\frac{1}{2}}\frac{\arccos(s_2/s)}{s_2(s^2-s_2^2)^{\frac{1}{2}}}-\frac{2\pi}{3^{\frac{1}{2}}s^2}\right], \quad (112)$$

where $c(s)$ is defined by

$$c(s)\varphi(a)=3^{\frac{1}{2}}\frac{\arccos(a/s)}{a(s^2-a^2)^{\frac{1}{2}}}+\frac{2\pi}{3^{\frac{1}{2}}s^2} \quad (112a)$$

and φ is the solution of the wave equation of the deuteron (cf. 108a). This choice ensures that ψ is approximately a solution of the wave equation (109b) in region II. Moreover, (112) is chosen so that it joins smoothly to solution (111b) at $s_1=a$.

[44] Cf. Jahnke-Emde, Tables of Functions, p. 199. The function K_0 is also known as Macdonald's Bessel function: Watson, *Theory of Bessel Functions*, p. 78.

Region III: $s_2 < a$, $s > l$: Analogous to region II.
Region IV: $s < l$. We put

$$\psi(\mathbf{s}_1\mathbf{s}_2) = (s/l)\psi[(l/s)\mathbf{s}_1,\ (l/s)\mathbf{s}_2]. \quad (112b)$$

This assumption serves merely to keep the wave function small in region IV.

The integral (110) can then be carried out. The rather involved calculation is found in Thomas' paper. The contribution of region I (outside) may be reduced to an integral over the *surfaces* separating region I from II, III and IV. The contribution of region IV involves the kinetic energy of the particles when they are close together, and is therefore proportional to K, (108f). Regions II and III each contribute, of course, the same amount. These contributions partly cancel the surface integrals over the surface between II or III and I, which arise from the outside region I. The net result for the ratio of the binding energies of H^3 and H^2 is

$$\frac{|E|}{\epsilon} > \frac{24(2\pi \cdot 3^{-\frac{3}{2}} - 1)}{\pi} \frac{\mu^2}{\alpha^2} |\log \mu l|^3 \times\left[1 + 0\left(\frac{K}{(a/l)|\log \mu l|}\right) + 0\left(\frac{a}{l}\right)\right]. \quad (113)$$

Sign and magnitude of the second and third term are unknown so that it is necessary to make them small. This is always possible. We have only to choose

$$a/l = \delta \ll 1; \qquad \mu l = e^{-1/K\delta^2}. \quad (113a)$$

Then

$$|E|/\epsilon > 1.6(a\alpha)^{-2}e^{-2/K\delta^2}K^{-3}\delta^{-4}(1+0(\delta)). \quad (113b)$$

The factor $e^{-2/K\delta^2}K^{-3}\delta^{-4}$ may be rather small if the second and third term in (113) are large and necessitate the choice of a small δ. However, δ and K are both *independent* of a. Therefore it is always possible to make (113b) as large as one wishes by choosing a very small range a. This shows that an infinitely short range of the forces would lead to an infinite binding energy of the H^3 nucleus. Therefore the *range of the forces between neutron and proton must be finite* though possibly very small. Furthermore to any (experimentally given) finite ratio of the binding energies of H^3 and H^2, there must correspond a definite range a of the forces, when we assume the shape of the potential function $V(r)$.

Thomas has also shown that it is impossible to assume a "δ-function" for the potential energy. Such an assumption would mean that the wave function of a neutron and a proton would have a singularity when the two particles coincide, and would behave about in the following way:

$$\psi(\mathbf{r}_1\mathbf{r}_2) = A(\mathbf{r}_1+\mathbf{r}_2)(1/r_{12}-\lambda) + 0(r_{12}). \quad (114)$$

The proof is to a large extent analogous to that for a potential energy with finite but short range.

The argument is not changed if part of the binding energy of the deuteron is attributed to a Heisenberg force (§14). As we shall show in the next section, only *half* of such a Heisenberg force would be active in the binding of the H^3 nucleus, while the full force is active in the deuteron. This will make the H^3 nucleus relatively less stable. However, the only assumption necessary for Thomas' proof is that a stable state of the deuteron exists if the force between the neutron and proton in the deuteron is the same as the average force between the proton and each neutron in H^3. This is actually the case: It follows from the scattering of slow neutrons by protons (§15) that the deuteron would have practically *no* binding energy if the force acting were the Majorana *minus* the Heisenberg force, while Majorana *plus* Heisenberg force lead to the experimental binding energy. The Majorana force plus half the Heisenberg force would therefore give some binding energy in between, *viz.*, about 50 to 65 percent of the observed binding energy of the deuteron. This is quite sufficient for the validity of Thomas' arguments given in this section.

§20. Calculation of the Energy of H^3, He^3 and He^4 from the Variation Principle (F2, F3, P8, M9, W1)

The binding energies of the light nuclei H^2, H^3, He^3, He^4 are the most suitable quantities to deduce the nuclear forces from. The procedure must of course be indirect. A law of force is assumed, in which one or more parameters are left arbitrary. The binding energies of the light nuclei are calculated as functions of the parameters in the interaction force, and these parameters are then so determined that the calculated energies agree with the observed ones.

The binding energies of H^2, H^3, He^3, He^4 are sufficient to determine *three* parameters in the law of nuclear forces. It might seem four parameters could be determined from the four nuclear binding energies. However, the binding energies of the two nuclei H^3 and He^3 are closely related to each other. H^3 consists of two neutrons and one proton, He^3 of two protons and one neutron. Now we have assumed throughout this article that the nuclear forces are symmetrical in neutrons and protons, i.e., the force between two protons is exactly the same as between two neutrons[45] except for the Coulomb repulsion. This means that the energy of He^3 must differ from that of H^3 just by the Coulomb repulsion between the two protons. This prediction seems indeed to be true, although the experimental evidence is somewhat contradictory (see §22). Thus He^3 provides a check of our assumption of the symmetry of nuclear forces in protons and neutrons.

There remain then the three nuclei H^2, H^3 and He^4 to determine the nuclear forces. It would therefore be useless to assume a force containing more than three parameters. Accordingly, we assume that the range of the forces between like particles is the same as between neutrons and protons, but that the strengths of the forces are different. In formulae, we assume a neutron-proton interaction potential

$$J(r) = -Be^{-r^2/a^2} \tag{115}$$

and an interaction potential between like particles

$$K(r) = -Ce^{-r^2/a^2}. \tag{115a}$$

We have then to determine the common range a of the forces, and the strengths B and C, respectively, of the two kinds of interactions.

The neutron-proton interaction is assumed to be of the Majorana exchange type. The nature of the interaction between the like particles will be discussed in more detail in §24. For the present, it may be assumed to be an "ordinary" force; the force suggested in §24 will give the same result for all nuclei up to He^4.

With the potentials (115) (115a), the energy of any nucleus could in *principle* be calculated exactly from the Schrödinger equation. Actually the calculation can be carried out *explicitly* only for the deuteron. (§12.) Since the deuteron consists of one neutron and one proton, its binding energy gives us just one relation between the two constants B and a determining the neutron-proton force. This relation is given in Table III for two cases: (a) if the whole binding energy of the deuteron is due to the "Majorana force" (115), and (b) if part of the binding energy is due to a "Heisenberg force" depending on the relative spins of proton and neutron, as it seems to follow from the scattering of slow neutrons by protons (§14).

In the latter case, there is given an "effective" force $\bar{B}$, which is the strength B of the Majorana force plus one-half the strength of the Heisenberg force. It is this $\bar{B}$ which enters the energy of nuclei containing an even number of neutrons or an even number of protons or both, at least in first approximation, i.e., if the Heisenberg force is small compared to the Majorana force.

In that case, the complete eigenfunction of the H^3 nucleus may be written

$$\varphi(1,2,3) = \psi(x_1, x_2, x_3)\alpha(1)\cdot 2^{-\frac{1}{2}}[\alpha(2)\beta(3) - \beta(2)\alpha(3)]. \tag{116}$$

Here the first factor is the positional wave function. The second is the spin wave function of the proton (particle 1), whose spin is supposed to be parallel to the z axis. (The spin function α indicates a spin parallel to z, β a spin antiparallel to z, cf. 92a.) The last factor is the spin wave function of the two neutrons (particles 2, 3) whose spins are of course opposite to each other. The factor $2^{-\frac{1}{2}}$ stands for normalization.

Now consider a Majorana potential V_M and a Heisenberg potential V_H to act between the proton and the neutron 2. The potential energy operator operated on (116) gives then

$$\begin{aligned}(V_H + V_M)\varphi &= 2^{-\frac{1}{2}}\psi(x_2x_1x_3)\{J_M(r_{12})\alpha(1)[\alpha(2)\beta(3) - \beta(2)\alpha(3)] \\ &\quad + J_H(r_{12})\alpha(2)[\alpha(1)\beta(3) - \beta(1)\alpha(3)]\},\end{aligned} \tag{116a}$$

where J_H and J_M are the Heisenberg and Majorana potentials as functions of the distance r_{12}. The average value of the potential energy is then

$$\begin{aligned}\sum_\sigma \varphi(V_M + V_H)\varphi d\tau &= \int d\tau\psi(x_1x_2x_3)\psi(x_2x_1x_3)J_M(r_{12}) \\ &\quad \times \tfrac{1}{2}\sum_\sigma \alpha^2(1)[\alpha(2)\beta(3) - \beta(2)\alpha(3)]^2 \\ &\quad + \int d\tau\psi(x_1x_2x_3)\psi(x_2x_1x_3)J_H(r_{12}) \\ &\quad \times \tfrac{1}{2}\sum_\sigma[\alpha^2(1)\alpha^2(2)\beta^2(3) - \alpha^2(1)\alpha\beta(2)\alpha\beta(3) \\ &\quad - \alpha\beta(1)\alpha^2(2)\alpha\beta(3) - \alpha\beta(1)\alpha\beta(2)\alpha^2(3)].\end{aligned} \tag{116b}$$

When we carry out the summation over the spins of all

[45] It would be in accord with this assumption if the forces between like particles vanish identically.

three particles, we find two terms contributing unity to the first sum but only one term in the second, all other terms vanishing (cf. 92e). Therefore (116b) reduces to

$$\sum_{\sigma}\int\varphi(V_M+V_H)\varphi d\tau = \int d\tau\psi(x_1x_2x_3)\psi(x_2x_1x_3)(J_M+\tfrac{1}{2}J_H), \quad (116c)$$

which shows that half the Heisenberg interaction is to be added to the full Majorana term.

It might have been expected that the Heisenberg force would not contribute at all to the energy of nuclei like H^3, containing only pairs of neutrons. For one might think that of the two neutrons *one* would have spin parallel to the given spin of the proton, the other antiparallel so that the contributions of the Heisenberg forces due to the two neutrons would cancel each other. The fallacy is due to the use of the word antiparallel: Two spins are antiparallel only if the wave function is antisymmetrical in the respective spin coordinates. In the wave function $\alpha(1)\beta(2)$, however, merely the z components of the spins 1 and 2 are antiparallel, the spins themselves might still be just as well parallel as antiparallel, because we can form either of the two wave functions

$$\begin{aligned}&2^{-\frac{1}{2}}(\alpha(1)\beta(2)+\alpha(2)\beta(1)),\\&2^{-\frac{1}{2}}(\alpha(1)\beta(2)-\alpha(2)\beta(1)),\end{aligned} \quad (116d)$$

the first corresponding to parallel, the second to antiparallel spins. The correct reasoning is therefore as follows: If the components of two spins in a given direction are parallel, the spins are certainly parallel; if the components are opposite, it is just as likely that the spins are parallel as that they are antiparallel. Thus the neutron which has its spin *component* parallel to the proton will have just once the Heisenberg interaction with the latter, while the neutron whose spin is opposite the proton spin, has no Heisenberg interaction at all with the proton. Therefore we get two Majorana interactions and one Heisenberg interaction in the H^3 nucleus, or per neutron one Majorana and one-half Heisenberg term, as we found in (116c).

Accordingly, the value $\bar{B}$ of Table III has to be used for computations of the energies of H^3 and He^4.

We return to the problem of determining the binding energy from the given force. While the Schrödinger equation can be integrated explicitly for the deuteron, this is not the case for the nuclei H^3 and He^4. Approximate methods must be used. The method used most frequently for similar problems is the Ritz variational method.[46] It gives a definite upper bound for the energy of any quantum mechanical system, but this upper bound sometimes converges very slowly towards the correct energy. The energy of nuclei is, unfortunately, one of the cases in which this is true. The reason for this is the very short range of the nuclear forces. We know from the theory of the deuteron (§12) that the correct wave function (cf. 38a) changes very rapidly when the two particles have a distance smaller than the range a of their mutual forces, while for large distances $r>a$ the wave function varies very slowly. The same will be true for the correct wave function of more complicated nuclei: We shall find a comparatively *slow* variation of the wave function *in general*, and superposed on it a *rapid* change *whenever* two nuclear *particles* come close *together*. Such a combination of rapidly and slowly varying functions cannot be easily represented analytically. In fact, a rather complicated trial wave function had to be chosen by Thomas (§19) in order to prove that the binding energy of H^3 tends to infinity when the range of the forces goes to zero. The use of a wave function of similar complication for *quantitative* computations of the binding energy for a finite range of the forces would be prohibitively laborious, especially for more complicated nuclei than H^3. On the other hand, any reasonably *simple* analytical expression will be a poor approximation to the correct wave function, because the analytical expression will change too rapidly when the nuclear particles are far away from each other, and too slowly when they are close together. This will be particularly true if the range of the forces is very small. Therefore we cannot expect any too good agreement between the results of the Ritz variational method and the experimental binding energies, and we must expect the discrepancy to become worse with decreasing range of the forces. (See following, especially Table IV.)

The first thing to do in applying the variational method is to choose an approximate wave function for the system. A sufficiently simple function for the H^3 nucleus is

$$\psi=\exp\left[-\tfrac{1}{2}\nu(r_{12}^2+r_{13}^2)-\tfrac{1}{2}\mu r_{23}^2\right], \quad (117)$$

where the subscript 1 denotes the proton, 2 and 3 the two neutrons so that r_{12} is the distance of the first neutron from the proton and r_{23} the distance between the two neutrons. The function $e^{-\frac{1}{2}\nu r^2}$ "ties together" two unlike particles, $e^{-\frac{1}{2}\mu r^2}$ two like particles. Since the forces between unlike particles are larger, we expect ν to be larger than μ. For the α-particle, we choose analogously:

[46] Cf., e.g., *Handbuch der Physik*, Vol. 24, 1, p. 353.

$$\psi=\exp\left[-\tfrac{1}{2}\nu(r_{13}^2+r_{14}^2+r_{23}^2+r_{24}^2)-\tfrac{1}{2}\mu(r_{12}^2+r_{34}^2)\right], \quad (117a)$$

1 and 2 referring to the protons, 3 and 4 to the neutrons.

With the wave functions (117), (117a) all integrals occurring in the variational method can be carried out elementarily; e.g., for the H^3 nucleus we obtain for the kinetic energy

$$E_{\text{kin}}=-\frac{\hbar^2}{2M}\frac{\int\psi(\Delta_1+\Delta_2+\Delta_3)\psi d\tau_2 d\tau_3}{\int\psi^2 d\tau_2 d\tau_3}=3\hbar^2/2M(\mu+2\nu), \quad (117b)$$

where Δ_1 is the Laplacian with respect to the coordinates of the first particle, and the integrals have to be extended over all positions of the second and third particle relative to the first. The potential energy becomes

$$E_{\text{pot}}=-2\bar{B}\frac{\int e^{-r_{12}^2/a^2-\nu r_{12}^2-\frac{1}{2}(\mu+\nu)(r_{13}^2+r_{23}^2)}d\tau_2 d\tau_3}{\int e^{-\nu(r_{12}^2+r_{13}^2)-\mu r_{23}^2}d\tau_2 d\tau_3}-C\frac{\int e^{-r_{23}^2/a^2-\mu r_{23}^2-\nu(r_{12}^2+r_{13}^2)}d\tau_2 d\tau_3}{\int e^{-\nu(r_{12}^2+r_{13}^2)-\mu r_{23}^2}d\tau_2 d\tau_3} \quad (117c)$$

$$=-16\bar{B}\left[\frac{\nu(\nu+2\mu)}{(\nu+\mu)(5\nu+\mu+4a^{-2})}\right]^{\frac{3}{2}}-C\left[\frac{\nu+2\mu}{\nu+2\mu+2a^{-2}}\right]^{\frac{3}{2}}. \quad (117d)$$

The form of the integral in the first numerator in (117c) is due to the Majorana type of the forces: The particles 1 and 2 are supposed to interact, we have therefore to multiply the wave function (117) with the function obtained by interchanging particles 1 and 2 in (117) instead of taking the square of (117). The factor 2 in the first term of (117c) arises from the fact that we have *two* neutrons interacting with the proton.

The expressions (117b) (117d) may be simplified by introducing the abbreviations

$$\sigma=\tfrac{1}{4}(5\nu+\mu)a^2 \qquad p=(2\nu+4\mu)/(5\nu+\mu) \qquad T=\hbar^2/Ma^2. \quad (118a)$$

Then the upper bound for the energy of H^3 becomes

$$E^0(H^3)=(2+p)\sigma T-2\bar{B}[p(4-p)/(1+2p)]^{3/2}(\sigma/\sigma+1)^{3/2}-C(p\sigma/p\sigma+1)^{3/2}. \quad (118)$$

Similarly, we find for the α-particle

$$E^0(He^4)=\tfrac{3}{2}(2+p)\sigma T-4\bar{B}[p(2-p)]^{3/2}/(\sigma/\sigma+1)^{3/2}-2C(p\sigma/p\sigma+1)^{3/2} \quad (119)$$

with the abbreviations

$$\sigma=\tfrac{1}{2}(3\nu+\mu)a^2, \qquad p=(2\nu+2\mu)/(3\nu+\mu) \quad (119a)$$

(118) and (119) have to be minimized by varying the parameters p and σ.

We want to carry out the calculations first without assuming a force between like particles, in order to obtain an idea about the degree of approximation afforded by (118), (119). We know from the preceding section that the eigenvalue of H^3 must tend to minus infinity if we let the range a of the forces go to zero, i.e., if B and T in (118) go to infinity, and if at the same time the relation between B and T is kept such that the binding energy of the deuteron remains correct. This must hold *a fortiori* for He^4. We may reasonably expect the binding energy of both nuclei to decrease monotonously with decreasing a.—For very *long* range of the forces, the binding energy of H^3 will be just twice, that of He^4 just four times the binding energy of the deuteron.

Actually, the minimum of (118) and, to a lesser extent, that of (119), behave very differently from these expectations. While for long range of the forces the result is rather satisfactory, the minimum of $E^0(He^4)$ decreases only very slowly with decreasing range of the forces, and that of $E^0(H^3)$ even *increases* with decreasing a. This shows that the variational method becomes increasingly worse with decreasing range of the forces, which is to be expected (see above).

The procedure to determine the minimum of (118), (119) is the following. First of all, the minimum with respect to p is determined. The position of that minimum depends only slightly on the value of σ, and the dependence of the minimum energy on p is quite negligible.* The minimum lies at

* When the ratio of the potential energy of the H^3 nucleus to its kinetic changes from 1.0 to 1.2, p_0 changes from 0.735 to 0.765; the coefficient of T in (117c) from 2.735 to 2.765, i.e., about 1.1 percent; the coefficient of B increases by 1.0 percent; the ratio of the coefficients, which is the most important quantity determining the binding energy, rises by 0.12 percent which is quite negligible.

TABLE IV. *Energies of H^3 and He^4 as functions of the range of the neutron-proton forces.* (All energies in MV, a in 10^{-13} cm.)

			H^3 ($E_{obs}=-8.3$)			He^4($E_{obs}=-27.6$)				
T	a	$\bar{B}$	σ_0	E^0	E_{Feen}	σ_0	E_{kin}	E_{pot}	E^0	E_{Feen}
5	2.86	22.05	0.740	− 1.60	——	1.080	23.65	32.7	− 9.05	——
10	2.02	38.4	0.551	− 0.44	−2.8	0.891	38.95	49.15	−10.2	−15.6
20	1.43	69.4	0.389	+ 1.67	+1.0	0.760	66.5	77.9	−11.4	−17.4
40	1.01	129.2	(0.25)*	+ 5.30	+4	0.666	116.5	129.5	−13.0	−19.3
80	0.715	245	(0.25)*	+12.9	——	0.593	207.5	220.0	−12.5	——

* The expression (118c) has, for these values of T, no minimum. 0.25 is the position of the minimum when it just disappears.

$$p_0=0.75, \text{ corresponding to } \mu=0.54\nu \text{ for } H^3, \quad (118b)$$

$$p_0=0.92, \text{ corresponding to } \mu=0.7\nu \text{ for } He^4. \quad (119b)$$

Inserting these values, (118), (119) reduce to

$$E^0(H^3) = 2.750T\sigma-1.925_5\bar{B}(\sigma/\sigma+1)^{3/2}, \quad (118c)$$

$$E^0(He^4)=4.378T\sigma-3.961\,\bar{B}(\sigma/\sigma+1)^{3/2}. \quad (119c)$$

The minimum of these expressions as functions of σ is found for σ_0, which is given by

$$(\sigma_0+1)^{5/2}\sigma_0^{-\frac{1}{2}} = \begin{cases} (3/2)(1.925_5/2.750)\bar{B}/T & \text{for } H^3, \quad (118d) \\ (3/2)(3.961/4.378)\bar{B}/T & \text{for } He^4. \quad (119d) \end{cases}$$

From these equations, σ_0 can easily be determined for any given ratio

$$\lambda=\bar{B}/T=M\bar{B}a^2/\hbar^2. \quad (118e)$$

σ_0 is then inserted in (118c, 119c) and E^0 calculated.

The result is shown in Table IV. For 5 different ranges of the forces (column 2), the strength of the force, i.e., the constant $\bar{B}$ (column 3), is computed from the theory of the deuteron (Table III, §12), due account being taken that part of the binding of the deuteron arises from Heisenberg forces (cf. above, and §14). For each pair a, $\bar{B}$ (or T, $\bar{B}$), σ_0 and the total energy is calculated for H^3 and He^4. In addition, the values of the energy derived by Feenberg (F2) are given; they differ from ours because he assumed the deuteron binding energy to be entirely due to Majorana forces.[47] Moreover, for the α-particle the kinetic and potential energy are given separately, in order to show that both these quantities increase rapidly when the forces become stronger and of shorter range.

[47] Feenberg corrected for the Heisenberg forces at a later stage of his calculations.

The result is by no means satisfactory. Even for the α-particle, the lowest value of the energy obtainable is −13.0 MV, i.e., only one-half of the observed binding energy of 28 MV. (Feenberg's values are somewhat better, because he assumed a stronger force for any given value of the range, for the reason mentioned.) For H^3, the binding energy disappears completely for ranges below $2.6\cdot10^{-13}$ cm. The reason for the unsatisfactory result is, of course, that our wave functions (117), (117a) represent a very poor approximation to the exact eigenfunction if the range of the force is short.

Feenberg has tried to get a better approximation by choosing better wave functions: The best approximation was obtained with the wave function

$$\psi=\psi_0+\lambda H\psi_0, \quad (120)$$

where λ is a parameter to be varied, ψ_0 the wave function (117) and H the Hamiltonian operator, i.e., the sum of the operators of kinetic and potential energy. Wave functions of the type (120) were first introduced for use with the variational method by Hassé (H6). They often correct the original wave function ψ_0 to something very well approximating the correct one. In our case, (120) gives an energy of about $E^0+0.6(E'-E^0)$ where E^0 is the energy calculated from (117c) and E' that following from Feenberg's "method of the equivalent two-body-problem" (cf. §21).

A still better result for H^3 with a variational method was obtained by Present (P8). He chose a simple exponential potential $V_0e^{-r/a}$ between neutrons and protons, with $a=10^{-13}$ cm and $V_0=97$ MV. This value of V_0 follows from the theory of the deuteron (Table III), it is not corrected for Heisenberg forces. No forces are assumed to act between the two neutrons. Present assumed furthermore that the interaction was of the "ordinary" (Wigner) rather than the exchange (Majorana) type (this simplifies calculations and should give a somewhat too high result for the binding energy). For the wave function, Present chose a simple exponential times a power series in the distances between the particles, varying the coefficients of the terms in the series. With eight terms in the power series, he obtained a binding energy of 4.85 MV for the H^3 nucleus. The results for the binding energy obtained with increasing number of terms in the eigenfunction showed rapid convergence. He therefore concludes that the correct binding energy is 4.9 ± 0.05 MV. A similar value (about 4.7 MV for $a=10^{-13}$ cm, or, in their notation, $\mu=0.5\cdot 10^{13}$ cm^{-1}) was obtained by Massey and Mohr (M9) by a variation calculation using a different expression for the approximate wave function. Feenberg's method of the equivalent two-body problem which will be described in the following section, gives, according to Present, 4.5 MV, i.e., again almost the same binding energy.

In judging these figures, it should be borne in mind that a binding energy of twice the binding energy of the deuteron, i.e., 4.3 MV, would be expected for H^3 even in the most unfavorable case, *viz.*, for infinitely long range of the forces. The actual results found from the calculations are only very slightly larger than this value, at the best, by about 15 percent. On the other hand, the *observed* binding energy of 8.3 MV is nearly twice as large as our "elementary" figure 4.3 MV. This seems to show that a *very much shorter range of the forces would be required* in order to explain the observed mass defect of H^3, if we assume only forces between neutrons and protons and none between two neutrons. On the other hand, no range of the forces shorter than $1\cdot 10^{-13}$ cm seems to be reconcilable with the binding energies of α-particle and H^2 (Table V, §21). We must therefore *conclude that attractive forces between two neutrons must exist in order to explain the observed mass defects of H^2, H^3, He^4.*

This conclusion is strengthened by the fact that Present has purposely made some neglections in his calculations which will tend to make his result for the binding energy too big: Firstly, he assumed Wigner instead of Majorana forces. Secondly, he did not take into account the fact that part of the binding energy of H^2 is due to Heisenberg forces which will contribute relatively less to the binding energy of H^3 (see above). Thirdly, a range of the forces of $1\cdot 10^{-13}$ cm seems already somewhat shorter than can be reconciled with the theory of the α-particle (Table V). Present's conclusion that forces between neutrons must exist, is also strongly confirmed by the more extensive calculations presented in the next section.

The opposite conclusion has been reached by Dolch (quoted by Weizsäcker, W1) also on the grounds of a variational calculation. The details of Dolch's calculation are not yet available, but only some curves representing the values of B and a in the expression (115) which are necessary to obtain the correct binding energies of H^2, H^3 and He^4, respectively (Fig. 1 in Weizsäcker's paper). Apparently, the data for H^2 have not been obtained by an *exact* solution of the Schrödinger equation for that nucleus but also by a variational method. What the particular method was, seems rather doubtful. It seems from the data published that the wave function used was less good than the simple wave function (121); e.g., for $a=1.4\cdot 10^{-13}$ cm, the wave function (121) gives, according to (121a), the correct binding energy for the deuteron if B is chosen to be 87 MV. The value obtained from Dolch's calculation for $a=1.4\cdot 10^{-13}$ cm ($b=2.0$ in Weizsäcker's notation) is 0.100 mass unit = 93 MV, i.e., *more* than from (121).[48] Since the B required to give the observed value for the binding energy of the deuteron is larger in Dolch's calculations, his wave function represents a poorer approximation to the wave function of the deuteron than even the simple function (121). On the other hand, his results for H^3 and He^4 seem to be better approximations than those obtained from (117), (117a) and listed in Table IV. It does not seem consistent to us to compare a fairly good approximation for H^3 and He^4 with a very poor approximation for H^2. For these reasons, we cannot follow the conclusions drawn by Weizsäcker and Dolch from Dolch's calculations, *viz.*, that no forces between like particles need to be assumed to explain the mass defects of H^2, H^3, He^4. (Moreover, if Dolch's calculations are corrected for Heisenberg forces it becomes even more necessary to assume forces between like particles.)

From the variational calculations of Present and of Massey and Mohr we can also conclude that Feenberg's method of the equivalent two-body problem (§21) gives an almost correct result for the binding energy of H^3 (4.5 MV, compared to 4.9 MV according to the best variational calculation of Present). We shall therefore apply this method with some confidence to H^3 and He^4 in the next section.

[48] We are indebted to Dr. Feenberg for drawing our attention to this point.

§21. FEENBERG'S "EQUIVALENT TWO-BODY PROBLEM" (F2, F3)

Since the variational method gives, at least with simple choices for the approximate wave function, very poor results for the binding energy of H^3 and He^4, that method cannot be used for a determination of the nuclear force constants from the observed binding energies. A better method has to be devised. The method used thus far most extensively is Feenberg's method of the equivalent two-body problem. A similar method has been used by Wigner in his first calculation of the binding energies of He^4 (W12).

The method of the equivalent two-body problem cannot be founded rigorously. But it is about as likely that it gives too low a result for the binding energy for a given force, as too high one. Moreover, Present has shown that for H^3 the use of a better approximate wave function with the variation method, gives a binding energy very nearly equal to that obtained by the "equivalent two-body problem."[49]

The method proceeds as follows: The variation method is applied to the *deuteron*, with the same type of wave function which was used for H^3 and He^4 (cf. 121). The energy of the deuteron (cf. 121a) resulting from the variation method has a form exactly analogous to that found for energies of the nuclei H^3 and He^4 (cf. 118c, 119c), only the coefficients being different. The "variational energy" of H^2 would therefore have exactly the same value as that for H^3, if the constants T and B determining the force were replaced by some other constants T' and B' which are multiples of T and B chosen in such a way as to make up for the difference in the coefficients in formulae (118c) and (121a). Next the exact wave equation of the deuteron is solved with the new constants T' and B' which may be called the "equivalent force constants" for the H^3 nucleus. The result of the solution of the deuteron equation with the force constants $T'B'$ ("equivalent two-body problem") is assumed as the true energy of the H^3 nucleus.

The wave function chosen for the deuteron is

$$\psi = e^{-\frac{1}{2}\nu r^2} \tag{121}$$

with r the distance of proton and neutron. With this wave function, the energy of the deuteron resulting from the variation principle is

$$E^0(H^2) = \tfrac{3}{2}T\sigma - B(\sigma/\sigma+1)^{\frac{3}{2}}, \tag{121a}$$

where
$$\sigma = \nu a^2. \tag{121b}$$

We may write the energy (117c) of H^3 in a form analogous to (121a), *viz.*,

$$E^0(H^3) = \tfrac{3}{2}T'\sigma - B'(\sigma/\sigma+1)^{\frac{3}{2}} \tag{122}$$

if we put

$$T' = 1.833T, \qquad B' = 1.925\bar{B}. \tag{122a}$$

The method of the equivalent two-body problem assumes that the energy of the H^3 nucleus is equal to the eigenvalue E' in the wave equation (equivalent two-body equation)

$$(\hbar^2/M)\Delta\psi + (E' + B'e^{-r^2/a'^2})\psi = 0 \tag{122b}$$

where (cf. 118a)

$$a'^2 = \hbar^2/MT'. \tag{122c}$$

The eigenvalue E' in (122b) may be obtained, without any further calculation, from Table III in §12. Suppose we have two equations of the type (122b), one with the constants E', B', a', T' and another with the constants E'', B'', a'', T''. Suppose furthermore that

$$B'/T' = B''/T'' \tag{123}$$

or in other words

$$B'a'^2 = B''a''^2. \tag{123a}$$

Then the ratio of the energies is the same as the ratio of the potentials, *viz.*:

$$E'/B' = E''/B''. \tag{123b}$$

To see this, introduce into (122b) new coordinates which are a'/a'' times the old ones, thus:

$$r'' = ra''/a', \quad \Delta'' = (a'/a'')^2\Delta, \tag{123c}$$

where Δ'' denotes the Laplacian with respect to the new coordinates. Multiplying the resulting equation by $(a'/a'')^2$, we find

$$(\hbar^2/M)\Delta''\psi + (E'(a'/a'')^2 + B'(a'/a'')^2e^{-r''^2/a''^2})\psi = 0 \tag{123d}$$

which, with (123a) and (123b), is equivalent to

$$(\hbar^2/M)\Delta''\psi + (E'' + B''e^{-r''^2/a''^2})\psi = 0, \tag{123e}$$

showing that $E'' = E'B''/B'$ is the eigenvalue corresponding to the force field $B''T''$ if E' is the eigenvalue corresponding to the force $B'T'$. Now Table III gives the pairs of values $B''T''$ which lead to the binding energy $E'' = -2.14$ MV. From these, we can easily deduce the eigenvalue corresponding to a given pair of force constants $B'T'$ by use of (123b).

As an illustration, we want to calculate the energy of H^3 for the case $T = 10$ MV. Table III gives us as the value $\bar{B}$

[49] The deviation of the result of the equivalent two-body problem from that of Present's refined variational calculation is only about 6 percent of the difference between the simple variation method and Present's calculation.

corresponding to this T according to the theory of the deuteron, $\bar{B}=38.4$ MV. From (121a), we find for the force constants of the two-body problem equivalent to the H^3 problem, $T'=18.33$ MV, $B'=73.9$ MV. The ratio of the two is $B'/T'=4.03$. We seek in Table III this particular ratio B''/T'' and find listed above it the value $T''=12.5$ MV. This value of T'' would lead to a binding energy $-E''=2.14$ MV. The binding energy of H^3 therefore becomes $-E'=-E''T'/T''=2.14\cdot 18.3/12.5=3.13$ MV.

The analogous procedure can be carried out for He^4, only the relation between the equivalent force constants $B'T'$ and the given constants of the actual force, $\bar{B}\ T$, being changed to (cf. 118c, 121a)

$$T'(He^4)=2.919T, \quad B'(He^4)=3.961\bar{B}. \tag{124}$$

TABLE V. *Energies of H^3 and He^4 from the method of the equivalent two-body problem, as functions of the range of the forces.* (No forces between like particles.) (All energies in MV, range a in 10^{-13} cm. Values in parentheses estimated.)

			H^3 obs. $E=-8.3$		He^4 obs. $E=-27.6$			
T	a	$\bar{B}$	$-E'$	$-E_{Feen}$	$-E'$	C	$-E$	$-E_{Feen}$
5	2.84	22.05	2.82	(4.2)	11.05	0.56	10.5	(18.2)
10	2.02	38.4	3.00	4.5	14.05	0.72	13.3	24.3
15	1.65	54.0	3.12	4.7	16.7	0.83	15.9	26.8
20	1.43	69.4	3.29	4.9	19.15	0.93	18.2	29.0
30	1.16_5	99.3	3.46	5.2	23.65	1.10	22.55	(33.2)
40	1.01	129.2	3.70	5.4	28.25	1.24	27.0	—
50	0.905	158.3	3.86	—	32.55	1.35	31.2	—
60	0.825	187.	4.02	—	36.5	1.45	35.05	—
80	0.715	245.	4.30	—	44.3	1.63	42.7	—

Table V gives the result of the calculations for H^3 and He^4 for various ranges a of the force. The forces between like particles are still assumed to be zero. E' is the energy derived from the "equivalent two-body problem." C is the correction for the Coulomb repulsion of the two protons in the α-particle (cf. §22), $E=E'+C$ is the total calculated energy. In the table the energies calculated by Feenberg are included, the difference being again that Feenberg determined the force constant B for every range a so that the total binding energy of the deuteron is accounted for by the Majorana force while we considered part of it as due to a Heisenberg force. The differences are seen to be rather large, Feenberg's binding energies being, or course, much larger for any given range a of the forces.

We observe that the binding energy of the α-particle increases rapidly with decreasing range of the forces, the experimental value of 27.6 MV being obtained for

$$T=41.5 \text{ MV}, \quad a=0.99\cdot 10^{-13} \text{ cm}, \quad \bar{B}=133 \text{ MV}. \tag{125}$$

With Feenberg's assumptions about the origin of the binding energy of the deuteron, the potential hole would be much less deep and narrow, *viz.*,

$$T=16.8 \text{ MV}, \quad a=1.56\cdot 10^{-13} \text{ cm}, \quad \bar{B}=60 \text{ MV}. \tag{125F}$$

The binding energy of the H^3 nucleus turns out to be much smaller than the observed one (8.3 MV) and to increase only very slowly with decreasing range. For the range which yields the experimental value for the binding energy of the α-particle ($T=41.5$ MV) the binding energy of H^3 is only 3.7 MV, i.e., 45 percent of the observed value. (With Feenberg's assumptions, it would be markedly better, *viz.*, 4.8 MV at $T=17$ MV, which, however, is still unsatisfactory.) We must therefore conclude that either (a) the method of the equivalent two-body problem, while satisfactory for the α-particle, gives much too small binding energies for the H^3 nucleus, or (b) there are additional forces which depress the energy of H^3 relatively much more than that of the α-particle.

The first possibility can be excluded almost with certainty, in view of the variational calculations of Present (cf. end of §20). Therefore we adopt alternative (b). The additional forces which we assume are attractive forces between like particles. This assumption is the more preferable over assumption (a) since we have already found other evidence for the forces between like particles (§10, §18).

The energy of H^3 and He^4 resulting from the variation method and including forces between like particles, is given in (118), (119). These two expressions do not have exactly the same form as the "variational energy" of H^2 given in (121a). For the last term in (118) as well as (119) contains $(p\sigma/p\sigma+1)^{\frac{3}{2}}$ instead of $(\sigma/\sigma+1)^{\frac{3}{2}}$. Therefore the method of the equivalent two-body problem cannot be applied immediately.

Two procedures suggest themselves: (a) We put simply $p=1$. In this way, we lose one parameter in the variation principle and therefore impair the result somewhat. The error will, however, not be serious because the introduction of forces between like particles tends to equalize the constants μ and ν in the wave function and therefore to bring p closer to one. If the forces between like particles were equal to those between unlike ones, we would have exactly $\mu=\nu$ and $p=1$. For the ratio of the forces derived below (Eq. (128)), one finds $p=0.94$ for H^3 and $p=0.985$ for He^4 as the values giving the minimum of the energy.

(b) We introduce a new parameter σ' by putting

$$\sigma=\sigma'(1+c(1-p)). \qquad (126)$$

c can then be fixed so that the *sum* of the second and third term in (118) has the form $(2Bf(p)+C)(\sigma'/\sigma'+1)^{3/2}$. This is achieved by putting approximately

$$c=C/(2B+C). \qquad (126a)$$

The minimum of (118) with respect to p occurs then approximately for

$$p=1-(B-C)/2(2B+C) \qquad (126b)$$

and (118) reduces to

$$E(H^3)=\left[3-\left(\frac{B-C}{2B+C}\right)^2\right]T\sigma' + \left[2B+C-\tfrac{1}{4}B\left(\frac{B-C}{2B+C}\right)^2\right](\sigma'/\sigma'+1)^{3/2}. \qquad (126c)$$

The results obtained from this equation are almost the same as from method (a). The constant C, determined in the way described below from the binding energy of H^3 turns out 0.15 MV smaller by method (b) than by method (a) if we take the range of the forces finally chosen (cf. $T=8$ MV). For shorter range, the correction is larger, being 0.9 MV for $T=20$ and 1.7 MV for $T=30$. But these differences are, of course, very small compared to the accuracy of the method. For the α-particle the effect of applying method (b) would be still smaller. We have applied method (b) in the final calculations, but we shall now describe the procedure according to method (a).

Putting $p=1$ in (118), (119) and comparing the result to (121a), we obtain for the parameters in the equivalent two-body problems:

$$\text{for } H^3: \quad T'=2T, \quad B'=2B+C, \qquad (127)$$

$$\text{for } He^4: \quad T'=3T, \quad B'=2(2B+C). \qquad (127a)$$

Since we want now to fix *three* constants B, C and T (T is equivalent to the range of the forces a), we must use the exact binding energies of all three nuclei H^2, H^3 and He^4. Therefore we choose the following procedure: We take a given value of T. The value of $\bar{B}$ corresponding to it can be read directly from Table III, which is based on the theory of the deuteron. C can then be determined from the observed binding energy of the H^3 nucleus, *viz.*, $E'=8.28$ MV$=3.87$ times the binding energy of the deuteron. The method of determination is that described in Eqs. (123) to (123e): We know that the value $T'=2T$, together with the unknown value of B', must give the observed binding energy $E'=3.87E_0$. Therefore the force constants $T''=T'/3.87$, and $B''=B'/3.87$, would give exactly the binding energy of the deuteron. Therefore we have only to look in Table III for the value $T''=T'/3.87 =2T/3.87$, and to find the corresponding value of B''/T'' from the table: This value of B''/T'', multiplied by the given $T'=2T$, gives the required value B' for the H^3 nucleus. Subtracting $2\bar{B}$ from B', we find *that* value of C which gives the correct binding energy of H^3 (and H^2) for the given value of T; we may call it $C(H^3, T)$.

The same procedure is then carried out for the α-particle. In this case, a correction has first to be applied to the observed binding energy to allow for the Coulomb repulsion of the two protons (for its determination, see (129b)). The result of the calculation is another value for C, *viz.*, $C(He^4, T)$ which is the value necessary to give the correct binding energy of He^4 and H^2.

TABLE VI. *Strength of the force between like particles necessary to explain the binding energies of H^3 and He^4.* (Equivalent two-body problem. Energies in MV, a in 10^{-13} cm.)

T	a	$\bar{B}$	H^3 B'	H^3 C	He^4 COUL.*	He^4 1/2 B'	He^4 C
5	2.86	22.05	62.15	18.05	0.70	64.7	20.6
7½	2.33	30.4	81.75	20.95	0.79	81.85	21.05
10	2.02	38.4	100.3	23.5	0.86	97.5	20.7
15	1.65	54.0	135.8	27.8	0.96	127.7	19.7
20	1.43	69.4	169.3	30.5	1.05	156.7	17.9
30	1.165	99.3	233.8	35.2	1.18	211.4	12.8

* Coulomb energy of the two protons.

It is seen from Table VI that $C(H^3, T)$ increases rapidly with decreasing range of the forces while $C(He^4, T)$ stays constant down to about $1.5\cdot 10^{-13}$ cm range and then *decreases.* This is due to the fact that the binding energy of the α-particle may be explained without having recourse to like-particle forces if we only assume $a=1.00\cdot 10^{-13}$ cm while the corresponding calculations for H^3 (Table V) do not give a satisfactory result.

We now choose the point where the curves $C(H^3, T)$ and $C(He^4, T)$ cross. The constants of the force for this point are

$$T=7.6 \text{ MV}, \quad \bar{B}=30.7 \text{ MV}, \quad C=21.0 \text{ MV}, \quad a=2.32\cdot 10^{-13} \text{ cm}. \qquad (128)$$

These constants give, according to our method, the correct binding energies of the three light nuclei H^2, H^3 and He^4.

Feenberg obtained slightly different constants by assuming in the first part of his calculation that there are no Heisenberg forces and then correcting for them afterwards. His values are

$$\bar{B}=34 \text{ MV}, \quad C=21 \text{ MV}, \quad a=2.17\cdot 10^{-13} \text{ cm}, \tag{128F}$$

the difference arising from slightly different experimental values for the binding energies of $H^2H^3He^4$.

It must be emphasized that the result (128) is by no means accurate. There is no way of estimating the error of the method of the equivalent two-body problem. It seems that the value of C (strength of the forces between like particles) is not as sensitive to errors as the values of B (force between proton and neutron) and a (range of the forces). E.g., let us suppose that the C values for the α-particle are too low by 2 MV, and those for H^3 too high by the same amount. Then the "crossing point" would be shifted to $T=11.2$ MV, $\bar{B}=42$ MV, $C=22.5$ MV, $a=1.91\cdot 10^{-13}$ cm, i.e., $\bar{B}$ would increase by almost 40 percent, a would show a corresponding decrease by almost 20 percent, while C would change by only 7 percent. The present determination can therefore not be regarded as final. However, at least the existence of forces between like particles seems almost certain, and the order of magnitude should not be very different from (128).

Moreover, it is very gratifying that the values for the force between like particles obtained from our theory *agree almost perfectly with those following from the scattering of protons by protons* (§18). This agreement is very satisfactory and represents at present the *only real check on our fundamental assumptions about nuclear forces.* That this is so, can be seen as follows: We have five accurate experimental data at our disposal in order to fix the constants in the theory of nuclei, *viz.*, the binding energies of $H^2H^3He^4$, the scattering of slow neutrons by protons, and the scattering of protons by protons. The first four are needed to fix the four constants: range of the forces, strength of the Majorana and Heisenberg forces between neutron and proton, and strength of the proton-proton forces. Only the fifth experimental result is therefore available for a check of the consistency of our assumptions.

§22. Comparison of H^3 and He^3

We have assumed throughout that the nuclear forces are symmetrical in neutrons and protons (§6); i.e., the force between two neutrons is assumed to be the same as that between two protons, disregarding the Coulomb force between the latter. From this point of view, the difference between the binding energies of H^3 and He^3 should be due entirely to the Coulomb repulsion between the two protons in He^3. A computation of this effect will give the most direct check of the assumption of symmetry of the nuclear forces in neutrons and protons. At the same time, we may compute the Coulomb energy for the α-particle, which was needed as a correction to the calculations in the preceding section (Tables V and VI).

The Coulomb energy may be calculated very easily from the wave function (117); the result is for He^3

$$\text{C.E.}=e^2[2(\nu+2\mu)/\pi]^{\frac{1}{2}}. \tag{129}$$

Similarly, we obtain for the α-particle

$$\text{C.E.}=e^2[4(\nu+\mu)/\pi]^{\frac{1}{2}}. \tag{129a}$$

In order to compute these expressions, we have to calculate ν and μ. It seems sufficiently accurate to put $\nu=\mu$ which is very nearly true for our case (see above). Then we may use (118a) (119a) to replace $\nu=\mu$ by σ, giving

$$\text{C.E.}=2(e^2/a)(\sigma/\pi)^{\frac{1}{2}} \tag{129b}$$

for *both* H^3 and He^4. Now σ may be determined approximately by making the *variational* energy of H^3 and He^4 a minimum. For H^3 and $\mu=\nu$ this variational energy is (cf. 118, put $p=1$):

$$E=3T\sigma-(2B+C)(\sigma/\sigma+1)^{\frac{3}{2}}. \tag{129c}$$

The minimum occurs for

$$(\sigma+1)^5/\sigma=[(2B+C)/2T]^2. \tag{129d}$$

For our values of the constants (128), this

expression has the value 29.0. This gives

$$\sigma = 0.934, \qquad (129e)$$

$$\text{C.E.} = 0.68 \text{ MV}. \qquad (129f)$$

A somewhat larger value of the Coulomb energy may be expected in view of the fact that the two protons shall be more frequently close together than would be expected from the wave function (117), owing to their strong mutual attraction at close distances. However, it can be expected that this correction is not very large.

Therefore (129f) should be approximately equal to the observed difference between the binding energies of H^3 and He^3, He^3 being the heavier nucleus. This difference is given experimentally directly by the difference between the energies evolved in the two reactions

$$H^2 + H^2 = H^3 + H^1 + Q_1, \qquad (130)$$

$$H^2 + H^2 = He^3 + n^1 + Q_2. \qquad (130a)$$

For we may write:

Binding energy of H^3 = mass of two neutrons + one proton − H^3

$$= 2n^1 + 2H^1 - (H^3 + H^1).$$

Binding energy of He^3

$$= n^1 + 2H^1 - He^3 = 2n^1 + 2H^1 - (He^3 + n^1).$$

Therefore difference of binding energies equals

$$(He^3 + n^1) - (H^3 + H^1) = Q_1 - Q_2. \qquad (130b)$$

The reaction energy Q_1 is very well known from experiments of Oliphant, Kempton and Rutherford (O1), it is

$$Q_1 = 3.97 \pm 0.01_5 \text{ MV}. \qquad (130c)$$

Q_2 has been measured by Bonner and Brubaker (B20) and by Dee and Walton (D1). The latter authors used the range of recoil He^4 nuclei set into motion by the neutrons to determine the neutron energy, the value of Q_2 derived from their measurements is about 2.95 MV (chapter XVI). This value is rather uncertain because the ranges of the recoil α-particles are very short, and lie in a region where the range-energy relation is not very well established. Bonner and Brubaker used recoil *protons* whose range is much longer and falls into the region where the range-energy relation is best known. They find

$$Q_2 = 3.21 \pm 0.13 \text{ MV}. \qquad (130d)$$

If we decide for this value, the difference of the binding energies of He^3 and H^3 turns out to be

$$(He^3) - (H^3) = Q_2 - Q_1 = 0.76 \pm 0.14 \text{ MV} \qquad (130e)$$

in close agreement with the calculated value (129f).

§23. Excited States of the α-Particle (F4, C17)

Experimental evidence has been obtained by Crane and Lauritsen (C17) that the α-particle possesses excited states. The evidence is based on the fact that γ-rays are emitted when protons fall on Li^7 nuclei. The reaction taking place is probably (cf. chapter XVI)

$$Li^7 + H^1 = He^4 + He^{4*}; \quad He^{4*} \rightarrow He^4 + \gamma, \qquad (131)$$

where He^{4*} denotes an excited α-particle which, after the nuclear transmutation is completed, radiates its excitation energy as a γ-ray. (Cf. chapter XIII for arguments against the γ-ray being emitted *during* the nuclear transmutation itself.)

The γ-ray spectrum from reaction (131) appears to be complex. The maximum γ-energy observed is 16 MV, i.e., nearly the total energy available in the reaction (17 MV). In addition, there seem to be some γ-rays of smaller energy. A satisfactory interpretation would be to assume one excited level of the α-particle at 16 MV above the ground state, and one or two more excited states at lower energies, the lowest being perhaps 10 MV above ground. Then the low energy γ-rays would come from transitions between the high levels, while the energetic γ-rays would correspond to transitions from one of the high levels to the ground state.

α-particles with an excitation energy of 16 MV would be perfectly stable against disintegration. For the disintegration which would require least energy would be into $H^3 + H^1$, and this process would require 19.4 MV energy, as calculated from the masses of H^1H^3 and He^4.

The problem is now whether (one or more) stable excited states of the α-particle can reasonably be expected. Feenberg (F4) has shown that the answer is affirmative provided the values for the force constants derived in the preceding section are anywhere near correct. There are probably two or three stable excited states not very far apart.

One could try to calculate the energy of the excited states by the variation method. This method is, however, very troublesome for excited states[50] because their wave functions must be chosen orthogonal to that of the ground state. Moreover, the variation method did not give very good results even for the ground state. Feenberg therefore chose a different method for the treatment of the excited levels of the α-particle.[51]

The method is based on the so-called *sum rules* for the matrix elements of the coordinates. We introduce the following coordinates

$$\mathbf{u}=\mathbf{r}_1+\mathbf{r}_2-\mathbf{r}_3-\mathbf{r}_4;\quad \mathbf{v}=\mathbf{r}_1-\mathbf{r}_2;\quad \mathbf{w}=\mathbf{r}_3-\mathbf{r}_4;\quad \mathbf{S}=\tfrac{1}{4}(\mathbf{r}_1+\mathbf{r}_2+\mathbf{r}_3+\mathbf{r}_4), \tag{132}$$

where $\mathbf{r}_1\mathbf{r}_2$ are the positions of the protons and $\mathbf{r}_3\mathbf{r}_4$ those of the neutrons. $\frac{1}{2}\mathbf{u}$ is the vector from the center of gravity of the two neutrons to that of the two protons, $\mathbf{v}$ the vector from one proton to the other, $\mathbf{w}$ the corresponding vector for the neutrons, and $\mathbf{S}$ the position of the center of gravity of the α-particle. The 12 Cartesian components of these four vectors form an orthogonal set of coordinates in the configuration space of the four particles. Accordingly, the Laplacian operator (kinetic energy) transforms into a sum of Laplacian operators with respect to the coordinates (132) without any cross terms, thus:

$$\Delta_1+\Delta_2+\Delta_3+\Delta_4=4\Delta_u+2\Delta_v+2\Delta_w+\tfrac{1}{4}\Delta_S. \tag{132a}$$

We insert this expression into the wave equation, and then deduce the sum rule for the oscillator strength in the usual way.* If u' denotes any Cartesian component of the vector $\mathbf{u}$, and u_{mn}' its matrix element with respect to the two states mn of the α-particle, we have for any state m

$$\sum_n (E_n-E_m)\,|u_{mn}'|^2=2\hbar^2/M, \tag{133}$$

$$\sum_n (E_n-E_m)\,|v_{mn}'|^2=\sum_n (E_n-E_m)\,|w_{mn}'|^2=\hbar^2/M, \tag{133a}$$

the sums including, of course, integrations over the continuous spectrum. We apply (133) particularly to the ground state $m=0$. We denote by E_1 the energy of the lowest excited state for which the matrix element u_{0n}' does not vanish. Then we have certainly from (133)

$$(E_1-E_0)\sum_n |u_{0n}'|^2<2\hbar^2/M. \tag{133b}$$

The sum occurring here can be evaluated at once:

$$\sum |u_{0n}'|^2=(u'^2)_{00}, \tag{133c}$$

where $(u'^2)_{00}$ is the average of u'^2 over the eigenfunction of the ground state. We obtain therefore

$$E_1-E_0<2\hbar^2/M(u'^2)_{00}. \tag{134}$$

(134) gives an upper bound for the energy of the first excited state which involves only the knowledge of the wave function of the ground state. We take the wave function (117a) which may be rewritten

$$\psi_0=e^{-\frac{1}{2}\nu u^2-\frac{1}{2}(\mu+\nu)(v^2+w^2)}, \tag{134a}$$

where u is the length of the vector $\mathbf{u}$. Then we find easily

$$(u'^2)_{00}=\int e^{-\nu u'^2}u'^2du'/\int e^{-\nu u'^2}du'=1/(2\nu). \tag{134b}$$

From (119a) we find $\nu=\frac{1}{2}\sigma/a^2$ if we put $p=1$ [actual value $p=0.985$, cf. §21, above Eq. (126)] so that

$$E_1-E_0<2\hbar^2\sigma/Ma^2=2T\sigma \tag{134c}$$

(cf. 118a). σ may be determined by making the variational energy of the ground state of the α-particle, *viz.* (cf. (119) with $p=1$)

$$4\tfrac{1}{2}T\sigma-2(2B+C)(\sigma/\sigma+1)^{\frac{3}{2}} \tag{134d}$$

a minimum. Inserting the values of the constants (128), this gives

$$\sigma=1.32, \tag{134e}$$

$$E_1-E_0<20.1\text{ MV}. \tag{135}$$

Since the condition for stability against disintegration into H^3+H^1 is

$$E_1-E_0<19.4\text{ MV}, \tag{135a}$$

[50] Cf. the corresponding calculations for the helium *atom*, *Handbuch der Physik*, Vol. 24/1, p. 364.

[51] Actually, Feenberg has also carried out variation calculations for the excited levels. However, the calculations were only performed for an ordinary rather than an exchange interaction. Moreover, the resulting energy levels lay not appreciably lower than the upper limit derived from the sum rule.

* *Note added in proof:* Feenberg has pointed out that the sum rules (133), (133a) hold only for *ordinary* forces. For Majorana forces, it seems that the upper bound for the energy turns out much lower than (135). This would make the existence of stable excited states a certainty.

the level E_1 needs only to be slightly below the upper limit (135) in order to be stable. Thus the stability of the state E_1 seems almost certain.[52]

The properties of the level E_1 may be deduced from the fact that the coordinate u has a non-vanishing matrix element between the ground state (completely symmetrical wave function) and E_1. This has two consequences: Firstly, the eigenfunction of E_1 must be symmetrical in the space coordinates of the protons and in the space coordinates of the neutrons; this makes the state a singlet state. Secondly, the eigenfunction must change sign if the two neutrons are replaced by the two protons and *vice versa* which shows that the function has a nodal plane; it is therefore a P function. Consequently, the state E_1 is a ^1P state.

Two ^3P levels are obtained by making the eigenfunction antisymmetrical in the space coordinates of the two protons, or of the two neutrons. Calling these ^3P levels E_2 and E_3, it follows that the matrix elements v_{02}' and w_{03}', respectively, are different from zero. Upper bounds for the energies E_2 and E_3 can be obtained in a way analogous to E_1, *viz.*,

$$E_2-E_0<\hbar^2/M(v'^2)_{00}, \tag{135b}$$

$$(v'^2)_{00}=1/(2\mu+2\nu). \tag{135c}$$

With $\mu=\nu$, the upper bounds for E_2 and E_3 become identical to that for E_1. From this fact, we might conclude that both the levels E_2 and E_3 might be stable. However, since the two levels have the same symmetry properties, there will be some "interaction" between them which will depress one level and raise the other possibly above the stability limit.

With all reservation, we may therefore form the following tentative picture of the excited states of the α-particle: There will be a ^1P state, odd in the u coordinate, even in the v and w coordinates, less than 20 MV above the normal state. Furthermore, there will be a ^3P state even in u, odd in v or w, which we might expect to lie lower than the ^1P state because of interaction with the other ^3P state. This other ^3P state will, because of the same interaction, probably lie *higher* than the ^1P state and probably it will not be stable at all.

We propose to identify the ^1P state with the observed level at 16 MV, the lower ^3P state with a lower level at, perhaps, 10 MV. This choice seems plausible from the standpoint of selection rules: Only the coordinate u has an electric dipole moment attached to it. Allowed optical transitions will therefore lead from a state odd in u to a state even in u or *vice versa*. Both the transitions from ^1P to ^1S and to ^3P are therefore allowed. The fact that the latter transition is an "intercombination" should only moderately decrease the intensity, because the rather strong Heisenberg forces (§14) prevent the spin of the α-particle from being a true quantum number. The transition $^3P-{}^1S$ must then also occur, because it is the only way in which α-particles in the ^3P state can get rid of their energy. We therefore expect three lines, corresponding to the transitions $^1P-{}^1S$ (about 16 MV),[53] $^1P-{}^3P$ ($\approx$6 MV) and $^3P-{}^1S$ ($\approx$ 10 MV). This seems compatible with experiments, in view of the small number of observations made and the large statistical fluctuations to be expected accordingly.

V. Statistical Theory of Nuclei

§24. The Hartree Method

In this and the following chapter, we shall use an approximation to the nuclear problem in which each particle is, in first approximation, supposed to move independently of the others. This method has been introduced into *atomic* physics by Hartree (H4), and has been used in that domain with great success.

In the Hartree approximation, we assume certain wave functions $\psi_1, \psi_2\cdots\psi_Z$ for each individual proton, and $\varphi_1\cdots\varphi_N$ for the neutrons. Each of these wave functions is supposed to be a function of the position as well as of the spin

[52] Feenberg has pointed out that the eigenfunction (134a) falls off too rapidly for large values of u, the correct dependence being $e^{-\alpha u}$ rather than $e^{-\frac{1}{2}\mu u^2}$. This makes the calculated value of $(u'^2)_{00}$ too small. A correct wave function would therefore give a *lower* value for the upper bound (135).

[53] There is, of course, no theoretical foundation for the figures. They are chosen to fit the experiments as good as possible.

of the respective particles; p_i and n_i shall stand for all coordinates (positional and spin) of the ith proton and neutron, respectively. Then a wave function of the nucleus as a whole which satisfies Pauli's principle, is (cf., e.g., reference S22)

$$\Psi=\begin{vmatrix}\psi_1(p_1) & \psi_2(p_1) & \cdots & \psi_Z(p_1)\\ \psi_1(p_2) & \psi_2(p_2) & \cdots & \psi_Z(p_2)\\ \cdots & \cdots & \cdots & \cdots\\ \psi_1(p_Z) & \psi_2(p_Z) & \cdots & \psi_Z(p_Z)\end{vmatrix}\cdot\begin{vmatrix}\varphi_1(n_1) & \varphi_2(n_1) & \cdots & \varphi_N(n_1)\\ \varphi_1(n_2) & \varphi_2(n_2) & \cdots & \varphi_N(n_2)\\ \cdots & \cdots & \cdots & \cdots\\ \varphi_1(n_N) & \varphi_2(n_N) & \cdots & \varphi_N(n_N)\end{vmatrix}. \qquad (136)$$

The total energy of the nucleus is given, in first approximation, by

$$E=\left[\int\Psi^*\left[V-(\hbar^2/2M)\left(\sum_{i=1}^{N}\Delta_{ni}+\sum_{i=1}^{Z}\Delta_{pi}\right)\right]\Psi d\tau\right]\Big/\int|\Psi|^2d\tau, \qquad (137)$$

where V is the total potential energy as a function of the coordinates, Δ_{ni} and Δ_{pi} are the Laplacian operators with respect to the coordinates of the neutron i and proton i, and the integral has to be extended over all coordinates of all particles.

The expression (137) actually represents an upper bound for the energy of the system since it is well known that the right-hand side of (137) becomes an absolute minimum if we insert the correct wave function Ψ_0 instead of Ψ (Schrödinger's variation principle). Of course, if Ψ_0 is inserted in (137), that equation will give the *correct* energy E_0. Thus the correct energy is always lower than the E calculated from (137) with an *approximate* wave function Ψ. This has a very important consequence: If Ψ contains one or more parameters which are left arbitrary in the early stages of the calculation, the values of the parameters should finally be determined in such a way as to make (137) as small as possible. This will then make Ψ and E as close approximations to the correct wave function and energy as possible with the assumed form of the wave function (Ritz method). The parameter which we shall usually introduce into the wave function and then fix by this minimum condition, is the nuclear radius, but occasionally more parameters will be introduced.

Before we evaluate (137), we want to say a few words about the Hartree approximation in nuclear physics. It can be said at once that this approximation will not be as successful in nuclear theory as in the theory of atoms. The main reason for this is the saturation type of the nuclear forces: Any given nuclear particle interacts essentially only with two particles of the other kind (§7). Therefore the force between a given pair of particles will be of the same order of magnitude as the force exerted by the whole nucleus on one particle. This is contrary to the assumptions of the Hartree theory. These are that in first approximation the total action of the nucleus on one particle may be represented by an *average* field, corresponding to the average distribution of all other particles over the nucleus. The "correlations" between the different particles, i.e., the fact that the motion of one particle is influenced by the instantaneous position of the others, is supposed to cause only small perturbations in the Hartree theory. These assumptions of the Hartree theory are well satisfied in the atomic problem because the force due to the nucleus, and the force corresponding to the *average* charge distribution of the electrons, are very much stronger than the fluctuations of the force caused by, say, a close approach of one other electron to the electron considered. In nuclear physics, the force on one neutron changes by 100 percent or more according to whether a proton happens to be near the neutron or not. Therefore the correlations between the nuclear particles will be of extreme importance for any satisfactory calculation of nuclear energies, and the Hartree method will afford only a poor approximation. In spite of these serious objections against the Hartree method, we are forced to use it because no better method seems practicable at the moment.

Proceeding now to the evaluation of the energy (137), it is useful to assume the eigenfunctions of the individual particles to be orthogonal and normalized. This is most conveniently achieved by assuming the ψ's and φ's to be the solutions of certain one-particle wave equations, thus:

$$\begin{aligned}(\hbar^2/2M)\Delta\psi_i+(E_i-V_p)\psi_i=0,\\ (\hbar^2/2M)\Delta\varphi_i+(W_i-V_n)\varphi_i=0,\end{aligned} \tag{138}$$

where V_p and V_n are certain "auxiliary potentials" which may be chosen suitably so as to make the energy (137) a minimum.

We assume now the total potential energy V in (137) to consist of a number of terms corresponding to the interaction of pairs of particles:

$$V=\sum_{i=1}^{Z}\sum_{k=1}^{N}V_{ik}(p_i, n_k)+\tfrac{1}{2}\sum_{i=1}^{Z}\sum_{k\neq i}P_{ik}(p_i, p_k)$$
$$+\tfrac{1}{2}\sum_{i=1}^{N}\sum_{k\neq i}N_{ik}(n_i, n_k), \tag{139}$$

where V_{ik} represents the interaction of a proton and a neutron, P_{ik} that between two protons and N_{ik} that between two neutrons. The factors 1/2 stand in order to count each pair only once.

The evaluation of (137) is then straightforward and similar to the theory of complex atoms (C13). The result is

$$E=\sum_{i=1}^{Z}E_i+\sum_{i=1}^{N}W_i-\sum_{i=1}^{Z}\int|\psi_i|^2V_pd\tau-\sum_{i=1}^{N}\int|\varphi_i|^2V_nd\tau$$
$$+\sum_{i=1}^{Z}\sum_{k=1}^{N}\int\psi_i^*(p)\varphi_k^*(n)V(p, n)\psi_i(p)\varphi_k(n)d\tau_pd\tau_n$$
$$+\tfrac{1}{2}\sum_{i=1}^{Z}\sum_{k\neq i}^{Z}\Big[\int|\psi_i(1)|^2|\psi_k(2)|^2P(1, 2)d\tau_1d\tau_2$$
$$-\int\psi_k^*(1)\psi_i^*(2)\psi_i(1)\psi_k(2)P(1, 2)d\tau_1d\tau_2\Big]$$
$$+\tfrac{1}{2}\sum_{i=1}^{N}\sum_{k\neq i}^{N}\Big[\int|\varphi_i(1)|^2|\varphi_k(2)|^2N(1, 2)d\tau_1d\tau_2$$
$$-\int\varphi_k^*(1)\varphi_i^*(2)\varphi_i(1)\varphi_k(2)N(1, 2)d\tau_1d\tau_2\Big]. \tag{140}$$

The first line contains the kinetic energy of the particles. The second line represents the effect of the interaction between unlike particles, the next two lines the proton and the last two the neutron interaction. It is seen that exchange terms appear only in the four last lines which refer to the interaction between like particles, because the terms arise from the antisymmetry of the wave function with respect to particles of the same kind.

We now assume that $V(p, n)$ is an interaction of the Majorana type. In other words, if xs denote position and spin of the proton, $\xi\sigma$ the same quantities for the neutron, then

$$V(p, n)\psi_i(xs)\varphi_k(\xi\sigma)=J(x-\xi)\psi_i(\xi s)\varphi_k(x\sigma), \tag{141}$$

where J is an ordinary function of the distance between the particles, as treated in chapter III.

The integral signs in (140) imply, of course, a summation over the spin coordinates as well. This sum can be carried out if we assume that each of the functions ψ and φ is the product of a function which depends only on the space coordinates and one depending only upon spin, which assumption is always justified as long as the "auxiliary" potentials V_p and V_n (cf. 139) do not depend upon spin. Then the contribution of the interaction between proton i and neutron k to (140) becomes

$$V_{ik}=\int\psi_i^*(x)\varphi_k^*(\xi)\psi_i(\xi)\varphi_k(x)J(x-\xi)dxd\xi, \tag{141a}$$

the integral now extending over the space coordinates only.

The interaction between protons contains the Coulomb interaction and the specific interaction between like particles discussed in §§10, 18 and 21. Nothing definite is known about the type of these forces. Two types suggest themselves: The interaction may or may not depend on the relative orientation of the spins of the particles.[54] We decide in favor of the former choice because an interaction independent of the relative spins would be essentially equivalent to a Wigner type force and would therefore lead to difficulties, giving excessive binding energies for heavy nuclei[55] (§28, V1). We therefore assume

$$\begin{aligned}N(1, 2)&=-\tfrac{1}{3}K(r_{12})(\boldsymbol{\sigma}_1\cdot\boldsymbol{\sigma}_2),\\ P(1, 2)&=-\tfrac{1}{3}K(r_{12})(\boldsymbol{\sigma}_1\cdot\boldsymbol{\sigma}_2)+e^2/r_{12}.\end{aligned} \tag{142}$$

The forces neutron-neutron and proton-proton have been assumed equal, except for the Coulomb repulsion (cf. §6, 22). $\boldsymbol{\sigma}$ denotes the Pauli spin

[54] Majorana and Heisenberg forces may be expressed in terms of these two kinds of forces, owing to the antisymmetry of the wave function with respect to like particles, reference V1.

[55] A linear combination of a large force depending on spin, and a small force independent of spin, cannot be excluded.

operator. The factor $-\frac{1}{3}$ has been inserted in order to make the interaction of two neutrons (or protons) with opposite spins equal to $+K$; namely, for two such neutrons, we have (with $\boldsymbol{\sigma}$ denoting the resultant $\boldsymbol{\sigma}_1+\boldsymbol{\sigma}_2$, which is zero):[56]

$$2\boldsymbol{\sigma}_1\cdot\boldsymbol{\sigma}_2=\boldsymbol{\sigma}^2-\boldsymbol{\sigma}_1^2-\boldsymbol{\sigma}_2^2=-6. \tag{142a}$$

K is thus identical with the quantity denoted by K in (115a)

To evaluate the neutron-neutron interaction in (140), we first add formally a term $k=i$ referring to the interaction of a neutron with itself. This does not change the expression, because the "ordinary" and the "exchange" term cancel exactly for $k=i$. Then we write each wave function φ_i as the product of a spatial wave function u_i and a spin wave function, which is α or β according to whether the spin component in a given direction z is $+\frac{1}{2}$ or $-\frac{1}{2}$. We then take a particular state $\varphi_i=u_i\alpha$, and consider its interaction with the *two* states $\varphi_{k1}=u_k\alpha$ and $\varphi_{k2}=u_k\beta$. (u_k may or may not be equal to u_i.) Then we have for the "direct" part of the interaction (first integral in last line of (140)):

$$-\tfrac{1}{3}\int|u_i(1)|^2|u_k(2)|^2K(r_{12})d\tau_1d\tau_2$$
$$\sum_{s_1s_2}\alpha(s_1)\alpha(s_2)(\boldsymbol{\sigma}_1\cdot\boldsymbol{\sigma}_2)\alpha(s_1)\alpha(s_2) \tag{142b}$$

and

$$-\tfrac{1}{3}\int|u_i(1)|^2|u_k(2)|^2K(r_{12})d\tau_1d\tau_2$$
$$\sum_{s_1s_2}\alpha(s_1)\beta(s_2)(\boldsymbol{\sigma}_1\cdot\boldsymbol{\sigma}_2)\alpha(s_1)\beta(s_2) \tag{142c}$$

corresponding to the two states φ_{k1} and φ_{k2}. Now the spin wave functions obey the relations:

$$\sigma_z\alpha=\alpha,\quad \sigma_z\beta=-\beta,\quad \sigma_x\alpha=\beta,\quad \sigma_x\beta=\alpha,$$
$$\sigma_y\alpha=i\beta,\quad \sigma_y\beta=-i\alpha. \tag{142d}$$

Therefore

$$(\boldsymbol{\sigma}_1\cdot\boldsymbol{\sigma}_2)\alpha(1)\alpha(2)=\alpha(1)\alpha(2),$$
$$(\boldsymbol{\sigma}_1\cdot\boldsymbol{\sigma}_2)\alpha(1)\beta(2)=-\alpha(1)\beta(2)+2\beta(1)\alpha(2). \tag{142e}$$

Thus the sum over the spin coordinates s_1s_2 in (142b) gives $+1$, that in (142c) -1. The two expressions (142b) (142c) are thus equal and opposite; therefore the ordinary part of the interaction between the neutrons vanishes, if we take the interaction potential (142).[57]

The exchange part (second integral in last line of (140)) becomes

$$+\tfrac{1}{3}K_{ik}\sum_{s_1s_2}\alpha(s_1)\alpha(s_2)\cdot\alpha(s_1)\alpha(s_2) \tag{143}$$

and

$$+\tfrac{1}{3}K_{ik}\sum_{s_1s_2}\beta(s_1)\alpha(s_2)$$
$$\cdot[-\alpha(s_1)\beta(s_2)+2\beta(s_1)\alpha(s_2)] \tag{143a}$$

for the interaction of φ_i with φ_{k1} and φ_{k2}, respectively. K_{ik} is the integral

$$K_{ik}=\int u_k^*(r_1)u_i^*(r_2)u_i(r_1)u_k(r_2)$$
$$\times K(r_{12})d\tau_1d\tau_2. \tag{143b}$$

The summation over s_1s_2 gives 1 and 2, respectively, in (143) and (143a), so that both contributions together are just equal to K_{ik}. Thus K_{ik} is the interaction between one neutron i and a pair of neutrons k of opposite spin; or $\frac{1}{2}K_{ik}$ is the average interaction between two neutrons ik. Therefore the total interaction of all neutrons is $\frac{1}{4}\sum_{i=1}^{N}\sum_{k=1}^{N}K_{ik}$, since another factor $\frac{1}{2}$ comes from (140).

For the protons, the same result holds, but there is to be added the Coulomb repulsion which gives rise to a "direct" and to an "exchange" term.

Collecting our terms, we may now write for (140),

$$E=T+V+C-A+F \tag{144}$$

where T is the kinetic energy, V the contribution of the forces between protons and neutrons, C the direct Coulomb interaction between the protons, A the correction to the Coulomb interaction because of proton exchange, and F the contribution of the nuclear forces between like particles. Explicitly, we have

[56] $\boldsymbol{\sigma}_1$ is *twice* the spin angular momentum of particle 1 in units $\hbar$. The eigenvalue of the square of the angular momentum is $s(s+1)$ where $s=\frac{1}{2}$. Therefore $\boldsymbol{\sigma}_1^2=4\cdot\frac{3}{2}\cdot\frac{1}{2}=3$.

[57] This fact prevents the neutron-neutron interaction from giving excessive contributions to the binding energy of heavier nuclei. This is actually the reason for choosing the particular form (142) for the interaction potential.

$$T=\sum_{i=1}^{Z}(E_i-\int|\psi_i|^2V_pd\tau)+\sum_{i=1}^{N}(W_i-\int|\varphi_i|^2V_nd\tau), \tag{144a}$$

$$V=\sum_{i=1}^{Z}\sum_{k=1}^{N}\int\psi_i^*(x_1)\psi_i(x_2)\varphi_k^*(x_2)\varphi_k(x_1)J(r_{12})d\tau_1d\tau_2, \tag{144b}$$

$$C=\sum_{i=1}^{Z}\sum_{k=1}^{Z}\int|\psi_i(x_1)|^2|\psi_k(x_2)|^2(e^2/r_{12})d\tau_1d\tau_2, \tag{144c}$$

$$A=\sum_{i=1}^{Z}\sum_{k=1}^{Z}\int\psi_i^*(x_1)\psi_i(x_2)\psi_k^*(x_2)\psi_k(x_1)(e^2/r_{12})d\tau_1d\tau_2, \tag{144d}$$

$$F=\tfrac{1}{4}\sum_{i=1}^{Z}\sum_{k=1}^{Z}\int\psi_i^*(x_1)\psi_i(x_2)\psi_k^*(x_2)\psi_k(x_1)K(r_{12})d\tau_1d\tau_2$$
$$+\tfrac{1}{4}\sum_{i=1}^{N}\sum_{k=1}^{N}\int\varphi_i^*(x_1)\varphi_i(x_2)\varphi_k^*(x_2)\varphi_k(x_1)K(r_{12})d\tau_1d\tau_2. \tag{144e}$$

A formal simplification of these equations can be achieved by introducing the total density of protons and neutrons, *viz.*,

$$\rho_p(x)=\sum_{i=1}^{Z}|\psi_i(x)|^2,\quad \rho_n(x)=\sum_{i=1}^{N}|\varphi_i(x)|^2, \tag{145}$$

and Dirac's "mixed densities"

$$\rho_p(x_1,x_2)=\sum_{i=1}^{Z}\psi_i^*(x_1)\psi_i(x_2),$$
$$\rho_n(x_1,x_2)=\sum_{i=1}^{N}\varphi_i^*(x_1)\varphi_i(x_2). \tag{145a}$$

Obviously,

$$\rho_p(x_2,x_1)=\rho_p^*(x_1,x_2),\quad \rho_p(x)=\rho_p(x,x).$$

Interchanging the order of summations and integrations in (144a) to (144e), we have

$$T=\sum_{i=1}^{Z}E_i+\sum_{i=1}^{N}W_i-\int(V_p\rho_p+V_n\rho_n)d\tau, \tag{146a}$$

$$V=\int J(r_{12})\rho_p(x_1x_2)\rho_n^*(x_1x_2)d\tau_1d\tau_2, \tag{146b}$$

$$C=\tfrac{1}{2}\int(e^2/r_{12})\rho_p(x_1)\rho_p(x_2)d\tau_1d\tau_2, \tag{146c}$$

$$A=\tfrac{1}{4}\int(e^2/r_{12})|\rho_p(x_1x_2)|^2d\tau_1d\tau_2, \tag{146d}$$

$$F=\tfrac{1}{4}\int K(r_{12})(|\rho_p(x_1x_2)|^2+|\rho_n(x_1x_2)|^2)d\tau_1d\tau_2. \tag{146e}$$

§25. The Statistical Model. Qualitative Conclusions (M6)

In the statistical model the eigenfunction of each individual proton or neutron is supposed to be that of a free particle *viz.*, a plane wave. These eigenfunctions are inserted in (146) and the energy of the nucleus calculated.

Accordingly, the auxiliary potentials V_p and V_n are assumed to have certain constant, negative values inside a sphere of radius R (nuclear radius) and to be zero outside that sphere. The radius R is an arbitrary parameter which has to be determined in such a way as to make the nuclear energy a minimum (cf. §24, beginning). The value of the auxiliary potentials V_p and V_n inside the "nucleus" does not affect the nuclear energy materially because only the kinetic energy of the particles and the form of the wave functions enter the formula (146) neither of which quantities depend upon the value of the auxiliary potentials except for surface effects (§29).

For the sake of the simplicity of the eigenfunctions it may be allowable to replace the spherical "box" for the nuclear particles by a cubical one having the same volume. Denoting the length of the cube by l we have

$$l^3=4\pi R^3/3,\quad l=R(4\pi/3)^{\frac{1}{3}}. \tag{147}$$

The wave functions in such a box are plane waves

$$\psi_i=l^{-\frac{3}{2}}e^{i(\mathbf{k}_i\cdot\mathbf{r})},\quad \varphi_i=l^{-\frac{3}{2}}e^{i(\mathbf{k}_i\cdot\mathbf{r})}. \tag{147a}$$

The factor $l^{-\frac{3}{2}}$ serves to normalize the functions.

The components of the wave vectors $\mathbf{k}_i$ are determined by the boundary condition for the eigenfunctions at the boundary of the cube. Neglecting the details of surface effects we may impose the usual condition

$$k_{ix}=2\pi\kappa_x/l, \quad k_{iy}=2\pi\kappa_y/l, \quad k_{iz}=2\pi\kappa_z/l, \tag{148}$$

$\kappa_x\kappa_y\kappa_z$ being three integers. For each triple of integers $\kappa_x\kappa_y\kappa_z$ there is just one proton state and one neutron state of either spin direction, so that the total number of neutron (or proton) states having wave vector components between k_x and k_x+dk_x, k_y and k_y+dk_y, k_z and k_z+dk_z is

$$ndk_xdk_ydk_z=2(l/2\pi)^3dk_xdk_ydk_z, \tag{148a}$$

the factor 2 arising from the two possible spin directions. The number of states for which the absolute value of the wave vector is between k and $k+dk$, is (cf. 147)

$$2(l/2\pi)^3 4\pi k^2dk=(4/3\pi)R^3k^2dk. \tag{148b}$$

The kinetic energy of a particle of wave number k is

$$T(k)=\hbar^2k^2/2M, \tag{149}$$

so that the number of neutrons with kinetic energy between T and $T+dT$ becomes

$$N(T)dT=(2^{5/2}/3\pi)M^{\frac{3}{2}}\hbar^{-3}R^3T^{\frac{1}{2}}dT. \tag{149a}$$

The states of lowest energy will be those for which the kinetic energy, and therefore the wave number, is smallest. The number of neutron states having wave numbers below k_0, or kinetic energies below T_0, is

$$N=(4/3\pi)R^3\textstyle\int_0^{k_0}k^2dk=(4/9\pi)R^3k_0^3, \tag{150}$$

$$N=(2^{7/2}/9\pi)M^{\frac{3}{2}}\hbar^{-3}R^3T^{\frac{3}{2}}. \tag{150a}$$

From these equations we may find the maximum wave number and kinetic energy of the neutrons in the nucleus if the total number of neutrons is N, *viz.*,

$$k_0=(9\pi N/4)^{\frac{1}{3}}R^{-1}, \tag{150b}$$

$$T_0=(9\pi N/4)^{\frac{2}{3}}\hbar^2/2MR^2. \tag{150c}$$

The average kinetic energy per neutron follows from (149a)

$$\bar{T}=\textstyle\int N(T)TdT/\int N(T)dT=\frac{3}{5}T_0. \tag{150d}$$

$\bar{T}$ is thus proportional to the 2/3 power of the number of neutrons, just as T_0. The total kinetic energy is therefore proportional to $N\bar{T}\propto N^{5/3}R^{-2}$. All these formulae are familiar from Fermi statistics. They apply to protons as well as to neutrons, with only N to be replaced by Z.

We may now proceed to calculate the interaction of neutrons and protons (146b), using our free particle eigenfunctions (147a). We have first to calculate the mixed density

$$\rho(\mathbf{r}_1\mathbf{r}_2)=\sum_{i=1}^{N}\psi_i^*(\mathbf{r}_1)\psi_i(\mathbf{r}_2)=l^{-3}\sum_{i=1}^{N}e^{i\mathbf{k}_i\cdot(\mathbf{r}_2-\mathbf{r}_1)}$$

$$=l^{-3}(l^3/4\pi^3)\int_0^{k_0}dk_xdk_ydk_ze^{i\mathbf{k}\cdot\mathbf{r}}, \tag{151}$$

where $\mathbf{r}=\mathbf{r}_2-\mathbf{r}_1$. Introducing a polar coordinate system in k space with the polar axis parallel to $\mathbf{r}$, and using (147) again, we have

$$\rho(\mathbf{r}_1\mathbf{r}_2)=(1/4\pi^3)\textstyle\int_0^{k_0}2\pi\sin\theta d\theta k^2dke^{ikr\cos\theta}$$

$$=(1/\pi^2r)\textstyle\int_0^{k_0}kdk\sin kr$$

$$=(\sin k_0r-k_0r\cos k_0r)/\pi^2r^3. \tag{151a}$$

This function has a pronounced maximum for $r=0$, *viz.*,

$$\rho(\mathbf{r}_1\mathbf{r}_1)=k_0^3/3\pi^2=N/(4\pi R^3/3) \tag{151b}$$

[cf. (150)]. $\rho(\mathbf{r}_1\mathbf{r}_2)$ falls off rapidly when k_0r becomes larger than unity. This means that there is practically no correlation between the wave functions at two points whose distance is considerably larger than $1/k_0=\lambda_0/2\pi$, where λ_0 is the shortest wave-length of any neutron in the nucleus. Because of (150), $1/k_0\approx\frac{1}{2}RN^{-\frac{1}{3}}$ which is small compared to the radius of the nucleus if N is large. The mixed density is therefore only large if the two points $\mathbf{r}_1$, $\mathbf{r}_2$ are very close together compared to the radius of the nucleus. This is very essential for the validity of the statistical model: One may reasonably hope that the actual wave functions of the nuclear particles resemble plane waves at least over a limited region of the nucleus even if they differ widely from plane waves if considered over the whole nucleus.

A number of important conclusions can be drawn by inserting (151a) into (146b), even without assuming a special form of the interaction potential $J(r_{12})$. We only introduce for convenience a certain length a which determines the width of the potential hole J, and we put

$$J(r)=-V_0f(r/a), \tag{152}$$

where V_0 is a constant determining the strength of the interaction. f may be any function; e.g., we may take any of the following:

rectangular hole: $f(x)=1$, for $x<1$,
$f(x)=0$ for $x>1$,

exponential function: $f(x)=e^{-x}$,

Gaussian function: $f(x)=e^{-x^2}$,

or a more complicated function. The only assumptions we make are

(1) $J(r)$ vanishes at infinity sufficiently rapidly
(2) $J(r)$ does not become infinite at $r=0$ more strongly than $1/r$
(3) a is chosen in such a way that the main drop of the function $f(x)$ occurs near $x=1$.

The total potential energy of the neutron-proton interaction becomes now, according to (146b) and (151a)

$$V=-\frac{4\pi}{3}R^3V_0\int_0^\infty 4\pi r_{12}^2dr_{12}f(r_{12}/a)\times\frac{\sin k_Nr_{12}-k_Nr_{12}\cos k_Nr_{12}}{\pi^2r_{12}^3}\times\frac{\sin k_Pr_{12}-k_Pr_{12}\cos k_Pr_{12}}{\pi^2r_{12}^3},\quad(152\text{a})$$

where k_N and k_P are the maximum wave numbers for neutrons and protons, as given by (150b). (In case of k_P, we have to replace N in (150b) by Z.) The integral over $d\tau_1d\tau_2$ in (146b) has been replaced by an integral over $d\tau_1d\tau_{12}$, i.e., we integrate first over the coordinates of particle 2, keeping the position of the first constant, which is equivalent to integrating over the relative coordinate $\mathbf{r}_{12}$ (volume element $4\pi r_{12}^2dr_{12}$). Since the integrand falls off rapidly for large distances r_{12}, the result of this first integration will not be materially changed if we extend the range of integration with respect to r_{12} to infinity. This extension corresponds to the neglection of surface effects. Then the result will obviously not depend upon the position of the first particle $\mathbf{r}_1$ so that the integration over $d\tau_1$ gives simply a factor equal to the volume of the nucleus, *viz.*, $(4\pi/3)R^3$.

It is convenient to introduce the radius of a sphere corresponding to the volume occupied by one particle, *viz.*,

$$r_0=R(N+Z)^{-\frac{1}{3}}\quad(153)$$

and to put accordingly

$$r_{12}=r_0\rho;\quad r_0/a=\alpha,\quad(153\text{a})$$

$$k_Nr_{12}=\kappa_N\rho,$$
$$k_Pr_{12}=\kappa_P\rho,\quad(153\text{b})$$

with

$$\kappa_P=[9\pi Z/4(N+Z)]^{\frac{1}{3}};\qquad \kappa_N=[9\pi N/4(N+Z)]^{\frac{1}{3}}.\quad(153\text{c})$$

Then ρ is our new integration variable while α is a parameter determining the radius of the nucleus. α has to be varied such that the total energy becomes a minimum. (152a) may now be rewritten:

$$V=-\frac{16}{3}(N+Z)V_0\pi^{-2}\int_0^\infty f(\alpha\rho)\times(\sin\kappa_N\rho-\kappa_P\rho\cos\kappa_N\rho)\times(\sin\kappa_P\rho-\kappa_P\rho\cos\kappa_P\rho)\rho^{-4}d\rho.\quad(154)$$

Since κ_N and κ_P depend only upon the ratio N/Z but not upon the magnitude of either N or Z, the integral is a function of α and N/Z only. Similarly, if we introduce the notations (153), etc., into (150c, d), we have

$$T=T_N+T_P=\frac{3}{5}\frac{\hbar^2}{2M}(Nk_N^2+Zk_P^2)=\frac{3}{5}\frac{\hbar^2}{2Ma^2}\left(\frac{9\pi}{4}\right)^{2/3}(N+Z)\frac{N^{5/3}+Z^{5/3}}{(N+Z)^{5/3}\alpha^2}.\quad(154\text{a})$$

In this formula, the kinetic energy has been written as the number of particles, $N+Z$, times a function of N/Z and α only. Therefore, from (154) and (154a)

$$E=T+V=(N+Z)F(N/Z,\alpha),\quad(155)$$

where the function F can be determined when the interaction potential J, i.e., the function f, is given.

Making (155) a minimum by varying α, we see: (1) The value of α corresponding to the minimum will only depend upon the *ratio* N/Z but not upon the number of particles $N+Z$. This means, according to (153, 153a) that the radius of the nucleus R is, for a given ratio N/Z, proportional to $(N+Z)^{\frac{1}{3}}$; in other words that *the volume of the nucleus* is, again for given N/Z, *proportional to the number of particles* contained in it. (2) *The binding energy* $-E$ of the nucleus is, for a given ratio N/Z, proportional to the number of *particles* contained in the nucleus.

These two conclusions are in accord with experiment. In fact, we have assumed exchange forces to act between neutrons and protons just in order to account for the two experimental facts mentioned. (§7.) Thus our calculation merely shows that *exchange forces are really suitable for obtaining the desired result.*

In making this statement we must bear in mind that actually we have only shown that the *statistical model* applied to a nucleus held together by exchange forces, leads to a binding energy proportional to the number of particles. Apart from the neglection of surface effects, we have made two assumptions:

(a) the "auxiliary potential energy" V_N and V_P was supposed to be constant over the nucleus. This led to a constant density of the particles (cf. (151b)).

(b) the wave functions were supposed to be representable by plane waves over a region of the order of one wave-length (cf. the remarks after (151b)).

The first assumption is very plausible indeed: There is no force establishing a correlation between the positions of the particles and a *fixed* point, in contrast, e.g., to the case of atomic electrons. There also is no long-distance force between the particles which might establish differences between the density at different points. The force upon one particle actually depends only on the density in its immediate neighborhood; one given value of the density will lead to minimum energy; any fluctuations of the density from point to point in the nucleus will obviously lead to an increase in energy. A mathematical proof of the constancy of the density for a particular case was given by Majorana (M6).

The second point will, of course, not actually be fulfilled because of correlations between different particles (§24). However, it can be shown by a group theoretical argument that, even with quite general assumptions about the eigenfunctions, the total energy of a nucleus is proportional to the number of particles if exchange forces act between them.[58]

We now discuss the quantitative implications of (154), (154a), (155). The behavior of the integral in (154) can be easily found for the two limiting cases $\alpha \ll 1$ and $\alpha \gg 1$. In the first case, we may replace $f(\alpha\rho)$ by $f(0)$. Then the integral may be carried out elementarily and gives

$$V = \begin{matrix} -2J(0)Z & \text{if } Z<N \\ -2J(0)N & \text{if } Z>N \end{matrix} \qquad \alpha \ll 1. \qquad (156)$$

[58] We are indebted to Professor Wigner for this communication.

If, therefore, the radius of the nucleus is so small that the average distance of neighboring particles, r_0, is small compared to the range of the forces, a, then the total potential energy can be thought of as due to the interaction of each particle of the sort of which there are fewer in the nucleus, with two particles of the other sort (one of each spin direction).[59] The total potential energy (156) is independent of α while the kinetic energy (154a) increases as α^{-2}. Therefore the total energy is certainly positive for sufficiently small α, and the most favorable value of α is certainly not $\alpha=0$.

Similarly, for α very large, we may replace all factors *except* $f(\alpha\rho)$, by their value for very small ρ, and obtain

$$V = -3NZV_0(N+Z)^{-1}\int f(\alpha\rho)\rho^2 d\rho$$

$$= -\frac{3NZ}{N+Z}\frac{V_0}{\alpha^3}\int f\left(\frac{r}{a}\right)\left(\frac{r}{a}\right)^2 d\left(\frac{r}{a}\right)$$

$$= \text{const}\cdot\alpha^{-3}. \quad \alpha \gg 1. \qquad (156a)$$

Thus, for large α, the potential energy decreases more rapidly than the kinetic. Therefore the total energy must be positive for very large α. A negative total energy, i.e., a resultant binding energy, can only be obtained for intermediate values of α, and since all constants are of the order of magnitude unity, the minimum may be expected to lie at a value of α near unity. This means that r_0 is of the same order as a. Now r_0 is the radius of a sphere whose volume is equal to the volume occupied by one nuclear particle. r_0 is thus of the same order as the distance between neighboring particles in the nucleus. We conclude that this distance is about as large as the radius of action of the nuclear forces. This conclusion corresponds to reality: The radius of radioactive nuclei is about 8 to $9\cdot 10^{-13}$ cm (chapter IX), their atomic weight about 220, therefore $r_0 = 9\cdot 10^{-13}\cdot 220^{-\frac{1}{3}} \approx 1.5\cdot 10^{-13}$ cm which is indeed of the same order of magnitude as the range of nuclear forces (§21).

[59] Formula (156) has, of course, only a meaning if $J(0)$ is finite. If this is not the case, $J(\alpha\rho)$ cannot be replaced by $J(0)$ however small α.

§26. Quantitative Results and Limitations of the Statistical Model (H9, W2, W10)

In order actually to calculate the binding energy and the nuclear radius, we must make a special assumption about the form of the potential $J(r)$. We take (cf. 115)

$$J(r) = -Be^{-r^2/a^2}. \qquad (157)$$

This potential has been assumed in the extensive calculations of Feenberg about light nuclei; therefore the constants B and a are well known. We have to insert $f(\alpha\rho) = e^{-\alpha^2\rho^2}$ into (154). The integrations can then be carried out explicitly and elementarily, giving

$$\begin{aligned} V = -B(N+Z)\pi^{-\frac{1}{2}}x^{-3}\{&[2-x^2(n^{\frac{2}{3}}+n^{\frac{1}{3}}(2-n)^{\frac{1}{3}} \\ &+(2-n)^{\frac{2}{3}})]\cdot e^{-\frac{1}{4}x^2[n^{\frac{1}{3}}-(2-n)^{\frac{1}{3}}]^2} \\ &-[2-x^2(n^{\frac{2}{3}}-n^{\frac{1}{3}}(2-n)^{\frac{1}{3}}+(2-n)^{\frac{2}{3}})] \\ &\times e^{-\frac{1}{4}x^2[n^{\frac{1}{3}}+(2-n)^{\frac{1}{3}}]^2}+\pi^{\frac{1}{2}}x^3\Phi(\tfrac{1}{2}x[n^{\frac{1}{3}}+(2-n)^{\frac{1}{3}}]) \\ &-\pi^{\frac{1}{2}}(n-1)x^3\Phi(\tfrac{1}{2}x[n^{\frac{1}{3}}-(2-n)^{\frac{1}{3}}])\}, \end{aligned} \qquad (158)$$

where[60]

$$n = 2N/(N+Z), \quad 2-n = 2Z/(N+Z), \qquad (158a)$$

$$x = \frac{3}{2}\left(\frac{\pi}{3}\right)^{\frac{1}{3}}\frac{a}{r_0} = \frac{3}{2}\left(\frac{\pi}{3}\right)^{\frac{1}{3}}\frac{1}{\alpha}, \qquad (158b)$$

and $\Phi(y)$ is the Gaussian error integral (cf., e.g., Jahnke-Emde, Tables of Functions, p. 97).

If the nucleus contains equally many neutrons and protons, we have $n=1$ so that (158) reduces to[61]

$$\begin{aligned} V = -\pi^{-\frac{1}{2}}x^{-3}B(N+Z)\{&2-3x^2 \\ &+(x^2-2)e^{-x^2}+\pi^{\frac{1}{2}}x^3\Phi(x)\}, \end{aligned} \qquad (159)$$

while the kinetic energy (154a) may be rewritten:

$$T = \frac{3}{10}\frac{\hbar^2}{Ma^2}x^2(N+Z). \qquad (159a)$$

The most satisfactory procedure would now be to insert the values for B and a derived from the theory of light nuclei, and to determine the minimum of $V+T$ and the corresponding value of x. Unfortunately, the binding energy obtained in this way is much too small; it is, indeed, entirely wiped out when due corrections are made for the Coulomb repulsion of the protons and the "surface effect" (§29).

The values for B and a derived from the theory of light nuclei are:

$$\text{(a)}\quad B = 133 \text{ MV}, \quad a = 0.99\cdot 10^{-13} \text{ cm}, \quad T = \hbar^2/Ma^2 = 42 \text{ MV}, \qquad (160)$$

if no interaction between like particles is assumed (cf. 125),

$$\text{and (b)}\quad B = 41 \text{ MV},^{62} \quad a = 2.32\cdot 10^{-13} \text{ cm}, \quad T = 7.6 \text{ MV}, \qquad (160a)$$

if the interaction between like particles has the value derived in (128). According to Table VII, we get with these constants the following results in the statistical model:

(a) without interaction between like particles we have $D = MBa^2/\hbar^2 = 3.17$. For this value of D, the energy $V+T$ has *no minimum at all* but is positive for any value of x.

(b) with interaction between like particles: $D = MBa^2/\hbar^2 = 5.3_5$. Minimum of energy for $x = 1.99$ $\quad r_0 = 0.79a = 1.82\cdot 10^{-13}$ cm.

$$\begin{aligned} \text{Minimum energy} &= -0.067B(N+Z) \\ &= -2.7(N+Z) \text{ MV}. \end{aligned} \qquad (160b)$$

With the constants under (a), we thus obtain no binding energy at all, even without correcting for Coulomb repulsion and surface effect. With the constants under (b), we find a small binding energy, about one-quarter of the observed binding energy for medium weight nuclei. However, even this small binding disappears when we correct for Coulomb repulsion and surface effect (see below).

We must therefore conclude that *the statistical model is quite inadequate for the treatment of nuclear binding energies*. This is not very surprising in view of the objections raised in §24 against the application of the Hartree method to

[60] The old constants κ_N and κ_P are related to n by

$\kappa_N = (9\pi/8)^{\frac{1}{3}}(2N/N+Z)^{\frac{1}{3}} = (3/2)\cdot(\pi n/3)^{\frac{1}{3}}$
$\kappa_P = \frac{3}{2}(\pi(2-n)/3)^{\frac{1}{3}}$.

[61] This formula was first calculated by Heisenberg (quoted by Flügge, F12).

[62] The figure given is the force between a proton and a neutron, plus one-half the force between like particles ($B+\frac{1}{2}C$, cf. §28). This combination enters the energy of a heavy nucleus containing approximately equal numbers of protons and neutrons (cf. 174).

the nuclear problem which hold *a fortiori* for the statistical method. In view of these objections we would even go so far as to say that any expression for the nuclear forces which *would* make the nuclear binding energy following from the statistical model equal to the observed binding, must certainly give too large values for the nuclear forces.

It is of some interest to note that the statistical model becomes the worse the shorter the range of the forces. This is due to the fact that the wave functions will change very rapidly whenever the distance between two particles is smaller than the range of the nuclear forces (cf. §20). The fact is shown by a comparison of our cases (a) and (b). We may add that for *zero* range of the forces the potential energy (159) does never exceed 0.73 times the kinetic energy (159a), if the constant $D=MBa^2/\hbar^2$ is taken from the theory of the deuteron (2.70, cf. Table III).

The value of the statistical model can, according to the foregoing, not lie in quantitative calculations of nuclear binding energies, but lies, in our opinion, in qualitative results such as those obtained in the preceding section, and in the possibility of setting *upper* limits to the nuclear forces (§27). Furthermore, we might try to deduce the dependence of the binding energy on the atomic weight, on the ratio of the numbers of neutrons and protons, etc.; but all these conclusions should be considered as tentative only.

Obviously, if we want to make use of the statistical model, we must fix the constants B and a *ad hoc* and must not take the values derived from the theory of light nuclei. The constants B and a derived from the statistical model are, then, of course not correct; but it may seem more consistent in drawing conclusions from that model to fix the constants entering it also from the statistical model.

To fix the constants B and a, we need *two* experimental data. Various data have been used by different authors; e.g., Heisenberg (H9) used the mass defects of medium-sized and heavy nuclei, Wick (W10) used mass defect and radius of the oxygen nucleus, v. Weizsäcker (W2) the mass defects of O^{16} and Hg^{200}. Wick's method seems preferable to us because, to a first approximation, the mass defect per particle should be the same for all nuclei, according to the statistical model. Only the Coulomb repulsion of the protons and the surface effects should produce some differences in the mass defects per particle. Since the surface effect cannot be treated as satisfactorily as the "volume" energy, one should not base the determination of the constants on quantities which depend sensitively on the surface effect, as does the difference of the mass defects of O^{16} and Hg^{200}. The only objection which could be raised against Wick's procedure is that the radius of O^{16} is not well known experimentally. This can be avoided by taking radius and binding energy of a heavy atom as standards. In doing so, we get the further advantage of minimizing the surface effect.

We therefore take radius and binding energy of Hg^{200} as standards. Since this nucleus has an atomic weight just below that of radioactive nuclei, it seems reasonable to assume a radius slightly smaller than theirs. We take

$$R_{\mathrm{Hg}^{200}}=8\cdot 10^{-13}\ \mathrm{cm}. \tag{161}$$

The binding energy follows from Aston's determination of the atomic weight of Hg, *viz.*, 200.016, and from the atomic weights of proton and neutron which are 1.00807 and 1.00846, respectively (cf. (75a)). This gives

$$\begin{aligned} E_{\mathrm{Hg}^{200}} &= 200.016-(80\cdot 1.00807+120\cdot 1.00846) \\ &= 200.016-201.661 \\ &= -1.645\ \text{mass units} = -1530\ \text{MV}. \end{aligned} \tag{161a}$$

Per nuclear particle, the binding energy is -7.6_5 MV.

From the nuclear radius (161), we can immediately calculate the kinetic energy of all neutrons and protons together, which becomes, according to (150c), (150d), with $N=120$ and $Z=80$:

$$T=\frac{3}{5}(ZT_Z+NT_N)=\frac{3}{5}\left(\frac{9\pi}{4}\right)^{\frac{2}{3}}\frac{\hbar^2}{2MR^2}\times(Z^{5/3}+N^{5/3})=2980\ \text{MV}. \tag{161b}$$

Here T_Z and T_N denote the maximum kinetic energies for protons and neutrons, which are, for Hg, 22.2 and 28.9 MV, respectively. The average kinetic energy of each of the 200 particles is, according to (161b), 14.9 MV.

The potential energy of the nuclear forces, V, is obtained from

$$E=T+V+C+S, \tag{162}$$

where C is the potential energy of the Coulomb repulsion of the protons which is (Eq. (16))

$$C=\tfrac{3}{5}Z^2e^2/R=670 \text{ MV} \tag{162a}$$

and S is the surface energy which we estimate (cf. 184)

$$S=450 \text{ MV}. \tag{162b}$$

Therefore
$$\begin{aligned} V&=-(1530+2980+670+450)\\ &=-5630 \text{ MV}, \end{aligned} \tag{163}$$

$$|V|/T=1.89. \tag{163a}$$

The procedure for determining B and a is then the following: For any value of $D=MBa^2/\hbar^2$, we may determine the minimum of $E'=V+T$ as a function of x, using (159) and (159a). The value $x_{\min}$ for which that minimum occurs, is given as a function of D in Table VII. (n has been put equal to 1, corresponding to an equal number of protons and neutrons.) Inserting $x_{\min}$ into (159), (159a) we may find the ratio V/T as a function of D (fourth line of Table VII). That value of D which gives the observed ratio V/T, is the "correct" value of D for the statistical model; for $|V|/T=1.89$ we find

$$D=MBa^2/\hbar^2=9.6. \tag{164}$$

The corresponding value of $x_{\min}$ is 2.70 corresponding to (cf. 158b)

$$a=0.64\cdot 2.70r_0=1.73r_0 \tag{164a}$$

and with the observed value $R=8\cdot 10^{-13}$ cm:

$$r_0=8\cdot 10^{-13}\cdot 200^{-\frac{1}{3}}=1.37\cdot 10^{-13} \text{ cm}, \tag{164b}$$

$$a=2.37\cdot 10^{-13} \text{ cm}, \tag{164c}$$

$$B=\hbar^2D/Ma^2=70 \text{ MV}. \tag{164d}$$

Thus the range of the nuclear forces turns out almost identical with that derived by Feenberg from light nuclei, whereas the depth of the potential hole must be chosen about 70 percent larger than the actual depth in order to bring the "statistical energy" into agreement with the observed nuclear binding energy.

From the table we note that the depth of the potential hole (fifth line) depends practically only on the nuclear radius and not on the observed binding energy. The latter determines only the range a of the forces.

We may now use our constants to deduce the dependence of the "volume energy" $V+T$ on the ratio of the numbers of neutrons and protons. Expanding (158), (158a) in a power series in

$$n-1=(N-Z)/(N+Z) \tag{165}$$

we have

$$V(n)=V(1)+\frac{B}{3(\pi)^{\frac{1}{2}}}\frac{(N-Z)^2}{N+Z}\frac{1}{x}\times(x^2-1+e^{-x^2}), \tag{165a}$$

$$T(n)=T(1)+\frac{\hbar^2}{6Ma^2}\frac{(N-Z)^2}{N+Z}x^2, \tag{165b}$$

where $V(1)$ and $T(1)$ are the values of potential and kinetic energy for $n=1$, i.e., $N=Z$. Adding (165a) and (165b), we have for the total energy

$$E(n)=E(1)+\beta(N-Z)^2/(N+Z). \tag{166}$$

Inserting the values for x, B and a derived above for the statistical model, we find

$$\beta=39 \text{ MV} \quad \text{(statistical)}. \tag{166a}$$

Inserting, on the other hand, the constants B and a derived from light nuclei (cf. 160a), and determining x from that value of a and the observed r_0 (cf. 164b) with the help of (158b), we have

$$\beta=26 \text{ MV} \quad \text{(force constants from light nuclei)}. \tag{166b}$$

In both cases, the increase of the kinetic energy

TABLE VII. *Relation of constants in the statistical model.* Potential $J(r)=-Be^{-r^2/a^2}$.

$D=MBa^2/\hbar^2$	3.5	4	5	6	7	8	9	10	11
$x_{\min}$	1.22	1.56	1.92	2.14	2.34	2.48	2.62	2.76	2.90
a/r_0	0.78	0.99	1.22	1.36	1.48	1.57	1.66	1.75	1.84
$\|V\|/T$	0.90	1.04	1.25	1.40	1.54	1.67	1.81	1.94	2.08
$B/[\hbar^2/Mr_0^2]$	5.8	4.1	3.35	3.25	3.20	3.23	3.25	3.26	3.27
$E_{\min}/B(N+Z)$	+0.013	−0.008	−0.053	−0.093	−0.127	−0.159	−0.187	−0.213	−0.234

contributes about 8 MV to the constant β while the main contribution arises from a decrease in magnitude of the potential energy when the numbers of neutrons and protons become different.

The value of β has to be compared to the value derived in §8 from the observed ratio of the numbers of neutrons and protons in heavy nuclei (then denoted by $\alpha\epsilon$, cf. (15), (19b)), *viz.*,

$$\beta = 20 \text{ MV} \quad \text{(semiempirical)}, \tag{166c}$$

which is somewhat less than the two theoretical values (166a), (166b). The agreement is improved by introducing forces between like particles (§28).

The nuclear radius is also somewhat altered when the numbers of protons and neutrons are unequal, because the minimum of $V(n)+T(n)$ (cf. 165a, b) occurs for a slightly different value of x than the minimum of $V(1)+T(1)$. The nuclear radius becomes

$$R = r_0(N+Z)^{1/3}[1+0.665(N-Z/N+Z)^2]. \tag{167}$$

The correction term containing $N-Z$ is very small, amounting to only $3\frac{1}{2}$ percent even in the case of uranium ($Z=92$, $N=146$).

In concluding this section, we want to mention that quite analogous results are obtained for the simple exponential law of force

$$J(r) = -Be^{-r/a}, \tag{168}$$

which has been the basis of previous investigations of the statistical model for nuclei by Heisenberg (H9), Wick (W10) and v. Weizsäcker (W2). Instead of (158), we find

$$\begin{aligned} V = -\pi^{-1}x^{-3}B(N+Z)\{&n^{1/3}(2-n)^{1/3}x^2 \\ &-(\tfrac{1}{4}+\tfrac{3}{4}x^2[n^{2/3}+(2-n)^{2/3}]) \log \frac{1+x^2[n^{1/3}+(2-n)^{1/3}]^2}{1+x^2[n^{1/3}-(2-n)^{1/3}]^2} \\ &+2 \arctan (x[n^{1/3}+(2-n)^{1/3}]) \\ &-2(n-1) \arctan (x[n^{1/3}-(2-n)^{1/3}])\}, \end{aligned} \tag{168a}$$

which for $n=1$ ($N=Z$) reduces to

$$\begin{aligned} V(n=1) = -\pi^{-1}x^{-3}B(N+Z) \\ \times\{x^2-(\tfrac{1}{4}+\tfrac{3}{4}x^2) \log (1+4x^2) \\ +\arctan 2x\} \end{aligned} \tag{168b}$$

a formula first derived by Heisenberg.[63] Again, we obtain practically no binding energy if we insert the constants B and a derived from the theory of light nuclei. The constants necessary to give the observed binding energy and radius of the Hg^{200} atom, are

$$D = MBa^2/\hbar^2 = 6.7, \tag{169}$$

$$x_{\min} = 1.75; \quad a = 0.64 \cdot 1.75 r_0 = 1.53 \cdot 10^{-13} \text{ cm}, \tag{169a}$$

$$B = 118 \text{ MV}. \tag{169b}$$

[63] Heisenberg (H9) gives the energy per unit volume, i.e. (168b) divided by the nuclear volume $(4\pi/3)R^3 \cdot (N+Z)$.

These constants are not very different from those obtained by Wick ($B=88$ MV, $a=1.47 \cdot 10^{-13}$ cm), as is to be expected since the experimental data used are similar. (The difference in B arises from the fact that Wick did not correct for the surface energy which is rather large for his standard nucleus O^{16}.) Weizsäcker found $B=184$ MV, $a=1.03 \cdot 10^{-13}$ cm, $D=4.7_5$, i.e., a very short range and very deep hole. Heisenberg's results deviate from ours to the other side, *viz.*, $B=25$ MV, $a=8 \cdot 10^{-13}$ cm.

The dependence of the energy on the "isotopic number" $I=N-Z$ becomes with the exponential potential:

$$V = V(1) + \frac{B}{6\pi x}\frac{(N-Z)^2}{N+Z}(4x^2 - \log(1+4x^2)), \tag{168c}$$

giving for the constant β in (166) a value 42 MV, i.e., almost the same as that obtained from the e^{-r^2/a^2}-potential.

§27. Disproof of Ordinary Forces

In this section we want to prove that the assumption of ordinary forces between neutrons and protons would lead to binding energies of heavy nuclei far larger than those observed. To prove this, we use the statistical model which certainly gives a lower bound for the binding energy.

If we replace the "exchange" (Majorana) by "ordinary" (Wigner) forces the binding energy of the ground state of the deuteron would not be influenced at all, and also that of the α-particle would remain almost unchanged. Thus we obtain practically the same force constants B and a as for Majorana forces.

The expressions (144b), (146b) have to be replaced by

$$V' = \sum_{i=1}^{Z}\sum_{k=1}^{N} \int |\psi_i(x_1)|^2|\varphi_k(x_2)|^2 J(r_{12})d\tau_1 d\tau_2 \tag{170}$$

$$= \int J(r_{12})\rho_p(x_1)\rho_n(x_2)d\tau_1 d\tau_2, \tag{170a}$$

where ρ_p is the *ordinary* proton density as defined in (145). With the plane wave functions used in the statistical model, we have obviously

$$\begin{aligned} &\rho_p(r) = \text{const.} = Z/(4\pi R^3/3) \quad \rho_n = N/(4\pi R^3/3) \quad (\text{for } r<R), \\ &\rho_p = \rho_n = 0 \quad (\text{for } r>R). \end{aligned} \tag{170b}$$

We must carry out the integral in (170a) exactly, i.e., we must not consider R to be large compared to the range a of the forces, because this assumption will be contradicted by the result.

With

$$J(r) = -B'e^{-r^2/a^2} \tag{171}$$

the elementary calculation gives

$$\begin{aligned} V' &= -B'ZN(3/4\pi)^2 8\pi^2 R^{-6} \\ &\quad \times \int_0^R r_1 dr_1 \int_0^R r_2 dr_2 \int_{|r_1-r_2|}^{r_1+r_2} r_{12} dr_{12} e^{-r_{12}^2/a^2} \\ &= -6B'ZNy^{-6}\{\pi^{1/2}y^3\Phi(y)+2-3y^2+e^{-y^2}(y^2-2)], \end{aligned} \tag{171a}$$

where

$$y = 2R/a. \tag{171b}$$

For simplicity, we assume an equal number of neutrons and protons, $N=Z$. We divide (171a) by the total kinetic energy $T=(6/5)NT_{0N}$ where T_{0N} is the maximum kinetic energy per particle as given by (150c), and obtain

$$|V'|/T=(5/3\pi)(\pi/3)^{\frac{1}{3}}A^{\frac{1}{3}}(Ma^2B'/\hbar^2)y^{-4} \times\{\pi^{\frac{1}{2}}y^3\Phi(y)+\cdots\}. \quad (171c)$$

Inserting for $Ma^2B'/\hbar^2$ the smallest possible value, *viz.*, 2.70 (cf. Table III, §12) we have

$$|V'|/T>1.50A^{\frac{1}{3}}[\pi^{\frac{1}{2}}y^{-1}\Phi(y)+2y^{-4} -3y^{-2}+e^{-y^2}(y^{-2}-2y^{-4})]. \quad (172)$$

Now we remember that R, and therefore y, is an arbitrary parameter which has to be chosen so that the total energy is a minimum. If we therefore choose *that* value for y which makes the square bracket of (172) a minimum, we can only obtain too *low* a value for the binding energy. The maximum of the square bracket is obtained for $y=2.80$ and has the value 0.288. Thus

$$|V'|/T>0.433A^{\frac{1}{3}}. \quad (172a)$$

Thus for $A>12$ the potential energy becomes larger than the kinetic, even in the poor approximation obtained by the statistical model. If A is much larger than 12, the kinetic energy is only a small fraction of the potential. But the potential energy (171a) is of the order of magnitude A^2B'. In other words, the binding energy of nuclei of atomic weight greater than, say, 30, would increase as the *square* of the atomic weight, in contradiction to experience (§7). Moreover, the total binding energy would attain huge values; e.g., if we insert $a=2.3\cdot10^{-13}$ cm (cf. 128) and $B'=41$ MV (cf. 160a), the minimum of the total energy for uranium is obtained for $y=0.62$ and has the value

$$E_U=(V'+T)_U=-366{,}000 \text{ MV}. \quad (172b)$$

This would correspond to a mass defect of the uranium nucleus of about 380 mass units, i.e., about one and a half times the mass of uranium itself! Since our theoretical value is a lower limit, the impossibility of the assumption of "ordinary" forces which are attractive over their whole range, has been established strikingly enough.

In addition to the very large binding energies, the model gives much too small nuclear radii. For y is approximately constant for all values of A; indeed it even decreases somewhat with increasing A. The nuclear radius becomes therefore independent of A, and is of the same order as the range of the forces; e.g., for uranium the radius would turn out to be about $1.3\cdot10^{-13}$ cm, one-seventh of the observed radius.

The Coulomb forces are, of course, unimportant compared to the huge binding energies resulting from our model and can therefore not alter the conclusions.

Our statement that ordinary forces are impossible, must, however, be qualified in two respects. Firstly, "ordinary" forces which are repulsive at very small distances and attractive at somewhat larger distances can, of course, not be disproved. Plausibility is the only argument against the assumption of such forces between *elementary* particles. (Cf. §31 where such forces are *derived* for the (complex) α-particle.)

Secondly, a *small* "ordinary" force in addition to a large "exchange" force cannot be excluded. We denote by the "strength" of the "ordinary" force in MV and assume the range to be the same for ordinary and exchange force. If we accept the values B and a given in (164c, d) for the exchange force, we find:

Change of nuclear properties caused by a small additional ordinary force

Atomic weight $A=$	10	30	100	238
Relative change of nuclear radius (percent)	$-1.70B'$	$-2.85B'$	$-4.0\ B'$	$-4.6\ B'$
Change of binding energy per particle (MV)	$0.63B'$	$0.87B'$	$1.07B'$	$1.19B'$

An "ordinary" force of $B'=5$ MV might be just tolerable. It would correspond to a difference between the binding energies per particle for U^{238} and Si^{30} of $0.32B'=1.6$ MV, the uranium nucleus having the stronger binding. Such a difference seems about the highest reconcilable with the experimental facts. B' might, of course, be negative and of about the same magnitude. In any case the "ordinary" forces must be very small (not more than about 7 percent) compared to the exchange forces.

§28. Forces Between Like Particles

The total interaction energy due to the forces between like particles has, according to (146e), practically the same form as that due to the interaction between protons and neutrons (146b). As in §21, we assume the shape of the interaction potential K for like particles to be the same as for unlike ones, and the range of the forces to be the same in both cases, so that

$$K(r)=-Ce^{-r^2/a^2}. \quad (173)$$

Then we find in analogy to (158):

$$F=-\tfrac{1}{4}\pi^{-\frac{1}{2}}x^{-3}C(N+Z)\{[4-3x^2(n^{\frac{2}{3}}+(2-n)^{\frac{2}{3}})] -[2-x^2n^{\frac{2}{3}}]e^{-x^2n^{\frac{2}{3}}}-[2-x^2(2-n)^{\frac{2}{3}}] \times e^{-x^2(2-n)^{\frac{2}{3}}}+\pi^{\frac{1}{2}}x^3[n\Phi(xn^{\frac{1}{3}}) +(2-n)\Phi(x(2-n)^{\frac{1}{3}})]\}, \quad (173a)$$

where x has the same meaning as in (158b).

The value of x which makes $T+V+F$ a minimum, depends only on $n=2N/(N+Z)$, and upon the force constants $MBa^2/\hbar^2$ and $MCa^2/\hbar^2$, but *not* upon the total number of particles. This makes, as in §26, the volume of the nucleus and the total binding energy proportional to the number of particles, in agreement with experience. Quantitatively, for $n=\frac{1}{2}$,

$$F/V=C/2B. \tag{174}$$

The total interaction, due to interaction between like *and* unlike particles, is therefore proportional[62] to $B+\frac{1}{2}C$. From the theory of light nuclei (cf. 128), we obtain

$$C/2B=0.35. \tag{174a}$$

Using the values for B, C and a derived in (128), we found the result (160b) for the binding energy which is much too small, although the addition of the interaction between like particles helps somewhat to increase the binding energy following from the statistical model (cf. the result obtained without forces between like particles, above Eq. (160b)).

We again determine the constants *ad hoc* from the statistical model, as in §26, Eq. (164). The only alteration necessary is that we must now make $B+\frac{1}{2}C$ as large as B was in §26. Keeping the ratio $B : C$ as given by the theory of light nuclei, *viz.* $\frac{1}{2}C=0.35B$, the B of §26, Eq. (164d) should be reduced by a factor 1/1.35, giving

$$\begin{aligned} B&=70/1.35=52 \text{ MV}\\ C&=0.70\cdot 52=36 \text{ MV}\\ a&=2.37\cdot 10^{-13} \text{ cm}\end{aligned} \tag{175}$$

(from statistical model, with forces between like particles equal to 0.70 times the neutron-proton forces).

The interaction of like particles has some influence on the dependence of the nuclear energy on $N-Z$. As might be expected, the potential energy of this interaction decreases (i.e., its absolute magnitude becomes *larger*) when the numbers of neutrons and protons become different. The constant β defined in (166) consists now of three parts, arising respectively from the forces between protons and neutrons, the forces between like particles, and the kinetic energy. For the values of the constants B, a, C chosen in (175) we have

$$\begin{aligned}\beta&=0.44B+0.39\hbar^2/Mr_0^2-0.070C\\ &=23+8-1=30 \text{ MV},\end{aligned} \tag{176}$$

while the constants (128) would give

$$\beta=22 \text{ MV} \tag{176a}$$

as compared to 39 and 28 MV, respectively, when no forces between like particles are assumed (cf. 166a, 166b). The semiempirical value of §8, *viz.*, 20, is almost identical with (176a).

In addition to the "specific nuclear forces" between like particles we have also the electrostatic repulsion between the protons which we found responsible for the increase of the ratio N/Z with increasing atomic weight (§8). We assume the protons to be uniformly distributed over the nucleus so that their density is $Z/(4\pi R^3/3)$; then their interaction, without taking into account exchange, is (cf. 146c)

$$\begin{aligned}C&=\tfrac{1}{2}Z^2\int(e^2/r_{12})(4\pi R^3/3)^{-2}d\tau_1 d\tau_2\\ &=\tfrac{3}{5}e^2Z^2/R,\end{aligned} \tag{177}$$

each of the space integrals $d\tau_1 d\tau_2$ extending over a sphere of radius R. The result has to be corrected for exchange (term A, cf. (146d)). With the expression (151a) for the mixed density, and with the assumption $k_0R\gg 1$ (i.e., large atomic weight), we obtain[64]

$$\begin{aligned}A&=\frac{1}{4}\frac{4\pi}{3}R^3\int_0^\infty \frac{e^2}{r_{12}}\frac{4\pi r_{12}{}^2dr_{12}}{(\pi^2 r_{12}{}^3)^2}\\ &\qquad\times(\sin k_Pr_{12}-k_Pr_{12}\cos k_Pr_{12})^2\\ &=e^2R^3k_P{}^4/3\pi^2=3^{5/3}2^{-8/3}\pi^{-2/3}e^2Z^{4/3}R^{-1}.\end{aligned} \tag{177a}$$

Expressing R in terms of r_0 (cf. 153) we have for the total effect of the electrostatic forces

$$\begin{aligned}C-A&=(e^2/r_0)(Z/N+Z)^{1/3}\\ &\qquad\times(\tfrac{3}{5}Z^{5/3}-3^{5/3}\pi^{-2/3}2^{-8/3}Z)\end{aligned} \tag{177b}$$

$$=(e^2/r_0)(Z/N+Z)^{1/3}(0.600Z^{5/3}-0.460Z). \tag{177c}$$

With our value $r_0=1.37\cdot 10^{-13}$ cm, deduced from experiment, we have

$$e^2/r_0=1.08 \text{ MV}. \tag{177d}$$

[64] Weizsäcker's expression (W2, Eq. (50)) is too large by a factor $(4\pi/3)^{1/3}$.

The exchange term is proportional to the nuclear charge, and amounts to only about 0.18 MV per nuclear particle. The first term, which increases as $Z^{5/3}$, reaches the value 890 MV for uranium, i.e., $3\frac{3}{4}$ MV per particle in the U nucleus, which reduces the binding energy of that nucleus by about one-third (cf. the deduction of the nuclear forces from the mass defect of Hg in §26).

The Coulomb forces have also some effect on the nuclear radius. The relative change of the latter due to Coulomb forces is, using (159), (159a) and putting $B=70$ MV:

$$\delta R/R = +0.60(e^2/Br_0)Z^2A^{-4/3} = 0.0092Z^2A^{-4/3}. \quad (177e)$$

For uranium, this amounts to 5.3 percent, for Fe^{56} to 2.8 percent. We found in (167) that the radius of uranium should be increased by 3.5 percent because the numbers of neutrons and protons are not equal. Altogether, we should thus expect that the radius of U is $(3.5+5.3-2.8)$ percent $=6$ percent larger than would be expected from the radius of the Fe nucleus, assuming the nuclear volume to be proportional to the number of particles. However, it must be borne in mind that the nuclear radius cannot easily be defined to such an accuracy.

§29. The Surface Effect (W2, W10)

Wick (W10) has first pointed out that the nuclear binding energy will be reduced, especially for small nuclei, by the existence of a surface of the nucleus. Particles at the surface interact, in the average, only with half as many other particles as do particles in the interior of the nucleus. The situation is, of course, quite analogous to the surface tension of liquids.

Weizsäcker (W2) has calculated the surface effect quantitatively. He has shown that the effect consists of a "classical" and a "quantum mechanical" part. Classically, we may assume the nucleus to have a sharp boundary at a certain distance R from the center. Those particles which are nearer to the boundary than the range of the forces a, will then not contribute their full share to the binding energy. This leads to an increase of the total energy by an amount of the relative order a/R (classical surface effect).

In quantum theory, the boundary can never be sharp, because this would mean an infinite derivative of all particle wave functions and consequently an infinite kinetic energy. The surface layer must thus be spread out over a certain region of the order of magnitude of one particle wave-length. The narrower the surface region, the larger will be the additional kinetic energy; the broader the region, the more will the total potential energy be reduced in magnitude. Therefore, there will be an optimum breadth of the surface layer.

Weizsäcker (W2) has calculated the surface effect by an extension of the statistical model. If we use the statistical model in its ordinary form we find no surface effect at all. To see this, we assume that there is a certain "auxiliary potential energy" U which has a given negative value $-U_0$ inside a sphere of radius R (nuclear radius) and then rises gradually to zero outside that sphere. We assume the rise of the potential to be gradual enough so that the statistical method is applicable; i.e., we suppose that we may choose volume elements $d\tau$ so small that the potential energy U is sufficiently nearly constant inside $d\tau$, and on the other hand so large that we may apply the considerations of §25 to each volume element. This condition means essentially that the thickness of the surface layer, i.e., the region in which the auxiliary potential U changes from $-U_0$ to 0, must be *large compared to the wave-length of the particles in the nucleus.*

If this condition is fulfilled, the particle density at any point $\mathbf{r}$ (or in any volume element $d\tau$), as well as the maximum and the average kinetic energy of the particles at that point, and the contribution of $d\tau$ to the total potential energy are all completely determined by the value of the "auxiliary potential" U at that point. We have (cf. (150))

$$\rho(\mathbf{r}) = N/(4\pi R^3/3) = k_{\max}{}^3(\mathbf{r})/3\pi^2 = (2M[E_0 - U(\mathbf{r})]\hbar^{-2})^{3/2}/3\pi^2, \quad (178)$$

where

$$(\hbar^2/2M)k_{\max}{}^2 = E_0 - U(\mathbf{r}) \quad (178a)$$

is the maximum kinetic energy of any particle at the point $\mathbf{r}$, E_0 being the total energy of the most energetic particle. The average kinetic energy of the particles at $\mathbf{r}$ is

$$\frac{3}{5}\frac{\hbar^2}{2M}k_{\max}{}^2 = \frac{3^{5/3}\pi^{4/3}}{10}\frac{\hbar^2}{M}\rho^{2/3}. \quad (178b)$$

Multiplying this expression by ρ, we obtain the kinetic energy per unit volume.

The mixed density is given by (151a), with only k_0 to be replaced by $k_{\max}$, provided again

k_{max} does not change appreciably between the two points $\mathbf{r}_1$, $\mathbf{r}_2$, i.e., provided the auxiliary potential changes sufficiently slowly. The contribution to the potential energy per unit volume can then be calculated similarly to (158) or (159) if we assume the density of neutrons and protons to be equal at every point. We have only to divide (159) by the total nuclear volume $(4\pi/3)(N+Z)r_0^3$, and to express, in the definition (158b) of x, the quantity r_0 by the density ρ of protons or neutrons. Since r_0 was defined as the radius of a sphere containing one particle, i.e., in the average one-half neutron and one-half proton, we have

$$(4\pi/3)r_0^3 = \tfrac{1}{2}\rho^{-1}, \tag{178c}$$

$$x = (3\pi^2\rho)^{1/3}a. \tag{178d}$$

The *total* energy per volume element $d\tau$ becomes

$$dE = -\tfrac{2}{3}\pi^{-5/2}a^{-3}B\{2-3x^2-(2-x^2)e^{-x^2} + \pi^{1/2}x^3\Phi(x)\}d\tau + 3^{5/3}\pi^{4/3}(\hbar^2/10M)\rho^{5/3}d\tau. \tag{178e}$$

The total energy *per particle* in the volume element $d\tau$, *viz.*, $dE/(\rho d\tau)$, is, as we know, a minimum if x has the value x_{min} derived in §26 (Table VII). The value of ρ corresponding to x_{min} will be called the standard density ρ_0; it will be the density in the interior of nuclei. Any region of the nucleus in which the density ρ falls short of its standard value ρ_0 will increase the total energy of the nucleus over its value derived in §26.

From this point of view, the total energy of the nucleus would attain its minimum value if the density is ρ_0 throughout the nucleus, falling to zero suddenly at the boundary. This minimum value would be exactly equal to the energy derived in §26; thus there would be no surface effect at all, as we mentioned before. However, in order to make the statistical method applicable, we had to assume that the density varies *slowly* at the boundary of the nucleus, more accurately that is does not drop from ρ_0 to 0 in a distance shorter than the wave-length of the nuclear particles, which is of the order of r_0. If the density changes too rapidly, the kinetic energy becomes much larger than its value in the statistical model.

We therefore assume that the *drop of the density from* ρ_0 *to 0* occurs in a layer of a thickness λ, of the order of magnitude r_0. Furthermore, we assume the "auxiliary potential" U to fall off linearly in the surface region. This means, according to (178), that ρ varies as the 3/2-power of the distance from the surface. Denoting by y the coordinate perpendicular to the surface, we thus assume

$$\begin{aligned} &\rho = 0 && \text{for } y<0, \\ &\rho = \rho_0(y/\lambda)^{3/2} && \text{for } 0<y<\lambda, \\ &\rho = \rho_0 && \text{for } \lambda<y. \end{aligned} \tag{179}$$

It is then easy to calculate the change of the nuclear energy due to the surface effect. We denote by S the total surface of the nucleus, by Ω its volume, where Ω is defined by

$$\Omega\rho_0 = N, \tag{179a}$$

N being the total number of neutrons. T_0 is the total kinetic energy without surface effect, given in §26. Then we obtain a decrease of the kinetic energy[65] by

$$\delta T = -(4/35)T_0 S\lambda/\Omega \tag{179b}$$

and an increase of the potential energy by

$$\delta V = \tfrac{3}{5}\pi^{-1/2}B(N+Z)(S\lambda/\Omega)x_0^{-3}(\tfrac{1}{2}x_0^2 - 2 + 3x_0^{-2} - (3x_0^{-2}+1)e^{-x_0^2}), \tag{179c}$$

where x_0 is the value of x making the nuclear energy a minimum. We insert (cf. 164d, 161b)

$$S = 4\pi R^2, \quad \Omega = 4\pi R^3/3, \tag{179d}$$

$$B = 70 \text{ MV}, \quad T_0 = 15(N+Z) \text{ MV}, \quad x_0 = 2.70 \tag{179e}$$

and obtain

$$\begin{aligned} \delta T &= -5.1(N+Z)\lambda/R \text{ MV}, \\ \delta V &= +8.3(N+Z)\lambda/R \text{ MV}, \\ \delta E &= +3.2(N+Z)\lambda/R \text{ MV}, \end{aligned} \tag{180}$$

or, using (153)

$$\delta E = 3.2(N+Z)^{2/3}\lambda/r_0 \text{ MV}. \tag{180a}$$

λ/r_0 should, according to our assumptions, be of the order of magnitude unity. A considerably larger value seems necessary to obtain agreement with experiment (cf. §30, especially (185a)).

Weizsäcker has tried to determine λ theoretically. He supposes that each individual wave function contains an exponentially decreasing factor near the surface of the nucleus, but behaves otherwise in the same way as the

[65] This is due to the reduced *density* in the surface layer.

statistical method assumes. The exponential decay introduces an additional term in the kinetic energy. It is then assumed that the density corresponding to each individual state decreases in the same way as the total density. Obviously the additional term in the kinetic energy will then be the larger the thinner the surface layer. On the other hand, the term (179c) is the larger the thicker the surface layer. The condition that the sum of the two terms shall be a minimum, leads immediately to a determination of the thickness of the surface layer and of the additional surface energy. However, the basic assumption that the decrease in density at the surface is due to a uniform decrease of the density due to each individual state, does not seem to be justified: In reality when we approach the surface one state after the other "dies out" because its total energy becomes less than the potential energy at the given point; and this dying out accounts for the decrease in density without any exponential decay of the individual wave functions being necessary. (The exponential tail of each wave function can be neglected in the statistical method.)

Flügge (F12) has used Weizsäcker's method to treat light nuclei, for which the surface layer cannot be considered as thin compared to the nuclear radius so that the distinction of the "interior" of the nucleus and the "surface layer" is no longer justified. He finds that the mass defects of all light nuclei from He^4 to Si^{28} can be well represented by Weizsäcker's extension of the statistical method described above, the potential energy of the interaction between and proton being assumed as

$$J(r) = -Be^{-r^2/a^2}$$

with $$B = 85 \text{ MV}, \quad a = 1.46 \cdot 10^{-13} \text{ cm}. \tag{181}$$

The density of neutrons and protons is supposed to depend like a Gaussian function on the distance r from the center of the nucleus, *viz.*,

$$\rho(r) = e^{-2r^2/R^2}, \tag{181a}$$

where R is the "nuclear radius." R is not exactly proportional to the cube root of the atomic weight, as it is for heavy nuclei, but is somewhat larger for light nuclei, *viz.*,

$$R = (0.67A^{\frac{1}{3}} + 0.93A^{-\frac{1}{3}}) \cdot 10^{-13} \text{ cm}. \tag{181b}$$

Accordingly, the volume of light nuclei is larger than would be expected if the volume were exactly proportional to the number of particles.

Flügge has also carried through calculations using a simple exponential potential, with similar results.

§30. Weizsäcker's Semiempirical Formulae (W2)

Since the statistical model does not give satisfactory results, Weizsäcker has proposed a semi-empirical method for calculating nuclear energies. The nuclear energy is assumed to have a *form* such as is indicated by the statistical method, and, indeed, by very simple qualitative considerations. However, in the formula certain *constants* are left arbitrary and are determined from experimental data.

We choose the following form for the *total mass* (energy) of an *atom* which is slightly simpler than Weizsäcker's:

$$M = NM_n + ZM_p - \alpha A + \beta(N-Z)^2/A + \gamma A^{\frac{2}{3}} + \tfrac{3}{5}(e^2/r_0)Z^2A^{-\frac{1}{3}}, \tag{182}$$

where A is the atomic weight, N and Z the numbers of neutrons and protons, $r_0A^{\frac{1}{3}}$ the nuclear radius, M_n and M_p the masses of neutron and hydrogen atom and $\alpha\beta\gamma r_0$ empirical constants. The first two terms in (182) are self-evident. The third represents the main binding energy which we know to be proportional to the number of particles in the nucleus A, the constant α to be determined empirically.

The fourth term in (182) represents the decrease of the binding energy when the numbers of protons and neutrons become different; it has the form derived in §26 from the statistical model. This form holds, of course, only if $N - Z \ll A$; but this condition is fulfilled for all existing nuclei.

The fifth term is the surface effect, the last term the Coulomb repulsion of the protons. Both these terms have again the form suggested by the statistical model. The exchange correction to the electrostatic repulsion (177a) may be considered as contained in the first term since it is proportional to Z.

To determine the constants, we proceed in the following way:

1. We determine r_0 from the empirical radii of radioactive nuclei this gives (cf. 17a, b)

$$r_0 = 1.48 \cdot 10^{-13} \text{ cm} \tag{182a}$$

$$\tfrac{3}{5}e^2/r_0 = 0.58 \text{ MV}. \tag{182b}$$

2. β is determined so that the most stable nucleus of atomic weight 200 has the nuclear charge 80 (Hg^{200}). This gives

$$\beta = \left[\frac{3}{5}\frac{e^2}{r_0}\frac{Z}{A^{\frac{1}{3}}} + \frac{1}{2}(M_p - M_n)\right]\frac{A}{2(N-Z)}$$

$$= \left(0.58 \cdot \frac{80}{200^{\frac{1}{3}}} - 0.20\right)\frac{200}{80} \text{ MV},$$

$$\beta = 19.5 \text{ MV}. \tag{183}$$

With this value for β, the most stable nucleus of atomic weight A has the "isotopic number"

$$I_A=(N-Z)_A=A\frac{0.3(e^2/r_0)A^{2/3}+\frac{1}{2}(M_p-M_n)}{2\beta+0.3(e^2/r_0)A^{2/3}},$$

$$I_A=A(A^{2/3}-0.7)/(134+A^{2/3}), \qquad (183a)$$

and the mass

$$M_{\min}(A)=\tfrac{1}{2}A(M_n+M_p)-\alpha A+\gamma A^{2/3}$$
$$+\frac{3}{20}(e^2/r_0)A^{5/3}-A\frac{[0.3(e^2/r_0)A^{2/3}-\frac{1}{2}(M_n-M_p)]^2}{2(2\beta+0.3(e^2/r_0)A^{2/3})}$$
$$=\tfrac{1}{2}A(M_n+M_p)-\alpha A+\gamma A^{2/3} \qquad (183b)$$
$$+0.145A^{5/3}\frac{135}{134+A^{2/3}}.$$

In the last transformation, some very small terms involving the difference of the masses of neutron and hydrogen atom have been neglected. The figures represent energies in MV.

3 and 4. We determine the coefficients α and γ so that the masses of O^{16} and Hg^{200} are correct. We have for O^{16}: atomic weight 16.0000; $8M_n+8M_p=16.1322$.

Difference	-0.1322 mass unit $=-123.1$ MV,
Coulomb energy	$0.58\cdot 8^2\cdot 16^{-1/3}=14.7$ MV,
The term	$\beta(N-Z)^2/A$ vanishes.

Therefore

$$-16\alpha+16^{2/3}\gamma=-(123.1+14.7)=-137.8\text{ MV}. \qquad (184a)$$

For Hg^{200}: atomic weight 200.016; $120M_n+80M_p=201.661$.

Difference	-1.645 mass units $=-1532$ MV,
Coulomb energy	$0.58\cdot 80^2\cdot 200^{-1/3}=633$ MV,
	$\beta(N-Z)^2/A=19.5\cdot 40^2/200=156$ MV.

Therefore

$$-200\alpha+200^{2/3}\gamma=-(1532+633+156)$$
$$=-2321\text{ MV}. \qquad (184b)$$

From (184a), (184b) we find

$$\alpha=13.86\text{ MV},\quad \gamma=13.2\text{ MV}. \qquad (184)$$

We convert all energies into thousandths of a mass unit, and insert the constants into (182). Then the excess of the exact atomic weight of an atom (A,Z) over the "mass number" A is, in thousandths of a mass unit:

$$E=1000(M-A)=8.0_5Z+8.4_5N-14.9A$$
$$+21(N-Z)^2/A+14.2A^{2/3}+0.625Z^2A^{-1/3} \qquad (185)$$

or

$$E=-6.6_5A+0.4I+21I^2/A+14.2A^{2/3}$$
$$+0.625Z^2A^{-1/3}, \qquad (185a)$$

where $I=N-Z=A-2Z$ is the isotopic number.

As Weizsäcker has pointed out, this formula is immediately applicable only to nuclei with even numbers of neutrons and protons. Nuclei containing an odd number of either neutrons or protons have higher mass (less binding energy). This can be seen by the following argument. The energy (185a) is, for a given number of protons Z, very nearly a quadratic function of the number of neutrons N. Therefore, any further neutron which may be bound to a given nucleus, would be bound less strongly than the preceding neutron. Actually, however, if we have a nucleus containing an even number of neutrons and protons, it will always bind *two* additional neutrons with the *same* binding energy, because they both can be bound in the same state. Therefore, the energy of a nucleus containing an *odd* number of neutrons is to be computed by taking the arithmetic mean between the energies of the two adjacent nuclei with even numbers of neutrons. The same is true for nuclei containing an odd number of protons. The energy of nuclei with both N and Z odd are to be obtained by double interpolation.

Table VIII gives the mass excess for some nuclei throughout the periodic table in thousandths of mass units. It shows the relative importance of main energy, surface energy and Coulomb repulsion. The agreement with experiment is rather satisfactory. The incomplete agreement for the standard O^{16} is due to rounding off in (185).

Table IX gives the mass excess of the known light nuclei, calculated and observed. The values derived by Weizsäcker from his semiempirical formula are also given. They agree somewhat better with the observed values than ours. The reason is that Weizsäcker adjusted his constants so that the masses of light nuclei are represented as well as possible. Weizsäcker's formula is

$$E=[-(\alpha^2+\beta^2)^{1/2}+(\alpha^2+\beta^2((Z-N)/A)^2)^{1/2}]$$
$$\times[A-1-\gamma(A-1)^{2/3}]$$
$$+(3e^2/r_0A^{1/3})(\tfrac{1}{5}Z^2-2^{-4/3}Z^{4/3}) \quad (185b)$$

with $\alpha=1.6$, $\beta=13.9$, $\gamma=0.6$ thousandths of a mass unit, and $r_0=1.26\cdot 10^{-13}$ cm.

The general trend of the nuclear masses seems to be represented fairly well by the theoretical formulae. Notable exceptions are the lightest nuclei, for which the formulae cannot be expected to hold, and a marked break near oxygen. While the experimental energies are in the average about equal to the theoretical ones up to O^{16}, and in some instances (Be^8 C^{12}) even lower, they are very much higher than the theoretical energies for the nuclei between O and Al. The differences reach 6 to 8 milli-mass units. For still heavier nuclei, the agreement improves again, and is almost complete for sulphur. The reason for all these facts seems to be the completion of a "closed neutron shell" at O^{16} (cf. §33).

The differences between the energies of isobars such as $C^{13}N^{13}$, $N^{15}O^{15}$, $O^{17}F^{17}$, etc., seem to agree well with the experimental values deduced from the upper limit of β-ray spectra (§39). In agreement with experiment, B^{10} turns out to be more stable than Be^{10}, N^{14} more stable than C^{14}, but F^{18} less stable than O^{18}.[66] However, there are also notable discrepancies, especially for heavier nuclei of even atomic weight; e.g., the nuclei Si^{28} and Al^{28} should be equally heavy according to our semiempirical formula, while actually Al^{28} is 5 units heavier, P^{30} should be 4 units heavier than Si^{30} and is actually 6 units heavier, P^{32} should be 1 unit lighter than S^{32} and is 2 units heavier. This shows that our method does not give a big enough difference between nuclei of even mass and odd charge and such of even mass and even charge especially for heavier nuclei. It therefore points to the necessity of introducing forces between like particles (§10, 18, 21, 28).

For the lightest nuclei maximum stability is found for those which contain an integral number of α-particles (Be^8, C^{12}, O^{16}). For higher atomic weight these differences in stability gradually disappear, and the maximum of stability is shifted towards nuclei containing more neutrons than protons (compare Be^8, Be^9, Be^{10} to S^{32}, S^{33}, S^{34}). It should be noted that the exceptional stability of light nuclei containing an integral number of α-particles is obtained without the assumption that there are actually α-particles as secondary units in the nucleus. We have only made the rather obvious assumption that neutrons and protons form shells of two, each shell containing two particles of opposite spin and equal spatial wave function.

[66] This seems to show that the break in the isotope pattern near O (cf. §34) is *not* connected with the completion of a closed shell at O^{16}. For the latter fact is *not* represented in our calculation, while the former comes out automatically since the greater stability of O^{18} compared to F^{18} is sufficient to explain the change in the isotope pattern.

TABLE VIII. *Masses of some nuclei calculated from the semi-empirical formula (185), in thousandths of a mass unit.*

Atom	Mass Excess of Consti.*	Main Energy	Term in $(N-Z)^2$	Surface Energy	Coulomb Energy	Total Theor.	Total Exp.
O^{26}	+132.0	−238.4	0	90.1	15.8	−0.5	0.0
A^{40}	329	−596	8	166	59	−33	−29
K^{82}	329	−1222	25	268	186	−78	−73
Xe^{134}	1101	−1997	106	372	354	−62	−71
Hg^{200}	1658	−2980	168	486	680	+16	+16
U^{238}	1953	−3547	257	545	838	+52	+99**

* $8.0_5Z+8.4_5N$, i.e., the excess of the masses of the neutrons and protons contained in the nucleus over the "mass number" $Z+N$.
** Computed from the energies of the α-particles emitted in the uranium series, and an assumed mass of Pb^{206} of 206.020.

We want to use our empirical formula for a redetermination of the limit of stability against α-disintegration (cf. §8). We take the most stable nucleus of given atomic weight A, whose energy is given by (183b). Inserting the values of α and γ from (185), we find for the mass excess of the most stable nucleus of atomic weight A in thousandths of a mass unit

$$E_{\min}(A)=-6.65A+14.2A^{2/3}$$
$$+0.156A^{5/3}135/(134+A^{2/3}). \quad (186)$$

The condition for α-instability is

$$E_{\min}(A)-E_{\min}(A-4)>3.35 \quad (186a)$$

because 3.35 thousandths of a mass unit is the mass excess of the helium atom. This gives

$$-26.6+37.9A^{-1/3}+0.625A^{2/3}$$
$$\times\frac{135\cdot((5/3)\cdot 134+A^{2/3})}{(134+A^{2/3})^2}>3.35. \quad (186b)$$

This condition is fulfilled for

$$A>147, \quad (186c)$$

TABLE IX. *Mass excesses of light nuclei calculated from the semiempirical formula (185), in thousandths of a mass unit.*

Nucleus	Theor.	Weizsäcker	Exp.	Nucl.	Theor.	Weiz.	Exp.	Nucl.	Theor.	Weiz.	Exp.	Nucl.	Theor.	Weiz.	Exp.
He^3	29.3	13.3	16.4	He^4	10.8	9.6	3.4	He^5	16.8	14.1	—	He^6	22.8	18.6	—
Li^5	17.5	15.3	—	Li^6	16.5	13.9	16.1	Li^7	15.5	12.7	16.9	Li^8	25.1	17.6	18.3
Be^7	16.5	13.7	—	Be^8	8.3	7.5	7.0	Be^9	10.4	10.1	13.9	Be^{10}	12.5	13.6	15.4
B^9	12.5	12.1	15.5	B^{10}	10.5	10.4	14.6	B^{11}	8.5	8.5	11.1	B^{12}	13.2	12.8	16.6
C^{11}	10.6	10.3	14.2	C^{12}	4.6	3.5	3.7	C^{13}	4.7	6.0	6.9	C^{14}	4.9	8.5	7.8
N^{13}	7.5	8.7	10.0	N^{14}	4.8	6.5	7.6	N^{15}	2.2	4.3	5.3	N^{16}	4.6	7.9	7.5
O^{15}	5.0	7.4	8.6	O^{16}	−0.4	0.4	0.0	O^{17}	−1.0	1.8	4.0	O^{18}	−1.7	3.4	4.5
F^{17}	2.4	5.5	7.8	F^{18}	−0.5	3.1	>5.5	F^{19}	−3.4	0.4	+4.0	F^{20}	−2.6		5.4
Ne^{19}	0.0	—	—	Ne^{20}	−5.2	−2.3	−0.2	Ne^{21}	−6.7		−0.5	Ne^{22}	−8.3		−2.2
				Na^{22}	−5.9		−0.3	Na^{23}	−9.2		−2	Na^{24}	−9.2		−0.5
				Mg^{24}	−10.2		−6.1	Mg^{25}	−12.2		−6.5	Mg^{26}	−14.2		−7.5
				Al^{26}	−10.8		−3.5	Al^{27}	−14.5		−10.5	Al^{28}	−15.2		−10.0
				Si^{28}	−14.8		−15.	Si^{29}	−17.2		−16	Si^{30}	−19.6		−17.5
				P^{30}	−15.7		−11.5	P^{31}	−19.5		−20	P^{32}	−21.0		−20.5
				S^{32}	−19.5		−22.5	S^{33}	−22.4		−23.5	S^{34}	−25.3		−25.5

Note: All experimental data on nuclei of atomic weight greater than 17, with the exception of F^{19} and Ne^{22}, are tentative only. They are based on the scarce scattered data about transmutations of these heavier elements, and partly only on interpolation. Errors up to about 3 units in the difference between neighboring elements, and maybe 10 units in the absolute values, seem possible

i.e., we obtain practically the same condition for α-instability as in §8 when we neglected the surface effect. An estimate of the average kinetic energy of α-particles emitted by radioactive nuclei is obtained by inserting into the left-hand side of (186a) an average atomic weight for radioactive nuclei, say, 226 (radium). Then we have in satisfactory agreement with the experimental average energy of radioactive α-particles

$$Q_\alpha = E_{\min}(226) - E_{\min}(222) - 3.35 = 3.8 \text{ MV}. \quad (186d)$$

VI. More Detailed Theory of Heavier Nuclei

Not many definite results concerning the details of the structure of heavier nuclei have yet been obtained. We shall discuss in this chapter only a few of the attempts to obtain such a theory, and only those which we consider likely to become starting points for future development.

§31. α-Particles as Subunits of Heavier Nuclei

The following arguments have been given for the assumption that α-particles exist in heavier nuclei as subunits:

1. The mass defect per particle is for all heavier nuclei of the same order of magnitude as for the α-particle. In other words, when a heavy nucleus (atomic weight between 20 and 200) is built up, most of the binding energy is set free when groups of two neutrons and two protons are combined into α-particles (27 MV per α-particle) and only a relatively small additional energy (about 7 MV per α-particle) is gained when the α-particles are put together in the heavy nucleus.

2. The assumption of α-particles as subunits therefore seems to offer a straightforward method for a theoretical calculation of the binding energy of heavier nuclei: Already in "zero approximation," the binding energy of the heavy nucleus would be the sum of the binding energies of the α-particles contained in it; and if it can be shown that α-particles attract each other, there will be justified hope to arrive at a theoretical binding energy reasonably near the observed one in the next approximation. In contrast to this, the "Hartree" approximation which assumes the elementary particles to move independently in the nucleus (§§32 to 35) fails to lead to satisfactory results for the binding energy, whether it is used in the crude form of the statistical method (chapter V) or in the more elaborate one described in this chapter (§§35, 36).

3. Among the light nuclei, those which can be regarded as consisting exclusively of α-particles, i.e., Be^8, C^{12}, O^{16}, Ne^{20}, etc., have higher binding energies per particle than any of their neighbors.

4. Radioactive nuclei emit α-particles.

However, the last two arguments can easily be refuted, and in the course of refuting argument 3 we shall come across some strong arguments against the existence of α-particles as subunits.

Ad 4: This argument is not at all conclusive, because it can be shown by very simple considerations involving energy and probability only, that *α-particles are the only particles which can be emitted from heavy radioactive nuclei.* Firstly, the internal energy of the α-particle is much lower than that of the preceding nuclei H^1, H^2, H^3, He^3. Therefore a given nucleus Z^A (Z=nuclear charge, A=atomic weight) may have higher energy than the nuclei $(Z-2)^{A-4}$ and He^4 together, but will in general have lower energy than, say, $(Z-1)^{A-3}$ plus H^3. Thus it is energetically unstable against emission of an α-particle, but stable against emission of any of the lighter particles. Of course, the nucleus Z^A will in general be energetically unstable against the breaking-up into $(Z-6)^{A-12}$ and C^{12}, or $(Z-8)^{A-16}$ and O^{16} if it is unstable against α-emission. But here the second point, *viz.*, the probability considerations, set in: It is almost impossible that such a heavy particle as C^{12} "leaks through" the high and broad potential barrier existing between it and the residual nucleus (chapter IX), while the comparatively light α-particle with its comparatively small charge may leak through quite easily.

Ad 3: This rule may be explained without assuming α-particles as subunits. For there are two main principles governing the structure of nuclei: Firstly, the overwhelming strength of the neutron-proton interaction which, for small atomic weight, makes those nuclei most stable which contain equally many neutrons and protons (§6). Secondly, the "even-odd rule" (§10) stating that nuclei are most stable if they contain even numbers of neutrons and protons, the reason being the Pauli principle (§30) and probably in addition attractive forces between like particles (§10, 18, 21). The nuclei containing exclusively α-particles are favored by both these points which explains their particular stability. In fact, we could even account quantitatively for the difference between the binding energies of these nuclei and their neighbors without assuming α-particles as subunits (§30).

After having disposed of arguments 4 and 3, we shall give a more general argument against α-particles as nuclear subunits. For nuclei heavier than about 30, the preference for nuclei composed exclusively of α-particles ceases to exist. The reason is of course the Coulomb repulsion of the α-particles; this repulsion makes it necessary that stable heavy nuclei contain some extra neutrons as "mortar" keeping the α-particles together.

This fact in itself does not speak against the existence of α-particles as subunits. However, as far as the rather scarce experimental evidence goes, it seems that the binding energy of these additional neutrons is materially the same as the interior binding energy of an α-particle per particle, i.e., 7 to 8 MV. If the concept of α-particles as nuclear subunits were a good approximation, we would expect that all interactions between α-particles and additional neutrons, or between pairs of α-particles, must be small compared to the internal binding energy of the α-particle. A binding energy of 8 MV per additional neutron must correspond to a large perturbation of the α-particles, so that it becomes very doubtful to what extent one may speak of their existence as subunits in nuclei at all. This holds at least for nuclei which contain a large number of extra neutrons.

Another argument which points in the same direction may be drawn from Heisenberg's attempt to calculate the interaction between two α-particles. This interaction will, similarly to chemical interactions, consist of two parts, the exchange interaction and the van der Waals interaction. The exchange interaction is obtained by averaging the mutual potential energy of all the individual particles over the unperturbed eigenfunction of the interacting systems (molecules or α-particles), taking due account of the Pauli principle. The van der Waals interaction is connected with a mutual polarization of the two interacting systems. Now the α-particle is a "closed-shell" system, analogous to the rare gas atoms in chemistry [i.e., protons and neutrons in the α-particle fill all places in the lowest quantum state (1*s* state)]. Accordingly, the exchange interaction between two α-particles, or between an α-particle and an elementary particle, must be repulsive, just as between a rare gas atom and another atom (rare gas or otherwise). The reason is that the eigenfunction must be antisymmetric with respect to interchange of a neutron or proton in the α-particle, and one in the system interacting with it, because of the Pauli principle. This introduces nodes into the wave function which lead to increased total energy. The van der Waals forces are always attractive.

In molecular theory, the repulsive exchange forces are very much stronger than the van der Waals forces at close distances, making molecules practically impenetrable for each other. On the other hand, the van der Waals forces extend to much larger distances. For the exchange forces exist only if the wave functions of the two interacting molecules overlap, while the van der Waals forces are not subject to this condition. The exchange forces between molecules therefore fall off exponentially with increasing distance of the molecules, whereas the van der Waals forces behave as a *power* of the distance r, usually r^{-6}. The reason for the slow decrease of the van der Waals forces is the very slow decrease of the force between two individual electrons, *viz.*, e^2/r.

In the case of nuclei, the force between elementary particles falls off very rapidly. We shall therefore expect the van der Waals forces to have a range only slightly larger than the exchange forces. Roughly, we may expect the range of the exchange forces between two α-particles to be about equal to the diameter of the α-particle, while the van der Waals forces will extend over a distance about equal to that diameter *plus* the range of the forces between neutron and proton. Since the latter is certainly not larger and probably considerably *smaller* than the diameter of the α-particle, the difference will not be great.

We shall thus obtain an interaction potential between two α-particles which has the following characteristics: There will be a strong repulsion at close distances, a not quite so strong attraction at slightly larger distances, and the Coulomb repulsion at great distances. Such a shape of the potential seems to agree qualitatively with the scattering of α-particles by α-particles (chapter X).

Heisenberg has computed the interaction between two α-particles, assuming a potential Be^{-r^2/a^2} between a neutron and a proton and no interaction between like particles. For each α-particle a wave function of the form (117a) was assumed. The result of Heisenberg's calculation is very surprising: The exchange force acts only if the two α-particles coincide exactly,[67] and the van der Waals force has the same range as the neutron-proton force. It does not seem clear whether this peculiar result is due to the particular form of interaction potential and wave functions (Gaussian functions!) or to the approximations made in Heisenberg's derivation. Detailed investigation will be needed to clear up this point.

However, it seems at least certain from Heisenberg's calculations that both exchange and van der Waals forces become of the same order of magnitude as the binding energy of the α-particle when the two α-particles are close together. This again seems to be a serious objection against the α-particle approximation.

It may be asked why the binding between α-particles is so small if the forces between them are large. The reason is probably the small region of space in which there is a large attraction between them.

It may be mentioned that this peculiar shape of the interaction between two α-particles as a function of their distance, would make the structure of nuclei composed of α-particles very simple provided the approximation from α-particles has any sense: Two neighboring α-particles in a nucleus would in general have a mutual distance falling inside the "trough" of the interaction potential, i.e., very near a given value. This means that the structure of nuclei containing exclusively α-particles could be considered from a purely geometrical point of view: The 3 α-particles of C^{12} would be arranged in an equilateral triangle, the 4 in O^{16} in a tetrahedron, etc. The binding energy (energy of the respective nucleus compared to the energy of the corresponding number of free α-particles) would then in first approximation be proportional to the number of pairs of neighboring α-particles which is 1 for Be^8, 3 for C^{12} (triangle), 6 for O^{16} (tetrahedron), and 3 more for each additional α-particle. Experimentally, the mass of Be^8 is almost exactly that of two α-particles; thus the mutual attraction of one pair of α's is not sufficient to overcome the kinetic energy associated with their relative motion, the situation being similar to that found in the deuteron

[67] Heisenberg assumes that the centers of gravity of the α-particles are not exactly localized. The interaction between two α-particles at a fixed distance s is obtained from his formulae (14), (15), (17) by letting η go to infinity. Then the exchange interaction (15) becomes the Dirac δ-function, while the van der Waals interaction is proportional to e^{-2r^2/a^2}.

(cf. Massey and Mohr, M10). The binding energy of the next α-particle is 0.0067 mass unit (mass of $Be^8=8.0070$, $He^4=4.0034$, $C^{12}=12.0037$, binding energy $Be^8+He^4-C^{12}$), of the following 0.0071 mass unit ($C^{12}+He^4-O^{16}$). From our simple picture, we would expect the binding energy of the fourth α-particle (leading to O^{16}) to be about 50 percent larger than that of the third, because three new "bonds" are created when the fourth, and only two when the third α-particle is added. On the other hand, it seems to be correct that the addition of every further α-particle sets about the same energy free as the addition of the fourth, in agreement with expectations from our simple approximation.

Summarizing, we must say that it can at present not be decided how much truth is in the assumption of α-particles as nuclear subunits. Certainly, this assumption must not be taken literally, and the α-particles undergo considerable deformations (polarizations) in the nucleus. On the other hand, the approximation assuming the elementary particles to move independently (Hartree approximation) is certainly not correct either, but must be supplemented by introducing correlations between the particles (end of §36). Such correlations would lead at least in the direction towards the α-particle approximation. The truth will therefore probably lie between the two extremes, as Heisenberg (H10) has pointed out. However, it seems to us that at present the Hartree approximation offers more prospects for being perfected.

§32. Quantum States of Individual Particles (Neutron and Proton "Shells") (H10, B9, E3, G13)

The opposite extreme to the assumption of α-particles as nuclear subunits is that of independent motion of the individual protons and neutrons. This assumption can certainly not claim more than moderate success as regards the calculation of nuclear binding energies (§35). However, it is the basis for a prediction of certain periodicities in nuclear structure for which there is considerable experimental evidence (§33, 34). Also, the individual-particle-approximation seems to offer some hope for the development of a rational theory of nuclear spins in the future (§36).

The procedure in the individual particle-scheme is very simple: To start with, we assume a certain "auxiliary potential" which we suppose to act on each proton and neutron. (The auxiliary potentials may be chosen different for protons and neutrons, to account roughly for the electrostatic repulsion between the protons.) We calculate the wave functions of the individual particles in the auxiliary potential. Then we compute the total kinetic and potential energy of the nucleus from the wave functions, with the help of (144), (146).

In "zero-order" approximation, the energy of the nucleus will be given by the sum of the eigenvalues of the individual particles in the auxiliary potential, provided the latter has been chosen suitably. This zero-order approximation will be studied in this section and will lead us to the prediction of periodicities in nuclear structure. In the following two sections we shall discuss the experimental evidence concerning the periodicity. In §35 and 36, we shall then proceed to the "first approximation," in which the energy of the nucleus is calculated as the average of kinetic plus potential energy, averaged over the wave function of the nucleus. We shall also indicate the probable influence of a second approximation (§36).

For the zero-order approximation, we need only know the eigenvalues of the individual particles in a given auxiliary potential. The first problem is therefore the suitable choice of an auxiliary potential. The potential suggesting itself immediately is the simple potential hole: The potential is assumed to be $-U_0$ inside a sphere of radius R (nuclear radius) and zero outside. Such a potential will represent the actual state of affairs fairly accurately since we know that the density of nuclear matter is practically constant inside heavy nuclei; thus the potential energy of one particle in the field of the nucleus will also be practically constant.[68] A still better approximation may be obtained by letting the potential go to zero gradually at the edge of the nucleus, but such refinements seem hardly worth while at present. Only for light nuclei the simple potential hole will be unsatisfactory because the thickness of the "surface layer" (§29) in which the potential goes gradually to zero, is of the same order as the nuclear radius (Flügge, F12). Therefore it seems more appropriate actually to represent the gradual change in the auxiliary

[68] We shall take, in this section, the *exact* solutions of the Schrödinger equation in a spherically symmetrical potential hole. The result will differ considerably from the result of the statistical method, in which the eigenfunctions were approximated by plane waves (chapter V).

potential, which can be done by choosing, e.g., an "oscillator" potential

$$U=-U_0+\tfrac{1}{2}Cr^2=-U_0+\tfrac{1}{2}M\omega^2r^2 \quad (187)$$

as the auxiliary potential (Heisenberg, H10). (ω is the frequency of a "classical oscillator" of mass M in the potential U.) The oscillator potential has the additional advantage of giving very simple wave functions (Hermitian functions).

(a) We shall first discuss the quantum states of the individual particles in the *oscillator potential* (187). The quantum states may be described by three quantum numbers n_1, n_2, n_3, which are connected to the energies of the vibrations along the x, y, and z axis, respectively. The total energy of a particle is

$$E=-U_0+\hbar\omega(N+\tfrac{1}{2}) \quad (187a)$$

with

$$N=n_1+n_2+n_3+1. \quad (187b)$$

The eigenfunction of the state n_1, n_2, n_3 is

$$\psi=e^{-\frac{1}{2}\rho^2}(x^{n_1}+\cdots)(y^{n_2}+\cdots)(z^{n_3}+\cdots) \quad (187c)$$

with

$$\rho^2=M\omega r^2/\hbar, \quad (187d)$$

the dots denoting lower powers in x, y and z, respectively. In order to compare the results for the "oscillator" potential with those for other central fields, e.g., the potential hole, the wave function must be written as a function of r times a spherical harmonic. This is always possible by suitable linear combinations of the wave functions (187c); e.g., we have:

$N=1: \psi=\psi_{000}=e^{-\frac{1}{2}\rho^2}$, $l=0$, 1s level,

$N=2: \psi=\psi_{001}=e^{-\frac{1}{2}\rho^2}\rho\cos\theta$, $l=1$, 2p level,

and two similar functions ψ_{010} and ψ_{100}

$N=3: \psi_1=\psi_{200}+\psi_{020}+\psi_{002}$

$=e^{-\frac{1}{2}\rho^2}(\rho^2-(3/2))$, $l=0$, 2s level,

$$\left.\begin{aligned}\psi_2&=\psi_{002}-\tfrac{1}{2}(\psi_{200}+\psi_{020})\\&=e^{-\frac{1}{2}\rho^2}\rho^2((3/2)\cos^2\theta-\tfrac{1}{2}),\\ \psi_3&=\psi_{110}=e^{-\frac{1}{2}\rho^2}\rho^2\sin^2\theta\\&\quad\times\sin 2\varphi,\end{aligned}\right\} l=2,\ 3d\ \text{level},$$

and their similar functions $\psi_{011}\psi_{101}$ and $\psi_{200}-\psi_{020}$.

$$\left.\begin{aligned}N=4: &\ \psi_{003}+\psi_{201}+\psi_{021}\\&=e^{-\frac{1}{2}\rho^2}(\rho^3-(5/2)\rho)\cos\theta)\end{aligned}\right\} l=1,\ 3p\ \text{level},$$

and two similar functions

$$\left.\begin{aligned}&\psi_{003}-(3/2)(\psi_{201}+\psi_{021})\\&=e^{-\frac{1}{2}\rho^2}\rho^3((5/2)\cos^3\theta\\&\qquad-(3/2)\cos\theta)\end{aligned}\right\} l=3,\ 4f\ \text{level}.$$

and six similar functions.

We see first of all, that oscillator levels with even N correspond to odd azimuthal quantum numbers l because the wave functions are odd functions of the coordinates xyz, and *vice versa*. There is considerable degeneracy of levels, the second s level coinciding with the first d level, etc. The levels are designated by the usual spectroscopic notation, giving the lowest level of azimuthal quantum number l the principal quantum number $n=l+1$ and numbering consecutive levels of the same l by successive values of n. Then $n-1$ is the total number of nodes of the wave function, radial and angular together.[69]

The general relation between the principal quantum number n in polar coordinates, the azimuthal quantum number l and the "energy quantum number" N is

$$N=2n-l-1. \quad (187e)$$

This follows from the examples given above, and can be shown generally.—For a given energy (given N), we have $\frac{1}{2}N$ or $\frac{1}{2}(N+1)$ different quantum levels in the nl scheme, according to whether N is even or odd. These levels have the azimuthal quantum numbers $l=N-1$, $N-3$, $N-5$, etc., and the principal quantum numbers $n=N$, $N-1$, $N-2$, etc., respectively. The total statistical weight of the energy level N is $N(N+1)$, taking account of the spin (factor 2 in statistical weight). Thus the weight of the levels $N=1, 2, 3, 4, 5, 6, 7$, is 2, 6, 12, 20, 30, 42, 56, respectively. The total number of quantum states having an N smaller or equal to N_0, is 2, 8, 20, 40, 70, 112, 168 for $N_0=1$ to 7.

[69] In some theoretical papers on nuclei the lowest level of *any l* has been given the principal quantum number 1. This seems an unhappy choice, in view of the analogy to atomic spectroscopy.

(*b*) If we take a *simple potential hole* as auxiliary potential, the wave functions are spherical harmonics, multiplied by Bessel functions of order $l+\frac{1}{2}$ of the radius, l being the order of the spherical harmonics (azimuthal quantum number). If the walls of the hole are infinitely high, the Bessel functions must vanish for $r=R$ (nuclear radius). If the height of the walls is finite, this boundary condition has to be relaced by a more complicated one, involving the wave function and its derivative.

The order of the energy levels has been worked out by Elsasser (E3) for infinitely high walls, by Margenau (M7) for finite walls of a certain height. The arrangement of the energy levels is in both cases similar to that for the oscillator potential, but the "accidental degeneracy" of levels with different l and the same N which we found for the oscillator potential, is of course removed. The oscillator level N splits into levels with given l and n in such a way that the level of highest l lies lowest. The arrangement of the levels is shown in Fig. 8 for the oscillator potential, the potential hole with infinitely high walls, and the potential hole of finite depth, just sufficient to take 58 particles (this is the case considered by Margenau). The figure shows all levels below $100\hbar^2/MR^2$ in a potential hole of radius R with infinitely high walls. These levels are also given in Table X. According to our

TABLE X. *Energy levels in potential hole with infinite walls.* Energy in units $\hbar^2/MR^2$ where R = radius of hole.

l	1st Level of Azimuthal Quantum Number l Des.*	En.**	2nd Des.	En.	3rd Des.	En.	4th Des.	En.	l	1st Level Des.	En.
0	1*s*	4.93	2*s*	19.74	3*s*	44.42	4*s*	78.96	6	7*i*	55.27
1	2*p*	10.12	3*p*	29.85	4*p*	59.45	5*p*	98.92	7	8*j*	67.98
2	3*d*	16.61	4*d*	41.35	5*d*	75.96			8	9*k*	81.79
3	4*f*	24.40	5*f*	54.25	6*f*	93.83			9	10*l*	96.74
4	5*g*	33.51	6*g*	68.49							
5	6*h*	43.76	7*h*	83.98							

* Spectroscopic designation.
** Energy.

scheme, we should expect a successive filling-up of the quantum states with neutrons and protons. The first two neutrons (or protons) will go into the 1*s* shell, the next six into the 2*p* shell, etc. The shells are tabulated in Table XI in the order of their energy; below each shell the number of quantum states in it is given (n_i); in the third

TABLE XI. *Successive filling of neutron (or proton) shells in potential hole with infinitely high walls.*

Shell	1*s*	2*p*	3*d*	2*s*	4*f*	3*p*	5*g*	4*d*	6*h*	3*s*	5*f*	7*i*	4*p*	8*j*	6*g*
n_i	2	6	10	2	14	6	18	10	22	2	14	26	6	30	18
N_i	2	6	12		14	6	18	34			40		6	48	
$S_i=\sum_{k=1}^{i} N_k$	2	8	20		34	40	58	92			132		138	186	

line the n_i's of shells with nearly identical energy are added (N_i); in the last line the N_i's of all shells up to the one considered are added: The figures in the last line (S_i) therefore represent the numbers of neutrons (or protons) for which we would expect a shell (or group of shells of nearly identical energy) to be completed.

Whenever a shell is completed, we should expect a nucleus of particular stability. When a new shell is begun, the binding energy of the newly added particles should be less than that of the preceding particles which served to complete the preceding shell. We should thus expect that the 3rd, 9th, 21st, etc., neutron or proton is less strongly bound than the 2nd, 8th, 20th.

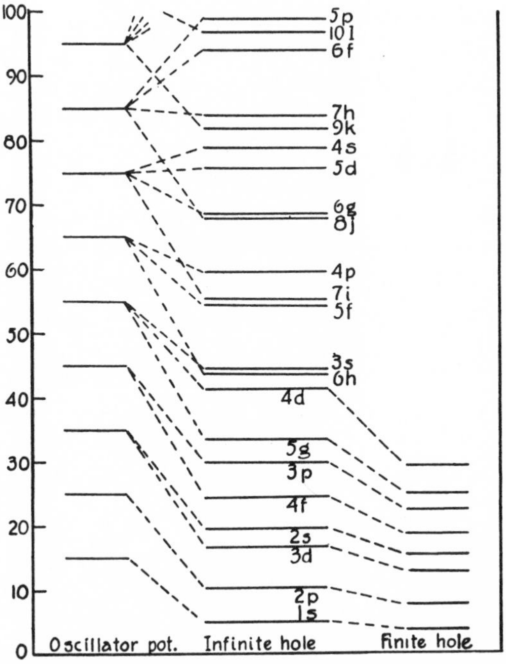

FIG. 8. Energy levels in an oscillator potential, in a potential hole with infinitely high walls, and in a hole with finite walls. The levels in the infinite hole are drawn to scale.

§33. Evidence for Periodicities from the Energy of Nuclei

We have seen in the preceding section that the concept of individual quantum states for the elementary particles leads to a particular stability of nuclei containing 2, 8, and 20 neutrons (or protons), because these numbers of neutrons are just sufficient to fill the $1s$, $2p$ and $(3d+2s)$-shell. The first number mentioned (2 neutrons, 2 protons) corresponds to the α-particle whose particular stability is well known but cannot be used as an argument for our scheme because it follows from any approximation whatever. The last number (20 neutrons, 20 protons) leads to the nucleus Ca^{40}. Unfortunately no exact data about nuclear masses are available for such high atomic weights so that no direct check is possible concerning the special stability of Ca^{40}. Some indirect evidence will be mentioned in §34. There remains thus the nucleus containing 8 neutrons and 8 protons, i.e., O^{16}, to test the "shell structure" hypothesis by means of nuclear energies. It seems in fact that there is ample evidence for a particular stability of O^{16}, and thus for the individual-particle approximation.

To be free from other fluctuations of the binding energy, we shall compare the analogous nuclei He^4, Be^8, C^{12}, O^{16}, Ne^{20}, etc., all of which can be considered as containing exclusively α-particles, and the nuclei which are produced by adding neutrons and protons to these "standard" nuclei. In Table XII, the masses of analogous nuclei are given (part (a)), together with the increase in mass connected with the addition of one or more particles to the standard nucleus (part (b)).

By comparing the figures in any one column of Table XIIb, it is seen immediately that the *figures decrease steadily as we go down the column, with the sole exception of a marked increase when going from the C^{12} to the O^{16} line.* This feature repeats itself in each column of the table. It means that a *given particle* or group of particles is *bound* the *more strongly* to a "standard nucleus" *the heavier* that *standard nucleus* is, *but* that the binding *to* the nucleus O^{16} is *less strong than to* C^{12}. This is exactly what we must expect if O^{16} marks the completion of a neutron and proton "shell": *All particles added to O^{16} must go into the next outer shell ($3d$) and will therefore be bound much less strongly than the preceding particles.* The fact that the decrease in binding energy from C^{12} to O^{16} occurs whatever particle or particles are added to the "standard nucleus," constitutes very strong evidence for the shell-structure indeed.

It seems worth while to discuss the reliability of the data underlying the Table XII, from which we drew our conclusion. The most reliable data are those for the addition of 2 neutrons and one proton to the standard nucleus. The general downward trend of the mass increase connected with that addition is unmistakable. The increase from C^{12} to O^{16} depends on the mass differences $N^{15}-C^{12}$ and $F^{19}-O^{16}$. These differences are very well established from transmutation data, and if there is any error, it can only be in the direction that $N^{15}-C^{12}$ is actually smaller, or $F^{19}-O^{16}$ actually larger than the figures given in our table, changes which would strengthen our point. The mass differences are based on the reactions (L4, L5)

$$N^{14}+H^2=C^{12}+He^4 \qquad (A)$$

together with $$N^{14}+H^2=N^{15}+H^1 \qquad (B)$$

and on $$F^{19}+H^1=O^{16}+He^4. \qquad (C)$$

It is extremely unlikely that a γ-ray is connected with (A). On one hand, C^{12} probably has no excited states less than 5.5 MV above the ground state;[70] on the other hand, the assumption that the α-particles observed in reaction (A) are associated with a γ-ray of as much as 5.5 MV energy, would conflict severely with a great number of other transmutation, and with mass spectroscopic data. Thus γ-ray emission, if any, can only be associated with processes (B) and (C). In case of (B), this would *lower* the mass of N^{15}, in case of (C), it would *raise* the mass of F^{19}. The figures given in Table XII are therefore an upper limit for $N^{15}-C^{12}$, and a lower limit for $F^{19}-O^{16}$, which makes our conclusion valid *a fortiori.*

The next most reliable column in Table XII is that referring to the addition of one neutron. The

[70] No lower excited state has been observed. Also it is very unlikely theoretically that a nucleus of so high binding energy as C^{12} should have any low-lying excited states (cf. excited states of the α-particle, §23).

situation is similar to the foregoing case. The crucial mass differences are $C^{13}-C^{12}$ and $O^{17}-O^{16}$. Of these, the latter is derived independently from three different reactions: (C9, N2)

$$O^{16}+H^2=O^{17}+H^1, \quad \text{(D)}$$

$O^{16}+H^2=F^{17}+n^1$, together with

$$F^{17}\rightarrow O^{17}+\epsilon^+, \quad \text{(E)}$$

$O^{16}+H^2=N^{14}+He^4$, together with

$$N^{14}+He^4\rightarrow O^{17}+H^1. \quad \text{(F)}$$

All three figures check closely, at least if Haxel's data are used for the second reaction in (F). It is therefore almost impossible that the data can be invalidated by γ-emission because it is exceedingly improbable that all three reactions (D), (E), (F) lead to the same excited state of O^{17}. (Reaction (D) was used for the actual determination of the mass difference.)

The difference $C^{13}-C^{12}$ may be obtained in two ways, *viz.*; (C9, C10, T10)

$$C^{12}+H^2=C^{13}+H^1, \quad \text{(G)}$$

$C^{12}+H^2=N^{13}+n^1$, together with

$$N^{13}\rightarrow C^{13}+\epsilon^+. \quad \text{(H)}$$

(G) leads to the figure given in the table, assuming no γ-rays. (H) leads to a value about 1 MV *less* for the difference $C^{13}-C^{12}$, indicating that there may be a γ-ray of 1 MV associated with the proton group in (G). This would again be more favorable to our argument than the values given in Table XII.

It may thus be said safely that the completion

TABLE XII. (*a*) *Masses of analogous light nuclei.*

Standard Nucleus	Nucleus Obtained by Addition to the Standard Nucleus of: 1 Neutron	1 Proton	2 Neutrons	1 Neu.+1 Pro.	2 Neu.+1 Pro.
He^4= 4.0034	He^5> 5.013 *J*	(Li^5)	(He^6)	Li^6= 6.0161	Li^7= 7.0170
Be^8= 8.0070	Be^9= 9.0139	B^9= 9.0155*F*	Be^{10}=10.0154	B^{10}=10.0152	B^{11}=11.0117
C^{12}=12.0037	C^{13}=13.0069	N^{13}=13.0100	C^{14}=14.0077*E*	N^{14}=14.0076	N^{15}=15.0053
O^{16}=16.0000	O^{17}=17.0040	F^{17}=17.0078	O^{18}=18.0065*H*	F^{18}>18.0065	F^{19}=19.0040
Ne^{20}=19.9994*G*	Ne^{21}=20.999 *J*	(Na^{21})	Ne^{22}=21.9977	Na^{22}=21.9996*F*	Na^{23}=22.9980*E*

(*b*) *Mass increase connected with the addition to the standard nucleus of*

Standard	1 Neutron	1 Proton	2 Neutrons	1 Neu.+1 Pro.	2 Neu.+1 Pro.	α-Particle
He^4	1.009*D*	——	——	2.0127*A*	3.0136*A*	4.0036*A*
Be^8	1.0069*A*	1.0085*C*	2.0084*B*	2.0082*A*	3.0047*A*	3.9967*A*
C^{12}	1.0032*A*	1.0063*B*	2.0040*B*	2.0039*A*	3.0016*A*	3.9963*A*
O^{16}	1.0040*A*	1.0078*C*	2.0065*C*	>2.0065*C*	3.0040*A*	3.9994C
Ne^{20}	1.000*D*	——	1.998*C*	2.000*D*	2.999*C*	3.995*D*

Explanation of signs used in Table XII: A, B, C, D denote decreasing grades of certainty of the values given. Figures denoted by *A* are deduced from reliable transmutation data, *B* from transmutations whose interpretation is not absolutely certain. *C* means that at least one of the masses used is based on mass spectroscopic or band spectroscopic data or such β-disintegrations for which the upper limit of the electron energy is not exactly known. *D* data are based on estimates. *E, F, G, H, J* refer to various ways of obtaining nuclear masses: *E*=transmutation whose interpretation is not quite certain, *F*=β-disintegration with unaccurately known upper limit, *G*=mass spectroscopic, *H*=band spectroscopic data, *J*=estimate from analogous nuclei. Masses without letter in the first table are well established by transmutation data.

Special remarks: He^5 is estimated according to Atkinson (A6), by comparison with analogous nuclei. C^{14} is obtained from the reaction N^{14}+slow neutrons=C^{14}+protons, Na^{23} from the reaction $Ne^{20}+He^4=Na^{23}+H^1$. The last reaction is subject to some doubt, firstly because the measurements are very old, secondly because Ne consists of three isotopes and it is not known from which isotope the observed protons arise. But firstly Ne^{20} is the most abundant isotope, and secondly there are reasons to believe that the longest range protons are emitted from Ne^{20}, rather than Ne^{21} or Ne^{22}.—The mass of Ne^{20} itself is obtained by assuming the ratio of the masses of Ne^{20} and Ne^{22} as determined mass spectroscopically by Bainbridge, to be correct. The mass of Ne^{21} is then estimated by assuming the difference $Ne^{21}-Ne^{20}$ to be about 0.001 mass unit larger than $Ne^{22}-Ne^{21}$, in analogy to $O^{16,\,17,\,18}$ and $C^{12,\,13,\,14}$. For details of the determination of nuclear masses and references to experimental papers, cf. chapter XVII.

Example of calculation: The increase in mass when one neutron and one proton are added to C^{12}, is equal to the mass of N^{14}, i.e., of the nucleus which is formed by that addition, minus the mass of C^{12}. Thus increase = 14.0076 − 12.0037 = 2.0039.

of a neutron-proton shell at O^{16} is established beyond doubt from the data about nuclear masses.

§34. Periodicities in the Existing Isotopes (M11, E3, G13, G3)

If all the isotopes found in nature are represented in a diagram, preferably plotting the "isotopic number" $I=N-Z$ against the atomic weight A (Fig. 2), unmistakable breaks are apparent; e.g., the maximum isotopic number I does not increase quite smoothly with increasing A, but seems to "refrain from increasing" up to a certain A, and then to jump suddenly by several units. Bartlett (B9) has first suggested a connection between these irregularities and the neutron and proton shells discussed in the preceding section, while Elsasser (E3) and Guggenheimer (G13) have worked out some details.

Before discussing the experimental results, it is necessary to give a strong warning against taking the neutron and proton shells too literally. This has been done very frequently in the past with the effect of discrediting the whole concept of neutron and proton shells among physicists.

First of all, the concept of quantum states of the individual particles is to be regarded as a zero-order approximation only, which has to be completed by a consideration of at least one, preferably two more approximations (§36). This fact alone shows that the effects connected with the completion of a shell cannot be too well marked, and it seems reasonable to expect them to be the less well marked the greater the number of particles already in the nucleus. Therefore, apparent *deviations from* the *simple shell structure* expected should of course be *attributed* to the *crude approximation* used. *Under no circumstances* do such deviations justify far reaching *ad-hoc assumptions* such as the introduction of *negative protons* as building stones of nuclei. There is at present not a single piece of reasonably well-founded evidence for the existence of negative protons in nuclei, but several grave reasons against this existence.

Secondly, it should be borne in mind that the filling of individual quantum states is not the only thing determining nuclear energies. In fact, other points have a much stronger influence on nuclear energies and stability. These points are:

(a) The general trend of the isotopic number as a function of the nuclear charge (§8).

(b) The odd-even rule, i.e., the rule that there are no stable nuclei of odd nuclear charge greater than 7 and even atomic weight (§10), and that nuclei are most stable when they contain even numbers of neutrons and protons.

(c) The isobar rule, stating that there exist almost no pairs of neighboring isobars (§43).

(d) The general trend of the average number of isotopes per element to increase from light to medium atomic weight (cf. end of §10) and to decrease again for the heaviest elements (because of instability against α-decay).

Thirdly, great caution should be applied in drawing conclusions from the mere existence or nonexistence of isotopes because there may be some rare isotopes yet undiscovered, and on the other hand, some spurious ones among the reported isotopes (cf. Mattauch, M11).

We shall now discuss the experimental evidence.

(a) *Nuclei of odd atomic weight*

These nuclei seem to be more suitable for the detection of irregularities in the increase of the isotopic number than nuclei of even weight, because there exists in general only one stable nucleus for a given odd atomic weight, so that we may give a definite isobaric number $I=A-2Z$ for any value of A. Moreover, there is no theoretical reason for any preference for even or odd nuclear charge for these nuclei, and no such preference seems actually to exist.

Table XIII gives, for each isotopic number from 1 to 43, the nucleus of maximum atomic weight observed for the given I. Column 2 gives the chemical symbol of the nucleus, column 3 its atomic weight $A(I)$. In column 4 we have calculated the atomic weight $A_0(I+1)$ which would correspond to the isotopic number $I+1$ according to formula (18b). If the increase of the isotopic number were quite regular, we should expect that all nuclei of odd atomic weight *smaller* than A_0 should have the isotopic number I, all nuclei of odd weight *larger* than A_0 the isotopic number $I+2$. Therefore, the greatest odd number smaller than A_0, let us say (A_0), should be the heaviest nucleus of isotopic number I. The difference $\delta A=A(I)-(A_0)$ (column 5) measures the deviation of the isotope scheme from regularity: A *positive* value δA indicates that relatively too many protons, a negative value that too many neutrons are contained in the respective nucleus.

Leaving out small fluctuations ($\delta A = \pm 2$), we observe two major and two minor deviations from a regular increase. The first major fluctuation is an excess of protons (positive δA) in all nuclei between $A = 75$ and about 110, the second an excess of neutrons (negative δA) in the nuclei immediately following, *viz.*, from 110 to 140. The minor deviations are the single nucleus K^{39} (too many protons) and the group from 150 to 180 (slightly too many neutrons, the difference δA being scarcely significant).

The theoretical sequence of levels of the individual particles is given in Table XI (§32). According to that table, we should expect "closed shell structures" for nuclei containing 2, 8, 20, 34, 40, 58, 92, 132 neutrons or protons.

Of all the fluctuations found experimentally in the isotope scheme, there is only one which can readily be explained on the grounds of these "closed shells," *viz.*, the case of K^{39}. This nucleus constitutes actually quite a strong piece of evidence for the completion of a neutron shell with 20 neutrons. For the isotopic number moves actually *against* the general trend, being 3 for the atomic weight 37 (nucleus Cl^{37}) and dropping back to 1 for K^{39}. The explanation in the neutron-shell scheme is that for $A = 39$ the nucleus with isotopic number 3 (A^{39}) would contain 18 protons and 21 neutrons, i.e., one neutron outside the closed shell, which makes the nucleus less stable than K^{39} which contains 19 protons and 20 neutrons, all of them *inside* the (3*d*, 2*s*) shell. For $A = 37$ there is no influence of the completion of the shell; both the nuclei of isotopic number 1 (A^{37}) and 3 (Cl^{37}) would contain only neutrons in the inner shell, *viz.*, 19 and 20, respectively. Therefore $A = 37$ can be considered as "regular" showing that for "regular" nuclei of this atomic weight the isotopic number should be 3, and that the stability of K^{39} rather than A^{39} really is to be attributed to an irregularity, *viz.*, the completion of a neutron shell.

On the other hand, the two "long periods" in the isotope scheme do not fit at all to the simple shell concept. For $A = 110$, i.e., the end of the first period, we have about 48 protons and 62 neutrons. The latter number is near to 58, so that we may expect a closed *neutron* (5*g*) shell. The number of protons is midway in the 5*g* shell. Thus we would expect, from the naive standpoint, to find an excess of *neutrons* in the nuclei below $A = 110$, exactly the contrary of the experimental result.

TABLE XIII. *The isotopic number of nuclei of odd atomic weight.* (Explanation in text.)

I	ELEM.	$A(I)$	$A_0(I+1)$	δA	I	ELEM.	$A(I)$	A_0	δA
1	K	39	31	+8	25	La	139**	151	−12
3	Ti	47	46.5	+2	23(!)	Sm	147**	143.5	+4
5	Cu	63	60	+4	25	Eu	151**	151	0
7	Ga	69	72	−2	27	Gd	155	158.5	−2
9	Br	79	77.5	+2	29	Dy	161	166.5	−4
11	Ru	99	92	+8	31	Yb	171	173	−2
13	Pd	105	101.5	+4	33	Hf	177	179	−2
15	Sn	115*	110	+6	35	Re	185	186	0
17	Sn	117	119.5	−2	37	Os	189	193	−4
19	Te	123	128.5	−4	39	Hg	199	200	0
21	Kr	129	136	−6	41	Tl	203	206	−2
23	Ba	135**	143.5	−8	43	Bi	209	212	−2

* *Remarks:* There are two isobars for $A = 115$, *viz.*, Sn and In. Assuming In to have the smaller energy (it is more abundant!), Sn^{115} would only exist because the transition $Sn^{115} + \epsilon^- = In^{115} + n^0$ is forbidden. For $A = 113$ there are also two isobars, In and Cd. Thus 111 (Cd) is the heaviest nucleus of isotopic number 15 which is certainly energetically stable. This would make $\delta A = 2$ for $I = 15$, more in line with the general trend of δA.

** There is a pronounced irregularity for $I = 23$ and 25: The isotopic number 25 appears already in Ba^{137} and La^{139}. Then the isotopic number drops back to 23 in Pr^{141}, Nd^{143} and Sm^{147}, to reach 25 again in Nd^{145}, Sm^{149} and Eu^{151}. We had therefore to include the values $I = 23$ and 25 *twice* in our table.

Similarly, the end of the experimental period of an excessive number of neutrons ends with La^{139}, i.e., $Z = 57$ and $N = 82$. This corresponds to a closed shell of protons, but does not represent any particular point with regard to the neutrons.[72]

Therefore it seems that the naive theory of neutron-proton shells fails for higher atomic number. The reason may be the following: Because of the interaction of the particles, there will be a great number of energy levels of the nucleus as a whole, for a given distribution of the protons and neutrons over the individual quantum states. Let us call such a distribution of the particles a "configuration" and the levels of the nucleus, corresponding to a given configuration, a "level system." Then the lowest level of a system will lie much lower than the average energy of the level system, the difference being largest when the outermost shells of neutrons and protons are just half-filled, because this state of affairs corresponds

[72] Bartlett, Elsasser and Guggenheimer have left out the shells 2*s*, 3*p* and 3*s*, without giving any reason for such a procedure. According to them, the 5*g* shell should be filled when there are 50 particles, the (4*d* 6*h*) shell with 82 particles. These numbers would agree with the experimental result, but they lack theoretical foundation.

to the largest number of levels in the system. Accordingly, we shall have two effects counteracting each other: The energy of zero approximation (average of the energies of a system of levels) decreases when a shell approaches completion, but the difference between the average energy and the lowest level of a system decreases as well. The minimum energy will therefore lie between the middle and the end of a shell. The larger the number of places in the shell, the larger will be the number of levels in a "system," and the more will the minimum energy be shifted towards half-complete shells. Thus it may happen that 48 rather than 58 protons, and 82 rather than 92 neutrons, correspond to minimum energy.

This explanation of the discrepancy between the observed isotope scheme and the naive theory is tentative only and somewhat *ad hoc*. However, it seems certainly necessary to include higher approximations to the energy than the zero approximation, in nuclear physics much more than in atomic physics, and the direction of the deviation from the naive theory due to higher approximations is correctly given by our argument.

(*b*) *Even atomic weight*

For even atomic weight, even nuclear charge Z corresponds to greater stability than odd charge (§10, and rule 2 above). This means that stable nuclei of atomic weight $4n$ (n an integer) have isotopic numbers $I=A-2Z$ divisible by 4, while nuclei of atomic weight $4n+2$ have in general isotopic numbers of the form $4m+2$.

The first rule holds without exception, the second rule for all nuclei of the type except the very lightest ones (H^2, Li^6, B^{10}, N^{14}). For these four nuclei, the "even-odd rule" conflicts too severely with the rule that among light nuclei those with isotopic number 0 are by far the most stable, because the forces between neutrons and protons are the strongest forces in the nucleus. It seems significant that this conflict is decided in favor of the even-odd rule (isotopic number 2) as early as for $A=18$, for which atomic weight we have the stable nucleus O^{18} containing 8 protons and 10 neutrons while F^{18}, which would contain an equal number of neutrons and protons, is unstable. We do not consider it significant that the change from isotopic number 0 to 2 occurs just after the completion of the $2p$ shells of neutrons and protons (cf. §30).

The change of the isotopic number of the nuclei $4n+2$ from 0 to 2, which is such a natural consequence of general principles, is quite sufficient to explain the striking change of the isotope pattern at $Z=8$: For $Z\leq 7$, each element has two isotopes of isotopic number 0 and 1, whether its nuclear charge Z is even or odd [*Exceptions:* For $Z=1$, there is in addition the proton, with isotopic number -1. For $Z=2$, the isotope He^5 is unknown, and there are very strong reasons that it does not exist at all (Atkinson, A6), being unstable against disintegration into an α-particle and a neutron. For $Z=8$, Be^8 is unknown because it can disintegrate into two α-particles. The exceptions are not serious for our point: We are now interested in the most stable nucleus for any given atomic weight; thus it suffices that He^5 is more stable than Li^5, and Be^8 more so than Li^8 which is certainly true]. For $Z\geq 8$, each element of even Z has three isotopes, of isotopic number 0, 1, 2; while the elements of odd Z have only one isotope each ($I=1$). This change of the pattern means only that the nuclei of weight $4n+2$ have odd nuclear charge for $A\leq 14$, even charge for $A\geq 18$. There is therefore no profound reason behind the change of the isotope pattern at oxygen.

Turning now to the nuclei of atomic weight $4n$, we remark that for low atomic weight these nuclei have isotopic number 0 throughout. This fact, together with the fact that they contain even numbers of neutrons and protons, makes them the most stable nuclei in the region (§30). (The special stability follows also from the picture of α-particles as subunits. §31.) The heaviest nucleus of this kind is Ca^{40}. It may be significant that this nucleus contains just 20 neutrons and protons, corresponding to complete $1s$, $2p$, and ($3d$, $2s$) shells.

The behavior of heavier nuclei of even weight shows fluctuations analogous to those of the nuclei of odd weight. In fact, the nuclei of odd atomic weight lie always in the center of the broad band filled by nuclei of even weight (Fig. 2). There are therefore the same difficulties in explaining the observed periodicities as for the odd nuclei.

We shall now mention a few other points connected with the shell-structure of nuclei.

(a) *Radioactivity.* A feature giving some support to our ideas about shell-structure is the start of α-radioactivity. The lightest α-radioactive atom (except for the odd case of Sm) is Po^{210}. For this element, $Z=84$ and $N=126$. This is fairly near the completion of the group of proton shells ($4d$, $6h$, $3s$) and the group of neutron shells ($5f$, $6i$). The shells would be complete for $Z=92$ and $N=132$, but we expect minimum energy for somewhat smaller values of Z and N.

(b) *Nuclear spins and magnetic moments.* Nothing definite can be said in this respect until a better method of attack has been found, or at least until the first and second approximation mentioned in §36 have been calculated. However, it seems significant that large spins appear for the first time after the $4f$ shell is begun (scandium, $Z=21$, spin 7/2).

With all reserve, we may be allowed to infer from our considerations the probable spin of K^{40} which is supposedly the radioactive isotope of K (Klemperer, K4). K^{40} contains 19 protons and 21 neutrons, i.e., complete 20-shells minus a proton in the $2s$ shell plus a neutron in the $4f$ shell. The orbital momentum of K^{40} is therefore most probably 3, while we cannot say whether the spins of proton and neutron are parallel or antiparallel to each other and to the orbital momentum. In any case, a total momentum of 2, 3 or 4 would result. 4 would be amply sufficient, 3 just sufficient to explain the long life of radioactive K^{40} (§43).

(c) *Another type of shell-structure* was suggested by Landé (L2). He assumed that as many neutrons and protons as possible are combined in α-particles while the remaining neutrons are arranged in shells. It was necessary to assume that some inner neutron shells are left incomplete while outer shells are being built up, and that the inner shells are completed afterwards. The capacity of the successive neutron shells was assumed to be 2, 6, 8, 12. It seems hard to attach any theoretical significance to these numbers.

In conclusion, we want to emphasize again that *reliable* conclusions about the shell-structure of nuclei can only be drawn when atomic weight determinations will be available which are guaranteed to be accurate to at least three decimals, i.e., 1 part in 100,000 for atomic weights of the order 100.

§35. Energy of O^{16} and Ca^{40} in the Hartree Approximation (H10)

Heisenberg (H10) has calculated the energy of the closed-shell nuclei He^4, O^{16}, Ca^{40}, using the individual-particle approximation and assuming oscillator wave functions (cf. §32). He assumes the "Gaussian" potential

$$J(r_{12}) = -Be^{-r^2/a^2} \tag{188}$$

to act between a neutron and a proton, and no force between like particles. The wave function of the neutrons and protons in the $1s$ shell is

$$\psi_{1s} = (\mu/\pi)^{\frac{3}{4}} e^{-\mu r^2}, \tag{188a}$$

where r is the distance from the center of the nucleus and μ a constant characteristic for the auxiliary oscillator potential used (cf. §32). If this potential is

$$U = \tfrac{1}{2}Cr^2 = \tfrac{1}{2}M\omega^2 r^2, \tag{188b}$$

where M is the mass of the proton (or neutron) and $\omega/2\pi$ its "classical" frequency in the oscillator potential, then

$$\mu = (MC)^{\frac{1}{2}}/\hbar = M\omega/\hbar. \tag{188c}$$

The wave functions of the other states ($2p$, $3d$, etc.) are similar to (188a). C (or ω, or μ) has to be regarded as an arbitrary parameter which must be fixed in such a way as to make the total energy a minimum.

For the details of the calculation we refer to Heisenberg's paper. The procedure is the following: The kinetic energy is equal to half the eigenvalues of the particles in the oscillator potential, i.e., $(3/4)\hbar\omega$, $(5/4)\hbar\omega$ and $(7/4)\hbar\omega$, respectively, for particles in the $1s$, $2p$, and ($3d$, $2s$) shell. The potential energy consists of the neutron-proton interaction and the Coulomb repulsion between the protons: Both terms can be calculated from the wave functions, but the final expressions are somewhat complicated. Potential and kinetic energy are added, and the sum minimized as a function of $x=\mu a^2$ (cf. (188), (188c)). The result represents the binding energy as a function of the force constants B and a. Instead of calculating the mass defect from the constants derived from the theory of light nuclei (chapter IV), Heisenberg uses the inverse procedure, i.e., to calculate the force constants from the observed mass defect. (This procedure has been used by us in connection with the statistical model, chapter V.) These constants may be compared to those necessary to give the correct eigenvalue of the α-particle, with the *variation principle* and the eigenfunction

$$\psi = e^{-\lambda(r_{12}^2+r_{34}^2)-\nu(r_{13}^2+r_{14}^2+r_{23}^2+r_{24}^2)}. \tag{188d}$$

For any given range a of the forces, the constant B must be chosen about 25 percent larger to obtain the correct binding energy of He^4 with the oscillator wave functions, than with the wave function (188d). This means that the oscillator wave functions are less good approximations than (188d). The reason is, of course, that in (188d) the interacting particles are linked directly to each other while in the oscillator approximation they are linked to the center of the nucleus, and only indirectly to each other.

The approximation to the energy of the nuclei O^{16} and Ca^{40} afforded by the oscillator potential is slightly worse than for He^4, the values of B necessary to obtain agreement with the observed binding energy being about 8 percent and 15 percent larger for O^{16} and Ca^{40}, than for He^4.

Nothing is changed if forces between like particles are assumed, provided the force between two like particles is of the form (cf. 142)

$$K(r_{12})\boldsymbol{\sigma}_1\cdot\boldsymbol{\sigma}_2. \tag{189}$$

For in this case the total interaction between all neutrons will be of the form

$$\tfrac{1}{4}\int K(r_{12})\,|\rho_n(\mathbf{r}_1\mathbf{r}_2)|^2 d\tau_1 d\tau_2 \tag{189a}$$

(cf. 146e), while the interaction between neutrons and protons is

$$\int J(r_{12})\rho_n^*(\mathbf{r}_1\mathbf{r}_2)\rho_p(\mathbf{r}_1\mathbf{r}_2)d\tau_1 d\tau_2, \tag{189b}$$

where J is the interaction potential between neutrons and protons, and $\rho_n\rho_p$ the mixed densities of neutrons and protons (§24). Since the numbers of neutrons and protons and their wave functions are equal, we have $\rho_n=\rho_p$ and all interactions together give an integral of the form (189b), with only J being replaced by $J+\frac{1}{2}K$. Thus the result for the binding energy will be the same as without like-particle forces, only B means now the sum of the interaction between two unlike particles, and half the interaction between two like particles. This will not change the comparison between the results for He^4, O^{16}, Ca^{40}, with each other and with the result of the variational method applied to He^4. (It would change the comparison with the theory of the deuteron.)

§36. The Coupling Scheme in Nuclei

For nuclei with incomplete shells, it is of great interest to find out how the orbital momenta and the spins of the individual particles are coupled to the resultant nuclear momentum I. Up to the present, practically no calculations concerning this problem have been carried out. But it seems as if the Hartree approximation discussed in the preceding sections, might in the future lead to a rational theory of the nuclear spins I and the associated magnetic moments μ, at least for light nuclei. One must, however, be prepared that higher order perturbations may seriously affect the picture, at least as regards the magnetic moments.

It seems reasonable to assume Russell-Saunders coupling to hold at least approximately in the nucleus, the Heisenberg forces being small compared to the Majorana forces (§13, 14). We shall thus introduce a total orbital momentum Λ of the nucleus and a total spin momentum Σ whose resultant is the "nuclear spin" (total angular momentum of the nucleus) I. The momenta Λ and Σ are the resultants of the λ's and σ's of the individual particles. A level of the nucleus shall be denoted similarly to the usual spectroscopic way, giving first the configuration of the protons, then that of the neutrons, then the characteristics of the level of the nucleus as a whole; e.g., $(1s^2\,2p)\,(1s^2\,2p^2)^2P_{3/2}$ means that there are two protons in the $1s$ shell, one in the $2p$ shell, 2 neutrons in $1s$, two in $2p$, and that the resultant orbital momentum of the nucleus as a whole is 1 (P term), the resultant spin 1/2 (doublet term), the total nuclear moment 3/2.

The energy of the various levels corresponding to a given configuration of neutrons and protons, can be calculated by a method similar to that used in the theory of optical spectra. The calculation is, however, much more involved because there are twice as many particles in each shell (neutrons and protons!) and the particles of different kind are not equivalent. This makes the number of levels extremely high. The terms expected from all possible neutron and proton configurations in the p shell are listed in Table XIV. (6 ^3P means that the configuration leads, among others, to 6 different ^3P terms of the nucleus.)

The calculation of the energy levels is simple only in the case of one neutron and one proton in the p shell. Since there are no restrictions due to the Pauli principle in this case, we may disregard the spins entirely in first approximation. We denote by 1 the coordinates of the proton, by 2 those of the neutron, and by M the component of the total orbital momentum in a given direction (axis of the polar coordinate system). We leave out the wave functions of the particles in the closed 1*s* shell. Putting furthermore

$$f=(8/3)\pi^{-\frac{1}{2}}e^{-(\rho_1^2+\rho_2^2)}\rho_1\rho_2, \quad (190)$$

where ρ is defined as in (187d), we have:

For $M=2$ one wave function

$$\psi_2=(3/8\pi)f\sin\theta_1 e^{i\varphi_1}\sin\theta_2 e^{i\varphi_2}. \quad (190a)$$

For $M=1$ two wave functions

$$\begin{aligned}\psi_{11}&=(3/8\pi)\sqrt{2}f\sin\theta_1 e^{i\varphi_1}\cos\theta_2,\\ \psi_{12}&=(3/8\pi)\sqrt{2}f\cos\theta_1\sin\theta_2 e^{i\varphi_2}.\end{aligned} \quad (190b)$$

For $M=0$ three wave functions

$$\begin{aligned}\psi_{01}&=(3/8\pi)f\sin\theta_1 e^{i\varphi_1}\sin\theta_2 e^{-i\varphi_2},\\ \psi_{02}&=(3/8\pi)f\sin\theta_1 e^{-i\varphi_1}\sin\theta_2 e^{i\varphi_2}, \quad (190c)\\ \psi_{03}&=(3/4\pi)f\cos\theta_1\cos\theta_2.\end{aligned}$$

Following Slater's "method of sums,"[73] the energy of the *D* term is given by the diagonal matrix element of the interaction between neutron and proton, corresponding to the wave function (190a) for $M=2$. The *P* term is found by adding the diagonal matrix elements corresponding to the two functions (190b), and subtracting the *D* term from the sum. The *S* term is equal to the sum of the three diagonal elements corresponding to (190c), *minus* the sum of the two diagonal elements corresponding to (190b). In every case, an additional constant has been left out containing the interaction of the 1*s* neutrons and protons among themselves and with the 2*p* neutrons and protons, and also the kinetic energy of the particles. Since this constant is the same for all three terms *SPD*, it is irrelevant for the question of which term is the lowest.

[73] J. C. Slater, Phys. Rev. **34**, 1293 (1929).

The interaction between neutron and proton may be expanded in spherical harmonics of the angle Θ between the radius vectors $\mathbf{r}_1$ and $\mathbf{r}_2$ of the two particles, *viz.*,

$$V(r_{12})=-V_0-3V_1P_1(\cos\Theta)-5V_2P_2(\cos\Theta)-\cdots, \quad (191)$$

where the minus sign has been chosen in order to make V_0, V_1, etc., *positive*. If we take

$$V(r_{12})=-Be^{-r_{12}^2/a^2}, \quad (191a)$$

then

$$V_l=(-i)^l(\pi a/4ir_1r_2)^{\frac{1}{2}}B\times J_{l+\frac{1}{2}}(2ir_1r_2/a^2)e^{-(r_1^2+r_2^2)/a^2}, \quad (191b)$$

where J is the Bessel function. In particular,

$$V_0=(B/2x)(e^x-e^{-x})e^{-(r_1^2+r_2^2)/a^2}, \quad (191c)$$

$$V_1=(B/2x^2)[e^x(x-1)-e^{-x}(x+1)]e^{-(r_1^2+r_2^2)/a^2}, \quad (191d)$$

$$V_2=(B/2x^3)[e^x(x^2-3x+3)-e^{-x}(x^2+3x+3)]e^{-(r_1^2+r_2^2)/a^2} \quad (191e)$$

with

$$x=2r_1r_2/a^2. \quad (191f)$$

Every V_l is positive for any positive value of x.

Considering that the interaction is of the Majorana type, we have now, e.g., for the

TABLE XIV. *Nuclear levels expected from various neutron and proton configurations in the p shell.*

Configurations	Terms	Total Number
$(p^1)(p^1)$; $(p^1)(p^5)$; $(p^5)(p^5)$	^3D ^3P ^3S ^1D ^1P ^1S	6
$(p^2)(p^1)$; $(p^4)(p^1)$; $(p^2)(p^5)$; $(p^4)(p^5)$	^4D ^4P ^4S ^2F 2^2D 3^2P ^2S	10
$(p^2)(p^2)$; $(p^2)(p^4)$; $(p^4)(p^4)$	^5D ^5P ^5S 2^3F 3^3D 5^3P ^3S ^1G ^1F 4^1D 2^1P 3^1S	25
$(p^3)(p^2)$; $(p^3)(p^4)$	^6P ^4F 3^4D 3^4P 2^4S ^2G 3^2F 5^2D 6^2P 2^2S	27
$(p^3)(p^3)$	^7S 2^5D 2^5P ^5S ^3G 3^3F 6^3D 6^3P 3^3S ^1G 3^1F 4^1D 4^1P ^1S	38

average value of the interaction over the wave function ψ_{11} (190b):

$$V^{(11)}=\int\psi_{11}^*(1,2)V(r_{12})\psi_{11}(2,1)d\tau_1d\tau_2$$
$$=(9/32\pi^2)\int f^2(r_1r_2)V(r_{12})\sin\theta_1e^{-i\varphi_1}$$
$$\cos\theta_2\sin\theta_2e^{i\varphi_2}\cos\theta_1d\tau_1d\tau_2. \quad (192)$$

We see that the Majorana force makes the diagonal matrix element very different from its usual form for "ordinary forces." To evaluate (192), we put[74]

$$\sin\theta_2\cos\theta_2e^{i\varphi_2}=(8\pi/15)^{\frac{1}{2}}Y_{21}(\theta_2\varphi_2), \quad (192a)$$

where Y is a normalized spherical harmonic. Moreover, we insert (191) for $V(r_{12})$ and expand the spherical harmonics of Θ according to the addition theorem[75]

$$(2l+1)P_l(\cos\Theta)$$
$$=4\pi\sum_{m=-l}^{l}Y_{lm}(\theta_1\varphi_1)Y_{lm}^*(\theta_2\varphi_2). \quad (192b)$$

Only the term $Y_{21}(\theta_1\varphi_1)Y_{21}^*(\theta_2\varphi_2)$ in the expansion (191), (192b) contributes to (192), since all other terms vanish upon integration over the angles. We thus have

$$V^{(11)}=-(9/32\pi^2)(8\pi/15)4\pi\int f^2(r_1r_2)$$
$$V_2(r_1r_2)r_1^2dr_1r_2^2dr_2=-\tfrac{3}{5}L, \quad (192c)$$

where L denotes the integral. With the value (190) for f, and (191e) for V_2, the integral can be evaluated; the result is

$$L=\int f^2V_2r_1^2dr_1r_2^2dr_2=(5/3)\alpha^{3/2}(2+\alpha)^{-7/2}. \quad (193)$$

In a similar manner, we may evaluate the other diagonal elements of V; e.g., for the state ψ_2 we obtain

$$V^{(2)}=(9/64\pi^2)\int f^2V\sin^2\theta_1\sin^2\theta_2d\tau_1d\tau_2$$
$$=(1/16\pi^2)\int f^2V[1-P_2(\theta_1)][1-P_2(\theta_2)]$$
$$\times d\tau_1d\tau_2=-(K+\tfrac{1}{5}L), \quad (192d)$$

where

$$K=\int f^2V_0r_1^2dr_1r_2^2dr_2$$
$$=\tfrac{1}{3}\alpha^{3/2}(2+\alpha)^{-7/2}(5+6\alpha+3\alpha^2). \quad (193a)$$

[74] Cf. *Handbuch der Physik*, Vol. 24/1, p. 275, Eq. (1.8).
[75] *Handbuch der Physik*, Vol. 24/1, p. 559, (65.59), also p. 554, (65.21), (65.22).

The diagonal matrix elements of V are:

$$V^{(2)}=-(K+\tfrac{1}{5}L), \qquad V^{(11)}=V^{(12)}=-(3/5)L, \quad (194)$$
$$V^{(01)}=V^{(02)}=-(6/5)L,\ V^{(03)}=-(K+(4/5)L).$$

Therefore the energy levels:

$$D=V^{(2)}=-K-\tfrac{1}{5}L, \quad (195)$$
$$P=2V^{(11)}-V^{(2)}=+K-L, \quad (195a)$$
$$S=2V^{(01)}+V^{(03)}-2V^{(11)}=-K-2L. \quad (195b)$$

Since both K and L are positive and $K>L$ (cf. 193, 193a), the lowest level is the S level, the next D, and the highest P. This order is opposite to the order of levels in atomic spectra. The reason for this reversion of the order is that the forces between the particles are attractive in nuclei, repulsive in atoms (Coulomb force between the electrons).

The ratio of the intervals SD : DP *cannot* be predicted on general grounds as in atomic spectra, but depends on the radial wave functions and on the form of the interaction potential. The reason is the Majorana type of the forces: This makes the integral K appear with *different sign* in the expressions for the energy of the different levels, while K would appear with the same sign throughout in atomic theory. The interval (DP) therefore depends on K as well as L, while in atomic spectra both intervals (DP) and (SD) would only depend on L.

Now we have to consider the spin interaction. We know (§14) that there are Heisenberg forces between neutron and proton which increase the Majorana forces if the two particles have parallel spin, and decrease them if the spins are antiparallel. Thus the spins of the neutron and the proton in the $2p$ shell will be *parallel* in the lowest state of the nucleus Li^6. This lowest state is thus a ^3S state, i.e., the total angular momentum of the nucleus must be 1 unit, and the magnetic moment must be the same as for the deuteron, save for perturbations. Both these predictions seem to agree with the experiments of Fox and Rabi, (F13).

In view of the extreme complication of the level scheme for more than 2 particles in the $2p$ shell (Table XIV), and of the fundamental difference between atomic and nuclear theory due to the Majorana forces, it does not seem possible to make any safe predictions regarding the spins of other nuclei, without actually carrying the calculations through. However, it seems as if the order of levels is roughly opposite to that in atoms: Thus we might expect to find always a level of *low* orbital momentum Λ and comparatively low resultant spin[76] Σ as the lowest level of nuclei. It is conceivable that one might explain in this way the fact that all nuclei containing even numbers of neutrons and protons seem to have zero total momentum I.

The calculations carried out in this section represent only the first approximation to the energy. This will probably be insufficient in many cases. A second approximation may be obtained as follows: We first determine the configuration of neutrons and protons which has lowest energy in zero approximation. Then we determine which of the energy levels corresponding to the given configuration lies lowest. (Thus far the procedure corresponds to the procedure in this section.) Then we look for the next higher neutron-proton configurations and take those terms arising from them which have the same symmetry as the lowest level which we have determined just before. We may calculate the perturbation of the ground level due to these higher levels of the same symmetry according to the usual methods, with regard to energy as well as magnetic moment, etc. In some cases, it may happen that the lowest level does not have the angular momentum deduced from the first approximation but a different angular momentum which is more favored by the interaction with the higher levels arising from other proton-neutron-configurations.

§37. Van Vleck's Potential (V1)

It is of great interest to investigate whether the forces exerted by a nucleus on one of its particles, can adequately be represented by a potential, i.e., whether the wave equation for a given particle can be written in the form

$$(\hbar^2/2M)\Delta\psi+(W-U(r))\psi=0. \quad (196)$$

[76] Two factors governing the spin should be distinguished: The Pauli principle, and the actual forces acting on the spin (Heisenberg forces). The Pauli principle will probably require *low* resultant spin of all neutrons, and low resultant spin of all protons, as far as these terms have any meaning. Inasmuch as the Pauli principle does not yet determine the resultant spin, the Heisenberg forces will tend to make it as *large* as possible.

This is not obvious in the case of Majorana forces, in fact, the wave equation which one obtains at first does not have the form (196) at all.

To find the appropriate one-particle wave equation in the Majorana theory, we start from the general Majorana equation (29b), leave out the spins,[77] and write the positional wave function as a product of wave functions of all individual particles,[77] *viz.*,

$$\Psi(x_1x_2\cdots x_Z\xi_1\cdots\xi_N)=\prod_{i=1}^{Z}\psi_i(x_i)\prod_{k=1}^{N}\varphi_k(\xi_k). \quad (196a)$$

We then integrate over the positions of all protons and neutrons except one proton, let us say the jth, after having multiplied Eq. (29b) by

$$\prod_{i\neq j}^{Z}\psi_i^*(x_i)\prod_{k=1}^{N}\varphi_k(\xi_k):$$

The wave function and coordinate of the proton j shall simply be denoted by ψ and x. Then we find

$$(\hbar^2/2M)\Delta\psi+W\psi=\sum_k\int d\boldsymbol{\xi}\varphi_k^*(\boldsymbol{\xi})J(\mathbf{x}-\boldsymbol{\xi})\psi(\boldsymbol{\xi})\varphi_k(\mathbf{x}). \quad (196b)$$

W is a constant connected to the total energy of the nucleus E and certain integrals over the wave functions. The right-hand side contains a sum over all proton wave functions and has not at all the familiar form $U(r)\psi(x)$.

Van Vleck has shown that (196b) nevertheless can be reduced to the form (196) with a suitably chosen potential function $U(r)$. This is at least true if the wave functions φ and ψ are solutions of a wave equation of the type (196) with a simple-hole potential; i.e., U is supposed to be constant and equal to $-U_0$ inside a sphere of radius R, and to be zero outside. From this assumption it cannot only be derived that Eq. (196b) can be reduced to the form (196) but moreover that the potential U acting on the neutron is again a simple potential hole of radius R. Thus the scheme is consistent.

In the simple potential hole, we may represent the wave functions φ_k and ψ_i by plane waves (statistical method, chapter V) or by spherical waves (§32) which can be considered as superpositions of plane waves. Since $\sum_k\varphi_k^*(\xi)\varphi_k(x)$ is the mixed density (§24), we have (cf. 151)

$$\rho(\boldsymbol{\xi}\mathbf{x})=\sum_k\varphi_k^*(\xi)\varphi_k(x)$$
$$=(2/h^3)\int_0^P d\mathbf{p}\exp\left[i(\mathbf{x}-\boldsymbol{\xi})\cdot\mathbf{p}/\hbar\right], \quad (197)$$

where $P=\hbar k_0$ is the maximum momentum of the neutrons (cf. 150b). For ψ, we write

$$\psi(\boldsymbol{\xi})=\exp\left(i\mathbf{p}_0\cdot\boldsymbol{\xi}/\hbar\right). \quad (197a)$$

J may be expanded in a Fourier series

$$J(\mathbf{x}-\boldsymbol{\xi})=\int d\mathbf{q}a(\mathbf{q})\exp\left(i\mathbf{q}\cdot(\mathbf{x}-\boldsymbol{\xi})/\hbar\right), \quad (197b)$$

where the coefficients $a(\mathbf{q})$ are given by

$$a(\mathbf{q})=h^{-3}\int d\mathbf{x}J(x)\exp\left(-i\mathbf{q}\cdot\mathbf{x}/\hbar\right). \quad (197c)$$

[77] Spin and symmetry of the wave function do not enter in first approximation.

Then the right-hand side of (196b) becomes

$$Q=2h^{-3}\int_{\Omega}d\boldsymbol{\xi}\int_0^{\infty}d\mathbf{q}\int_0^{P}d\mathbf{p}a(\mathbf{q})\exp\left[i(\mathbf{p}+\mathbf{q})\cdot\mathbf{x}/\hbar+i(\mathbf{p}_0-\mathbf{p}-\mathbf{q})\cdot\boldsymbol{\xi}/\hbar\right]. \quad (198)$$

In this equation, we invert the order of integration, integrating first over $\boldsymbol{\xi}$ (i.e., the volume of the nucleus), then over $\mathbf{q}$, finally over $\mathbf{p}$. If the nucleus is very large compared to the wave-lengths $\hbar/p$ of the particles and the range of the forces (which is approximately $\hbar/q$), we have

$$\int d\boldsymbol{\xi}\exp\left[i(\mathbf{p}_0-\mathbf{p}-\mathbf{q})\cdot\boldsymbol{\xi}/\hbar\right]=h^3\delta(\mathbf{p}_0-\mathbf{p}-\mathbf{q}), \quad (198a)$$

δ being Dirac's δ-function. Then the integration over $\mathbf{q}$ yields (cf. 197a)

$$\int a(\mathbf{q})d\mathbf{q}\exp\left[i(\mathbf{p}+\mathbf{q})\cdot\mathbf{x}/\hbar\right]\delta(\mathbf{p}_0-\mathbf{p}-\mathbf{q})=a(\mathbf{p}_0-\mathbf{p})\exp\left(i(\mathbf{p}_0\cdot\mathbf{x})/\hbar\right)=a(\mathbf{p}_0-\mathbf{p})\psi(\mathbf{x}) \quad (198b)$$

since the δ-function makes the integrand vanish except for $\mathbf{q}=\mathbf{p}_0-\mathbf{p}$. On the right-hand side of (198b) we have now the wave function of the proton $\psi(\mathbf{x})$, as required. Inserting into (198), we find

$$Q=2\psi(\mathbf{x})\int_0^{P}d\mathbf{p}a(\mathbf{p}_0-\mathbf{p}). \quad (198c)$$

Thus (196b) reduces to the form (196) with

$$U=2\int_0^{P}d\mathbf{p}a(\mathbf{p}_0-\mathbf{p}). \quad (199)$$

Thus we may indeed use an ordinary Schrödinger equation for each individual particle. Moreover, U does not depend on the direction of p_0, since $a(\mathbf{p}_0-\mathbf{p})$ depends only on $|\mathbf{p}_0-\mathbf{p}|$ and the integral in (199) goes over a sphere in p space. The form (196) is thus independent of our assumption in (197a) that ψ is a plane wave: It is valid for any linear combination of plane waves of given wave number $p_0/\hbar$ (or given energy W), e.g., for spherical waves, etc. (Van Vleck treated first the case of spherically symmetrical waves.)

However, there is still a serious flaw in our considerations: U obviously depends on the value of p_0, i.e., on the energy of the particle (it decreases with increasing energy). Thus we do not obtain the same wave equation for individual-particle states of different energy. Therefore the solutions of (196), each taken with the potential U appropriate for its particular energy parameter W, do not form an orthogonal set. Thus this method is not applicable for the construction of a set of individual wave functions to be used for the calculation of nuclear energies according to the scheme of §§35, 36. All wave functions of such a set must be derived from one and the same "auxiliary potential" which may be chosen as some average of the U's derived from (199).

The scheme of this section may, however, be useful in deriving an approximate wave function for one particular particle, and for this purpose (neutron scattering) the method was originally devised by Van Vleck.

The value of U can easily be obtained from (199) if the range of the nuclear forces is *long* compared to the wave-length of the particles ($\hbar/P$ or $\hbar/p_0$). In this case, $a(q)$ will only be large for small q's so that

$$U\approx\begin{cases}2\int_0^{\infty}d\mathbf{p}a(\mathbf{p}_0-\mathbf{p})=2J(0) & \text{if } p_0<P\\ 0 & \text{if } p_0>P.\end{cases} \quad (199a)$$

This would mean that only such neutrons would be bound whose momentum is less than the maximum momentum of the protons.

Actually, the range of the forces is of the same order as the wave-lengths. In this case, U will decrease slowly with increasing p_0. It may be expressed in terms of the mixed density of the protons (197), with the help of (197c), (199):

$$U=2h^{-3}\int_0^{\infty}d\mathbf{x}\int d\mathbf{p}J(x)\exp\left[i(\mathbf{p}-\mathbf{p}_0)\cdot\mathbf{x}/\hbar\right]$$

$$=\int_0^{\infty}d\mathbf{x}J(x)e^{-i\mathbf{p}_0\cdot\mathbf{x}/\hbar}\rho(\mathbf{x}) \quad (199b)$$

$$=(4\pi\hbar/p_0)\int_0^{\infty}rdr\rho(r)J(r)\sin(p_0r/\hbar). \quad (199c)$$

With the "Gaussian potential"

$$J(r)=-Be^{-r^2/a} \quad (199d)$$

U can easily be evaluated, the result being

$$U=-B\left[\Phi(x-y)+\Phi(x+y)+\pi^{-\frac{1}{2}}y^{-1}\left(e^{-(x+y)^2}-e^{-(x-y)^2}\right)\right] \quad (199e)$$

with $x=Pa/2\hbar,\quad y=p_0a/2\hbar,$ (199f)

and Φ the error integral (cf. Jahnke-Emde, Table of Functions).

VII. β-Disintegration and Nuclear Forces

§38. Disproof of the Existence of Electrons in Nuclei

It is now generally believed that no electrons exist inside nuclei. The main reasons are the following:

1. *The statistics of nuclei.* Nuclei of even atomic weight generally obey Bose, such of odd weight Fermi statistics. This is to be expected (§4) if the nucleus contains only neutrons and protons. If we would, however, replace one neutron by a proton plus an electron, there would be an increase of the number of elementary particles by one, and therefore a change from Bose to Fermi statistics and *vice versa*.

2. *The nuclear spin.* The corresponding argument holds for the nuclear spins which are integer or half-integer according to whether the atomic weight is even or odd (§5).

3. *The nuclear magnetic moments.* They are all

of the order of the proton magneton $e\hbar/2Mc$, while they should be of the order of the Bohr magneton $e\hbar/2mc$ if electrons existed in the nucleus.

4. *The size of the electron wave function.* The wave-length of an electron with a kinetic energy of the order of a few MV (energy of most β-rays !) is much larger than the nuclear radius (§3, end).

5. *The impossibility of a potential barrier sufficient to keep the electrons inside the nucleus.* This argument is the strongest of all, and we shall therefore discuss it in detail.

The nuclei emitting β-particles have mean lifetimes from 1/50 of a second up to about 10^8 years. There must, therefore, be some force keeping the β-particles inside the nucleus for that length of time, in spite of the fact that they have amply enough energy to escape. It might be tried to assume a potential barrier keeping the electrons from leaving the nucleus, in analogy to α-particles (chapter IX). There are three grave reasons in the way of such an assumption:

(a) To all our knowledge, a nucleus *attracts* an electron at any distance. This is certainly true at large distances (Coulomb force) and at very small distance (owing to the very assumption that there are electrons bound in the nucleus). In order to provide a potential barrier for the electrons, there would have to be a strong repulsion at intermediate distances (a few times nuclear radius, say).

(b) In relativistic theory it is nearly impossible to devise *any* potential barrier which would keep high energy electrons inside the nucleus. To see this, it is sufficient to consider the relativistic *Schrödinger* equation (without spin), *viz.*,

$$\hbar^2c^2\Delta\psi+[(E-V)^2-m^2c^4]\psi=0. \qquad (200)$$

E is the total energy of the electron, i.e., its kinetic energy at infinite distance from the nucleus plus mc^2. This equation has a solution of exponential character only if

$$|E-V|<mc^2. \qquad (200a)$$

However, it is necessary that the solution is of exponential type in the region of the potential barrier, because only in this case the potential barrier prevents the particles from immediate escape. This means that the potential energy V inside the potential barrier must not differ from the total energy E of the electron by more than mc^2. It is obvious that such a requirement is very unlikely to be fulfilled by a given potential barrier, especially for such nuclei for which the energy E of the β-particles is very large. There is one case (B^{12}) in which $E=24\ mc^2$. Then V would have to be between 23 and 25 mc^2, a very improbable assumption indeed.[78]

(c) Granted that V has really a value satisfying (200a), there would be extremely large perturbations of the optical electrons due to that potential barrier which absolutely contradict experiment. The most favorable assumption is that $V=E$ inside the barrier, let us say for r between R and $R+b$ (R=nuclear radius, b=breadth of barrier, r=distance between nucleus and electron). In this case, the solution of (200) is

$$\psi=Ae^{-mcr/\hbar}\quad \text{for } R<r<R+b, \qquad (200b)$$

where A is a constant. The lifetime is then, similar to that of nuclei emitting α-particles (cf. chapter IX)

$$\tau=(R/c)e^{2mcb/\hbar}. \qquad (200c)$$

Putting $R=8\cdot10^{-13}$ cm, $c=3\cdot10^{10}$ cm/sec., and assuming a lifetime $\tau=1$ sec., we have

$$2mcb/\hbar=\log(4\cdot10^{22})=52, \quad b=26\hbar/mc; \qquad (200d)$$

i.e., the breadth of the potential barrier would have to be much larger than the Compton wave-length $\hbar/mc$. Since the radius of the K shell of heavy atoms is less than twice the Compton wave-length, the potential barrier assumed would change the potential acting on the K electrons, and even on more distant electrons, beyond recognition, and would have tremendous effects on the energies of all these electrons. This disproves completely the assumption of a potential barrier keeping the electrons in the nucleus.

Therefore we are forced to assume that the *electrons observed in β-disintegration did not pre-exist in the emitting nucleus. We suppose that they are formed in the same moment when they are actually emitted*, and that it is this process of formation which is so improbable that it accounts for the long lifetime of β-emitting nuclei.

The process of β-disintegration should therefore be compared not to α-disintegration, but to the emission of light by atoms (or nuclei). This comparison to light seems quite advantageous to explain what is meant by non-pre-existence and formation in the moment of emission: Nobody would say that a hydrogen atom in the third state contains the light quanta corresponding to the spectral lines it may emit, *viz.*, the $H\alpha$-line and the two first lines of the Lyman series. (The second Lyman line may be emitted immediately, or the first after the emission of $H\alpha$.) Still, the hydrogen atom is capable of emitting these light quanta, and it is generally accepted that the

[78] The fact that the β-particles coming from a given nucleus have a continuous energy spectrum, would make the situation quite untenable, because we would have to assume a different height of the barrier for different nuclei of the same species, corresponding to the energy of the β-particle emitted. But the continuous energy distribution of the β-rays cannot be properly understood without the neutrino hypothesis anyway, so we prefer not to use it here as an argument.

quanta are produced in the moment of their emission. The emission of electrons by nuclei is entirely analogous, we have just to substitute "electron" throughout for "light quantum."

§39. The Neutrino (F7, B17, N1, C14, C15, C16)

The assumption that the β-particles which are emitted by radioactive nuclei, did not exist in the nucleus before the emission but are "created" in the moment of emission, solves the difficulties 3, 4 and 5 pointed out in §38. However, there remain the difficulties about statistics and spin, and the still graver difficulty of the continuous β-spectra. These difficulties can only be solved by introducing a new, hypothetical particle, having no charge, very small mass (electron mass or less), spin $\frac{1}{2}\hbar$, and Fermi statistics. This particle is called the neutrino.

The main evidence for the neutrino is the continuous character of the β-spectra. The β-particles emitted by radioactive nuclei do not all have the same energy, but have energies distributed over the whole range from zero to a certain upper limit which we shall denote by E_0. This is in violent contradiction to the fact that the parent nucleus before the emission of the β-ray, and the product nucleus after the emission, have quite exactly determined energies. This follows for natural radioactive nuclei from the fact that the α-particles emitted have definite energies for each transformation. For the artificial radioactive nuclei the proof is even more conclusive, since the masses of the radioactive nucleus as well as of the product nucleus can be determined with very great accuracy, by means of the energy balance of nuclear transmutations involving only heavy particles, or by mass-spectrographic measurements.

We are thus confronted with the following situation: A parent nucleus which is in a quantized state of definite energy, emits an electron and leaves a residual product nucleus, again in a quantized state of definite energy. However, the energy of the emitted electron is not equal to the difference ΔE between the energies of the nucleus before and after emission, but may have *any* amount between 0 and E_0, the energy in each particular case being apparently determined by chance.

There are only two ways of accounting for this situation: either (a) we have to give up the conservation of energy for β-disintegration or (b) we have to assume that, simultaneously with the electron, another particle is emitted which ordinarily is not detected. Such an assumption would immediately account for the experimental facts: The total available energy ΔE will be distributed among the electron and the second, unobservable, particle (neutrino). The electron would therefore only receive a part of ΔE which will vary from case to case. The maximum kinetic energy the electron may receive is

$$E_0 = \Delta E - (m+\mu)c^2, \tag{201}$$

where m and μ are the masses of electron and neutrino. E_0 will thus be the upper limit of the β-spectrum.

It seems that the hypothesis (a) i.e., nonconservation of energy, should not be made if it can possibly be avoided. Not only in classical physics and in all branches of atomic physics has the principle of conservation of energy proved extremely successful, but also in the transmutations of nuclei it holds perfectly as long as only heavy particles (of at least proton mass) are involved. This success seems to justify the retention of the principle by all means.

Moreover, there seems to be direct experimental proof against the nonconservation hypothesis, at least if one accepts that energy is conserved *statistically*, in the average over a great number of β-processes. Such an assumption seems necessary; if it were not made, it would be possible to construct a machine for perpetual motion, either using β-processes or their inverse. Assuming now statistical conservation of energy, we have

$$\bar{E} = \Delta E - mc^2, \tag{201a}$$

$\bar{E}$ being the average kinetic energy of the β-particles. This equation is contradicted by experiments, most violently for the artificial radioactive nuclei Li^8 and B^{12} (C14, 15, 16).

The nucleus Li^8 disintegrates into $Be^8+\epsilon^-$, the β-particles having an average energy (C16)

$$\bar{E}_{Li} = 3.8 \text{ MV}. \tag{201b}$$

The difference between the masses of Li^8 and Be^8 can be derived from the data

$$Li^7+H^2=Li^8+H^1+Q, \qquad (201c)$$

$$Li^7+H^1=2He^4+17.2 \text{ MV}, \qquad (201d)$$

$$Be^8=2He^4+0.3 \text{ MV}, \qquad (201e)$$

$$H^2=2H^1-1.2 \text{ MV}. \qquad (201f)$$

The first transmutation is the one used to produce the radioactive Li^8. The energy Q evolved in it is not known with any certainty. However, it is known that ordinary lithium (mixture of Li^6 and Li^7) bombarded by deuterons yields one and only one proton group of a range of about 30 cm, corresponding to an energy evolution around 5 MV. This group has been attributed to the reaction

$$Li^6+H^2=Li^7+H^1 \qquad (201g)$$

since at the time the existence of Li^8 was not suspected, and since the separated isotope Li^6 showed the group. However, no search for the proton group was made with pure Li^7 as target. Therefore we consider it most plausible that *both* isotopes Li^6 and Li^7 contribute to the group, the energies evolved in the two reactions being accidentally the same. In any case, it is absolutely certain that the energy evolved in (201c) is ***not more*** than 5 MV, it might be equal to that figure or less.—The energy evolved in (201d) is very accurately measured; the difference of the energies of Be^8 and 2α-particles is deduced from the transformation of B^{11} by protons and is certainly accurate to ± 0.5 MV, the difference between the deuteron and two protons is based on the mass spectrographic determination of Bainbridge and also certainly correct to ± 0.5 MV. Thus we find

$$\Delta E=17.2-0.3-1.2-Q \geqslant 10.7 \text{ MV}. \qquad (201h)$$

This is more than 6 MV greater than the average energy of the β-particles emitted by Li^8 which is irreconcilable with the assumption of statistical conservation of energy in β-disintegration.

On the other hand, the figure in (201h) agrees perfectly with the upper limit of the β-spectrum emitted by Li^8 which is

$$E_0=10.5 \text{ MV}. \qquad (201i)$$

Therefore Eq. (201) holds, which gives strong support to the neutrino hypothesis, designated as alternative (b) above.

The evidence from B^{12} (C15) is similar: B^{12} is formed in the reaction

$$B^{11}+H^2=B^{12}+H^1. \qquad (202)$$

From the bombardment of boron by deuterons several proton groups arise, the longest having a range of 92 cm. This bombardment may, besides (202), lead to the reaction

$$B^{10}+H^2=B^{11}+H^1. \qquad (202a)$$

The decision between (202) and (202a) is possible by observing the number of the protons of each group and the number of the β-particles emitted by B^{12}. It is found that the β-particles are much more numerous (20 times) than the protons in any group of range longer than 10 cm. Therefrom it follows that the energy evolved in (202) is less than 2.5 MV. With the masses $B^{11}=11.0111$, $H^2=2.0142$, $H^1=1.0081$, we have thus

$$B^{12} \geqslant 11.0111+2.0142-1.0081 -0.0027=12.0145 \qquad (202b)$$

and, with the mass 12.0037 for C^{12}, we find

$$\Delta E-mc^2 \geqslant 0.0108 \text{ mass unit}=10.1 \text{ MV}. \qquad (202c)$$

This is to be compared to the observed average energy of the β-particles

$$\bar{E}=4 \text{ MV} \qquad (202d)$$

and to the maximum energy

$$E_0=11.5 \text{ MV}. \qquad (202e)$$

Again, the experiments definitely contradict the assumption of only statistical conservation of energy, and are in good agreement with the neutrino hypothesis.

Further examples are the artificial radioactive nuclei F^{17}, P^{30}, etc. We can thus say that the proof against the only statistical conservation of energy is conclusive, and that the idea of non-conservation of energy has to be abandoned altogether, in favor of the neutrino hypothesis.

Further support for the neutrino hypothesis is derived from the difficulty about the statistics and the spin of nuclei (1 and 2, §38). The assumption that only neutrons and protons are present in nuclei, solves this difficulty for stationary states of nuclei. However, there remains a difficulty for the β-transformation. For the atomic weight of the nucleus remains unchanged in the β-transformation, therefore its statistics does not change and its spin remains integer (half-integer) if it was integer (half-integer) before. On the other hand, the emitted electron has Fermi statistics and spin $\frac{1}{2}$: Consequently, the total spin of the system cannot be conserved in the β-disintegration, being, for a nucleus of even weight, integral for the parent nucleus, and half-integral for the product nucleus and the electron together. Such a nonconservation of total spin, and a similar nonconservation of the statistics of the system, is almost as bad a contradiction against very well-established laws of nature as the nonconservation of energy would be. Therefore we are again forced to assume the emission of a second particle (neutrino) in the β-disintegration. Doing this, the difficulty is removed if we assume the neutrino to have the spin $\frac{1}{2}\hbar$ and Fermi statistics, as every other elementary particle (electron, positron, proton, neutron). Then the resultant of the spins of electron and neutrino is integral (1 or 0), and the resultant of the spins of all particles left after the β-disintegration, *viz.*, product nucleus, electron and neutrino, is integral or half-integral according to whether the parent nucleus has integral or half-integral spin which allows the total angular momentum to be conserved. Similarly, the statistics remains conserved because now the total number of particles (protons and neutrons in the nucleus, electron and neutrino) increases by 2 in the β-transformation which leaves the statistics of the system unchanged (§4).

A further point to support the neutrino hypothesis is the success of the theory of the β-decay, especially as regards the energy distribution of the electrons (K5, and §40), and the dependence of the lifetime on the maximum energy (§41).

There is thus considerable evidence for the neutrino hypothesis. Unfortunately, all this evidence is indirect; and more unfortunately, there seems at present to be no way of getting any direct evidence. At least, it seems practically impossible to detect neutrinos in the *free state*, i.e., *after* they have been emitted by the radioactive atom. There is only *one* process which neutrinos can *certainly* cause. That is the inverse β-process, consisting of the capture of a neutrino by a nucleus together with the emission of an electron (or positron). This process is, however, so extremely rare (§42) that a neutrino has to go, in the average, through 10^{16} km of solid matter before it causes such a process. The present methods of detection must be improved at least by a factor 10^{13} in sensitivity before such a process could be detected.

Whether there are other processes by which a free neutrino may be detected, depends entirely on its properties. We know for certain that the neutrino has no charge, because the charge of the electron alone accounts for the change of the charge of the radioactive nucleus in β-emission (increase by one unit). The absence of charge precludes any strong ionization due to neutrinos. However, it is theoretically quite conceivable that the neutrino might have a magnetic moment associated with its spin. The ionization due to such a magnetic moment has been calculated (B14) and was found to be about $100n^2$ ions per km path in air, n being the magnetic moment expressed in Bohr magnetons. Nahmias (N1) has searched for ionization produced by neutrinos, using strong radioactive sources shielded by large amounts (about 1 meter) of Pb in order to absorb α-, β- and γ-rays and leave only the neutrinos. No ionization was found larger than the fluctuations of the ionization due to cosmic rays, in spite of the latter's intensity having been cut down by performing the experiment in an underground railway of London. The evaluation shows that neutrinos cannot form more than 1 ion in about 500,000 km path in air, which means that their magnetic moment, if any, must be smaller than 1/7000 Bohr magneton. It seems therefore probable that the neutrino does not have any magnetic moment at all. This makes it futile to search for ionization produced by neutrinos.

Therefore the only hope of getting more direct evidence for the neutrino is from the radioactive decay itself. The recoil of the product nucleus, which can be observed in principle, will decide definitely between the hypothesis of nonconservation of energy and the neutrino hypothesis. According to the neutrino hypothesis, the momentum of the recoil nucleus should be equal and opposite to the resultant of the momenta of the electron and the neutrino. Therefore, if the momentum of the recoil nucleus and the emitted electron can be observed simultaneously as to magnitude and relative direction, the momentum of the neutrino can be inferred. On the other hand, the energy of the neutrino is directly given as the difference between the upper limit of the β-spectrum and the energy of the β-particle actually observed in a particular experiment. Now the neutrino momentum p and its kinetic energy E must be related by an equation of the form

$$(E+\mu c^2)^2=p^2c^2+\mu^2c^4, \qquad (203)$$

μ being the neutrino mass. All observations must be representable by the same value of μ—a severe test to the neutrino hypothesis if the experiments can be performed. It is seen that such experiments would lead to a direct determination of the neutrino mass as well as to a more direct proof for its existence.

The difficulty of the experiments lies in the smallness of the kinetic energy of the recoil nucleus. If we assume that all the energy available (E_0) is given to the electron, the recoil energy of a nucleus of mass M is easily found to be

$$E_r=E_0(E_0+2mc^2)/2Mc^2$$
$$=540E_0^{\mathrm{MV}}(E_0^{\mathrm{MV}}+1)/A \text{ volts}, \qquad (203a)$$

where E_0^{MV} is the upper limit of the β-spectrum in MV, and A the atomic weight of the radioactive nucleus. For $E_0=2$ MV, which is about average for artificial radioactive nuclei, and $A=20$, we have $E_r=160$ volts. The most favorable case would be Li^8, with $E_0=10.5$ MV, $A=8$ and therefore $E_r=8000$ volts; unfortunately, this element has a very short life ($\frac{1}{2}$ sec.).

The present evidence about neutrinos can be summarized as follows:

No charge
Very small mass, probably zero, at least small compared to electron mass (from β-spectra, §40)
Spin $\frac{1}{2}\hbar$
Fermi statistics
Magnetic moment less than 1/7000 Bohr magneton, if any
No detectable effects in free state

In concluding this section, a word must be said about antineutrinos. It seems probable that neutrinos obey a wave equation similar to the Dirac equation, only the charge (and possibly also the mass) being zero. This wave equation will allow solutions with both positive and negative energy. Just as in the case of electrons, it must be assumed that all states of negative energy are ordinarily filled, in order to avoid the serious difficulties connected with the possibility of transitions from positive to negative energy. A state of negative energy which happens to be empty, is equivalent to a particle analogous to the positron. This particle is called an antineutrino. Since the neutrino has no charge and probably no magnetic moment, the antineutrino cannot be distinguished from the neutrino in any way. There is thus no need of distinguishing neutrinos and antineutrinos, except for the mathematical formalism.

§40. Theory of β-Disintegration

If a nucleus emits a β-particle, its charge increases by one unit while its weight remains unchanged. In other words, the number of protons in the nucleus increases by one, while the number of neutrons decreases by one. Thus the β-transformation can be considered as consisting of the transformation of one neutron into one proton, one electron and one neutrino:

$$n^1 \rightarrow H^1+\epsilon^-+n^0. \qquad (204)$$

Similarly, a radioactive process in which a positron is emitted, is to be considered as

$$H^1 \rightarrow n^1+\epsilon^++n^0. \qquad (204a)$$

It need hardly be mentioned after the discussions of §38 that the *neutron* should *not* be considered as *composed of a proton, an electron and a neutrino, but is only able of transforming into these three particles*, and similarly for the proton.

The problem of the theory of β-disintegration is to calculate the *probability* of the processes (204), (204a). Of course, this cannot be done on the grounds of any existing theory, but an entirely new "force" has to be introduced which produces just the transitions (204), (204a), i.e., which converts a neutron into a proton (or *vice versa*) and at the same time produces a (negative or positive) electron and a neutrino. Such a force has been introduced by Fermi (F7), using the analogy to the emission of light discussed at the end of §38.

The probability that a charged particle emits light and at the same time goes over from the state m to the state n, is given by the well-known formula (relativistic theory)

$$w = C\left|\int u_n^*(\mathbf{r})\mathbf{A}(\mathbf{r})\cdot\boldsymbol{\alpha}u_m(\mathbf{r})d\mathbf{r}\right|^2, \qquad (205)$$

where u_n and u_m are the wave functions of the particle, $\boldsymbol{\alpha}$ the Dirac operator, C a certain constant and $\mathbf{A}(\mathbf{r})$ the vector potential of a light wave of the correct frequency $(E_m - E_n)/h$ and unit intensity, at the place of the charged particle. (205) can also be expressed by saying that there is a certain term in the Hamiltonian of the charged particle which corresponds to the spontaneous emission of radiation and which has the form

$$H = C'\mathbf{A}(\mathbf{r})\cdot\boldsymbol{\alpha}, \qquad (205a)$$

where the transitions of the particle under the influence of this Hamiltonian have to be calculated according to the ordinary methods of the perturbation theory. The vector potential $\mathbf{A}$ may be regarded as a sort of wave function of the emitted light quantum: Thus in the Hamiltonian there appears the wave function of the emitted particle at the place of the emitting particle.

It is reasonable to assume a similar expression for the interaction between a heavy particle, an electron and a neutrino. There are only two differences. Firstly, two particles are produced rather than one, therefore both the wave functions of electron and neutrino have to appear in the Hamiltonian. Secondly, the emission of the two particles changes the character of the heavy particle, converting a neutron into a proton and *vice versa*. Let us introduce an operator Q which corresponds to the conversion of a neutron into a proton, and Q^* corresponding to the opposite conversion. Then a plausible expression for the Hamiltonian of β-emission would be

$$H = g(\psi^*\varphi^*Q + \psi\varphi Q^*), \qquad (206)$$

where ψ is the wave function of the electron, φ that of the neutrino, both taken at the place of the heavy particle. The first term corresponds to the creation of an electron and a neutrino, together with the conversion of a neutron into a proton, the second term to the absorption of an electron and a neutrino, or the emission of a positron and an antineutrino, together with the conversion of a proton into a neutron. g is a constant to be derived from experiment.

The mathematical treatment is simplified, and the physical ideas and results not changed (K5) if we let one particle be created and one absorbed in each process, rather than two created or two absorbed. This can be done by assuming that the emission of a negative electron is associated with that of an antineutrino (or with the absorption of a neutrino), while the emission of a positron (or absorption of an electron) is accompanied by the emission of a neutrino. This is equivalent to our previous assumptions, because of the equivalence of neutrino and antineutrino. The Hamiltonian (206) is then to be replaced by

$$H = g(\psi^*\varphi Q + \psi\varphi^* Q^*). \qquad (206a)$$

The probability of a β-transformation is given by the ordinary nonstationary perturbation theory. If u_m and u_n are the eigenfunctions of the heavy particle before and after emission, and G_eG_n the number of states of electron and neutrino per unit energy interval, the probability of a β-emission in which the electron receives an energy between E and $E+dE$, is per unit time:

$$w = (2\pi/\hbar)g^2\left|\int d\mathbf{r}u_n^*(\mathbf{r})u_m(\mathbf{r})\psi^*(\mathbf{r})\varphi(\mathbf{r})\right|^2 \times G_nG_e dE. \qquad (206b)$$

Thus far, we have not considered relativity. The introduction of relativistic wave functions for the light particles is absolutely essential because their energies are much larger than their "rest" energy, mass $\cdot c^2$. The introduction of relativity for the heavy particles would not be necessary, except for the calculation of forbidden transitions (§§41 to 43) and for symmetry.

To set up the relativistic analog to (206b), we start from the requirement that the integrand in (206b) is relativistically invariant (F8, K5). From two functions ψ and φ, we may build up five quantities, which behave under Lorentz transformations, respectively, like a scalar, a vector, a tensor, a pseudovector and a pseudoscalar,[79] *viz.*,

Scalar: $i(\psi\dagger\varphi) = (\psi^*\beta\varphi)$, (207a)

Four vector:

$$-i(\psi\dagger\boldsymbol{\gamma}\varphi) = \begin{cases} i(\psi^*\boldsymbol{\alpha}\varphi) & \text{(space components),} \\ (\psi^*\varphi) & \text{(time component),} \end{cases} \qquad (207b)$$

Tensor:

$$(\psi\dagger\gamma_i\gamma_k\varphi) = \begin{cases} (\psi^*\beta\boldsymbol{\sigma}\varphi) & \text{(if } i \text{ and } k = 1, 2, 3), \\ (\psi^*\beta\boldsymbol{\alpha}\varphi) & \text{(if either } i \text{ or } k = 4), \end{cases} \qquad (207c)$$

Pseudovector:

$$(\psi\dagger\gamma_i\gamma_k\gamma_l\varphi) = \begin{cases} (\psi^*\boldsymbol{\sigma}\varphi) & \text{("space" components, } i = 4,\ kl = 1, 2, 3), \\ i(\psi^*\gamma_5\varphi) & \text{("time" component, } ikl = 1, 2, 3), \end{cases} \qquad (207d)$$

Pseudoscalar: $(\psi\dagger\gamma_1\gamma_2\gamma_3\gamma_4\varphi) = (\psi^*\beta\gamma_5\varphi)$. (207e)

[79] Pauli, *Handbuch der Physik*, Vol. 24/1, p. 220, etc.

Here $\boldsymbol{\alpha}=(\alpha_x, \alpha_y, \alpha_z)$ and β are the ordinary Dirac matrices, $\boldsymbol{\gamma}$ is the "matrix vector" with the components

$$\gamma_k=-i\beta\alpha_k \quad \text{for } k=1, 2, 3, \quad \gamma_4=-\beta. \tag{207f}$$

$\boldsymbol{\sigma}$ is the Pauli spin operator, *viz.*,

$\sigma_l=-i\alpha_i\alpha_k$, where the indices ikl follow cyclically upon each other, each being one of the numbers 1, 2, 3 (207g)

$\psi^\dagger$ is the "conjugate" to the Dirac wave function, *viz.*,

$$\psi^\dagger=-i\psi^*\beta \tag{207h}$$

and finally

$$\gamma_5=\gamma_1\gamma_2\gamma_3\gamma_4. \tag{207i}$$

The factors i and $-i$ on the left-hand sides of (207a) to (207e) are chosen so as to make the main components on the right-hand side real.

Five quantities analogous to (207a) to (207e) can be formed from the wave functions of the heavy particles $u_m u_n$. Multiplying any of the quantities (207a) to (207e) with the corresponding quantity formed from $u_m u_n$, we obtain an invariant. Thus we have five different possibilities to replace the integrand in (206b) in relativity theory:

Scalar: $$(u_n^*\beta u_m)(\psi^*\beta\varphi), \tag{208a}$$

Vector: $$(u_n^* u_m)(\psi^*\varphi)-(u_n^*\boldsymbol{\alpha} u_m)\cdot(\psi^*\boldsymbol{\alpha}\varphi), \tag{208b}$$

Tensor: $$(u_n^*\beta\boldsymbol{\sigma} u_m)\cdot(\psi^*\beta\boldsymbol{\sigma}\varphi) + (u_n^*\beta\boldsymbol{\alpha} u_m)\cdot(\psi^*\beta\boldsymbol{\alpha}\varphi), \tag{208c}$$

Pseudovector: $$(u_n^*\boldsymbol{\sigma} u_m)\cdot(\psi^*\boldsymbol{\sigma}\varphi) - (u_n^*\gamma_5 u_m)(\psi^*\gamma_5\varphi), \tag{208d}$$

Pseudoscalar: $$(u_n^*\beta\gamma_5 u_m)(\psi^*\beta\gamma_5\varphi). \tag{208e}$$

Fermi chose originally an expression similar to (208b). From the standpoint of the general theory of nuclear forces, the "tensor" or the "pseudovector" expression (208c), (208d) are preferable (§44). (The dot means scalar product.)

For the heavy particles, the operator β practically does not change the wave functions, the operator $\boldsymbol{\sigma}$ acts on the spin part of the wave function but leaves the order of magnitude practically unchanged, while $(u_n^*\boldsymbol{\alpha} u_m)$ and $(u_n^*\gamma_5 u_m)$ are small compared to $(u_n^* u_m)$, *viz.*, of the relative order v/c where v is the velocity of the heavy particles.[80] Therefore the second terms in (208b, c, d) can practically be neglected (except for forbidden transitions, §§41 to 43). This makes the results from expressions (208a) and (208b), and from (208c) and (208d), very nearly identical.

The energy distribution of the β-particles can easily be calculated from (206b) after a definite one of the expressions (208a) to (208e) has been chosen to replace the integrand in (206b). One must simply insert a plane wave for the neutrino wave function, while the electron wave function ψ is to be taken in the Coulomb field of the disintegrated nucleus. For light nuclei, it is allowable to neglect the effect of the Coulomb field and thus to replace the electron wave functions by plane waves as well. Since the wave-lengths of electron and neutrino are large compared to the nuclear radius for all known β-transformations, ψ and φ may be regarded as constant and taken outside the integral. With these approximations, the energy distribution of the β-particles turns out almost identical whichever of the expressions (208a) to (208e) is accepted. The result for w is

$$w=\frac{1}{2\pi^3}\frac{mc^2}{\hbar}\left(\frac{g}{mc^2(\hbar/mc)^3}\right)^2\frac{E_e}{mc^2}\frac{p_e}{mc} \times\frac{E_n}{mc^2}\frac{p_n}{mc}\frac{dE}{mc^2}|G|^2, \tag{209}$$

where $E_e E_n p_e p_n$ are energy and momentum of electron and neutrino. (The energies are supposed to include the terms mc^2 and μc^2, resp.) G is the matrix element

$$G=\int d\tau u_n^* u_m \quad \text{for (208a) or (208b)}, \tag{209a}$$

$$G=\int d\tau u_n^*\boldsymbol{\sigma} u_m \quad \text{for (208c) or (208d)}. \tag{209d}$$

A small term of the relative order $mc^2\mu c^2/E_e E_n$ has been neglected in (209).

From the shape of the β-spectrum near the maximum energy of the electrons, the mass of the neutrino can be deduced (F8). The experimental evidence points to a mass very small compared to the electron mass, probably zero. This conclusion is reached as follows: If E_e is near its upper limit E_0, the factors E_e and p_e in (209) may be regarded as constant. If the neutrino mass is not zero, E_n may also be considered constant, *viz.*, equal to μc^2, as long as $E_0-E_e\ll\mu c^2$. Under the same condition, we may insert for p_n the nonrelativistic expression $p_n=(2\mu)^{\frac{1}{2}}(E_n-\mu c^2)^{\frac{1}{2}}=(2\mu)^{\frac{1}{2}}(E_0-E_e)^{\frac{1}{2}}$. Thus at the upper limit of the energy spectrum (209) would go to zero as $(E_0-E_e)^{\frac{1}{2}}$, i.e., with *vertical tangent.* Actually, the observations show that the number of β-particles per unit energy goes to zero with *horizontal* tangent near the upper limit of the β-spectrum. This can only be understood if μ is assumed to be zero: Then $E_n=cp_n=E_0-E_e$, and (209) becomes proportional to $(E_0-E_e)^2$, if the electron energy E_e is near the upper limit E_0. A very

[80] Cf. *Handbuch der Physik*, Vol. 24/1, p. 301, etc.

small mass of the neutrino, up to about $\frac{1}{5}$ of the electron mass, would however seem tolerable in the light of the present evidence.

In the following, we shall put the neutrino mass equal to zero. Then (209) can be written

$$w=\frac{mc^2}{2\pi^3\hbar}\left(\frac{gm^2c}{\hbar^3}\right)^2 |G|^2\epsilon(\epsilon^2-1)^{\frac{1}{2}}(\epsilon_0-\epsilon)^2 d\epsilon \quad (210)$$

with the abbreviations

$$\epsilon=E_e/mc^2, \quad \epsilon_0=E_0/mc^2. \quad (210a)$$

If $\epsilon_0 \gg 1$, which is the case for many of the radioactive substances, the 1 in $(\epsilon^2-1)^{\frac{1}{2}}$ may be neglected compared to ϵ over the larger part of the energy spectrum. Then w becomes proportional to $\epsilon^2(\epsilon_0-\epsilon)^2$, i.e., there is a maximum of the probability for equal distribution of the energy among electron and neutrino ($\epsilon=\frac{1}{2}\epsilon_0$) and the distribution is symmetrical with respect to the two particles (P5). This is in contradiction to experiments: It is found generally that the electron receives, in the average, much less than half the maximum energy E_0. In other words, the neutrino energy is in the average larger than the electron energy.[81]

This shows that the theory in the form hitherto used does not account for the experimental facts. It is necessary to correct it in such a way that the emission of neutrinos of high energy becomes theoretically more probable. This has been done by Konopinski and Uhlenbeck (K5), by introducing the derivative of the neutrino wave function with respect to time instead of the wave function itself. There are three possible expressions involving the first derivative of φ which correspond to the expressions (208b, c, d), *viz.*,

Vector: $(u_n^* u_m)(\psi^*\beta\partial\varphi/\partial t)$

$$-c(u_n^*\boldsymbol{\alpha} u_m)(\psi^*\beta \text{ grad } \varphi), \quad (211b)$$

Tensor: $(u_n^*\beta\boldsymbol{\sigma} u_m)\cdot(\psi^*[\alpha\times\text{grad } \varphi])$

$$+(u_n^*\beta\boldsymbol{\alpha} u_m)(\psi^*\boldsymbol{\alpha}(\partial\varphi/c\partial t)-\psi^* \text{ grad } \varphi), \quad (211c)$$

Pseudovector:

$$(u_n^*\boldsymbol{\sigma} u_m)\cdot(\psi^*\beta\{[\alpha\times\text{grad } \varphi]-i\boldsymbol{\sigma}(\partial\varphi/c\partial t)\})$$
$$+i(u_n^*\gamma_5 u_m)(\psi^*\beta(\boldsymbol{\sigma} \text{ grad } \varphi)). \quad (211d)$$

(211b) is the expression chosen by Uhlenbeck, while a linear combination of (211b) and (211c) or (211d) must be taken if one wants to connect β-emission and general nuclear forces (§44). Making the same assumptions as when deriving (209), *viz.*, small nuclear charge and zero neutrino mass, we obtain from (211)

$$wd\epsilon=\frac{mc^2}{2\pi^3\hbar}\left(\frac{g}{mc^2(\hbar/mc)^4}\right)^2 |G|^2\epsilon(\epsilon^2-1)^{\frac{1}{2}}(\epsilon_0-\epsilon)^4 d\epsilon, \quad (212)$$

where G has the same meaning as in (209a) if the expression (211b) is chosen, while it is $(\frac{2}{3})^{\frac{1}{2}}$ and $(\frac{1}{3})^{\frac{1}{2}}$ times (209b), respectively, if (211c) or (211d) is accepted for the interaction between heavy particle, electron and neutrino. (212) differs from (210) by containing the *fourth* power of the neutrino energy $\epsilon_0-\epsilon$, rather than the second. This difference arises from the *derivatives* contained in (211): Since $\partial\varphi/\partial t=-iE_n\varphi/\hbar$, the introduction of the derivative introduces a factor E_n in the integral in (206b), and therefore a factor $E_n^2=(E_0-E_e)^2$ in the transition probability w. This additional factor $(E_0-E_e)^2$ is just what is required to bring about agreement with the experimental energy distribution in β-spectra: The factor makes for an increase in the probability of emission of slow electrons and fast neutrinos, compared to that of fast electrons and slow neutrinos. The most probable electron energy is shifted to $\frac{1}{3}E_0$ for large E_0.

A more quantitative comparison between formula (212) and the experimental energy distributions for all well-investigated β-spectra, has been carried out by Konopinski and Uhlenbeck (K5). The result is very satisfactory. Moreover, the *total* disintegration probability (integral of (212) over ϵ), i.e., the reciprocal lifetime of the radioactive nucleus, is also well represented by (212) in its dependence on the maximum energy (§41). It seems therefore that one of the expressions (211) must be very nearly correct. We shall therefore accept the Konopinski-Uhlenbeck theory as the basis of our future discussions.

[81] This is true of the light radioactive nuclei as well as of the heavy ones. The discrepancy can therefore not be attributed to the neglection of the electrostatic action of the nucleus on the electrons.

§41. Lifetime *vs.* Maximum Energy in β-Disintegration (F8, S1, K16)

The total probability of β-decay can be obtained easily by integrating (212) over all electron energies from $\epsilon=1$ to $\epsilon=\epsilon_0$. The result is

$$(\log 2)/\tau=\int_1^{\epsilon_0} w d\epsilon=|G|^2 f(\epsilon_0)/\tau_0, \quad (213)$$

where τ is the half-life of the β-disintegrating nucleus,

$$\tau_0=(2\pi^3\hbar/mc^2)(mc^2)^2(\hbar/mc)^8 g^{-2} \quad (213a)$$

is a time characteristic for β-decay, G is the matrix element referring to the transition of the heavy particles (cf. 209a, b), and

$$f(\epsilon_0)=(\epsilon_0^2-1)^{\frac{1}{2}}\left(\frac{1}{105}\epsilon_0^6-\frac{2}{21}\epsilon_0^4-\frac{247}{420}\epsilon_0^2-\frac{8}{105}\right)$$
$$+\tfrac{1}{2}\epsilon_0(\epsilon_0^2+\tfrac{1}{2})\log(\epsilon_0+(\epsilon_0^2-1)^{\frac{1}{2}}) \quad (213b)$$

is a function of the upper limit $E_0=mc^2\epsilon_0$ of the β-spectrum. If the kinetic energy of the β-particles is small, i.e., if ϵ_0-1 is small compared to unity, (213b) can conveniently be expanded in powers of ϵ_0-1, with the result

$$f(\epsilon_0)=(256/5\cdot 7\cdot 9\cdot 11)\sqrt{2}(\epsilon_0-1)^{5\frac{1}{2}}$$
$$+0(\epsilon_0-1)^{6\frac{1}{2}}. \quad (213c)$$

The lifetime τ of β-disintegrating nuclei is, according to (213), inversely proportional to $f(\epsilon_0)$, and therefore decreases rapidly with increasing kinetic energy of the β-particles. For small kinetic energy, $\tau\propto(\epsilon_0-1)^{-5\frac{1}{2}}$ (cf. 213c), for large kinetic energy, $\tau\propto\epsilon_0^{-7}$ (cf. 213b). This behavior agrees, qualitatively and quantitatively, with experiment. This can be seen from Table XV, in which the product $f(\epsilon_0)\tau$ is listed for a number of radioactive nuclei, for which the upper limit of the β-spectrum ϵ_0 is well known.[82] Now according to (213),

$$\tau f(\epsilon_0)=\tau_0\log 2|G|^{-2}. \quad (213d)$$

Here τ_0 is a universal constant. The matrix element G will be nearly unity if the neutron before the disintegration is in almost the same state as the proton after the disintegration. This will be true at least for a large number of light radioactive nuclei. If there is considerable difference between the states of neutron and proton, G will be smaller than unity. G may even vanish, in this case we have a "forbidden transition" which will be discussed below.

Since we expect the matrix element G to be nearly unity for a considerable number of radioactive nuclei, the product τf should have approximately the same value for all these nuclei. This is actually true for the first group of nuclei in Table XV (group 0*A*), for all of which τf has a value between $0.4\cdot 10^5$ and $3\cdot 10^5$, in spite of considerable differences between the lifetimes τ of the various nuclei. Absolute agreement of the values τf is of course not to be expected because of the differences in the matrix element G. However, the agreement is good enough to allow the determination of a rough value of the universal constant τ_0. This constant must lie at least in the neighborhood of

$$\tau_0=0.7\cdot 10^5 \text{ sec.}, \quad (214)$$

i.e., approximately one day. Using this value, and the value of the "characteristic electronic time"

$$\hbar/mc^2=1.3\cdot 10^{-21} \text{ sec.} \quad (214a)$$

we find from (213a) for the constant g of the β-decay

$$g=1.1\cdot 10^{-13}mc^2(\hbar/mc)^4 \quad (214b)$$
$$=6.5\cdot 10^{-3}Mc^2(\hbar/Mc)^4 \quad (214c)$$
$$=1.9\cdot 10^{-60} \text{ erg cm}^4. \quad (214d)$$

Corresponding to the fact that the lifetime of β-disintegrating nuclei (order of some seconds) is extremely long compared to nuclear times ($\sim 10^{-21}$ sec.), the constant g turns out to be extremely small if mc^2 and $\hbar/mc$ are chosen as the unit of energy and length, respectively (cf. 214b). Since the β-decay is a property of heavy particles, it may seem more appropriate to choose Mc^2 and $\hbar/Mc$ as units: In these units, g is about 1/150. The smallness of g causes some difficulties if one tries to connect the neutron-proton forces and the β-decay (§44).

Formula (213), (213b) is only true for light nuclei for which the influence of the nuclear field

[82] A very useful method for determining this upper limit has been suggested by Kurie, Richardson and Paxton (K16).

on the wave function of the β-particle may be neglected. For heavy nuclei the wave functions of the electron in the Coulomb field must be used. Then (212) must be replaced by (cf. K5)

$$w d\epsilon = \frac{mc^2}{2\pi^3\hbar} \frac{g^2}{(mc^2)^2(\hbar/mc)^8} |G|^2 \times \epsilon(\epsilon^2-1)^{\frac{1}{2}}(\epsilon_0-\epsilon)^4 e^{\pi\gamma\epsilon(\epsilon^2-1)^{-\frac{1}{2}}} \times |\Gamma(s+i\gamma\epsilon(\epsilon^2-1)^{-\frac{1}{2}})/\Gamma(2s+1)|^2 \times (2pR/\hbar)^{2s-2} d\epsilon, \quad (215)$$

where

$$\gamma = Z/137, \qquad s=(1-\gamma^2)^{\frac{1}{2}},$$

$$p = \text{electron momentum}, \quad (215a)$$

$$R = \text{nuclear radius}, \quad \Gamma = \Gamma\text{-function}.$$

(215) may be approximated as follows

$$w d\epsilon = \frac{|G|^2}{\tau_0} 2\pi\gamma\epsilon(\epsilon_0-\epsilon)^4 \frac{\epsilon}{1-e^{-2\pi\gamma\epsilon(\epsilon^2-1)^{-\frac{1}{2}}}} \times [\epsilon^2(1+4\gamma^2)-1]^{s-1}\left(\frac{Rmc}{\hbar}\right)^{2(s-1)} \frac{4}{(2s)!^2}. \quad (215b)$$

The expression $e^{-2\pi\gamma\epsilon(\epsilon^2-1)^{-\frac{1}{2}}}$ may be neglected, even for very high energy and medium large nuclear charge. The main differences between (212) and (215b) are: (1) an additional factor $2\pi\gamma$, which, for $Z=88$ (radium), is equal to 4, and (2) the four last factors in (215b) which stand instead of $(\epsilon^2-1)^{\frac{1}{2}}$, giving, for $Z=88$, $R=9\cdot10^{-13}$ and $\epsilon=3$, an increase by a factor 6. The first factor would be present in non-relativistic wave mechanics as well; it is due to the influence of the Coulomb field which increases the probability of the electron being near the nucleus. The other factors are characteristic of the relativistic wave mechanics. Both (1) and (2) tend to increase the probability of β-disintegration for heavy nuclei, altogether by a factor of about 24. *The lifetime of a heavy β-radioactive nucleus should therefore be considerably shorter than that of a light radioactive nucleus, if the upper limit of the β-spectrum is the same in both cases.*[83]

[83] It is therefore not correct to plot lifetimes *vs.* energy in the *same* diagram for heavy and light nuclei (Sargent curves). If heavy and light radioactive nuclei are to be compared, the factors due to the Coulomb field and to relativity have to be taken into account.

From (215b), the reciprocal half-life may easily be found by integrating over ϵ. The square bracket may be regarded as constant for this purpose. We find a formula of the type (213), with

$$f_0(\epsilon_0) = 2\pi\gamma[\bar{\epsilon}^2(1+4\gamma^2)-1]^{s-1}\left(\frac{2}{5}\frac{Rmc}{\hbar}\right)^{2(s-1)} \times \frac{4}{(2s!)^2}\left[\frac{1}{105}\epsilon_0^7 - \left(\frac{1}{5}\epsilon_0^5 - \frac{1}{3}\epsilon_0^3 + \frac{1}{7}\epsilon_0\right)\right]. \quad (216)$$

Here $\bar{\epsilon}$ is a suitably chosen average energy of the electrons. The dependence on the maximum energy is mainly contained in the last bracket; in this bracket the first term is much larger than the others unless ϵ_0 is very small. For large energies ϵ_0 the dependence on energy is the same for light (cf. 213b) and heavy nuclei (216).

The values of $\tau f(\epsilon_0)$ for some heavy radioactive nuclei are tabulated in Table XV under $0B$. While they agree fairly well among themselves, they are appreciably higher than the values for light nuclei (group $0A$ in table). This may indicate that the matrix elements G for heavy nuclei are in the average smaller than for light ones, which would be quite plausible because neutron and proton wave functions are certainly very different in heavy, and very similar in light nuclei. Probably, a β-disintegration of a heavy nucleus is always connected with a complete rearrangement, which should reduce the value of G. (Changes in the fundamental expression for the β-disintegration do not affect the ratio of the disintegration probability of light and heavy nuclei appreciably.)

We shall now discuss the *forbidden transitions.* We call a β-disintegration forbidden if the matrix element G vanishes for the transition. The most common cause for this will be a change of the total angular momentum I of the nucleus. We may distinguish forbidden transitions of the first, second, third . . . kind according to changes of I by $L=1, 2, 3\cdots$. The "forbidden" transitions will, of course, not be completely forbidden but will only be much less probable than the "allowed" transitions with $L=0$.

In order to calculate the probability of forbidden transitions, we must not make certain approximations which we made thus far (cf.

paragraph above 209). The dependence of the electron and neutrino wave function on the coordinates inside the nucleus, and the "small" second term in the fundamental expression (211b) for the interaction, must be taken into account. The small term in the interaction does in no case yield a bigger transition probability than the large term, and can therefore be neglected even now, provided we want only to know the order of magnitude of the effects.

If the nuclear moment changes by L, the product of the eigenfunctions u_n and u_m will contain a spherical harmonic of order L,[84] besides a factor depending on the radius r. The transition probability will then depend on the integral (cf. (211b), $\partial\varphi/\partial t=-iE_n\varphi/\hbar=\text{const}\cdot\varphi$)

$$\int P_L(\theta)F(r)[\psi_1{}^*(r\theta)-\psi_3{}^*(r\theta)]e^{-iE_nz/\hbar c}d\tau, \qquad (217)$$

where $F(r)P_L(\theta)$ represents the product $u_m{}^*u_n$, the exponential is the neutrino wave function, and ψ_1 and ψ_3 are the first and third Dirac component of the electron wave function.[85] The direction of motion of the neutrino is assumed parallel to z, and without loss of generality z may be chosen as axis of the polar coordinate system, so that $z=r\cos\theta$. The exponential in (217) may be expanded; any term in the expansion is much smaller than the preceding, the ratio being about $E_nR/\hbar c$ (R=nuclear radius), i.e., about 1/20 for neutrino energies of about 1 MV. The nth term in the expansion contains a factor $(\cos\theta)^n$. Of the electron wave functions ψ_1 and ψ_3, *one* contains a spherical harmonic $P_k(\theta)$, the other $P_{k-1}(\theta)$ if $k-\frac{1}{2}=j$ is the angular momentum of the electron. The function containing P_k is somewhat smaller than the other, but only by a factor of the order γ, i.e., about 1/2 for heavy nuclei and still 1/7 for a nucleus as light as K.

In order that (217) does not vanish we must take such a term in the expansion of the neutrino wave function that its product with ψ_1 or ψ_3, contains $(\cos\theta)^L$. For this purpose, we have to take the $L-k$th term, multiplied by the smaller electron function, or the $L-k+1$st term, multiplied by the larger electron function. According to the foregoing, the former choice will give much the larger contribution to the integral. The contribution from different values of k is about the same, that from $k=1$ (s and $p_{\frac{1}{2}}$ electrons) being possibly slightly larger than that from higher values of $k(2, 3, \cdots, L)$. The contribution of $k=1$ to the transition probability will therefore be of the same order as the total transition probability.[86] The lifetime in the case of forbidden transitions then becomes again of the form (213), but now with $f(\epsilon_0)$ being replaced by

$$f_L(\epsilon_0)=\frac{4\pi}{9}\gamma^3\epsilon_0{}^{2L+5}\left(\frac{mcR}{\hbar}\right)^{2L-4+2s}$$
$$\times\frac{L}{(2L+3)(2L+4)(2L+5)}\cdot\frac{1}{1^2\cdot 3^2\cdot 5^2\cdot\cdots\cdot(2L-1)^2}$$
$$\times\frac{4}{(2s!)^2}[\bar{\epsilon}^2(1+4\gamma^2)-1]^{s-1} \qquad (218)$$

and G being replaced by

$$G_L=\int u_n{}^*u_mY_{LM}(r/R)^Ld\tau, \qquad (218a)$$

where Y_{LM} is a spherical harmonic and M is the difference between the magnetic quantum numbers of the states u_m and u_n. G_L will again be of the order of magnitude unity, but rather smaller.

[84] There may be terms containing higher spherical harmonics, but their contribution to the transition probability is negligible.

[85] The neutrino spin has been assumed parallel to z. If it is opposite, ψ_2 and ψ_4 appear instead of $\psi_1\psi_3$. This does not change the result.

[86] The addition of the contributions of higher k's will increase the transition probability, the consideration of the second term in the interaction expression (211b) will decrease it somewhat. The result seems to be about within a factor 2 equal to (218).

TABLE XV. *Lifetimes of β-radioactive nuclei.**

Nucleus	τ(sec.)	ϵ_0-1	$f(\epsilon_0)$	$\tau f(\epsilon_0)$	Nucleus	τ	ϵ_0-1	$f(\epsilon_0)$	$\tau f(\epsilon_0)$
Group 0A					Group 1A				
C^{11}	1200	2.5	30	$0.4\cdot10^5$	Li^8	1/2	21.8	$3\cdot10^7$	$1.5\cdot10^7$
N^{13}	660	2.83	60	$0.4\cdot10^5$	B^{12}	1/50	25.8	$12\cdot10^7$	$0.6\cdot10^7$
O^{15}	150	3.9	450	$0.7\cdot10^5$	N^{16}	9	(13)	10^6	$0.9\cdot10^7$
F^{17}	70	4.9	1900	$1.3\cdot10^5$	F^{20}	40	(10)	$2\cdot10^5$	$0.8\cdot10^7$
Si^{27}	150	3.9	450	$0.7\cdot10^5$	Na^{22}	$1.5\cdot10^7$	(1.2)	0.4	$0.6\cdot10^7$
Mg^{27}	620	(3.9)	450	$2.8\cdot10^5$	Na^{24}	54000	4.1	600	$3.4\cdot10^7$
					Al^{28}	180	7.8	$3.6\cdot10^4$	$0.7\cdot10^7$
Group 0B			f_0	τf_0	P^{30}	195	9.6	$1.3\cdot10^5$	$2.5\cdot10^7$
UX_2	94	5.5	19700	$18\cdot10^5$	Si^{31}	9600	4.1	600	$0.6\cdot10^7$
Ra B	2300	2.27	21.9	$5\cdot10^5$	Cl^{38}	2200	12.5	10^4	$2.2\cdot10^7$
Th B	55000	1.70	1.1	$6\cdot10^5$	K^{42}	4850	9.0	10^5	$4.5\cdot10^7$
Th C″	275	4.5	5700	$16\cdot10^5$					
Ac C″	410	3.73	1410	$6\cdot10^5$	Group 1B			f_0	τf_0
					Ra C	1700	7.14	140000	$2.4\cdot10^8$
Group 2 or Higher					Ra E	$6\cdot10^6$	3.38	650	$4\cdot10^9$
P^{32}	$1.3\cdot10^6$	4.1	600	$0.8\cdot10^9$	Th C	8500	5.29	18000	$1.5\cdot10^8$
K^{40}	$3\cdot10^{15}$	1.4	1.0	$3\cdot10^{15}$	$MsTh_2$	32000	6.21	55000	$1.7\cdot10^9$

* Data on maximum energy mostly from Kurie, Richardson and Paxton (K16) and from Fowler, Delsasso and Lauritsen (F15).

Comparing (218) to (216), we find

$$f_L(\epsilon_0)/f_0(\epsilon_0)=\frac{\gamma^2}{9}\left(\frac{E_0R}{\hbar c}\right)^{2L-2}\times\frac{7\cdot 6\cdot 5}{(2L+5)(2L+4)(2L+3)}\,\frac{L}{1^2\cdot 3^2\cdot 5^2\cdot\cdots(2L-1)^2}. \quad (218b)$$

Thus the probabilities of the first forbidden ($L=1$) and the allowed transitions are in the ratio $\gamma^2/9$, which is about 1 : 40 for heavy nuclei and considerably less for light nuclei. However, for very light nuclei and high energy E_0 this does not hold, the ratio being then $(E_0R/\hbar c)^2$ rather than $\gamma^2/9$ whenever the first quantity is larger than the second. Since the forbidden transitions of the first order have, for light nuclei, been observed mainly in cases of high maximum energy E_0, a ratio 1 : 100 is in the average to be expected for the probabilities of first forbidden and allowed transitions. In Table XV, we have listed a number of nuclei for which the β-disintegration is apparently of the first forbidden type: Group 1A contains light nuclei, 1B heavy nuclei of this type. For simplicity, we have again calculated $\tau f_0(\epsilon_0)$ for each nucleus. Since actually $\tau f_1(\epsilon_0)$ should be equal to $\tau_0(\log 2)|G|^{-2}$, we expect τf_0 to be about 100 times as large. Indeed, the values of τf_0 are about 100 times larger for group 1 than for group 0. The difference between heavy and light nuclei is again found for the forbidden transitions.

The probability of forbidden transitions of higher order decreases, according to (218b), by a factor $(E_0R/\hbar c)^2$ per order. Besides, there is another factor (last factor in (218b)) which also decreases rapidly with increasing L. Thus the lifetime for β-active nuclei becomes very long if the β-disintegration corresponds to a forbidden transition. There are three β-disintegrations known which have exceedingly long lifetimes: K, Sr and Nd. The isotope concerned in the case of K is probably K^{40}, according to a suggestion of Klemperer (K4). The maximum energy of the β-rays of K is about 0.7 MV$=1.4mc^2$, the nuclear radius is probably about $4.5\cdot 10^{-13}$ cm, therefore $E_0R/\hbar c=0.016$. *A change of the angular momentum by $L=3$ will be amply sufficient to account* for the observed lifetime. We have for $L=3$ from (218b)

$$f_3/f_0=\frac{1}{9}\cdot\frac{1}{7^2}\cdot 0.016^4\cdot\frac{7\cdot 6\cdot 5}{11\cdot 10\cdot 9}\cdot\frac{3}{3^2\cdot 5^2}=4\cdot 10^{-13} \quad (218c)$$

corresponding to a life $2.5\cdot 10^{12}$ times longer than for an allowed transition. Actually, the quantity $\tau f(\epsilon_0)$ is only $3\cdot 10^{10}$ times larger for K than for most allowed disintegrations. Thus the assumption of a change of the nuclear moment by 3 units is more than sufficient to account for the long life of K^{40}. Such a change seems likely from general considerations (§34).

§42. The Inverse β-Processes: Capture of Orbital Electrons by Nuclei, Disintegration of Nuclei by Electrons and Neutrinos

The following three "inversions" of the ordinary β-process seem of interest

(1) The capture of an orbital electron of the atom by a nucleus, with the emission of a neutrino.

(2) The capture of an incident free electron by a nucleus, with the emission of a neutrino.

(3) The capture of an incident neutrino by a nucleus, with the emission of a (positive or negative) electron.

Processes (1) and (2) lead to a decrease of the nuclear charge by one unit, (3) to an increase, if a negative, a decrease, if a positive electron is emitted.

All three processes are, of course, only possible if the necessary energy is available; e.g., the condition for process (1) is, if the binding energy of the orbital electron is neglected compared to the nuclear energies:

$$Z^A+\epsilon^- > (Z-1)^A+n^0. \quad (219)$$

Here Z^A denotes the mass of the original nucleus, of charge Z and mass number A, $(Z-1)^A$ is the mass of the product nucleus, ϵ^- that of the captured electron and n^0 that of the emitted neutrino. The condition is certainly fulfilled for all positron-emitters; in fact, for these the more stringent condition

$$Z^A > (Z-1)^A+\epsilon^+ + n^0 \quad (219a)$$

must be fulfilled. However, the process of capture of an electron is of no great interest for positron emitters, because for these nuclei the emission of a positron will in general be much more probable than the capture of an orbital electron.

There will, however, certainly be some nuclei Z^A for which (219) but not (219a) is fulfilled. One of these nuclei is He^3, provided the mass of the neutrino is zero (or very small), as we have assumed in §40. F^{18} might be another, judging from the general trend of the masses of analogous nuclei in its neighborhood. Finally, it is probable that In^{113}, Sn^{115} and Te^{123} (or perhaps one of them) belong in this category, being isobaric with the "neighboring" nuclei Cd^{113}, In^{115} and Sb^{123}, respectively (cf. §43).

The probability of capture is, of course, greatest for the K electrons, since they are most frequently inside the nucleus. The probability can easily be calculated from the general theory of β-decay. It is a function of the kinetic energy $E_0=\epsilon_0 mc^2$ given to the neutrino in the process, i.e., of the excess of the energy on the left-hand side of (219) over than on the right-hand side. The lifetime may again be expressed in the form (213), with G having the form (213b) and (218a) for allowed and forbidden transitions, respectively, and with

$$f_K(\epsilon_0)=2\pi\gamma^3\epsilon_0^4\left(\frac{2R}{a_K}\right)^{2s-2}\frac{1+s}{2s!} \tag{220}$$

for allowed transitions ($L=0$) and

$$f_K{}^L(\epsilon_0)=\frac{2\pi}{9}\gamma^5\frac{\epsilon_0^2(L-1+\epsilon_0)^2}{1^2\cdot 3^2\cdot\cdots(2L-1)^2L} \times\left(\frac{E_0R}{\hbar c}\right)^{2L-2}\left(\frac{2R}{a_K}\right)^{2s-2}\frac{1+s}{2s!} \tag{220a}$$

for forbidden transitions in which the nuclear moment changes by L. a_K denotes the Bohr radius of the K shell, for s see (215a).

For the nucleus He^3 we have $\gamma=2/137$, and therefore $s\approx 1$. Thus

$$f_K=2\cdot 10^{-5}(E_0/mc^2)^4 \tag{221}$$

and, with $G=1$ and (213):

$$\tau=2.5\cdot 10^9(mc^2/E_0)^4 \text{ sec.} =100(mc^2/E_0)^4 \text{ years.} \tag{221a}$$

The energy set free in the capture of an electron by He^3 is equal to the difference of the masses of He^3 and H^3. This difference seems to be somewhere in the neighborhood of 0.0002 mass unit $=0.4mc^2$ (cf. chapter XVII, or §22). With this value, He^3 would have a lifetime of about 5,000 years. This would mean that He^3 cannot be found in nature since it would have decayed long ago; but artificially produced He^3 would not change over into H^3 in any time allowing easy observations. It is to be noted that the capture of an electron by a nucleus is not observable as a β-process but could only be deduced from the fact that the product substance (in our case H^3) slowly accumulates in a material which originally contained only the parent substance (He^3).

For $Z=50$, i.e., in the region in which pairs of neighboring isobars are found (§43), we have $\gamma=0.365, s=0.93, R=7\cdot 10^{-13}$ cm, $a_K=1.05\cdot 10^{-10}$ cm, and therefore

$$f_K{}^0=0.62\epsilon_0^4, \tag{222}$$

$$f_K{}^L=0.009\epsilon_0^2\frac{(L-1+\epsilon_0)^2}{1^2\cdot 3^2\cdot\cdots(2L-1)^2L}\left(\frac{E_0R}{\hbar c}\right)^{2L-2}. \tag{222a}$$

Therefore the lifetime becomes

$$\tau=0.8\cdot 10^5(mc^2/E_0)^4 \text{ sec.}=1(mc^2/E_0)^4 \text{ days} \quad \text{for } L=0, \tag{222b}$$

$$\tau=0.15(mc^2/E_0)^4 \text{ years} \quad \text{for } L=1, \tag{222c}$$

$$\tau=0.8\cdot 10^4(mc^2/E_0)^4[mc^2/(E_0+mc^2)]^2 \text{ years} \quad \text{for } L=2, \tag{222d}$$

$$\tau=0.9\cdot 10^9(mc^2/E_0)^6[mc^2/(E_0+2mc^2)]^2 \text{ years} \quad \text{for } L=3. \tag{222e}$$

The lifetime thus increases very rapidly with increasing order of the forbidden transition, a fact which is very important for the question of stability of isobars (§43).

We shall now turn to the processes (2) and (3) mentioned at the beginning of this section, i.e., the disintegration of nuclei by free electrons or neutrinos. Both these processes are about equally probable for equal energy of the incident particle, because the β-theory is almost symmetrical in electrons and neutrinos. Both processes are exceedingly rare, because of the small value of the characteristic constant g. Their probability can be estimated very easily from the probability of capture of a β-particle: The probability of "allowed" capture processes is proportional to the probability of the incident particle being at the nucleus. For the capture from the K shell, this probability is equal to the nuclear volume divided by the volume of the K shell, which, for small nuclear charge, is πa_K^3. For a free electron moving through a material containing per cm³ N atoms whose nuclei can be disintegrated, the probability of being in the nucleus is the nuclear volume divided by the volume per atom, the latter being equal to $1/N$. The ratio of the probabilities is thus $\pi N a_K^3$. Since $a_K = \hbar/mc\gamma$, the function $f(\epsilon_0)$ becomes for the capture of a free electron by light nuclei (cf. 220)

$$f_e(\epsilon_0) = 2\pi^2 N(\hbar/mc)^3 \epsilon_0^4. \qquad (223)$$

The time until a free electron is captured, is thus (cf. 213) $\tau_0 \log 2/|G|^2 f_e(\epsilon_0)$. Assuming that the electrons travel with the velocity of light, the path traveled before causing a disintegration, would be c times the "lifetime," i.e.,

$$l_e = c\tau_0 \log 2 \frac{1}{2\pi^2|G|^2}\left(\frac{mc^2}{E_0}\right)^4\left(\frac{mc}{\hbar}\right)^3 N^{-1}. \qquad (223a)$$

For a solid material all of whose atoms may be disintegrated by the electron, the number N may be estimated as $6\cdot 10^{22}$. Putting $G=1$, we thus obtain

$$l_e = 2\cdot 10^{22}(mc^2/E_0)^4 \text{ cm}. \qquad (223b)$$

Thus, even if the disintegration of the nuclei is energetically possible and corresponds to an "allowed" transition, and even if the energy E_0 given to the neutrino is very big, the probability of the process if entirely negligible. In other words, the disintegration of nuclei by the capture of free electrons and emission of neutrinos is practically unobservable.[87]

The probability of disintegration of nuclei by free neutrinos is almost the same as that by electrons. The path which a neutrino of energy E_0 has to travel in a solid material containing $6\cdot 10^{22}$ nuclei per cm³, is

$$l_n = 2\cdot 10^{22}(mc^2/E_0)^2(mc^2/E)(mc/p) \text{ cm} \qquad (223c)$$

if E and p are energy and momentum of the electron which would be emitted in the nuclear disintegration, and if every nucleus in the material can be disintegrated by the neutrinos of the given energy. It is indeed very unfortunate that the probability of the disintegration of nuclei by neutrinos is so unobservably small, because this disintegration is the only action of free neutrinos which can be predicted with certainty.

§43. Stability of Isobars and Forbidden β-Processes

We have found in §10 that two nuclei of the same mass number and nuclear charge differing by one unit (neighboring isobars) cannot both be stable. In fact, pairs of neighboring isobars practically do not occur at all in nature. However, a few such pairs seem well established experimentally. These are Cd^{113} In^{113}; In^{115} Sn^{115}; Sb^{123} Te^{123}; K^{40} Ca^{40}; and Rb^{87} Sr^{87}. All the isotopes in the first three pairs have been confirmed recently by Bainbridge[88] using hydrogen-free sources, thus excluding spurious "isotopes" due to hydrides. Other rare isotopes which would be isobaric with "neighboring" well-established nuclei could not be confirmed by Bainbridge. These are Cd^{115} (would be isobaric to In^{115}), $Sn^{121}(Sb^{121})$, $Hg^{197}(Au^{197})$, $Hg^{203}(Tl^{203})$, $Pb^{203}(Tl^{203})$, $Pb^{205}(Tl^{205})$, $Pb^{209}(Bi^{209})$, Pb^{210} (this nucleus can be said to be spurious almost with certainty, because the radioactive element Ra D is identical with Pb^{210}, thus Pb^{210} cannot occur in nature as a stable isotope). All these isotopes seem to be, at least, much rarer than was originally claimed; however, Bainbridge believes that his results need rechecking in order to be sure that the doubtful isotopes really do not exist.

The last two of the well-established pairs of isobars do not enter our present discussion, because K and Rb are known to be radioactive. In the case of K, it seems highly probable

[87] Electrons can, however, disintegrate nuclei by virtue of their electric field. This process is analogous to the disintegration by γ-rays (cf. §16, and B17) and might be observable; the disintegrating electron is *not* captured in the process.—The disintegration by *capture* of electrons may also become observable if the expression for the probability of β-disintegration is modified in such a way as is necessary to explain the nuclear forces (§44) and if at the same time the energy of the incident electron and the emitted neutrino are of the order 137 mc^2.

[88] We are indebted to Dr. Bainbridge for communicating his results to us before publication.

that K^{40} is responsible for the radioactivity. In fact, K^{40} was discovered by Nier (N3) after Klemperer (K4) had given good reasons for its being the radioactive isotope of K (cf. §33, 40). In the case of Rb, it seems reasonable to assume that Rb^{87} is the radioactive isotope, just because it is isobaric with Sr^{87}. This view is confirmed by the fact that no other isotopes of Rb than Rb^{85} and Rb^{87} could be detected (N4).

There remain three pairs of well-established neighboring isobars, none of which shows any observable β-activity. The number of these pairs is very small indeed, compared to the number of "allowed" isobaric pairs with nuclear charges differing by two units, which is over 50 (cf. Table I). However, the fact that three "forbidden" pairs of isobars exist, is significant enough and must be explained.

There appear to be two possible explanations. Either (a) the mass of the neutrino is not zero or (b) the β-transformation of one isobar into the other is highly forbidden.

If we accept alternative (a), the conditions for the energetic stability of two isobars Z^A and $(Z-1)^A$ are the following: The nucleus $(Z-1)^A$ must not be capable of β-disintegration. This is certainly the case if its mass is smaller than the sum of the masses of the particles which would be formed in such a disintegration, i.e., the nucleus Z^A, an electron and a neutrino. We thus have the condition for the masses

$$(Z-1)^A < Z^A + \epsilon^- + n^0. \qquad (224)$$

Similarly, Z^A must be incapable of capturing one of the orbital electrons attached to it (§42) and emitting a neutrino. This condition will be fulfilled if the sum of the masses of the original particles is smaller than the masses of the produced particles, *viz.*,

$$Z^A + \epsilon^- < (Z-1)^A + n^0. \qquad (224a)$$

The nuclei Z^A and $(Z-1)^A$ will therefore *both* be energetically stable if

$$(Z-1)^A - n^0 < Z^A + \epsilon^- < (Z-1)^A + n^0. \qquad (224b)$$

In other words, if the mass of the nucleus Z^A, plus an electron is identical with the mass of the nucleus $(Z-1)^A$ within an accuracy of one neutrino mass, then both isobars Z^A and $(Z-1)^A$ will be energetically stable against β-transformations.

We may try to obtain an estimate of the neutrino mass from this condition, assuming that there are just 3 pairs of neighboring isobars of atomic weight below 150. We have shown in §9 and 10 (cf. (21)) that, in a very rough approximation, the exact weight of the atoms of mass number A can be represented as a function of the charge Z as follows:

$$E(Z) = B + \kappa(Z - Z_A)^2 = B + C(Z - Z_A)^2/A, \qquad (225)$$

where B and C are certain constants and Z_A is the "most favorable" charge for the atomic weight (cf. (19a)) A. C depends only slightly on the atomic weight A, and has, for A in the neighborhood of 120, the value 100 MV (cf. (21), (22)). If we have two isobaric nuclei of charges Z and $Z-1$, we define a quantity β by putting

$$Z_A = Z - \tfrac{1}{2} + \beta. \qquad (225a)$$

β is supposed to lie between $-\frac{1}{2}$ and $+\frac{1}{2}$. We assume now that any value of β in this range is equally probable, i.e., that the values of β are distributed perfectly at random. The weights of the two isobaric atoms, as functions of β, are

$$\left.\begin{aligned} &B + (C/A)(\tfrac{1}{2} - \beta)^2 \quad \text{for the nucleus of charge } Z, \\ &B + (C/A)(\tfrac{1}{2} + \beta)^2 \quad \text{for the nucleus of charge } Z-1, \end{aligned}\right\} \qquad (225b)$$

so that the difference is

$$\Delta E = 2C\beta/A. \qquad (225c)$$

This difference is supposed to be smaller than the neutrino mass n_0. Thus the two isobars will both be stable if

$$|\beta| < n_0 A/2C. \qquad (225d)$$

The probability for this is, for random distribution of the β's:

$$p = n_0 A/C. \qquad (225e)$$

The total number of isobaric pairs of odd atomic weight smaller than 150 should therefore be

$$P = \sum_{\text{all odd } A\text{'s}}^{150} n_0 A/C = 150^2 n_0/4C = 5600 n_0/C. \qquad (225f)$$

Putting this equal to three, and inserting $C = 100$

MV, we find

$$n_0 = 0.05 \text{ MV}. \quad (225g)$$

Thus the mass of the neutrino must be one-tenth of the electron mass in order to explain the observed number of pairs of isobars on the basis of energetic stability.[89] Such a mass would seem just reconcilable with the data about the energy distribution in β-spectra. However, it does not seem very plausible to assume a neutrino mass which is so small, and still not zero. The assumption of a zero mass would seem much more satisfactory.

We are thus led to alternative (b) (see above) which assumes that one of the neighboring isobars may be energetically unstable, but does not transform into the other because the corresponding β-transformation is highly forbidden.

We must distinguish two cases: Either the isobar of larger nuclear charge has higher energy, or that of smaller charge. In the first case, there will be an "inverse" β-process, i.e., a capture of K electrons by the unstable nucleus; in the second case, there will be an ordinary β-process. Only in the second case there will be a radioactivity which is observable in principle; whereas the "inverse" β-process would manifest itself only in the gradual disappearance of the more energetic isobar. Accordingly, we have to make the following requirements regarding the lifetime of the unstable isobar:

(a) If the nucleus of higher charge is the unstable one, its lifetime must be of the order of the age of the earth (about 10^9 years). Otherwise, this nucleus could no longer be found on the earth.

(b) If the nucleus of *lower* charge has higher energy, its lifetime must be such that radioactivity becomes unobservable.[90] In order to compute the necessary lifetime, we may assume that a radioactivity of one β-particle per hour from one cm² area of a solid target could just be observed. Furthermore, let us assume an upper limit of the β-spectrum of 200,000 volts, corresponding to an average energy slightly below 100,000 volts. If τ is the half-life in years, the number of β-particles observed per cm² per hour is

$$\beta = Nx_0 \log 2/24 \cdot 365\tau, \quad (226)$$

where N is the number of atoms of the disintegrating isotope per cm³ in the material and x_0 the average depth from which β-particles will escape. The latter is influenced by scattering and stopping and can be calculated (B14, formulae (32), (30), (29)). If the disintegrating isotope constitutes almost 100 percent of the material, which would be the case for In^{115}, one finds $Nx_0 = 5 \cdot 10^{19}$ cm^{-2} for $Z=50$ and an electron energy of 100,000 volts. Then (226) becomes

$$\beta = 4 \cdot 10^{15}/\tau \quad \text{particles/hour}. \quad (226a)$$

Thus a lifetime of 10^{15} years would be required in order to make the radioactivity unobservable.

When we find a pair of neighboring isobars in nature, without observing any β-radioactivity in the isobar of smaller nuclear charge, it is *more likely* that the *isobar of larger nuclear charge is the energetically unstable* one. We shall therefore assume in the following that case (a) is realized in all three observed isobaric pairs, so that In^{113}, Sn^{115} and Te^{123} are the energetically unstable isobars.

For these isobars, the lifetime is given by (222b, c, d, e). It is seen that a change of the nuclear moment by $L=2$ units would lead to the required lifetime of 10^9 years or more, only if the energy E_0 of the emitted neutrino is less than about 25,000 volts. It is rather improbable that the energies of the two isobars coincide to that accuracy. However, a change of the nuclear moment by 3 units will lead to a lifetime of 10^9 years even if E_0 is as large as 500,000 volts which is certainly a conservative estimate of the energy difference between the isobars. Thus the existence of pairs of neighboring isobars is possible, if

1. The difference of the nuclear spins of the two isobars is at least 3 units.

2. The isobar of the larger charge has the higher energy.

[89] This figure is based on the assumption that the dependence of the energy of isobaric nuclei on the nuclear charge is perfectly regular, as given by (225). If there are irregularities, a larger value would be required for the neutrino mass. Also, it should be pointed out that it is not very satisfactory to base statistics on only 3 pairs.

[90] In making this statement, it is assumed that no radioactivity can be observed from the elements Cd, In and Sb. It would, of course, be very interesting to search for such radioactivity. The β-particles may be expected to have very small energy.

Differences of 3 units in total spin seem very plausible, in view of the high angular momenta of the individual particles in the nucleus (§32), and the high total momenta observed for a number of nuclei (§48). Thus the explanation of isobaric pairs on the grounds of forbidden β-transitions seems satisfactory. There is, then, no objection to assuming the neutrino mass to be zero.

In concluding this section, we want to mention the most fundamental pair of isobars, *viz.*, neutron and proton. Of these particles, the neutron must be unstable, its weight being about $E_0 = 350{,}000$ volts higher than the weight of a proton and an electron together. This figure is based on the binding energy of the deuteron (2.14 MV, §16) and Bainbridge's determination of the masses of proton and deuteron ($H^2 = 2.01423$, if $H^1 = 1.00807$). There seems to be some evidence (A5) that the deuteron is actually heavier than 2.01423; this would make the neutron even more unstable.

The transition $n^1 \rightarrow H^1 + \epsilon^- + n^0$ is certainly "allowed," therefore its probability is given by (213), (213b). Inserting $\epsilon_0 = 1.7$ into (213b), we obtain $f(\epsilon_0) = 0.02$, and with (213), (214), and $G = 1$:

$$\tau_{\text{Neutron}} = 2.5 \cdot 10^6 \text{ sec.} = 1 \text{ month.} \quad (227)$$

A similar value has been deduced by Motz and Schwinger (M17). The lifetime (227) is too long to allow observation of the β-decay of neutrons. This would even be true if we take Aston's value for the mass of H^2 (2.0148) which would lead to an upper limit of the β-spectrum of the neutron of about 0.9 MV, therefore $\epsilon_0 = 2.8$, $f(\epsilon_0) = 4$, and

$$\tau = 1.2 \cdot 10^4 \text{ sec.} = 3 \text{ hours.} \quad (227a)$$

It might also be worth while to estimate the lifetime of the nuclei Be^{10} and C^{14} which are formed in certain transmutations. Since the nuclei B^{10} and N^{14} are known to be stable, their isobaric neighbors Be^{10} and C^{14} must be unstable, if we assume the neutrino mass to be zero. The difference in energy seems to be very small in both cases, probably about 100,000 to 200,000 volts. Assuming the β-transitions to be allowed, the lifetimes would be between $\frac{1}{2}$ and 20 years.

§44. Nuclear Forces and β-Disintegration

It was first suggested by Heisenberg[91] that there may be a connection between the "Fermi field" corresponding to β-disintegration, and the forces between neutrons and protons. This connection may be thought of as analogous to the connection between the emission of light (electromagnetic field) and the Coulomb interaction between charged particles. In quantum electrodynamics, the Coulomb interaction between two particles is not introduced as a separate assumption, but each particle is only assumed to interact with the electromagnetic field. Only because *both* charged particles interact with the field, there is also some interaction between them. The Coulomb interaction is thus regarded as a second approximation of the interaction between field and individual particles.

The same program may be carried out for Fermi's β-field and the nuclear forces. Let us suppose we have two particles 1 and 2 and want to investigate the interaction energy corresponding to the transformation of the first particle from a neutron into a proton, simultaneously with the inverse transformation of the second particle. Given is the Hamiltonian describing the interaction of both particles with the Fermi field, which we assume in the Konopinski-Uhlenbeck form (206a), (211b), leaving out the small term containing the α operator of the heavy particles:

$$\begin{aligned} H &= H_1 + H_1^* + H_2 + H_2^*, \\ H_1 &= (gE_n/\hbar c)(\psi^*(\mathbf{r}_1)\beta\varphi(\mathbf{r}_1))Q_1. \end{aligned} \quad (228)$$

Here E_n is the neutrino energy, β the Dirac operator, $\varphi(\mathbf{r}_1)$ and $\psi(\mathbf{r}_1)$ neutrino and electron wave function at the place of the heavy particle, Q the operator transmuting a neutron into a proton and Q^* the inverse operator. The indices 1 and 2 refer to first and second heavy particle, and H_1^* is the complex conjugate of H_1.

The Hamiltonian (228) will, in second approximation, lead automatically to a simultaneous transformation of particle 1 from a neutron into a proton, and of 2 from a proton into a neutron. The Hamiltonian connected with this transformation is, according to the ordinary Schrödinger

[91] Heisenberg, Lectures at the Cavendish Laboratory, Cambridge, 1934. Unpublished.

perturbation theory

$$W=-\sum_b \frac{H_2{}^{*bd}H_1{}^{ab}}{E_b-E_a}-\sum_c \frac{H_1{}^{cd}H_2{}^{*ac}}{E_c-E_a}. \quad (228a)$$

Here a, b, c, d denote initial, two intermediate and final state, $H_1{}^{ab}$, etc., the respective matrix elements and E_aE_b the total energies. The states are describable as follows:

Initial state (a): Particle 1 = Neutron, 2 = Proton, all electron and neutrino states of negative energy occupied, all states of positive energy empty.

Intermediate state (b): Particle 1 = Proton, 2 = Proton, one neutrino state of negative energy $-E_n$ empty, one electron state of positive energy E_e occupied

Intermediate state (c): Particle 1 = Neutron, 2 = Neutron, one electron state of negative energy $-E_e$ empty, one neutrino state of positive energy E_n occupied

Final state (d): Particle 1 = Proton, 2 = Neutron, all negative states occupied, all positive states empty.

Thus in both cases

$$E_b-E_a=E_c-E_a=E_n+E_e. \quad (228b)$$

We now insert for φ and ψ plane waves, normalized per unit momentum:

$$\varphi(\mathbf{r}_1)=h^{-3}\bar{\varphi}(\mathbf{p}_n)\exp[i\mathbf{p}_n\cdot\mathbf{r}_1/\hbar], \quad (228c)$$

etc., where $\mathbf{p}_n$ is the momentum of the neutrino and $\bar{\varphi}$ a constant spinor. After a simple calculation involving these spinors, (228a) reduces to

$$W=-2\left(\frac{g}{\hbar c}\right)^2 h^{-6}\int\frac{d\mathbf{p}_n d\mathbf{p}_e}{E_n+E_e}E_n{}^2\exp[i(\mathbf{p}_n-\mathbf{p}_e)\cdot\mathbf{r}/\hbar] \quad (228d)$$

if the neutrino mass is assumed to be zero. $\mathbf{r}=\mathbf{r}_1-\mathbf{r}_2$ is the distance of the two particles, and the integral extends over all momentum space of electron and neutrino. Introducing polar coordinates in $\mathbf{p}_n$ and $\mathbf{p}_e$-space, and integrating over the angles, we obtain

$$W=-32\pi^2\left(\frac{g}{\hbar c}\right)^2(2\pi\hbar)^{-6}\left(\frac{\hbar}{r}\right)^2 \times\int\frac{p_n dp_n p_e dp_e E_n{}^2}{E_n+E_e}\sin(p_n r/\hbar)\sin(p_e r/\hbar). \quad (228e)$$

The main contribution to the integral clearly comes from high energies of electron and neutrino. We therefore put

$$p_n=E_n/c, \quad p_e=E_e/c. \quad (228f)$$

Furthermore, we introduce the abbreviations

$$x=E_n r/\hbar c, \quad y=E_e r/\hbar c. \quad (228g)$$

Then (228e) becomes

$$W=-\tfrac{1}{2}\pi^{-4}\frac{g^2}{\hbar c}r^{-7}\int_0^\infty\frac{x^3dxdy}{x+y}\sin x\sin y. \quad (228h)$$

The integral diverges. It can, however, be calculated as

$$\left(\frac{\partial^4 F(p,q)}{\partial p^3\partial q}\right)_{p=q=1} \quad (228i)$$

with

$$F=\int_0^\infty\int\frac{dxdy}{x+y}\cos px\cos qy. \quad (228j)$$

F has the value[92] $\pi/2(p+q)$; therefore (228i) becomes $4!\pi/2(1+1)^5=3\pi/8$.

Thus $\quad W=-(3/16)\pi^{-3}(g^2/\hbar c)r^{-7}. \quad (229)$

The interaction of a neutron and a proton is thus proportional to the inverse seventh power of their distance, i.e., it increases very rapidly with decreasing distance, as we have always assumed. The interaction would become infinite for $r=0$, and the binding energy of all nuclei would become infinite, if (229) held down to $r=0$. We must therefore assume that, for some reason as yet unknown, (229) breaks down at small distances. To define the breakdown radius a more accurately, we put

$$\begin{aligned} W&=-W_0=-(3/16)\pi^{-3}(g^2/\hbar c)a^{-7} && \text{for } r<a,\\ W&=\text{expression (229)} && \text{for } r>a. \end{aligned} \quad (229a)$$

This would correspond practically to a potential hole of depth W_0 and radius a. By choosing a suitable value for the breakdown distance a, we could make the interaction (229a), and the corresponding binding energies, as large as we please; e.g., if we want to obtain the correct

---

[92] To calculate F, we introduce $s=x+y$, $t=x-y$, $\alpha=(p+q)/2$, $\beta=(p-q)/2$. Then

$$F=\frac{1}{4}\int_0^\infty\frac{ds}{s}\int_{-s}^{s}dt[\cos(\alpha s+\beta t)+\cos(\beta s+\alpha t)] \quad (A)$$

$$=\frac{1}{4}\int_0^\infty\frac{ds}{s}\frac{1}{\beta}[\sin(\alpha+\beta)s-\sin(\alpha-\beta)s] +\frac{1}{4}\int_0^\infty\frac{ds}{s}\frac{1}{\alpha}[\sin(\beta+\alpha)s-\sin(\beta-\alpha)s]. \quad (B)$$

Now $$\int_0^\infty\frac{ds}{s}\sin\kappa s=\begin{cases}+\pi/2, & \text{if } \kappa>0,\\ -\pi/2, & \text{if } \kappa<0.\end{cases} \quad (C)$$

Since $\alpha+\beta=p>0$, $\alpha-\beta=q>0$, the first integral in (B) is zero, the second is equal to $\pi/4\alpha=\pi/2(p+q)$.

value for the binding energy of the deuteron, we must choose (cf. (40))

$$W_0a^2 = (\pi^2/4)\hbar^2/M, \tag{229b}$$

$$\text{i.e., } (\pi a)^5 = \frac{3}{4}g^2M/\hbar^3c = \frac{0.75\cdot 3.6\cdot 10^{-120}\cdot 1.7\cdot 10^{-24}}{1.1\cdot 10^{-81}\cdot 3\cdot 10^{10}} \times 1.3\cdot 10^{-73}\text{ cm}^5,$$

$$a = 0.85\cdot 10^{-15}\text{ cm},\quad W_0 = 1.3\cdot 10^{12}\text{ volts.} \tag{230}$$

However, the assumption of such a short range and such a large magnitude of the forces between neutron and proton is quite unsatisfactory. It would lead to an extremely big binding energy of the α-particle (chapter IV). In order to obtain agreement with the empirical facts about nuclear forces, we must assume that the "cut-off" of the interaction (229) occurs in the neighborhood of $r=2\cdot 10^{-13}$ cm (§21). This, however, leads to an interaction energy which is negligibly small compared to the empirical value. The quantity W_0a^2, which determines the binding energy of the deuteron, is proportional to a^{-5}, and is therefore decreased by a factor $250^{-5}=10^{-12}$ if the range of the forces is increased by a factor 250, from $0.8\cdot 10^{-15}$ to $2\cdot 10^{-13}$ cm. *Thus the interaction comes out to be too weak by a factor* 10^{12}, *if we cut the potential (229) off at the observed range of the nuclear forces.*

This highly unsatisfactory result is, of course, due to the extremely small value of the constant g which governs the β-emission. However, the general idea of a connection between β-emission and nuclear forces is so attractive that one would be very reluctant to give it up. In principle, several ways seem open:

(a) The interaction leading to β-emission is only part of a more general interaction. The other "components" of that general interaction are larger. This hypothesis was suggested by Heisenberg, in analogy to electrodynamics, where also the Hamiltonian leading to emission and absorption of light (transverse electromagnetic waves) is small compared to that connected with "longitudinal" electromagnetic waves. The latter cause most of the Coulomb interaction. How such a modification could be introduced into the β-theory, is of course not clear.

(b) The fundamental expression for the β-emission contains actually higher derivatives of the electron and neutrino wave function. This would mean a more rapid increase of the β-disintegration probability with increasing energy. Now the constant g is derived from the given lifetimes of nuclei emitting β-rays of a few MV energy. On the other hand, the energies which contribute most to the interaction of a neutron and a proton at a distance a, are those for which the electron and neutrino wave-length is of the order a. The energy corresponding to $a=2\cdot 10^{-13}$ cm, is $E=\hbar c/a\approx 100$ MV, i.e., about 100 times the energy of most β-particles. Each additional derivative introduced into the expression (211), multiplies the Hamiltonian corresponding to β-decay by a factor E_e or E_n, according to whether it is introduced into the electron or the neutrino wave function. Since W contains (cf. 228a) the square of the β-Hamiltonian, a factor E_e^{2k} (or E_n^{2k}) is introduced if k more derivatives are introduced in (211). Since the constant g must be redetermined in such a way as to make the lifetimes of β-decaying nuclei agree with experiment, W is multiplied by the $2k$th power of the ratio of the energies occurring in (228a), and in β-decay. This ratio being about 100, W is multiplied by 10^{4k}. The introduction of *three* more derivatives (i.e., altogether four) into (211) would, therefore, bring about agreement with the observed nuclear forces.

However, such a change of the fundamental assumptions of the theory would also lead to a considerable decrease of the lifetime of nuclei emitting high energy β-rays, compared to those emitting less energetic ones. This would destroy the agreement obtained in §41, Table XV. The only way out would be to assume that all the observed high energy β-transformations are forbidden at least of the second order, which does not seem plausible at all.

(c) It may be that the behavior of electrons of wave-length near $e^2/mc^2 (=2.8\cdot 10^{-13}$ cm) is so completely different from the usual one, that these electrons contribute much more to (228a) than we anticipate, without the β-interaction differing appreciably from the Konopinski-Uhlenbeck expression for lower energies. This idea is sufficiently vague to make it hard to disprove it.

Assuming for the present that the problem of the magnitude of the nuclear forces, as compared to the β-interaction, will be solved in some way or other in the future, we may inquire into the *nature* of the forces between neutron and proton following from the β-theory. The force is clearly an exchange force, being connected with a change of roles of neutron and proton (see above). This is in agreement with our empirical knowledge about neutron-proton forces. However, it turns out to be a *Heisenberg* force, because the spins remain unchanged in the process (228) if we consider a particle at a given *point*, $\mathbf{r}_1$ or $\mathbf{r}_2$. Only the "charge passes over" from point $\mathbf{r}_2$ to $\mathbf{r}_1$, so that the neutron formed in the process, which is situated at $\mathbf{r}_2$, has the same spin as the proton previously situated at that place. This is the characteristic of a Heisenberg force, which is not admissible for the fundamental force of nuclear physics (§7, §11).

However, it is easy to change the fundamental expression for the β-disintegration so as to obtain a Majorana rather than a Heisenberg force. It is only necessary to accept one of the interactions (211c), (211d), rather than the Konopinski-Uhlenbeck interaction (211b). Besides irrelevant changes in the magnitude of the forces, this will introduce a factor $\boldsymbol{\sigma}_{\text{heavy}}$ in the Hamiltonians (228), and therefore a factor $\boldsymbol{\sigma}_1\cdot\boldsymbol{\sigma}_2$ in the expression (228a) for the interaction energy of neutron and proton. The product $\boldsymbol{\sigma}_1\cdot\boldsymbol{\sigma}_2$ is positive when proton and neutron spin are parallel, in this case we obtain the same result as before. $\boldsymbol{\sigma}_1\cdot\boldsymbol{\sigma}_2$ is negative if the spins are antiparallel; in this case we therefore find the sign of the interaction reversed. This is exactly the difference between a Heisenberg and a Majorana force (Eq. (30a, b)). Their signs are *equal* if the spins of proton and neutron are parallel, opposite for opposite spins.

However, the numerical factor is not correct. $\boldsymbol{\sigma}_1\cdot\boldsymbol{\sigma}_2$ has the value 1 for parallel, -3 for antiparallel spins. The force between neutron and proton would thus be 3 times as large for antiparallel than for parallel spins, whereas actually (§14) the forces are almost equal, and the force for *parallel* spins is even somewhat bigger. Thus we ought to take a suitable linear combination of the forces (211b) and (211c, d) to obtain a neutron-proton interaction of the desired form. This again is somewhat unsatisfactory.

The replacement of (211b) by (211c, d) does not change our discussion about β-spectra materially. The only difference is that changes of the *spin* angular momentum of the nucleus by one unit are now allowed transitions. If the coupling between spin and orbital momentum is weak, we have, then, just to refer to changes of the orbital momentum of the nucleus rather than to its total momentum, in discussing forbidden β-transitions. If the spin-orbit coupling in the nucleus is strong, which is probably the case, changes of the total momentum by $L+1$ will be about as probable as changes by L were in our previous discussion. Then we must require a change by 4 units for radioactive K^{40}, and for the isobaric pairs (§43).

The "β-hypothesis of nuclear forces" gives, in first approximation, only forces between neutrons and protons. In second approximation forces between like particles would appear, the mechanism being about as follows: Each of two neutrons emits "virtually" (intermediate state, cf. 228a) an electron and a neutrino, and then absorbs the particles emitted by the other neutron. It should be expected that this second approximation is *not* small compared to the first. For if the β-theory is to lead to the observed magnitude and range of neutron-proton forces, we must assume that for electron and neutrino energies of about 100 MV the β-interaction H (228) is also of the order 100 MV: Electrons of energy $\sim$100 MV should give the main contribution to W, because their wave-length is of the order of the range of the nuclear forces. On the other hand, W is of the order $H^2/(E_e+E_n)$, according to (228a, d), and is about 30 MV empirically (§21). This requires the matrix elements of H, corresponding to the emission of electrons and neutrinos of 100 MV, to be also of the order 100 MV. But if this is true, we should expect the second approximation to be nearly as big as the first, i.e., the forces between like particles should be not much smaller than those between neutron and proton, in agreement with the conclusions from nuclear binding energies (§21).

§45. The Magnetic Moments of Proton and Neutron (W11)

It was first suggested by Wick (W11) that the anomalous values of the magnetic moments of proton and neutron (§5) may be explained on the grounds of the β-theory. According to that theory, a neutron can never be regarded as entirely isolated, but is always associated with a "β-particle field" surrounding it. In other words, if we observe a neutron at a given moment, we shall not always find a neutron, but sometimes we shall find a proton, an electron and a neutrino instead. During the short intervals of time when the neutron is replaced by a proton, an electron and a neutrino, an external magnetic field will find the spin of the electron to act upon. The interaction energy between a magnetic field and a neutron, will therefore be equal to the interaction energy between the field and an electron, times the probability of finding the neutron temporarily "dissolved" into proton, electron and neutrino at any given instant, times the probability that the spin of the electron is parallel to the spin of the neutron rather than antiparallel.

The same argument holds for the proton, with the only difference that "positron" should be inserted instead of "electron." Moreover, the *proton* will interact with the magnetic field even if it is not "dissolved"; during these times, its interaction will correspond to its "normal" magnetic moment $\hbar e/2Mc$, which follows from the Dirac theory.

The probability that a neutron is found to be temporarily dissolved into a proton, an electron of energy E_e and a neutrino of energy E_n is, according to the Schrödinger perturbation theory,

$$|H(E_e, E_n)|^2/(E_e+E_n)^2, \qquad (231)$$

where H is the matrix element of the β-interaction (211) which refers to the emission of an electron and a neutrino of the respective energies E_e, E_n.

If we insert in (231) the ordinary interaction derived from the probability of β-disintegration itself, we are faced with the same difficulties as in the preceding section when trying to account for the nuclear forces: If we accept the β-interaction as it stands, the expression for the magnetic moment of the neutron will diverge. If we avoid the divergence by "cutting off" the β-interaction for high energies of electron and neutrino, we shall obtain much too small a value for the magnetic moment.

We shall therefore assume, just as in the preceding section, that the present β-ray theory is not adequate in the region of high electron and neutrino energies, but that a future correct theory will give a higher disintegration probability for electron energies of the order 100 MV. We shall furthermore assume that this probability will turn out to be such as to give the correct magnitude of the nuclear forces. This means (end of §44) that the matrix elements H in (231) are nearly as large as the denominator E_e+E_n, if E_e and E_n are of the order 100 MV. Then (231) becomes almost unity; may be of the order 1/10. In other words, during a considerable fraction of the time, the neutron will be found dissolved into proton, electron and neutrino.

The magnetic moment of a high energy electron is one-third Bohr magneton.[93] The correlation between the spin directions of the neutron and the emitted electron is not easy to estimate; it depends on the particular linear combination of the expressions (211b, c, d) which represents the correct β-ray interaction.[94] A correlation of a few percent would be sufficient to account for the observed magnetic moment of the neutron, *viz.*, $2.0\hbar/2Mc = 1/900$ Bohr magneton. According to the foregoing, the probability that an

---

[93] This can be seen easily from the Dirac wave function of a free electron with spin parallel to the Z axis, whose components are

$$\psi_1 = -Acp_z \quad \psi_2 = -Ac(p_x+ip_y) \quad \psi_3 = -A(E+mc^2) \quad \psi_4 = 0,$$

$$A = [2E(E+mc^2)]^{-\frac{1}{2}} \exp[i(\mathbf{p}\cdot\mathbf{r}-Et)/\hbar].$$

The magnetic moment, in Bohr magnetons, is given by

$$(\psi^*\sigma_z\psi) = \psi_1^*\psi_1 - \psi_2^*\psi_2 + \psi_3^*\psi_3 - \psi_4^*\psi_4$$
$$= [c^2p_z^2 - c^2(p_x^2+p_y^2) + (E+mc^2)^2]/2E(E+mc^2).$$

Averaging all over directions of motion of the electron, we obtain

$$p_x^2 = p_y^2 = p_z^2 = \tfrac{1}{3}p^2 = (E+mc^2)(E-mc^2)/3c^2.$$

We have therefore

$$(\psi^*\sigma_z\psi) = [E+mc^2 - \tfrac{1}{3}(E-mc^2)]/2E = \tfrac{1}{3} + \tfrac{2}{3}mc^2/E.$$

For small E, of the order of mc^2, this expression has the value 1; for large E ($\gg mc^2$), it is only one-third.

[94] If the β-ray interaction is represented by one of the expressions (211b, c, d) *alone*, there will be no correlation between the spins of neutron and electron.

electron is present, is of the order 1/10; the magnetic moment of the electron, *if* present, is 1/3 Bohr magneton; thus the magnetic moment of the neutron would be 1/30 Bohr magneton if the spin of the emitted electron were *always* parallel to the neutron spin.

It appears thus that our theory gives rather *too large* a value for the magnetic moment of the neutron if we deduce the β-interaction from the forces between neutron and proton.

The magnetic moment of the proton can be calculated in the same way. Since the β-ray theory is perfectly symmetrical in neutrons and protons, the *additional* moment of the proton should be equal and opposite to the magnetic moment of the neutron. By additional moment we understand the excess of the actual magnetic moment of the proton over its "elementary" moment $\hbar/2Mc$ which follows from the Dirac theory. We should thus conclude that the *sum* of the magnetic moments of neutron and proton, i.e., the magnetic moment of the deuteron, is equal to the elementary magnetic moment of the proton, since the additional moment of the proton, and the magnetic moment of the neutron, cancel each other. The observed value of the magnetic moment of the deuteron is about $0.85\hbar/2Mc$. This is nearly, but not exactly, equal to the value following from our considerations. The difference may either (a) be connected with the difference in mass between neutron and proton, or (b) with the fact that the proton is actually dissolved into a neutron, a positron and a neutrino during a considerable fraction of the time, and does not possess its "elementary moment" during that time.

VIII. Nuclear Moments

The essential features of atomic and molecular structure can be accounted for by the quantum theory on the assumption of an atom consisting of a small massive nucleus surrounded by electrons which are held in the nuclear field. It is quite satisfactory for most of the purposes of atomic and molecular structure to consider the field only at distances from the nucleus sufficiently large that its field is a Coulomb field. As far as the main features of atomic and molecular spectra are concerned, therefore, the nucleus does not enter except as the center of this Coulomb field. There are, however, certain facts of both atomic and molecular spectra which lead directly to information concerning the atomic nucleus.

For atoms it is well known that the totality of energy states found from the usual analyses of spectra can be accounted for both as to number and, with some difficulty, as to position by the quantum theory treatment of the electrons moving in the central Coulomb field. These states are characterized by quantum numbers of the electrons, by the total angular momentum of the electrons and usually by the spin and orbital angular momenta as well. The most closely adjacent states are those (at least for the case of Russell-Saunders coupling) which have only different total angular momenta. Such states compose a "multiplet" and are referred to as fine structure because of their frequent close spacing.

A more detailed examination of the spectral lines involved frequently has indicated that states considered above are not themselves single but are actually composed of a group of states. This multiplicity is called hyperfine structure. The hyperfine structure of spectral lines cannot be accounted for on the basis of the assumptions mentioned above. It was first suggested by Pauli (P4) that hyperfine structure is due to the action of the electrons in the field of a nuclear magnetic dipole. That the interaction is essentially a magnetic one is immediately seen by a comparison of hyperfine structure groups with ordinary fine-structure multiplets. Such a comparison shows striking similarities. The nuclear origin of hyperfine structure is confirmed on many sides but it seems sufficient to mention only one such confirmation at this time. Those states which have electrons with a higher probability of being near the nucleus show the structure while those with a very low probability show no structure. It is now completely certain that the hyperfine structure of spectral lines is for the greatest part due to the interaction of a nuclear magnetic moment with the electrons not appearing in closed shells. The origin of the nuclear magnetic dipole whose presence gives rise to the hyperfine structure is probably due to the motion of electrical charges and it is to be expected that the nucleus possesses a certain angular momentum (spin). This same conclusion is of course attained since the nucleus is considered to be built of protons and neutrons which themselves have intrinsic angular momenta and magnetic moments. A detailed study of the hyperfine structure for a particular element allows us to determine the angular momentum of the nucleus in question and, with somewhat more difficulty in interpretation, the magnitude of the nuclear magnetic moment as well. Complications arise when the element has several isotopes but these can frequently be overcome.

For diatomic molecules composed of like atoms it is found that the presence of a nuclear angular momentum changes the statistical weight of the rotational states. For zero nuclear angular momentum we find that alternate lines of the bands are missing. With a nuclear angular momentum it is found that successive lines of the bands have an intensity ratio which depends on the magnitude of the nuclear spin. It is thus possible by studying the intensities in such molecular spectra to determine the nuclear angular momentum.

There are other indications of the presence of nuclear spins and magnetic moments. In hydrogen for example, since the proton is known to have a spin, the hydrogen molecule may consist of two hydrogen atoms with their nuclear spins in the same (orthohydrogen) or opposite directions (parahydrogen). The presence of these two sorts of hydrogen is known theoretically from the behavior of the specific heat at low temperatures (D2) and indeed the two sorts of hydrogen have been separated experimentally. In the presence of a paramagnetic gas such as oxygen the rate of conversion from para- into orthohydrogen depends on the size of the nuclear magnetic moment. It has thus been possible (F1) to determine the ratios of the nuclear magnetic moments for hydrogen and deuterium from this dependence. A more detailed account of these atomic and molecular effects will be found in the following paragraphs.

§46. The Interaction of the Nuclear Moment with the Electrons

If the interaction which gives rise to the hyperfine structure is due to the presence of a nuclear magnetic dipole in the field of the electrons it should be possible to deduce certain properties of it without calculation. Let us denote by **I** and **J** the angular momenta of the nucleus and the electrons, respectively, in units of $\hbar$. The interaction term which must be added to the potential energy of the system will be proportional to the cosine of the angle between **I** and **J**, and may be written

$$V=A'(\mathbf{I}\cdot\mathbf{J})=\tfrac{1}{2}A'(\mathbf{F}^2-\mathbf{I}^2-\mathbf{J}^2), \qquad (232)$$

where **F** is the vector sum of **I** and **J** or the total angular momentum of the system in units $\hbar$. The addition to the energy which such a perturbing potential will give can be found immediately from the characteristic values for the squares of angular momenta.

$$W=\tfrac{1}{2}A\{F(F+1)-I(I+1)-J(J+1)\}. \qquad (233)$$

All the states of the hyperfine structure group have the same values of the quantum numbers I and J. F takes on a series of values from $I+J$ down to $|I-J|$ as determined by the quantum theory treatment of vectors. W then has a series of values $A[IJ]$; $A[IJ-(I+J)]$; $A[(IJ+1)-2(I+J)]$; $A[(IJ+3)-3(I+J)]$ . . . and the energy differences of successive states are $A(I+J)$; $A(I+J-1)$; $A(I+J-2)$. . . . For the energy differences the part in brackets is just the larger F value for the two states considered. This regularity is called the interval rule and the factor A the interval factor. (233) above leads directly to this regularity and may therefore be regarded as a statement of the interval rule. Since (233) follows directly from the cosine form of the interaction it is expected that the interval rule will hold exactly (G9). The only exception to this should occur when two atomic states with different J are separated by an amount which is not large compared to the hyperfine structure separations. The hyperfine structure separations are rarely more than a few cm^{-1} so this exception occurs very infrequently. Recently deviations from the interval rule have been found which do not come under this exception and which therefore mean that the form of the interaction term must be slightly modified. This can be done by the assumption that the nucleus has a small electric quadrupole moment and will be discussed somewhat further in §50.

It is apparent from (233) that the relative spacing and number of the hyperfine structure states is fixed as soon as the quantum numbers F, I and J are known. In case $I<J$, then I can be determined directly from the number of hyperfine states which would be $2I+1$. For $I>J$ however, the number of states is determined by J. In either case the hyperfine quantum number F can be determined from exact measurements of the relative separations by the use of the interval rule (except for $J=\frac{1}{2}$). With F thus determined and J known, the value of I can be found. It is thus possible to determine the nuclear angular momentum without any further knowledge of the nature of the interaction.

The interaction constant A will contain the nuclear magnetic moment μ and factors which relate to the electrons and the probability of their being near the nucleus. The constant μ is related to the absolute size of the hyperfine structure separations. In order to determine μ it will be necessary to make a determination of the other factors in A and then use the experimental size to find μ. Let us consider the case of a single electron in the field of the nuclear magnetic dipole with charge Ze. In order to have the calculation apply to s electrons as well as others it is necessary to use the Dirac equations. We wish then to write the interaction term in terms of the vector potential

$$A=[\boldsymbol{\mu}\mathbf{r}]/r^3=g\mu_0[\mathbf{Ir}]/r^3, \tag{234}$$

where $\boldsymbol{\mu}=g\mu_0\mathbf{I}$ is the nuclear magnetic moment and $\mu_0=e\hbar/2mc$. If this vector potential is introduced into the Dirac equations and if the two "small" wave functions are eliminated to obtain an equation in the two "large" wave functions, the term representing the interaction of the nucleus with the electron can be written (B23)(F5)(B12).

$$H_1=g\mu_0(\mathbf{A}\cdot\mathbf{I}), \tag{235}$$

where

$$\mathbf{A}=2\mu_0[r^{-3}\mathbf{L}-r^{-3}\boldsymbol{\sigma}+3r^{-5}(\mathbf{r}\cdot\boldsymbol{\sigma})\mathbf{r}]$$
$$\times(1+eA_0/2mc^2)^{-1}-2\mu_0[r^{-2}\boldsymbol{\sigma}-r^{-4}(\mathbf{r}\cdot\boldsymbol{\sigma})\mathbf{r}]$$
$$\times\frac{d}{dr}\left(1+\frac{eA_0}{2mc^2}\right)^{-1}. \tag{235a}$$

Here $\mathbf{L}\hbar$ is the angular momentum vector, $\boldsymbol{\sigma}\hbar$ is the spin angular momentum and $\boldsymbol{\sigma}$ is $\frac{1}{2}$ times the Pauli spin matrix, A_0 is the scalar potential of the nuclear electrostatic field and E has been replaced by mc^2. With (235) for the interaction and using the properties of angular momenta and the fact that $\mathbf{A}$ contains only electron variables the perturbed energy can be found (B23) to be

$$W_1=g\mu_0(\mathbf{AJ})_J[F(F+1)-I(I+1)$$
$$-J(J+1)][2J(J+1)]^{-1}. \tag{236}$$

$(\mathbf{A\ J})_J$ is the diagonal element in the matrix $(\mathbf{A\ J})$ for state J. $(\mathbf{A\ J})$ can be found from (235a) using $\mathbf{J}=\mathbf{L}+\boldsymbol{\sigma}$. If the second term of (235a) is neglected in comparison to the first and if $eA_0/2mc^2$ is neglected compared to 1, we obtain

$$(\mathbf{A\ J})=2\mu_0[r^{-3}\mathbf{L}^2-r^{-3}\boldsymbol{\sigma}^2+3r^{-5}(r\cdot\boldsymbol{\sigma})^2$$
$$+3r^{-5}(\mathbf{r}\cdot\boldsymbol{\sigma})(\mathbf{r}\cdot\mathbf{L})]. \tag{237}$$

Here $(\mathbf{r}\cdot\mathbf{L})=0$ since $\mathbf{L}=\hbar[\mathbf{rp}]$ and $3r^{-5}(\mathbf{r}\cdot\boldsymbol{\sigma})^2-r^{-3}\boldsymbol{\sigma}^2=0$ due to the properties of the Pauli spin matrix $2\boldsymbol{\sigma}$.

Using (237) which holds for electrons other than s electrons (236) now becomes

$$W_1=g\mu_0{}^2\frac{L(L+1)}{J(J+1)}\overline{(r^{-3})}$$
$$\times[F(F+1)-I(I+1)-J(J+1)]. \tag{238}$$

For s electrons the first term of (235a) vanishes and we have

$$(\mathbf{A\ J})^s=-2\mu_0[r^{-2}\boldsymbol{\sigma}^2-r^{-4}(\mathbf{r}\cdot\boldsymbol{\sigma})^2]$$
$$\times\frac{d}{dr}\left(1+\frac{eA_0}{2mc^2}\right)^{-1}=-\mu_0r^{-2}\frac{d}{dr}\left(1+\frac{eA_0}{2mc^2}\right)^{-1}$$

The diagonal element for J becomes

$$(\mathbf{A}\,\mathbf{J})_{J^s} = -\mu_0 \int_0^\infty R^2(r) r^{-2} \frac{d}{dr}\left(1+\frac{eA_0}{2mc^2}\right)^{-1} r^2 dr$$

$$= \mu_0 \int_0^\infty \left(1+\frac{eA_0}{2mc^2}\right)^{-1} \frac{d}{dr} R^2(r) dr$$

$$\cong \mu_0 R^2(0). \quad (239)$$

$R(r)$ is the radial function and $R(0)$ its value at the origin. $eA_0/2mc^2$ is neglected in comparison to 1. From (238), (236) and (239) it is possible by comparison with (233) to obtain expressions for the separation factors. For single electrons these will be called a_s or a_l and the quantum numbers l, j, etc.

$$a_s = (4/3) g(I) \mu_0{}^2 R^2(0), \quad (240)$$

$$a_l = 2g(I)\mu_0{}^2 \overline{r^{-3}} (l(l+1)/j(j+1)). \quad (241)$$

Since $\mu = g\mu_0 I$, equations (240) or (241) determine the size of the nuclear magnetic moment[95] if the separation factors as well as the quantum numbers are known from experiment and if it is possible to determine $R^2(0)$ or $\overline{r^{-3}}$. For any particular atom a knowledge of the wave functions is necessary therefore to determine the magnetic moment. Furthermore it is necessary to have wave functions which are quite accurate in the neighborhood of the nucleus if μ is to be determined accurately. At present such wave functions are not available. Calculations by Wills and Breit (W15) and by Shoupp, Bartlett and Dunn (S21) for Na $3s$, $3p_{3/2}$ and $4p_{3/2}$ states using Hartree functions give $\mu = 5.85$, 22 and 10.4 nuclear magnetons, i.e., $\mu_0/1838$, respectively, while $\mu = 2.5$, 5.1, 3.1 n.m., respectively, using Fock functions. Both of these sorts of wave functions give the spin doublet separations too high. Empirical correction assuming that the wave functions give the correct ratio of doublet separation to hyperfine structure separation leads to values for μ of 2.5–3.0 n.m. While Na may not be a favorable example, it is apparent that calculations using present wave functions are rather dangerous. It is very desirable, therefore, to have some approximation method which leads to consistent results and which could be applied even where wave functions are not available.

[95] It is customary to give μ/μ_0, i.e., to measure μ in units of μ_0, the Bohr magneton. Nuclear moments are of the order $10^{-3}\mu_0$ and it is customary to express them in the nuclear magneton (n.m) unit $\mu_0/1838$.

Let us consider the case of hydrogenic wave functions. For this case $R^2(0) = 4Z^3/n^3 a_0{}^3$ and we find that (240) can be rewritten in the form

$$a_s = \frac{8}{3}\frac{R\alpha^2 Z^3}{n^3} g(I)\ \mathrm{cm}^{-1}, \quad (242)$$

where $R = me^4/4\pi\hbar^3 c$ and $\alpha = e^2/\hbar c$. For hydrogenic functions $\overline{r^{-3}} = Z^3/n^3 a_0{}^3(l+1)(l+\frac{1}{2})l$ and using this value in (241) we find

$$a_l = g(I)\frac{R\alpha^2 Z^3}{n^3 j(j+1)(l+\frac{1}{2})}\ \mathrm{cm}^{-1}. \quad (243)$$

For this case (243) reduces to (242) for $j = \frac{1}{2}$ and $l = 0$.

For atoms consisting of a single electron outside a closed shell (243) would be a poor approximation if Z denotes the atomic number, since the screening of the other electrons which plays an important role, would be neglected. A way in which the screening can be taken into account is indicated by the fact that (243) is similar to the expression for the spin doublet separation.

$$a(\Delta\nu) = \frac{\Delta\nu}{l+\frac{1}{2}} = \frac{R\alpha^2 Z^4}{n^3(l+\frac{1}{2})l(l+1)}\ \mathrm{cm}^{-1}.$$

Approximations to the spin doublet separation which replace Z^4 by $Z_i{}^2(1+z)^2$ and n by n^* have been found to hold for alkali type atoms. Here z is the degree of ionization, n^* the effective total quantum number and $Z_i = Z$ for s electrons while $Z_i = (Z-4)$ seems to fit the data for p electrons. For higher l the difference $Z - Z_i$ becomes greater and the necessary approximation more difficult to obtain. This approximation for deeply penetrating states was derived first by Landé (L1) using the Bohr theory and can now be justified on the basis of the wave mechanics.

By a similar argument Goudsmit (G6) and Fermi and Segrè (F11) have obtained an approximation for the hyperfine structure using the observed doublet separation

$$a_l = g(I)\frac{\Delta\nu\, l(l+1)}{j(j+1)(l+\frac{1}{2})Z_i}\ \mathrm{cm}^{-1}. \quad (244)$$

This approximate expression can be used with

$\Delta\nu$ and Z_i to calculate $g(I)$ from the observed separation factor. For s electrons, an approximation[96] similar to that made in the doublet separation gives

$$a_s = g(I)\frac{R\alpha^2 Z_i(1+z)^2}{n^{*3}j(j+1)(l+\frac{1}{2})}\ \mathrm{cm}^{-1}. \quad (245)$$

With Z_i fixed as above (244) and (245) become semiempirical formulas whose main virtues are that they apply to a considerable range of examples and lead to values of μ which, although determined from different states of the electron, are nearly the same.

Certain approximations ($E \cong mc^2$; $eA_0 \ll 2mc^2$) were made to obtain (240) and (241) which hold for light elements but are not valid for heavier ones. More exact calculations by Breit (B22) and Racah (R5) which do not use these approximations show that (244) or (245) should be multiplied by a factor

$$\kappa(j, Z) = 4j(j+1)(j+\tfrac{1}{2})/(4\rho^2-1)\rho \quad (246)$$

where $$\rho = [(j+\tfrac{1}{2})^2 - Z^2\alpha^2]^{\frac{1}{2}}.$$

This factor becomes important (~ 1.20) for $Z=40$ for $j=\frac{1}{2}$ and for $Z=80$ for $j=3/2$. This relativistic correction factor can be carried over directly to the approximations (244) and (245) using Z_i as determined before for Z above. The doublet separation appearing in (244) is subject to the same sort of correction. In this case it is found (B22, R5) that the doublet separation should be multiplied by a factor

$$\lambda(l, Z) = [2l(l+1)/Z^2\alpha^2][\{(l+1)^2 - Z^2\alpha^2\}^{\frac{1}{2}} - 1 - (l^2 - Z^2\alpha^2)^{\frac{1}{2}}]. \quad (247)$$

Using these correction factors (245) and (244) become

$$\mu' = \frac{aIn^{*3}j(j+1)(l+\frac{1}{2})}{R\alpha^2 Z_i(1+z)^2\kappa(l, Z_i)}1838, \quad (248)$$

$$\mu' = \frac{aIj(j+1)(l+\frac{1}{2})Z_i\lambda(l, Z_i)}{\Delta\nu\, l(l+1)\kappa(l, Z_i)}1838. \quad (249)$$

Here μ' is the nuclear magnetic moment in nuclear magnetons ($\mu_0/1838$) and $\Delta\nu$ is the observed doublet separation. Table XVI gives the values (G6) of κ and λ for several Z_i. If the doublet separation is large as it is for the heavy elements a further correction must be made (B24).

A comparison with (238) for example shows that for g (or μ) positive the hyperfine structure levels will be arranged with that state with smallest F as lowest. In such a case the levels are said to be regular, the a factor positive and μ positive. Similarly for the case where the state with largest F is lowest the levels are said to be inverted, the a factor negative and μ negative.

TABLE XVI.

Z_i	κ $j=1/2$	κ $j=3/2$	λ $l=1$	Z_i	κ $j=1/2$	κ $j=3/2$	λ $l=1$
10	1.01	1.00	1.00	70	1.78	1.10	1.12
20	1.04	1.01	1.00	80	2.25	1.15	1.17
30	1.09	1.02	1.01	85	2.61	1.17	1.20
40	1.18	1.03	1.03	90	3.10	1.20	1.24
50	1.30	1.05	1.05	92	3.36	1.21	1.27
60	1.49	1.07	1.08				

Table XVI gives the correction factor κ for the hyperfine separation and the correction factor λ for the doublet separation for various values of Z_i.

If the atom considered has more than one valence electron the above relations cannot be applied directly. It is frequently the case that the interaction is due to the presence of one penetrating s electron in the group of valence electrons. In such a case the separation for a given state can be obtained simply in terms of the separation constant of the s electron (G7). For many configurations all of the valence electrons will have a considerable interaction with the nucleus. It is possible in such cases to find relations which give the hyperfine structure size in terms of the separation constants of the various electrons involved.[97] Having thus determined the one-electron separation factors, the nuclear magnetic moments can be found from (248) and (249). If these equations are to yield

[96] It has been pointed out by Fermi and Segrè that (255) should be multiplied by a factor $(1-ds/dn)$ where $n-s=n^*$. This factor would increase the values of μ calculated from the low s states of Na I, Cs I and Tl III by 2 percent, 5 percent and 10 percent, respectively. This factor is certainly negligible for light elements and for heavy elements is probably smaller than errors due to perturbation effects.

[97] The hyperfine structure for several valence electrons is discussed in the following papers: Goudsmit (G5); Guttinger and Pauli (G16); Racah (R5); Breit and Wills (B28) (sp, sd, sf, p^2, $p \cdot p$, p^3, p^2s); Crawford (C18) (d^2s); Crawford and Wills (C22) (p^3s). The more recent work of Breit, Crawford and Wills give the separations for the indicated electron configuration in intermediate coupling as well, in terms of the one electron separation factors.

constant values of μ from different atomic states, we see that the hyperfine separation must differ greatly for the various states. This is indeed found to be the case.

It is well known that perturbations between states are very prevalent in atomic spectra; i.e., the wave function representing a particular atomic state contains the pure wave function not only of that state but of others which are thus said to perturb it. These perturbations frequently become very large when the states involved are close together, since the real wave function can be written

$$\psi_0' = \psi_0 + \sum_i \psi_i V_{0i}/(E_0 - E_i), \qquad (250)$$

where V_{0i} is the matrix element of the electrostatic energy between the two states. The factor $(E_0 - E_i)^{-1}$ insures that the correction to the wave function usually will not be great. A case of very small perturbation would be if $V_{01} = 1$ ev and $(E_0 - E_1) = 10$ ev. The perturbing state might well be above the ionization potential with such a value of $E_0 - E_1$. ψ_0' would contain $0.1\psi_1$, and the ψ_1^2 would make only a 1-percent contribution[98] in $\psi_0'^2$. It has been pointed out by Fermi and Segrè (F11) that even in such unfavorable cases the hyperfine structure perturbation may be large. Suppose that the hyperfine structure of the perturbing state is 50 times that of state 0, which might well be the case if the former had an unpaired s electron and the latter did not. The hyperfine structure for the latter state would be increased by $50 \cdot 1$ percent due to the perturbation and it would thus be half again as large as without the perturbation. It is very difficult to calculate such perturbations exactly and since they may have a great influence in the determination of μ it is desirable to avoid using states in the determination of μ which are subject to great change. States with small hyperfine structure will have the greatest percentage change. It is therefore desirable in the determination of μ to use states which have large hyperfine structure and are not subject to violent perturbation. States which perturb each other are those with the same value of J and the same parity. For strict L–S coupling there is the additional restriction that the states must have the same resultant L and S.

[98] In some cases it is also possible to have cross terms contributing. If the two electron configurations for ψ_0 and ψ_1 differ in only a single electron and are furthermore of the same parity, there will be a term $\int \psi_0 H' \psi_1 d\tau$ if H' is the hyperfine interaction term. The most frequent perturbations are between configurations which differ in two electrons but this effect might be expected between ps and pd for example.

§47. Methods Used to Determine the Nuclear Angular Momentum and the Hyperfine Structure Separations

I. Direct observation of the hyperfine structure

There are a considerable number of methods which are used to determine the nuclear spin and the hyperfine separations. For the greater part the methods are best applicable to different cases; i.e., they are very largely supplementary in scope. The greatest amount of information has come from the direct study of the hyperfine structure of the spectral lines. It is possible from such a study to get information from both normal and excited states and for almost any atom whose spectrum can be excited. Information about excited states is important in order to obtain independent determinations of μ. The main limitations are those arising from the complexity of the patterns and the smallness of the separations.

A. Number of components. If each of the two states between which a radiative transition takes place, has hyperfine structure then the resulting spectral line will show structure. This line structure will be more complex than the state structure because the selection rules for the hyperfine quantum number F may be shown to be the same as for the total electronic angular momentum J, namely, $F \rightarrow F$, $F \pm 1$, $0 \rightarrow 0$ being forbidden. It is thus not infrequent for line patterns to consist of fifteen or twenty components which, because of their close spacing, overtax the present possibilities of resolution. There are many cases, however, where the experimental possibilities are sufficient. If the hyperfine levels of one state are very close and are considerably larger for the other state, the resulting pattern will show essentially only the larger separations. Because of the interval rule such a line has a characteristic "flag" appearance, i.e., the separations decrease uniformly across the pattern. It has been pointed out in §46 that the number of hyperfine states into which an atomic level is split by the presence of a nuclear moment is $2I+1$ for $J>I$ or $2J+1$ for $I>J$. It is apparent that the complete analysis of the hyperfine structure for any state which is split into a number of components less than $2J+1$ is a conclusive determination of I. For this purpose the presence of flag-type patterns is very useful. For small values of the nuclear spin the hyperfine structure of the spectral line is very frequently resolved completely. Exact values of the hyperfine structure separations will come for the most part from those patterns which can be completely resolved.

B. Relative separations. It is frequently the case in simple spectra of the one-electron type that no state of sufficiently large J can be found which has any appreciable hyperfine structure. In such a case the interval rule (233) can be used to determine the F values for the hyperfine states and thus the value of the nuclear spin I if J is known. The relative separations are not sensitive to I if I is large so that in such a case extremely accurate values of the hyperfine separations must be used. A determination of I by the use of the interval rule relies absolutely on its validity and therefore upon the cosine law of interaction. In view of deviations from this law which have been found recently and which seem to be due to the presence of an electric quadrupole moment, it can be considered as safe to use the interval rule to determine I only for those cases which would show no quadrupole effects (B15) namely, those states involving only s or $p_{\frac{1}{2}}$ electrons. Since the interval rule cannot be applied to a state with $J=\frac{1}{2}$, it can now be used with safety only for certain states involving more than a single electron.

C. Relative intensities. The relative intensities of the components of a hyperfine structure multiplet have been shown to obey the intensity relations (H14) which hold for multiplets in Russell-Saunders coupling when the quantum numbers L, J and S are replaced by J, F, and I, respectively. With J known for both initial and final state it is possible from an accurate knowledge of the relative intensities to deduce I since the F value can be written in terms of I and J. This method is very useful where the spin cannot be determined directly from the number of components.

D. Zeeman effect. There is one further method for determination of the nuclear spin from the direct observation of the hyperfine structure and that comes in the study of the Zeeman effect. The Zeeman effect of an atomic level leads to a displacement $M_J g(J)\mu_0 H$ where H is the magnetic field strength, M_J a magnetic quantum number for the state J and $Jg(J)\mu_0$ is the magnetic moment of the atom for the state J. If a nuclear moment is included two additional terms must be added to the energy (B3).

$$W_1 = M_J g(J)\mu_0 H + M_I g(I)\mu_0 H + AM_I M_J.$$

M_I refers to one of the magnetic substates of I and, $g(I)\mu_0 I$ is the nuclear magnetic moment and hence the second term is negligible compared to the first. The third term represents the interaction between the nuclear magnetic moment and the outside electrons, A being the separation constants. The above expression for the energy holds only for strong fields, i.e., for those fields in which the Zeeman-effect separation is large compared to the hyperfine separations. The usual field strengths which are used in Zeeman effects give separations which generally fulfill this condition and we therefore usually have a Paschen-Back effect of the hyperfine structure multiplet. Due to the third term above, each M_J state is split into $2I+1$ (the possible number of values of M_I) "hyperfine" states. In a spectral line each transition $M_J \rightarrow M_J'$ will consist of $2I+1$ components since in the strong field case changes in M_I will not be allowed. It is thus possible to determine the nuclear spin directly from the number of components in the Zeeman effect. The hyperfine Zeeman effect is also known experimentally and theoretically for field strengths which are not "strong."

II. Polarization of resonance radiation

If we have atoms in a weak magnetic field and excite them with their own radiation incident in the direction of the field, it is known that the radiation produced will show polarization effects. It is found that the percentage polarization changes with the strength of the magnetic field. This change with field strength comes about because the intensities of the Zeeman components may be shown for particular states to depend on H/A where H is the field strength and A is the hyperfine separation constant.. Different components of the radiation are thus emitted by atoms in the field with varying intensities and even if the components cannot be resolved this change in intensity is evident in the polarization of the emitted radiation. The polarization for a particular field also depends on the nuclear spin. The important application of this method as it has been developed by Ellet and Heydenburg (E2)(H12)(H13)(L3) is not in the determination of the nuclear spin but in the determination of very small hyperfine structure separations which cannot be studied directly. It has been possible by this method to determine hyperfine separations which are less than 10^{-3} cm^{-1}. The small separations of the excited states of Na and Cs have been obtained in this way.

III. Molecular and atomic beams

The principle of the Stern-Gerlach effect has been applied directly to a beam of hydrogen molecules by Stern Estermann and Frisch (F14)(E4). The beam of hydrogen molecules is passed through a strong magnetic field which has a steep gradient at right angles to the direction of the beam. The beam is separated into a number of components depending on the number of magnetic substates and the magnetic moment can be determined from the amount of the splitting. For hydrogen molecules there is, in addition to the magnetic moment which may come from the two nuclei, a magnetic moment arising from the molecular rotation. For parahydrogen the two nuclei have their nuclear moments in opposite directions so that for pure parahydrogen any observed magnetic splitting is entirely of rotational origin. It is possible to determine the rotational magnetic moment from observations on parahydrogen and then to make the necessary correction on the moment observed for orthohydrogen when ordinary hydrogen is used for the beam. The orthohydrogen will be split into three components (total spin 1) each of which will consist of three components due to the rotational moment. The central one of these coincides with the parahydrogen position but the magnetic moment can be calculated from the separation of the outer two components, corrections being made for the rotational moment. Measurements made by this method constitute the only direct determination of the nuclear magnetic moment.

A beam of atoms passed through a strong magnetic field with gradient perpendicular to the beam shows a separation into $(2J+1)$ components (Stern-Gerlach). If a nuclear moment is present each of these components consists of $2I+1$ individual components which for strong fields all fall together. Breit and Rabi (B26) have pointed out that as the field strength approaches zero these individual components no longer all fall together. In the case of the hydrogen atom $(I=\frac{1}{2})$ in its normal state there are four magnetic substates which for weak fields are all separate in the deflection pattern. Furthermore the spacing of the components changes as the field decreases until the two central ones come together at zero field. The deflection in the field will depend on the component of the magnetic moment in the field direction. For the two states with $M=M_I+M_S=0$ this component will depend on H/A, where H is the magnetic field strength and A is the hyperfine separation factor. The components of the magnetic moment in the field direction are

$$\mu_z(M_S,\, M_I)=\mu_z(\tfrac{1}{2},\, \tfrac{1}{2})=\mu_0;$$

$$\mu_z(\tfrac{1}{2},\, -\tfrac{1}{2})=\frac{x\mu_0}{(1+x^2)^{\frac{1}{2}}};\quad \mu_z(-\tfrac{1}{2},\, \tfrac{1}{2})=\frac{-x\mu_0}{(1+x^2)^{\frac{1}{2}}};$$

$$\mu_z(-\tfrac{1}{2},\, -\tfrac{1}{2})=-\mu_0,$$

where $x=\mu_0 H/\pi\hbar c\Delta\nu$. An accurate measurement of the deflection pattern, together with knowledge of the field gradient, allows the determination of $\mu_z(\frac{1}{2},\, -\frac{1}{2})$ and $\mu_z(-\frac{1}{2},\, \frac{1}{2})$ and hence a determination of $\Delta\nu$ the hyperfine separation. The nuclear spin is obtained directly from the number of components $(2I+1)$ which have the same M_S and which thus fall together in strong fields. The method of atomic beams has been extensively developed by Rabi and his co-workers (R2)(R3)(R1)(M13)(F13) and has recently been extended by these workers (R4)(K3) so that the regularity or inversion of the hyperfine structure can also be detected. This allows one to say whether the magnetic moment is positive or negative. It has been applied to hydrogen and the various alkalis, its limitations coming chiefly from the fact that a beam of atoms must be produced and detected. This method gives the nuclear spin and the hyperfine structure separation of the normal state only. It thus gives just the information which cannot be obtained from the study of the polarization of resonance radiation. The atomic beams method can be applied in cases where the hyperfine structure separation is too small to be measured directly. It has been applied successfully to H, D, Li, Na, K, Rb and Cs.

IV. Band spectra

The presence of a nuclear spin causes a change in the statistical weight associated with a given rotational state of a homonuclear diatomic molecule and thus causes a change in the expected intensities found in the band spectrum. If there were no nuclear spin, the statistical weight of any state for which the total angular momentum is J, would be $2J+1$. The presence of a nuclear spin further increases the degeneracy and changes the statistical weight of the states. It is found that the states whose wave functions are symmetrical in the position coordinates of the two nuclei are not affected in the same way as those whose wave functions are antisymmetrical. Let us suppose that a wave function symmetrical in the position coordinates is multiplied by a weight factor g_s while one antisymmetrical is multiplied by a factor g_a. We know (see paragraph 4) that a given molecule will have only those states which have wave functions which are totally antisymmetric (symmetric) if the nuclei obey the Fermi (Bose) statistics. If the particular nucleus has even atomic weight it will follow the Bose statistics, for odd atomic weight the Fermi statistics. For nuclear spin zero therefore we see that either g_s or g_a must be zero since the symmetry is entirely determined by the position coordinates.

If I is the nuclear spin of each of the two nuclei there will be $2I+1$ spin orientations and thus $(2I+1)^2$ spin functions representing the components of the two nuclear spins. Of these, $(2I+1)$ have the same component for both spins and will thus be symmetrical. For the remaining $2I(2I+1)$ functions, half can be built up as symmetrical combinations and half as antisymmetrical combinations. There are thus $(I+1)(2I+1)$ symmetrical and $I(2I+1)$ antisymmetrical spin functions. For a molecule containing nuclei which obey the Fermi statistics we can make the total wave function antisymmetrical from either a symmetrical or antisymmetrical position function by making the spin function either antisymmetrical or symmetrical. Because of the unequal weighting of the states corresponding to these latter, those states for which the position functions are antisymmetrical will have the greater weight. The ratio will be

$$g_a/g_s=(I+1)/I.$$

For a molecule containing nuclei which obey the Bose statistics the total wave function can be made symmetrical from either a symmetrical or antisymmetrical position function with a spin function which is either symmetrical or antisymmetrical. This leads to a ratio of the weight factors

$$g_s/g_a=(I+1)/I.$$

Successive rotational states will show alternate symmetry in the position coordinates and thus will have different statistical weights. This leads to a band structure in which successive lines have an intensity ratio $(I+1)/I$, no matter whether the Fermi or the Bose statistics are obeyed. A determination of the positional symmetry characteristics for the rotational states allows one to decide the symmetry of those states of greater weight and hence whether the nuclei follow the Fermi or the Bose statistics. H^1, Li^7, Na^{23}, P^{31}, Cl^{35}, and K^{39} are found to obey the Fermi statistics while H^2, He^4, N^{14}, O^{16} and S^{32} are found to obey the Bose statistics.

The above method is very useful in the determination of nuclear spins, particularly for light nuclei. The main advantage of the band spectrum determination is that since it depends only on the nuclear angular momentum, it is possible to obtain the spin of nuclei even in case the spin is zero. Methods which determine the hyperfine structure separation are unable to distinguish between zero magnetic moment and zero spin. On the other hand, no information about the nuclear magnetic moment can be obtained by this method of alternating intensities. Nuclei whose angular momenta have been determined from band spectra include H^1, H^2, He^4, Li^7, C^{12}, N^{14}, O^{16}, F^{19}, Na^{23}, P^{31}, S^{32} and Cl^{35}.

V. Other methods

The specific heat of a gas will depend on the distribution of molecules over the various low states. At ordinary temperatures, kT is large compared to the distance between the rotational states and any possible weighting of these states is irrelevant. For very low temperatures this is not the case and the specific heat of hydrogen, for example is well known at temperatures where kT is not large. For the hydrogen molecule the alternate rotational states are symmetrical in the nuclei beginning with the lowest with $J=0$, while those with J odd are antisymmetrical. From above this means that since the nuclei are expected to obey the Fermi statistics, $g_s/g_a = I/(I+1)$. Under ordinary conditions there will be no transitions from the symmetrical to the antisymmetrical states so that if the gas were run down to a very low temperature the molecules would not concentrate in the lowest state. This lack of transitions means that hydrogen must be treated effectively as a mixture of two gases and the specific heat determined accordingly. It was only after this had been done by Dennison (D2), that it was possible to get an explanation of the behavior of the specific heat at low temperatures. Furthermore it was found necessary to weight the even and odd states in the ratio $\frac{1}{3}$ in order to fit the specific heat and hence, from above, $I=\frac{1}{2}$.

While there are no transitions between symmetrical and antisymmetrical states under ordinary conditions, it is well known that it is possible to separate parahydrogen and also orthodeuterium by adsorption of the hydrogen or deuterium on charcoal at the temperature of liquid hydrogen. These preparations can be made with considerable purity and are very stable. It is, however, possible to induce *para-ortho* transitions by the presence of an inhomogeneous magnetic field such as that due to the paramagnetic oxygen molecule. A. Farkas and L. Farkas (F1) have found that both parahydrogen and orthodeuterium will slowly reach the equilibrium condition in the presence of oxygen. It is possible to compare the rates of conversion of parahydrogen and orthodeuterium under suitable conditions and Kalckar and Teller (K1) have found that the relative speed of conversion depends only on the spins and magnetic moments and on the equilibrium concentrations. It is possible by measuring the relative speeds of conversion to make an accurate determination of the ratio of magnetic moments for hydrogen and deuterium. Farkas and Farkas find $\mu_P/\mu_D = 3.96 \pm 0.11$.

§48. Values of Nuclear Spins and Magnetic Moments

There are a considerable number of elements which have been investigated by one or more of the methods described in §46 and nuclear spins determined. These values of the nuclear spins are gathered in Tables XVII, XVIII and XIX. The first of these contains those nuclear spins which it is believed are known with certainty. Those known with somewhat less certainty are given in Table XVIII where a grade of A, B or C has been appended to indicate decreasing certainty. Such a division into groups of this sort is necessarily somewhat arbitrary because there are no sharp divisions in the degree of certainty of the nuclear spin. It is apparent from §47, however, that not all methods used to determine the nuclear spin are equally sure and it is because of this that the above division has been made. The

Table XVII. *Nuclear spins.*

Element	Source	Z	A	I	Element	Source	Z	A	I
H	*	1	1	1/2	Rb	*	37	85	5/2
H	*	1	2	1				87	3/2
He	B	2	4	0	Cd	*	48	111	1/2
Li	*	3	7	3/2				113	1/2
C	B	6	12	0	In	*	49	115	9/2
N	*	7	14	1	Sn	*	50	117	1/2
O	B	8	16	0				119	1/2
F	*	9	19	1/2	Sb	*	51	121	5/2
Na	*	11	23	3/2				123	7/2
Al	*	13	27	1/2	Cs	*	55	133	7/2
P	B	15	31	1/2	Pr	H(W5)	59	141	5/2
S	B	16	32	0	Eu	*	63	151	5/2
K	*	19	39	3/2				153	5/2
	*	19	41	3/2	Ta	H(M5)	73	181	7/2
Mn	*	25	55	5/2	Re	H(Z1)	75	185	5/2
Cu	*	29	63	3/2		(G14)(M12)		187	5/2
			65	3/2	Hg	*	80	199	1/2
Ga	*	31	69	3/2				201	3/2
			71	3/2	Tl	*	81	203	1/2
As	*	33	75	3/2				205	1/2
Br	*	35	79	3/2	Pb	*	82	207	1/2
			81	3/2	Bi	*	83	209	9/2

Table XVII contains those nuclear spins which are believed to be certain. In the "Source" column, B means determined from band spectra and H from hyperfine structure, while an asterisk indicates that it is discussed in the text.

TABLE XVIII. *Additional probable nuclear spins.*

ELEMENT	SOURCE	Z	A	I	ELEMENT	SOURCE	Z	A	I
Li	*	3	6	1C	Xe	*	54	129	1/2A
Cl	B	17	35	5/2C				131	3/2B
K	*	19	41	>1/2	Ba	*	56	135	5/2C
Sc	*	21	45	7/2A				137	5/2C
V	*	23	51	7/2B	La	*	57	139	7/2A
Co	*	27	59	7/2A	Tb	H(S6)	65	159	3/2B
Zn	*	30	67	3/2B	Ho	H(S14)	67	165	7/2A
Kr	*	36	83	9/2C	Tu	H(S13)	69	169	1/2B
Sr	*	38	87	≧3/2C	Lu	*	71	175	7/2B
Cb	*	41	93	9/2B	Hf	H(R8)	72	177	≦3/2C
Ag	*	47	107	3/2C				179	≦3/2C
			109	3/2C	Ir	H(S15)	77	191	1/2C
I	*	53	127	5/2A	Pt	H(V2)	78	195	1/2C
					Au	*	79	197	3/2C
					Pa	H(S5)	91	231	3/2A

Table XVIII contains additional nuclear spins which are considered probable. Decreasing probability is indicated by the letters A, B, C, with those marked A being nearly certain.

tables give in the second column an indication of the source of the information in some cases and in other cases simply an asterisk to indicate that the particular element is discussed briefly below.

The nuclear spins are in every case assigned to particular isotopes. In general this can be done simply, because the element is either single or consists of only two isotopes whose relative abundances are well known. In such cases there is no confusion in assigning the nuclear spins. In other cases where there are a large number of isotopes some further interpretation is necessary. It is first noticed that the majority of the isotopes represented in the tables of spins have odd mass numbers and that all of these isotopes with odd mass numbers show half-integer spins. The isotopes with even mass numbers show integer spins, mostly zero. Furthermore it should be stated that no measurable hyperfine structure has been found for any isotope with even mass number A, and even atomic number Z (meaning either small magnetic moment or zero spin). Now the cases mentioned above for which there exist a considerable number of isotopes have all been studied by the direct observation of the hyperfine structure. In the interpretation of the resulting patterns it has been assumed that the even isotopes show no structure. This assumption is very reasonable in view of the above regularity and indeed has a striking confirmation in the cases of Pb and Hg. For the first of these Kopfermann (K7) has been able to designate the isotopes by using uranium and thorium lead. For Hg the evidence that the strong central component is formed by the even isotopes which do not have hyperfine structure is already well indicated by the intensities but is confirmed by the presence of one line for which the central

TABLE XIX. *Nuclear magnetic moments.*

ELEMENT	Z	A	I	μ	CLASS	OBS. RATIO
H	1	1	1/2	2.9	I	3.96
		2	1	0.85	III	
Li	3	6	1C	∼0.8	III	
		7	3/2	3.2	I	
N	7	14	1	∼0.2	III	
F	9	19	1/2	3	I	
Na	11	23	3/2	2.0	I	
Al	13	27	1/2	2.2	I	
K	19	39	3/2	0.40	I	1.81
		41	3/2	±0.22	I	
Sc	21	45	7/2A	3.6	I	
Cu	29	63	3/2	2.5	I	1.00
		65	3/2	2.5	I	
Zn	30	67	3/2B	−1.7	II	
Ga	31	69	3/2	2.1	I	0.79
		71	3/2	2.7	I	
As	33	75	3/2	1.5	I	
Kr	36	83	9/2C	−1	II	
Rb	37	85	5/2	1.4	I	0.494
		87	3/2	2.8	I	
Sr	38	87	≧3/2C	∼−0.8	II	
Ag	47	107	3/2C	0.2	I	
		109	3/2C	0.2	I	

ELEMENT	Z	A	I	μ	CLASS	OBS. RATIO
Cd	48	111	1/2	−0.65	II	1.00
		113	1/2	−0.65	II	
In	49	115	9/2	5.7	I	
Sn	50	117	1/2	−0.89	II	1.00
		119	1/2	−0.89	II	
Sb	51	121	5/2	3.7	I	1.32
		123	7/2	2.8	I	
Xe	54	129	1/2A	−0.9	II	−1.11
		131	3/2B	0.8	II	
Cs	55	133	7/2	2.5	I	
Ba	56	135	5/2C	1.0	II	
		137	5/2C	1.0	II	
La	57	139	7/2A	2.8	I	
Eu	63	151	5/2		I	2.2
		153	5/2		I	
Au	79	197	3/2C	0.3	I	
Hg	80	199	1/2	0.5	II	−0.90
		201	3/2	−0.6	II	
Tl	81	203	1/2	1.4	I	0.98
		205	1/2	1.4	I	
Pb	82	207	1/2	0.6	II	
Bi	83	209	9/2	4.0	I	

Table XIX contains the magnetic moments (μ) of the various elements in units $eh/2Mc$. A number given in the ratio column is the ratio of the magnetic moment of the isotope with smaller A (mass number) to that with larger A. All elements given here are discussed briefly in the text.

components are absent (G8). This line is ordinarily forbidden but arises because two atomic states are not separated by an amount large compared to the hyperfine structure. The "forbidden" line should thus occur only for those isotopes possessing structure. The other cases in which a number of even isotopes exist in addition to the odd ones mentioned here are Zn, Kr, Sr, Cd, Sn, Xe, Ba and Hf. These even isotopes either have spin zero or very small magnetic moments.

Most of the known nuclear spins are of isotopes with Z and A both odd (class I) and these spins are half-integer. There are fewer with Z even and A odd (class II) and these are also half-integer. There are only three (this type of nucleus is very rare) with Z odd and A even (class III) and they have spin unity. There are four with Z and A both even (class IV) and these spins are zero.

The nuclear magnetic moments are gathered in Table XIX. These are mostly derived from the hyperfine structure separation by the use of Goudsmit's equations (248) and (249). The magnetic moment is given in the column headed μ and is in units $\mu_0/1838$. All elements for which magnetic moments are given, are discussed briefly in the text. The nuclear magnetic moments roughly follow certain regularities. In general those nuclei of class I have magnetic moments which are large (>1) and positive. The nuclei of class II all seem to have small magnetic moments and are mostly negative. The three of class III are all small, that of H^2 being positive. No magnetic moments are known for elements of class IV, since none of them show any hyperfine structure. As mentioned before this may mean zero angular momentum or small magnetic moment. Since the few nuclei of this type for which spins are known all have $I=0$, perhaps the former possibility is the more probable.

In view of the approximations which are necessary in order to determine nuclear magnetic moments, there is one sort of information which is of great interest because it does not depend on these approximations. In cases where there are two isotopes the ratio of the magnetic moments can be determined directly from the hyperfine separations and the values of the spins by making use of the fact that wave functions of the two isotopes are the same in the neighborhood of the nucleus. There are twelve such ratios which are known and they are listed in Table XIX, giving the ratio of the magnetic moment of the isotope with smaller A to that of the isotope with larger A. Since they should have an accuracy which is limited only by the accuracy of the hyperfine measurements these ratios constitute the most accurate information about nuclear magnetic moments.

There is really no satisfactory way of ascertaining the errors in the values of the nuclear magnetic moments themselves. Calculations using wave functions have been found to give widely varying results in the case of Na (see p. 209). The better the wave functions were corrected, the more nearly did the values of the magnetic moment obtained agree with that obtained by the use of the Goudsmit-Fermi-Segrè approximate formulas. Probably the best indication of the accuracy comes from the consistency of the values obtained from different atomic states, i.e., by using different individual electron hyperfine separation constants. In determining the magnetic moment care must be exercised not to use any atomic states which are subject to large perturbations (see p. 211). It is nearly impossible to avoid such perturbations in atoms which have complicated configurations of valence electrons, though it is sometimes possible to find cases where they are not important. In general the most accurate information will come from atoms or ions which have simple valence configurations. There are numerous nuclei whose external electron structure is too complicated to allow a determination of the nuclear moments at present (for example Eu).

H

The nuclei H^1 and H^2 are known to have spin $\frac{1}{2}$ and 1, respectively. These values have been determined from band spectra (H16)(K2)(M19) and in the first case also by the atomic beams method (unpublished) and from the specific heat at low temperatures (D2). The value of the magnetic moment ($\mu=2.9$) of H^1 is that determined by Rabi, Kellogg and Zacharias from the determination of the hyperfine structure separation of the normal state (unpublished and (R2)(R4)). The magnetic moment of H^2 is also the value given by Rabi, Kellogg and Zacharias

(R3). For hydrogen the calculations of the magnetic moment use the exact wave functions of course; the errors arise from the difficulties of measurement. The value 0.7 for H^2 is in agreement with the ratio $\mu(H^1)/\mu(H^2)=3.96$ determined by Farkas and Farkas (F1) which is more accurately known than either of the magnetic moments.*

Li

The work of Fox and Rabi (F13) using the atomic beam method indicates that Li^6 has a spin $\geqq 1$. It seems probable that the spin is 1 (see §36). The ratio of the magnetic moments can be determined directly and gives $\mu(Li^6)/\mu(Li^7)\leqq 0.25$. This means that the magnetic moment of the Li^6 is about the same as for H^2. For Li^7 the nuclear spin has been determined from band spectra by Harvey and Jenkins (H5), from hyperfine structure by the work of Guttinger (G15), Schuler (S2); Guttinger and Pauli (G16) and Granath (G10) and by atomic beams by Fox and Rabi (F13). The magnetic moment has been determined from the hyperfine structure separations of Li II $1s2s\,^3S_1$ (G10) and from the hyperfine separation of the ground state of Li I (F13). The two values thus obtained are practically identical.

N

The N^{14} nucleus is known to have a spin of 1 from the alternations in intensity in band spectra as observed by Ornstein and Van Wijk (O2). From the study of lines in the N I spectrum expected to have large hyperfine structure by reason of the large nuclear interaction, it has been concluded that the magnetic moment must be in the neighborhood of 0.2 or smaller (B1). Because of the importance of the magnetic moments of light elements it would be very desirable to have a measurement on the normal state by the atomic beams method. The nitrogen nucleus follows the Bose statistics (H11)(R7).

F

The fluorine nucleus has a spin of $\frac{1}{2}$, determined from band spectra by Gale and Monk (G1) and confirmed from hyperfine structure measurements by Campbell (C2). The nuclear magnetic moment has been obtained from these latter measurements by Brown and Bartlett (B30). They have carried out calculations using Hartree functions and have obtained values which are in the neighborhood of 3. There is quite a variation in the values obtained from different levels, at least part of which is due to the fact that the hyperfine structure patterns are not completely resolved.

Na

The nuclear spin of Na^{23} has been determined from atomic beam deflection (R1), from band spectra (J5), from hyperfine structure intensities (G12), and from the polarization of resonance radiation (E2)(L3), and the values thus found are all $I=3/2$. The hyperfine structure separations of several states have been measured (R1)(E2)(L3)(G12)(J3)(F13) and the magnetic moments have been calculated (W15)(S21) by using the G.F.S. relations and with various sorts of wave functions (see p. 209). The value given in these tables is that obtained from the G.F.S. relations for the states $3s\,^2S_{1/2}$ and $4p\,^2P_{3/2}$. The values for $3p\,^2P_{3/2}$ are between 2.25 and 2.6 depending on the value of the separation used.

Al

The nuclear spin $I=\frac{1}{2}$ has been determined by Ritschl (R11) from observations on the number of components in the hyperfine structure. The measurements give only a single one-electron separation factor $a(3s)$ since the separation observed in the Al II and Al I terms are attributed to this electron (R11)(P3)(G8). The value of the magnetic moment is determined from this single separation. Brown and Cook (B31) have obtained magnetic moment 2.4 from the same separation by using Hartree functions.

K

The nuclear spin of K^{39} has been determined by the atomic beam method by Millman (M13) and is 3/2. The magnetic moment is determined from the separation of the normal state $4s\,^2S_{1/2}$ which has been accurately determined by the deflection method (F13) as well as by the hyperfine structure measurements of Jackson and

* *Note added to proof:* More recently Rabi, Kellogg and Zacharias have found that $\mu(H^2)=0.85\pm0.03$. This is not in agreement with the ratio determined by Farkas and Farkas and this latter is therefore subject to doubt.

Kuhn (J3)(J4). Recent measurements (Rabi and co-workers, unpublished) with atomic beams indicate that the hyperfine structure of the normal state of K^{39} is regular and the magnetic moment therefore positive. This result is in disagreement with the observed intensities (J4) of the hyperfine structure, but this may be due to reversal in the hyperfine lines. The value -0.40 for the magnetic moment is given from the G.F.S. relation while Gibbons and Bartlett (G4) get $\mu=1.2$ using Hartree functions. This is a wide discrepancy but in view of similar difficulties with Hartree functions in Na, the former value is taken. The nuclear spin of K^{41} has been found by Manley (unpublished) to be 3/2 using atomic beams. The magnetic moment determined from the separation of the normal state is ± 0.22.

Sc

Schuler and Schmidt (S11) and Kopfermann and Rasmussen (K10) have found the nuclear spin of Sc^{45} to be 7/2, and the first workers have determined it from the number of components. The magnetic moment has been determined by the second workers from the $ds^2\,{}^2D_{5/2}$ and ${}^2D_{3/2}$ separations. Since these separations are not resolved directly but are inferred from an unresolved pattern and since it is doubtful what the correct value of Z_i should be (they use $Z_i=8$), the resulting magnetic moment is very approximate. The two separations give the same μ, however.

V

The hyperfine structure of the V I spectrum has been investigated by Kopfermann and Rasmussen (K12) who assign a nuclear spin of 7/2. Because of the extreme complexity of the unresolved patterns this value cannot be considered as certain. The interactions are too complicated to determine a magnetic moment from the observed separations.

Mn

The nuclear spin of 5/2 for Mn^{55} has been determined by White and Ritschl (W7). The complexity of the electron configurations do not allow a determination of a value for μ.

Co

From the work of More (M16) and Kopfermann and Rasmussen (K11) the nuclear spin 7/2 of Co^{59} is practically certain. No value of the magnetic moment can be obtained from the measured separations because of the complexity of the electron configurations.

Cu

The hyperfine structure of Cu I has been studied by Ritschl (R10) who found the spins of both Cu^{63} and Cu^{65} to be 3/2. Though the lines in the resultant patterns are not coincident for the two isotopes they both have the same spin and the same hyperfine separations. This means that they both have the same magnetic moment. The magnetic moment 2.5 is determined from the $d^{10}s\,{}^2S_{1/2}$ and the $d^9s^2\,{}^2D_{3/2,\,5/2}$. These give, respectively, 2.5, 2.5 and 2.1 with the value $Z_i=19.6$ determined by Fermi and Segrè for the d electron. These terms are selected because they are expected to show the smallest perturbation effects from terms with much larger structure. Perturbation effects may, however, be present for both of these terms and this adds to the uncertainty of the magnetic moment. The approximate agreement of the values given above indicates that these perturbations are probably not serious.

Zn

Hyperfine structure has been found by Schuler and Westmeyer (S19) for Zn II. The observed components are very weak compared to the strong lines assigned to the even isotopes in accordance with expectations for Zn^{67}. They conclude that the spin is 3/2 though this conclusion is not certain because the pattern is not completely resolved. By using the observed separation for $d^9s^2\,{}^2D_{5/2}$ the magnetic moment would be about -1.7. This value has no other confirmation.

Ga

The nuclear spins of Ga^{69} and Ga^{71} have been determined by Jackson (J1) and Campbell (C1) from a study of the hyperfine structure. The two nuclei have the same spin $I=3/2$ but different magnetic moments as found by Campbell. The ratio $\mu^{69}/\mu^{71}=0.79$ is quite exact. The individual magnetic moments are determined from the separation constants for the $5s\ ^2S_{1/2}$ and $4p\ ^2P_{1/2}$ states in Ga I and from the separation constant $a(4s)=0.43\ \text{cm}^{-1}$ as determined from numerous separations in Ga II. These give 2.24, 2.36 and 2.07 for μ^{69}, respectively. Perhaps none of these values is very accurate.

As

The nuclear spin of As^{75} has been determined by Tolansky (T4) and by Crawford and Crooker (C20) from interval measurements on the hyperfine structure. Interval measurements may be influenced by the presence of an electric quadrupole moment for the nucleus. Crawford and Crooker's determination on the intervals of $4s5s\ ^3S_1$ of As IV is free from this possibility, however. The magnetic moment $\mu=1.5$ is calculated from the separation factor $a(4s)$ determined from the As IV measurements. The separations in As II are subject to large perturbations so cannot be used directly to determine μ.

Br

The nuclear spins of Br^{79} and Br^{81} seem to be quite certainly 3/2 from the work of Tolansky (T3). Though none of the spectral lines investigated were completely resolved, the appearance of almost identical structures having four components for several lines having $4p^45s\ ^4P_{5/2}$ as a final state indicated $I=3/2$. Both isotopes have the same abundance and the structures are superimposed. There are several separations known but the interaction with the nucleus is through all the five electrons and perturbations are also very probable. No magnetic moment is determined.

Kr

Krypton has a number of even isotopes and one odd one, Kr^{83}. The spectral lines have been studied by Kopfermann and Wieth-Knudsen (K13) and are found to have a very strong central component corresponding to the even isotopes and a weak structure attributed to Kr^{83}. It is possible to conclude that $I>\frac{1}{2}$ and the value 9/2 makes the weighted center of the fine structure coincide with the even isotopes. This value is uncertain. By using Goudsmit's sum relations (G5) it is possible to obtain a value for μ from $a(5s)$. This value is negative and roughly unity.

Rb

The spins of Rb^{85} and Rb^{87} have been determined by Kopfermann (K8) from the hyperfine structure of the Rb II lines and also by Fox and Millman (unpublished) using the atomic beams method. Both methods give $I(85)=5/2$ and $I(87)=3/2$. The magnetic moments have been determined from the separations for $5s\ ^2S$, the normal state of Rb I. The ratio of the magnetic moments (0.494) is believed to be quite accurate ($\sim$1 percent).

Sr

Strontium is known to have hyperfine structure (M18)(S17)(S3)(W4) and it is attributed to Sr^{87}. The spin is very uncertain but using the value $I=3/2$ a value for the magnetic moment can be obtained from the separation of the normal state of Sr II. ($\Delta\nu=-0.15\ \text{cm}^{-1}$.)

Cb

The spin of Cb^{93} has been determined by Ballard (B7) from the hyperfine structure. The lines are not completely resolved but a careful study has yielded the value $I=9/2$. The value of the magnetic moment (B7), $\mu=3.7$ is very rough at best.

Ag

A doubling of the resonance lines of silver has been observed by Hill (H15). Because of the intensity ratio of the components it is concluded that the structure is hyperfine structure and not isotope shift, since the two isotopes have nearly the same abundances and the observed structure shows a weak and a stronger component. A tentative value of 3/2 is suggested for the spins of both isotopes. If this is correct then the two isotopes would have the same magnetic moments. They would be about +0.2.

Cd

Cadmium has six isotopes, 4 even and 2 odd whose mass numbers are 111 and 113. The lines of the spectrum have been studied by Schuler and co-workers (S4)(S9) and are found to have a strong central component attributed to the even isotopes and a simple hyperfine structure attributed to 111 and 113 together. The spins of the two odd isotopes are certain under these assumptions which are made almost mandatory by the detection of isotope shift (S20) in Cd II. The magnetic moments for the odd isotopes are determined from the separation factor $a(5s)$ for the $5s$ electron, which gives rise to most of the structure in Cd I, and that of the $6s$ electron (J6). These give $\mu=-0.66$ and -0.63, respectively, for both isotopes.

In

The nuclear spin of In^{115} has been determined by Jackson (J2) and Paschen and Campbell (P1)(P2). Though no lines have been found where the large spin could be determined directly from the number of components the accurate measurement of the separations of $5s6s\ {}^3S_1$ which would not be expected to show any quadrupole effect, makes $I=9/2$ quite certain. The unusually large magnetic moment is determined from the separations observed in In I, $a(6s)=0.056$ cm^{-1} and $a(6p_{\frac{1}{2}})=0.076$ cm^{-1} and deduced from In II, $a(5s)=0.70$ cm^{-1}. These give $\mu=5.2$, 5.9 and 6.0, respectively, for the magnetic moment.

Sn

Tin has a large number of isotopes but only two of these Sn^{117} and Sn^{119} are odd and present in any considerable amount. A study of the hyperfine structure by Tolansky (T5) and by Schuler and Westmeyer (S18) shows that the lines can be interpreted in a manner similar to Cd, assigning the weak structure to the odd isotopes. If this structure is due to both odd isotopes then it is quite certain that the spin of each is 1/2. From the hyperfine separations of $6s\ {}^2S_{1/2}$ and $6p\ {}^2P_{1/2}$ in Sn II the magnetic moments are $\mu=-0.90$, -0.87, respectively.

Sb

The nuclear spins of the two antimony isotopes Sb^{121} and Sb^{123} have been determined by the work of Crawford and Bateson (C19) Badami (B4) and Tolansky (T6). Crawford and Bateson have completely resolved the line $5s6p\ {}^3P_0 \rightarrow 5s6s\ {}^3S_1$ of Sb IV and interval measurements lead uniquely to the spins $I(121)=5/2$ and $I(123)=7/2$. The interval rule is expected to hold very well here as a possible quadrupole moment would show no effect for these states. The magnetic moments are determined from the separation factor $a(5s)=1.4$ cm^{-1} for Sb^{121} determined from these measurements. A determination of the same $a(5s)$ from the sp^3 configuration of Sb II leads to a lower value of $a(5s)=1.04$ but this might be expected since this configuration is undoubtedly perturbed and its hyperfine structure made smaller thereby. By using the former value, $\mu(121)=3.7$. The ratio $\mu(121)/\mu(123)=1.32$ can be determined directly from the observed separations.

I

The hyperfine structure of the iodine spectrum has been studied by Tolansky (T7)(T8)(T9) who has observed a large number of lines. Interval measurements on lines showing structure predominantly from one level indicate that the spin is very probably 5/2. The large number of possible energy states and the existence of large perturbations do not permit an evaluation of the magnetic moment of I^{127} at present.

Xe

There are a considerable number of Xe isotopes but only two, Xe^{129} and Xe^{131} with odd mass numbers. Kopfermann and Kindal (K9) and Jones (J7) find that the observed hyperfine structures can be accounted for by assigning nuclear moments to these odd isotopes. The spin of Xe^{129} is almost surely 1/2 and that of Xe^{131} probably 3/2. With the sum relations of Goudsmit (G5) it is possible to find a separation factor $a(6s)=-0.164$ cm^{-1} for Xe^{129}. This gives a rough value for the magnetic moment $\mu=-0.9$. The ratio of the moments is known from the observed separations, and the spins $\mu(129)/\mu(131)=-1.11$. In this case the two magnetic moments have opposite signs.

Cs

The nuclear spin of Cs^{133} has been determined from hyperfine structure by Kopfermann (K6) and by the method of atomic beams by Cohen (C11). The value $I=7/2$ is quite certain. Accurate measurements of the hyperfine separations of several states have been made by Granath and Stranathan (G11) and Heydenburg (H13).

The separation factors and magnetic moments obtained from them are as follows: $a(6^2S_{1/2}) = 0.0767$ cm^{-1}, $\mu = 2.70$; $a(6^2P_{1/2}) = 0.00925$ cm^{-1}, $\mu = 2.45$; $a(6^2P_{3/2}) = 0.00142$ cm^{-1}, $\mu = 2.37$; $a(7^2P_{1/2}) = 0.00329$ cm^{-1}, $\mu = 2.67$; $a(7^2P_{3/2}) = 0.000486$ cm^{-1}, $\mu = 2.48$. The values $Z_i = Z$ and $Z_i = Z-4$ have been used for s and p electrons as usual. Slightly different values of Z_i determined either from the doublet separation or from the observed ratio of the hyperfine separations from the two members of the doublet lead to small changes in the above values (G11)(H13).

Ba

Kruger, Gibbs and Williams (K14) have concluded from a study of the intensities of the Ba II hyperfine structure that the spin of Ba^{135} and Ba^{137} is probably 5/2. The presence of several isotopes makes this value rather uncertain. All the lines show strong central components corresponding to the even isotopes. Using the separations (K14)(R12) of the $6\,^2S$ and $6\,^2P_{1/2}$ states the magnetic moments are respectively $\mu = 1.06$ and 0.82 if $I = 5/2$.

La

The angular momentum of La^{139} is very probably $7/2\ \hbar$ from the hyperfine structure measurements of Anderson (A1)(A2). The magnetic moment has been determined by Crawford and Grace (C21) from the separations observed in La III. They find for the states $6s\,^2S$ and $6p\,^2P_{1/2}$ the values $\mu = 2.84$ and 2.87. From a study of the d^2s configuration Crawford (C18) gets for the separation factor a somewhat different value which gives $\mu = 2.5$.

Eu

The two europium isotopes Eu^{151} and Eu^{153} have been found by Schuler and Schmidt (S15) to have angular momentum $5/2\ \hbar$ for each type of nucleus. These values are established directly by the number of components observed. The individual magnetic moments cannot be found easily due to the complicated electronic structure, but their ratio is found directly. There is evidence in Eu that the interval rule is not followed equally well for both isotopes. This is a deviation which cannot be explained by a perturbation effect (see §50).

Lu(Cp)

From the relative intensities of the hyperfine components, Schuler and Schmidt (S16) find a value $7/2\ \hbar$ for the angular momentum of the Lu^{175} nucleus. Evidence of deviations from interval rule are also found.

Au

Hyperfine structure found for gold by Ritschl (R9) and Wulff (W16) indicates that Au^{197} has a nuclear spin which is probably 3/2. By using this value and the observed separation of the normal state, $\mu = 0.2$.

Hg

Mercury has several isotopes of even mass number and two odd ones Hg^{199} and Hg^{201} to which the observed hyperfine structure is attributed. These two isotopes have been found by Schuler, Keyston and Jones (S8)(S7) to have spins 1/2 and 3/2, respectively. Numerous lines of the Hg I spectrum have been investigated and the separations are well known. These separations, however, are due mainly to the $6s$ electron which has a large separation factor. The values of $a(6s)$ for Hg^{199} determined from various configurations do not agree particularly well (from $6s\,9s$, $a(6s) = 1.37$ neglecting $a(9s)$; from $6s\,6p$, $a(6s) = 1.15$). These give $\mu(199) = 0.52$, 0.43, respectively. The ratio of the magnetic moments gives $\mu(199)/\mu(201) = -0.90$.

Tl

Hyperfine structure has been found in the Tl I, II and III spectra (S10)(M3)(M2). All of these patterns show conclusively that $I = \frac{1}{2}$ for Tl^{203} and Tl^{205}. The separations in Tl II are due almost entirely to the $6s$ electron so that while it is possible to make several estimations of this separation constant this leads to only one determination of μ. By using (248) and (249) the following values of μ are obtained from the various separations: Tl III, $a(7s) = 1.37$ cm^{-1}, $\mu = 0.9$; $a(8s) = 0.606$ cm^{-1}, $\mu = 1.5$; $a(6p_{1/2}) = 1.21$ cm^{-1}, $\mu = 1.8$; $a(7p_{1/2}) = 0.375$, $\mu = 1.4$; Tl II $a(6s) = 5.8$ cm^{-1}, $\mu = 1.7$; Tl I $a(7s) = 0.402$ cm^{-1}, $\mu = 1.1$; $a(6p_{1/2}) = 0.710$ cm^{-1}, $\mu = 2.0$. These values vary rather widely and are probably not very trustworthy. Calculations by Breit (B24) and Wills (W14) give $\mu(6p_{1/2}) = 1.45$ and $\mu(7s)$

$=1.35$ from the Tl I separations. There are rather large perturbations in the thallium spectra which make some of the separations very irregular and those expected to be particularly bad are not included above. Though the components due to the two isotopes are usually not separated there is evidence (S10) that Tl^{205} may have slightly larger separations, though this small effect may be due to the presence of a quadrupole moment for one of the isotopes.

Pb

The lead isotopes, with $A=204$, 206, 207, 208, are all evident in the hyperfine structure. Kopfermann (K7) has shown, by using samples of uranium lead (206) and thorium lead (208), just which single components should be assigned to these isotopes. Pb^{207} shows a structure which is due to a nuclear spin $I=\frac{1}{2}$. The presence of perturbations makes the determination of μ somewhat unreliable. Using the separation (M4) of $6s\ 7s\ ^3S_1$ from Pb III and neglecting the value $a(7s)$ we find from (248), $\mu=0.64$. This value would be smaller had $a(7s)$ not been neglected. Using $a(6s)=1.89$ as determined by Rose (R13) from Pb II sp^2 we find $\mu=0.40$. This is probably too small due to perturbations. Breit and Wills (B28) find $\mu=0.75$ and 0.67 from considering the $6p^2$ and $6p\ 7s$ configurations of Pb I.

Bi

For Bi^{209}, $I=9/2$. This value has been determined from the study of the Zeeman effect for the hyperfine structure by Back and Goudsmit (B3). Hyperfine structure has been observed and studied in the spectra of Bi I–V. Of these Bi II and III may be expected to show large perturbation effects due to excitation of inner electrons. For Bi IV, there are also perturbations but it is possible from some of the unperturbed levels to find (M1), $a(6s)=2.3\ \text{cm}^{-1}$. This value is practically the same as that determined directly from $6s6p_{3/2}$ which should not be greatly perturbed. From Bi V the separation of the normal state (A3) gives $a(6s)=2.6\ \text{cm}^{-1}$. By using this value in (248), $\mu=3.5$. It is possible that the p^3 configuration of Bi I is not badly perturbed and we may obtain the separation constant, $a(6p_{1/2})=0.375\ \text{cm}^{-1}$. By using (249) this gives $\mu=3.5$. This agreement does not mean much since the second value should certainly be larger due to the screening of the other p electrons. Breit and Wills find $\mu=5.4$ from p^3 of Bi I. The value 3.5 given in Table XIX, probably is not very reliable.

§49. Isotope Shift in Atomic Spectra

The spectral lines which are due to the different isotopes of an element usually do not have the same wave-length and the energy states of the various isotopes must therefore be spaced differently. If this effect occurs alone, there are as many components of a given spectral line as there are isotopes and their intensities are proportional to the relative abundances. If we consider the energy states for a particular isotope and compare them to those of another, we cannot say from the isotope shift alone how the energy levels of one isotope are placed with respect to those of another; we can detect only differences in the separations of the energy levels. Accordingly we expect such an isotope effect to be detectable only by means which study the transitions between energy states.

In the preceding paragraphs the presence of several isotopes was a complication which made the interpretation of the hyperfine separations more difficult. We found that isotopes having even mass number A show no hyperfine separations except those few peculiar nuclei of class III ($H^2\ Li^6\ B^{10}\ N^{14}$). Those nuclei with A odd generally show hyperfine structure. For elements which have a number of isotopes we may expect isotope shift *alone* for those with even A and isotope shift plus hyperfine structure for those with odd A. In the first case we can determine the relative displacement of the levels directly from the observed shifts. In the second case it is necessary to allow for the hyperfine structure and find the position of the hypothetical level without hyperfine structure. Because of the regular spacing and known weighting of the levels, this can be done quite easily and it is thus possible to determine the isotope shift for all the isotopes whether they show hyperfine structure or not though the interpretation in case they do is much more involved.

In order to ascertain the cause of isotope shift it is perhaps best to examine first of all the effect of different mass for the different isotopes. It is well known that a correction to the energy must be made in a hydrogen-like atom if the nucleus has been regarded as fixed. It is also well known that the final energy depends upon the nuclear mass and that the H^1 and H^2 spectral lines are not coincident but are displaced from each other by an amount which is just that expected by their mass difference.

In the case of an atom with several electrons it is found that the shift in the energy levels consists of two parts (H17)(B11), one of which is just like the hydrogen case and is called the normal effect and a new part which is characteristic of the many electron problem and is called the specific effect. If the center of mass is regarded as fixed the Schrödinger equation for N electrons of mass m and a nucleus of mass M can be written

$$-\hbar^2\left[\frac{1}{2\mu}\sum_{\kappa=1}^{N}\nabla_\kappa^2+\frac{1}{M}\sum_{\kappa<i}\nabla_\kappa\cdot\nabla_i\right]+(V(x)-W)\psi=0, \quad (251)$$

where $\mu=\dfrac{mM}{M+m}$, $\nabla_\kappa=\mathbf{i}\dfrac{\partial}{\partial x_\kappa}+\mathbf{j}\dfrac{\partial}{\partial y_\kappa}+\mathbf{k}\dfrac{\partial}{\partial z_\kappa}$,

and x_κ, y_κ and z_κ are the coordinates of the κth electron with respect to the nucleus. Let us consider that the characteristic values $W(m)$ and the solutions $\psi(m)$ are known for the case of the stationary nucleus. In order to compare with these values let us neglect the second term in (251) momentarily. It may then be seen by introducing coordinates with respect to the center of mass that the energy W' is related to $W(m)$, the energy for a stationary nucleus simply by

$$W'=\alpha W(m), \quad \text{where} \quad \alpha=\mu/m. \quad (251a)$$

Next if we consider the second term in (251) as a perturbation, it gives a contribution to the energy

$$\Delta W=-\frac{\hbar^2}{M}\int\psi^*(m)[\sum_{\kappa<i}\nabla_\kappa\cdot\nabla_i]\psi(m)d\tau. \quad (251b)$$

We now have the difference between the energy in (251) and that for the problem of the stationary nucleus, separated into two parts (251a) the normal effect and (251b) the specific effect. In order to evaluate (251b) it is necessary to use wave functions for the particular atom in question. Hughes and Eckart (H17) carried this out for the case of Li and found agreement with experimental data of Li II and Li I. Bartlett and Gibbons (B11) have made the calculation for the case of Ne. In this case the agreement is not satisfactory since some of the lines show much larger specific shifts than expected.

The shift due to the specific effect (251b) is inversely proportional to the mass M or to A if we replace M by A the mass number of the particular isotope considered. We have then for the normal effect.

$$W'=\frac{A}{A+m}W(m)=\left(1-\frac{m}{A}+\cdots\right)W(m), \quad (252a)$$

$$W'-W(m)=-(m/A)W(m).$$

Since the specific effect is also inversely proportional to the mass number, the total energy displacement $D=W'-W(m)+\Delta W$ is proportional to $1/A$. If we consider an element with several isotopes with mass numbers A, $A+\delta_1$, $A+\delta_2$, etc., the relative displacement for these various isotopes can be written

$$D_0-D_1=C/A-C/(A+\delta_1)\cong C\delta_1/A^2,$$
$$D_0-D_2\cong C\delta_2/A^2. \quad (252b)$$

Since δ_1, δ_2, etc., are practically always small compared to A this is a good approximation. The relative displacement is thus proportional to the differences of the masses of the two isotopes. For several isotopes with successive mass numbers we expect the components due to the various isotopes to be equally spaced. This affords a means of examining the isotope shift without making the calculations in each case. The first element in the periodic table which has more than two isotopes all of which have sufficient abundance, is Mg. There are three isotopes with $A=24$, 25, and 26, present with relative abundances 7, 1, 1. According to (252b) we should expect any energy state to have relative displacements proportional to the mass differences and hence the same must be true of the relative shift in the spectral lines. Some of the spectral lines of Mg I do show isotope shift (B2) but instead of showing three equally spaced components, show only two components.* This means that the observed shift cannot be explained by mass effect alone. For other elements it has been found that the elements with even A are usually spaced approximately according to (252b) but the odd isotopes do not occupy the required positions.[99] Schuler and Schmidt (S12) have found in Sm that the even isotopes also do not seem to be regularly spaced. It must be concluded that mass effect alone is not sufficient to explain the observed shifts for any but the light elements.

If the isotope displacement of a state for which the outer electrons are seldom in the neighborhood of the nucleus (large l) is arbitrarily called zero, then it is found that states with electrons having small l and particularly with s electrons, have large isotope displacement. The amount of the displacement increases with the number of penetrating electrons and the degree of penetration. This means that perturbation effects are very important for isotope displacement as well as for hyperfine structure. A simple example (B2) of this is found in the case of Mg where most of the observed shifts are directly due to perturbation effects and are found to be quantitatively related to the amount of the perturbation. The importance of states showing penetration indicates that calculations (D3) particularly for heavy atoms, which assume Russell-Saunders coupling and neglect perturbations must be considered as unreliable.

The presence of large displacements where penetrating electrons are involved indicates that isotope shift may be due to some difference in the field in the neighborhood of the nucleus. It has been pointed out by Bartlett (B8) that such a difference is to be expected if one assumes constant nuclear density. The heavier isotopes have greater radii since $R\sim A^{\frac{1}{3}}$, and will bind a penetrating electron more loosely than the lighter nuclei. Calculations have been carried out by Breit and Rosenthal (B25)(R14) and Racah (R6) who find that the change in the energy of an s state due to a change in nuclear radius Δy_0 is

$$\delta W=\frac{Ra_0^2}{Z}R^2(0)\frac{1+s}{[\Gamma(2s+1)]^2}\left[y_0^{2s}+\frac{v}{2\gamma^2}y_0^{2s+1}\right]\frac{\Delta y_0}{y_0},$$

where $s=(1-\gamma^2)^{\frac{1}{2}}$; $y_0=\dfrac{2Zr_0}{a_0}$; $v=\dfrac{V}{mc^2}$; $\gamma=\dfrac{Z}{137}$.

r_0 is the nuclear radius in cm inside of which V, the potential energy is considered constant; a_0 is the Bohr radius and R the Rydberg constant; $R(0)$ is the value of the Schrödinger radial function at $r=0$ and it may be obtained in terms of the observed hyperfine separation from (240) if the magnetic moment has been previously determined from (248) or (249). Breit has found that, with $R^2(0)$ determined in this way, there is general agreement for Hg, Tl and Pb. Since it is expected that r_0 will be larger than normal for nuclei containing an odd neutron or proton, the shift for a nucleus with Z even and A odd will be more nearly like that of nucleus Z, $(A+1)$ than like Z, $(A-1)$. For heavy nuclei, however, it is known that the displacement for nucleus Z, A is more nearly like Z, $A-1$. It is possible that the extra neutron may interact with the electrons to cause this effect.[99] Detailed comparisons have not yet been made for other elements, but the light elements are in disagreement with the "radius" effect since for Ne, Mg, Cu and Zn it would be necessary to ascribe smaller radii to the nuclei of larger mass number. Although no entirely satisfactory explanation of the observed isotope shifts has been found, the variation for different atomic states indicates a difference in the fields of the various isotopes in the neighborhood of the nucleus.

§50. Quadrupole Moments

It has been found by Schuler and Schmidt (S15) that certain elements show deviations from the interval rule. Such deviations are to be expected when two atomic states are very close together but otherwise the interval rule should

* The possibility that Mg^{25} shows hyperfine structure can be eliminated because of the nature of the states involved.

[99] It was suggested by Breit and Condon at the New York meeting of the American Physical Society, February 1936, that a small neutron-electron interaction might, though insignificant elsewhere, show itself in the isotope shift. If the odd neutron which nuclei of even Z and odd A contain, (Mg^{25} for example) would show such interaction it might explain why these isotopes do not appear centrally between those of mass $A-1$ and $A+1$.

be obeyed if the interaction between the electrons and the nuclear moment follows a cosine law. The deviations found by Schuler and Schmidt (S15) for Eu and by Mintz and Granath (M14) for Bi cannot be explained by perturbation effects. In the first case the two isotopes Eu^{151} and Eu^{153} show different deviations from the interval rule and in the second case deviations are found for the $p^3\ {}^2D_{3/2}$ state of Bi which is not sufficiently close to any other state to allow an explanation on the basis of perturbations.

Schuler and Schmidt observed that the deviations could be satisfactorily accounted for by the presence of an interaction term which is proportional to the square of the cosine of the angle between the nuclear spin and the extranuclear angular momentum. Such an interaction, they pointed out, would be expected if the nucleus has an electric quadrupole moment. It is indeed reasonable that nuclei should have small electric quadrupole moments. Such a moment arises if the protons are not distributed over the nucleus on the average with spherical symmetry.

The calculation of the contribution to the energy due to the presence of an electric quadrupole moment for the nucleus has been carried out by Bethe (B15) and Casimir (C3). It is found that for the case of a single electron outside closed shells the change in the energy of a hyperfine state F associated with the atomic state $j=l-\frac{1}{2}$, can be written

$$W_F=-\tfrac{1}{2}\overline{R^2}e^2\overline{r^{-3}}\times\frac{3C(C+1)-4I(I+1)j(j+1)}{(2l+1)(2l+3)(2I-1)(2I+3)}, \quad (253)$$

where $C=F(F+1)-I(I+1)-j(j+1)$. $\overline{R^2}$ is a measure of the nuclear quadrupole moment since R is the coordinate of a nuclear proton with respect to the center of gravity of the nucleus. It is possible to determine $\overline{r^{-3}}$ from (241). It is also possible to replace $\overline{r^{-3}}$ by $\overline{Z(r)r^{-3}}/Z_i$ and to determine this quantity from the fine structure separation since $\overline{Z(r)r^{-3}}=\Delta\nu/R\alpha^2a_0^3(l+\frac{1}{2})$. To find the corresponding expression for W_F when $j=l+\frac{1}{2}$, (253) should be multiplied by $(2l+3)/(2l-1)$.

From (253) it is found that for $j=\frac{1}{2}$, W_F vanishes. There is, therefore, no quadrupole effect for s and $p_{1/2}$ electrons. Other states will be expected to show effects which are roughly proportional to the fine structure doublet separation so that large effects will be expected for low $p_{3/2}$ and d electrons for the heavy elements.

For the case of several electrons the situation is somewhat more complicated. In the case of Eu, Casimir found that the observed deviations could be accounted for by quadrupole moments $\overline{R^2}=5.1$, $2.4\cdot10^{-24}$ cm² for Eu^{151} and Eu^{153}, respectively. For Bi, Bethe found a quadrupole moment $\overline{R^2}=0.61\cdot10^{-24}$ cm². This moment for Bi is about the size which would be expected if a single proton were unsymmetrically distributed in the nucleus.

References

The following list of references is not intended to be a complete list It does contain those papers referred to explicitly in the text. Certain papers from which experimental facts have been used are not referred to here but will be given in Part B where the results of experiments on nuclear disintegrations will be presented.

The references are arranged alphabetically and are indicated by a letter and a number representing the position in the alphabetical list. The paragraph in which the reference appears is indicated.

A1. Anderson, O. E., Phys. Rev. **45**, 685 (1934). (§48)

A2. Anderson, O. E., Phys. Rev. **46**, 473 (1934). (§48)

A3. Arvidsson, Nature **126**, 565 (1931). (§48)

A4. Aston, *Mass Spectra and Isotopes* (London, 1933). (§1, 2)

A5. Aston, Nature **135**, 541 (1935). (§16, 43)

A6. Atkinson, Phys. Rev. **48**, 382 (1935). (§33)

B1. Bacher, Phys. Rev. **43**, 1001 (1933). (§48)

B2. Bacher and Sawyer, Phys. Rev. **47**, 587 (1935). (§49)

B3. Back and Goudsmit, Zeits. f. Physik **47**, 174 (1928). (§47, 48)

B4. Badami, Zeits. f. Physik **79**, 206, 224 (1932). (§48)

B5. Bainbridge, Phys. Rev. **42**, 1 (1932); **43**, 103, 378, 424; **44**, 56, 57, 123 (1933). (§2)

B6. Bainbridge and Jordan, Phys. Rev. **49**, (1936). (§43)

B7. Ballard, Phys. Rev. **46**, 806 (1934). (§48)

B8. Bartlett, Nature **128**, 408 (1931). (§49)

B9. Bartlett, Phys. Rev. **41**, 370 (1932); **42**, 145 (1932). (§32)

B10. Bartlett, Phys. Rev. **49**, 102 (1936). (§11)

B11. Bartlett and Gibbons, Phys. Rev. **44**, 538 (1933). (§49)

B12. Bethe, *Handbuch der Physik* Vol. **24**/1, p. 385. (§46)

B13. Bethe, Phys. Rev. **47**, 633 (1935). (§2, 16, 33)
B14. Bethe, Proc. Camb. Phil. Soc. **31**, 108 (1935). (§39)
B15. Bethe, Phys. Rev. (1936). (§47, 50)
B16. Bethe and Peierls, Proc. Roy. Soc. **A148**, 146 (1935). (§11, 12, 16)
B17. Bethe and Peierls, Nature **133**, 532, 689 (1934). (§39, 42)
B18. Bethe and Peierls, Proc. Roy. Soc. **A149**, 176 (1935). (§14, 15)
B19. Bhabha, Nature **134**, 934 (1934). (§15)
B20. Bonner and Brubaker, Phys. Rev. **49**, 19 (1936). (§22)
B21. Brasch, Lange, Waly, Banks, Chalmers, Szilard and Hopwood, Nature **134**, 880 (1934). (§16)
B22. Breit, Phys. Rev. **35**, 1447 (1930). (§46)
B23. Breit, Phys. Rev. **37**, 51 (1931). (§46)
B24. Breit, Phys. Rev. **38**, 463 (1931). (§46, 48)
B25. Breit, Phys. Rev. **42**, 348 (1932). (§49)
B26. Breit and Rabi, Phys. Rev. **38**, 2082 (1931). (§47)
B27. Breit and Wigner, Phys. Rev. **48**, 918 (1935). (§11)
B28. Breit and Wills, Phys. Rev. **44**, 470 (1933). (§46, 48)
B29. Briscoe, J. Am. Chem. Soc. **127**, 696, 1925; 282, 1927. (§1)
B30. Brown and Bartlett, Phys. Rev. **45**, 527 (1934). (§48)
B31. Brown and Cook, Phys. Rev. **45**, 731 (1934). (§48)
C1. Campbell, Nature **131**, 204 (1933); also unpublished. (§48)
C2. Campbell, Zeits. f. Physik **84**, 393 (1933). (§48)
C3. Casimir, Physica **2**, 719 (1935). (§50)
C4. Chadwick, Phil. Mag. **40**, 734 (1920). (§1).
C5. Chadwick, Proc. Roy. Soc. **A136**, 705 (1932). (§1)
C6. Chadwick, Proc. Roy. Soc. **A142**, 1 (1933). (§14, 15)
C7. Chadwick and Goldhaber, Nature **134**, 237 (1934). (§16)
C8. Chadwick and Goldhaber, Proc. Roy. Soc. **A151**, 479 (1935). (§16)
C9. Cockcroft, Inter. Conf. on Physics (London, 1934). (§33)
C10. Cockcroft and Walton, Proc. Roy. Soc. **A144**, 704 (1934). (§33)
C11. Cohen, Phys. Rev. **46**, 713 (1934). (§48)
C12. Condon and Breit, Phys. Rev. **49**, 229 (1936)
C13. Condon and Shortley, *Theory of Atomic Spectra* (Cambridge Univ. Press, and Macmillan Co.). (§24)
C14. Crane, Delsasso, Fowler and Lauritsen, Phys. Rev. **47**, 971 (1935). (§39)
C15. Crane, Delsasso, Fowler and Lauritsen, Phys. Rev. **47**, 887 (1935). (§39)
C16. Crane, Delsasso, Fowler and Lauritsen, Phys. Rev. **48**, 484 (1935). (§39)
C17. Crane, Delsasso, Fowler and Lauritsen, Phys. Rev. **48**, 125 (1935). (§23)
C18. Crawford, Phys. Rev. **47**, 768 (1935). (§46, 48)
C19. Crawford and Bateson, Can. J. Research **10**, 693 (1934). (§48)
C20. Crawford and Crooker, Nature **131**, 655 (1933). (§48)
C21. Crawford and Grace, Phys. Rev. **47**, 536 (1935). (§48)
C22. Crawford and Wills, Phys. Rev. **48**, 69 (1935). (§46)
D1. Dee and Walton, Proc. Roy. Soc. **A148**, 623 (1935). (§23)
D2. Dennison, Proc. Roy. Soc. **A115**, 483 (1927). (§47, 48)
D3. Dickinson, Phys. Rev. **46**, 598 (1934). (§49)
D4. Dunning, Pegram, Fink and Mitchell, Phys. Rev. **48**, 265 (1935). (§14, 17)
E1. Ehrenfest and Oppenheimer, Phys. Rev. **37**, 333 (1931). (§4)
E2. Ellett and Heydenburg, Phys. Rev. **46**, 583 (1934). (§47, 48)
E3. Elsasser, J. de phys. et rad. **4**, 549 (1933); **5**, 389, 635 (1934); Comptes rendus **199**, 46, 1213 (1934). (§32, 34)
E4. Estermann and Stern, Zeits. f. Physik **85**, 17 (1933). (§47)
F1. Farkas and Farkas, Proc. Roy. Soc. **A152**, 152 (1935). (§5, 47, 48)
F2. Feenberg, Phys. Rev. **47**, 850 (1935). (§12, 20, 21)
F3. Feenberg, Phys. Rev. **48**, 906 (1935). (§12, 20, 21)
F4. Feenberg, Phys. Rev. **49**, 328 (1936). (§23)
F5. Fermi, Zeits. f. Physik **60**, 320 (1930). (§46)
F6. Fermi, Nature **133**, 898 (1934). (§1)
F7. Fermi, Zeits. f. Physik **88**, 161 (1934). (§39, 40, 41)
F8. Fermi, Phys. Rev. **48**, 570 (1935). (§17)
F9. Fermi, Amaldi, d'Agostino, Rasetti and Segrè, Proc. Roy. Soc. **A146**, 483 (1934). (§1)
F10. Fermi and Amaldi, Ricerca Scientifica **1**, 56 (1936).
F11. Fermi and Segrè, Zeits. f. Physik **82**, 729 (1933). (§46)
F12. Flugge, Zeits. f. Physik **96**, 459 (1935). (§26, 29)
F13. Fox and Rabi, Phys. Rev. **48**, 746 (1935). (§47, 48)
F14. Frisch and Stern, Zeits. f. Physik **85**, 4 (1933). (§47)
F15. Fowler, Delsasso and Lauritsen, Phys. Rev. **49**, 561 (1936)
G1. Gale and Monk, Astrophys. J. **69**, 77 (1929). (§48)
G2. Gamow, Comptes rendus **197**, 1620 (1933). (§3)
G3. Gamow, Inter. Conf. on Physics (London, 1934). (§31, 34)
G4. Gibbons and Bartlett, Phys. Rev. **47**, 692 (1935). (§48)
G5. Goudsmit, Phys. Rev. **37**, 663 (1931). (§46, 48)
G6. Goudsmit, Phys. Rev. **44**, 636 (1933). (§46)
G7. Goudsmit and Bacher, Phys. Rev. **34**, 1501 (1929). (§46)
G8. Goudsmit and Bacher, Phys. Rev. **43**, 894 (1933). (§48)
G9. Goudsmit and Back, Zeits. f. Physik **43**, 321 (1927). (§46)
G10. Granath, Phys. Rev. **42**, 44 (1932). (§48)
G11. Granath and Stranathan, Phys. Rev. **48**, 725 (1935). (§48)
G12. Granath and Van Atta, Phys. Rev. **44**, 935 (1934). (§48)

G13. Guggenheimer, J. de phys. et rad. **5**, 253, 475 (1934). (§34)
G14. Gremmer and Ritschl, Zeits. f. Instrumentenk. **51**, 170 (1931). (§48)
G15. Guttinger, Zeits. f. Physik **64**, 749 (1930). (§48)
G16. Guttinger and Pauli, Zeits. f. Physik **67**, 743 (1931). (§46, 48)
H1. Hall, Phys. Rev. **49**, 401 (1936). (§16)
H2. Harkins, Phys. Rev. **19**, 135 (1922). (§1)
H3. Harkins, Gans, Kamen and Newson, Phys. Rev. **47**, 511 (1935). (§15)
H4. Hartree, Proc. Camb. Phil. Soc. **24**, 89 (1928). (§24)
H5. Harvey and Jenkins, Phys. Rev. **58**, 789 (1930). (§48)
H6. Hassé, Proc. Camb. Phil. Soc. **26**, 542 (1930). (§20)
H7. Heisenberg, Zeits. f. Physik **77**, 1 (1932). (§1, 6, 7, 8, 11)
H8. Heisenberg, Zeits. f. Physik **78**, 156; **80**, 587 (1932). (§6, 7, 8)
H9. Heisenberg, Rapport du VIIme Congrès Solvay (Paris, 1934). (§26)
H10. Heisenberg, Zeits. f. Physik **96**, 473 (1935). (§26, 32, 35)
H11. Heitler and Herzberg, Naturwiss. **17**, 673 (1929). (§48)
H12. Heydenburg, Phys. Rev. **43**, 640 (1933). (§47)
H13. Heydenburg, Phys. Rev. **46**, 802 (1934). (§47, 48)
H14. Hill, E. L., Proc. Nat. Acad. Sci. **15**, 779 (1929). (§47)
H15. Hill, H., Phys. Rev. **48**, 233 (1935). (§48)
H16. Hori, Zeits. f. Physik **44**, 834 (1927). (§48)
H17. Hughes and Eckart, Phys. Rev. **36**, 694 (1930). (§49)
I1. Iwanenko, Comptes rendus **195**, 439 (1932); Nature **129**, 798 (1932). (§1)
J1. Jackson, Zeits. f. Physik **75**, 229 (1932). (§48)
J2. Jackson, Zeits. f. Physik **80**, 59 (1933). (§48)
J3. Jackson and Kuhn, Proc. Roy. Soc. **A148**, 335 (1935). (§48)
J4. Jackson and Kuhn, Nature **137**, 107 (1936). (§48)
J5. Joffe and Urey, Phys. Rev. **43**, 761 (1933). (§48)
J6. Jones, Proc. Phys. Soc. **45**, 625 (1933). (§48)
J7. Jones, Proc. Roy. Soc. **A144**, 587 (1934). (§48)
K1. Kalckar and Teller, Proc. Roy. Soc. **A150**, 520 (1935). (§5, 47)
K2. Kapuschinski and Eymers, Proc. Roy. Soc. **A122**, 58 (1929). (§48)
K3. Kellogg, Rabi and Zacharias, Phys. Rev. **49** (1936). New York meeting. (§47)
K4. Klemperer, Proc. Roy. Soc. **A148**, 638 (1935). (§34, 41, 43)
K5. Konopinski and Uhlenbeck, Phys. Rev. **48**, 7 (1935). (§40 to 45)
K6. Kopfermann, Zeits. f. Physik **73**, 437 (1931). (§48)
K7. Kopfermann, Zeits. f. Physik **75**, 363 (1932). (§48)
K8. Kopfermann, Zeits. f. Physik **83**, 417 (1933). (§48)
K9. Kopfermann and Kindal, Zeits. f. Physik **87**, 460 (1934). (§48)
K10. Kopfermann and Rasmussen, Zeits. f. Physik **92**, 82 (1934). (§48)
K11. Kopfermann and Rasmussen, Zeits. f. Physik **94**, 58 (1935). (§48)
K12. Kopfermann and Rasmussen, Zeits. f. Physik **98**, 624 (1936). (§48)
K13. Kopfermann and Wieth-Knudsen, Zeits. f. Physik **85**, 353 (1933). (§48)
K14. Kruger, Gibbs and Williams, Phys. Rev. **41**, 322 (1932). (§48)
K15. Kurie, Phys. Rev. **44**, 463 (1933). (§15)
K16. Kurie, Richardson and Paxton, Phys. Rev. **48**, 167 (1935); **49**, 203 (1936). (§41)
L1. Landé, Zeits. f. Physik **25**, 46 (1924). (§46)
L2. Landé, Phys. Rev. **43**, 620 (1933). (§34)
L3. Larrick, Phys. Rev. **46**, 581 (1934). (§47, 48)
L4. Lawrence and Livingston, Phys. Rev. **44**, 316 (1933). (§33)
L5. Lawrence, MacMillan and Henderson, Phys. Rev. **47**, 254 (1935). (§33)
M1. McLay and Crawford, Phys. Rev. **44**, 986 (1933). (§48)
M2. McLennan and Allin, Proc. Roy. Soc. **A129**, 43 (1930). (§48)
M3. McLennan and Crawford, Proc. Roy. Soc. **A132**, 10 (1931). (§48)
M4. McLennan, Crawford and Leppard, Nature **128**, 301 (1931). (§48)
M5. McMillan and Grace, Phys. Rev. **44**, 949 (1933). (§48)
M6. Majorana, Zeits. f. Physik **82**, 137 (1933). (§7, 11, 25)
M7. Margenau, Phys. Rev. **46**, 613 (1934). (§32)
M8. Massey and Mohr, Proc. Roy. Soc. **A148**, 206 (1935). (§14)
M9. Massey and Mohr, Proc. Roy. Soc. **A152**, 693 (1935). (§20)
M10. Massey and Mohr, Nature **136**, 141 (1935). (§31)
M11. Mattauch, Zeits. f. Physik **91**, 361 (1934). (§1, 34)
M12. Meggers, King and Bacher, Phys. Rev. **38**, 1258 (1931). (§48)
M13. Millman, Phys. Rev. **47**, 739 (1935). (§47, 48)
M14. Mintz and Granath, Phys. Rev. **49**, 196 (1936). (§50)
M15. Monod-Herzen, J. de phys. et rad. **5**, 95 (1934). (§15)
M16. More, Phys. Rev. **46**, 470 (1934). (§48)
M17. Motz and Schwinger, Phys. Rev. **48**, 704 (1935). (§43)
M18. Murakawa, Tokyo Inst. Phys. Chem. Res. **18**, 305 (1932). (§48)
M19. Murphy and Johnston, Phys. Rev. **46**, 95 (1934). (§48)
N1. Nahmias, Proc. Camb. Phil. Soc. **31**, 99 (1935). (§39)
N2. Newson, Phys. Rev. **48**, 790 (1935). (§33)
N3. Nier, Phys. Rev. **48**, 283 (1935). (§43)
N4. Nier, Phys. Rev. **49**, 272 (1936). (§43)
O1. Oliphant, Kempton and Rutherford, Proc. Roy. Soc. **A149**, 406 (1935). (§22)
O2. Ornstein and Van Wijk, Zeits. f. Physik **49**, 315 (1928). (§48)

P1. Paschen, Sitzungsber. d. Preuss. Akad. **24**, 430 (1935). (§48)
P2. Paschen and Campbell, Naturwiss. **22**, 136 (1934). (§48)
P3. Paschen and Ritschl, Ann. d. Physik **18**, 867 (1933). (§48)
P4. Pauli, Naturwiss. **12**, 741 (1924), Chapter 8.
P5. Perrin, Comptes rendus **197**, 1625 (1933); **198**, 2086 (1934). (§39)
P6. Plesset, Phys. Rev. **49** (1936). New York Meeting. (§11)
P7. Present, Phys. Rev. **48**, 919 (1935). (§18)
P8. Present, Phys. Rev. **49** (1936). New York Meeting. (§20)
R1. Rabi and Cohen, Phys. Rev. **46**, 707 (1934). (§47, 48)
R2. Rabi, Kellogg and Zacharias, Phys. Rev. **46**, 157 (1934). (§47, 48)
R3. Rabi, Kellogg and Zacharias, Phys. Rev. **46**, 163 (1934). (§47, 48)
R4. Rabi, Kellogg and Zacharias, Phys. Rev. **49**, 421(A) (1936). (§47, 48)
R5. Racah, Zeits. f. Physik **71**, 431 (1931). (§46)
R6. Racah, Nature **129**, 723 (1932). (§49)
R7. Rasetti, Zeits. f. Physik **61**, 598 (1930). (§48)
R8. Rasmussen, Naturwiss. **23**, 69 (1935). (§48)
R9. Ritschl, Naturwiss. **19**, 690 (1931). (§48)
R10. Ritschl, Zeits. f. Physik **79**, 1 (1932). (§48)
R11. Ritschl, Nature **131**, 58 (1933). (§48)
R12. Ritschl and Sawyer, Zeits. f. Physik **72**, 36 (1931). (§48)
R13. Rose, Phys. Rev. **47**, 122 (1935). (§48)
R14. Rosenthal and Breit, Phys. Rev. **41**, 459 (1932). (§49)
R15. Rutherford, Phil. Mag. **21**, 669 (1911). (§1)
S1. Sargent, Proc. Roy. Soc. **A139**, 659 (1933). (§40)
S2. Schuler, Zeits. f. Physik **66**, 431 (1930). (§48)
S3. Schuler, Zeits. f. Physik **88**, 323 (1934). (§48)
S4. Schuler and Bruck, Zeits. f. Physik **56**, 291 (1929). (§48)
S5. Schuler and Gollnow, Naturwiss. **22**, 511 (1934). (§48)
S6. Schuler and Gollnow, Naturwiss. **22**, 714 (1934). (§48)
S7. Schuler and Jones, Zeits. f. Physik **74**, 631 (1932). (§48)
S8. Schuler and Keyston, Zeits. f. Physik **72**, 423 (1932). (§48)
S9. Schuler and Keyston, Zeits. f. Physik **71**, 413 (1931). (§48)
S10. Schuler and Keyston, Zeits. f. Physik **70**, 1 (1931). (§48)
S11. Schuler and Schmidt, Naturwiss. **22**, 758 (1934). (§48)
S12. Schuler and Schmidt, Zeits. f. Physik **92**, 148 (1934). (§49)
S13. Schuler and Schmidt, Naturwiss. **22**, 838 (1934). (§48)
S14. Schuler and Schmidt, Naturwiss. **23**, 69 (1935). (§48)
S15. Schuler and Schmidt, Zeits. f. Physik **94**, 457 (1935). (§48, 50)
S16. Schuler and Schmidt, Zeits. f. Physik **95**, 265 (1935). (§48)
S17. Schuler and Westmeyer, Naturwiss. **21**, 561 (1933). (§48)
S18. Schuler and Westmeyer, Naturwiss. **21**, 660 (1933). (§48)
S19. Schuler and Westmeyer, Zeits. f. Physik **81**, 565 (1933). (§48)
S20. Schuler and Westmeyer, Zeits. f. Physik **82**, 585 (1933). (§48)
S21. Shoupp, Bartlett and Dunn, Phys. Rev. **47**, 705 (1935). (§46)
S22. Slater, Phys. Rev. **34**, 1293 (1929). (§24)
S23. Sommerfeld, Zeits. f. Physik **80**, 415 (1933). (§7)
T1. Teller, Phys. Rev. **49**, 421(A) (1936). (§14)
T2. Thomas, Phys. Rev. **47**, 903 (1935). (§9, 19)
T3. Tolansky, Proc. Roy. Soc. **A136**, 585 (1932). (§48)
T4. Tolansky, Proc. Roy. Soc. **A137**, 541 (1932). (§48)
T5. Tolansky, Proc. Roy. Soc. **A144**, 574 (1934). (§48)
T6. Tolansky, Proc. Roy. Soc. **A146**, 182 (1934). (§48)
T7. Tolansky, Proc. Roy. Soc. **A149**, 269 (1935). (§48)
T8. Tolansky, Proc. Roy. Soc. **A152**, 663 (1935). (§48)
T9. Tolansky, Proc. Phys. Soc. London **48**, 49 (1936). (§48)
T10. Tuve and Hafstad, Phys. Rev. **48**, 106 (1935). (§33)
T11. Tuve, Heydenburg and Hafstad, Phys. Rev. **49**, 432 (1936). (§18)
V1. Van Vleck, Phys. Rev. **48**, 367 (1935). (§24, 37)
V2. Venkatesachar and Sibaiya, Nature **136**, 65 (1935). (§48)
W1. Weizsäcker, Physik. Zeits. **36**, 779 (1935). (§20)
W2. Weizsäcker, Zeits. f. Physik **96**, 431 (1935). (§8, 26, 29, 30)
W3. Westcott and Bjerge, Proc. Camb. Phil. Soc. **31**, 145 (1935). (§17)
W4. Westmeyer, Zeits. f. Physik **94**, 590 (1935). (§48)
W5. White, H. E., Phys. Rev. **34**, 1397 (1929). (§48)
W6. White, H. E. and Eliason, Phys. Rev. **44**, 753 (1933). (§47)
W7. White, H. E. and Ritschl, Phys. Rev. **35**, 146 (1930). (§48)
W8. White, M. G., Phys. Rev. **49**, 309 (1936). (§18)
W9. Wick, Zeits. f. Physik **84**, 799 (1933). (§12, 15)
W10. Wick, Nuovo Cim. **11**, 227 (1934). (§26)
W11. Wick, Att. Acad. Lincei **21**, 170 (1935). (§45)
W12. Wigner, Phys. Rev. **43**, 252 (1933). (§9, 11, 19, 21)
W13. Wigner, Zeits. f. Physik **83**, 253 (1933). (§14, 15)
W14. Wills, Phys. Rev. **45**, 883 (1934). (§48)
W15. Wills and Breit, Phys. Rev. **47**, 704 (1935). (§46)
W16. Wolff, Phys. Rev. **44**, 512 (1933). (§48)
Y1. Young, Phys. Rev. **48**, 913 (1935). (§10)
Z1. Zeeman, Gisolf and de Bruin, Nature **128**, 637 (1931). (§48)

NUCLEAR PHYSICS

Part B

Nuclear Dynamics, Theoretical

REVIEWS OF MODERN PHYSICS

VOLUME 9 | APRIL, 1937 | NUMBER 2

Nuclear Physics

B. Nuclear Dynamics, Theoretical*

H. A. BETHE†
Cornell University

TABLE OF CONTENTS

* Part A of this report ("Stationary States of Nuclei," by H. A. Bethe and R. F. Bacher) appeared in this journal, April, 1936 (Vol. 8, pp. 82–229). The final part C ("Nuclear Dynamics, Experimental," by M. Stanley Livingston and H. A. Bethe) will appear in the issue of July, 1937.

† The author is very greatly indebted to Drs. E. J. Konopinski and M. E. Rose for extensive help in computing tables, drawing figures and especially for a critical revision of the whole manuscript.

IX. Nuclear Processes as Many-Body Problems

§51. Bohr's General Theory (B32[1])

BOHR (B32) was the first to point out that every nuclear process (disintegration, capture of particles, etc.) must be treated as a many-body problem. It is not at all permissible to use a one-body (Hartree) approximation, particularly not in the case of heavy nuclei.

Previously, the one-body approximation had been used extensively for the treatment of nuclear processes (cf. the detailed discussions in §73). The method was taken over from the theory of atomic collisions where it had proved highly successful. Let us consider, as a typical atomic collision, the ionization of an atom by an electron. To treat this process, it is customary to consider the incident as well as the ejected electron as moving in the average field of the residual ion. In this way, "Hartree wave functions" are obtained for the two electrons. Then the interaction between the two electrons is considered as a perturbation, and the matrix element of this perturbation is calculated with the Hartree wave functions for the electrons. The square of the matrix element gives the probability of the collision. This method, known as the "method of the distorted wave functions," gives a good approximation whenever the interaction between the two electrons is small compared to the interaction between one of the electrons and the atom or small compared to the kinetic energy of the incident electron. This is generally true in atomic collisions, and therefore the method described is quite satisfactory for them[2] except for very small energies of the incident electron. It is, in the case of atoms, far superior to the often used Born approximation in which the average interaction between atom and incident electron is neglected.

However, the method of the distorted wave functions is not at all applicable in nuclear physics. As we know (§25[3]), the interaction between one particle and a whole nucleus is only of the same order of magnitude as that between two nuclear particles. Therefore, it has in general no sense to consider the "average potential" acting on the incident particle, in an earlier approximation than the interaction causing the nuclear process itself. Furthermore, the interaction is, in general, large compared to the energy of the incident particle so that the second condition for the applicability of the method of the distorted wave functions is also not fulfilled.

The difference between an atomic and a nuclear collision may be described in a variety of ways. The descriptions are all equivalent but differ in being more or less intuitive and in making use of more classical or more quantum-mechanical concepts.

In an atomic collision, the interaction between the incident particle and the individual electrons of the atom is small, as we have already mentioned. Therefore it is comparatively seldom that a particle in passing through an atom imparts energy to an atomic electron; in other words, inelastic collisions are rare.[4] In most cases, the incident particle will go through the atom without interacting with any particular atomic electron and without losing any energy; it will only be affected by the average field of force of the atom and will be deflected thereby (elastic scattering). The time which the particle spends in the atom is of the order of the atomic dimensions divided by the velocity of the particle. (For electrons, the time is even smaller because the electron is accelerated inside the atom.) Because of this short time spent inside the atom, it is highly improbable that, e.g., radiation is emitted in the collision. Thus atomic collisions are characterized by a very large elastic scattering, a smaller inelastic scattering,[4] and a very small probability of the emission of radiation.

If a particle falls on a *nucleus*, it cannot possibly pass through it without interacting with

[1] A letter and a number, e.g. B32, refer to an original paper. A list of references is given at the end of this article.

[2] It has been applied to the problem of the ionization of the K shell (S17, M9) and to the elastic scattering of electrons by heavy atoms (H25) with considerable success. A wider application has thus far been prevented by mathematical difficulties.

[3] The sections §1 to §50 are contained in part A of this report, Rev. Mod. Phys. **8**, 82 (1936).

[4] This is correct for the collisions in which the incident particle actually passes through the atom (close collisions) and is, in general, deflected by a large angle. In addition, there are numerous inelastic collisions due to particles passing the atom at a distance (B16, W20). These "distant collisions" are due to the long range of the forces between incident particle and atomic electrons (Coulomb forces). Since the nuclear forces are short range forces, there is no analog to the "distant collisions" in nuclear physics.

the individual nuclear particles, for the average distance between them is of the same order as the range of the nuclear forces (§25) due to the peculiar character (short range) of these forces. An incident particle which passes between two nuclear particles must therefore necessarily interact strongly with them; we know that the nuclear forces are very strong (over 10 MV) inside their range of action. The incident particle will therefore lose part of its energy as soon as it strikes the surface of the nucleus, by transferring it to the nuclear particles. As the particle goes on, there will be a further dissipation of its energy among the nuclear particles. As a result, the energy which is initially concentrated in the incident particle, will very soon be distributed among all the particles of the system consisting of the original nucleus and the incident particle (compound nucleus). Each of the particles of the "compound nucleus" will have some energy, but none will have sufficient energy to escape from the rest. Only after a comparatively long time, the energy may "by accident" again be concentrated on one particle so that this particle can escape. Even then, the escaping particle need not be of the same kind as the incident, and even if it is of the same kind, the energy may (and will in general) be divided between the escaping particle and the remaining nucleus in a way different from the initial state of affairs; in other words, the residual nucleus may remain in an excited state. Only if the escaping particle is of the same kind as the incident *and* the internal state of the nucleus is not changed may we speak of an elastic collision. It is obvious that an elastic collision is only a very special case and must therefore be quite a rare event compared to the many kinds of possible inelastic collisions, i.e., such in which either the nucleus is excited (ordinary inelastic collision) and such in which a particle of a different kind is emitted (transmutation). Furthermore it is evident that the time spent by the particle inside the nucleus will in general be very long; it may be several orders of magnitude larger than the time which the particle would need to traverse the nucleus on a straight path without loss of velocity. This provides ample opportunity for the emission of radiation during the collision. Therefore the characteristics of nuclear collisions are: a quite small probability of elastic scattering, a much larger probability of inelastic scattering including transmutations, and a comparatively high probability of the emission of radiation. In some cases radiative processes are even the most probable of all (cf. §58, 61, 62).

The description of the nuclear collisions given in the foregoing shows that the situation during the greater part of the collision may be described as a quasi-stationary state of the "compound nucleus." This compound nucleus will exist during a time which is very large compared with the "characteristic nuclear time" which may be defined as the nuclear radius divided by some average velocity of the nuclear particles, i.e., about[5] $12\cdot 10^{-13}/4\cdot 10^{+9}=3\cdot 10^{-22}$ sec. A theory of nuclear collisions is therefore primarily a theory of the (quasi-stationary) states of the compound nucleus, and of the transitions from these states to states in which one particle or other of the compound nucleus is separated from the rest. The collision may be described by the scheme:

Initial nucleus+incident particle→
compound nucleus→final nucleus
+outgoing particle.

It is therefore characterized by a double transition, whereas atomic collisions consist in direct transitions from initial to final state.

We have arrived at the idea of the compound nucleus from an almost geometrical argument, *viz.* from the fact that the incident particle cannot find its way through the nucleus in between the nuclear particles but must interact with them. The nucleus is, in Bohr's terminology, a "closed system." An additional particle cannot go through the system but can only be amalgamated with it. In contrast to this, an atom is an "open system" which may be traversed by an external particle without any difficulty, and without the particle's becoming incorporated in the system.

The existence of quasi-stationary states of the compound nucleus, and the nonexistence of an analog in atomic collisions, may also be discussed from an *energetic* rather than geometric point of view. We start from the energy levels of the individual particles (nuclear particles and electrons,

[5] For the value of the nuclear radius, cf. §68. The velocity of $4\cdot 10^{+9}$ corresponds to a kinetic energy of about 10 MV, cf. §26.

respectively) in the two cases. It is true that this method of approach, while satisfactory for atoms, does not give a good approximation for nuclei. However, it has the virtue of starting from the extreme opposite to the many-body concept so that the fact that it leads finally to the concept of quasi-stationary states of the compound nucleus is still more convincing.

As we know, the potential energy acting on one nuclear particle (proton or neutron) may, with sufficient accuracy, be represented by a simple "rectangular" hole. In such a hole, the energy levels for the individual particles are almost uniformly distributed from the bottom of the hole to the top. (Strictly speaking, the density of energy levels increases as $E^{\frac{1}{2}}$ where E is the energy above the bottom; cf. §25 and 53.) The depth of the hole is about[6] 18 MV. Since the binding energy of the most loosely bound particle is of the order 8 MV, the potential hole is normally filled with particles up to 8 MV below the top, i.e. to a little more than half its height. There is a great number of empty energy levels between this energy and the top; in fact, the number of empty levels is greater than the number of occupied ones. Let us now consider a state of the nucleus which has, say, 6 MV more energy than the ground state. Such a state may be obtained by exciting *one* of the nuclear particles by 6 MV and leaving the rest unexcited. But equally well, the total excitation energy may be shared between two, three or more particles. It is easily possible that as many as ten particles share in the excitation because the levels of the individual particles lie fairly close together and it is therefore only necessary to supply quite a small amount of energy to each excited particle. (For a quantitative discussion, cf. §53.) Thus it is obvious that there is an enormous variety of ways in which the given total excitation energy (6 MV) may be shared between several particles whereas the configuration in which all the energy is concentrated on one particle represents only a single possibility.

If there were no interaction between the particles, we would thus obtain a very large number of energy levels of the complete nucleus corresponding to the various distributions of the excitation energy among the particles. Besides, we would get a few energy levels corresponding to the excitation of a single particle. In reality, there is interaction between the particles which causes a mixing of the eigenfunctions of the various levels. Therefore, the levels do not fall into different classes corresponding to the excitation of one, two, three . . . particles, but each level is of mixed character. If the nucleus is in any arbitrary state, its excitation energy is part of the time concentrated on one particle, part of the time shared between two, three etc., particles. The exact value of the probability of the concentration of the energy on one particle may be deduced from the eigenfunction of the nuclear state; it is certainly very small because there are so few configurations in which the energy is concentrated and so many in which it is distributed. The number of energy levels, on the other hand, is evidently very great, i.e., the spacing between neighboring energy levels of the nucleus as a whole is extremely small, very much smaller than the spacing between the energy levels of an individual particle would be.

We have chosen the excitation energy of the nucleus lower than the energy necessary to dissociate it into a free particle plus a residual nucleus, in order to avoid dealing with a continuous spectrum for the particle. However, we see now that this restriction is irrelevant because the configuration in which the energy is concentrated on one particle is quite unimportant for the description of a nuclear state. Therefore, the nuclear states which are excited by more than the dissociation energy (about 8 MV) have essentially the same character as those of lower energy; their eigenfunctions are not very different. The only difference is that these more highly excited states may with a certain, quite small, probability disintegrate into a free particle plus a residual nucleus. This probability is given by the coefficient with which the function representing the excitation of a single particle is contained in the eigenfunction of the nuclear state in question. But apart from this comparatively rare event of disintegration, *a nucleus has a series of closely spaced energy levels which have essentially the same*

[6] Cf. (150c) §25. The nuclear radius is about $R=r_0A^{\frac{1}{3}}$ with $r_0=2.05\cdot10^{-13}$ cm (cf. §68) and $N=\frac{1}{2}A$. Therefore the kinetic energy of the most energetic particle $T_0=(9\pi/8)^{\frac{2}{3}}(\hbar^2/2Mr_0^2)=10.3$ MV. The binding energy of this particle is about 8 MV, therefore the depth of the hole $10+8=18$ MV.

character below and above the dissociation energy. The character of these levels will only change decidedly when the excitation energy *per particle* becomes of the order of a few MV, i.e., the excitation energy of the nucleus 100 MV or more.

For atoms, the situation is entirely different. The binding energies of the individual electrons differ by very large amounts: e.g., in the uranium atom the binding energies of successive electrons are approximately 110,000, 20,000, 4000, 800, 200, 40, and 6 volts. The reason for these large differences is, of course, the character of the atomic field, *viz.* a screened Coulomb field which is extremely large near the nucleus and falls off very rapidly at greater distances. The energy region in which there are empty electronic levels, extends over only a few volts. Accordingly, if we excite only outer electrons, we can only obtain excitation energies of a few tens of volts even if we excite many electrons simultaneously. Moreover, since there are only very few electrons in the outer shell as compared to the many particles in a nucleus, there will be comparatively few ways in which a given amount of excitation energy may be shared between the electrons. This makes the configuration in which the whole energy is concentrated on one electron, relatively more probable. The atomic states whose energy is sufficient for a dissociation, will "disintegrate" much more easily into an ion plus an electron than nuclei in corresponding states. (In spite of all this, some states of atoms above the ionization potential, in which two electrons are excited, are known in spectroscopy, especially for the alkaline earths.)

If we excite an inner electron, we obtain, of course, a rather large excitation energy. However, the energy levels of the atom which are obtained in this way, are still restricted to a minute energy region. E.g., if we excite the K electron of uranium and some outer electrons, we obtain a number of energy levels all of which lie between, say, 110,000 and 110,050 volts (above the ground state of the atom). The spectrum of the energy levels of the atom as a whole contains therefore only a few very narrow regions in which there are discrete energy levels, and these regions are separated by immense "empty" spaces in which there is no level at all. The reason is the small region in which there are empty levels for the electrons, a region which is for uranium about 20,000 times smaller than the binding energy of the K electrons, as compared to one-half the binding energy of the most strongly bound particle in nuclei (see above). If we consider the "compound system" formed by the addition of an electron to an atom, i.e., a negative ion, the situation becomes even more extreme because the binding energy of the last electron in a negative ion is considerably less than in an atom. Then the bands in which the compound system possesses energy levels, will extend only over a few volts each. In general, the energy of an incident electron will therefore not fall into one of these regions, and no quasi-stationary state of the compound system will be found in the collision. (It can be shown that the states of the compound system are unimportant even if the energy of the incident electron falls just into one of the energy bands, except for very slow electrons.) Therefore an atomic compound system has no quasi-stationary states of any importance for atomic collisions.

We have seen that the stationary states of the compound nucleus are responsible for quite a different relative magnitude of the various kinds of nuclear collisions (elastic, inelastic, transmutation, radiation) as compared to atomic collisions. But they have another, even more striking consequence: the phenomenon of *resonance.* If the energy of the incident particle is such that the total energy of the system is just equal, or nearly equal, to one of the energy levels of the compound nucleus, the probability of the formation of the compound nucleus will obviously be much greater than if the energy of the particle falls in the region between two resonance levels. Therefore we shall find characteristic fluctuations of the yield of every nuclear process with the energy, from high values at the resonance energies to low values between resonance levels. These resonance phenomena are most pronounced with slow neutrons (Chapter X) but have also been observed in the radiative capture of protons and in transmutations caused by α-particles (Chapter XIII).

The study of the resonance phenomena in nuclear processes is of paramount importance for nuclear physics. First of all, the spacing between neighboring levels of the compound nucleus may be deduced from the resonances. The determination of the spacing as a function of the mass number and of the excitation energy of the nucleus will enable us to check our theoretical ideas about nuclear structure (cf. §53).

Secondly, the width of the resonance levels is of great interest. Just as in the theory of atoms the width of an excited level is given by the probability of its emitting radiation (W6), so the width of the level of a compound nucleus is given by the total probability of the emission of particles of any kind—neutrons, protons, α-particles, γ-rays etc., by the compound nucleus. Thus the width of the levels enables us to determine the probability of the concentration of energy on any one of the particles in the compound nucleus. This may be supplemented by a study of the relative probability of the emission of various kinds of particles.

The experimental data on nuclear levels are not yet very extensive. The spacing of the energy levels seems to be of the order of 10 volts for nuclei of atomic weight 100 or more and excitation energies just sufficient for a dissociation into a neutron and a residual nucleus (about 8 MV excitation). (These figures are obtained from the slow neutron experiments (§60).) For light nuclei, spacings of a few hundred thousand volts seem to prevail (proton capture (§81), γ-ray spectra (§89), resonances in α-particle disintegrations (§82)). The same order of magnitude has been found for heavy nuclei just above the ground state (from γ-ray evidence and long range α-particles).

A survey of the experimental results for the widths of levels will be given at the end of the next section.

§52. The Dispersion Formula (B51, B15)

The probability of a nuclear process as a function of the energy of the incident particle is given by a formula similar to the ordinary dispersion formula for the scattering of light. This dispersion formula was first derived by Breit and Wigner (B51) for the case of one resonance level of the compound nucleus and then generalized by Bethe and Placzek (B15) for an arbitrary number of resonance levels. For a general derivation, see §55.

Any nuclear process may be described as follows: a particle P falls on a nucleus A which is in a state p. A compound nucleus C is formed which may have a number of energy levels W_r. A particle Q is emitted by the compound nucleus, leaving a residual nucleus B in a state q. The emitted particle Q may or may not be identical with the incident particle P; in the first case, we speak of scattering, in the second case of transmutation. In the special case that Q is a light quantum, we have radiative capture, while we deal with a photoelectric effect, if P is a light quantum and Q a material particle. For the initial and final state energy must be conserved

$$W_{Ap}+W_P+E_P=W_{Bq}+W_Q+E_Q, \quad (254)$$

where W_{Ap} denotes the internal energy of nucleus A in state p, W_P the internal energy and E_P the kinetic energy of the incident particle, etc.[7] For the intermediate state (compound nucleus), of course, conservation of energy is unnecessary.

The probability of the nuclear process

$$A+P\rightarrow C\rightarrow B+Q \quad (254a)$$

may be calculated using the ordinary methods of the Dirac perturbation theory. Since we are dealing with a double process, a second-order perturbation calculation is necessary. The result for the cross section of the process is (B15, and §55)

$$\sigma=4\pi^3\lambda\!\!\!^{-2}\left|\sum_r \frac{H^{App}{}_{Cr}\,H^{Cr}{}_{BQq}}{W_{Ap}+W_P+E_P-W_{Cr}+\frac{1}{2}i\gamma_r}\right|^2. \quad (255)$$

Here $\lambda\!\!\!^{-}$ is the "wave-length" of the incident particle,

$$\lambda\!\!\!^{-2}=\hbar^2/2ME_P, \quad (256)$$

$$M=M_AM_P/(M_A+M_P). \quad (256a)$$

H is the Hamiltonian of the interaction between the particles concerned, $H^{APp}{}_{Cr}$ its matrix element corresponding to the transition from the initial state (nucleus A in state p+particle P of energy E_P) to the intermediate state (compound nucleus C in state r). To calculate the matrix element, the wave function of the particle P must be normalized per unit energy. γ_r is the total effective width of the level r of the compound nucleus. It is a sum of the partial effective widths due to the emission of various kinds of particles by the compound nucleus:

$$\gamma_r=\sum_{Q'}\gamma^r{}_{Q'}. \quad (257)$$

[7] Throughout this report, we denote by W the energies of nuclear quantum states and by E the kinetic energies of particles.

Among the particles Q' which may be emitted there are the incident particle P, the outgoing particle Q and possibly others. The width $\gamma^r{}_{Q'}$ again may be written in the form

$$\gamma^r{}_{Q'} = \sum_{q'} \gamma^r{}_{Q'q'}, \tag{257a}$$

where $$\gamma^r{}_{Q'q'} = 2\pi |H^{Cr}{}_{B'Q'q'}|^2 \tag{257b}$$

is that part of the width of level r which is due to the disintegration of the compound nucleus C into a nucleus B' in state q' and a particle Q' with an energy given by the conservation law (254).

It must be kept in mind that for all quantities occurring in (255) the energy of the corresponding particle (P, Q or Q') must be chosen according to the conservation law (254). Therefore all quantities depend on the energy E_P of the incident particle. This holds also for the effective width γ_r which is therefore in general not the true width of the level r. This true width Γ_r is obtained when the resonance energy

$$E_{Pr} = W_{Cr} - W_{Ap} - W_P \tag{258}$$

is inserted for E_P, so that

$$\Gamma_r = \gamma_r(E_{Pr}). \tag{258a}$$

The cross section (255) shows resonance phenomena. Pronounced maxima occur whenever the energy of the incident particle E_P is near one of the resonance energies (258). In this case, generally one term in (255) will be predominant, namely the one referring to the particular resonance level r in question. Then (255) simplifies to the "one level formula"

$$\sigma^{Pp}{}_{Qq} = \pi\lambda^2 \frac{\gamma^r{}_{Pp}\gamma^r{}_{Qq}}{(E_P - E_{Pr})^2 + \frac{1}{4}\gamma_r^2}, \tag{259}$$

which was first derived by Breit and Wigner. This formula is most useful for applications and sets the resonances in evidence very clearly. The total probability of the production of particles of kind Q summed over all possible levels q in which the nucleus B may be left, is

$$\sigma^{Pp}{}_Q = \sum_q \sigma^{Pp}{}_{Qq} = \pi\lambda^2 \frac{\gamma^r{}_{Pp}\gamma^r{}_Q}{(E_P - E_{Pr})^2 + \frac{1}{4}\gamma_r^2} \tag{259a}$$

if only one level of C is important.

Special considerations are necessary if some of the nuclear states involved are degenerate. The calculations which have been carried out by Bethe and Placzek replace (255) by the formula

$$\sigma^{Pp}{}_{Qq} = \frac{\pi\lambda^2}{(2i+1)(2s+1)} \sum_{ljl'j'J} (2J+1) \times \left| \sum_r \frac{u^{CrJ}{}_{ApPlj}\, u^{CrJ}{}_{BqQl'j'}}{W_{Ap} + W_P + E_P - W_{CrJ} + \frac{1}{2}i\gamma_{rJ}} \right|^2. \tag{260}$$

Here and in the following we denote by i, J and i' the angular momenta of initial, compound and final nucleus; s and s' are the spins of incident and outgoing particle; ll' and jj' their orbital and total angular momenta. p, r, q mean all quantum numbers of initial, compound and final nucleus other than the angular momenta. The sums over l and l' extend over all values of these quantities from 0 to ∞, the sum over j from $|l-s|$ to $l+s$ and the sum over j' correspondingly; the sums over rJ extend over all states of the compound nucleus. The u's are matrix elements similar to the H's; they are defined so that the partial width of a level is simply equal to the square of u, *viz.*

$$\gamma^{rJ}{}_{Pplj} = (u^{CrJ}{}_{ApPlj})^2, \tag{261}$$

$$\gamma^{rJ}{}_{Pp} = \sum_{lj} \gamma^{rJ}{}_{Pplj}. \tag{261a}$$

Furthermore, the u's are real but may be positive or negative.

If only one level is important, (260) reduces again to the very simple formula

$$\sigma^{Pp}{}_{Qq} = \pi\lambda^2 \frac{2J+1}{(2s+1)(2i+1)} \frac{\gamma^{rJ}{}_{Pp}\gamma^{rJ}{}_{Qq}}{(E_P - E_r)^2 + \frac{1}{4}\gamma_{rJ}^2}, \tag{262}$$

which is the generalization of the Breit-Wigner formula (259) for nonvanishing angular momenta of the nuclei concerned.

The quantities H, u and γ depend on the energy of the particle involved, because the matrix element H (or u) contains the wave function of the incident particle. As we have mentioned, this wave function must be normalized per unit energy. If we assume the particle to be free and to have an orbital momentum l, its normalized wave function has the form (B15, Eq. (15))

$$\psi=\left(\frac{2}{\pi}\frac{dk}{dE}\right)^{\frac{1}{2}}\frac{1}{r}\chi_l(kr)\,Y_{lm}(\vartheta\varphi), \qquad (263)$$

where $k=1/\lambda\!\!\!^{-}$ is the wave number of the particle,

$$E=(\hbar^2/2m)k^2 \qquad (263a)$$

its energy, Y_{lm} a normalized spherical harmonic and χ_l *that* solution of the radial wave equation,

$$\frac{d^2\chi_l}{dr^2}+\left(k^2-\frac{l(l+1)}{r^2}\right)\chi_l=0, \qquad (263b)$$

which behaves asymptotically as sin $(kr-\frac{1}{2}l\pi)$. At small r,

$$\chi_l=\frac{1}{1\cdot 3\cdots(2l-1)(2l+1)}(kr)^{l+1}. \qquad (263c)$$

Now for the matrix element only the behavior of the wave function ψ *inside* the nucleus is important. Therefore, if the wave-length $\lambda\!\!\!^{-}=1/k$ is large compared to nuclear dimensions, we are only interested in the values of ψ for small values of kr. Then, according to (263) (263c), the matrix element will depend on k as

$$(dk/dE)^{\frac{1}{2}}k^{l+1}. \qquad (264a)$$

Since dE/dk is proportional to k (cf. 263a)), the matrix element will be proportional to $k^{l+\frac{1}{2}}$. Thus we may write

$$u^{rJ}{}_{Pp}=b^{rJ}{}_{Pp}\,k^{l+\frac{1}{2}}=b^{rJ}{}_{Pp}\,\lambda\!\!\!^{-}{}^{-(l+\frac{1}{2})} \qquad (264)$$

and, because of (261),

$$\gamma^{rJ}{}_{Pp}=(b^{rJ}{}_{Pp})^2\lambda\!\!\!^{-}{}^{-(2l+1)}, \qquad (264b)$$

where b is a constant. Introducing the true width Γ (cf. (258a)) instead of the effective width γ, and similarly the value of the matrix element at resonance, U, instead of u, we have

$$u^{rJ}{}_{Pp}=U^{rJ}{}_{Pp}(\lambda\!\!\!^{-}{}_{Pr}/\lambda\!\!\!^{-})^{l+\frac{1}{2}}=U^{rJ}{}_{Pp}(E/E_{Pr})^{\frac{1}{2}l+\frac{1}{4}} \qquad (265)$$

$$\gamma^{rJ}{}_{Pp}=\Gamma^{rJ}{}_{Pp}(\lambda\!\!\!^{-}{}_{Pr}/\lambda\!\!\!^{-})^{2l+1}=\Gamma^{rJ}{}_{Pp}(E/E_{Pr})^{l+\frac{1}{2}}, \qquad (265a)$$

where $\lambda\!\!\!^{-}{}_{Pr}$ is the wave-length corresponding to the resonance energy E_{Pr}. It can easily be shown (see B15, p. 454) that the formulae (264) (265) hold also if the particle is not free.

If, as we have assumed, the wave-length of the particle is large compared to nuclear dimensions, then only the partial wave $l=0$ will have appreciable amplitude in the nucleus, as can be seen directly from (263c). Therefore for slow particles (264) (265) become

$$u^{rJ}{}_{Pp}=b^{rJ}{}_{Pp}\lambda\!\!\!^{-}{}^{-\frac{1}{2}}=a^{rJ}{}_{Pp}v^{\frac{1}{2}}=c^{rJ}{}_{Pp}E^{\frac{1}{4}}$$
$$=U^{rJ}{}_{Pp}(v/v_{Pr})^{\frac{1}{2}}, \qquad (266)$$

$$\gamma^{rJ}{}_{Pp}=(c^{rJ}{}_{Pp})^2E^{\frac{1}{2}}=\text{etc.}, \qquad (266a)$$

where v is the velocity of the particle and a, b, c are certain constants whose relation is obvious.

For γ-rays, there is no partial wave $l=0$, so that $l=1$ (dipole radiation) becomes the most important part. The dependence of the matrix element for γ-rays on the wave-length is again given by (264) (cf. B15). However, in contrast e.g., to slow neutrons, the energy of γ-rays is ordinarily very much larger than the spacing between energy levels; therefore the ratio E/E_{Pr} of the actual to the resonance energy will, for γ-rays, be very near unity for all those levels rJ which contribute appreciably to the cross section. Thus we may, for γ-rays, replace the matrix element u by its value at resonance U, and the actual width $\gamma=u^2$ by the true width $\Gamma=U^2$, independent of l.

The same arguments hold for fast particles. This is particularly welcome because the exact behavior of the matrix element as a function of energy cannot be obtained without assuming a special nuclear model, as soon as the wave-length of the particle becomes comparable with or smaller than the nuclear dimensions. But we do not need to know this behavior because we wish to know the matrix elements u only in the neighborhood of the resonance energy, let us say in an energy region of the extension of a few times the spacing between resonance levels. Inasmuch as the spacing of levels is very small compared to the energy of the fast particle, the matrix element u and the effective width γ may be regarded as constant over the energy region considered, and be replaced by their values at resonance, U and Γ. Or, alternatively, formulae (265) (265a) may be used for fast particles: Even though they are not strictly true for this case, their use is permitted over energy regions small compared to the particle energy itself.

Inserting (264) into (260), we have

$$\sigma^{Ppl}{}_{Qql'}=\frac{\pi}{(2s+1)(2i+1)}\lambda\!\!\!^{-1-2l}\lambda\!\!\!^{-\prime-(1+2l')}\sum_{jj'J}(2J+1)$$

$$\times\left|\sum_r\frac{b^{rJ}{}_{ApPlj}\,b^{rJ}{}_{BqQl'j'}}{E_P-E_{CrJ}+\frac{1}{2}i\gamma_{rJ}}\right|^2 \tag{267}$$

with

$$b^{rJ}{}_{ApPlj}=U^{rJ}{}_{ApPlj}\lambda\!\!\!^{-}{}_{Pr}{}^{l+\frac{1}{2}}, \tag{267a}$$

$$\sigma^{Pp}{}_{Qq}=\sum_{ll'}\sigma^{Ppl}{}_{Qql'}. \tag{267b}$$

$\lambda\!\!\!^{-}$ and $\lambda\!\!\!^{-}{}'$ are the wave-lengths of incident and outgoing particle; all the other quantities have the same meaning as before. If only one level rJ is important, (267) reduces to

$$\sigma^{Ppl}{}_{Qql'}=\frac{\pi(2J+1)}{(2s+1)(2i+1)}\lambda\!\!\!^{-1-2l}\lambda\!\!\!^{-}{}_{Pr}{}^{2l+1}$$

$$\times\left(\frac{\lambda\!\!\!^{-}{}_{Qr}}{\lambda\!\!\!^{-}{}'}\right)^{2l'+1}\frac{\Gamma^{rJ}{}_{Ppl}\Gamma^{rJ}{}_{Qql'}}{(E_P-E_{Pr})^2+\frac{1}{4}\gamma_{rJ}{}^2}. \tag{268}$$

In this formula, only the factors $\lambda\!\!\!^{-}$, $\lambda\!\!\!^{-}{}'$ and E_P itself change with the energy E_P of the incident particle,[8] all the other quantities are constants characterizing the nuclear level.

In order to discuss formulae (267) (268), it is useful to distinguish between fast and slow incident and outgoing particles. A particle is called slow if its wave-length is large compared to nuclear dimensions. This is true for energies up to a few hundred thousand volts for neutrons or protons interacting with heavy nuclei and up to about a million volts if they interact with light nuclei. For α-particles, the limits are one-quarter of these figures. γ-rays can always be regarded as slow or fast at ones discretion, it being simpler to regard them as fast. We have then the following cases:

[8] The effective width γ_{rJ} will in general consist of contributions from fast particles Q and from slow particles which may be emitted by the compound nucleus rJ. The contributions of the fast particles (including γ-rays) will not depend sensitively on the energy of the incident particle, at least not over energy regions of a few thousand or even a few hundred thousand volts. The contributions of the slow emitted particles *will* depend strongly on the energy, according to (265a). But these contributions will in general be small because of the small factor $E^{l+\frac{1}{2}}$. Thus in general, γ_{rJ} will be approximately constant and equal to the true width Γ_{rJ} over fairly large energy regions; and only if the contribution of slow emitted particles happens to be large will there be a noticeable dependence of γ_{rJ} on energy.

A. Incident and outgoing particle fast

In this case the variation of the factors $\lambda\!\!\!^{-}$ and $\lambda\!\!\!^{-}{}'$ in (267) (268) with energy is irrelevant and may be disregarded (cf. above). The only factor depending on energy is then the resonance factor (last factor in (267) (268)).

B. Incident particle slow, outgoing particle fast

We may put $l=0$ for the incident particle and neglect the dependence on the wave-length $\lambda\!\!\!^{-}{}'$ of the outgoing particle. Then (267) reduces to

$$\sigma^{Pp}{}_{Qq}=\frac{\pi}{(2s+1)(2i+1)}\lambda\!\!\!^{-}\sum_{l'j'J}(2J+1)$$

$$\times\left|\sum_r\frac{U^{rJ}{}_{ApP0s}\lambda\!\!\!^{-}{}_{Pr}{}^{\frac{1}{2}}U^{rJ}{}_{BqQl'j'}}{E_P-E_{Pr}+\frac{1}{2}i\gamma_{rJ}}\right|^2 \tag{269}$$

and if only one level is important, we obtain from (268)

$$\sigma^{Pp}{}_{Qq}=\frac{\pi(2J+1)}{(2s+1)(2i+1)}\lambda\!\!\!^{-}\lambda\!\!\!^{-}{}_{Pr}\frac{\Gamma^{rJ}{}_{Pp}\Gamma^{rJ}{}_{Qq}}{(E_P-E_{Pr})^2+\frac{1}{4}\gamma_{rJ}{}^2}. \tag{270}$$

In both (269) and (270), there appears beside the resonance factor a factor $\lambda\!\!\!^{-}$ which indicates proportionality of the cross section with the reciprocal of the velocity v of the incident particle (again apart from resonance). This "$1/v$-factor" makes the cross section for phenomena produced by slow particles, especially slow neutrons, very large (cf. Chapter X). (About the dependence of γ_{rJ} on energy, cf. footnote 8)

C. Incident particle fast, outgoing particle slow

Following the same considerations as before, (267) reduces to

$$\sigma^{Pp}{}_{Qq}=\frac{\pi}{(2s+1)(2i+1)}\lambda\!\!\!^{-}{}'^{-1}\sum_{ljJ}(2J+1)$$

$$\times\left|\sum_r\frac{U^{rJ}{}_{ApPlj}U^{rJ}{}_{BqQ0s'}\lambda\!\!\!^{-}{}_{Pr}\lambda\!\!\!^{-}{}_{Qr}{}^{-\frac{1}{2}}}{E_P-E_{Pr}+\frac{1}{2}i\gamma_{rJ}}\right|^2. \tag{271}$$

The cross section is in this case inversely proportional to $\lambda\!\!\!^{-}{}'$, or directly proportional to the velocity v' of the outgoing particle. Thus slow particles are only rarely produced in nuclear processes.

D. Incident and outgoing particle slow

The cross section contains apart from the resonance factor, a factor $\lambda\!\!\!^{-}/\lambda\!\!\!^{-}{}' \propto v'/v$. In the particular case of the elastic scattering of slow particles, $v'=v$ so that the cross section depends on the energy only through the resonance factor while there is no general trend such as the $1/v$ or the v'-law. E.g., if only one level is important, the elastic scattering becomes

$$\sigma^{Pp}{}_{Pp}=\frac{\pi(2J+1)}{(2s+1)(2i+1)}\lambda\!\!\!^{-}{}_{Pr}{}^{2}\frac{(\Gamma^{rJ}{}_{Pp})^{2}}{(E_P-E_{Pr})^{2}+\frac{1}{4}\gamma_{rJ}{}^{2}} \tag{272}$$

with the same notation as in (270).

If one or both of the particles involved are slow, there are strict selection rules for the angular momenta. E.g., if the incident particle is slow, its orbital momentum will be zero and therefore its total momentum j equal to its spin s. We have then the selection rules

$$|i-s|\leq J\leq i+s; \qquad J+i\geq s, \tag{273}$$

which means that only a fraction of the levels of the compound nucleus will contribute to the cross section. If the outgoing particle is slow, there will be a similar selection rule determining the angular momentum i' of the final nucleus. γ-rays are to be considered as "slow" particles in this connection, with s being replaced by l, i.e., 1 and 2 for dipole and quadrupole radiation, respectively. For fast material particles no useful selection rule holds.

The application of the dispersion formulae developed in this section to the various nuclear processes will form the main content of this report. From the experimental data, we shall deduce the widths corresponding to the emission of various sorts of particles. The main results known thus far are the following:

The *γ-ray width* seems to be of the order of 0.1 to 1 volt for most of the "slow neutron levels" of medium heavy nuclei, i.e., for levels with an excitation energy around 8 MV. (§61.) For *light* nuclei, the γ-ray width is of the order of 1 to 10 volts for rather large excitation energies ($\sim$15 MV), according to the evidence from the capture of protons (§81).

The *neutron width* is for medium heavy nuclei ($A\sim100$) and excitation energies around 8 MV, of the order of 10^{-3} volts at a neutron energy of the order of a volt (§60). According to (266a), the neutron width is *cet. par.* proportional to the square root of the neutron energy. It will therefore be about 1 volt for neutrons of 1 MV energy, and near 10 volts if the neutron energy is of the order of the nuclear interaction potentials ($\sim$20 MV). There is not much evidence on the neutron width for light nuclei except that it is very much larger.

The *proton width* is known for some very light nuclei, e.g., Be^8 (§81). It contains one factor representing the transmission of protons through the potential barrier (§68) and one factor giving the proton width without barrier. The latter should be comparable to the neutron width; it is of the order of 50,000 volts for the resonance level of Be^8 at an excitation energy of 17 MV and a proton energy of 400 kv. Data from other light nuclei are similar.

The widths corresponding to the emission of deuterons, α-particles, etc., from the compound nucleus are, apart from the different transmission of the potential barrier, of the same order as for protons and neutrons (Chapter XIII).

§53. The Distribution of Nuclear Energy Levels. (B13, B4, B33, F30, O12)

We have already found in our qualitative discussions in §51 that the energy levels of a heavy nucleus will be very closely spaced and that their spacing will decrease rapidly with increasing energy. In order to find theoretical expressions for the magnitude of the spacing and its dependence on the energy we must, of course, use some model of the nucleus.

Two models suggest themselves which may be considered as opposite extremes:

(a) We may start with *free* individual particles and consider the total energy of the nucleus as equal to the sum of the energies of the individual particles. This amounts to assuming the interaction between the particles to be small; the nucleus would then be comparable to a gas.

(b) We may consider the interaction to be large, more accurately, large compared to the kinetic energy of the particles. Then the nucleus will in first approximation correspond to a drop of liquid, the distance between neighboring particles remaining almost constant all the time. It seems that assumption (b) will come nearer the truth.

A calculation using assumption (a) was given by Bethe (B13) and also by Oppenheimer (O12), while assumption (b) was proposed by Bohr and Kalckar (B33). An intermediate model, treating the nuclear particles like the electrons in metals, i.e., as half-free particles with correlations between their positions, was treated by Bardeen (B4).

In all cases, we consider the nuclear problem as a statistical one. E.g., if we take assumption (b), we shall first find out the normal modes of vibrations of the particles in a drop of liquid, and then assume that each normal mode has an excitation energy given by the Planck formula. Summing the excitation energies of all normal modes, we obtain the total excitation energy of the nucleus as a function of the "temperature" introduced in the Planck formula. This relation between total excitation energy and temperature should, of course, be considered as a definition of the latter.

Since we are only dealing with excitation energies very small compared to the total binding energy of the nucleus (10 as compared to $\sim$1000 MV), the temperature is always "low." From the relation between energy and temperature we may obtain the specific heat and the entropy of the nucleus using the ordinary thermodynamical relations. The entropy S is then by definition related to the number of states of the nucleus $\rho \sim e^S$. Thus the number of states (per energy interval) can be expressed in terms of the excitation energy of the nucleus.

If we take a different model for the nucleus, e.g., the free particle model, we shall only change the relation between energy and temperature, and therefore between the density of levels and the excitation energy. The principle of the calculation remains unchanged. We shall therefore carry out the calculations first assuming a general relation between energy and temperature, and shall specify the nuclear model only at the end.

Our object is to obtain the average spacing of the nuclear levels

$$D(U) = 1/\rho(U), \tag{274}$$

where $\rho(U)dU$ is the number of levels with energy between U and $U+dU$. To obtain the "density of levels" $\rho(U)$, we consider the expression familiar from statistics:

$$\sum_k e^{-E_k/\tau} = e^{-F/\tau}, \tag{275}$$

in which the sum is to be extended over all levels E_k of the nucleus. τ is an arbitrary parameter, and $F(\tau)$ a certain function of τ defined by (275). If the levels are very dense, the expression (275) can be written as an integral

$$\int \rho(E) e^{-E/\tau} dE = e^{-F/\tau}. \tag{275a}$$

It will be possible to find $\tau(U)$ from this equation if we can (1) choose the parameter τ in such a way that only energies near U contribute appreciably to the integral in (275a), (2) determine F as a function of τ and both these quantities as functions of U.

Provided τ *can* be determined so that condition (1) is fulfilled, $\rho(E)e^{-E/\tau}$ will have a sharp maximum near $E=U$. Then we have obviously

$$\frac{\sum E_k e^{-E_k/\tau}}{\sum e^{-E_k/\tau}} = \frac{\int E\rho(E)e^{-E/\tau}dE}{\int \rho(E)e^{-E/\tau}dE} = U. \tag{276}$$

This relation *determines* the parameter τ, as a function of U, if the energy spectrum of the system is known. Conversely, it expresses U as a function of τ. The two functions $U(\tau)$ and $F(\tau)$ are connected by

$$U = \frac{d(F/\tau)}{d(1/\tau)} = F - \tau\frac{dF}{d\tau}, \tag{276a}$$

which follows immediately from (275a) (276). τ, U and F correspond, in ordinary statistical theory, respectively, to kT, to the energy of the system and to the free energy.

As is shown in classical statistics, the condition (1) will be fulfilled by the τ as defined in (276), provided the system contains sufficiently many particles. Then the main contribution to the integral (275a) will come from energy levels E_k near U. Thus we may write instead of (275a)

$$e^{-F/\tau} = \rho(U)e^{-U/\tau}\lambda(U), \tag{277}$$

where $\lambda(U)$ is a quantity of the dimensions of an

energy which is to be calculated by evaluating the integral in (275a). It is a slowly varying function of the excitation energy U as compared to the rapidly varying functions $e^{-F/\tau}$, $e^{-U/\tau}$ and $\rho(U)$. We rewrite (277)

$$\rho(U)=e^{(U-F)/\tau}/\lambda(U). \tag{277a}$$

According to the definition (276), τ is that temperature for which the average energy of the system is U, and F is the corresponding free energy. We introduce the abbreviation

$$(U-F)/\tau=S. \tag{278}$$

S is then the entropy divided by Boltzmann's constant k. Then we have

$$\rho(U)=e^{S(U)}/\lambda(U) \tag{278a}$$

and it remains only to obtain, in specific terms, $S(U)$ and $\lambda(U)$.

For S we have, according to (276a)

$$S=(U-F)/\tau=-dF/d\tau. \tag{278b}$$

Also from (276a) we find

$$\frac{dU}{d\tau}=-\tau\frac{d^2F}{d\tau^2} \tag{278c}$$

so that

$$S=\int\frac{d\tau}{\tau}\frac{dU}{d\tau}=\int\frac{Cd\tau}{\tau}, \tag{279}$$

where C is the specific heat. From (279) the well-known relation

$$\frac{dS}{d\tau}=\frac{1}{\tau}\frac{dU}{d\tau} \tag{279a}$$

follows immediately, and, since τ and U are uniquely related to each other,

$$\frac{dS}{dU}=\frac{1}{\tau}, \tag{279b}$$

expressing the elementary definition of the entropy. S is therefore known as a function of U if the relation between U and τ is known.

To determine λ, we insert (278a) in (275a) and have

$$e^{-F/\tau}=\int e^{S(E)-E/\tau}dE/\lambda(E), \tag{280}$$

where $S(E)$ is the entropy corresponding to the energy E. Since λ varies slowly with the energy, and since the integrand is very small except for $E\approx U$, we may replace $\lambda(E)$ by $\lambda(U)$ and have, using (277) again:

$$\lambda(U)=\int e^{S(E)-S(U)+(U-E)/\tau}dE. \tag{280a}$$

With (279b) the exponent becomes

$$\frac{dS}{dU}(E-U)+\frac{1}{2}\frac{d^2S}{dU^2}(E-U)^2+\frac{U-E}{\tau}$$
$$=-\frac{1}{2\tau^2}\frac{d\tau}{dU}(E-U)^2. \tag{280b}$$

The integration gives then immediately

$$\lambda(U)=(2\pi)^{\frac{1}{2}}\tau(dU/d\tau)^{\frac{1}{2}}. \tag{280c}$$

With (274) and (278a), the average spacing between levels becomes thus

$$D(U)=1/\rho(U)=\lambda(U)e^{-S(U)}$$
$$=\tau(2\pi dU/d\tau)^{\frac{1}{2}}e^{-S}. \tag{281}$$

We make now the more special assumption that U depends on the temperature according to a power law

$$U=\alpha\tau^n. \tag{282}$$

This is true for the model (a) proposed above with n equal to 2. For model (b) it is fulfilled for low temperature (low excitation energies) with $n=7/3$, and for high temperature with $n=4$ (see below). With the law (282) we find, using (279),

$$S=\frac{\alpha n}{n-1}\tau^{n-1}=\frac{n}{n-1}(\alpha U^{n-1})^{1/n} \tag{283}$$

and inserting in (281)

$$D(U)=(2\pi n)^{\frac{1}{2}}$$
$$\left(\frac{U^{n+1}}{\alpha}\right)^{\frac{1}{2}n}\exp\left(-\frac{n}{n-1}\alpha^{1/n}U^{(n-1)/n}\right). \tag{284}$$

The spacing decreases, according to (284), exponentially with increasing energy. The exponent contains $U^{\frac{1}{2}}$ for the Fermi gas model (model a, $n=2$), $U^{4/7}$ for the liquid drop model at low and $U^{3/4}$ at high temperatures. The "liquid" model thus gives a somewhat more rapid decrease of the spacing.

The exact value of α can, of course, only be obtained from a special model. Quite generally α increases with the number of particles A in the nucleus, so that the spacing of nuclear levels *decreases* with increasing A, for constant excitation energy U. The decrease with A is in general more rapid the slower the decrease with increasing energy; e.g., it is more rapid for the free particle model (a) than for the "liquid drop" model (b).

A. Free particle model

In this model, the nucleus is considered as a mixture of two Fermi gases of neutrons and protons. At zero temperature, the particles occupy all the lowest levels while the higher levels are empty. At higher temperatures, there will be some particles in the higher levels and some of the lower levels will be empty.

The statistical treatment given above is not directly applicable to this case because it presupposes that the energy levels of the given system containing, say, N neutrons and Z protons, are known. In the customary treatment of Fermi statistics (and any gas statistics) not only systems of the correct number of particles are considered but also such for which the number of particles is slightly different.

The only way in which the given number of particles is taken into account is by making the statistical probability a maximum for that number of particles. This is achieved by suitably choosing a certain parameter ζ, the Fermi energy. But even with the proper choice of ζ, there remains the fact that the distribution function derived in Fermi statistics gives a finite probability also for the states in which the total number of particles does not exactly have the correct value, so that we must multiply the density $\rho(U)$ obtained in (281) by the probability that the number of particles of each kind has the correct value.

In addition to this, we must remember that, owing to the selection rules, only compound states with one or a few values of the angular momentum are important for a given nuclear process. Therefore, we are interested in the states of the nucleus with a given angular momentum J. This constant of motion should be treated in the same way as N and Z so that we have altogether three constants of motion besides the energy, *viz.* N, Z and J.

For generality, we assume that an arbitrary number m of constants of motion $N_1 N_2 \cdots N_m$, is given. Then there will be an equal number of parameters $\zeta_1 \cdots \zeta_m$ in the distribution function which must be chosen in such a way that in the average $N_1 N_2 \cdots$ have their correct values. Then we have instead of (275)

$$e^{-\Phi/\tau} = \sum_{k,\, N_1' \ldots N_m'} e^{(-E_k + N_1'\zeta_1 + \ldots + N_m'\zeta_m)/\tau}. \tag{285}$$

The sum extends over all possible values of $N_1' \cdots N_m'$ including their correct values $N_1 \cdots N_m$. (285) may be rewritten in integral form

$$e^{-\Phi/\tau} = \int \rho(E, N_1' \cdots N_m') e^{-E/\tau + \alpha_1 N_1' + \ldots + \alpha_m N_m'} \tag{285a}$$

with

$$\alpha_i = \zeta_i / \tau. \tag{285b}$$

Φ must be regarded as a function of the parameters τ, $\alpha_1 \cdots \alpha_m$. We have

$$\left(\frac{\partial(\Phi/\tau)}{\partial(1/\tau)}\right) = U, \tag{286}$$

$$\frac{1}{\tau}\left(\frac{\partial \Phi}{\partial \alpha_i}\right) = -N_i, \tag{286a}$$

where the partial derivative with respect to any parameter α_i is to be taken with the other parameters α_k $(k \neq i)$ and τ being kept constant.

The density of levels may again be written

$$\rho(U, N_1 \cdots N_m) = e^S / \lambda \tag{287}$$

with

$$S(U, N_1 \cdots N_m) = (U - \Phi)/\tau - \alpha_1 N_1 - \cdots - \alpha_m N_m. \tag{287a}$$

S is a function of the energy U and the "constants of motion" $N_1 \cdots N_m$ only but does not depend explicitly[9] on the parameters τ and α. The derivatives of the entropy with respect to the "constants of motion" are

$$\partial S / \partial U = 1/\tau, \tag{288}$$

$$\partial S / \partial N_i = -\alpha_i. \tag{288a}$$

(288) means that formula (279b) and therefore (283) remains true in our more complicated case provided the number of particles is constant. We assume now again that λ is a slowly varying function of its arguments, and find, similarly to (280a):

$$\lambda(U) = \int e^{S(E) - S(U) + (U-E)/\tau + \alpha_1 (N_1' - N_1) + \ldots} \, dE dN_1' \cdots dN_m'. \tag{289}$$

With (288, 288a) the exponent becomes

$$\frac{1}{2}\sum_{i=0}^{m}\sum_{j=0}^{m} \frac{\partial^2 S}{\partial N_i \partial N_j}(N_i' - N_i)(N_j' - N_j)$$

$$= -\frac{1}{2}\sum_{ij} \frac{\partial \alpha_i}{\partial N_j}(N_i' - N_i)(N_j' - N_j), \tag{289a}$$

---

[9] This follows immediately by differentiating (287a) with respect to any of the parameters, keeping the other parameters and U, $N_1 \cdots N_m$ constant. E.g.,

$$\partial S / \partial \alpha_i = -(1/\tau)(\partial \Phi / \partial \alpha_i) - N_i = 0 \quad \text{(cf. (286a))}.$$

S and Φ correspond formally to Hamiltonian and Lagrangian in mechanics.

where we have put

$$E=N_0', \quad U=N_0, \quad 1/\tau=-\alpha_0. \tag{289b}$$

The integration of (289) is then straightforward and gives

$$\lambda(U)=(2\pi)^{\frac{1}{2}(m+1)}/\Delta^{\frac{1}{2}}, \tag{290}$$

where Δ is the Jacobian

$$\Delta=\left\|\frac{\partial\alpha_i}{\partial N_j}\right\|=\left\|-\frac{\partial^2 S}{\partial N_i\partial N_j}\right\|. \tag{290a}$$

If we are treating a mixture of degenerate Fermi gases, then the Fermi energy ζ_i for the ith kind of particles is determined solely by the number N_i of these particles, except for a very small dependence on the excitation energy U which we neglect. Since $\alpha_i=\zeta_i/\tau$, we have (cf. 289b)

$$\frac{\partial\alpha_0}{\partial N_0}=-\frac{\partial(1/\tau)}{\partial U}, \quad \frac{\partial\alpha_0}{\partial N_i}=-\frac{\partial(1/\tau)}{\partial N_i},$$

$$\frac{\partial\alpha_i}{\partial N_0}=\zeta_i\frac{\partial(1/\tau)}{\partial U}, \quad \frac{\partial\alpha_i}{\partial N_i}=\frac{1}{\tau}\frac{\partial\zeta_i}{\partial N_i}+\zeta_i\frac{\partial(1/\tau)}{\partial N_i},$$

$$\frac{\partial\alpha_i}{\partial N_k}=\zeta_i\frac{\partial(1/\tau)}{\partial N_k}. \tag{290b}$$

We add ζ_i times the zeroth row to the ith row of the determinant; then there remains in that row only the diagonal term $\tau^{-1}d\zeta_i/dN_i$. Thus the value of the Jacobian is

$$\Delta=-\left(\frac{\partial(1/\tau)}{\partial U}\right)_{N_i}\prod_{i=1}^{m}\frac{1}{\tau}\frac{d\zeta_i}{dN_i} \tag{291}$$

and the spacing between neighboring levels becomes (cf. (287))

$$D(U)=\frac{1}{\rho(U)}=(2\pi)^{\frac{1}{2}(m+1)}\tau^{\frac{1}{2}m+1} \times\left(\frac{dU}{d\tau}\prod_{i=1}^{m}\frac{dN_i}{d\zeta_i}\right)^{\frac{1}{2}}e^{-S(U)}. \tag{292}$$

We now turn to the special features of the free particle model. We assume that the particles move in a spherical potential hole of a certain depth, and that their interaction is small (except inasmuch as it is expressed by the potential hole). Neutrons and protons will obey Fermi statistics, and the number of neutrons having a kinetic energy between ϵ and $\epsilon+d\epsilon$ will be given by the usual formula of Fermi statistics (cf. B13)

$$n(\epsilon)d\epsilon=\tfrac{3}{2}Cf(\epsilon)\epsilon^{\frac{1}{2}}d\epsilon \tag{293}$$

with

$$C=(2^{7/2}/9\pi)(MR^2/\hbar^2)^{\frac{3}{2}}, \tag{293a}$$

$$f(\epsilon)=1/e^{(\epsilon-\zeta_1)/\tau}+1, \tag{293b}$$

R being the nuclear radius and ζ_1 the Fermi energy for neutrons. Integration gives for $\zeta_1\gg\tau$

$$N=\int n(\epsilon)d\epsilon=C\zeta_1^{\frac{3}{2}}(1+(\pi^2/8)(\tau/\zeta_1)^2+\cdots) \tag{294}$$

and for the energy

$$U_1=\tfrac{3}{5}C\zeta_1^{5/2}(1+(5\pi^2/8)(\tau/\zeta_1)^2+\cdots). \tag{295}$$

From (294), we have in sufficient approximation

$$\frac{dN}{d\zeta_1}\approx\frac{3}{2}C\zeta_1^{\frac{1}{2}}=\frac{3}{2}C^{\frac{2}{3}}N^{\frac{1}{3}}=\frac{3}{2}\frac{N}{\zeta_1} \tag{294a}$$

and a corresponding relation for the protons. From (295), we find after a short calculation (cf. B13)

$$U=U_1+U_2-U_0=\tfrac{1}{4}\pi^2C\tau^2(\zeta_1^{\frac{1}{2}}+\zeta_2^{\frac{1}{2}}), \tag{295a}$$

where U_1 is the total energy of the neutrons, U_2 that of the protons, U_0 the zero-point energy which depends only on the numbers of protons and neutrons but not on the temperature τ, and is therefore irrelevant for the determination of the specific heat $dU/d\tau$ and the entropy (cf. 278). Since the numbers of neutrons and protons are not very different, we may put $\zeta_1=\zeta_2=\zeta_0$ where ζ_0 is an average Fermi energy for protons and neutrons. Then

$$U=\tfrac{1}{2}\pi^2C\zeta^{\frac{1}{2}}\tau^2=\tfrac{1}{4}\pi^2(A/\zeta_0)\tau^2, \tag{296}$$

using (294) and putting $N=\frac{1}{2}A$, in analogy with our approximation[10] $\zeta_1=\zeta_2=\zeta_0$. From (294) (296) we find immediately the entropy S, and the quantities $dU/d\tau$, $dN/d\zeta_1$, $dZ/d\zeta_2$ occurring in (292).

We must now introduce a parameter γ analogous to the ζ's which makes a certain value J of the *angular momentum* most probable. It is more convenient to work with the component M of J in a given direction Z because M is the sum of

[10] The justification of these approximations is proved in B13.

the angular momentum components of the individual particles, *viz.*

$$M=\sum_{i=1}^{A} m_i. \tag{297}$$

We write for the probability to find a neutron state of energy and angular momentum m occupied

$$f(\epsilon, m)=1/(e^{(\epsilon-\zeta_1-\gamma m)/\tau}+1). \tag{297a}$$

The expressions (294) (296) for the total number of particles and for the total energy will be changed only by quantities of the order γ^2. The average value of the total angular momentum is

$$M=\int mg(\epsilon, m)\left(\frac{1}{e^{(\epsilon-\zeta_1-\gamma m)/\tau}+1}+\frac{1}{e^{(\epsilon-\zeta_2-\gamma m)/\tau}+1}\right)d\epsilon dm, \tag{297b}$$

where $g(\epsilon, m)$ is the number of states of an individual particle with given ϵ and m.

The number $g(\epsilon, m)$ of states with given ϵ and m of an individual particle in a spherical potential hole of radius R, can be found as follows. Leaving out the spin, we describe each particle state by the three quantum numbers n, l and m where l is the total angular momentum, m its component in the Z direction and n the radial quantum number. According to the WKB method, n is given by

$$n\pi=\int_{\min}^{R}\left(\frac{2M\epsilon}{\hbar^2}-\frac{(l+\frac{1}{2})^2}{r^2}\right)^{\frac{1}{2}}dr, \tag{298}$$

where $r_{\min}$ is the value of r for which the integrand vanishes. Since there is just one state for each integral value nlm, the number of states of given l and m per unit energy is

$$g(\epsilon, l, m)=\frac{dn}{d\epsilon}=\frac{1}{\pi}\frac{M}{\hbar^2}\int\left(\frac{2M\epsilon}{\hbar^2}-\frac{(l+\frac{1}{2})^2}{r^2}\right)^{-\frac{1}{2}}dr$$

$$=\frac{1}{2\pi\epsilon}\left(\frac{2M\epsilon R^2}{\hbar^2}-(l+\tfrac{1}{2})^2\right)^{\frac{1}{2}} \tag{298a}$$

The total number of states of given m and energy between ϵ and $\epsilon+d\epsilon$ becomes

$$g(\epsilon, m)d\epsilon=d\epsilon\int_{|m|}^{l_0} g(\epsilon, l, m)dl \tag{298b}$$

where l_0 is the value of l for which (298a) vanishes. This expression should be multiplied by 2 for spin. Putting $\zeta_1=\zeta_2=\zeta_0$ and expanding in powers of γ, (297b) becomes then

$$M=-4\gamma\int d\epsilon\frac{d}{d\epsilon}(e^{(\epsilon-\zeta_0)/\tau}+1)^{-1}\int dl\, g(\epsilon, l, m)\int_{-(l+\frac{1}{2})}^{l+\frac{1}{2}} m^2dm. \tag{299}$$

The elementary evaluation of this integral, using (298a), gives

$$M=c\gamma \tag{299a}$$

with

$$c=\tfrac{2}{5}(MR^2/\hbar^2)A. \tag{299b}$$

The entropy for a given N, Z, U may thus be written (cf. (288a))

$$S(N, Z, U, M)=S(N, Z, U, 0)-(M^2/2c\tau) \tag{300}$$

and the density of states, for given NZU, depends on M as (cf. (287))

$$\rho(N, Z, U, M)=\rho(N, Z, U, 0)\exp(-M^2/2c\tau). \tag{300a}$$

We must now determine the number of states of given J. This is equal to the difference between the number of quantum states with $M=J$, and the number of levels with $M=J+1$ (cf. B13) and therefore (cf. (300a))

$$\rho(N, Z, U, J)=\rho(N, Z, U, 0)\times[(J+\tfrac{1}{2})/c\tau]\exp[-(J+\tfrac{1}{2})^2/2c\tau]. \tag{301}$$

It can be shown (see (299b)) that $c\tau$ is rather large (≈ 1000) so that, for the important values of J, the exponential in (301) may be neglected. Therefore we obtain for the spacing of levels of angular momentum J (cf. (292))

$$D(N, Z, U, J)=D_0/(2J+1), \tag{302}$$

$$D_0=\frac{2c\tau}{\rho(N, Z, U, 0)}=2c(2\pi)^2\tau^{7/2}\times\left(\frac{dU}{d\tau}\frac{dN}{d\zeta_1}\frac{dZ}{d\zeta_2}\frac{dM}{d\gamma}\right)^{\frac{1}{2}}e^{-S(N, Z, U, 0)}. \tag{302a}$$

Inserting (296) (294a) (279) (299a) (299b) we have

$$D_0 = 3\sqrt{2}\pi^3\tau^4(Ac/\zeta_0)^{3/2}e^{-\frac{1}{2}\pi^2 A\tau/\zeta_0}, \qquad (302b)$$

$$D = (18/5)^{3/2}(A^2U^2/\zeta_0)e^{-\pi(AU/\zeta_0)^{1/2}}. \qquad (302c)^{11}$$

Here the Fermi energy ζ_0 is according to (293a) (294) and with $N=\frac{1}{2}A$:

$$\zeta_0 = (A/2C)^{2/3} = 1.15_5 P, \qquad (303)$$

where

$$P = \hbar^2/Mr_0^2, \qquad (303a)$$

$$r_0 = RA^{-1/3}. \qquad (303b)$$

This radius r_0 is independent of A if the nuclear volume is proportional to the number of particles. From the radii of α-radioactive nuclei (Chapter XI)

$$r_0 = 2.0_5 \cdot 10^{-13}\text{ cm}, \qquad (303c)$$

so that

$$P = 10\text{ MV}, \qquad (303d)$$

$$\zeta_0 = 11.5\text{ MV}, \qquad (303e)$$

which is of the same order as the binding energy of neutrons or protons in nuclei

We introduce now the abbreviation

$$x = \pi(AU/\zeta_0)^{1/2} = (AU/1.17)^{1/2}\ (U\text{ in MV}). \qquad (304a)$$

Then the spacing becomes (cf. (302c))

$$D_0 = 8\cdot 10^5 x^4 e^{-x}\text{ volts.} \qquad (304)$$

For $U=8$ MV and $A=120$, we have $x=28.7$ and $D_0=0.2$ volt. This result for the spacing is very small indeed[12] and is certainly smaller than the spacing observed in slow neutron experiments (cf. §60) which may be of the order of 10 volts for the atomic weight and excitation energy considered. This shows that our assumptions are inadequate, particularly the assumption of free particles with small interaction.

B. Free particle model with correlations

Bardeen (B4) has pointed out that the free particle model in the form used in A, is not in accord with the assumption of exchange forces between the particles. As Van Vleck has shown (§36 of this report, ref. V 1 of part A), the potential acting on one particle depends strongly on the wave number of that particle, and its absolute value decreases appreciably with increasing wave number, i.e., with increasing kinetic energy of the particle. This means that the total energy of an individual particle,

$$E = E_{\text{kin}} + E_{\text{pot}} \qquad (305)$$

depends much more strongly on the wave number k than the kinetic energy itself.

We may put

$$(dE/dk)_{E=\zeta} = \gamma(dE_{\text{kin}}/dk)_{E=\zeta}. \qquad (305a)$$

With the range and magnitude of the nuclear forces derived from the binding energy of light nuclei, Bardeen finds

$$\gamma \approx 2. \qquad (305b)$$

Now the number of individual particles states per unit energy is proportional to dk/dE, and is therefore reduced by the "correlations" to about one-half of its value. A reduction in the density of individual particle states means a much larger reduction in the density of states of the nucleus as a whole. Quantitatively, we can take account of the change of the density of individual states as follows: The excitation energy U is proportional to the density of individual quantum states times the square of the temperature τ. This follows from (296) if we consider that, with kinetic energy alone, the density of individual states of energy near ζ, is $\frac{3}{2}C\zeta^{1/2}$ (cf. (293)); it also follows from elementary considerations on Fermi statistics.

Thus the constant α in

$$U = \alpha\tau^2 \qquad (306)$$

must be multiplied by $1/\gamma$. But, according to (283),

$$S = 2(\alpha U)^{1/2} \qquad (306a)$$

Therefore, if U is kept fixed, the entropy "with correlation" is connected to that with completely free particles by

$$S_{\text{corr}} = S_{\text{fr}}\gamma^{-1/2} \qquad (306b)$$

and the quantity x in (304) has to be replaced by

$$x' = x\gamma^{-1/2}. \qquad (306c)$$

[11] This formula differs somewhat from that given by Bethe (B13, Eqs. (49) (50a)). Part of the difference *viz.* a factor $(\log 4)^{1/2}$ is due to the improved treatment of the angular momentum, another part (factor $2^{3/4}$) to a numerical error in Bethe's paper. These errors have also been pointed out by Bardeen.

[12] It seems somewhat surprising that the free particle model gives *too small* a spacing.

TABLE XX. *Spacing of nuclear energy levels according to Bardeen's method. The figures give the spacing D_0 (in volts) of levels of zero angular momentum, at an energy equal to the average neutron dissociation energy Q (cf. Table XXI, row O).*

A	20	50	100	200
Q (MV)	9.5	9.1	8.2	6.9
$D_0(R_0^*=9\cdot 10^{-13}$ cm)	$1.1\cdot 10^6$	$4.2\cdot 10^4$	1000	10
$D_0(R_0^*=12.5\cdot 10^{-13}$ cm)	$6\cdot 10^4$	300	1.2	$1.2\cdot 10^{-3}$

* R_0 = nuclear radius for $A_0=230$. The radius for any nucleus is assumed to be $R_0(A/A_0)^{1/3}$.

The resulting spacing of energy levels, as a function of the atomic number, is given in Table XX for various nuclear radii and is compared to the spacing obtained from the free particle model without correlations.

C. Liquid drop model

It was pointed out by Bohr and Kalckar (B33) that a nucleus should be considered as a drop of liquid rather than as a gas because the interaction between the nuclear particles is large and therefore fluctuations of the density very improbable. A liquid drop is capable of two types of vibrations, namely surface waves and volume waves. The surface waves are not connected with any changes in volume; the potential energy is then only due to the surface tension and is comparatively small. Therefore the frequency of the surface waves is rather small, and this type of waves will most easily be excited at "low temperatures" i.e., in all cases relevant for nuclear disintegrations. The volume waves involve a compression of the nuclear liquid which is connected with a large increase in potential energy, a high frequency and therefore small excitation probability at "low" temperatures.

1. The surface waves.—For simplicity, assume first that the surface of the liquid is plane when there are no vibrations. Let the xy plane be the equilibrium surface, and let $\zeta(x, y)$ be the (vertical) displacement at a point x, y of the surface. Then the increase in surface area is for small displacements

$$\tfrac{1}{2}\int dxdy[(\partial\zeta/\partial x)^2+(\partial\zeta/\partial y)^2] \qquad (307)$$

integrated over the surface. If G is the total surface energy of the liquid in *equilibrium* and $S=\int dxdy$ the total surface area, the increase in potential energy is

$$V=(G/2S)\int dxdy[(\partial\zeta/\partial x)^2+(\partial\zeta/\partial y)^2]. \qquad (307a)$$

In order to find the kinetic energy we must know the velocity **u** at every point in the liquid. We assume that the motion of the liquid is irrotational; then **u** may be written as the gradient of a velocity potential which we call $\partial\psi/\partial t$, *viz.*

$$\mathbf{u}=-\operatorname{grad}\partial\psi/\partial t. \qquad (308)$$

Since the surface waves are not connected with any compression of the liquid, we have

$$\operatorname{div}\mathbf{u}=-\Delta(\partial\psi/\partial t)=0. \qquad (308a)$$

Then the kinetic energy becomes

$$T=\tfrac{1}{2}\rho\int u^2d\tau=\tfrac{1}{2}\rho\int(\operatorname{grad}\partial\psi/\partial t)^2d\tau$$

$$=\tfrac{1}{2}\rho\int\frac{\partial\psi}{\partial t}\frac{\partial\psi}{\partial z\partial t}dxdy, \qquad (308b)$$

where ρ is the density. The last integral is to be extended over the surface $z=0$ of the liquid and the vanishing of the Laplacian has been used.

From (308) it follows that the displacement may be written

$$\boldsymbol{\varrho}=-\operatorname{grad}\psi \qquad (309)$$

and in particular the displacement of the particles on the surface of the liquid

$$\zeta=-(\partial\psi/\partial z)_{z=0}. \qquad (309a)$$

Because of the absence of volume changes we have again

$$\Delta\psi=0. \qquad (309b)$$

We assume now that ψ is a periodic function of the coordinates in the surface plane, x and y, and of t. Then from (309b) it follows that

$$\psi=a\cos(k_xx+k_yy)e^{kz}\cos\omega t, \qquad (309c)$$

where
$$k=(k_x^2+k_y^2)^{\frac{1}{2}}, \qquad (309d)$$

while the relation between ω and k must be calculated. The positive sign of the term kz in the exponent of (309c) makes the displacement and the velocity vanish for large negative z.

The total potential energy (307a) becomes now, with (309a) and (309c)

$$V=(G/2S)a^2\cos^2\omega t\,k^4\int\sin^2(k_xx+k_yy)dxdy$$

$$=\tfrac{1}{4}Ga^2\cos^2\omega t\,k^4, \qquad (310)$$

since the average of $\sin^2$ over the surface is $\tfrac{1}{2}$. Similarly, the kinetic energy (308b) becomes

$$T=\tfrac{1}{4}\rho Sa^2\omega^2\sin^2\omega t\,k. \qquad (310a)$$

If the sum of potential and kinetic energy is to be

constant, we must have

$$\omega = k^{3/2}(G/\rho S)^{1/2}. \tag{311}$$

From this expression, we can easily find the number of normal modes per unit frequency ω, a number which is needed for calculating the energy of the liquid drop as a function of the temperature. Following the usual method, we obtain for the number of normal modes per $dk_x dk_y$

$$(S/4\pi^2)dk_x dk_y. \tag{312}$$

Therefore the number of vibrations with wave numbers between k and $k+dk$ is

$$(S/4\pi^2)2\pi k dk \tag{312a}$$

and that with frequencies between ω and $\omega+d\omega$, according to (311)

$$p(\omega)d\omega = (S/2\pi)(\rho S/G)^{2/3}\tfrac{2}{3}\omega^{1/3}d\omega. \tag{312b}$$

Essentially the same formula can also be derived for a spherical drop by a more detailed study of its surface vibrations, using an analysis of the deformation in spherical harmonics.

For a spherical nucleus, the surface area is

$$S = 4\pi R^2 \tag{313}$$

and the density $\rho = 3AM/4\pi R^3$. (313a)

The surface energy G can be deduced from empirical data on nuclear binding energies using the method outlined in §30; we find

$$G = \Gamma A^{2/3}, \tag{314}$$

where $\Gamma = 9.6 \text{ MV} \approx 10 \text{ MV}$ (314a)

if the nuclear radius has the value $2.05 \cdot 10^{-13}A^{1/3}$ (§68). If we introduce, instead of ω, the quantum energy

$$\epsilon = \hbar\omega \tag{315}$$

(312b) may be written

$$p_S(\epsilon)d\epsilon = 4 \cdot 3^{-1/3}(\Gamma P)^{-2/3}A^{2/3}\epsilon^{1/3}d\epsilon \tag{315a}$$

with P as defined in (303a, d). For a heavy nucleus ($A = 200$), this gives about 3 normal modes of surface vibrations[12a] with quantum energies below 1 MV. This seems reasonable in connection with the fact that the low excitation levels of radioactive nuclei lie at a few hundred kilovolts.

[12a] From a direct analysis of the vibrations of the spherical drop, it follows that the lowest frequency lies actually at about 1.1 MV. Our distribution function (315a) will therefore give a good approximation only for nuclear temperatures of 1 MV or higher. For lighter nuclei, the minimum temperature required increases as $A^{-1/3}$.

2. *Volume waves.*—An ideal liquid will admit only longitudinal volume waves. The number of normal modes is

$$\Omega(2\pi)^{-3} \cdot 4\pi k^2 dk = (2/3\pi)R^3\omega^2 d\omega/u_0^3, \tag{316}$$

where u_0 is the velocity of sound. This quantity is connected to the compressibility of the substance, i.e., to the second derivative of the energy E with respect to the density σ,

$$u_0^2 = \frac{\sigma^2}{MA}\frac{d^2E}{d\sigma^2} = \frac{R^2}{9MA}\frac{d^2E}{dR^2}, \tag{317}$$

where MA is the mass, E the energy, R the radius and $\sigma \propto R^{-3}$ the density of the nucleus ($dE/dR = 0$). d^2E/dR^2 may be estimated from the statistical formula for the energy as a function of the radius, (159) (159a):

$$\frac{E}{AT_0} = \frac{3}{10}x^2 - \frac{B}{\pi^{1/2}T_0}\{2x^{-3} - 3x^{-1} + (x^{-1} - 2x^{-3})e^{-x^2} + \pi^{1/2}\Phi(x)\}, \tag{318}$$

where $T_0 = \hbar^2/Ma^2$, $x = \frac{3}{2}\left(\frac{\pi}{3}\right)^{1/3}\frac{a}{r_0}$ (318a)

and Be^{-r^2/a^2} is the interaction between two particles. We use for B, T_0 and x the values derived from the new nuclear radius according to the same method as used in §26, *viz.*

$$B = 34 \text{ MV}, \quad T_0 = 1.65 \text{ MV}, \quad a = 5.0 \cdot 10^{-13} \text{ cm}, \quad x = 3.66. \tag{318b}$$

We obtain for the second derivative of E:

$$R^2\frac{d^2E}{dR^2} = x^2\frac{d^2E}{dx^2} = 6AT_0\left[\frac{1}{10}x^2 + \frac{B}{T_0\pi^{1/2}}x^{-3}\{x^2 - 4 + (x^4 + 3x^2 + 4)e^{-x^2}\}\right]. \tag{318c}$$

Inserting the values (318b), this becomes

$$R^2(d^2E/dR^2) = 35A \text{ MV}. \tag{319}$$

Therefore, according to (317)

$$K = Mu_0^2 = (R^2/9A)(d^2E/dR^2) \approx 4 \text{ MV}. \tag{319a}$$

Inserting (319a) into (316), we obtain for the number of longitudinal volume waves with a quantum energy between ϵ and $\epsilon+d\epsilon$

$$p_l(\epsilon)d\epsilon = (2/3\pi)(KP)^{-3/2}A\epsilon^2 d\epsilon. \tag{320}$$

The total number of "longitudinal" modes of vibration is equal to the number of particles, A. The number of modes of surface waves is of the order of the number of particles in the surface, i.e., about $A^{2/3}$. For large A, this is negligible compared to A. Thus longitudinal waves and surface waves together do not give all possible modes of vibration whose number must be $3A$ (number of degrees of freedom). The additional modes must apparently be "transverse" volume waves. If such waves have fairly short wave-length, their frequency will be of the same order as that of longitudinal waves of the same length: For short waves, only the relative displacement of close neighbors matters, but the arrangement and interaction of close neighbors is about the same in a liquid as in a crystal, and in a crystal there is no great difference between the frequencies of waves of the same wave-length and different polarization provided the wave-length is short. Thus, for short waves, we shall get approximately the correct number of modes of vibration per $d\epsilon$ if we multiply (320) by three, i.e. the number of possible polarizations is

$$p_v(\epsilon)d\epsilon=(2/\pi)(KP)^{-3/2}A\epsilon^2 d\epsilon. \qquad (320a)$$

The frequency of long transverse volume waves in a liquid is an unsolved problem. We may only hope that such waves are not very essential for the problem of the specific heat. This may perhaps be justified by the fact that to a certain extent, the long transverse waves will be replaced by surface waves. Moreover, at the low temperatures which are important for nuclear reactions, the influence of the surface waves is much more important than that of the volume waves. In the absence of any correct treatment, we shall therefore assume (320a) to give the correct number of volume waves for low as well as for high quantum energy.

3. Thermal properties of the liquid drop.—The total number of normal modes of quantum energy between ϵ and $\epsilon+d\epsilon$ is (cf. (315a), (320a))

$$p(\epsilon)d\epsilon=(4\cdot 3^{-1/3}(\Gamma P)^{-2/3}A^{2/3}\epsilon^{1/3}$$
$$+(2/\pi)(KP)^{-3/2}A\epsilon^2)d\epsilon. \qquad (321)$$

The energy at the "temperature" τ is then, according to Planck's formula

$$U=\int_0^\infty \frac{p(\epsilon)d\epsilon}{e^{\epsilon/\tau}-1}=\frac{4}{3^{1/3}}C_{4/3}\frac{A^{2/3}\tau^{7/3}}{(\Gamma P)^{2/3}}+\frac{2}{\pi}C_3\frac{A\tau^4}{(KP)^{3/2}}, \qquad (322)$$

where $$C_n=\int_0^\infty \frac{x^n dx}{e^x-1}=n!\zeta(n+1), \qquad (322a)$$

$$\zeta(n+1)=1^{-(n+1)}+2^{-(n+1)}+\cdots, \qquad (322b)$$

$$C_{4/3}=1.694, \quad C_3=6.50. \qquad (322c)$$

We introduce a "critical temperature" τ_0 as that temperature at which the contributions of surface and volume waves to the energy become equal. According to (322),

$$\tau_0=(2\pi C_{4/3}/C_3)^{3/5}3^{-1/5}(K/\Gamma)^{2/5}(KP)^{1/2}A^{-1/5}$$
$$=1.080(K/\Gamma)^{2/5}(KP)^{1/2}A^{-1/5}. \qquad (323)$$

At this temperature, the contributions of surface and volume waves to U each have the value

$$U_0=2^{17/5}\pi^{7/5}3^{-4/5}C_{4/3}{}^{12/5}C_3{}^{-7/5}(K/\Gamma)^{8/5}(KP)^{1/2}A^{1/5}$$
$$=5.60(K/\Gamma)^{8/5}(KP)^{1/2}A^{1/5}. \qquad (324)$$

If we insert the values of P, Γ and K (cf. (303d), (314a), (319a)), we have

$$U_0=8.2A^{1/5}\text{ MV}, \qquad (324a)$$

which, for $A=100$, has the value 20.6 MV. In most practical cases, the excitation energy will therefore be considerably less than $2U_0$, and therefore the temperature less than τ_0. This means that surface waves are in general more important than volume waves.

We introduce the abbreviation

$$T=\tau/\tau_0. \qquad (325)$$

Then (322) reduces to (cf. (323), (324))

$$U/U_0=T^{7/3}+T^4. \qquad (325a)$$

The entropy is

$$S=\int\frac{d\tau}{\tau}\frac{dU}{d\tau}=\frac{U_0}{\tau_0}\left(\frac{7}{4}T^{4/3}+\frac{4}{3}T^4\right)$$
$$=S_0\left(\frac{21}{16}T^{4/3}+T^4\right) \qquad (326)$$

with $$S_0=4U_0/3\tau_0=6.93(K/\Gamma)^{6/5}A^{2/5}. \qquad (326a)$$

With $P=\Gamma=10$ MV and $K=4$ MV, we have

$$S_0=2.31A^{2/5}, \qquad (326b)$$

$$\tau_0=4.75A^{-1/5}\text{ MV}. \qquad (326c)$$

Finally, the quantity λ in (281) has the value

$$\lambda=(2\pi dU/\partial\tau)^{1/2}\tau=\lambda_0\left(\frac{7}{12}T^{4/3}+T^3\right)^{1/2}T \qquad (327)$$

with

$$\lambda_0=(8\pi U_0\tau_0)^{1/2}=12.3(K/\Gamma)(KP)^{1/2}=31\text{ MV}. \qquad (327a)$$

The temperature, T, the entropy S/S_0 and the quantity λ/λ_0 are plotted in Fig. 9 as functions of the energy U/U_0. The plots are almost straight lines, especially for the entropy. The values of U_0S_0 and τ_0 for various values of the nuclear mass are given in lines B to D of Table XXI. From these values and from the curves of Fig. 9, the temperature, entropy etc. can be obtained for any excitation energy of the nucleus. The temperatures τ so deduced are listed in lines E to G of Table XXI, for excitation energies of 5, 10 and 20 MV. They are seen to be much smaller than

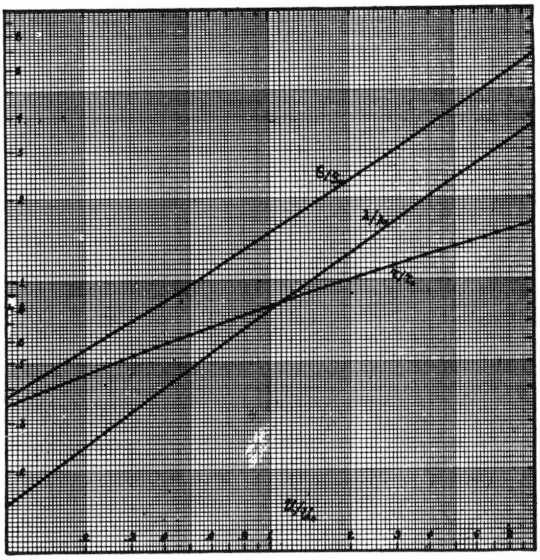

FIG. 9. Nuclear temperature τ, entropy S, and λ (cf. (281)) as functions of the excitation energy U in the liquid drop model. The constants U_0, S_0, τ_0, λ_0 are given in formulae ((324a), (326b), (326c), (327a)).

the excitation energies U, *viz.* about 1 to 2 MV for 10 MV excitation. This is, of course, due to the fact that the excitation energy is, even at low temperatures, distributed among many normal modes of vibration.

From temperature, entropy and λ, the spacing of nuclear levels can be deduced with the help of (281). The result is given in lines H to K of Table XXI, for four different excitation energies (5 to 20 MV).

Table XXI shows clearly the very rapid decrease of the spacing with increasing excitation energy and increasing size of the nucleus which is expected from qualitative considerations (B32). Excitation energies between 15 and 20 MV are easily obtained if fast deuterons (kinetic energy ~5 MV) are used as projectiles, since the large internal energy of the deuteron forms part of the excitation energy of the compound nucleus. With such energies and heavy nuclei, the spacing is seen to be only of the order of a millivolt.

The dependence of the results on the assumptions made in the calculation is shown in lines M and N of Table XXI. In the upper of these lines, we have tabulated the spacing obtained when only surface waves are considered, for $U=10$ MV. The results differ only by a factor 2.5 to 3 from that found with surface and volume waves for the same excitation energy. This is quite welcome because our treatment of volume waves was much less satisfactory than that of the surface waves. This concerns the general procedure, especially the treatment of the transverse volume waves, and also the determination of the constant K which was only possible with the help of the statistical model of nuclei whereas Γ and P could be deduced from empirical data. Therefore the small influence of the volume waves increases our confidence in the results. Of course, a larger influence is to be expected for higher excitation.

In the N line of Table XXI, the spacing as calculated, assuming the old nuclear radius derived by Gamow from the one-body model of the

TABLE XXI. *Spacing of nuclear energy levels in the "liquid drop" model.*

	Atomic Weight A		10	20	50	100	200
B	Characteristic energy U_0 (MV)		13.0	14.9	17.9	20.6	23.7
C	Characteristic temperature τ_0 (MV)		3.00	2.61	2.17	1.89	1.64
D	Characteristic entropy S_0		5.8	7.65	11.05	14.6	19.3
E	Tempera-	for $U=5$ MV	1.72	1.43	1.11	0.92	0.75
F	ture τ	10	2.18	1.82	1.41	1.17	0.97
G	(MV)	20	2.78	2.31	1.81	1.49	1.24
H		for $U=5$ MV	110,000	36,000	5,900	1,100	150
I	Spacing	$U=10$ MV	13,000	2,700	180	16	0.9
J	$D(U)$ in	$U=15$ MV	2,000	240	9.5	0.35	0.007
K	volts	$U=20$ MV	400	33	0.55	0.015	$1.5\cdot10^{-4}$
L		$U=Q$	16,000	3,300	310	60	15
M	Spacing for	Without volume waves	30,000	9,000	500	40	2.7
N	$U=10$ MV	With $r_0=1.48\cdot10^{-13}$	100,000	30,000	4,000	500	60
O	Q = neutron dissociation energy (MV)		9.3	9.5	9.1	8.2	6.9

α-particle disintegration, is also given. The corresponding values of the spacing are, as is to be expected, very much (up to 70 times) larger than those derived with the larger radius. A spacing of as much as 500 volts for $A=100$ seems quite irreconcilable with the evidence from slow neutron experiments which may be taken as an argument for the larger nuclear radius independent of the α-particle disintegration (Chapter XI).

The greatest interest of Table XXI lies, of course, in comparisons with experimental data. The most extensive of these are the data obtained from slow neutron experiments which indicate spacings of the order of 10–20 volts for $A=100$ and 200, and 100 to a few hundred volts for $A=50$ (§60, especially Table XXII). The levels of the compound nucleus responsible for neutron capture are those whose excitation energy is about equal to the dissociation energy Q of the compound nucleus into a neutron and a residual nucleus. This dissociation energy Q is estimated in the last line of Table XXI from the semi-empirical formula for mass defects (cf. §30, and B13); the corresponding spacing is given in line L. The values thus obtained for $A=50$, 100 and 200 compare very favorably with the average spacing of neutron levels estimated from experimental data.

Less satisfactory is the agreement with data from proton capture and α-particle disintegrations of very light nuclei. E.g., the levels of P^{31} are well known from the α-particle disintegration of Al^{27}. The spacing is about 0.3 to 0.5 MV for α-particle energies of the order of 5 MV, corresponding to excitation energies of about 12 MV (cf. §82, Table XXXX). Table XXI would indicate a spacing of less than 1000 volts for this case. Similarly, 3 levels below 1 MV are known in the capture of protons by F^{19} (§81), corresponding to an average spacing of about 0.3 MV. The corresponding excitation energy of the compound nucleus, Ne^{20}, is 12 MV so that, from Table XXI, we should expect a spacing of about 1000 volts.

Part of the discrepancy in the latter case may be attributed to the fact that the normal state of Ne^{20} lies particularly low compared to neighboring nuclei. The reasons for this are well known (even number of neutrons as well as protons), but these reasons will only depress the ground state and not the excited states of the nucleus. Therefore the excitation relative to the ground state should not be taken as significant for the "temperature" of the nucleus. Tentatively, we may replace in such a case the ground state by the average of the ground states of two neighboring nuclei, in our case perhaps F^{19} and Ne^{21}. This would reduce the excitation energy by about 4 MV so that a spacing of 10 kv might be expected from Table XXI. But even this spacing is much smaller than the observed one.

It seems therefore that our calculations fail for very light nuclei. This is not very surprising in view of the assumptions made; especially the distinction between surface and volume waves would seem objectionable when almost all the particles in the nucleus are surface particles. Moreover, it might be possible that the nuclear volume increases faster than the number of particles owing to some repulsive interactions; this would make light nuclei smaller and therefore the spacing of their energy levels wider. However, it seems more plausible that the discrepancies are due to a failure of the liquid drop model; the Bardeen model (section B) seems to give much more reasonable values (Table XX) for the spacing of the levels of light nuclei. It is difficult to decide what atomic weight is required for the validity of the liquid drop model; we shall assume it to be valid for $A \geqslant 50$.

In conclusion, we give here the formulae for the temperature, entropy etc. when only surface waves are important. This assumption is a good approximation as is shown by Table XXI. From (322) we have then

$$U=4\cdot 3^{-\frac{1}{3}}C_{4/3}(\Gamma P)^{-\frac{2}{3}}A^{\frac{2}{3}}\tau^{7/3}$$

$$=4.7(\Gamma P)^{-\frac{2}{3}}A^{\frac{2}{3}}\tau^{7/3}, \tag{328a}$$

$$\tau=0.515(\Gamma P)^{2/7}A^{-2/7}U^{3/7}, \tag{328b}$$

$$S=3.39(A\,U^2/\Gamma P)^{2/7}, \tag{328c}$$

$$\lambda=2.75(\Gamma P/A)^{1/7}U^{5/7}, \tag{328d}$$

$$D=\lambda e^{-S}. \tag{328e}$$

With $\Gamma=P=10$ MV, and U in MV:

$$\tau=1.92A^{-2/7}U^{3/7}\text{ MV}, \tag{329a}$$

$$S=0.91A^{2/7}U^{4/7}, \tag{329b}$$

$$\lambda=5.3A^{-1/7}U^{5/7}\text{ MV}. \tag{329c}$$

§54. The Widths of Nuclear Levels and the Evaporation Model (B33, W7, B15, F30)

There are two ways of calculating the widths of nuclear levels: Either we may start from the assumption that a particle falling on the nucleus will, in a large percentage of all cases, amalgamate with it and form a compound nucleus. This assumption seems plausible because of the large interaction between nuclear particles and is confirmed by slow neutron measurements (cf. Section D); it forms the basis of the "evaporation model" (Sections D, E) which is especially useful for high energy particles. The other, more ambitious method is to try to calculate the matrix elements (widths) from the wave functions (Sections A to C). Although quantitative results have not been obtained from this method, the formalism seems useful especially for the treatment of the widths of low compound states which occur, e.g., in natural radioactivity (§67).

A. Wave function of the incident particle and nuclear potential

The matrix element is defined as

$$H_r^{np}=\int \psi_{Cr}^* H \psi_{An}\psi_{Pp} d\tau, \tag{330}$$

where ψ_{An}, ψ_{Pp} and ψ_{Cr} are the wave functions of initial nucleus, incident particle and compound nucleus, respectively, the subscripts $n\ p\ r$ specifying the states of the three systems. H is the Hamiltonian of the interaction between incident particle and initial nucleus.

In the calculation of this matrix element, the main uncertainty arises from the function ψ_{Pp} representing the incident particle. All the other factors, *viz.* the Hamiltonian H and the wave functions ψ_{An} and ψ_{Cr} of the initial and compound nucleus, can either be considered as known (Hamiltonian) or can be calculated in principle from the Schrödinger equation. It is true that this calculation would be impracticable in the case of complicated nuclei; but at least the functions ψ_{An} and ψ_{Cr} are uniquely determined by the respective Hamiltonians.

This is not the case for the particle wave function ψ_{Pp}. Here we can make any of three choices:

(a) We can take ψ_{Pp} to be a plane wave,

(b) We can take ψ_{Pp} as the solution of the one-particle Schrödinger equation in an *attractive* potential extending approximately over the volume of the nucleus,

(c) We can take a *repulsive* potential of the same extension.

This ambiguity refers, of course, only to the determination of a "zero-order potential" and of the zero-order wave functions of the particle. It is not an ambiguity in the problem of nuclear disintegration itself but only in our method of solving that problem. If we were able to solve the problem rigorously, we would not need to speak of the particle wave function, and of transitions between compound and dissociated state at all; we would only have to solve the Schrödinger equation for the complete system, and the solution would automatically represent the compound state as long as all particles are close together, and behave asymptotically like the wave function of the initial nucleus, times a spherical wave of the particle P, when this particle is far away from the rest of the system. There would then be no question as to the choice of the phase of the particle wave function because this, among other things, would follow directly from the solution of the wave equation.

As it is, we cannot solve our problem rigorously but must apply a method of successive approximations, the Schrödinger-Dirac perturbation method. Even so, we shall certainly arrive ultimately at the correct result (provided the method converges) from whatever "zero-order wave functions" we start. But it will take a great number of approximations before we arrive at the correct result if we choose an unsuitable zero-order wave function. Therefore we want a particle wave function ψ_{Pp} which makes the perturbation theory converge as rapidly as possible; if possible, we want it such that the second approximation written down in (255) already represents the probability of the process fairly well.

It is evident that the convergence of the successive approximations of the perturbation theory must be very different for the three different choices of the particle potential mentioned above. This follows simply from the fact that, although all zero-order wave functions must ultimately lead to the same result, the value of the matrix element (330) is quite different for different "zero-order" potentials. If we choose an attractive potential, (330) will in the average be much larger than if we start from plane waves, mainly because with an attractive potential there is a chance of having a resonance already in the one-particle wave function ψ_{Pp}, which increases the value of ψ_{Pp} inside the nucleus and hence the matrix element (330). A repulsive potential, on the other hand, will make (330) (in the average) smaller than a zero potential because in a (sufficiently strong) repulsive potential, ψ_{Pp} will decrease exponentially from the surface of the nucleus inwards, and will be smaller than the plane wave function even at the surface itself. The matrix element would decrease indefinitely with increasing repulsive potential except for the fact that the perturbing potential H tends to increase; H is the difference between the actual interaction between particle P and nucleus A, and the "zero-order potential" V used to calculate ψ_{Pp}. Consequently, there will be a certain medium value of the repulsive potential V for which (330) will be a minimum (in the average over the states r).

It seems likely that a repulsive potential which just makes the average matrix elements a minimum will be the best choice for the zero-order potential. Our requirement for a good zero-order wave function is that it shall make the higher approximations of the perturbation theory small compared to the first. This will be the case if the individual matrix elements are small, particularly for high levels r of the compound nucleus. It will be shown below that indeed a repulsive potential will make the matrix elements sufficiently small so that, at least in a particular case, the higher approximations of the perturbation theory are negligible. In contrast to this, we shall find that plane waves make the higher approximations more important than the first, and this will hold *a fortiori* for particle wave functions in an attractive potential.

Our result that we have to choose a repulsive potential to calculate the zero-order wave functions for the incident particle seems very reasonable from the standpoint of the many-body picture of the nucleus. As we have seen in §51, the incident particle will give part of its energy to the nuclear particles as soon as it strikes the surface of the nucleus; it will thus lose energy as it goes inside and its wave function will change to such an extent that we cannot reasonably speak of the free particle inside the nucleus at all. It is therefore reasonable to choose a wave function for the incident particle which "dies out" inside the nucleus.

When it had not yet been realized that the problem of nuclear dynamics is essentially a many-body problem, an attractive potential was used to represent the action of the initial nucleus on the incident particle (B12, A7). This attractive potential was taken as the average interaction energy between nucleus and particle. The use of the average potential seemed very appropriate in the one-body picture when the incident particle was supposed to traverse the nucleus more or less undisturbed. However, in the many-body picture there is no undisturbed traversing of the nucleus, and there appears to be no room for the attractive average potential.

In the theory of radiation, the wave functions of the light quantum which are used in the dispersion formula, are taken simply as plane waves. This appears to be the most straightforward procedure, more logical than to take part of the perturbation already into account when calculating the zero order wave functions. However, the case of radiation is particularly favorable because the interaction between light and matter is very small; in our problem we must apply special tricks to enforce convergence in spite of the large interaction. It is, of course, not clear whether our wave functions are actually suitable to this end, but at any rate they are more likely to be suitable than the other possible functions.

To specify the "zero-order" potential more precisely, we have to define the magnitude of the repulsive potential V_0 and its radius of action. The latter must obviously be taken approximately equal to the nuclear radius R. If the perturbation theory is to have any meaning, the wave functions of the compound states, *plus* those of the dissociated states, should form as nearly a complete set as possible; this obviously requires that the one group of wave functions dies out where the other goes into effect, i.e. that the repulsive potential acts inside a sphere of approximately the nuclear radius.

The height V_0 of the potential barrier must be of the order of nuclear energies (10 MV). A smaller height would be unfavorable because it would not make the particle wave function ψ_{Pp} small enough inside the nucleus, a larger V_0 would make the perturbing potential H too large, as already mentioned above. If V_0 is chosen of the same order as the actual interaction between nucleus A and particle P, its influence will be moderate.

In order to investigate the influence of the higher approximations of the perturbation theory, we calculate the elastic scattering of slow particles (wave-length $\lambda\!\!\!^{-}$ large compared to the nuclear radius R). The scattering may be divided into three parts:

(a) The "zero-order" scattering contained in the zero-order wave function itself. Since the particle is assumed to be slow, the repulsive potential is practically impenetrable for it so that the nucleus acts like a hard sphere of radius R. The scattering cross section of zero order is therefore

$$\sigma_1 = 4\pi R^2. \tag{331}$$

(b) The terms in the dispersion formula (255) (second-order scattering) which arise from energy levels close to the particle energy. This "resonance scattering" σ_2 may be larger or smaller than the zero-order scattering according to whether the particle energy is near a resonance level or far away. It has no bearing on the question of the convergence of the perturbation calculation.

(c) The terms in the dispersion formula arising from distant energy levels. These terms do not change appreciably with the particle energy and may therefore be regarded as part of the potential scattering. Their ratio to the cross section σ_1 will give us the relative importance of the second as compared to the zero-order approximation which we want to calculate. There is no difficulty in separating "close" and "distant" levels since the contribution of "medium distant" levels can be shown to be very small.

The dispersion formula (255) gives for elastic scattering

$$\sigma_2+\sigma_3=\pi\lambda\!\!\!^{-2}\left|\sum_r \frac{\gamma^r{}_{Pp}}{E_P-E_{Pr}+\frac{1}{2}i\gamma_r}\right|^2. \tag{332}$$

For given states of initial nucleus and incident particle, the matrix elements $\gamma^r{}_{Pp}$ will be in the average independent of the energy E_r of the compound nucleus, except that for very high compound states $\gamma^r{}_{Pp}$ will fall off (see below). Therefore the contributions to the sum over r in (332) will come mainly from energy regions in which there are many levels, i.e. from high energies. Since for very high energies γ falls off, the main contribution will come from the energy region just before γ starts decreasing.[13] Let us denote the particle energy at which this occurs by E_C. Then we can replace the resonance denominator in (332) by $-E_C$ for all important terms r. The sum over r may be evaluated using the completeness relation:

$$\sum_r \gamma^r{}_{Pp}/(-E_C) = -2\pi\sum_r\left|\int \psi_{Cr}{}^*H\psi_{An}\psi_{Pp}d\tau\right|^2/E_C$$

$$= -2\pi\int |H\psi_{An}\psi_{Pp}|^2 d\tau/E_C. \tag{333}$$

The volume element $d\tau$ may be written as $d\tau_A d\tau_P$ where $d\tau_A$ is the volume element in the configuration space of all the particles in the original nucleus A, and $d\tau_P$ the volume element for the incident particle. Integrating over $d\tau_A$ and considering the normalization of ψ_{An}, we have

$$\int |H(r_A, r_P)|^2|\psi_{An}(r_A)|^2 d\tau_A = U^2(r_P), \tag{333a}$$

where U is an irregularly varying function of the coordinate r_P of the particle, having the magnitude of a nuclear energy. Inserting (333a) in (333) (332), we have

$$\sigma_3 = (4\pi^3\lambda\!\!\!^{-2}/E_C{}^2)\left(\int U^2\psi_P{}^2 d\tau_P\right)^2. \tag{334}$$

The wave function ψ_P is normalized per unit energy, corresponds to the energy E_P and to zero angular momentum of the particle, and satisfies the one-particle Schrödinger equation in a potential which has the value V_0 for $r<R$ and 0 for $r>R$. (The latter assumption is equivalent to assuming the incident particle to be a neutron.) The usual methods give

$$\psi = \begin{cases} \dfrac{1}{\pi(2\hbar v)^{\frac{1}{2}}r}\left(\dfrac{E}{V_0}\right)^{\frac{1}{2}} e^{\kappa(r-R)} & r<R \\[2ex] \dfrac{1}{\pi(2\hbar v)^{\frac{1}{2}}r}\sin k(r-R+1/\kappa) & r>R \end{cases} \tag{335}$$

[13] Actually, the condition is that the product of γ and the density of energy levels must have a maximum.

with $k=(2ME)^{\frac{1}{2}}/\hbar, \quad \kappa=(2MV_0)^{\frac{1}{2}}/\hbar.$ (335a)

Therefore

$$4\pi\int_0^R U^2\psi_P{}^2r^2dr=(U^2)_{\text{Av}}\frac{E}{V_0}\frac{4\pi}{2\kappa}\frac{1}{2\pi^2\hbar v}$$
$$=(U^2)_{\text{Av}}E^{\frac{1}{2}}/2\pi V_0{}^{\frac{3}{2}} \qquad (336)$$

and the second-order cross section becomes

$$\sigma_3=(\pi\hbar^2/2M)(U^2)_{\text{Av}}{}^2/V_0{}^3E_C{}^2. \qquad (337)$$

The energies U, V_0 and E_C are all of the same order of magnitude. U must be somewhat larger than V_0 because it contains V_0 besides the nuclear potential. We estimate $U\sim 15$ MV, $V_0\sim 10$ MV and (cf. (347a)) $E_C\sim 8$ MV. Therefore the last factor in (337) will be about 0.8 MV^{-1} and σ_3 is about $5\cdot 10^{-25}$ cm². It is therefore much smaller than σ_1 which is of the order of 10^{-23} cm². This indicates that with our choice of the wave functions, the perturbation calculation actually converges quite rapidly, at least in the case of the elastic scattering.

If we had chosen plane waves, the wave function inside the nucleus would have had the constant value $k/\pi(2\hbar v)^{\frac{1}{2}}$, so that

$$4\pi\int U^2\psi^2r^2dr=(2/3\pi)(U^2)_{\text{Av}}k^2R^3/\hbar v$$
$$=(2M/3\pi\hbar^2)(U^2)_{\text{Av}}R^3/\lambda\!\!\!^-, \qquad (336a)$$

$$\sigma_3=\frac{16\pi}{9}R^2\frac{(U^2)_{\text{Av}}{}^2}{E_C{}^2(\hbar^2/MR^2)^2}. \qquad (337a)$$

With $R=10^{-12}$ cm, $\hbar^2/MR^2$ is about 0.4 MV. For U we take 10 MV instead of 15 MV, because it no longer includes $-V_0$. The critical energy E_C must be chosen higher than before;[14] we take $E_C=17$ MV. Then the last factor in (337a) is about 200, and σ_3 becomes approximately $1000R^2=80\sigma_1$, which shows that the perturbation calculation would diverge with plane waves.

For attractive potentials the perturbation calculation will diverge even more seriously. If $\kappa=(2M|V_0|)^{\frac{1}{2}}/\hbar$ is the wave number inside the nucleus, we have in this case

$$\psi=\frac{k}{\pi(2\hbar v)^{\frac{1}{2}}r(k^2\sin^2\kappa R+\kappa^2\cos^2\kappa R)^{\frac{1}{2}}}\sin\kappa r \qquad (335b)$$

for $r<R$. Then

$$4\pi\int U^2\psi_P{}^2r^2dr=\frac{(U^2)_{\text{Av}}R}{\pi\hbar v}\frac{k^2}{k^2\sin^2\kappa R+\kappa^2\cos^2\kappa R}, \qquad (336b)$$

$$\sigma_3=\pi R^2\frac{(U^2)_{\text{Av}}{}^2}{E_C{}^2}\frac{1}{(E_C\sin^2\kappa R+V_0\cos^2\kappa R)^2}. \qquad (337b)$$

This cross section becomes extremely large in the case of "one-particle resonance," i.e. if

$$\cos\kappa R\approx 0. \qquad (337c)$$

In this case there can, of course, be no convergence of the perturbation calculation. We may try to save matters by averaging over all possible phases κR of the particle wave function. Then we find

$$(\sigma_3)_{\text{Av}}=\frac{\pi}{2}R^2\frac{(U^2)_{\text{Av}}{}^2}{E_C{}^2V_0{}^{\frac{1}{2}}E^{\frac{1}{2}}}, \qquad (337d)$$

which still becomes extremely large for small energies of the incident particle.

Our considerations show that only with a repulsive zero-order potential can we expect a reasonable convergence of the perturbation expansion.

B. Potential scattering

The two contributions to the potential scattering σ_1 and σ_3 have opposite phase: The phase of the wave scattered from a hard sphere (σ_1) is certainly opposite to the phase of the incident wave (the two waves must just cancel each other at the surface of the sphere). On the other hand, the scattered wave σ_3 due to the dispersion effect of highly excited nuclear states has the same phase as the incident wave, just as light scattered by an atom has when the frequency of the light is smaller than the characteristic frequency of the atom. Thus the total potential scattering cross section is

$$\sigma_{\text{pot}}=(\sqrt{\sigma_1}-\sqrt{\sigma_3})^2. \qquad (338)$$

If we estimate σ_3 to be about 5 percent of σ_1 (see above), the total potential scattering will be decreased by 40 percent due to the influence of the high levels of the compound nucleus.

In order to find the interference effects between potential scattering and resonance scattering, we must separate the former into contributions of

---

[14] The estimate of the critical energy E_C in (344) to (347) is based on a comparison of the experimental width of slow neutron levels with the sum of the neutron widths of all levels as derived from the completeness relation (336). If a different wave function is taken for the neutron (particle P), the completeness relation will give a different result for the sum of all neutron widths. On the other hand, the experimental width of one level is given; therefore a different number of levels below E_C will be deduced. With plane waves, we have according to (336a) instead of (344):

$$N(E_C)=\frac{\Sigma\gamma^r{}_{Pp}}{(\gamma^r{}_{Pp})_{\text{Av}}}=\frac{4}{3}\frac{MR^3}{\hbar^2}\frac{(U^2)_{\text{Av}}}{\lambda\!\!\!^-\gamma_{\text{Av}}}=5\cdot 10^9, \qquad (344a)$$

taking $U=10$ MV, $\hbar^2/MR^2=0.4$ MV, $R=10^{-12}$ cm, $\lambda\!\!\!^-=3.10^{-10}$ cm, $\gamma=2\cdot 10^{-4}$ volts. This number of levels is very much larger than that derived from the repulsive potential wave functions in (344), corresponding to the large amplitude of the plane waves inside the nucleus. With (344a) and the assumptions leading to (347a), we find

$$U_C\approx 25 \text{ MV}, \quad E_C\approx 17 \text{ MV}.$$

various angular momenta J of the whole system. Neglecting σ_3 and assuming the orbital momentum of the incident particle to be zero the contribution of the angular momentum J to the cross section of the potential scattering is

$$\sigma_{1J}=4\pi\frac{2J+1}{(2i+1)(2s+1)}R^2. \quad (339)$$

Considering that only states of the same J interfere, and considering the phase of the potential scattering, the total elastic scattering of slow neutrons becomes

$$\sigma=\frac{\pi}{(2i+1)(2s+1)}\sum_J(2J+1)\times\left|2R+\lambda\sum_r\frac{\gamma^{rJ}{}_{Pp}}{E_P-E_{Pr}+\frac{1}{2}i\gamma_{rJ}}\right|. \quad (340)$$

If only one resonance level is important, this reduces to

$$\sigma=4\pi R^2+\frac{2J+1}{(2i+1)(2s+1)}\frac{\pi\lambda\gamma^{rJ}{}_{Pp}}{(E_P-E_{Pr})^2+\frac{1}{4}\gamma_{rJ}{}^2}\times[4R(E_P-E_{Pr})+\lambda\gamma^{rJ}{}_{Pp}]. \quad (341)$$

Near each resonance level, the cross section has a minimum on the side of low energies and a maximum at an energy slightly higher than the resonance energy E_{Pr}. If we use the fact that $\gamma^{rJ}{}_{Pp}$ varies inversely as the wave-length for slow particles (cf. 266), and if we assume γ_{rJ} to be constant and equal to the true width Γ_{rJ}, we have

$$\sigma=4\pi R^2+\frac{2J+1}{(2i+1)(2s+1)}\frac{\pi\lambda_{Pr}\Gamma^{rJ}{}_{Pp}}{(E_P-E_{Pr})^2+\frac{1}{4}\Gamma_{rJ}{}^2}\times[4R(E_P-E_{Pr})+\lambda_{Pr}\Gamma^{rJ}{}_{Pp}], \quad (341a)$$

$$\sigma/\sigma_1=1+\frac{2J+1}{(2s+1)(2i+1)}\frac{\tau(\tau+4x)}{1+4x^2} \quad (342)$$

with $\tau=\lambda_{Pr}\Gamma^{rJ}{}_{Pp}/R\Gamma_{rJ}$,

$$x=(E_P-E_{Pr})/\Gamma_{rJ}. \quad (342a)$$

If the resonance scattering is large compared to the potential scattering ($\tau\gg1$), the minimum cross section occurs at $x=-\frac{1}{2}\tau$ and the minimum at $x=1/2\tau$, the respective values of the cross section being

$$\sigma_{\min}=\left(1-\frac{2J+1}{(2s+1)(2i+1)}\right)\sigma_1, \quad (343)$$

$$\sigma_{\max}=\frac{2J+1}{(2s+1)(2i+1)}\tau^2\sigma_1. \quad (343a)$$

C. Sum rules and dependence of the matrix element on the energy of the compound state

We now want to discuss the average behavior of the matrix element (330) as a function of the energy of the compound nucleus W_{Cr} and the energy of the particle E_P. The wave functions of the various states of the compound nucleus will certainly not be fundamentally different. Consequently, the matrix elements will also not show any particular trend with the energy of the compound state, the state of the particle being kept fixed. This is another point in which nuclear theory is in contrast to atomic theory: Consider, e.g., the matrix elements of the optical transitions from the ground state of an atom to its excited states. (The excited states correspond to the states of the compound nucleus, the ground state to the initial nucleus, the light quantum to the incident particle.) These matrix elements fall off as the inverse third power of the principal quantum number n of the excited state of the atom (cf. B16, p. 442). The reason is simply that in the excited states the valency electron gets farther and farther away from the core of the atom, and its probability of being in the region occupied by the ground state decreases as n^{-3}. On the other hand, a nucleus in an excited state will be just as concentrated as in its ground state, and therefore the wave function of an excited state of the "compound nucleus" will overlap just as much with that of the initial nucleus as the wave function of the ground state of the compound nucleus.

This will, of course, not be true for extremely high excited states of the compound nucleus. The width of such states will be much more accurately given by the evaporation model (Section D) according to which the width of high compound levels decreases as their spacing. This means a decrease in the partial width at high energy and, in addition, it describes the approximate quantitative way in which the decrease takes place.

In order to complete our calculations in A, we want to estimate the critical excitation energy U_C above which the widths fall off, *not* with the correct dependence of the widths on the energy of the compound state but with the dependence assumed in A, *viz.* constant width up to U_C and width zero at higher energies. Such an estimate may be obtained by combining the experimental results about neutron width with the completeness relation and with some assumptions about the density of nuclear energy levels. From the experiments about neutron capture, we may deduce a neutron width of the order of about $3\cdot 10^{-4}$ volts for a neutron energy of about 1 volt and an atomic weight around 100. On the other hand, we have from the completeness relation according to (333) (333a) (336):

$$\sum_r \gamma^r{}_{Pp} = 2\pi \sum_r \left| \int \psi_{Cr}{}^* H \psi_{An} \psi_{Pp} d\tau \right|^2$$

$$= 2\pi \int U^2(r_P)\psi_P{}^2 d\tau_P = U^2{}_{\mathrm{Av}} E^{\frac{1}{2}}/V_0{}^{\frac{3}{2}}. \quad (344)$$

Since we have assumed $\gamma^r{}_{Pp}$ to be in the average independent of the energy up to the critical excitation energy U_C, the sum on the left of (344) represents just the value of a single matrix element, times the number of states $N(U_C)$ contributing, i.e., the number of states with energy below U_C. We have therefore

$$N(U_C) = (U^2{}_{\mathrm{Av}} E^{\frac{1}{2}}/V_0{}^{\frac{3}{2}} \gamma_{\mathrm{Av}}) = (10^{7/2}/3\cdot 10^{-4}) = 1\cdot 10^7, \quad (345)$$

assuming U and V_0 to be about 10^7 volts. There will thus be about ten million energy levels below the energy U_C.

The critical excitation energy U_C may then be calculated using the values for the spacing of the levels derived from the liquid drop model. Using the general formula (281), we have for the number of levels below U_C

$$N(U_C) = \int_0^{U_C} e^{S(U)} \frac{dU}{\tau} \left(2\pi \frac{dU}{d\tau} \right)^{-\frac{1}{2}}. \quad (346)$$

According to the general thermodynamical relation (279b)

$$dU/\tau = dS. \quad (346a)$$

The last factor in (346) changes slowly compared to the exponential and may therefore, without appreciable error, be replaced by its value for $U = U_C$. Then we have

$$N(U_C) = (2\pi dU/d\tau)_C{}^{-\frac{1}{2}} e^{S(U_C)}. \quad (346b)$$

Using again the general formula for the spacing, this gives

$$N(U_C) = \tau(U_C)/D(U_C). \quad (347)$$

The solution of this equation, for $N = 10^7$ and $A = 100$, is about (cf. Table XXI)

$$U_C = 16 \text{ MV}. \quad (347a)$$

Thus U_C is of the order of a nuclear energy, and $E_C = U_C - Q$ is of the order of the dissociation energy Q itself.

D. The evaporation model, general theory

Frenkel (F30) and Bohr (B33) have proposed to calculate the width of nuclear energy levels, and in particular its dependence on the energy of the compound and final nucleus, by analogy with the process of evaporation. Weisskopf (W7) has given a more detailed treatment of this "evaporation model." The method is especially valuable in predicting the energy distribution of the particles emitted (cf. Section E).

As in the theory of the rate of vaporization, we shall express the probability of emission of a particle by the compound nucleus (vaporization) in terms of the probability of the reverse process (condensation). For this purpose, we shall consider a large box of volume Ω containing nuclear particles (neutrons, protons) in equilibrium with complex nuclei of various kinds. Let $g_{Ak}F(A, W_{Ak}, \mathbf{p}_A)d\mathbf{p}_A$ be the number of nuclei of kind A, whose internal quantum state is W_{Ak} and whose center of gravity moves with the momentum $\mathbf{p}_A$. g_{Ak} is the statistical weight of state k, and $d\mathbf{p}_A$ is the volume element in momentum space. Then, in thermal equilibrium

$$g_{Ak}F(A, W_{Ak}, \mathbf{p}_A)d\mathbf{p}_A = g_{Ak} \frac{\Omega d\mathbf{p}_A}{(2\pi\hbar)^3} e^{-(W_{Ak}+E_A)/kT}, \quad (348)$$

where E_A is the kinetic energy

$$E_A = p_A{}^2/2M_A. \quad (348a)$$

The internal energy W_{Ak} is most conveniently defined as the total energy minus the energy of the free particles constituting nucleus A. The number of nuclei A having an internal energy between W_A and $W_A + dW_A$ and a momentum in the interval $d\mathbf{p}_A$ is

$$F(A, W_A, \mathbf{p}_A)\rho(W_A)dW_A d\mathbf{p}_A, \quad (348b)$$

where $\rho(W_A)$ is the density of energy levels of nucleus A, counting each level as many times as it is degenerate.

We now consider the disintegration of the "compound" nucleus C into two nuclei A and P. The internal states of the latter nuclei are fixed and may be denoted by k and n respectively. The compound nucleus may be in a state r in the interval between W_C and $W_C + dW_C$. We denote the probability of disintegration of C into A and P by $w_{AP}{}^C$; it will depend on W_C and on the states k and n of the final nuclei A and P. $w_{AP}{}^C$ should be considered as the average of the disintegration probabilities of the compound states in the energy interval dW_C. The probability of the reverse process, i.e., the recombination of A and P to form C, will be denoted by $w_C{}^{AP}$. This recombination will, of course, only be possible when momentum and energy are conserved.

The conservation of momentum requires

$$\mathbf{p}_A+\mathbf{p}_P=\mathbf{p}_C, \quad (349)$$

the conservation of energy

$$W_{Cr}+E_C=W_{Ak}+W_{Pn}+E_A+E_P, \quad (349a)$$

where the E's denote kinetic, the W's internal energies. We introduce, besides $\mathbf{p}_C$, the momentum of the relative motion of nuclei A and P, *viz.*

$$\mathbf{p}=(M_A\mathbf{p}_P-M_P\mathbf{p}_A)/(M_A+M_P). \quad (349b)$$

Then we have

$$E_A+E_P=E+E_C, \quad (349c)$$

where E is the kinetic energy of the relative motion

$$E=p^2/2M, \quad (349d)$$

M = reduced mass = $M_AM_P/(M_A+M_P)$. The conservation of energy becomes then

$$W_{Cr}=W_{Ak}+W_{Pn}+E. \quad (350)$$

The probabilities $w_{AP}{}^C$ and $w_C{}^{AP}$ are connected by the principle of detailed balancing. The number of disintegration processes $C\rightarrow A+P$ must be equal to the number of recombinations $A+P\rightarrow C$. This gives

$$\rho(W_C)F(C,\,W_C,\,\mathbf{p}_C)d\mathbf{p}_CdW_C\cdot w_{AP}{}^C = \int_\omega g_{Ak}g_{Pn}F(A,\,W_{Ak},\,\mathbf{p}_A)F(P,\,W_{Pn},\,\mathbf{p}_P) \times d\mathbf{p}_Ad\mathbf{p}_Pw_C{}^{AP}. \quad (351)$$

The integral on the right-hand side extends over all *directions* of the relative motion $\mathbf{p}$, keeping W_C and $\mathbf{p}_C$ constant. Because of (349) (349b), we may write

$$d\mathbf{p}_Ad\mathbf{p}_P=d\mathbf{p}_Cd\mathbf{p} \quad (351a)$$

and divide right and left-hand side of (351) by $d\mathbf{p}_C$. Understanding $w_C{}^{AP}$ to be an *average* over all directions of $\mathbf{p}$, we may then carry out the integration over the angle ω and obtain, instead of $d\mathbf{p}$,

$$4\pi p^2dp=4\pi p^2(dp/dE)dW_C, \quad (351b)$$

using the conservation of energy, (350). (351) divided by dW_C becomes

$$\rho(W_C)F(C,\,W_C,\,\mathbf{p}_C)w_{AP}{}^C=g_{Ak}g_{Pn}F(A,\,W_{Ak},\,\mathbf{p}_A) \times F(P,\,W_{Pn},\,\mathbf{p}_P)\cdot 4\pi pM\cdot w_C{}^{AP}. \quad (352)$$

We now insert in (352) the distribution function (348). Because of the conservation of energy (349a), the Boltzmann factors on the left and right-hand side of (352) cancel and we obtain

$$w_{AP}{}^C=w_C{}^{AP}\frac{g_{Ak}g_{Pn}}{\rho(W_C)}\frac{4\pi Mp\Omega}{(2\pi\hbar)^3}. \quad (353)$$

Here we may replace $\hbar w_{AP}{}^C$ by the partial width $\Gamma^{Cr}{}_{AkPn}$ of the compound levels referring to the disintegration into nuclei A and P in the states k and n, respectively (see the definition of Γ in §52). Γ means here the energy value of the width, averaged over all compound states r of energy near W_C. Furthermore, we write

$$\rho(W_C)=1/D'(W_C)=1/D_r', \quad (353a)$$

where D' is the average spacing of the compound levels. Then we have

$$\Gamma^{Cr}{}_{AkPn}/D_r'=\Omega w_C{}^{AP}g_{Ak}g_{Pn}(M^2v/2\pi^2\hbar^2). \quad (354)$$

The probability of recombination $w_C{}^{AP}$ is proportional to the number of collisions between nuclei A and P. Let R be the "collision radius" of the nuclei A and P. If quantum effects can be neglected, i.e., for short de Broglie wave-length $\hbar/Mv$, this radius may be taken as the sum of the actual radii of nuclei A and P. Then we must calculate the probability that a particle P moving with the velocity v hits a stationary sphere A of radius R if the particle P is distributed uniformly over the volume Ω. This probability is, per unit time,

$$w_C{}^{AP\prime}=4\pi R^2v_n/\Omega, \quad (355)$$

where v_n is the average value of the component of the velocity v normal to the surface of the sphere. As is shown easily, $v_n=\frac{1}{4}v$. We assume now that the average probability for recombination in a collision is ξ. Then the probability of recombination becomes

$$w_C{}^{AP}=\pi R^2v\xi/\Omega \quad (355a)$$

and the partial width

$$\Gamma^{Cr}{}_{AkPn}/D_r'=\frac{g_{Ak}g_{Pn}}{\pi}\cdot\frac{MR^2}{\hbar^2}\cdot\tfrac{1}{2}Mv^2\xi. \quad (355b)$$

Here we introduce

$$P'=\hbar^2/MR^2=PA^{-\frac{2}{3}}, \quad (356a)$$

where P is (cf. (303d)) about 10 MV and A the atomic weight of the compound nucleus. Then we have

$$\Gamma^{Cr}{}_{AkPn}/D_r'=(g_{Ak}g_{Pn}/\pi)(E/P')\xi. \quad (356)$$

The only unknown quantity in (356) is the "sticking probability." Bohr (B33) has suggested that ξ should be almost unity, which would mean that practically every collision between a particle P and a nucleus A would lead to an amalgamation of the two nuclei, and none of the collisions to an elastic reflection. In this case (356) gives a partial width of the compound levels of the same order as their spacing provided that the kinetic energy E is of order P', i.e., about 1 MV. The total width of the compound levels would be very much larger because in each case a great number of states k are possible for the "final nucleus" A.

The assumption $\xi \approx 1$ is very tempting because of the very large interaction between nuclear particles. Moreover, similar results have been obtained for the probability of condensation of molecules on liquids or solids of the same substance by Polanyi and Wigner (P6a). However, these calculations were carried out assuming equipartition of the energy, i.e., at temperatures high compared to the Debye temperature of the liquid or solid. In the nuclear problem, we are dealing with "low" temperatures. In this case, the transfer of energy from the incident particle (molecule) to the vibrations of the nucleus (crystal lattice) might be impeded. As pointed out by Polanyi and Wigner, this should give rise to smaller condensation probabilities. However, no calculations on this problem have been carried out.

Experimentally, the "sticking probability" ξ seems indeed to be considerably less than unity. From the experiments on the capture of slow neutrons, we can find the partial widths of a fairly large number of compound levels in the low energy region. We may then assume that the width increases as the square root of the neutron energy (§52) as long as the neutron wave-length remains smaller than the nuclear radius, or, in other words, as long as E is smaller than P' (cf. (356a)). Then we find for the neutron width corresponding to the kinetic energy P'

$$\Gamma(P')=\Gamma(E)(P'/E)^{\frac{1}{2}}=\Gamma' P'^{\frac{1}{2}}. \qquad (357)$$

$\Gamma'=\Gamma E^{-\frac{1}{2}}$ is given in Table XXVI for various levels of various elements; it is about 0.2–2 millivolts if E is measured in volts. P' is about $\frac{1}{4}$ MV for $A=100$ (cf. (356a)). Therefore $\Gamma(P')=0.1$ to 1 volt. On the other hand, the spacing of nuclear levels near the neutron dissociation energy is perhaps 5–20 volts for atomic weights of 100 to 200, both experimentally (§60) and theoretically (§53C). Thus

$$\Gamma(P')/D' \approx 0.005-0.2. \qquad (357a)$$

Taking 0.03 as an average, and putting $g_{Ak}=g_{Pn}=1$ (see below), this gives

$$\xi \approx 0.1, \qquad (357b)$$

i.e., a "sticking probability" of about ten percent. It seems that in some cases the "sticking probability" of slow neutrons approaches unity while in others it is as low as one percent.

Since ξ is expected to increase with the energy, and is not too far (about 1/10) from its high energy value (unity) even at moderate energy, we may *expect this quantity to change slowly compared to others such as the density of levels.* This seems to be the result of the "evaporation model" which is most valuable in predicting the probability of nuclear disintegrations and the energy distribution of the emitted particles. Quantitatively, we may expect ξ to be proportional to a low power of the temperature, perhaps directly proportional to the total excitation energy; but in the absence of a direct calculation this dependence is merely a guess, and it is also not very important for applications.

A word may be said about the statistical weight factors in (356). Since accidental degeneracies are highly improbable, these weights are given by the spins of the nuclei, i.e.,

$$g_{Ak}=2i+1, \quad g_{Pn}=2s+1. \qquad (358)$$

We shall now consider particularly the compound levels r with given angular momentum J. The spacing D_r' was defined not taking into account the degeneracy, and is therefore $2J+1$ times smaller than the true spacing D_{rJ}. Therefore

$$\Gamma^{C_rJ}{}_{AkPn}/D_{rJ}=\frac{(2i+1)(2s+1)}{2J+1}\cdot\frac{1}{\pi}\cdot\frac{E}{P'}\xi_J, \qquad (358a)$$

where ξ_J is the probability that the particles A and P, when colliding, stick together and form a compound nucleus of angular momentum J. If the orbital momentum is zero or small (slow particles), J may have any value between $|i-s|$ and $i+s$. If the probability ξ_J is proportional to the statistical weight, then

$$\xi_J=(2J+1)\xi/(2i+1)(2s+1), \qquad (358b)$$

where ξ is the total "sticking probability." Then (358a) reduces to

$$\Gamma^{C_rJ}{}_{AkPn}/D_{rJ}=\xi(E/\pi P'). \qquad (359)$$

This equation was used for computing ξ in (357b). The computation becomes more complicated when the orbital momentum l of the particle P becomes comparable to the spins s and i (medium fast particles). For very fast particles, it is again easy to compute Γ. Then J can take all values up to

$$l_0 = R/\lambda\!\!\!^{-} = MvR/\hbar \tag{360a}$$

and approximately

$$\xi_J = \xi(2J+1)/l_0^2 = (2J+1)\xi P'/2E, \tag{359b}$$

$$\Gamma^{CrJ}{}_{AkPn}/D_{rJ} = \xi(2i+1)(2s+1)/2\pi. \tag{360}$$

The relations given in this section are, of course, only true statistically, i.e., in the average over a sufficiently great number of levels of compound (or final) nucleus. For individual levels, large fluctuations of the probability ξ must be expected. Moreover, all the formulae derived are only valid if the wave-length of the emitted particle is smaller than the nuclear radius R, in other words if the kinetic energy E is greater than the characteristic energy P'. For smaller energies, quantum effects become important; the partial width is then proportional to the square root of the energy (cf. (266a)) and may, according to (359), be written

$$\Gamma^{CrJ}{}_{AkPn}/D_{rJ} = (\xi/\pi)(E/P')^{\frac{1}{2}}, \qquad E \ll P',\ J < i+s. \tag{361}$$

This formula is again only valid in the average over many levels.

E. The energy distribution of the particles produced in nuclear reactions

When the energy levels of the compound nucleus are sufficiently dense (§56), the number of nuclear processes leading to a final nucleus A in state k is simply proportional to the average partial width of the compound levels corresponding to level k of the final nucleus. According to (356), this width is proportional to

$$\Gamma \sim (2i+1)E\xi_k, \tag{362}$$

where i is the nuclear spin in state k, E the kinetic energy of the particles P emitted when nucleus A is left in state k, and ξ_k the "sticking probability" between particles of energy E and nuclei A in state k. In first approximation, we may consider $E\xi_k$ to be in the average independent of state k; then the probability of formation of state k is proportional to its statistical weight. The same is true of the number of particles P of energy $E = W_C - W_P - W_{Ak}$ emitted, i.e., of the number of particles forming the "group" corresponding to state k.

The assumption made ($E\xi_k$ independent of k) will be best justified for states k having not too different energy and, if possible, otherwise similar properties. A good example is the two lowest states of Li^7 which are supposed to form a "doublet" which, in spectroscopic notation, is to be designated as $^2P_{1/2,\,3/2}$ (F10, R10). These two states are formed in the reaction

$$Li^6 + H^2 = Li^7 + H^1.$$

The intensities of the two proton groups are very nearly in the ratio 2 : 1 which is the ratio of the statistical weights of the two states of Li^7. The ground state of Li^7 has $i = \frac{3}{2}$; accordingly, the more energetic proton group has the higher intensity.

The dependence of ξ on E over large energy intervals is unknown, even statistically. It might be expected that ξ decreases somewhat with decreasing energy of the state k of the final nucleus (i.e., for given compound state, with *increasing* kinetic energy E of the emitted particle) because the transfer of the kinetic energy of the particle to the nuclear vibrations in the condensation process may become increasingly difficult.

In many cases, especially for heavier nuclei, we are not interested in the number of nuclear processes leading to a *definite* final state k but only in the number of processes giving particles of kinetic energy between E and $E+dE$. This probability is proportional to (362), times the number of states of the final nucleus in the energy interval dE, i.e.,

$$w_{AP}{}^C dE \sim \rho(W_A)\xi E dE. \tag{363}$$

The most important factor here is the density of states

$$\rho(W_A) = \lambda^{-1} e^{S_A}, \tag{363a}$$

where S is the entropy of nucleus A. We know that the density of states increases with the excitation energy W_A of the final nucleus; therefore the number of particles emitted with a kinetic energy between E and $E+dE$ will increase rapidly with *decreasing* kinetic energy E of the particle. In other words, *most of the particles emitted in nuclear reactions will be comparatively slow*, leaving the residual nucleus as highly excited as possible.

The quantities ξ and λ in (363), (363a) are known to vary slowly compared to the exponential. Thus, approximately,

$$w_{AP}{}^{C} \sim E e^{S(W_{A0}-E)}, \tag{364}$$

where
$$W_{A0} = W_{Cr} - W_P \tag{364a}$$

is the excitation energy of the residual nucleus corresponding to zero kinetic energy of the emitted particle, and $S(W)$ is the entropy of nucleus A corresponding to the energy W. (364) is statistically correct (i.e., correct disregarding fluctuations due to the discrete quantum levels of the final nucleus) no matter what the excitation energy W_{A0}.

We now make an approximation which is only justified if the excitation energy of the residual nucleus, W_{A0}, is sufficiently high. Then we may expand the entropy in (364), *viz.*

$$S(W_{A0}-E) = S(W_{A0}) - E(dS/dW) + \cdots$$
$$= S(W_{A0}) - E/\tau + \cdots, \tag{365}$$

where τ is the "temperature" of nucleus A corresponding to the excitation energy W_{A0}. Since $S(W_{A0})$ is independent of E, we may then write for the distribution function of the emitted particles

$$w_{AP}{}^{C} dE \sim e^{-E/\tau} E dE, \tag{366}$$

a distribution strikingly similar to the Maxwell distribution and, indeed, closely connected to it. The distribution was derived by Weisskopf (W7) and also in a simpler but somewhat less rigorous way than used here by Frenkel (F30) and Bohr (B33).

According to (366), the energy distribution of the outgoing particles should have a maximum at $E=\tau$. Since the "temperatures" of nuclei with 5 to 20 MV excitation energy are only of the order of 1 to 4 MV, the temperature τ will always be quite small compared to the available excitation energy W_{A0}. Thus in most nuclear processes, only a small fraction of the available energy will in general be taken up by the outgoing particle, while the larger part remains in the residual nucleus as excitation energy. In some cases, this excitation energy may be large enough for the emission of a second particle: Then we have a so-called three-particle reaction, i.e., a reaction leading to three final products (cf. §85). Examples are $B^{11}+H^1=Be^{8*}+He^4=3\ He^4$; $B^{10}+H^2$ leading to the same products; Be^9+He^4 or $C^{12}+H^1$ or $B^{11}+H^2 \rightarrow Be^{9*}+He^4=3\ He^4+n^1$; $N^{14}+H^2=4\ He^4$, etc. In the other cases, the excitation energy of the residual nucleus will be given off as γ-radiation (Chapter XIV).

The energy distribution (366) is qualitatively confirmed in many nuclear reactions. A well-known example is the reaction

$$Be^9+He^4=C^{12}+n^1.$$

With α-particles from radon, the available energy W_{A0} is 13 MV (cf. Table LIII). However, very few neutrons of kinetic energy 13 MV are observed (D21, and §99, Fig. 40) while most of them have energies of about 4 MV which is of the same order as the temperature of a C^{12} nucleus with an excitation energy of 13 MV (cf. Table XX). Similar features are shown by other reactions with large energy evolution, such as

$$B^{10}+H^2=B^{11}+H^1, \quad N^{14}+H^2=C^{12}+He^4,$$

and other reactions caused by deuterons, and also reactions caused by α-particles and giving protons such as $P^{31}+He^4=S^{34}+H^1$ (cf. §99, 101).

According to the distribution formula (366), it is rather improbable that the emitted particles carry away the whole available energy as kinetic energy, leaving the nucleus in its ground state. Therefore it is often difficult to observe the "group" of disintegration particles corresponding to the ground state of the final nucleus (full energy group) and thus to obtain the total energy evolution in the transmutation which is important for the determination of nuclear masses. Quantitatively, the fraction of nuclei left in the ground state should be roughly equal to one over the total number of excitation levels of the residual nucleus with an excitation energy below the total available energy W_{A0}. The *relative* intensity of the full energy group of particles will therefore decrease when the available energy (i.e., also the energy of the incident particle) increases. This is again in agreement with experimental results for disintegrations produced by α-particles: In the reaction $Na^{23}+He^4=Mg^{26}+H^1$, the fastest proton group could be observed by König (K21) with the relatively slow α-particles from polonium (energy 5.3 MV) but not by May and Vadyanathan (M11) with Ra C′ alphas (7.7 MV). The danger of missing the full energy group of disintegration particles will be greater for heavier nuclei which have a greater density of levels.

For charged particles, the energy distribution (366) will be modified by the penetration through the potential barrier which favors the emission of high energy particles. In general, the probability of penetration through the potential barrier will increase faster with the particle energy than the Boltzmann factor $e^{-E/\tau}$ in (366) decreases. Therefore the most probable energy of the outgoing particle should be about equal to the height of the potential barrier while the emission of particles of lower and higher energy should be less probable. Fig. 10 gives the approximate theoretical distribution function for protons when the residual nucleus is Hg^{200} and the total available energy W_{A0} is 20 MV.

The influence of the potential barrier will be considered in more detail in Chapter XIII. The total probability of nuclear processes will be discussed in §56. The γ-ray width will be calculated in Chapter XIV.

§55. Derivation of the Dispersion Formula (B51, B33)

In this section, we shall give the derivation of the dispersion formula (255) for nuclear processes. We do this primarily in order to show the limitations to its validity. The *stationary* method of perturbation theory will be used in our treatment. This seems slightly simpler than the nonstationary method, especially when many levels of the compound nucleus are involved. A proof using the nonstationary (Dirac) method of perturbation theory was given in the original paper of Breit and Wigner.

A. One compound state

1. Notation and fundamental equations.—For simplicity, we treat first the case of only *one* compound state of wave function χ_C, and only

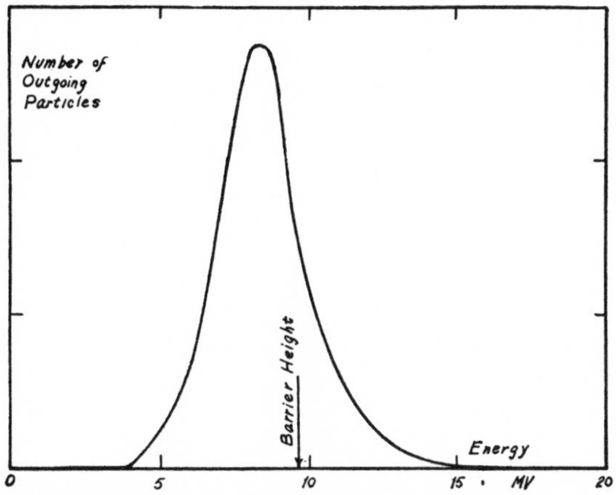

Fig. 10. Energy distribution of protons from a disintegration of Hg^{200}. Total available energy 20 MV, barrier height 9.6 MV.

two kinds of particles which can be emitted, P and Q. Let χ_A and χ_B be the wave functions of the residual nuclei A and B remaining after emission of the particles P and Q, respectively. χ_A, χ_B, χ_C are supposed to be normalized. If the particles P and Q are complex rather than elementary, there will be two further wave functions χ_P and χ_Q describing the internal motion of the elementary particles inside P and Q. Furthermore, there will be two wave functions ψ_P and ψ_Q describing the motion of the centers of gravity of the particles P and Q, relative to the centers of gravity of the respective residual nuclei. These functions ψ_P and ψ_Q we shall not assume to be normalized. They will be calculated, in the course of our discussion, as to form as well as to normalization. For the moment, we mention only that ψ_P consists of an incident plane wave of given amplitude, plus an outgoing spherical wave (scattering) whose amplitude we wish to calculate. ψ_Q is solely an outgoing wave whose amplitude determines the probability of the disintegration process $A+P\rightarrow B+Q$.

The total Hamiltonian H may be split in the following two ways:

$$H=H_A+H_P+T_P+U_P+V_{AP}, \quad (367)$$

$$H=H_B+H_Q+T_Q+U_Q+V_{BQ}. \quad (367a)$$

Here H_A contains the kinetic energies of and the interactions between the particles contained in nucleus A, while H_P contains the internal energies of the particles inside the incident nucleus P (If P is an elementary particle, $H_P=0$ and $\chi_P=1$).

T_P is the kinetic energy of the motion of the center of gravity of particle P with respect to nucleus A. U_P is an "effective potential energy of the particle P" which depends only on the distance between particle P and nucleus A, and not on the internal coordinates of A and P. At large distances between A and P, this potential U_P will be the Coulomb interaction between nucleus A and particle P, if the latter is charged, while U_P will be zero, if P is neutral (neutron or light quantum). If P is inside the nucleus, a repulsive potential (positive U_P) has to be chosen according to the considerations of §54A.—The last term in (367), V_{AP}, represents the interaction between nucleus A and particle P as far as it is not contained in the effective potential U_P. V_{AP} will depend on the internal coordinates of nuclei A and P as well as on the distance between them.

The internal wave functions χ introduced above satisfy equations such as

$$H_A\chi_A = W_A\chi_A. \tag{368}$$

Similar equations hold for χ_B, χ_P, χ_Q. The wave function of the compound nucleus satisfies

$$H\chi_C = W_C\chi_C \tag{368a}$$

except at the boundary of the compound nucleus: χ_C is supposed to be restricted to a finite region of space, inside of the "nuclear radius," whereas the correct wave function has (see below) continuations extending to infinity.

We write for the wave function of the complete system

$$\Psi = \chi_A\chi_P\psi_P + \chi_B\chi_Q\psi_Q + c\chi_C \tag{369}$$

and try to satisfy the Schrödinger equation

$$H\Psi = W\Psi \tag{369a}$$

as accurately as is possible with this form of Ψ. The energy W is

$$W = W_A + W_P + E_P, \tag{369b}$$

where W_A and W_P are the internal energies of the initial nuclei A and P and E_P the kinetic energy of their relative motion. In order to satisfy (369a) as well as possible we shall fulfill the three equations

$$\int \chi_C^*(H-W)\Psi d\tau_C = 0, \tag{370}$$

$$\int \chi_A^*\chi_P^*(H-W)\Psi d\tau_A d\tau_P = 0, \tag{370a}$$

$$\int \chi_B^*\chi_Q^*(H-W)\Psi d\tau_B d\tau_Q = 0. \tag{370b}$$

The first of these is a scalar equation which will serve to determine the coefficient c of the compound wave function χ_C in (369). The two others are differential equations for the as yet undetermined wave functions $\psi_P\psi_Q$ of incident and outgoing particle. In contrast to (370), the integrations in (370a, b) do not extend over all coordinates but only over the internal ones; the result is therefore a function of the coordinate $\mathbf{r}_{AP}$ of particle P with respect to nucleus A (or of $\mathbf{r}_{BQ}$ in case of 370b).

Using the expressions (367) (367a) for H, the Schrödinger equations (368) (368a) for the χ's and the fact that the χ's are normalized, we find

$$c(W-W_C) = \int \chi_C^*(V_{AP}-L_P)\chi_A\chi_P\psi_P d\tau_C + \int \chi_C^*(V_{BQ}-L_Q)\chi_B\chi_Q\psi_Q d\tau_C, \tag{371}$$

$$\begin{aligned} L_P\psi_P = {} & c\int \chi_A^*\chi_P^*(V_{AP}-L_P)\chi_C d\tau_A d\tau_P \\ & + \int |\chi_A|^2|\chi_P|^2 V_{AP}\psi_P d\tau_A d\tau_P \\ & + \int \chi_A^*\chi_P^*(V_{AP}-L_P)\chi_B\chi_Q\psi_Q d\tau_A d\tau_P \end{aligned} \tag{372}$$

and a similar equation for ψ_Q. L_P is the operator

$$\begin{aligned} L_P &= E_P - T_P - U_P \\ &= E_P + (\hbar^2/2M_P)\Delta_P - U_P \end{aligned} \tag{372a}$$

with M_P the reduced mass of P and A, and Δ_P the Laplacian operator with respect to the coordinates of P relative to A.

The second term in (372) gives an additional contribution to the potential scattering. This contribution may cause more irregular variations of the potential scattering from nucleus to nucleus than those predicted by the theory developed in §54. It is obvious how to calculate the effect of this term in first approximation. However, in our considerations we shall disregard the second as well as the last integral in (372): This corresponds to the assumption that most of the nuclear processes $A+P\rightarrow B+Q$ occur by way of formation and decay of the compound state C and only a small correction is supplied by direct transitions. This is probably true in most practical cases.

2. Solution of equation for particle wave functions.—Thus (372) reduces to

$$L_P\psi_P = c\int \chi_A{}^*\chi_P{}^*(V_{AP}-L_P)\chi_C d\tau_A d\tau_P. \quad (373)$$

This represents, according to (372a), an inhomogeneous differential equation for ψ_P. Since L_P is a spherically symmetrical operator, (373) may be separated in polar coordinates by putting

$$\psi_P = \sum_{lm}\psi_{Plm}(r_P)\,Y_{lm}(\vartheta_P\varphi_P)/r_P, \quad (373a)$$

where ψ_{Plm} satisfies the differential equation

$$(\hbar^2/2M_P)d^2\psi_{Plm}/dr^2 + (E_P - U_P - \hbar^2 l(l+1)/2M_P r^2)\psi_{Plm} = cC_{Plm}(r) \quad (374)$$

with

$$C_{Plm}(r_P) = \int r_P Y_{lm}{}^*(\vartheta_P\varphi_P)\chi_A{}^*\chi_P{}^*(V_{AP}-L_P)\chi_C \times d\tau_A d\tau_P d\omega_P. \quad (374a)$$

$d\omega_P$ is the element of solid angle in the coordinates of particle P. Thus the integration in (374a) goes over all coordinates of all particles in the system except the distance r_P between P and A.

The general solution of the ordinary inhomogeneous differential equation (374) is given by an arbitrary solution of this equation plus any multiple of the regular solution of the homogeneous equation. The right-hand side (inhomogeneity) will be appreciable only as long as $r_P < R$, i.e., as long as particle P is inside or near the nucleus A. For large r_P, the solution of (374) will therefore be identical with a certain solution of the homogeneous equation. But this will in general not be the regular solution, i.e., that solution which remains finite at the origin. Therefore we have for the asymptotic behavior of ψ_{Plm}:

$$\psi_{Plm} \rightarrow \alpha_{Plm} f_{Pl}(r_P) + \beta_{Plm} g_{Pl}(r_P), \quad (375)$$

where f and g are the regular and irregular solution of the homogeneous equation

$$(\hbar^2/2M_P)d^2 f, g/dr^2 + (E_P - U_P - \hbar^2 l(l+1)/2M_P r^2)f, g = 0. \quad (375a)$$

At the origin, the regular solution behaves as

$$f \sim r^{l+1} \quad (r \text{ small}), \quad (375b)$$

the irregular one as

$$g \sim r^{-l}. \quad (375c)$$

Asymptotically for large r, the behavior of f is

$$f = \sin(kr - \tfrac{1}{2}l\pi + \delta_l), \quad r \rightarrow \infty, \quad (376)$$

where

$$k = (2M_P E_P)^{\frac{1}{2}}/\hbar = M_P v/\hbar \quad (376a)$$

is the wave number, v the velocity of particle P (relative to nucleus A) and δ_l the phase shift of the wave (compared to a free particle wave function, with $U_P = 0$ in (375a)) due to the particle potential U_P. (376) fixes the normalization of f and therefore the magnitude of α and β in (375). For the irregular function g, we may choose any solution of (375a) which is linearly independent of f; the most convenient choice is

$$g = \cos(kr - \tfrac{1}{2}l\pi + \delta_l), \quad r \rightarrow \infty, \quad (376b)$$

i.e., the function shifted in phase by $\pi/2$ compared to the regular solution. The two solutions f and g satisfy for *any* r and any potential U_P the relation

$$g\,df/dr - f\,dg/dr = \text{const} = k, \quad (377)$$

which follows easily from (375a) by multiplying the equation for f by g, that for g by f, subtracting and integrating.

Of the constants α and β in (375), α_{Plm} is arbitrary (see above) while β_{Plm} is uniquely determined by the differential equation (374) with the additional condition that ψ must be finite at the origin. A convenient way for calculating β is to use the same procedure which led to (377): Multiply Eq. (375a) for f by ψ, Eq. (374) for ψ by f, subtract and integrate up to a very large distance r. Then we obtain

$$\lim_{r\rightarrow\infty}\left(f\frac{d\psi}{dr} - \psi\frac{df}{dr}\right) = \frac{2M_P}{\hbar^2}c\int_0^\infty f_{Pl}C_{Plm}dr. \quad (378)$$

Using the asymptotic expression (375) for ψ and the relation (377) between f and g, we have

$$\beta_{Plm} = -\frac{2M_P}{\hbar^2 k}c\int_0^\infty f_{Pl}C_{Plm}dr. \quad (379)$$

With (374a) and (376a), this may be rewritten

$$\beta_{Plm} = -(2/\hbar v)c\int f_{Pl}(r_P)\,Y_{lm}{}^*(\vartheta_P\varphi_P)\chi_A{}^*\chi_P{}^* \times (V_{AP}-L_P)\chi_C d\tau_A d\tau_P\, d\Omega_P/r_P, \quad (379a)$$

where $d\Omega_P = r_P^2 dr_P d\omega_P$ is the volume element in the space of particle P.[15] It is convenient to introduce in (379a) the function

$$F_{Plm} = f_{Pl}(r_P) Y_{lm}(\vartheta_P \varphi_P)/r_P. \quad (379b)$$

According to (372a) (375a), F satisfies the differential equation

$$L_P F_{Plm} = 0. \quad (379c)$$

Since the operator L_P is self-adjoint, its contribution to (379a) obviously vanishes and we find the final result:

$$\beta_{Plm} = -(2/\hbar v) c V^C{}_{APlm}{}^*, \quad (380)$$

where the matrix element $V^C{}_{APlm}$ is defined by

$$V^C{}_{APlm} = \int \chi_C^* V_{AP} \chi_A \chi_P F_{Plm} d\tau_C, \quad (380a)$$

$d\tau_C$ being the total volume element

$$d\tau_C = d\tau_A d\tau_P d\Omega_P = d\tau_B d\tau_Q d\Omega_Q. \quad (380b)$$

As already mentioned, the coefficient α in (375) may have any value as far as the differential equation (374) is concerned. In order to fix α, we must consider the required asymptotic behavior of ψ_{Plm}. We know that all ψ_{Qlm}'s (Q=produced particle) and all ψ_{Plm}'s with $m \neq 0$ must behave asymptotically as outgoing waves:

$$\psi_{Plm} \sim \exp(ikr_P), \quad \psi_{Qlm} \sim \exp(ik'r_Q) \quad (381)$$

without any term behaving as e^{-ikr}. The only exceptions are the partial waves $m=0$ for the incident particle which contain a term due to the incident plane wave. If we normalize the incident plane wave per unit current, it may be written in the familiar form (B15)

$$\psi_P{}^{\text{inc}} = v^{-\frac{1}{2}} e^{ikz} = (4\pi/v)^{\frac{1}{2}} (2ikr)^{-1} \sum_l (2l+1)^{\frac{1}{2}}$$
$$\times Y_{l0}(\vartheta)(e^{i(kr-\frac{1}{2}l\pi)} - e^{-i(kr-\frac{1}{2}l\pi)}). \quad (381a)$$

Therefore we have for particle P and $m=0$:

$$\psi_{Pl0} = \frac{i\pi^{\frac{1}{2}}(2l+1)^{\frac{1}{2}}}{kv^{\frac{1}{2}}} e^{-i(kr-\frac{1}{2}l\pi)} + \gamma e^{ikr}, \quad (382)$$

where γ is an arbitrary coefficient.

We compare the required asymptotic behavior (381) (382) to the known asymptotic behavior of the solutions f and g of the homogeneous equation. (a) In all cases except for particle P and $m=0$, we must have the asymptotic behavior

$$\psi \sim g + if = \text{const} \cdot e^{ikr}. \quad (383)$$

According to (375), this means

$$\alpha_{Plm} = i\beta_{Plm} \quad (383a)$$

or

$$\psi_{Plm} = \beta_{Plm}(g+if) \quad (383b)$$

The number of outgoing particles of the kind P and quantum numbers lm is then

$$\sigma_{Plm} = v_P |\beta_{Plm}|^2. \quad (384)$$

(b) If P is the incident particle and $m=0$, we may put

$$\alpha_{Pl0} = i\beta_{Pl0} + A_l, \quad (385)$$

so that, according to (375) (376) (376b),

$$\psi_{Pl0} = (\beta_{Pl0} - \tfrac{1}{2}iA_l) e^{i(kr-\frac{1}{2}l\pi+\delta_l)}$$
$$+ \tfrac{1}{2}iA_l e^{-i(kr-\frac{1}{2}l\pi+\delta_l)}. \quad (385a)$$

Comparing the coefficient of e^{-ikr} with (382), we find

$$A_l = \frac{2\pi^{\frac{1}{2}}(2l+1)^{\frac{1}{2}}}{kv^{\frac{1}{2}}} e^{i\delta_l}. \quad (386)$$

The outgoing spherical wave is obtained by subtracting the term l of (381a) from (385a):

$$\psi_{Pl0}{}^{\text{out}} = [\beta_{Pl0} e^{i\delta_l} + A_l \sin \delta_l] e^{i(kr-\frac{1}{2}l\pi)}. \quad (386a)$$

Therefore the scattering intensity (number of outgoing particles per second) is

$$\sigma_{Pl0} = |\beta_{Pl0} e^{i\delta_l} + A_l \sin \delta_l|^2 v_P. \quad (387)$$

This formula represents the interference between potential scattering (term $A_l \sin \delta_l$) and scattering by way of the compound state (term β_{Pl0}).

We may now write for any value of r:

$$\psi_{Plm}(r) = \alpha_{Plm} f_{Pl}(r) + \beta_{Plm} \gamma_{Pl}(r), \quad (388)$$

where $\beta_{Plm}\gamma_{Pl}{}^{(r)}$ is a particular solution of the inhomogeneous equation (374) which is chosen in such a way that it goes over asymptotically into βg, *without* any term containing f. It is important to note for the following that γ as well as g and f is *real*, since the differential operators in

[15] In contrast to $d\tau_P$ which refers to the coordinates of the particles *inside* P.

(375a) and (374) and the asymptotic expressions for f, g and γ are real. (The coefficient β may be complex, if C is complex. However, it follows from general considerations of spatial symmetry, etc. that C must have the same complex phase for all values of r, which is sufficient to make γ real.) We define now, in analogy to (379b),

$$G_{Plm}=\gamma_{Pl}(r_P)\,Y_{lm}(\vartheta_P\varphi_P)/r_P. \qquad (388a)$$

G_{Plm} is regular everywhere. Then we may write (373a):

$$\psi_P=\sum_{lm}\alpha_{Plm}F_{Plm}+\beta_{Plm}G_{Plm}. \qquad (388b)$$

Inserting (380) (383a) (385) this becomes

$$\psi_P=\sum_l A_lF_{Pl0}-2(c/\hbar v)\sum_{lm}V^C{}_{APlm}{}^*(G_{Plm}+iF_{Plm}). \qquad (389)$$

3. Determination of the coefficient of the compound wave function.—(389) expresses the particle wave function ψ_P in terms of calculable functions and of the constants A_l and c. It remains to determine c from Eq. (371) which has not been used yet. We insert (389) and obtain

$$c(W-W_C)=\sum_{lm}(A_l\delta_{m0}-i(2c/\hbar v)\,V^C{}_{APlm}{}^*)$$
$$\times\int\chi_C{}^*(V_{AP}-L_P)\chi_A\chi_PF_{Plm}d\tau_C$$
$$-(2c/\hbar v)\sum_{lm}V^C{}_{APlm}{}^*\int\chi_C{}^*(V_{AP}-L_P)\chi_A\chi_P$$
$$\times G_{Plm}d\tau_C+\text{similar terms with } BQ. \qquad (390)$$

The first integral here is by definition (380a) just $V^C{}_{APlm}$. The second integral is not easy to evaluate, but, as we shall show, it is fortunately unimportant for the theory. This is due to the fact that the integral, if it is complex at all, has the same complex phase as $V^C{}_{APlm}$; the complex character of both quantities arises solely from the dependence of the functions $\chi_C\,\chi_A\,\chi_P\,F$ and G on the angular variables, and this dependence is the same in both cases. Therefore the second sum is real; let us put its value equal to $\frac{1}{2}\hbar v\epsilon_{CP}$. We can then combine the second term on the right of (390), and the corresponding term due to the particles Q, with the term cW_C on the left, by simply putting

$$W_C{}'=W_C-\epsilon_{CP}-\epsilon_{CQ}. \qquad (390a)$$

Then the left-hand side will become $c(W-W_C{}')$. Thus the second term in (390) causes simply a shift of the resonance energy from W_C to $W_C{}'$. Such a shift is known from the theory of dispersion of light (Dirac frequency shift). It is wholly irrelevant for our considerations because we have no way of calculating the correct position W_C of the resonance levels, and therefore no way of telling whether the resonance energy has been shifted by the small amount $\epsilon_{CP}+\epsilon_{CQ}$. We shall, in the following, drop the prime in $W_C{}'$.

Inserting $V^C{}_{APlm}$ for the first integral in (390), and taking the term arising from particle Q into account, we obtain:

$$c=\sum_l A_lV^C{}_{APl0}/(W-W_C+\tfrac{1}{2}i\gamma_C) \qquad (391)$$

with

$$\gamma_C=\frac{4}{\hbar v_P}\sum_{lm}|V^C{}_{APlm}|^2+\frac{4}{\hbar v_Q}\sum_{lm}|V^C{}_{BQlm}|^2. \qquad (391a)$$

This can immediately be reduced to the form (257b) used in §52. We need only consider the normalization of the radial function f_{Pl} which enters the matrix element V (cf. (379a)). If φ_{Pl} is the corresponding wave function normalized per unit energy, we have (cf. B15, (15))

$$\varphi=(2/\pi)^{\frac{1}{2}}(dk/dE)^{\frac{1}{2}}f=(2/\pi\hbar v)^{\frac{1}{2}}f \qquad (391b)$$

and therefore

$$\gamma_C=2\pi\sum_{P'=P,\,Q}\ \sum_{lm}|H^C{}_{AP'lm}|^2, \qquad (392)$$

where $H^C{}_{APlm}$ is the same matrix element as $V^C{}_{APlm}$, only taken with wave functions normalized per unit energy, *viz.*:

$$H^C{}_{APlm}=(2/\pi\hbar v)^{\frac{1}{2}}V^C{}_{APlm}. \qquad (392a)$$

Inserting (391) and (386) in (380) (384), we find for the partial cross section corresponding to the production of particle Q with orbital momenta lm:

$$\sigma^P{}_{Qlm}=v_Q|\beta_{Qlm}|^2=\frac{4v_Q}{(\hbar v_Q)^2}\frac{4\pi}{v_Pk_P{}^2}$$
$$\times\left|\sum_{l'}(2l'+1)^{\frac{1}{2}}e^{i\delta_{l'}}\frac{V^C{}_{APl'0}V^C{}_{BQlm}{}^*}{W-W_C+\frac{1}{2}i\gamma_C}\right|^2 \qquad (393)$$
$$=4\pi^3\lambda\!\!\!^{-}{}_P{}^2\left|\sum_{l'}(2l'+1)^{\frac{1}{2}}e^{i\delta_{l'}}\frac{H^C{}_{APl'0}H^C{}_{BQlm}{}^*}{W-W_C+\frac{1}{2}i\gamma_C}\right|^2,$$

where $\lambda\!\!\!^{-}_P = 1/k_P$ is the wave-length of the incident particle divided by 2π. (393) is similar to (255). The interference terms between the various values of l' for the incident particle disappear when the cross section is summed over m (B15, appendix). If this summation is carried out, as well as the summation over all possible magnetic substates of the compound state C, the final nucleus B and the spin of the outgoing particle Q, and if the average is taken over all orientations of the spins of the initial nucleus A and the incident particle P, the one-level formula (259) is obtained for the total cross section. A much more complicated problem which has not yet been treated generally is the question of the angular distribution of the outgoing particles.

The *scattering* cross section will contain one term analogous to (393), and in addition a term representing the potential scattering and the interference between potential and resonance scattering. According to (386a), this term is for a *given angle* ϑ

$$\sigma^P{}_P{}^{\mathrm{pot}}(\vartheta) = v_P \left| \sum_l (A_l \sin\delta_l + \beta_{Pl0} e^{i\delta_l}) Y_{l0}(\vartheta) \right|^2 - v_P \left| \sum_l \beta_{Pl0} e^{i\delta_l} Y_{l0} \right|^2. \quad (393a)$$

Integrated over all angles, this gives

$$\sigma^P{}_P{}^{\mathrm{pot}} = v_P \sum_l \{ |A_l|^2 \sin^2\delta_l + (A_l \beta_{Pl0}{}^* e^{-i\delta_l} + A_l{}^* \beta_{Pl0} e^{i\delta_l}) \sin\delta_l \}$$

$$= 4\pi\lambda\!\!\!^{-}_P{}^2 \left\{ \sum_l (2l+1)\sin^2\delta_l - 2\pi\,\mathrm{Re} \frac{\sum_{l'} (2l'+1)^{\frac{1}{2}} e^{-i\delta_{l'}} H^C{}_{APl'0}{}^* \sum_l (2l+1)^{\frac{1}{2}} \sin\delta_l H^C{}_{APl0}}{W - W_C - \frac{1}{2} i\gamma_C} \right\}, \quad (394)$$

where Re denotes the real part. The first term in (394) represents the potential scattering in the strict sense, the second the interference between potential and resonance scattering.

B. Many compound states

We shall now discuss the more general case of many resonance levels which we distinguish by an index r. Moreover, we shall consider any number of different kinds Q of outgoing particles, and also the possibility of various states q of the final nucleus B. Then (369) is replaced by

$$\Psi = \sum_{Qq} \chi_{Bq} \chi_Q \psi_{Qq} + \sum_r c_r \chi_{Cr}, \quad (395)$$

where the sum over Q includes the incident particle P, and the sum over r goes over all states of the compound nucleus. Instead of (371), we obtain one equation for each compound state r:

$$c(W - W_r) = \sum_{Qq} \int \chi_{Cr}{}^* (V_{BQ} - L_Q) \chi_{Bq} \chi_Q \psi_{Qq} d\tau_C. \quad (396)$$

(373) is replaced by one equation for each possible emitted particle Q and each possible state q of the corresponding residual nucleus B, *viz.*

$$L_Q \psi_{Qq} = \sum_r c_r \int \chi_{Bq}{}^* \chi_Q{}^* (V_{BQ} - L_Q) \chi_{Cr} d\tau_B d\tau_Q. \quad (397)$$

The solution of Eqs. (397) is exactly analogous to that of (373). Defining F_{Qqlm} and G_{Qqlm} in exact analogy to (379b) (388a), we have, analogously to (388b),

$$\psi_{Qq} = \sum_{lm} \alpha_{Qqlm} F_{Qqlm} + \beta_{Qqlm} G_{Qqlm}. \quad (397a)$$

The coefficients β are obtained in analogy to (380):

$$\beta_{Qqlm} = -(2/\hbar v_{Qq}) \sum_r c_r V^{Cr}{}_{BQqlm}{}^*, \quad (398)$$

where v_{Qq} is the velocity of particle Q if nucleus B is left in state q. The coefficients α are determined from the asymptotic behavior of the wave functions, as in (383a) (385):

$$\alpha_{Qqlm} = i\beta_{Qqlm} + A_l \delta_{QP} \delta_{qp} \delta_{m0} \quad (398a)$$

with A_l given by (386). The cross section for the production of particle Q with orbital quantum numbers lm, and with the nucleus B being left in state q, is given by $v_{Qq}|\beta_{Qqlm}|^2$ as in (384): Only for the elastic scattering ($P=Q$, $p=q$, $m=0$) a term has to be added for the potential scattering, as in (387) (394).

The main difference compared to the one level case arises when we want to calculate the coefficients c_r from the equations (396). We insert (397a) in (396) and obtain

$$c_r(W-W_r)=$$

$$\sum_{Qqlm} i\beta_{Qqlm}\int \chi_{Cr}^*(V_{BQ}-L_Q)\chi_{Bq}\chi_Q F_{Qqlm}d\tau_C$$

$$+\sum_{Qqlm}\beta_{Qqlm}\int \chi_{Cr}^*(V_{BQ}-L_Q)\chi_{Bq}\chi_Q G_{Qqlm}d\tau_C$$

$$+\sum_l A_l\int \chi_{Cr}^*(V_{AP}-L_P)\chi_{Ap}\chi_P F_{Ppl0}d\tau_C. \qquad (399)$$

Here the second term is again unimportant inasmuch as it only shifts the resonance levels (cf. (390a)).[16] In the first and third terms, L_Q gives no contribution and the integrals are by definition the matrix elements (380a). Thus we have

$$c_r(W-W_r)=i\sum_{Qqlm}\beta_{Qqlm}V^{Cr}{}_{Qqlm}+\sum_l A_l V^{Cr}{}_{Ppl0}. \qquad (399a)$$

Here we insert β from (398), A_l from (386) and express V by H (cf. (392a)):

$$c_r(W-W_r)=-i\sum_s C_{rs}c_s+a_r, \qquad (400)$$

where the coefficients C are given in terms of the matrix elements by

$$C_{rs}=\pi\sum_{Qqlm}H^{Cr}{}_{Qqlm}H^{Cs}{}_{Qqlm}{}^* \qquad (400a)$$

and a_r is a constant connected to the amplitude of the incident wave by:

$$a_r=\sqrt{2}\pi\lambda\!\!\!^{-}_P\hbar^{\frac{1}{2}}\sum_l(2l+1)^{\frac{1}{2}}H^{Cr}{}_{Ppl0}e^{i\delta_l}. \qquad (400b)$$

Whereas (390) was a linear equation for the single unknown c, our result (400) represents a system of linear equations for the (infinitely many) unknowns c_r. Although it is always possible to solve this system rigorously the result will not in general be simple. However, a simple result is *always* obtainable when the spacing between adjacent energy levels W_r is large compared to the widths of the levels which are of the order of the coefficients C_{rs}. Then two cases are possible:

(a) The energy W coincides very nearly with one of the energy levels, let us say, W_n. Then c_n will be very much larger than all the other c_r's. This makes it permissible to neglect the other c_r's in the equation for c_n which gives (cf. (400))

$$c_n=a_n/(W-W_n+\tfrac{1}{2}i\gamma_n) \qquad (401)$$

with $$\tfrac{1}{2}\gamma_n=C_{nn}. \qquad (401a)$$

The other c_r's are then unimportant for the determination of the amplitudes β of the outgoing wave and may therefore be neglected.[17]

(b) The energy W lies in between two resonance levels, and sufficiently far from each. Then all the c_r's are small, and the sum over s in (400) may be left out entirely. Then

$$c_r=a_r/(W-W_r) \qquad (402)$$

for *all* r's.

The two formulae (401) (402) may be approximated by the same formula:

$$c_r=a_r/(W-W_r+\tfrac{1}{2}i\gamma_r) \quad \text{(all } r\text{)}. \qquad (403)$$

The error in this formula is small in all cases:

1. In case (b) the energy differences $W-W_r$ are all very large compared to the widths of the levels so that the addition of $i\gamma_r$ makes no appreciable difference.

2. In case (a) for $r=n$ formula (403) is identical with the correct formula (401).

3. The terms $r\neq n$ in case (a) are negligible compared to the term n, and likewise the coefficients c_r given by (403) are negligible compared to c_n given by the same formula. (The c_r $(r\neq n)$ are, of course, *not* correctly given by (403), cf. footnote 17.)

---

[16] Disregarding the first and third terms on the right-hand side of (399), and inserting β from (398), we have

$$c_r(W-W_r)=\sum_s c_sB_{rs}, \qquad (A)$$

where the B_{rs} are certain real coefficients. This system of equations is identical with that found in a simple perturbation problem. Its solution leads to a new set of eigenvalues W_r', and to a corresponding new set of wave functions $\chi_r'=\sum_s\gamma_{rs}\chi_s$ where the coefficients γ_{rs} follow from (A). The coefficients B_{rs} do not depend sensitively on the energy W of the system. Therefore the corrected resonance energies W_r' and the corrected wave functions χ_{Cr}' will be characteristics of the compound nucleus just as the old resonance energies W_r and wave functions χ_{Cr}. Therefore we may use the set of wave functions χ_{Cr}' just as well as χ_{Cr}, and thus eliminate the second term in (399) entirely.

[17] The value of c_r $(r\neq n)$ may be obtained from (400) by neglecting in the sum all terms except $s=n$. This gives, with (401),

$$c_r=\frac{a_r-iC_{rn}a_n/(W-W_n+iC_{nn})}{W-W_r},$$

which differs from the value (403) by the second term in the numerator. *The resonance for level n influences the amplitudes of all other compound states as well.* However, because of the large denominator $W-W_r$ (as compared to the denominator in (401)) c_r is irrelevant.

Inserting (403) in (398) (384), the dispersion formula (255) for the cross section is obtained. *This formula is therefore proved provided the width of the levels is small compared to their spacing.* For large width, the solution of the system (400) is much more complicated; it will be discussed for some special cases in the next section.

The case of degenerate levels offers no difficulties: Because of the selection rules for the magnetic quantum number, $H^{C_r}{}_{Qqlm}$ and $H^{C_s}{}_{Qqlm}$ cannot simultaneously be different from zero if the magnetic quantum numbers M_r and M_s of the two compound states are different. Therefore C_{rs} is zero if $M_r \neq M_s$. Similarly it can be shown that C_{rs} is only different from zero if r and s have the same *total* angular momentum J (cf. B15, appendix). Thus the system of Eqs. (400) falls into subsets each corresponding to a given J and M. Moreover, it follows from general considerations that the coefficients C_{rs} have the same value for all the magnetic substates M of the given states C_{rs}. Therefore the systems of Eqs. (400) belonging to the same J and different M's will all be equivalent.

§56. Fast Particles: Average Over the Resonances, Wide Levels

If the incident particle is fast, it will in general not be possible to define its energy accurately enough to observe resonance effects. This will be particularly true if the bombarded nucleus is heavy and has therefore closely spaced energy levels. Even the best sources (H6) of nuclear projectiles give particles whose energies fluctuate by about 1 percent, which means 10 kv at 1 MV particle energy. On the other hand (cf. §53) the spacing between neighboring energy levels is less than 10 kv probably for all nuclei containing more than about 50 particles, and it is only a few volts for heavy nuclei ($A=100$ or more, cf. §53). For fast particles and heavy nuclei, it will therefore only be possible to observe the *average* cross section of a nuclear reaction, averaged over a large number of resonance regions. It is the purpose of this section to calculate this average.

A. Width small compared to spacing

We assume first that the width of the levels of the compound states is small compared to their average spacing. Then the general formula (260) is valid and we obtain[18] by averaging over an energy interval ϵ:

$$(\sigma^{Pp}{}_{Qq})_{\text{Av}} = \frac{\pi\lambda\!\!\!^{-2}}{(2s+1)(2i+1)} \frac{1}{\epsilon} \int_\epsilon dE \sum_{Jrr'} (2J+1) \times \frac{U^{rJ}{}_{Pp} U^{r'J}{}_{Pp} U^{rJ}{}_{Qq} U^{r'J}{}_{Qq}}{(E-E_{rJ}+\frac{1}{2}i\Gamma_{rJ})(E-E_{r'J}-\frac{1}{2}i\Gamma_{r'J})}. \quad (404)$$

(The u's and γ's have been replaced by the U's and Γ's because each term rJ gives an appreciable contribution only near resonance, cf. §52.) The integration is to be extended over an energy interval ϵ large compared to the spacing of the levels D but small compared to the particle energy E itself. The integration can be carried out immediately and gives

$$(\sigma^{Pp}{}_{Qq})_{\text{Av}} = \frac{\pi^2\lambda\!\!\!^{-2}}{(2s+1)(2i+1)} \frac{1}{\epsilon} \sum_J (2J+1) \times \sum_{rr'(\text{in } \epsilon)} \frac{U^{rJ}{}_{Pp} U^{rJ}{}_{Qq} U^{r'J}{}_{Pp} U^{r'J}{}_{Qq} (\Gamma_{rJ}+\Gamma_{r'J})}{(E_{rJ}-E_{r'J})^2+\frac{1}{4}(\Gamma_{rJ}+\Gamma_{r'J})^2}. \quad (404a)$$

The sum has to be extended over all levels rr' in the energy interval ϵ. Here the mixed terms $r'\neq r$ can be neglected for two reasons. Firstly, they contain in the denominator the energy differences $E_{rJ}-E_{r'J}$ which are at least of the order of the spacing of the levels, while the terms $r'=r$ contain the width Γ_{rJ} instead; but we have assumed from the beginning that the widths are smaller than the spacing. Secondly, the numerator in the mixed terms will be positive for about as many terms as negative which makes the mixed terms cancel each other. Thus (404a) reduces to (cf. (261))

$$(\sigma^{Pp}{}_{Qq})_{\text{Av}} = \frac{2\pi^2\lambda\!\!\!^{-2}}{(2s+1)(2i+1)} \frac{1}{\epsilon} \sum_J (2J+1) \times \sum_{r\,(\text{in } \epsilon)} \frac{\Gamma^{rJ}{}_{Pp} \Gamma^{rJ}{}_{Qq}}{\Gamma_{rJ}}. \quad (404b)$$

[18] We use in this section the kinetic energy $E=E_P$ of the incident particle rather than the energy of the system $W=E+W_A+W_P$, and the resonance energies E_r as defined in (258) rather than the total energy W_{Cr} of the compound nucleus in state r. Moreover, we replace the effective widths γ by the true widths Γ (width at resonance) which is always allowed when dealing with fast particles (cf. §52). The indices $ljl'j'$ are unimportant for our considerations and are therefore left out.

Since the energy interval ϵ is supposed to contain many levels, we may replace the sum over r by the number of levels of angular momentum J times the average value of each term. Now the number of levels is ϵ/D_J, where D_J is the average spacing of levels of angular momentum J in the given energy region, so that we find

$$(\sigma^{Pp}{}_{Qq})_{\text{Av}} = \frac{2\pi^2\lambda\!\!\!^{-2}}{(2s+1)(2i+1)} \sum_J (2J+1) \times \left(\frac{\Gamma^J{}_{Pp}\Gamma^J{}_{Qq}}{\Gamma_J D_J} \right)_{\text{Av}}, \tag{405}$$

where the index r has been dropped because we are no longer referring to a specific level of the compound nucleus.

B. General formula for high energies

(405) may be simplified if we admit that the average in (405) may be taken by averaging $\Gamma^J{}_{Qq}/\Gamma_J$ and $\Gamma^J{}_{Pp}/D_J$ separately. This amounts to the assumption that the partial widths $\Gamma^J{}_{Pp}$ and $\Gamma_{Qq}{}^J$ have no correlation, i.e., that the levels of the compound nucleus cannot be classified into levels which easily emit a particle P and others which preferably emit Q. This is actually the basic assumption of Bohr's model of nuclear disintegrations and is equivalent to the statement that the nuclear wave functions are very different from "Hartree" wave functions and correspond to "ideal random motion" of the nuclear particles.

The expression $(\Gamma^J{}_{Pp}/D_J)_{\text{Av}}$ which now occurs in (405) may be expressed in terms of the "sticking probability" ξ_{PpJ} (cf. (358a)), i.e., the probability that a particle P colliding with nucleus A in state p will stick and form a compound nucleus of angular momentum J. We find (cf. (358a))

$$(\sigma^{Pp}{}_{Qq})_{\text{Av}} = 2\pi\lambda\!\!\!^{-2}(E/P')\sum_J \xi_{PpJ}(\Gamma^J{}_{Qq}/\Gamma_J)_{\text{Av}} \tag{406}$$

or, if we assume further that $\Gamma^J{}_{Qq}/\Gamma_J$ is independent of J, and use the definition of P' in (356a) and $\lambda\!\!\!^{-2} = \hbar^2/2ME$:

$$(\sigma^{Pp}{}_{Qq})_{\text{Av}} = \pi R^2 \xi_{Pp}(\Gamma_{Qq}/\Gamma)_{\text{Av}}, \tag{407}$$

where

$$\xi_{Pp} = \sum_J \xi_{PpJ}$$

is the total sticking probability of particle P.

The very simple result (407) is actually not surprising. It can be expressed as follows: The probability of any nuclear process is equal to the probability of formation of the compound nucleus from the initial particles, times the probability of its disintegration into the particular final particles in question. The probability of formation is

$$\sigma_{Pp} = \pi R^2 \xi_{Pp}, \tag{408}$$

i.e., the geometrical cross section of the nucleus times the "sticking probability." In fact, (408) may be considered as the definition of the sticking probability ξ.

The compound nucleus, once formed, must disintegrate in *some* way; therefore (408) must represent the total cross section for all nuclear processes together. The probability of a *particular* mode of disintegration is given by the partial width for that mode divided by the total width, averaged over a sufficiently large number of levels. Expressed in a formula,

$$\sigma^{Pp}{}_{Qq}/\sigma_{Pp} = (\Gamma_{Qq}/\Gamma)_{\text{Av}}. \tag{409}$$

This equation expresses the statement that the disintegration of the compound nucleus is independent of the way in which it has been formed, i.e., independent of the nature and the quantum state of the initial particles. This statement may be considered the simplest expression of the ideas of the Bohr theory of nuclear disintegrations. It is equivalent to the absence of correlations between the partial widths Γ_{Pp} and Γ_{Qq} which was used above in the proof of (407).

C. Limitations; angular momentum

The selection rules for the angular momentum may cause "correlations" between the partial widths which invalidate (407). For simplicity, we may suppose that initial nucleus, incident particle, outgoing particle and final nucleus all have zero spin. Then the total angular momentum J of the compound nucleus is equal to the orbital momentum of incident and outgoing particle. These orbital momenta must be smaller than

$$l_{Pp} = R/\lambda\!\!\!^{-}_{Pp} \quad \text{and} \quad l_{Qq} = R/\lambda\!\!\!^{-}_{Qq}, \tag{410}$$

respectively. Therefore the compound nucleus may be formed in any state with $J < l_{Pp}$, and, according to our assumptions, the probability of a given $J(<l_{Pp})$ is simply proportional to the statistical weight $2J+1$. On the other hand, a com-

pound state J can disintegrate into a particle Q and a residual nucleus in state q only if $J<l_{Qq}$. If $l_{Qq}<l_{Pp}$ the compound states with J between l_{Qq} and l_{Pp} cannot disintegrate in this way. Therefore, if Γ_{Qq} denotes the average of Γ_{QqJ} over all levels for which this quantity is not zero (i.e., with $J<l_{Qq}$), the cross section (407) will be reduced by a factor

$$\eta_{Qq}=(l_{Qq}/l_{Pp})^2=M_QE_{Qq}/M_PE_{Pp} \quad (\text{if } \eta<1). \tag{411}$$

Under these circumstances, the relative probability of two modes of disintegration will *not* be independent of the way in which the compound nucleus is formed. Compare, e.g., the disintegration products Pp and Qq. If $l_{Pp}>l_{Qq}$, the probabilities of emission of Pp as well as Qq are given directly by (407), if the incident particle is Qq. But if the incident particle is Pp, the probability of emission of Qq is reduced by the factor η which is smaller than unity. Thus the selection rule for the angular momentum introduces a certain correlation (in contrast to the fundamental assumption made in B) which favors elastic scattering as compared to other processes, i.e., favors the reemission of a particle of the same kind as the incident one.

This correlation due to the angular momentum can easily be treated in the way indicated in (411). Moreover, the influence of this correlation on the probability of reactions is comparatively small. It will be even smaller if the spins of the nuclei concerned are not zero. However, the fact remains that the existence of a constant of motion such as the angular momentum will necessarily cause a more or less serious breakdown of the assumption of "random motion" of the nuclear particles and of random distribution of the partial widths. According to all our knowledge, there is no other constant of motion besides the energy, the angular momentum and the parity (behavior of wave function with respect to a change of sign of all coordinates). If there were, we should find a more pronounced failure of the "random" assumption.

D. Total width large compared to spacing of levels, but partial width small

As we have shown in §54, the total width of highly excited nuclear states will in general be large compared to their spacing. We shall show that this does not invalidate the formulae derived in A, B provided that (1) the *partial* widths are all (or practically all) small compared to the spacing, and that (2) the partial widths of various nuclear states have no correlation. The second condition is in line with our general assumptions; the first amounts to assuming the sticking probability to be small compared to unity.

Mathematically, assumption (2) means that of the terms in the sum (400a) some will be positive and some negative if $s\neq r$, while they are, of course, all positive for $s=r$. Therefore it is reasonable to expect that the nondiagonal elements C_{rs} are in the average much smaller than the diagonal elements C_{rr}, the ratio being approximately equal to one divided by the square root of the number of terms in the sum (i.e., the number of different modes of disintegration). We shall therefore neglect the "nondiagonal" coefficients C_{rs} entirely. This can be shown to make no appreciable difference in some special cases.

With this assumption, (400) reduces to the form (390) familiar from the one-level case and has the solution

$$c_r=a_r/(E-E_r+\tfrac{1}{2}i\Gamma_r). \tag{412}$$

Then, according to (398) (392a),

$$\sigma^{Pp}{}_{Qqlm}=\frac{2\pi}{\hbar}\sum_{rs}\frac{a_ra_s{}^*H^r{}_{Qqlm}{}^*H^s{}_{Qqlm}}{(E-E_r+\tfrac{1}{2}i\Gamma_r)(E-E_s-\tfrac{1}{2}i\Gamma_s)}. \tag{413}$$

The averaging over energy gives, similarly to (404a),

$$(\sigma^{Pp}{}_{Qqlm})_{\mathrm{Av}}=\frac{2\pi^2}{\hbar\epsilon}\sum_{rs\,(\mathrm{in}\,\epsilon)}\frac{a_ra_s{}^*H^r{}_{Qqlm}{}^*H^s{}_{Qqlm}(\Gamma_r+\Gamma_s)}{(E_r-E_s)^2+\tfrac{1}{4}(\Gamma_r+\Gamma_s)^2}. \tag{413a}$$

In distinction from (404a), the resonance denominators are *not* much smaller for $r\neq s$ than for $r=s$. But, as we have assumed, H^r and H^s will have opposite signs for about as many pairs rs as for which they have the same sign, while H^rH^{r*} is always positive. Consequently, we expect that the contribution of the terms $r=s$ will be approximately proportional to the number of terms r, i.e., to the number of levels in the energy interval ϵ, which is ϵ/D. On the other hand, the contribution of the terms $r\neq s$ will be approximately proportional to the square root of the number of pairs rs which give an appreciable contribution. Since the contributions are negligible if $E_r-E_s\gg(\Gamma_r)_{\mathrm{Av}}$, there are, for each r, about Γ/D contributing levels s, and therefore altogether $\epsilon\Gamma/D^2$ contributing pairs rs. The contribution of these pairs is, according to the foregoing, proportional to $(\epsilon\Gamma)^{\frac{1}{2}}/D$ as compared to ϵ/D from the diagonal terms. Therefore, if we only average over an energy interval large compared to the average width of the levels, only the diagonal terms will be important and (413a) will reduce to

$$(\sigma^{Pp}{}_{Qqlm})_{\mathrm{Av}}=(2\pi/\hbar D)\,|a_r|^2(\Gamma^r{}_{Qqlm})_{\mathrm{Av}}/(\Gamma_r)_{\mathrm{Av}}, \tag{413b}$$

using the relation (257b) and replacing the sum over r in (413a) by the number of terms times the average of each term. This result is the same as for total width small compared to the spacing.

E. Partial width large compared to the spacing of levels

The case of large partial width can be treated rigorously if only one sort of particles can be emitted by a given compound state. The system of linear equations (400) has the general solution

$$c_1=\frac{\begin{Vmatrix} a_1 & a_2 & a_3 & \cdots \\ -iC_{12} & -iC_{22}+E_2-E & -iC_{32} & \cdots \\ \cdot & \cdot & \cdot & \cdots \\ \cdot & \cdot & \cdot & \cdots \end{Vmatrix}}{\Vert -iC_{sr}+\delta_{rs}(E_r-E) \Vert}=\frac{M_1}{N} \tag{414}$$

and similar expressions for the other c_r. If only one kind of particles P may be emitted and only one value of the angular momentum l is possible ($m=0$), then (400a) reduces to a single term, *viz.*

$$C_{rs}=\pi H_r H_s^*, \tag{414a}$$

where H_r is an abbreviation for $H^{C_r}{}_{Ppl0}$. Similarly, according to (400b):

$$a_r=AH_r, \tag{414b}$$

where A is a constant.

To evaluate the determinant N in the denominator of (414), we take out of the rth row the common factor $-i\pi H_r^*$, and out of the rth column the factor H_r. Then we have

$$N=(-i\pi|H_1|^2)(-i\pi|H_2|^2)\cdots\begin{Vmatrix} 1+i(E_1-E)/\pi|H_1|^2 & 1 & 1 & \cdots \\ 1 & 1+i(E_2-E)/\pi|H_2|^2 & 1 & \cdots \\ \cdots & \cdots & \cdots & \cdots \end{Vmatrix}. \tag{415}$$

We subtract the first row from each of the other rows which does not change the value of the determinant; then we have

$$N=\prod_r(-i\pi|H_r|^2)\begin{Vmatrix} 1+i(E_1-E)/\pi|H_1|^2 & 1 & 1 & \cdots \\ -i(E_1-E)/\pi|H_1|^2 & i(E_2-E)/\pi|H_2|^2 & 0 & \cdots \\ \cdots & \cdots & \cdots & \cdots \end{Vmatrix}. \tag{415a}$$

The determinant contains now zeros everywhere except in the diagonal, where the general element is $i(E_r-E)/\pi|H_r|^2$, in the first row which contains unity everywhere except for the first element, and in the first column whose elements (except for the first) are all equal to $-i(E_1-E)/\pi|H_1|^2$. The determinant can now be evaluated easily, giving

$$N=\prod_r(-i\pi|H_r|^2)\prod_r[i(E_r-E)/\pi|H_r|^2]$$
$$\times\left(1+i\sum_s\frac{\pi|H_s|^2}{E_s-E}\right). \tag{415b}$$

With
$$2\pi|H_r|^2=\Gamma_r \tag{416a}$$
this yields

$$N=\prod_r(E_r-E)\left(1+\tfrac{1}{2}i\sum_s\frac{\Gamma_s}{E_s-E}\right). \tag{416}$$

The determinant M in the numerator can be evaluated even more easily. We take out the same factors as before except from the first row, from which we take only the factor A. Then

$$M_1=AH_1\prod_{r=2}^{\infty}(-i\pi|H_r|^2)$$
$$\times\begin{Vmatrix} 1 & 1 & 1 & \cdots \\ 1 & 1+i(E_2-E)/\pi|H_2|^2 & 1 & \cdots \\ \cdots & \cdots & \cdots & \cdots \end{Vmatrix}. \tag{417}$$

Subtracting the first row from each of the others, we obtain a diagonal matrix with $i(E_r-E)/\pi|H_r|^2$ as the general diagonal term (except for $r=1$). Therefore

$$M_1=AH_1\prod_{r=2}^{\infty}(E_r-E). \tag{417a}$$

Dividing (417a) by (416), we find

$$c_1=\frac{AH_1}{E_1-E}\left(1+\tfrac{1}{2}i\sum_r\frac{\Gamma_r}{E_r-E}\right)^{-1}. \tag{418}$$

From (418) we obtain immediately the scattering cross section (exclusive of potential scattering), using (393) (398) (392a) (414b) (400b):

$$\sigma=4\pi^3\lambda\!\!\!^{-2}(2l+1) \times\left|\sum_r \frac{|H_r|^2}{E_r-E}\left(1+\tfrac{1}{2}i\sum_s\frac{\Gamma_s}{E_s-E}\right)^{-1}\right|^2. \quad (418a)$$

Using (416a) and considering that the parenthesis is independent of r:

$$\sigma=4\pi\lambda\!\!\!^{-2}(2l+1)\frac{\frac{1}{4}(\sum\Gamma_r/E_r-E)^2}{1+\frac{1}{4}(\sum\Gamma_r/E_r-E)^2}. \quad (419)$$

This shows that the cross section is always smaller than $4\pi\lambda\!\!\!^{-2}(2l+1)$ which represents the area of the incident beam corresponding to the angular momentum l. However, if the Γ_r are large, the difference between the cross section and $4\pi\lambda\!\!\!^{-2}(2l+1)$ is, in general, very small. This means that practically every particle of the given angular momentum is scattered if the widths of the levels are large.

There are, however, certain energies at which the scattering will still be small. Between any two energy levels, there is one value of the energy for which $\Sigma\Gamma_r/(E_r-E)$ vanishes. For this sum has the value $-\infty$ if E is just larger than one of the resonance energies E_r, and it is $+\infty$ just below the next level E_{r+1}. In between, it rises gradually and must therefore be zero for some energy about midway between the levels. For this energy, the scattering will therefore vanish. Instead of resonance maxima and practically zero cross section in between, we find now "resonance minima" with practically constant cross section $4\pi\lambda\!\!\!^{-2}$ in between.

We are interested in the cross section averaged over an energy interval large compared to the spacing of the levels. Obviously, only the neighborhood of the minima of the cross section is of interest, because elsewhere the cross section is constant. Near a minimum, we may in first approximation neglect the contribution of all levels but the two neighboring ones, which we denote by 1 and 2. Then we have for the position E_0 of the minimum the relation

$$\Gamma_1/(E_0-E_1)=\Gamma_2/(E_2-E_0). \quad (419a)$$

Near the minimum we find, using (419a),

$$\sum_r\frac{\Gamma_r}{E_r-E}\approx\left(\frac{\Gamma_1}{(E_0-E_1)^2}+\frac{\Gamma_2}{(E_2-E_0)^2}\right)(E-E_0)$$
$$=\frac{(\Gamma_1+\Gamma_2)^3}{D^2\Gamma_1\Gamma_2}(E-E_0), \quad (419b)$$

where $D=E_2-E_1$ is the spacing of the levels. The average of (419) over the energy interval from E_1 to E_2 is then

$$\sigma_{\text{Av}}=4\pi\lambda\!\!\!^{-2}(2l+1) \times\left(1-\frac{1}{D}\int_{E_1}^{E_2}\frac{dE}{1+\frac{1}{4}(\sum\Gamma_r/E_r-E)^2}\right) \quad (419c)$$
$$=4\pi\lambda\!\!\!^{-2}(2l+1)[1-2\pi D\Gamma_1\Gamma_2/(\Gamma_1+\Gamma_2)^3].$$

Replacing Γ_1 and Γ_2 by some average value Γ, we find finally

$$\sigma_{\text{Av}}=4\pi\lambda\!\!\!^{-2}(2l+1)(1-\tfrac{1}{4}\pi D/\Gamma). \quad (420)$$

The number of particles *not* scattered is thus proportional to D/Γ.

The theory given in this section applies to the elastic scattering of particles if no inelastic scattering and no transmutations are possible to any appreciable extent, and if furthermore scattered particle and scattering nucleus have zero spin. Then the orbital quantum numbers l and $m=0$ of the scattered particle are identical with the angular momentum quantum numbers J and $M=0$ of the compound nucleus. There will be a set of equations of the form (400) for each J, and each J will give a contribution of the size (420). The total cross section can be obtained by summing over all J up to $J=l_{Pp}$ (cf. (410)) similarly to section A of this §. The total cross section becomes then

$$\sigma_{\text{tot}}=4\pi R^2(1-\tfrac{1}{4}\pi D/\Gamma). \quad (420a)$$

The case of large partial width and many kinds of emissible particles seems rather more difficult to treat. However, the total cross section must obviously be again of the order πR^2, and the probability of emission of a given kind of particle will be related to its sticking probability in a way similar to the case of small partial width. This means that the emitted particles should have approximately the Maxwellian distribution discussed in §54D.

X. Neutrons

§57. Slow Neutron Processes. Classification and History. (F13, A7, A10, A11, D23, B12, B32, B51, etc.)

The phenomena produced by slow neutrons have been of the greatest importance for the development of the modern theory of nuclear processes. They supply the most detailed information yet available on the energy levels of heavy nuclei. Therefore in our discussion we shall treat the slow neutron processes first and later proceed to the processes produced by charged particles. In the treatment of the latter we shall make use of the results obtained from neutron experiments.

By "slow neutrons" we understand neutrons of energies up to a few thousand volts, including neutrons of "thermal" energy of the order kT. Slow neutrons are produced by surrounding a source of fast neutrons with paraffin, water or other substances containing hydrogen. The mechanism of the slowing down process and the energy distribution of neutrons in hydrogenic substances will be treated in §59. Sources of fast neutrons will be discussed in §92.

The processes which may occur when a slow neutron interacts with a nucleus are mainly of three types:

A. Elastic scattering

The largest elastic scattering cross section ($12\cdot 10^{-24}$ cm^2) has been observed in hydrogen (cf. §14); it is used for the production of slow neutrons (§59). The elastic scattering is also the most probable process in most of the other light nuclei (except Li^6, B^{10} and possibly N^{14}, cf. *C* and *D*) and a number of medium heavy nuclei such as Fe, Ni, Cu. The cross sections for these elements vary between about $2\cdot 10^{-24}$ and $10\cdot 10^{-24}$ cm^2. For nuclei which capture neutrons strongly (such as Rh, Ag, Cd, Hg) the scattering cross section is known to be small compared to the capture cross section (D23) but it may well be of the same absolute order of magnitude as for the elements mentioned before. No thorough investigation of the scattering of slow neutrons as a function of their energy has yet been made. A more detailed discussion of the scattering will be given in §63.

B. Simple capture of neutron with emission of γ-rays

Processes of the type[19]

$$Z^A + n^1 = Z^{A+1} + \gamma \tag{421}$$

are known for a great number of nuclei Z^A. In these "simple capture processes," the bombarded nucleus is transmuted into its isotope which is one mass unit heavier. This isotope may be either stable or radioactive. In the latter case, the capture process is most easily detected by the radioactivity produced. This production of artificial radioactivity was discovered by Fermi and his collaborators in 1934 (F13). Since then, 93 radioactive isotopes have been produced by simple capture of neutrons. About the production of stable isotopes by neutron capture, our information is necessarily less complete: We can only infer it from the absorption of the neutrons.

The γ-rays from capture processes have actually been found (A11) and their energy (F25, R3, K10, K11) and intensity (K9, G23) measured. They will be discussed in §90.

Not long after the discovery of the neutron-capture processes, Fermi, Amaldi, d'Agostino, Pontecorvo, Rasetti and Segrè (A7) found that the efficiency of the neutrons in producing radioactive isotopes increased greatly when the neutrons were slowed down by passing through paraffin or water. Cross sections up to $3000\cdot 10^{-24}$ cm^2 (Cd) were measured for the capture of these slow neutrons, i.e., cross sections more than a hundred times the geometrical cross section of the capturing nuclei.

Attempts were made by several authors (A7, B12, P6) to explain these large cross sections quantum mechanically using a one-body model. The neutron was assumed to move in a certain potential produced by the capturing nucleus. This model gave a capture cross section inversely proportional to the neutron velocity and could thus quite well explain the large increase of the cross section observed when slowing down the neutrons. Moreover, it gave the absolute cross section for slow neutrons reasonably well, and it was also capable of explaining the wide fluctua-

[19] Z^A denotes a nucleus of charge Z and mass number A.

tions of the capture cross section from element to element by differences in the neutron wave functions in the respective potentials.[20]

However, a serious difficulty soon arose in the ratio of scattering to capture cross section. The one-body theory predicted that the two cross sections should, for neutrons of thermal energy, be always of the same order of magnitude so that a large capture cross section would in every case imply large scattering as well. But experiments of Dunning, Pegram, Fink and Mitchell (D23) showed that in the strongly absorbing Cd the scattering cross section was less than 1 percent of the capture cross section.

The final blow to the one-body theory was the selective absorption of neutrons found by Moon and Tillman (M26), Amaldi and Fermi (A7) and Szilard (S29). These authors found that neutrons which made one substance radioactive had very little effect in activating another substance and *vice versa*. This was quite irreconcilable with the prediction of the one-body theory that the cross section should be inversely proportional to the neutron velocity for *any* capturing nucleus.

The phenomenon of selective absorption, while discrediting the one-body theory, led Bohr (B32) and Breit and Wigner (B51) to the correct theory of neutron capture based on the many-body concept of nuclear processes. The selective absorption of given "groups" of neutrons by given nuclei must be interpreted as a resonance effect of the neutrons with a virtual energy level of the "compound nucleus" formed by the temporary addition of neutron and capturing nucleus. Whenever the neutron has an energy coinciding with one of these resonance levels, there will be a large probability of neutron capture. As the resonance levels will be different from nucleus to nucleus, each nucleus will in general have its own characteristic group or groups of neutrons which are easily captured.

The Breit-Wigner theory of neutron capture will be discussed in the next section (§58), its application to the experiments in §§60–62. In these latter sections, we shall discuss the experimental results about the position and spacing of the resonance levels (§60, cf. the theoretical discussions in §53), the width of the levels (§§61, 62, cf. §54) and the absolute magnitude of the cross section (§62).

[20] There was, however, a difficulty about the dependence of the cross section on atomic number (V5).

C. Production of α-particles

The slow neutron is captured and an α-particle emitted, according to the scheme

$$Z^A+n^1\rightarrow Z^{A+1*}\rightarrow (Z-2)^{A-3}+\mathrm{He}^4. \quad (422)$$

Z^{A+1*} denotes the compound nucleus which is temporarily formed in an excited state (asterisk!). This type of reaction is known to occur with Li^6 and B^{10}. The reactions involved are

$$\mathrm{Li}^6+n^1=\mathrm{He}^4+\mathrm{H}^3, \quad (422a)$$

$$\mathrm{B}^{10}+n^1=\mathrm{Li}^7+\mathrm{He}^4. \quad (422b)$$

These reactions were discovered (A8, C12) by observing the heavy particles (H^3 and He^4) produced. The cross sections are very large, about $3000\cdot 10^{-24}$ and $900\cdot 10^{-24}\,\mathrm{cm}^2$ for the boron and the lithium reaction, respectively, with neutrons of thermal energy. The reactions are therefore used as sensitive methods for the detection of neutrons (cf. also §60).

D. Production of protons

This type which may be schematically written

$$Z^A+n^1=(Z-1)^A+\mathrm{H}^1 \quad (423)$$

has only been observed in a single instance, *viz.*

$$\mathrm{N}^{14}+n^1=\mathrm{C}^{14}+\mathrm{H}^1. \quad (423a)$$

It is likely that also the process

$$\mathrm{B}^{10}+n^1=\mathrm{Be}^{10}+\mathrm{H}^1 \quad (423b)$$

occurs; it has not yet been observed, but it is energetically possible. Other processes of this type are probably not possible energetically, which may be seen as follows: The bombarded nucleus Z^A is transmuted into an isobar with a nuclear charge smaller by one unit. Since the nucleus Z^A is known to be stable, $(Z-1)^A$ must have a larger atomic weight than Z^A. On the other hand, the neutron is heavier than the hydrogen atom only by 0.8 MV (cf. Chapter XVIII). Therefore the reaction (423) is only possible energetically if the mass difference $(Z-1)^A-Z^A$ is smaller than 0.8 MV. It is very improbable that this mass difference lies just between 0 and +0.8 MV, as it has to in order to make the process possible. In the cases (423a) and (423b) this condition happens to be fulfilled, but these are the only cases among the lighter nuclei. For heavier nuclei, a reaction of the type (423) would be extremely improbable even if energetically possible, because the protons produced have very low energy (<0.8 MV) and therefore cannot possibly penetrate the potential barrier of a heavy nucleus with any appreciable probability.

E. Other processes

Other processes produced by slow neutrons seem impossible. This is certainly true for ordinary inelastic collisions because "slow" neutrons do not have sufficient energy to excite even the first excited level of the bombarded nucleus which is usually several 10,000 volts above the ground state (cf. §88). Similarly, it seems energetically impossible that a deuteron could be emitted upon slow neutron bombardment.

Thus we are left, besides the elastic scattering, with only three possible types of slow neutron disintegrations, *viz.* simple capture, α-emission and proton emission. Of these, the γ-emission seems relatively very improbable for light nuclei, simply because the interaction between nuclear matter and the radiation field is so much smaller than between the nuclear particles themselves. On the other hand, for heavy nuclei the emission of charged particles after slow neutron bombardment is very improbable because of the high potential barriers. Thus we have, with a given nucleus, in general only *either* simple capture *or* emission of charged particles. (This is not true for reactions produced by *fast* neutrons, §65). Moreover, only in one case (B^{10}) can α-particles *and* protons be produced from the same initial nucleus. Thus we say that *as a rule slow neutrons can cause only a single type of transmutation,* either simple capture or α-emission or proton emission, *with a given nucleus.*

§58. The Breit-Wigner Theory of Slow Neutron Processes (B51, B15)

We apply our general formula (269) to the special case that the incident particle is a slow neutron. With $s=\frac{1}{2}$, we obtain

$$\sigma^{N0}{}_{Qq}=\frac{\pi\lambda}{2(2i+1)}\sum_{Jl'j'}(2J+1)\times\left|\frac{\lambda_r^{\frac{1}{2}}U^{rJ}{}_{N0}U^{rJ}{}_{Qql'j'}}{E-E_r+\frac{1}{2}i\gamma_r}\right|^2 \quad (424)$$

Here N denotes the neutron, the index 0 after N indicates that the initial nucleus is in its ground state, Q is the emitted particle, q denotes the state of the final nucleus, l' and j' the orbital and total angular momentum of the emitted particle. i and J are the angular momenta of initial and compound nucleus, r labels the states of the compound nucleus. E_r is the kinetic energy of a neutron which is in exact resonance with the state r of the compound nucleus (cf. (258)). E is the kinetic energy of the incident neutron, λ and λ_r the neutron wave-lengths (divided by 2π) corresponding to E and E_r. The U's are the matrix elements corresponding to the emission of neutron or particle Q from the state r of the compound nucleus, and γ_r is the effective width of state r. It has been assumed that the emitted particle Q has high energy compared to the neutron, so that $u^{rJ}{}_{Qq}$ does not change appreciably with the neutron energy and may therefore be replaced by $U^{rJ}{}_{Qq}$ (cf. §52, (265)). For the scattering of the slow neutrons themselves, we have according to ((340) (265a))

$$\sigma^{N0}{}_{N0}=\frac{\pi}{2(2i+1)}\sum_{J}(2J+1)\times\left|2R+\sum_r\frac{\lambda_r\Gamma^r{}_N}{E-E_r+\frac{1}{2}i\gamma_r}\right|^2, \quad (425)$$

where $$\Gamma^r{}_N=(U^r{}_{N0})^2 \quad (425a)$$

is the true neutron width of the nuclear level r.

From the formulae (424) (425), the cross section could be calculated as a function of the energy if the matrix elements and the energy levels of the compound nucleus were all known. The mathematical expressions are, however, rather complicated and not easy to interpret. Fortunately, they simplify very much in the cases which are of the greatest practical importance, namely when the cross section (424) is large. This is the case

(a) if one of the resonance denominators is small, i.e., if the energy is near a resonance level.

In this case, the sum over r in (424) reduces to a single term; we obtain a one-level formula similar to (270);

(b) if λ is large, i.e., if the velocity of the incident neutron is very small. In this case, the variation of the resonance factor with energy is negligible compared to the variation of λ. The cross section is then proportional to λ, i.e., inversely proportional to the neutron velocity v.

A. The resonance case

Formula (424) reduces to the Breit-Wigner one-level formula

$$\sigma^{N0}{}_{Qq}=\frac{\pi}{2}\left(1\pm\frac{1}{2i+1}\right)\frac{\lambda\lambda_r\Gamma_N\Gamma_{Qq}}{(E-E_r)^2+\frac{1}{4}\gamma_r{}^2}. \quad (426)$$

The $\pm$ sign stands according to whether the angular momentum J of the resonance level in question is $i+\frac{1}{2}$ or $i-\frac{1}{2}$; for $i=0$, only the $+$ sign is possible and the value of the parenthesis becomes 2.

The width γ in the denominator of (426) contains a term due to the possibility of emitting a neutron and a term due to emission of other particles. The neutron width varies as the square root of the neutron energy while the other term is practically independent of the neutron energy, if all the other particles are fast compared to the neutron. Thus we have

$$\gamma=\Gamma_N(E/E_r)^{\frac{1}{2}}+\sum_{Qq}\Gamma_{Qq}. \quad (427)$$

Because of the small neutron energy E, the neutron width is usually small compared to the other contributions to the width.[21] Then γ becomes independent of the energy and equal to the true width

$$\gamma\approx\Gamma\approx\sum_{Qq}\Gamma_{Qq}. \quad (427a)$$

In general, we are interested in the *total* cross section for the production of particles of the kind Q and of the corresponding final nucleus B, rather than in the *partial* cross section referring to a given state q of the final nucleus B. Therefore we sum (426) over all possible states q and obtain

[21] This follows directly from the experimental ratio of scattering to capture probability of slow neutrons, see below.

$$\sigma^N{}_Q=\sum_q\sigma^N{}_{Qq}=\frac{\pi}{2}\left(1\pm\frac{1}{2i+1}\right)\frac{\hbar^2}{2M(EE_r)^{\frac{1}{2}}}\times\frac{\Gamma_N\Gamma_Q}{(E-E_r)^2+\frac{1}{4}\Gamma^2}, \quad (428)$$

where
$$\Gamma_Q=\sum_q\Gamma_{Qq} \quad (428a)$$

is the total "particle Q width" of the nuclear level and Γ_N the total neutron width. Now we know from the general discussion in §57 that ordinarily only *one* kind of particles can be produced with appreciable probability under slow neutron bombardment, *viz.* either γ-rays or α-particles or protons. Furthermore, we know that the neutron width Γ_N is usually small compared to the contribution of the particles Q to the width (cf. (427) and the remarks after (432)). Therefore we may write in all practical cases

$$\Gamma_Q=\Gamma. \quad (429)$$

If we measure the cross section in cm^2 and all energies, including the widths, in volts, (428) may be rewritten thus:

$$\sigma_Q{}^N=(\sigma_0/(1+x^2))(E_r/E)^{\frac{1}{2}} \quad (430)$$

with

$$x=2(E-E_r)/\Gamma, \quad (430a)$$

$$\sigma_0=1.30_5\cdot 10^{-18}\left(1\pm\frac{1}{2i+1}\right)\frac{1}{E_r}\frac{\Gamma_N}{\Gamma}. \quad (430b)$$

(430b) shows that *the maximum cross section*, at exact resonance, *gives* a direct measure of the *ratio of the neutron width to the total width.* This is the basis of the experimental determinations of the neutron width.

In (430) the factor E_r/E may be put equal to unity over the whole resonance region provided the width Γ of the resonance level is small compared to the resonance energy E_r, which is true for most of the neutron levels known. In this case, the "shape" of the resonance level reduces, according to (430), exactly to the familiar shape of the optical lines. The cross section is one-half of its maximum value if the energy differs from the resonance energy by one-half of the "width" Γ. The case that E_r and Γ are of the same order will be discussed below.

The *elastic scattering* (425) becomes, if only one

level is important and the potential scattering is neglected,[22]

$$\sigma^N{}_N=\frac{\pi}{2}\left(1\pm\frac{1}{2i+1}\right)\lambda\!\!\!^-{}_r{}^2\frac{\Gamma_N{}^2}{(E-E_r)^2+\frac{1}{4}\Gamma^2},\tag{431}$$

making again the assumption (427). The ratio of elastic scattering to the capture (or disintegration) cross section (428) is

$$\sigma^N{}_N/\sigma^N{}_Q=(\Gamma_N/\Gamma_Q)(E/E_r)^{\frac{1}{2}}.\tag{432}$$

Experiments by Dunning, Pegram, Fink and D. P. Mitchell (D23) have shown that the scattering of slow neutrons by Cd is very small, probably less than one percent of the capture. Qualitatively the same seems to be true for other strongly absorbing substances such as Ag, Rh etc. (M19, M20). This shows that the *neutron width is generally much smaller than the γ-ray width for medium heavy nuclei.* This justifies our neglecting Γ_N compared to Γ_Q (cf. (427)). The result is furthermore of great importance in order to decide which values to deduce from the capture experiments for the neutron width Γ_N and the emitted-particle width Γ_Q: The capture cross section (428) contains only the product $\Gamma_N\Gamma_Q$ and the sum $\Gamma=\Gamma_N+\Gamma_Q$ of the two widths. From measurements of the capture cross section we can therefore deduce Γ_N and Γ_Q but cannot deduce which is which. The scattering experiments mentioned show that the larger of the two quantities must be identified with Γ_Q, the smaller one with Γ_N.

B. The $1/v$ region

If the energy of the incident neutron is small compared to all the resonance energies[23] E_r, the change of the resonance denominators in (424) with neutron energy is negligible compared to the change of the factor $\lambda\!\!\!^-$. Therefore, *for sufficiently small neutron energy, there will always be a region in which the capture cross section is inversely proportional to the neutron velocity.* Just how large this region is, depends on the position of the first resonance level (of positive or negative energy) and also to some extent on its width (see below).

By virtue of the large factor $\lambda\!\!\!^-$, the cross section will in general be quite large in the $1/v$ region. (Exceptions may occur if there is destructive interference between the contributions of various nuclear levels r for small neutron energies E.) The cross section of a given nucleus for slow neutrons will therefore, besides the resonance maxima, have one maximum at zero energy. While the resonance maxima lie, in general, at different neutron energies for different nuclei, the maximum at low energies is common for all nuclei affected by slow neutrons. In order to separate the characteristic resonance effects from the effect of the very slow neutrons, it is customary to absorb the latter by a suitable absorber. Such an absorber is Cd; it absorbs strongly all neutrons up to about 0.4 volt energy (cf. §61), a thickness of 0.3 mm being sufficient to reduce the intensity of very slow neutrons to less than one percent. The very slow neutrons strongly absorbed in Cd are known in the literature as the "C group" (C = cadmium). The activity produced in most detectors, under conditions specified in the next section, is in the average due in about equal parts to the C neutrons and to the neutrons in the resonance groups.

A closer examination of the validity of the $1/v$ law is possible if we accept the one-level formula (428) as valid in the $1/v$ region. This is in general *not* justified from experiment, as will be shown in §61. Only if the first resonance level lies very near zero[24] and if, in addition, this resonance level is "strong," i.e., has large widths Γ_N and Γ_Q for the neutrons as well as for the produced particles, may we expect the one-level formula to hold. This seems to be the case for Rh and Cd, but, e.g., not for Ag and I. However, the conditions for the validity of the $1/v$ law will not be very different whether the one-level or the many-level formula is to be used.

The one-level formula (428) may be written, after summing over q and putting $\gamma=\Gamma$ (cf. (427a))

$$\sigma^N{}_Q\sim\frac{E^{-\frac{1}{2}}}{(E-E_r)^2+\frac{1}{4}\Gamma^2}.\tag{433}$$

[22] The role of the potential scattering will be treated in §63.

[23] If the resonance energy E_r is *negative*, E must be small compared to $|E_r|$.

[24] "Near" must be understood in comparison with the average spacing of the levels. However, the first level will of course lie at an energy much *higher* than the limit of validity of the $1/v$ law. Cf. remark after (434).

The condition for the $1/v$ law is that the relative variation of $E^{-\frac{1}{2}}$ with energy shall be larger than that of the denominator in (433), *viz.:*

$$|(d/dE)(\log E^{-\frac{1}{2}})| \gg |(d/dE)\log[(E-E_r)^2+\tfrac{1}{4}\Gamma^2]|. \quad (433a)$$

This yields

$$4E|E_r-E| \ll (E_r-E)^2+\tfrac{1}{4}\Gamma^2. \quad (434)$$

This is certainly fulfilled if E is small compared to the larger of the two quantities[25] E_r and Γ. In words: *The 1/v law holds if the neutron energy is small compared to the first resonance energy, or is small compared to the width of this resonance level.*

From this rule, we must expect that the $1/v$ law holds only for a very small energy region in heavy nuclei whose levels are very dense (§53, 60) and narrow (§54, 61), but holds up to rather high energies for light nuclei, corresponding to the large spacing and width of their energy levels. In heavy nuclei, the first resonance level is often found at a neutron energy of about one volt (e.g., Rh, In, Ir, cf. §60) in which case the $1/v$ region will not extend much beyond thermal energies ($kT=1/40$ volt at room temperature). In some cases, such as Cd and Dy, the first level lies even lower (§61) so that even in the thermal region the cross section shows no proportionality with the reciprocal velocity. On the other hand, in light nuclei such as Li, B, N, the average spacing between nuclear levels may be expected to be of the order of hundred thousands of volts, and thus the first level will in general lie at an energy of this same order. But even if, by accident, the resonance level would lie very close to zero in one of these cases, its width would be very large: We know from experiments on the capture of protons by light nuclei that the width of the levels of light nuclei corresponding to the emission of particles is of the order of several ten thousands of volts if the particles are sufficiently fast (more than about 1 MV energy). This latter condition is certainly fulfilled for the slow neutron reactions in Li^6 and B^{10} in which α-particles are produced (cf. (422a) (422b), §64). For these reactions the $1/v$ law will therefore hold up to quite high neutron energies, of a few thousand volts at least. On the strength of these considerations, the absorption coefficient in boron has been used (S29, F33, W5, A7, H8, H9, G17, G18, cf. also §60) for determining the energy of the various resonance groups of neutrons activating various elements.

A very direct *experimental test of the 1/v law*, at least in the region of thermal energies, was made by Rasetti, Mitchell, Fink and Pegram (R5) for boron, and by Rasetti, Segrè, Fink, Dunning and Pegram (R4) for cadmium and silver. A steel or aluminum disk of 50 cm diameter was coated with the substance to be investigated and rotated at 6000 r.p.m. so that the linear velocity at the edge was about 140 meters/sec., comparable to the velocity of thermal neutrons (2200 m/sec.). A beam of slow neutrons was sent through the disk near the edge at an angle of about 65° with the normal to the wheel so that the relative velocity of the neutrons with respect to the absorbing nuclei in the wheel differed greatly according to whether the wheel was rotated in the direction or against the direction of the neutron beam. In spite of this change in relative velocity, no change in the absorption of the neutrons is expected when the $1/v$ law holds: For the time the neutrons need for traversing the wheel is independent of the velocity of the wheel, and the $1/v$ law just means that the capture probability per unit time is independent of the velocity. Any deviation from the $1/v$ law must, however, show in a different absorption of the neutrons for the two directions of rotation of the wheel.

The neutron intensity was measured by the number of disintegrations produced in Li with the help of an ionization chamber (cf. §94). No change in the absorption, outside the statistical error, was found for the B and Ag absorber, showing that they obey the $1/v$ law at least in the region of thermal energies. For Cd, however, an increase of 6.3 percent in the absorption coefficient was observed when the Cd disk was moved against the neutron beam. This corresponds under the conditions of the experiment to a cross section almost independent of the velocity (cf. §61).

C. Special cases of the one-level formula

For the applicability of the one-level formula to slow neutrons, it is only necessary that the

[25] This condition is only sufficient if $\Gamma_N \ll \Gamma_Q$. Otherwise, Γ will itself depend on the energy and a more stringent condition holds.

nearest level is very close as compared to the second nearest level, but it is not necessary that the nearest level is a real resonance level of positive energy. It is just as likely that large cross sections for slow neutrons are due to a "negative" level with a negative resonance energy E_r. In this case, of course, the substance in question would only abosorb the "C group" of neutrons and would not show any *characteristic* neutron absorption band, at least not at low energy. The cross section would be given by

$$\sigma^N{}_Q=\frac{\pi}{2}\left(1\pm\frac{1}{2i+1}\right)\frac{\hbar^2}{2M|E_r|^{\frac{1}{2}}}\frac{1}{E^{\frac{1}{2}}}\times\frac{\Gamma\Gamma_N}{(E+|E_r|)^2+\frac{1}{4}\Gamma^2}, \quad (428\text{b})$$

it would decrease monotonically with increasing energy, first as $E^{-\frac{1}{2}}$, later as $E^{-5/2}$.

Another case in which no characteristic absorption can be observed but only absorption of very slow neutrons, is the case of a resonance level very close to zero energy whose width is comparable to the (positive) resonance energy. Fig. 11 gives the behavior of the cross section as a

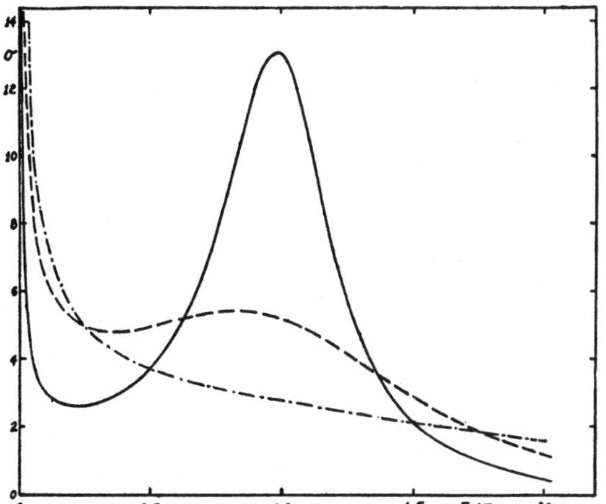

FIG. 11. The cross section for neutron capture as a function of energy for various ratios of width Γ to resonance energy E_r. ——— Width $\Gamma=\frac{1}{2}E_r$ (pronounced resonance), - - - - - $\Gamma=\sqrt{2}E_r$ (almost case of Cd), —·—·— $\Gamma=4E_r$ (almost $1/v$ law). The abscissa should be denoted E/E_r rather than E/E_0.

function of the neutron energy, for various values of the ratio $\epsilon=E_r/\Gamma$. There is a monotonic decrease of the cross section with energy for negative and very small positive ϵ. For $E_r=(5^{\frac{1}{2}}/4)\Gamma=0.56\Gamma$ the curve contains a point of inflection with horizontal tangent at $E=\frac{3}{5}E_r$. For larger ϵ, we find curves containing a minimum and a maximum at the energies

$$E_{\min}{}^{\max}=0.6E_r\pm(0.16E_r{}^2-0.05\Gamma^2)^{\frac{1}{2}}. \quad (435)$$

With increasing ratio E_r/Γ, the two extrema become more pronounced and more separated from each other, until for $E_r\gg\Gamma$ we have (cf. (430), (430a))

$$E_{\min}=\tfrac{1}{5}E_r,\quad E_{\max}=E_r, \quad (435\text{a})$$

$$\sigma_{\min}=(5^{5/2}/64)(\Gamma/E_r)^2\sigma_0=0.87(\Gamma/E_r)^2\sigma_0, \quad (435\text{b})$$

where $\sigma_0=\sigma_{\max}$ is the cross section at exact resonance (430b).

§59. DIFFUSION OF NEUTRONS (A11, F18, F17)

A. General

Before we can discuss the various methods of determining position, total width and neutron width of resonance levels, we must investigate the behavior of the neutrons in the hydrogen-containing substance (paraffin, water) etc. which is used to slow down the neutrons.

The questions to be answered are mainly the following:

(1) What is the energy distribution of the neutrons in a pure infinitely extended hydrogeneous substance?

(2) What is the distribution of the neutrons in space, and how does it depend on the neutron energy?

(3) How are the distributions affected if an absorbing substance is placed inside the hydrogeneous substance?

There are some further problems which are connected to those mentioned.

The approach to these problems is partly theoretical and partly experimental. We shall start from a few theoretical assumptions which seem reasonably certain, and rely on experimental determinations of the constants involved such as the mean free path of slow neutrons, their lifetime in hydrogeneous substances, etc. Since most experiments were made in paraffin, we speak of "paraffin" as representing hydrogeneous substances in general. We shall follow closely the theoretical treatment of Fermi.

The basic theoretical facts about the scattering of neutrons by protons are the following:

(*a*) In each collision with a proton the neutron will lose energy. As long as the neutron is fast compared to the proton, the probability that the neutron energy lies between E and $E+dE$ after the collision, is

$$w(E)dE=dE/E_0, \tag{436}$$

where E_0 is the neutron energy before the collision. This means that any value of the final energy of the neutron, between 0 and the initial energy E_0, is equally probable.

The proof follows immediately from the energy-momentum considerations given in §14, together with the fact that the scattering of neutrons by protons is practically spherically symmetrical[26] in a system of reference in which the center of gravity of the two particles is at rest (§15). If φ is the deflection of the neutron in the collision, the energy after collision is

$$E=E_0 \cos^2 \varphi \tag{436a}$$

(cf. (49d)), and the number of neutrons deflected between φ and $\varphi+d\varphi$

$$\text{const}\cdot \sin \varphi \cos \varphi d\varphi \tag{436b}$$

(cf. (72)). Combining (436a, b), we obtain (436).

From (436) it follows, that the arithmetical mean[27] of the energy of the neutron after one collision is $\frac{1}{2}E_0$, while the geometrical mean E_a is given by

$$\log (E_a/E_0)=\int_0^{E_0} \log (E/E_0)dE/E_0=-1 \tag{436c}$$

so that

$$E_a=E_0/e. \tag{437}$$

After n collisions, the geometric mean of the neutron energy will be

$$E_{an}=E_0e^{-n}. \tag{437a}$$

The actual energy distribution of the neutrons after a number of collisions with protons was calculated by Condon and Breit (C31).

(*b*) The mean free path between two collisions decreases very rapidly with decreasing neutron energy. In paraffin, it is about 5 cm for neutrons of 2 MV (theoretical value, cf. (62)) and decreases to about 1 cm (experimental value) for slow neutrons. The mean free path should be constant below, say, 10,000 volts (cf. (62)). For still slower neutrons whose energy is comparable to the vibrational energy of the hydrogen in paraffin ($\sim$0.4 volt) or smaller, the mean free path decreases further (see Sec. C) to about 0.3 cm for thermal neutrons (experimental value) and 0.2 cm for energy zero (theoretical, cf. (451), (464)).

Therefore the first part of the slowing down process will require a great amount of space, but

[26] Recently, additional experimental (H15) and some theoretical material has been advanced which is in contradiction with the spherical distribution of scattered fast neutrons. The theoretical calculations of Morse, Fisk, Schiff and Shockley (M28, M29, F22, F23) were made using a potential between neutron and proton of the form

$$V=-A(e^{-r/a}-\tfrac{1}{2}e^{-2r/a})$$

as suggested by Morse. It was found that such a potential indeed gives larger asymmetries in the scattering than a "rectangular hole" of equivalent radius. The reason is that the Morse potential (and similarly an exponential or a Gauss potential) is finite though small at fairly large distances, and therefore has a greater effect on the "p wave" ($l=1$) which causes the deviation from spherical symmetry. The actual calculations were made assuming a "range of the nuclear forces" which is about 50 percent larger than that expected from the binding energies of light nuclei; this will tend to increase the asymmetry. With this range, the pronounced asymmetry found experimentally by Kurie (K29) and by Harkins, Kamen and Newson (H15) could be duplicated theoretically if a neutron energy of 25 MV was assumed. The average energy of the Be+Rn neutrons used experimentally, is about 5 MV (§99B). Thus, even with the Morse potential which is particularly favorable for an asymmetry, only a few percent of the observed asymmetry would be expected if the correct neutron energy and range of the nuclear forces are used. In any case, an asymmetry in the scattering can certainly only exist for fairly high neutron energies and will be negligible below 1 MV, i.e., in the region in which we are primarily interested. *Added in proof:* The most recent experiments of Dee (private communication of Dr. Cockcroft) and of Kruger and Schoupp with 2.5 MV neutrons from the reaction $H^2+H^2=He^3+n^1$ give a spherical distribution of the recoil protons within the experimental error.

[27] There has been considerable controversy (A7, W15, G19, C31) about which average to use. It seems to us that the geometrical average originally suggested by Fermi (A7) is certainly preferable. This is already borne out by the fact that the arithmetical average of the logarithm of the energy (which is identical with the log of the geometric mean) decreases *linearly* with the number of collisions (cf. 437a) while the arithmetical average of the energy itself decreases exponentially, showing that $(\log E)_{Av}$ is the more suitable quantity. Moreover, the actual values of $\log E$ after n collisions form very nearly a Gaussian distribution around $(\log E)_{Av}$, whereas there is nothing as simple as a Gaussian distribution if we plot the probability against E itself. If any average other than the geometrical is taken, it seems to us that the average efficiency of the neutrons in producing radioactivity etc. should be chosen; assuming this efficiency to be roughly proportional to $1/v \propto E^{-\frac{1}{2}}$, we would have to define the "efficiency average" E^* after one collision by

$$E^{*-\frac{1}{2}}/E_0^{-\frac{1}{2}}=\int_0^{E_0}(E/E_0)^{-\frac{1}{2}}dE/E_0=2;$$

therefore $E^*=\frac{1}{4}E_0$.

when the neutrons have once been decelerated to about 100,000 volts, the remaining energy loss will occur in a relatively very small region of space. Ultimately, after 20 to 30 collisions, the neutron energy will be reduced to thermal energy. There will then, of course, be no further energy loss, but the neutrons will only diffuse in the paraffin. This will continue until they are absorbed, either by the protons in the paraffin themselves, or by some other absorber inserted in the paraffin.

The lifetime of the neutrons in pure paraffin may be calculated from the theoretical probability of the capture of neutrons by protons (cf. §17, (97a)) or from experimental data (see below, subsection E) and is found to be about 10^{-4} sec. This compares to a time of about $1.3\cdot 10^{-6}$ sec. which is necessary to slow the neutrons down to 1 volt energy. A lifetime of 10^{-4} sec. corresponds to somewhat over 100 collisions (cf. (485)) between neutron and protons after the neutrons have reached thermal equilibrium. Therefore a large fraction of the neutrons present in the paraffin will have thermal energies, and consequently a large part of the radioactivity produced in any detector placed in the paraffin will certainly be due to thermal neutrons.

In spite of the long time during which the neutrons have thermal energies, they will not diffuse over a very large distance in that time, owing to their very small mean free path. Measurements of Fermi give in the average a total "diffusion length" of about $3\frac{1}{2}$ cm, taken from the point at which the neutron first reaches thermal energies. This may be compared with an average path from the *source* to the point of absorption, of over 18 cm. Thus we can say that the neutrons spend most of their time as thermal neutrons, but that the region in which they move as thermal neutrons is small and its position is almost exclusively determined by the path of the neutron while it is fast. The spatial distribution of fast and slow neutrons will show only little difference. This has the practical consequence that it is impossible to shield a region of space against neutrons by just absorbing the slow ones, e.g., by Cd screens: The slow neutrons will be regenerated unless the fast ones are absorbed as well which is only possible by using sufficient amounts of paraffin. About 70 cm of water (50 cm of paraffin) are necessary to reduce the number of neutrons to one percent. (The density of neutrons decreases more rapidly because of the geometrical factor $1/r^2$.)

B. Energy distribution above 1 volt

The energy distribution of neutrons has been derived by Fermi (F18) under the assumptions that the energy distribution of the neutrons after *one* collision with a proton is uniform, from zero up to the initial energy E_0 (cf. (436)). This will be true as long as the neutron energy is large compared to the vibrational energy of the protons in paraffin (cf. Section C). Therefore, the validity of the formula to be derived will be restricted to energies above about one volt. The mean free path $l(E)$ and the probability of absorption (capture) of the neutron per unit time, $1/\tau$, may in the following be arbitrary functions of the energy.

We first derive the energy distribution for the case of ***negligible absorption***. This assumption is justified for pure paraffin or water, because the mean lifetime τ of a neutron with respect to capture by a proton, is extremely long compared to the total time required for the slowing down to thermal energies.

Let Q be the total number of neutrons per second emitted by the source, and $N(E)dE$ the number of neutrons of a given energy E present at any time in the whole paraffin. The decrease in the number of neutrons of energy E per second because of elastic collisions is, according to the definition of the mean free path,

$$N(E)vdE/l(E). \tag{438}$$

The number of neutrons thrown into the given energy interval dE by a collision, is

$$QdE/E_0+\int_E^{E_0} dE'N(E')[v'/l(E')]dE/E'. \tag{438a}$$

The first term represents the neutrons coming into the energy interval dE by the first collision after being emitted from the source, dE/E_0 being the probability that the neutron energy falls into the interval dE after the collision and E_0 being the initial energy of the neutron. The second term gives the neutrons of energy E produced by later collisions: $N(E')v'dE'/l(E')$ gives the total number of collisions of neutrons in the energy interval E' to $E'+dE'$, and dE/E' gives the probability that the energy of such a neutron is reduced from E' to E.

If the distribution $N(E)$ is to be stationary, (438) must be equal to (438a), i.e.,

$$\frac{N(E)v}{l(E)}=\frac{Q}{E_0}+\int_E^{E_0}\frac{N(E')v'}{l(E')}\frac{dE'}{E'}. \tag{439}$$

To solve this integral equation, we differentiate with respect to E:

$$\frac{d}{dE}\left(\frac{Nv}{l}\right)=-\frac{1}{E}\frac{Nv}{l} \tag{439a}$$

which integrates immediately to

$$Nv/l=\text{const}/E. \tag{439b}$$

Inserting this into (439) we determine the constant and find:

$$N(E)dE=Q\frac{l(E)}{v}\frac{dE}{E}. \tag{440}$$

In the region of slow neutrons (below 10,000 volts), $l(E)$ becomes independent of E, so that

$$N(E)\sim E^{-\frac{3}{2}}. \tag{440a}$$

In the more general case of finite absorption, we have instead of (439):

$$N(E)\left(\frac{v}{l(E)}+\frac{1}{\tau}\right)=\frac{Q}{E_0}+\int_E^{E_0}\frac{N'v'}{l'}\frac{dE'}{E'}, \tag{441}$$

where the second term on the left represents the neutrons of energy E absorbed per unit time. Differentiation gives

$$\frac{d}{dE}\left[N\left(\frac{v}{l}+\frac{1}{\tau}\right)\right]=-\frac{Nv}{lE}$$

$$=-N\left(\frac{v}{l}+\frac{1}{\tau}\right)\frac{v}{(v+l/\tau)E}. \tag{441a}$$

The integral of this is

$$N(E)=\frac{c}{v/l+1/\tau}\exp\left(\int_E^{E_0}\frac{dE}{E}\frac{v}{v+l/\tau}\right). \tag{441b}$$

Inserting this expression for $E=E_0$ into (441) fixes c:

$$c=Q/E_0. \tag{441c}$$

The distribution function (441b, c) simplifies considerably if, below a certain energy limit, l and τ are independent of the energy. This is almost certainly true in pure paraffin in the energy region from 1 to about 10,000 volts. It would also be true (with a smaller τ) if the paraffin (or water) contained a neutron absorber of small atomic weight such as Li and B. It would, however, not be true if an absorber with a large number of resonance absorption bands were contained in the paraffin—i.e., practically any substance of higher atomic weight. If we make the assumption of constant τ and l, the integral in (441b) can be evaluated and we obtain

$$N(E)=\frac{al}{(v+l/\tau)^3}. \tag{442}$$

The constant a can be determined by comparing (442) to (440) because there is a region in which l is practically independent of the energy and at the same time l/τ is negligible. Then we find

$$a=2Q/m, \tag{442a}$$

where m is the neutron mass.

C. Energies below 1 volt. Influence of chemical binding (F17)

1. General remarks.—The ordinary formulae for the scattering of neutrons by protons (§14, 15) are derived assuming that the proton is free. This assumption is justified as long as the neutron energy is very large compared to the "binding energy" of the proton which may be identified with $\hbar$ times the frequency of vibration of the hydrogen in the paraffin molecule. This frequency is about 3000 cm^{-1} for the CH bond, corresponding to about 0.4 volt. For slow neutrons, of energies less than about a volt, the binding of the protons must therefore be considered.

This binding has two effects: The first is that it is no longer possible to freely impart energy to the proton. The vibrations of the paraffin molecules may be divided into two groups. Firstly, vibrations in which a hydrogen atom moves relative to the rest of the molecule without any appreciable motion of other atoms, and secondly, vibrations of whole CH_2 groups with respect to the rest of the molecule. The transfer of energy to the first kind of vibrations will be similar to that to free protons inasmuch as there is a considerable probability that the neutron loses practically all its energy in one collision. However, such a transfer of energy is only possible as long as the neutron energy is larger than $\hbar$ times the frequency ω (number of vibrations per 2π sec.) of the respective vibrations which is rather high. A neutron of smaller energy can transfer energy only to the vibrations of the second kind; and the effect of the collisions will then be similar to that of collisions with a free[28] CH_2 molecule of mass 14. This fact will reduce greatly the average energy loss of the neutron per collision. Thus the *neutron energy will decrease more slowly once it has reached the "region of the chemical bond."*

The second effect of the proton binding is to change the probability of the collisions. We shall show that the cross section increases so that the *mean free path decreases with decreasing energy.* In the limit of very small energies, the mean free path is theoretically one-quarter of its value for neutrons above 1 volt energy (cf. (451)). As far as the efficiency of the slowing down is concerned, this effect will work in the opposite direction of the first, i.e., it makes the neutron lose energy faster.

Another effect of the binding is a change in the angular distribution of the neutrons after scattering: This distribution will be uniform (per unit solid angle) over the whole sphere at small energies while no neutrons are deflected through more than 90° by free protons (§15).

2. Validity of the Born approximation.—If there is an elastic force (natural frequency ω) on the proton and an interaction V between neutron (coordinate ξ) and proton (coordinate x), the wave equation of the system is

$$(\hbar^2/2m)(\Delta_x+\Delta_\xi)\Psi+(W-\tfrac{1}{2}m\omega^2x^2 - V(|x-\xi|))\Psi=0. \quad (443)$$

This wave equation looks, at first sight, rather unmanageable: On one hand, it is not separable, and on the other hand, it cannot be solved by regarding the interaction V as a small perturbation because V is of the order of 10 MV when neutron and proton are close together, and the Born perturbation method (cf. M32, Chapter VII) is only applicable when the perturbation is small compared to the energy of the particles. However, we can show that V may be replaced by an auxiliary potential U which may be so chosen that it gives the same results as V for the scattering, and at the same time can be treated as a small perturbation.

That this is possible is due to the fact that V is restricted to a very small region of space ($\sim 10^{-13}$ cm) while the oscillator potential $m\omega^2x^2$, the wave functions of the proton in that potential, and the plane wave representing the neutron, change only over very large distances ($\sim 10^{-9}$ cm). Therefore we may solve (443) for small distances $|x-\xi|$ between neutron and proton without paying any attention to the oscillator potential, and join the solution to a solution of (443) which is valid at large distances and in which the nuclear potential is neglected. For this joining on, which may be effected at some medium value s of $|x-\xi|$, it is immaterial how Ψ behaves for small $|x-\xi|<s$. Therefore the solution of (443) at large distances, and consequently the scattering, will remain unaltered if V is replaced by another potential U provided only the asymptotic behavior of the wave function for distances $|\xi-x|$ large compared to the range of the nuclear forces, is the same in the two potentials U and V.

To discuss this asymptotic behavior, it is sufficient to consider the wave equation of the relative motion of neutron and proton, neglecting the oscillator potential:

$$(\hbar^2/m)\Delta_\eta\psi+(E'-V(\eta))\psi=0, \quad (444)$$

where $\mathbf{n}=\boldsymbol{\xi}-\mathbf{x}$ and $\tfrac{1}{2}m$ is the reduced mass. If ψ is expanded in spherical harmonics, the radial wave functions of the partial waves $l\neq 0$ will be the same as for free particles, owing to the short range of the nuclear forces. This means spherical symmetry of the scattering (§15). For $l=0$ we have (cf. §14, Eq. (51))

$$r\psi=\sin K(r+r_0), \quad (445)$$

where
$$K=(mE')^{\frac{1}{2}}/\hbar. \quad (445a)$$

(445) corresponds to a scattered wave of amplitude r_0 in any direction.[29] The total scattering cross section σ_0 is

$$\sigma_0=4\pi r_0^2. \quad (445b)$$

(In the notation of §14, $r_0=1/\beta$ if the spins of neutron and proton are opposite, and $r_0=-1/\alpha$ if they are parallel.)

We compare the solution (445) with the solution of the differential equation

$$(\hbar^2/m)\Delta\psi'+(E'-U(\eta))\psi'=0, \quad (446)$$

where the "auxiliary potential" U is defined by

$$\begin{aligned} U&=-U_0 \quad \text{for} \quad \eta<R, \\ U&=0 \quad \text{for} \quad \eta>R. \end{aligned} \quad (446a)$$

If U is chosen so that the Born method is applicable, the amplitude of the wave scattered in the direction ϑ is

[28] Provided the neutron energy is still large compared to the quantum energies of all the oscillations of the CH_2 groups with respect to each other. For still lower energies of the neutron, the effective mass of the atomic groups with which the neutron collides, would be still larger.

[29] The scattering amplitude is generally (cf. M32, p. 24)

$$f(\vartheta)=\frac{1}{2iK}\sum_{l=0}^{\infty}(2l+1)(e^{2i\delta_l}-1)P_l(\cos\vartheta),$$

where δ_l is the phase shift of the partial wave of angular momentum l (cf. §14). In our case, $\delta_l=0$ for $l\neq 0$ and $\delta_0=Kr_0$ (cf. (51_0), (445)), so that

$$f(\vartheta)=2iKr_0/2iK=r_0.$$

(M32, p. 88, Eq. (5) and p. 87, Eq. (1))

$$f(\vartheta) = -(m/4\pi\hbar^2)\int U(\eta)\exp(iK(\mathbf{n}_0-\mathbf{n})\cdot\mathbf{n})d\tau, \quad (447)$$

where $\mathbf{n}_0$ and $\mathbf{n}$ are unit vectors in the direction of incident and scattered wave.

A. If we want this result to be identical with the scattering from the true nuclear potential, we have to fulfill the following conditions:

(1) $f(\vartheta)$ must be independent of ϑ (cf. above (445)). This means that the extension R of the "auxiliary potential" must be small compared to the wave-length $1/K$, i.e.,

$$KR \ll 1. \quad (447a)$$

(2) The amplitude (447) must be equal to r_0. Since the exponential reduces to unity, owing to (447a), we have with (446a)

$$f(\vartheta) = mU_0R^3/3\hbar^2 = r_0. \quad (448)$$

B. In order that the Born approximation is applicable, U_0 must be small compared to E':

$$U_0 \ll E'. \quad (448a)$$

This condition is reconcilable with (447a) and (448) because combination of (447a) and (448a) gives

$$mU_0R^3/3\hbar^2 \ll mE'/3\hbar^2K^3 = 1/3K, \quad (448b)$$

which is actually fulfilled according to (448) ($r_0 \approx 10^{-12}$ cm; $1/K$ = reciprocal neutron wave-length $\approx 10^{-9}$ cm or more). It is therefore possible to choose U_0 and R so that (447a) and (448a) are fulfilled simultaneously.

3. Calculation of cross section.—The Schrödinger Eq. (443) can now be solved by the Born method: In zero approximation, we take plane waves for the neutron, and oscillator wave functions for the proton. Let ψ_0 and ψ_n be the wave functions of the proton before and after the collision, $\mathbf{k}$ and $\mathbf{k}'$ the initial and final wave vector of the neutron. Then the differential cross section per unit solid angle $d\omega$ is, according to the Born formula

$$\sigma_n(\vartheta)d\omega = d\omega\frac{v'}{v}\left|\frac{2m}{4\pi\hbar^2}\int \exp[i(\mathbf{k}-\mathbf{k}')\cdot\boldsymbol{\xi}] \times U(|\mathbf{x}-\boldsymbol{\xi}|)\psi_0(\mathbf{x})\psi_n^*(\mathbf{x})d\mathbf{x}d\boldsymbol{\xi}\right|^2. \quad (449)$$

Now the potential U extends only over a region R small compared to the wave-length of the neutron, and small compared to the amplitude of the proton oscillator (cf. (447a)). Therefore in the exponential $\boldsymbol{\xi}$ may be replaced by $\mathbf{x}$. Then the integration over $\boldsymbol{\xi}$ can be carried out, and we obtain, using (448),

$$\sigma d\omega = (v'/v)(2r_0)^2\left|\int \exp[i(\mathbf{k}-\mathbf{k}')\cdot\mathbf{x}] \times\psi_0(\mathbf{x})\psi_n^*(\mathbf{x})d\mathbf{x}\right|^2 d\omega. \quad (450)$$

In the limiting case of *very small neutron energies*, only elastic collisions are possible so that $v' = v$. Also, k and k' are then very small so that the differential cross section per unit solid angle is $4r_0^2d\omega$, *independent of the angle.* The total cross section is (cf. (445b))

$$\sigma = 16\pi r_0^2 = 4\sigma_0, \quad (451)$$

i.e., *four times the cross section for free protons.* This factor 4 may be directly understood from the Born approximation, once the validity of this approximation has been established. According to (449), the scattering cross section in the Born approximation is proportional to the square of the mass m, for a given potential energy. Now if the protons are tightly bound, the neutron mass has to be inserted for m, while in the case of free protons we must use the reduced mass $m/2$.

For the calculation in the general case (neutron energy comparable to $\hbar\omega$) we assume that the proton is initially in the lowest vibrational state which will practically always be true. The wave functions of a one-dimensional oscillator are:

$$\psi_n = (2\pi)^{-\frac{1}{4}}n!^{-\frac{1}{2}}e^{-\frac{1}{4}\xi^2}H_n(\xi), \quad (452)$$

where

$$\xi = x(\hbar/2m\omega)^{-\frac{1}{2}} \quad (452a)$$

and the Hermitian functions are defined by (Jahnke-Emde, *Tables of Functions*, p. 105)

$$H_n(\xi) = e^{\frac{1}{4}\xi^2}\frac{d^n}{d\alpha^n}(e^{-\frac{1}{4}(\xi-\alpha)^2})_{\alpha=0}. \quad (452a)$$

The normalization of (452) is such that $\int\psi_n^2d\xi = 1$. Our matrix element (450) then is a product of three integrals referring to the x, y, and z coordinates. The x integral is

$$A_x = (2\pi)^{-\frac{1}{2}}n!^{-\frac{1}{2}}\left(\frac{d^n}{d\alpha^n}\int_{-\infty}^{\infty}e^{-\frac{1}{2}(\xi-\alpha)^2+iq_x\xi}d\xi\right)_{\alpha=0} \quad (453)$$

with

$$q_x = (k_x-k_x')(\hbar/2m\omega)^{\frac{1}{2}}. \quad (453a)$$

Elementary integration gives

$$A_x = (iq_x)^n n!^{-\frac{1}{2}}e^{-\frac{1}{2}q_x^2}. \quad (453b)$$

(450) reduces to

$$\sigma d\omega = 4r_0^2(v'/v)d\omega\frac{q_x^{2n_x}q_y^{2n_y}q_z^{2n_z}}{n_x!n_y!n_z!}e^{-q^2} \quad (454)$$

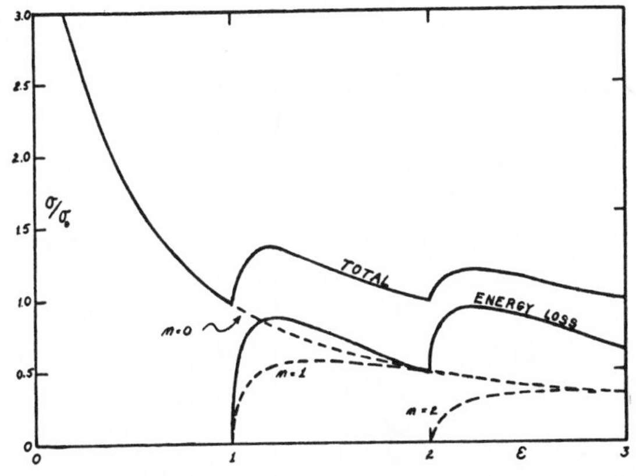

FIG. 12. Influence of the chemical binding of the protons on the slowing down of neutrons in a hydrogenous substance. Isotropic binding of hydrogen atoms assumed. Abscissa: ϵ=ratio of neutron energy to quantum energy of the vibration of the hydrogen in the molecule. Ordinate: Ratio of the cross section to that for free protons. Solid lines: total cross section and energy loss cross section. (The average energy loss per collision is one-half the neutron energy times the ratio of energy loss cross section to total cross section.) Broken lines: contributions of elastic collisions ($n=0$) and of one ($n=1$) and two ($n=2$) quantum excitations to the total cross section.

with

$$q^2 = q_x^2 + q_y^2 + q_z^2 = (\hbar/2m\omega)(\mathbf{k}-\mathbf{k}')^2$$
$$= (E+E'-2(EE')^{\frac{1}{2}}\cos\vartheta)/\hbar\omega. \quad (454a)$$

ϑ is the angle of deflection of the neutron and $n_x n_y n_z$ the vibrational quantum numbers of the excited state of the proton in the three directions of space. In (454a) it has been tacitly assumed that the vibrational frequencies are the same for the three directions of space, but (454) will be generally true even for anisotropic binding of the proton (cf. below).

4. *Isotropically bound protons.*—Under the assumption of *isotropic binding*, (454) may be summed over all the substates $n_x n_y n_z$ of the state of total vibrational quantum number

$$n = n_x + n_y + n_z, \quad (455a)$$

the summation giving

$$\sigma_n d\omega = 4r_0^2(E'/E)^{\frac{1}{2}} q^{2n} e^{-q^2} 2\pi \sin\vartheta d\vartheta/n!. \quad (455)$$

With (454a), we may write

$$dq^2 = 2(EE')^{\frac{1}{2}} \sin\vartheta d\vartheta/\hbar\omega. \quad (455b)$$

Furthermore, we introduce the abbreviation

$$\epsilon = E/\hbar\omega; \quad \epsilon - n = E'/\hbar\omega. \quad (456)$$

Then
$$\sigma_n(\vartheta)d\omega = \frac{\sigma_0}{\epsilon}\frac{q^{2n}}{n!}e^{-q^2}dq^2. \quad (456a)$$

The total cross section for excitation of the nth vibrational state can be obtained by integration between the limits

$$q_{\min} = \epsilon^{\frac{1}{2}} - (\epsilon - n)^{\frac{1}{2}}, \quad q_{\max} = \epsilon^{\frac{1}{2}} + (\epsilon - n)^{\frac{1}{2}}. \quad (456b)$$

The total cross section for excitation of level n becomes

$$\sigma_n = (\sigma_0/\epsilon)[f_n(\{\epsilon^{\frac{1}{2}} - (\epsilon-n)^{\frac{1}{2}}\}^2)$$
$$- f_n(\{\epsilon^{\frac{1}{2}} + (\epsilon-n)^{\frac{1}{2}}\}^2)] \quad (457)$$

with

$$f_n(x) = e^{-x}\left(1 + \frac{x}{1!} + \frac{x^2}{2!} + \cdots + \frac{x^n}{n!}\right). \quad (457a)$$

For $n=0$ (elastic collisions) σ reduces to

$$\sigma_{n=0} = (\sigma_0/\epsilon)(1 - e^{-4\epsilon}), \quad (458)$$

which decreases from $4\sigma_0$ at $\epsilon=0$ to $0.98\sigma_0$ at $\epsilon=1$.

For large energies ϵ and all values of n not very near ϵ, the second term in the square bracket in (457) is negligible while the first is practically unity. Then

$$\sigma_n = \sigma_0/\epsilon, \quad (\epsilon \gg 1) \quad (459)$$

independent of n. The maximum value of n is equal to ϵ by definition (456). Therefore the total cross section for large ϵ is σ_0, and any energy loss between zero and the total energy $\epsilon\hbar\omega$ is equally probable, just as we found assuming free protons.

Fig. 12 gives the cross sections for elastic scattering ($n=0$) and excitation of the first and second vibrational state for neutron energies between 0 and $3\hbar\omega$. The total cross section is given, and the "cross section for energy loss"

$$\sigma_l = (2/\epsilon)\sum_n n\sigma_n. \quad (460)$$

This definition of σ_l makes it equal to σ_0 for high energies. For the energies considered, σ_l (in contrast to σ_{total}) remains *smaller* than σ_0, but down to almost $\epsilon=1$ the difference is not large. This means that the efficiency of the paraffin in slowing down the neutrons remains practically the same down to a neutron energy equal to the energy $\hbar\omega$ of the proton vibration.

5. Quantitative considerations. Energy distribution below 1 volt.—The calculations given are not yet directly applicable to our problem because the protons in paraffin are not bound isotropically. The vibration frequency in the direction of the line joining the carbon and hydrogen atom is about $\nu_1 = 3000$ cm^{-1}, in the directions perpendicular to it, $\nu_2 = 800$ to 900 cm^{-1}. According to the foregoing, *a noticeable effect on the slowing down of the neutrons will not be found until the neutron energy has decreased to about* $\hbar\omega_2 = 2\pi\hbar c\nu_2$. This means that the neutron distribution law (440) should hold approximately down to the energy $E = \hbar\omega_2 \approx 0.1$ volt, with nearly constant $l(E)$. There will, of course, be a slight deviation from (440) in the direction of a larger number of neutrons for energies below, say, 1 volt, but, according to Fig. 12, this deviation will probably be less than a factor 2 for all energies above 1.25 $\hbar\omega_2 \approx 0.13$ volt.

Only for energies smaller than $\hbar\omega_2$, the transfer of energy to the C—H vibration will no longer be possible, and any further slowing down can only occur by transferring energy to the CH_2 groups as a whole. This is equivalent to an increase of the "effective mass" of the hydrogen from $M = 1$ to 14 (molecular weight of CH_2). Accordingly, the average energy loss of a neutron per collision is reduced to $2mM/(M+m)^2 \approx 1/8$ times the neutron energy. Therefore, if, e.g., we let neutrons pass from a block of hot paraffin to one of cold paraffin, it will take 8 collisions rather than one to "cool" the neutrons to the new temperature (more accurately: to reduce the difference between neutron and paraffin "temperature" to $1/e$ times its initial value).

It might be thought that the distribution function of neutrons at energies below $\hbar\omega_2$ will be affected accordingly, so that the number of neutrons of energy E would be roughly 8 times the number given by (440). This is not the case, simply because energies of this order are no longer in the domain of the validity of any formula similar to (440) but are already in the domain of the Maxwell distribution. The entire energy distribution of the neutrons can be divided into two parts: The high energy region where the neutrons lose energy continuously, and the low energy region where energy gains in a collision are about as frequent as losses. In the first region formula (440) will hold, with possible corrections for the efficiency of the decelerating process. In the second region, we shall have a Maxwell distribution. The total number of neutrons in the Maxwell region will be equal to the number produced per sec. Q, times the mean life τ of a neutron. Therefore the number having an energy between E and $E+dE$ is:

$$N(E)dE = 2\pi^{-\frac{1}{2}}Q\tau(kT)^{-\frac{3}{2}}E^{\frac{1}{2}}e^{-E/kT}dE. \quad (461)$$

The limit of validity between distributions (440) and (461) will be given by that energy $\epsilon_0 kT$ for which they give the same number of neutrons. We find

$$\epsilon_0^{-2}e^{\epsilon_0} = \tau(8kT/\pi m)^{\frac{1}{2}}/l(\epsilon_0 kT). \quad (461a)$$

The right-hand side is identical with the number N of collisions before absorption, introduced by Fermi (cf. (484)), except that the mean free path at energy $\epsilon_0 kT$ stands instead of that for energy kT. Assuming the ratio of the mean free paths to be 2.8 (cf. 464), the right-hand side becomes $N/2.8 = 53$, using Fermi's determination of N. Then (461a) has the solution

$$\epsilon_0 = 8, \quad (461b)$$

corresponding to about 1600 cm^{-1} at room temperature. *The Maxwell distribution thus extends up to about twice the energy of the weaker hydrogen bond, and joins directly to the distribution (440) with constant l* (of course, all these statements are approximate only).

6. Anisotropic binding of hydrogen.—We now consider the influence of the anisotropic hydrogen bond on the elastic scattering cross section at thermal energies. The differential cross section (455) becomes

$$\sigma_{el}d\omega \sin\chi d\chi = 4r_0^2 2\pi \sin\vartheta d\vartheta \tfrac{1}{2}\sin\chi d\chi$$
$$\times e^{-q_x^2 - q_y^2 - q_z^2} = \sigma_0 \sin\vartheta d\vartheta \sin\chi d\chi$$
$$\times e^{-2(1-\cos\vartheta)(\epsilon_1\cos^2\chi + \epsilon_2\sin^2\chi)}, \quad (462)$$

where ϑ is the deflection of the neutron (angle between $\mathbf{k}$ and $\mathbf{k}'$), χ the angle between the vector $\mathbf{k}-\mathbf{k}'$ and the direction of the strong bond, and

$$\epsilon_1 = E/\hbar\omega_1, \quad \epsilon_2 = E/\hbar\omega_2, \quad (462a)$$

so that

$$\epsilon_1 \approx \tfrac{1}{4}\epsilon_2. \quad (462b)$$

Since the energy is supposed to be small compared to both the bonds $\hbar\omega_1$ and $\hbar\omega_2$, we can expand the exponential in (462). Then the integration is straightforward and gives for the total cross section

$$\sigma_{el}=4\sigma_0\bigg[1-2\epsilon'+\frac{32}{15}(\epsilon'\epsilon_2+\tfrac{1}{4}\epsilon_1^2) -\frac{64}{35}\left(\epsilon'\epsilon_2^2+\tfrac{1}{4}\epsilon_1^2\epsilon_2+\frac{5}{24}\epsilon_1^3\right) +\frac{2048}{1575}(\epsilon'\epsilon_2^3+\cdots)+\cdots\bigg], \quad (463)$$

where $$\epsilon'=\tfrac{1}{3}\epsilon_1+\tfrac{2}{3}\epsilon_2. \quad (463a)$$

At room temperature, kT is about 200 cm^{-1}, so that $\epsilon_2\approx\frac{1}{4}$, $\epsilon_1\approx\frac{1}{16}$, $\epsilon'\approx\frac{3}{16}$. Then the elastic cross section becomes

$$\sigma_{el}=2.8\sigma_0. \quad (464)$$

Thus the mean free path of thermal neutrons will be about 2.8 times smaller than that of neutrons above one volt energy.[29a]

7. Conclusions. Mean free path of neutrons at and above thermal energies.—Amaldi and Fermi (A11) have measured the mean free path for thermal neutrons by determining the number of thermal neutrons which were scattered out of a neutron beam by varying thicknesses of paraffin. The number of thermal neutrons was measured by the difference in the radioactivities produced in a piece of rhodium with and without a screen of cadmium. Since Cd is known to absorb the thermal neutrons, such a measurement will indeed yield the number of thermal neutrons fairly accurately. The mean free path found in this way by Amaldi and Fermi was

$$l_{th}=0.3 \text{ cm}. \quad (465)$$

From this together with (464) we would conclude a mean free path for neutrons above one volt of about

$$l_0=0.85 \text{ cm}. \quad (465a)$$

Amaldi and Fermi also measured the mean free path of these faster neutrons directly. A detector for such neutrons, e.g., a sheet of Ag or Rh screened by Cd, was placed on top of a paraffin block. An absorber of the same material was inserted in the paraffin, at various depths x below the surface. The decrease in the activity of the detector δA was measured as a function of the depth x simultaneously with the activity A_0 of the lower surface of the absorber itself. Then $\delta A/A_0$ gives the probability that a neutron with an energy equal to the resonance energy of absorber and detector, is able to travel a distance x in paraffin without losing its property of being in resonance. Since all known resonance levels have a width small compared to the resonance energy, almost every collision will throw a neutron out of resonance. Therefore the measurements referred to will give directly the mean free path l_0 of the resonance neutrons. From both the measurements with Ag and Rh, Amaldi and Fermi deduced

$$l_0=1.1 \text{ cm}. \quad (465b)$$

It seems that the measurement of l_{th} is somewhat more accurate, and we accept therefore

$$l_0=0.9 \text{ cm} \quad (465c)$$

as the true value of the mean free path. Taking the density of paraffin as 0.90, the scattering cross section becomes

$$\sigma_0=\frac{14}{2\cdot 6.05\cdot 10^{23}\cdot 0.90\cdot 0.9}=14\cdot 10^{-24} \text{ cm}^2, \quad (466)$$

where 14 is the molecular weight of CH_2 and 2 the number of hydrogen atoms per CH_2 group. Inserting this into (62), we find for the energy of the virtual ^1S state of the deuteron

$$\epsilon'=105{,}000 \text{ volts}. \quad (467)$$

This figure replaces the figure of 40,000 volts given in (62a) which was erroneous because the effect of the chemical binding of the scattering proton had not been taken into account.

D. Spatial distribution of the neutrons in paraffin

The spatial distribution of the neutrons is a result of their multiple scattering during the process of being slowed down. Fig. 13 gives the distribution as observed by Amaldi and Fermi (A11) for two neutron energies (thermal and Ag resonance[30]). The quantity given is the number

[29a] Fermi (F17) gives 3.3 σ_0, assuming the hydrogen bond to be isotropic with $\nu=3000$ cm^{-1}.

[30] The distribution of the Rh and I resonance neutrons was also given by Amaldi and Fermi (A11, Fig. 7). These distributions are very similar to those of the Ag resonance neutrons.

of neutrons found in a spherical shell of radius r surrounding the source. The distribution curves show an increase at small r because of the increase in the area of the spherical shell, and then a decrease which is for large r practically exponential with a decay constant of 0.106 cm^{-1}. Since we know that the spatial distribution is primarily determined by the diffusion of the neutrons while fast (cf. A), the reciprocal of the decay constant, i.e., 9.4 cm, will be a measure of the mean free path for fast neutrons. The very complicated energy dependence (62) of the cross section for collisions with hydrogen atoms must be taken into account, and it must be considered that the neutrons which have suffered a collision have a smaller mean free path than those which have not. Furthermore, collisions with oxygen atoms are very important for fast neutrons, firstly because of the rapid decrease of the hydrogen cross section with increasing neutron energy, and secondly because a collision with an oxygen atom may cause a deflection of the neutron by a large angle while the deflection in collisions with hydrogen atoms can at most be 90°. The very different effect of oxygen and hydrogen collisions on the energy and the direction of the neutron makes the treatment of their combined effect rather difficult. Moreover, considerable uncertainty is introduced in the calculations because the magnitude of the cross section of oxygen for fast neutrons is known only approximately and its dependence on energy not at all. Finally, the whole problem is further complicated by the inhomogeneity of the neutrons from the source (Rn+Be).

Fermi has given a general formula for the mean square distance $(r^2)_{Av}$ of slow neutrons from the source, taking into account all the effects mentioned (F17). This rather complicated formula was applied by Horvay (H38) to the problem of neutron diffusion in water. Assuming for the cross section of oxygen the reasonable value $2\cdot 10^{-24}$ cm^2, Horvay finds that the observed value of $(r^2)_{Av}$ requires an initial average energy of the neutrons of about 3 MV which seems quite reasonable in comparison with the observed distribution of neutrons from a radon-beryllium source (D21).

Bethe (unpublished) has tried to obtain approximately the shape of the distribution curve,

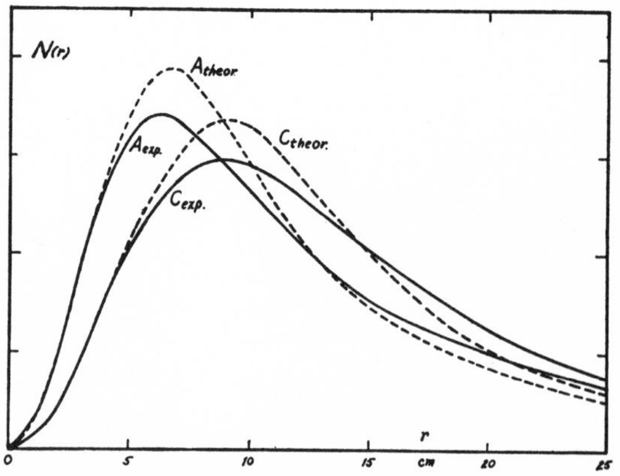

FIG. 13. Distribution of neutrons in a large volume of water, according to Amaldi and Fermi (A 11, Fig. 7). Abscissa: Distance r from source in cm. Ordinate: Number of neutrons of group A and C with distance between r and $r+dr$. Solid curves: observed. Broken curves: calculated with rough assumptions.

using very simplified assumptions about the diffusion process and determining a constant (essentially the mean distance traveled by fast neutrons) from the experimental data. The result of this admittedly rough calculation is also given in Fig. 13.

In this section, we shall limit ourselves to a calculation of the *difference* between the mean square distances of slow neutrons of different energies from the source. This difference can be used for a determination of neutron energies (A11, cf. below) and is comparatively easy to calculate because between 1 and 10,000 volts the mean free path is practically independent of the energy.

Let $\mathbf{r}_1$ be the distance traveled from the source by a neutron of final energy E_1 and $\mathbf{r}_2=\mathbf{r}_1+\mathbf{r}$ that for energy E_2. Then we want to calculate

$$(r_2^2)_{Av}-(r_1^2)_{Av}=(r^2)_{Av}+2(\mathbf{r}_1\cdot\mathbf{r})_{Av}. \quad (468)$$

The "correlation term" $(\mathbf{r}_1\cdot\mathbf{r})_{Av}$ may be neglected[31] if a large number of collisions is necessary[32] to slow the neutron down from E_1 to E_2. We now denote the number of these collisions by N, and the distance traveled between the $n-1$st and nth collisions by ρ_n. Then

[31] It can be shown that this term is actually compensated by other neglections made in the course of the calculation.

[32] This can always be achieved by choosing E_1 large. E.g., we may choose E_1 equal to 10,000 volts throughout.

$$(r^2)_{\text{Av}}=(\sum_{n=1}^{N}\varrho_n)^2{}_{\text{Av}}=\sum_n(\rho_n{}^2)_{\text{Av}}$$
$$+2\sum_{n<m}(\rho_m\rho_n\cos\vartheta_{mn})_{\text{Av}}, \quad (469)$$

where ϑ_{mn} is the angle between the vectors ϱ_m and ϱ_n. The three quantities in the last term are statistically independent of each other. Denoting the mean free path between two collisions by l, we have[33]

$$(\rho_n{}^2)_{\text{Av}}=2l^2, \quad (\rho_n)_{\text{Av}}=l, \quad (469\text{a})$$

and therefore

$$(r^2)_{\text{Av}}=2Nl^2(1+\sum_{m-n=1}^{\infty}(\cos\vartheta_{mn})_{\text{Av}}) \quad (470)$$

$$=2Nl^2\sum_{k=0}^{\infty}(\cos\vartheta_{n,\,n+k})_{\text{Av}}. \quad (470\text{a})$$

The sums have been extended to infinity because the terms decrease rapidly with k.

We shall first assume that only collisions with hydrogen atoms occur. Then the average number of collisions necessary is (cf. (437a))

$$N=\log(E_1/E_2). \quad (471)$$

The average cosine of the angle of deflection in a collision is (cf. (436b))

$$(\cos\vartheta_{n,\,n+1})_{\text{Av}}=\int_0^{\pi/2}\cos\vartheta\cos\vartheta\sin\vartheta d\vartheta\div$$
$$\int_0^{\pi/2}\cos\vartheta\sin\vartheta d\vartheta=\tfrac{2}{3}. \quad (472)$$

If φ is the angle between the plane of ϱ_{n+k-1}, ϱ_n and the plane of ϱ_{n+k-1}, ϱ_{n+k}, we have

$$\cos\vartheta_{n,\,n+k}=\cos\vartheta_{n,\,n+k-1}\cos\vartheta_{n+k-1,\,n+k}$$
$$+\sin\vartheta_{n,\,n+k-1}\sin\vartheta_{n+k-1,\,n+k}\cos\varphi. \quad (472\text{a})$$

The second term vanishes upon averaging over φ. Therefore we have, with (472),

$$(\cos\vartheta_{n,\,n+k})_{\text{Av}}=(\cos\vartheta_{n,\,n+k-1})_{\text{Av}}(\cos\vartheta_{n+k-1,\,n+k})_{\text{Av}}$$
$$=(\cos\vartheta_{n,\,n+1})_{\text{Av}}{}^k=(\tfrac{2}{3})^k. \quad (472\text{b})$$

Therefore the sum over k in (470a) gives 3, and

$$(r^2)_{\text{Av}}=6Nl^2=6l^2\log E_1/E_2. \quad (473)$$

This is the formula used by Amaldi and Fermi.

[33] The probability of a given value of ρ_n is $e^{-\rho_n/l}d\rho_n/l$ (definition of the mean free path).

Now we take the collisions with oxygen atoms into account. Let 2α be the ratio of the cross sections of oxygen and hydrogen for slow neutrons (1 to 10,000 volts). Then the number of collisions will now be (one oxygen atom per two hydrogen atoms!)

$$N=(1+\alpha)\log(E_1/E_2), \quad (474)$$

and the mean free path

$$l=l_0/(1+\alpha), \quad (474\text{a})$$

where l_0 is the mean free path with hydrogen only. After a collision with an oxygen atom, the direction of motion of the neutron will be practically random. Therefore the average cosine in (472) must be multiplied by the probability that the collision is with a hydrogen atom, giving

$$(\cos\vartheta_{n,\,n+1})_{\text{Av}}=2/3(1+\alpha). \quad (474\text{b})$$

(472b) will remain unchanged except for the last expression. Therefore we obtain instead of (473)

$$(r^2)_{\text{Av}}=[6l_0{}^2/(1+3\alpha)]\log(E_1/E_2). \quad (475)$$

l_0 is known from the measurements of Amaldi and Fermi, its value is 0.9 cm in paraffin (cf. (465c)). In water, it should be larger by the ratio of the number of hydrogen atoms per cm³ i.e., by the factor $0.9\cdot 18/1.0\cdot 14=1.16$, giving $l_0=1.04$ cm. The cross section of oxygen for slow neutrons was measured by Dunning and others (D23 and Table XXVII) and found to be $3.3\cdot 10^{-24}$ cm² so that (cf. (466))

$$\alpha=3.3/2\cdot 14=0.12. \quad (475\text{a})$$

Inserting in (475), we have

$$(r^2)_{\text{Av}}\approx 4.8\log(E_1/E_2)\text{ cm}^2. \quad (476)$$

The numerical factor 4.8 differs appreciably from the value used by Amaldi and Fermi (9.7). The difference is due partly to the neglection of the rather large effect of oxygen collisions, partly to a different value used for the mean free path in paraffin (1.1 instead of 0.9 cm). The value given in (476) agrees very much better with the measurements of Amaldi and Fermi (§60) than their old value.

At energies below one volt, (476) will fail. We may divide this low energy region into two parts: The region between 1 volt and thermal energies, and the Maxwell region. In the first region, the mean free path l, the energy loss, and the angular distribution of the neutrons will be complicated functions of the energy (cf. Section C), so that the increase in r^2 cannot be calculated in any way. In the Maxwell region we have complete diffusion and, according to (487b),

$$(r^2)_{\text{Maxwell}}=6L^2, \tag{477}$$

where L is the "diffusion length" of Maxwell neutrons. L has been measured by Amaldi and Fermi and found to be 2.1 cm in paraffin so that

$$(r^2)_{\text{Maxwell}}=26 \text{ cm}^2 \text{ paraffin}$$
$$=35 \text{ cm}^2 \text{ water.} \tag{477a}$$

On the other hand, Amaldi and Fermi have measured the difference in $(r^2)_{\text{Av}}$ between thermal neutrons and neutrons of about one volt energy (resonance group of Rh). In water, this difference is 50 cm². This would mean that $(r^2)_{\text{Av}}$ increases by $50-35=15$ cm² when the neutrons are slowed down from 1 volt to thermal energies. This figure is, of course, not very accurate.

E. Diffusion of thermal neutrons

When the neutrons have been slowed down to thermal energies, they will diffuse through the paraffin. Assuming that the mean free path is independent of the neutron velocity, and that the scattering of thermal neutrons by protons is isotropic, we have for the diffusion coefficient the well-known formula

$$D=\tfrac{1}{3}l_{\text{th}}v_a, \tag{478}$$

where
$$v_a=(8kT/\pi m)^{\frac{1}{2}} \tag{478a}$$

is the average velocity of thermal neutrons, and l the mean free path of thermal neutrons (cf. (465)).

For a stationary distribution, the diffusion equation takes the form

$$\partial F/\partial t=D\Delta F+q-F/\tau=0. \tag{479}$$

Here F is the number of neutrons per cm³. The first term on the right represents the neutrons entering a given volume element per unit time by diffusion, the last term the number disappearing from it by absorption, τ being the lifetime of a neutron with respect to capture. The second term, q, is the number of thermal neutrons produced in the volume element per second by the neutron source; in a homogeneous block of paraffin, q will be a slowly varying function of the coordinates whose variation we shall neglect. We shall treat the solution of (479) for various cases which are of practical importance:

1. Homogeneous paraffin, no absorber, homogeneous production of neutrons.—We have

$$q=\text{const.},\ \Delta F=0$$

and therefore

$$F=q\tau. \tag{480}$$

The total number of neutrons throughout the paraffin is

$$\int Fdv=Q\tau, \tag{480a}$$

where Q is the total number of neutrons emitted per second by the source.

2. Production of neutrons in a plane $x=0$.—We have $q=q'\delta(x)$ where δ is Dirac's δ function. The solution is

$$F=\tfrac{1}{2}q'(\tau/D)^{\frac{1}{2}}e^{-|x|(D\tau)^{-\frac{1}{2}}}. \tag{481}$$

It satisfies the condition

$$q=0,\ \text{i.e., } d^2F/dx^2=F/D\tau \text{ for } x\neq 0. \tag{481a}$$

The current starting from $x=0$ is equal to (cf. (488a))

$$-\lim_{\epsilon=0} 2D\left(\frac{dF}{dx}\right)_\epsilon=q'. \tag{481b}$$

(Factor 2 for the two sides of the plane.)

According to (481), the probability that a neutron emitted at $x=0$, reaches the plane x without being absorbed, is proportional to

$$p(x)=e^{-x/L}, \tag{482}$$

where
$$L=(D\tau)^{\frac{1}{2}} \tag{482a}$$

is the "diffusion length."

L has been measured by Amaldi and Fermi in the following way: An *absorber* of thermal neutrons (Cd) is placed at a certain plane $x=0$ in paraffin. Then the neutron distribution will be given by (480) *minus* (481) where q is the number of neutrons produced by the source per cm³ and sec., and q' the number absorbed in the absorber per cm² and sec. The activity in a detector placed at a distance x from the absorber, is then measured with and without the absorber. The difference turns out to be indeed an exponential function of x; its rate of decay measures the diffusion length L. The result of the measurements is

$$L=2.1 \text{ cm}. \tag{482b}$$

In the actual experiment, the detector was kept fixed at the surface of a paraffin block and the absorber placed at various depths x. Since the density of neutrons near the surface of a paraffin block is a function of x, it was necessary to measure the number q' of the neutrons absorbed by the absorber itself by measuring its own activity. The decrease in activity of the detector due to the insertion of the absorber, divided by the activity of the absorber, must then be proportional to $e^{-x/L}$.

3. Lifetime of neutron in paraffin.—Using the definition (478) of D and (482a) of L, we have

$$L^2=\tfrac{1}{3}l_{\text{th}}v_a\tau. \tag{483}$$

We define the "average number of collisions before absorption"

$$N=v_a\tau/l_{\text{th}}. \tag{484}$$

Then we have

$$L^2=\tfrac{1}{3}l_{\text{th}}{}^2N. \tag{484a}$$

With $L=2.1$, $l_{\text{th}}=0.30$ (cf. (465)), we find

$$N=150. \tag{485}$$

Considering the ratio 2.8 of the mean free paths for faster neutrons and thermal neutrons (cf. (464)), we find

$$N_0=v_a\tau/l_0=150/2.8=53. \tag{485a}$$

A theoretical calculation of τ on the lines of §17 gives, with the value $|\epsilon'|=105$ kv (cf. (467)) for the energy of the ^1S state of the deuteron:

$N_0=182$ if the ^1S state of the deuteron is stable, (486)

$N_0=75$ if the state is virtual.

The experimental value (485a) thus decides uniquely in favor of a *virtual* ^1S state of the deuteron. However, the agreement between the experimental and theoretical value of N_0 is not very good.

4. Point source of neutrons.—We assume now $q=0$ everywhere except at $r=0$. We find

$$D\,d^2(rF)/dr^2=rF/\tau, \tag{487}$$

$$rF=ce^{-r/L}, \tag{487a}$$

and therefore

$$(r^2)_{\text{Av}}=\frac{\int r^2Fr^2dr}{\int Fr^2dr}=6L^2=2Nl^2, \tag{487b}$$

a result which has been used in (477).

5. Absorption of neutrons by a thick plane absorber.—If there is an absorber at $x=0$ which absorbs *all* neutrons striking it, the neutron density will, in first approximation, be zero at $x=0$. If neutrons are produced at a rate q per cm³ per sec., the neutron density at a distance $+x$ from the absorber is (cf. (480), and (481) with q' chosen so as to make $F(0)=0$)

$$F=q\tau(1-e^{-x/L}). \tag{488}$$

The neutron current is, according to the definition of the diffusion coefficient

$$S=-D\text{ grad }F. \tag{488a}$$

Therefore the current of neutrons falling on the absorber from the right-hand side

$$S=D(\partial F/\partial x)_0=qD\tau/L=qL. \tag{489}$$

All the formulae given are valid only inside the paraffin. Near the surface of the paraffin, the density of neutrons of more than Maxwellian energy will be smaller than inside, and will decrease towards the surface. Therefore the rate of production of Maxwell neutrons q will vary fairly rapidly and in a complicated way with the coordinates so that the solutions given are no longer valid. All quantitative measurements of the absolute activity of detectors should therefore be made inside the paraffin rather than at its surface. The effect of an absorber for thermal neutrons, usually Cd, in the interior of the paraffin differs from that of an external surface in that the Cd does not disturb the distribution of the neutrons of more than thermal energy, and leaves therefore the rate of production of thermal neutrons constant all over space although their actual density will not be constant.

F. Angular distribution of the neutrons emerging from paraffin. Activity of detector in interior of paraffin and near absorber. Albedo (F17)

One of the most important practical questions about the distribution of thermal neutrons in paraffin is the problem of the activity produced in various conditions in a detector. The two most important cases are:

(1) A thin detector is placed in the interior of the paraffin, without any absorber near it.

(2) The detector is on one side covered by an absorber for thermal neutrons (Cd).

Case 1.—If F is the density of neutrons at some point in the interior of the paraffin, the number of neutrons crossing unit area of the detector at an angle ϑ with respect to its normal, is per unit time (from both sides together)

$$Fv_a \cos\vartheta \sin\vartheta d\vartheta. \quad (490)$$

The number absorbed per unit time in a thin detector of area s, thickness δ g/cm^2 and absorption coefficient K cm^2/g is accordingly

$$A = s\int_0^{\pi/2} Fv_a \cos\vartheta \sin\vartheta d\vartheta K\delta/\cos\vartheta = Fv_a Kw, \quad (490a)$$

where $w = s\delta$ is the weight of the detector. Thus we can *define* the density of neutrons by the activity produced in a detector

$$F = A/v_a Kw. \quad (491)$$

Case 2.—The activity produced in a detector placed on the front side of an absorber will be closely connected to the neutron current emerging from the paraffin near the absorber (cf. (489)). The connection will, however, involve the angular distribution of the neutrons emerging, which, near an absorber, will not be uniform as in the interior. Let us assume that the number of neutrons striking unit area of the absorber (and detector) in the direction ϑ per second is given by an expression of the form

$$f(\vartheta)\sin\vartheta d\vartheta = (a\cos\vartheta + b\cos^2\vartheta)\sin\vartheta d\vartheta, \quad 0<\vartheta<\pi/2 \quad (492)$$

$$f(\vartheta) = 0, \quad \pi/2<\vartheta<\pi.$$

We shall show below that the distribution actually has this form. If it does, the current is previously given by

$$S = \int_0^{\pi/2} f(\vartheta)\sin\vartheta d\vartheta = \tfrac{1}{2}a + \tfrac{1}{3}b \quad (492a)$$

and the activity by

$$A = Ks\int f(\vartheta)(\delta/\cos\vartheta)\sin\vartheta d\vartheta = Kw(a + \tfrac{1}{2}b) \quad (492b)$$

so that

$$A = SKw(a + \tfrac{1}{2}b)/(\tfrac{1}{2}a + \tfrac{1}{3}b). \quad (493)$$

We may also define the "effective density" from the activity by means of (491)

$$F_{\text{eff}}(0) = (S/v_a)(a + \tfrac{1}{2}b)/(\tfrac{1}{2}a + \tfrac{1}{3}b). \quad (493a)$$

According to (493a), the neutron density is not exactly zero at the absorber, as we have assumed in (488). On the other hand, it will still be true (cf. (488)) that F is approximately linear in the distance x from the absorber. We may thus write

$$F(x) = \alpha + \beta x \quad (494)$$

and have, according to (488a) (493a) (478)

$$\beta = S/D = 3S/lv_a, \quad (494a)$$

$$\alpha = (S/v_a)(a + \tfrac{1}{2}b)/(\tfrac{1}{2}a + \tfrac{1}{3}b). \quad (494b)$$

We may now use (494) in order to calculate the angular distribution of the emerging neutrons. A neutron which has a collision at the depth x and moves in the direction ϑ, has the probability $\exp(-x/l\cos\vartheta)$ of emerging from the paraffin without further collision. The number of neutron collisions per sec. at a depth between x and $x+dx$ is $dxF(x)v_a/l$, by definition of the mean free path l. Assuming isotropic collision, the fraction $\frac{1}{2}\sin\vartheta d\vartheta$ of the neutrons colliding will have the direction ϑ after collision. Thus the total number of neutrons emerging in the direction ϑ will be

$$f(\vartheta)\sin\vartheta d\vartheta = \tfrac{1}{2}\sin\vartheta d\vartheta\int dxF(x)(v_a/l)e^{-x/l\cos\vartheta}. \quad (495)$$

Inserting (494) and integrating over x, we obtain

$$f(\vartheta) = \tfrac{1}{2}v_a(\alpha\cos\vartheta + \beta l\cos^2\vartheta). \quad (495a)$$

This proves the distribution assumed in (492), with

$$a = \tfrac{1}{2}\alpha v_a, \quad b = \tfrac{1}{2}\beta l v_a. \quad (495b)$$

Inserting this into (494b), we find

$$\alpha=(S/v_a)(\alpha+\tfrac{1}{2}\beta l)/(\tfrac{1}{2}\alpha+\tfrac{1}{3}\beta l). \qquad (495c)$$

Comparison with (494a) gives

$$\alpha/\beta l=\tfrac{1}{3}(\alpha+\tfrac{1}{2}\beta l)/(\tfrac{1}{2}\alpha+\tfrac{1}{3}\beta l) \qquad (495d)$$

and therefore

$$\beta l=\sqrt{3}\alpha. \qquad (496)$$

Inserting this into (492) (495b), we find for the angular distribution of the neutrons

$$f(\vartheta)\sim\cos\vartheta+\sqrt{3}\cos^2\vartheta. \qquad (497)$$

This means that relatively more neutrons are emitted in the forward direction than would be expected according to the simple cosine law which is valid in the interior of the paraffin (cf. (490)). This agrees with observations of Fink (F20). The law (497) may be expected to hold not only for thermal neutrons emerging from paraffin near an absorber, but generally for all neutrons leaving a paraffin block at a surface (provided only the mean free path of the neutrons is small compared to the dimensions of the paraffin block).

The density distribution (494) becomes with (496)

$$F=\beta(x+l/\sqrt{3}). \qquad (498)$$

This means that the straight line representing the density as a function of the distance x from the absorber, would cut the x-axis at

$$x_0=-l/\sqrt{3}. \qquad (498a)$$

Amaldi and Fermi have observed the density distribution (A11, Fig. 8); it is sufficiently nearly linear and may be extrapolated to a cut with the abscissa at

$$x_0=-0.18 \text{ cm}. \qquad (498b)$$

This would mean $l=0.31$ cm, in very good agreement with the direct determination of the mean free path $l=0.30$ cm (cf. (465)).

As has already been mentioned, these considerations are also valid for neutrons of higher than thermal energy at the surface of paraffin. In calculating the effective mean free path, it must be considered that the directions of motion before and after a collision are correlated. By an argument similar to that in Section D, it may be shown that

$$l=3l_0/(1+3\alpha), \qquad (499)$$

where l_0 is the mean free path of the neutrons if only collisions with hydrogen are counted, and α is one-half the ratio of the cross sections of carbon and hydrogen, i.e., about 0.14 (cf. Table XXVII). Thus we should expect for neutrons above thermal energy in paraffin

$$-x_0=\sqrt{3}\cdot 0.9/1.42=1.1 \text{ cm}. \qquad (499a)$$

The activity of the detector becomes, if we insert (496) (495b) into (493),

$$A_a=\sqrt{3}SKw \qquad (500a)$$

and, with (489) (484a),

$$A_a=\sqrt{3}qLKw=ql_{\text{th}}N^{\frac{1}{2}}Kw. \qquad (500)$$

This may be compared with an activity in the interior of the paraffin of (cf. (490a), (480), (484))

$$A_i=q\tau v_a Kw=ql_{\text{th}}NKw \qquad (501)$$

so that

$$A_i/A_a=N^{\frac{1}{2}}. \qquad (502)$$

The activity of a thin detector in the interior of the paraffin should thus (cf. (485)) be about 12 times greater than near an absorber. Direct experiments by Amaldi and Fermi gave a ratio of about 11, again in excellent agreement with the value deduced from the ratio of diffusion path to mean free path.

The ratio (502) can be interpreted using the concept of the "albedo." A_a is the activity observed at the surface of a paraffin block. If we have paraffin on both sides of a detector, and if the two blocks of paraffin did not influence each other, the activity would be $2A_a$. In reality, each paraffin block reflects the neutrons coming from the other. If β is the reflection coefficient (albedo), then the number of neutrons once reflected will be $2A_a\beta$, that of the neutrons twice reflected $2A_a\beta^2$ etc., and the total activity

$$A_i=2A_a/(1-\beta) \qquad (502a)$$

so that the albedo has the value[33a]

$$\beta=1-2N^{-\frac{1}{2}}\approx 0.83. \qquad (503)$$

[33a] *Added in proof:* Halpern, Lueneburg and Clark (H11a) have given a vigorous solution of the albedo problem. They find

$$\beta=1-2.31N^{-\frac{1}{2}} \qquad (503a)$$

if the incident neutrons are distributed according to the cos ϑ law, and

$$\beta=1-2.48N^{-\frac{1}{2}} \qquad (503b)$$

for a $\cos^2\vartheta$ distribution.

§60. Neutron Resonance Energies (G17, G18, G18a, F33, A11, H8, H9, H10)

The most satisfactory method for determining neutron energies is a mechanical device which measures directly the neutron velocity. Such a mechanical velocity selector was constructed by Dunning, Pegram, Fink, Mitchell and Segrè (D25, D26, F19, F20) and used in order to determine the velocity distribution of the neutrons absorbed in Cd. The velocity selector consisted of four Duraluminum disks each of which bore 50 sectors of sheet cadmium, the spacing between the sectors being about the same (3.5°) as the angle subtended by each sector (3.7°). Two disks were mounted on a rapidly rotating shaft (up to about 5000 r.p.m.), a distance $d=54$ cm apart; the other two were fixed within 5 mm of the rotating disks. Each pair of disks, one rotating and the adjacent fixed one, represents a shutter for the neutrons absorbable in Cd, the shutter being opened and closed $50n/60$ times per second where n is the number of revolutions per minute. The two pairs of disks act as a velocity selector: Neutrons of velocity v are strongly absorbed by the shutters if

$$v=(100n/60)d. \qquad (504)$$

Thus, with $d=54$ cm, a speed of about $n=2500$ r.p.m. is necessary to absorb the "Maxwell neutrons" of $v=(2kT/m)^{\frac{1}{2}}=2.2\cdot 10^5$ cm/sec.

The experiments gave in fact a decrease in the number of transmitted neutrons, as detected by the disintegrations produced in boron with the help of an ionization chamber. The decrease was largest when the velocity selector rotated at a speed of about 2500 r.p.m. As a function of the speed of rotation, it followed closely the curve calculated from the assumption of Maxwellian distribution of the neutrons absorbed in cadmium (Fig. 14). This shows that (1) Cd is indeed a strong absorber for neutrons of thermal energies, (2) that boron is a good detector for such neutrons, (3) that thermal neutrons are present in large quantity in the interior of the paraffin and (4) that their distribution is practically Maxwellian, as we have concluded from our theoretical investigation of the diffusion of neutrons in paraffin (§59). A further confirmation of these points is found in experiments which show a shift of the maximum absorption to slower rotation upon cooling of the paraffin used for slowing down the neutrons.

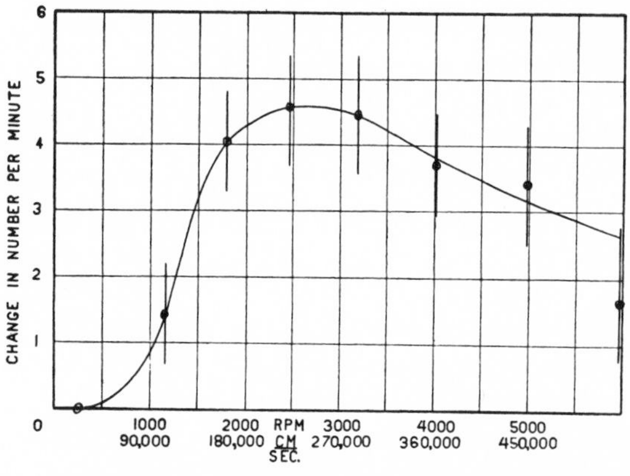

Fig. 14. Velocity selector of neutrons. Distribution of slow neutrons measured with the mechanical velocity selector by Dunning, Pegram, Fink, Mitchell and Segrè. The solid curve represents the Maxwell distribution at room temperature.

The mechanical velocity selector, while successful with thermal neutrons, does not seem applicable to the resonance groups of neutrons. As we shall see in this section, the energies of the neutrons causing resonance disintegration of nuclei, range from about 1 to several hundred volts, corresponding to velocities from 6 to 100 times thermal velocity. The rotational speed of the velocity selector would have to be increased in the same ratio, i.e., to between 15,000 and 250,000 r.p.m. which appears at present not to be feasible.

Another rather direct method which might in the future become usable is the diffraction of neutrons by a suitable scatterer which would serve as a measurement of the neutron wavelength. Thus far, however, only qualitative indications of neutron diffraction have been obtained (M23).

A. The boron absorption method

According to the foregoing, we are forced to make use of less direct methods for determining the neutron energy. The simplest of these methods is based on the assumption (§58) that the *absorption* coefficient of slow neutrons *in boron* and lithium is *inversely proportional to the neutron velocity*. This assumption is amply justified theoretically by the small density and large width to be expected for the energy levels of light nuclei,

(§58B). In addition, it has been confirmed by the following experiments.[34]

(1) Directly in the region of thermal energies by Rasetti, Mitchell, Fink and Pegram (R4, R5) using a mechanical arrangement which measures directly the deviation from the $1/v$ law (cf. §58B, end);

(2) Indirectly for higher energies by the fact that the absorption coefficients of Li and B change in the same way with energy (G17, H8);

(3) Approximately by the agreement between the observed ratios of the resonance energies of various elements derived by two different methods, *viz.* the absorption in boron and the spatial distribution of the resonance neutrons in paraffin (A11, cf. §60B);

(4) Qualitatively by comparison with experiments showing the order of the resonance levels of various nuclei in the energy scale (H10, cf. §60C).

Of the two light nuclei showing strong absorption of slow neutrons, boron is the more suitable because its absorption coefficient (per g/cm²) is about 10 times as high as that of Li. A high absorption coefficient is desirable in the first place in order to keep the required thickness of the absorber small and thus the geometry manageable. But even more important is the fact that scattering can be expected to be relatively smaller compared to the capture probability if the total absorption coefficient is higher because the scattering cross section will be of the same order for all light nuclei. Now the observed absorption is equal to the absorption by capture plus the absorption by scattering. The capture cross section decreases with increasing energy as $E^{-\frac{1}{2}}$ while the scattering cross section remains constant. Therefore the total absorption coefficient will cease to depend sensitively on the energy when the capture cross section becomes equal to or smaller than the scattering cross section. We estimate the scattering cross section to be about $4\cdot 10^{-24}$ cm² (average of other light elements, cf. D23 and Table XXVII). This must be corrected because in the experiments neutrons moving in all directions inside a hemisphere are counted. Therefore we assume the effective scattering cross section to be half the total, i.e., $2\cdot 10^{-24}$ cm². Then we find that capture and scattering will become equal at about 7 volts for Li ($\sigma_{\text{capt.}}=34\cdot 10^{-24}$ cm² at $E=kT\approx 1/40$ volt) and at about 2500 volts for $B(\sigma_{\text{capt.}}=600\cdot 10^{-24}$ cm² at $E=kT$). Lithium would therefore be useless for all resonance groups which lie at neutron energies higher than 7 volts (cf. H9). Even for boron, the observed absorption should be corrected for scattering which introduces quite an appreciable correction to the energy especially if the resonance energy lies high. At the same time, since the scattering cross section of boron can only be estimated roughly, a fairly large uncertainty will remain in the energy values given, even after the correction for scattering has been applied.

The principle of the experiment is very simple indeed: the absorption coefficient of boron must be measured, once for thermal neutrons and once for the resonance neutrons of the detector in question. As we know (§58), any neutron detector is made radioactive by thermal neutrons as well as by its particular "resonance neutrons." To separate the two activities, a sheet of Cd is placed between the neutron source and the detector: Cd is known to absorb all the thermal neutrons and to be practically transparent for the neutrons of higher energy ("resonance neutrons") (cf. §61G). Thus the difference of the activity in the detector with and without the Cd absorber will give the activity produced in it by thermal neutrons, while the activity with Cd shield must be attributed to the resonance neutrons. Both the activities, resonance and thermal, must be measured without the boron absorber and with boron, preferably as a function of the thickness of the latter. From these measurements, the absorption coefficients for the two groups of neutrons can be deduced. The energy of the resonance group is then

$$E_{\text{res}}=E_{\text{th}}(K_{\text{th}}/K_{\text{res}})^2. \qquad (505)$$

There is some question as to what to insert for the thermal energy. This problem is, however, well-defined if we remember that both the absorption in boron and the activity produced in the detector are proportional to $1/v$, i.e., to $E^{-\frac{1}{2}}$. Denoting by K_0 the absorption coefficient in

[34] A breakdown of the $1/v$ law for boron was announced by Arsenjewa-Heil, Heil and Westcott (A16) who maintained that all the resonance levels found experimentally lie actually inside the thermal region. This is due to a faulty interpretation of their experimental results, cf. footnote 37.

boron of neutrons of energy kT, and introducing

$$x=(E/kT)^{\frac{1}{2}} \tag{505a}$$

we have for the activity transmitted through a thickness l of boron:

$$I(l)=4\pi^{-\frac{1}{2}}I_0\int x^2e^{-x^2}dxe^{-(K_0/x)l}. \tag{506}$$

The average absorption coefficient K_{th} is defined by

$$I(l)=I_0e^{-K_{\text{th}}l}. \tag{506a}$$

The "effective thermal energy" to be inserted in (505) is then

$$E_{\text{th}}=kT(K_0/K_{\text{th}})^2. \tag{506b}$$

For very small thicknesses of boron, we obtain by a straightforward integration (B15)

$$E_{\text{th}}=(\pi/4)kT. \tag{507}$$

For larger thickness the integral must be evaluated numerically; the result is given in Fig. 15. Here the effective "thermal energy" E_{th} to be inserted in (505) is plotted against the (natural) logarithm of the transmission coefficient I/I_0 which follows directly from experiment. It is seen that E_{th} rises only very slowly with increasing thickness of the boron. For a transmission of 50 percent which is most convenient for experiments, we have E_{th} almost exactly equal kT as was already found by Goldsmith and Rasetti and used in the evaluation of their experiments.[34a]

Another point which requires some attention, is the oblique incidence of the neutrons on the absorber. If the experiments are made at the surface of the paraffin, the angular distribution of the emerging neutrons is approximately given by the law (497), $f(\vartheta)\sim\cos\vartheta+\sqrt{3}\cos^2\vartheta$, for both thermal and resonance neutrons. The absorption will thus not be exactly exponential, but will follow the absorption law

$$I=I_0c(Kl), \tag{508}$$

where the function

$$c(x)=\int_0^1 du(1+\sqrt{3}u)e^{-x/u}/(1+\tfrac{1}{2}\sqrt{3}) \tag{508a}$$

has been calculated by Fermi and is given in A11,

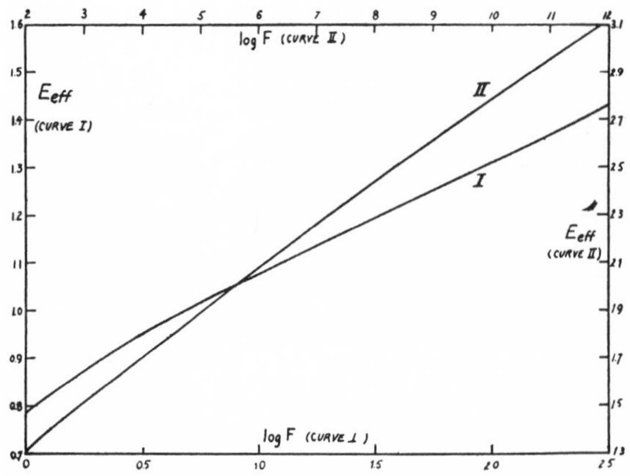

Fig. 15. Absorption of thermal neutrons in boron. The detector is assumed to obey the $1/v$ law in the thermal region. Abscissa: $|\log F|$ where $F=I/I_0$ is the transmission coefficient of the boron absorber. Ordinate: "Effective thermal energy" as a fraction of kT. In determining the energy of resonance neutrons by the boron method, the "effective energy" obtained from this figure should be inserted for the thermal energy.

Fig. 11 and in Fig. 16 of this report. From the given transmission ratio I/I_0, the value of $x=Kl$ can be read from the curve and therefrom the absorption coefficient K obtained. If the absorber thickness for each detector is chosen so that the transmission coefficient is the same for all detectors, the absorption coefficient in boron for the various resonance groups will be inversely proportional to the respective thickness of the boron absorbers used, independently of the angular distribution of the neutrons.

The most extensive measurements have been made by Goldsmith and Rasetti (G17). The results are listed in Table XXII. The values given by Goldsmith and Rasetti have been corrected for scattering, a cross section of $2\cdot10^{-24}$ cm^2 being assumed (cf. above), corresponding to an absorption coefficient of 0.11 cm^2/g due to scattering. This makes all the energy values slightly higher than those given by Goldsmith and Rasetti.

The top row in Table XXII refers to thermal neutrons,[35] measured with a Rh detector as the difference between the activities without and with Cd. The other figures refer to resonance groups. In general, one figure is given for each activity produced. This will represent the energy of the strongest resonance group, or some

[34a] More extensive calculations on the transmission of thermal neutrons by boron were made by Zahn (Z1a) and Laporte (L1b).

[35] According to Fig. 15, the "effective energy" of thermal neutrons corresponding to a boron transmission of 35 percent, is $1.09kT=0.028$ volt.

TABLE XXII. *Neutron resonance energies (from boron method).*

DETECTOR	FILTER	PERIOD	THICKNESS OF B ABSORBER g/cm²	TRANSMISSION %	ABS. COEFF. K (cm²/g)	K CORRECTED FOR SCATTERING	E (volt)
Rh[a]		44″	0.02	35.2	28	27.9	0.028
25 Mn	Cd	150′	0.58	43	0.73	0.62	60
29 Cu	Cd	5′	0.58	~50	0.57	0.46	100
33 As	Cd	26 hr.	0.58	54	0.49	0.38	150
35 Br	Cd	18′	0.58	50	0.57	0.46	100
45 Rh	Cd	44″	0.081	50	4.15	4.05	1.3
45 Rh	Cd	3.9′	0.054	~57	4.6	4.5	1.1
47 Ag	Cd	22″	0.108	63	2.8	2.7	3.0
47 Ag	Cd+Ag	22″[b]	0.108	60	2.1	2.0	5.5
47 Ag	Cd	2.3′	0.58	51.5	0.54	0.43	120
49 In	Cd	13″	0.108	~50	3	2.9	2.6
49 In	Cd	54′	0.081	53	3.8	3.7	1.6
53 I	Cd	25′	0.58	53	0.50	0.39	140
75 Re	Cd	20 hr.	0.23	64	0.83	0.72	40
77 Ir	Cd	19 hr.	0.108	46	3.5	3.4	1.9
79 Au	Cd	2.7 *d*	0.108	53	2.8	2.7	3.0

[a] Thermal neutrons.
[b] "*B* group."

average energy, if there are several resonance groups of about equal intensity.

Only in the case of the short period of Ag (22″), a separation into two resonance groups may easily be carried out, because part of the Ag "resonance neutrons" are very strongly absorbed in Ag (A group), another part (B group) much less strongly. This behavior suggests the presence of two resonance levels, a narrow one (A) and a broader one (B). The neutrons of the A group have an average absorption coefficient of 20 cm²/g in Ag; therefore a layer of 0.25 g/cm² Ag will absorb them almost completely while practically not affecting group B ($K\sim 0.3$ cm²/g in Ag). The activity due to neutrons passing through such a layer of Ag may thus be ascribed to the B group, the rest of the activity to A.

It is seen from Table XXII that there are several resonance levels near 1 volt, *viz.* Rh 44″, Rh 3.9′, In 54′, Ir 19 hr. These resonance levels can be shown to overlap (§61E) ("D group" of Amaldi and Fermi). The same is true for the "A" level of Ag 22″ and the level of Au (A group, energy near 3 volts). With the exception of In 13″ and of the "B" level of Ag 22″ (5 volts), all the remaining levels lie between 40 and 160 volts. The determination of their position is certainly not very accurate, partly because of the uncertainty of the scattering correction. It is plausible to assume (cf. G17) that these higher values actually represent averages over several levels[35a] while the lower values give the true positions of the resonance levels. If the first resonance level lies very low compared with the average spacing between levels, its effect will in general be much larger than that of higher levels, because the activity is cet. par. proportional to $E_r^{-\frac{3}{2}}$ (cf. (550a)). If the first resonance energy is of the order of the spacing of the levels, the second, third . . . levels will give contributions of the same order as the first.

From these considerations it seems that the average spacing between the energy levels of medium heavy and heavy nuclei ($Z>40$) is probably of the order of 10 to 30 volts, near zero kinetic energy of the neutron (cf. §53).

For the lighter nuclei (Mn to Br), the energy values observed are consistently high. It seems reasonable to assume that they represent the true position of the lowest level in spite of being "high," because lighter nuclei are expected to have a larger spacing between the levels (§53) and because no "low" values are observed for this lighter group. This spacing would then appear to be of the order of a few hundred volts, for Z near 30.

Other observations with the boron method have been made by Amaldi and Fermi (A11) and by v. Halban and Preiswerk (H8, H9). Amaldi and Fermi find values of the absorption coefficient in close agreement with Goldsmith and Rasetti, except for the thermal neutrons (38 cm²/g instead of 28 cm²/g). The value of Goldsmith and Rasetti was taken for the computation of Table XXII; it agrees with a value found by the same authors using an approximately parallel beam of neutrons, *viz.* 30 cm²/g, whereas Livingston and Hoffman (L33) find a value (36 cm²/g) near that of Amaldi and Fermi. The values found for the absorption coefficient of "resonance neutrons" in B are

GROUP	Rh 44″	Ag "A"	Ag "B"	I	Br
Goldsmith and Rasetti	4.15	2.8	2.1	0.5	0.57
Amaldi and Fermi	4.7	3	2.3	1.0	—
v. Halban and Preiswerk	6.3	3.4	—	0.7	0.35

The agreement is in general satisfactory. It is likely that for Br the determination of v.

[35a] This view is confirmed by experiments of Ruben and Libby (R17a) who show that the boron absorption coefficient of the neutrons producing activity in iodine decreases by about a factor 7 if the neutrons are first filtered by an iodine absorber.

Halban and Preiswerk is better, since from other evidence (cf. Section C) it seems that the Br level lies higher than the I level.

B. Method of the distribution of resonance neutrons in water

Amaldi and Fermi have also used an alternative method, based on the fact (§59D) that the mean square distance of the neutrons from the source increases with decreasing energy (cf. (476)). The activity of various detectors (Rh, Ag, I), screened by Cd against thermal neutrons, was measured at various points in a large tank of water which contained the neutron source at its center. The distributions are shown in A11, Fig. 7. They show a small but definite shift towards larger distances from the source in the order I, Ag (short period, A+B groups), Rh (short period, marked "D" in figure). The values of $(r^2)_{Av}$ measured are 262.2, 270.6 and 276.6 cm² water, respectively, for I, Ag and Rh. From (476) we have then for the ratio of the energies

$$I : Ag : Rh = 20 : 3.5 : 1. \qquad (509)$$

The agreement with the result of the boron method (160 : 3.5 : 1) is rather satisfactory considering the extreme sensitivity of the energy ratio to small errors in $(r^2)_{Av}$. The corrections suggested in §59D improve the agreement considerably as compared to the values given by Amaldi and Fermi (4.5 : 1.9 : 1).

C. Transformation of one resonance group into another

v. Halban and Preiswerk have pointed out that a neutron resonance group of high energy will be transformed into a group of lower energy by passing through paraffin, and have used this fact for establishing the *order* of various neutron resonance levels in the energy scale. The experimental arrangement is as follows: Neutrons emerging from a large paraffin block pass through seven thin paraffin plates of 3 mm thickness each. At the end of this series of plates, a neutron detector is placed, screened by cadmium. An absorber A can be interposed between any two of the paraffin plates, or at the end of the series right in front of the detector. The activity in the detector is measured as a function of the position of the absorber.

Immediately behind the absorber, the characteristic resonance group of the absorber will be missing in the neutron spectrum while otherwise the spectrum will be unaffected. However, when the neutrons pass through paraffin after traversing the absorber, the "hole" in the neutron distribution will shift to lower energies. Therefore, if the detector is made of the *same* material as the absorber, its activity will be strongly reduced if the absorber is placed right close to it, and will be much less reduced if paraffin is interposed between absorber and detector (v. Halban and Preiswerk, H8, H9). From such measurements, the mean free path of the resonance neutrons in paraffin may be deduced (A11, cf. §59 C7).

If the detector has a resonance level *higher* than that of the absorber, the result will be about the same except that the influence of the absorber is much smaller to start with, because it is only due to the overlapping of the neutron levels of absorber and detector. E.g., v. Halban and Preiswerk found that a certain Ag absorber reduced the activity of their I detector by only 5 percent but that of their Ag detector by 80 percent when there was no paraffin between absorber and detector.

However, when the resonance energy of the detector is *lower* than that of the absorber, the "hole" made by the absorber in the neutron distribution, will gradually cover the resonance region of the detector when the neutrons pass through paraffin. Therefore, in this case, the reduction of the activity of the detector by the absorber will *increase* with increasing thickness of the paraffin interposed between absorber and detector up to a certain maximum, and start decreasing only then. The position of the maximum will be an approximate measure of the ratio of the resonance energies (cf. method B) while its height will be connected to the width of the resonance level of the absorber.

With the method described, v. Halban and Preiswerk were able to determine the relative positions of the resonance levels of various elements. They found that the resonance energies could be arranged as follows:

$$Br\,(18') > I > Ag\,(22'') > Rh\,(44'') \sim In \sim Ir. \qquad (510)$$

This order agrees in general with the order found

by Amaldi and Fermi from the diffusion method B, and with that determined by the boron absorption method A.

The value of the method of transformation of neutron groups into each other lies firstly in the qualitative confirmation of the assumptions made about the boron absorption. Secondly, it is superior for establishing the order of high resonance energies which cannot be measured very well by the boron method (cf. above, scattering correction!). E.g., it shows that the Br level is higher than that of I while the evidence of the boron method is contradictory on this point. Moreover, it will indicate, at least qualitatively, when a nucleus has *several* resonance levels: If one level of nucleus A lies higher, another lower than the resonance level of a given other substance B, then we should expect an increase in the reduction of activity upon interposing paraffin *both* when A is used as absorber and B as detector, *and* when the arrangement is reversed. Applications of the method to the determination of the width of levels will be discussed in the following section.

D. Resonance levels of nuclei which are not activated

The determination of the position of resonance levels is much more difficult if the nucleus in question is not made radioactive by neutron capture. We can, in this case, only measure the absorption of the activity of other detectors in the substance in question. The information which can be obtained about nonactivated neutron absorbers is about the following:

(1) The absorption of thermal neutrons in the substance can be measured. The number of thermal neutrons can be measured either by the activity produced in any detector, or by the disintegration of Li or B. In each case, the difference between the number of disintegrations observed with and without a screen of Cd, gives the number of thermal neutrons. By this method, the absorption coefficients of almost all elements for thermal neutrons has been measured by Dunning, Pegram, Fink and Mitchell (D23) and for some additional elements (rare earths) by Hevesy and Levi (H31).

(2) The average absorption coefficient for faster neutrons can be measured, using a B or Li chamber shielded by Cd. The ratio of this absorption coefficient to that for thermal neutrons R_S is a characteristic of the substance S. It may be compared to the corresponding ratio R_B for boron. If R_S is approximately equal to R_B, the $1/v$-law holds approximately for the substance. Then the substance S will in general not possess any resonance levels at low energies. If $R_S > R_B$, we shall expect a resonance level of S at low energy but above thermal energy. The result $R_S < R_B$ indicates a resonance level in the thermal energy region itself or at small negative energy.

(3) The absorption coefficient for neutrons of certain known energies can be measured, using a detector whose resonance level is known. E.g., we may measure the absorption of one-volt neutrons by using Rh (44″), that of three-volt neutrons by using Ag (22″) as a detector, the detector being in each case screened by Cd. The interpretation is similar to, but more definite than in (2). For a number of absorbers, notably *Cd, Hg, Sm and Gd*, it has been found that the absorption coefficient for D neutrons (Rh resonance group) is almost negligibly small compared to that of the thermal neutrons. The ratio is about 1 : 100 in Cd, Sm and Gd (G17), as compared to 1 : 7 in boron. This shows that the absorbers mentioned must *have very low resonance levels*. Their exact position depends on the width assumed and can therefore be determined only in connection with other experiments measuring the width of the levels (cf. §61G).

It may happen that a given absorber shows very *strong* absorption for the neutrons of a known group. In this case, we would conclude that the absorbing substance has a resonance level coinciding with this group. No such case is known among the nonactivated absorbers, and it is indeed rather unlikely to find the level of an absorber in this way because the energy regions for which we possess specific detectors are very narrow and widely separated.

(4) The approximate position of the resonance levels of nonactivated absorbers may be obtained by the method described in C. A number of detectors must be used whose resonance energies form an increasing series. The absorber to be investigated is placed between the paraffin plates and the reduction in activity measured as

a function of the thickness of paraffin for the various detectors. Then it will be found that for the low-resonance detectors the intensity reduction will increase when a small amount of paraffin is interposed while it will decrease for the high-resonance detectors. The resonance energy of the absorber will lie between that of the last detector which shows the increase and the first one which does not. No experiments of this kind have yet been carried out.

(5) If the absorption of the substance is detectable by the boron or lithium detector, the energy of the resonance level can be estimated by interposing a boron absorber, in addition to the given substance, between neutron source and boron detector. The reduction of the neutron intensity by the substance can then be measured as a function of the thickness of the additional boron absorber, which determines the boron absorption coefficient of the resonance neutrons of the substance.

§61. Total (γ-Ray) Width of Neutron Resonance Levels (B15, A11, G18a, M2a, R4, L32)

According to the general formula (430), the capture cross section for neutrons of a given energy E may be written

$$\sigma_N(E)=\sigma_0/(1+(E-E_r/\tfrac{1}{2}\Gamma)^2) \qquad (511)$$

provided the energy of the neutron is not too different from the resonance energy E_r. σ_0 is the cross section at resonance (cf. 430b), Γ the total "natural width" of the level.

A. Doppler effect

In comparing this formula with experiment, it is in some cases necessary to take account of the *Doppler effect*. The velocities of the capturing nuclei are not negligible compared to the neutron velocities, and the Doppler width introduced in this way is sometimes comparable to, sometimes even larger than, the natural width.

If u_x is the velocity of the capturing nucleus in the direction of motion of the neutron, then $v-u_x$ will be the relative velocity of neutron and nucleus. The relative kinetic energy is then, neglecting terms of the order u^2/v,

$$E'=\tfrac{1}{2}m(v-u_x)^2=E-(2mE)^{\frac{1}{2}}u_x, \qquad (512)$$

where E is the absolute kinetic energy and m the mass of the neutron. Now the probability of finding a nucleus of velocity u_x is given by the Maxwell distribution function

$$w(u_x)du_x=(M/2\pi kT)^{\frac{1}{2}}e^{-Mu_x^2/2kT}du_x, \qquad (512a)$$

where M is the mass of the capturing nucleus. Inserting (512), we find for the probability of a given E' (for fixed E):

$$w(E')dE'=\pi^{-\frac{1}{2}}e^{-(E'-E)^2/\Delta^2}dE'/\Delta, \qquad (513)$$

where

$$\Delta=2(mE_rkT/M)^{\frac{1}{2}} \qquad (513a)$$

is the "Doppler width."[36] E_r has been written instead of E because the neutron energy is supposed to be near the resonance energy E_r.

The energy to be inserted in (511) is, of course, the relative energy E'. The effective cross section for neutrons of energy E is then

$$\sigma(\xi, E)=\int\sigma_N(E')w(E')dE'=\sigma_0\psi(\xi, x) \qquad (514)$$

with

$$x=(E-E_r)/\tfrac{1}{2}\Gamma, \qquad (514a)$$

$$\xi=\Gamma/\Delta, \qquad (514b)$$

$$\psi(\xi, x)=\frac{\xi}{2\pi^{\frac{1}{2}}}\int_{-\infty}^{\infty}\frac{e^{-\frac{1}{4}\xi^2(x-y)^2}}{1+y^2}dy, \qquad (515)$$

$$y=(E'-E_r)/\tfrac{1}{2}\Gamma. \qquad (515a)$$

$\psi(\xi, x)$ is in general a complicated function of x. Simple expressions are obtained

(a) for ξ very large (purely natural width)

$$\psi(\xi, x)=1/(1+x^2) \qquad (515b)$$

and (514) (515) reduce to (511);

(b) for ξ very small (pure Doppler width) and $x\ll\xi^{-2}$

$$\psi(\xi, x)=\tfrac{1}{2}\pi^{\frac{1}{2}}\xi e^{-\frac{1}{4}\xi^2x^2}; \qquad (515c)$$

(c) for very large $x\gg\xi^{-2}$ and any ξ

$$\psi(\xi, x)=1/(1+x^2); \qquad (515d)$$

(d) for $x=0$ (exact resonance) and any ξ

$$\psi(\xi, 0)=\tfrac{1}{2}\pi^{\frac{1}{2}}\xi e^{\frac{1}{4}\xi^2}[1-\Phi(\tfrac{1}{2}\xi)], \qquad (516)$$

where Φ is the Gaussian error function

$$\Phi(\tfrac{1}{2}\xi)=2\pi^{-\frac{1}{2}}\int_0^{\frac{1}{2}\xi}e^{-t^2}dt. \qquad (516a)$$

In the special case of large ξ (natural width), (516) reduces to unity, meaning (cf. (514)) that in this case the cross section at exact resonance is equal to σ_0. In the case of small ξ (Doppler width) we have $\psi(\xi, 0)=\frac{1}{2}\pi^{\frac{1}{2}}\xi$, meaning that the cross section at resonance is reduced to

$$\sigma_D(E_r)=\tfrac{1}{2}\pi^{\frac{1}{2}}\sigma_0\Gamma/\Delta. \qquad (516b)$$

Experimentally, the quantities measured most easily are the total activation, and the absorption of this total activity by thin absorbers of the same substance ("absorption coefficient for self-indication"). The total activation is proportional to

$$C=\int n(E)\sigma(\xi, E)dE, \qquad (517)$$

if $n(E)dE$ is the number of neutrons in the energy

[36] Δ is here defined as one-half of the quantity denoted by the same letter in the paper of Bethe and Placzek (B15).

interval dE. Since σ varies rapidly, n slowly with the energy, we have (cf. 514, 514a)

$$C=n(E_r)\sigma_0\tfrac{1}{2}\Gamma\int dx\psi(\xi,x). \qquad (517a)$$

As can easily be seen from the definition (515) of ψ, the integral has the value π independent of ξ so that we obtain

$$C=\tfrac{1}{2}\pi\Gamma n(E_r)\sigma_0. \qquad (518)$$

A thin absorber will absorb a fraction $c\sigma(\eta, E)$ of the activity produced by the neutrons of energy E, where c is a constant depending on the thickness of the absorber in g/cm^2, and $\eta=\Gamma/\Delta'$ where Δ' is the Doppler width for the absorber. This may be different from the Doppler width Δ for the detector because the temperatures may be different. The effective absorption "cross section for self-indication" is then

$$\sigma_s(\xi,\eta)=\frac{\int n(E)\sigma(\xi,E)\sigma(\eta,E)dE}{\int n(E)\sigma(\xi,E)dE}, \qquad (519)$$

The integral in the numerator may be evaluated (B15); then (519) becomes

$$\sigma_s(\xi,\eta)=\tfrac{1}{2}\sigma_0\psi(\zeta,0) \qquad (520)$$

with

$$\zeta=2\xi\eta(\xi^2+\eta^2)^{-\frac{1}{2}}. \qquad (520a)$$

In the usual case of equal temperature of absorber and emitter this becomes

$$\zeta=\xi\sqrt{2}. \qquad (520b)$$

According to the definition (514b) (513a) of ξ, this means that σ_s is one-half of the resonance cross section at one-half of the temperature of absorber and detector. If the temperatures of absorber and detector are not equal, one-half their arithmetical mean should be inserted, according to (520a).

From the total activity and the effective cross section for self-indication σ_s we can define the "effective width" of the neutron level

$$\Gamma_{\text{eff}}=C/n(E_r)\sigma_s, \qquad (521)$$

so that (cf. 518, 520)

$$\Gamma_{\text{eff}}=\pi\Gamma/\psi(\xi\sqrt{2},0). \qquad (521a)$$

For natural width, $\psi(\xi\sqrt{2},0)=1$ (cf. (516)) so that

$$\Gamma_{\text{eff}}=\pi\Gamma, \quad (\Gamma\gg\Delta), \qquad (522)$$

while for pure Doppler width $\psi(\xi\sqrt{2},0)=(\pi/2)^{\frac{1}{2}}\xi$ and (cf. (514b))

$$\Gamma_{\text{eff}}=(2\pi)^{\frac{1}{2}}\Delta, \quad (\Gamma\ll\Delta). \qquad (523)$$

For intermediate values of $\xi=\Gamma/\Delta$, a power series and a graph for Γ_{eff} have been given by Bethe and Placzek (B15, Fig. 2 and §5).

If the Doppler width is large compared to or of the same order as the natural width, the cross section for self-indication, and therefore also the effective width, will change with temperature. In the extreme case $\xi\ll1$ (pure Doppler width), σ_s will be inversely proportional to the Doppler width, i.e., proportional to $T^{-\frac{1}{2}}$. This behavior is exactly the same as would be expected for thermal neutrons assuming the validity of the $1/v$ law. A temperature effect of this type has actually been observed for the resonance activity of Ag by Arsenjewa-Heil, Heil and Westcott (A16).[37] If Doppler width and natural width are comparable, the temperature dependence of the cross section σ_s may serve for a determination of the natural width Γ from the known Doppler width (513a).[38]

B. Measurements of the effective width from activity and absorption coefficient

The effective width may be deduced (cf. (521)) from the total activity produced in a given detector by the resonance neutrons and the cross-section for self-indication if the number of neutrons per unit energy $n(E_r)$ is known. Let K be the effective absorption coefficient of the detector in cm^2/g for its resonance neutrons, so that

$$K_r=\sigma_s/M, \qquad (524)$$

where M is the mass of one atom of the detector. Then the total activity of a thin detector of weight w (in grams) is[39]

$$B_r=n(E_r)v_r\Gamma_{\text{eff}}K_rw, \qquad (524a)$$

where $n(E_r)dE$ is the number of neutrons per cm^3 in the energy interval dE about E_r, and therefore $n(E_r)v_rdE$ the number of neutrons in

[37] The effect was incorrectly interpreted by these authors as evidence that the Ag resonance neutrons had actually thermal energies.

[38] This experiment was first suggested and preliminary results reported by Frisch (Copenhagen Conference, June 1936).

[39] The relation between B_r and C (cf. (517), (518), 521)) is

$$B_r=(wv_r/M)C. \qquad (524b)$$

this energy interval striking unit area of the detector per sec. The "activity" is here defined as the number of neutrons captured per sec., or the number of β-rays emitted per sec. by the radioactive substance produced, immediately after an infinite time of irradiation. The detector must be so thin that it absorbs neither the resonance neutrons nor the β-electrons to any appreciable extent; otherwise corrections have to be applied (cf. A11).

The number of neutrons $n(E_r)$ can be expressed by the number emitted from the source with the help of (440):

$$n(E_r)v_r = ql_0/E_r, \tag{525}$$

where q is the number coming from the source per sec. and cm³ and l_0 the mean free path of neutrons above thermal energy. q is connected to the total number Q coming from the source per sec. by

$$Q = \int q d\Omega, \tag{525a}$$

the integral being extended over the whole volume of the paraffin. We should thus measure the activity of our detector as a function of its position, and integrate over the whole paraffin; then we find

$$\int B_r d\Omega = Ql_0 K_r w \Gamma_{\text{eff}}/E_r. \tag{526}$$

Q may now be obtained from measurements of the activity produced by *thermal* neutrons (in *any* detector). This activity is conveniently measured by determining the activity of the detector when it is covered by Cd from one side, and subtracting therefrom the activity obtained with Cd on both sides. With a Cd cover on one side, the activity of the detector due to thermal neutrons is, according to (500),

$$B_{\text{th}} = \sqrt{3}q'LK_{\text{th}}w, \tag{527}$$

where L is the diffusion length (§59E), K_{th} the absorption coefficient of the detector for thermal neutrons and q' the number of thermal neutrons produced per cm³ and sec. Owing to the different distribution of thermal and resonance neutrons in the paraffin (cf. §59D, Fig. 13), q' will in general be different from q. However, by integration we obtain

$$\int B_{\text{th}} d\Omega = QL\sqrt{3}K_{\text{th}}w, \tag{527a}$$

where Q has the same meaning as in (526). Combining (526) and (527a), we find[40]

$$\frac{\Gamma_{\text{eff}}}{E_r} = Y\frac{K_{\text{th}}}{K_r}\frac{L\sqrt{3}}{l_0}, \tag{528}$$

where

$$Y = \int B_r d\Omega / \int B_{\text{th}} d\Omega \tag{528a}$$

is the ratio of the activities. According to the measurements of Amaldi and Fermi, $L = 2.1$ cm (cf. (482b)), $l_0 = 0.9$ cm (cf. (465c)) so that

$$\sqrt{3}L/l_0 = 4.0, \tag{528b}$$

$$\Gamma_{\text{eff}}/E_r = 4.0\,YK_{\text{th}}/K_r. \tag{528c}$$

The actual measurements are complicated by the fact that the activities must be measured all over the paraffin. This can be avoided if only relative values of the "fractional width" Γ_{eff}/E_r are required, i.e., e.g., when Γ_{eff}/E_r has already been determined for one substance. The spatial distribution of various resonance groups is sensibly the same (cf. A11, Fig. 7). Therefore if B_1 and B_2 measure the activities of two detectors at any one place (e.g., near the maximum neutron density), we have according to (524a) (525)

$$\left(\frac{\Gamma_{\text{eff}}}{E_r}\right)_1 \div \left(\frac{\Gamma_{\text{eff}}}{E_r}\right)_2 = \left(\frac{B_r}{K_r w}\right)_1 \div \left(\frac{B_r}{K_r w}\right)_2. \tag{529}$$

A further complication is the finite absorption of neutrons and electrons in the actual target. This correction is somewhat ambiguous since the absorption coefficient for resonance neutrons decreases rather rapidly with increasing thickness of the absorber (cf. below, Section C) and this decrease depends on the ratio of natural to Doppler width (cf. above, Section A). Amaldi and Fermi corrected their results assuming a constant absorption coefficient which means an overcorrection, i.e., an overestimate of the true resonance activity without absorption and therefore an overestimate of the width Γ_{eff} (cf. (528)).

[40] It is, of course, not necessary that the activity of thermal and resonance neutrons is determined in the same detector. This fact was made use of in the determination of the width of the iodine level by Amaldi and Fermi. The activity produced by thermal neutrons in iodine is very small, therefore a Rh detector was used to determine the number of thermal neutrons. The numbers given for thermal neutrons in the I column of Table XXIII refer therefore to Rh, not to I.

TABLE XXIII. *Widths of neutron resonance levels.*

SUBSTANCE	Rh	Ag	I
$Y=\dfrac{\int B_{\mathrm{res}}d\Omega}{\int B_{\mathrm{th}}\,d\Omega}$	1.16	0.72	0.045
K_r (cm²/g)	19[40a]	20	0.38
K_{th} (cm²/g)	0.7	0.25	0.7
$\Gamma_{\mathrm{eff}}/E_r$	0.17	0.036	0.33
E_r(cf. Table XXII)	1.3	3.0	140
Γ_{eff} (volts)	0.22	0.11	50
$(2\pi)^{\frac{1}{2}}\Delta$(cf. (513a))	0.09	0.13	0.9
Γ	0.07	<0.03	15(?)

[40a] This large value for the resonance absorption coefficient of Rh was given recently by Manley, Goldsmith and Schwinger (M2a) on the basis of their own data and those of v. Halban and Preiswerk. It is almost 10 times the value originally given by Amaldi and Fermi (2.0 cm²/g) which causes large changes of all results about the Rh level (cf. B15).

We have recalculated the correction using the absorption curve for resonance neutrons given in Fig. 16, curve *w*. We find that the A group activity in the Ag detector of Amaldi and Fermi (thickness 0.057 g/cm²) is reduced by a factor of 0.45 by absorption (Amaldi and Fermi give 0.35). For Rh, we used the most recent value of the resonance absorption coefficient (M2a), *viz.* 19 cm²/g (A and F give 2.0) and find 0.079 for the absorption factor (A and F give 0.193).

The results of Amaldi and Fermi are given in Table XXIII. For Rh, the effective width obtained is 2.5 times larger than $(2\pi)^{\frac{1}{2}}$ times the Doppler width (cf. (523)), therefore the natural width is certainly much larger than the Doppler width and may be calculated from (522). For Ag, Γ_{eff} as obtained from the activity is exactly equal to $(2\pi)^{\frac{1}{2}}$ times the calculated Doppler width. Therefore the natural width must be very small compared to the Doppler width and cannot be calculated from Γ_{eff}. (Possibly measurements at lower temperatures would allow a determination of the natural width.) For iodine, the situation is the same as for Rh. However, the natural width of 15 volts obtained seems implausibly high (§87). The reason for this is, we believe, that the activity in iodine is due to a large number of resonance levels of various energies, the observed Γ_{eff} being the sum of the widths of all these levels. This assumption is plausible because the "effective" resonance energy of I lies very high (cf. the remarks after Table XXII).[40b]

Thus the measurement in Rh seems to be the only reliable figure for the natural width of nuclear resonance levels which is available at present. According to the considerations in §58, this width must be interpreted as the "γ-ray width," i.e., as giving the probability of γ-ray emission from the excited state of the compound nucleus. The lifetime of this state is $\hbar/\Gamma=1.0\cdot10^{-14}$ sec.

[40b] It is also confirmed by measurements of Ruben and Libby (R17a) (cf. reference 35a).

TABLE XXIV. *Test of the one-level formula.*

SUBSTANCE	Rh	Ag	I
Y	1.16	0.72	~ 3
ξ from (535)	7.5	13	3.3
ξ from Table XXIII	2	$\ll 1$	40

C. Test of the one-level formula

If we assume that the cross section is, from zero neutron energy up to the first resonance level E_r, determined by the first level alone, we have (cf. (430))

$$\sigma=\sigma_0\left(\frac{E_r}{E}\right)^{\frac{1}{2}}\frac{1}{1+(E_r-E/\frac{1}{2}\Gamma)^2}. \tag{530}$$

We may apply this formula to obtain the ratio of the activities due to thermal and resonance neutrons. For resonance neutrons we have, according to (518), (524b), (525):

$$B_r=ql_0\tfrac{1}{2}\pi\sigma_0 w\Gamma/ME_r, \tag{531}$$

independent of the ratio of Doppler width to natural width (cf. (518)). For thermal neutrons, we have simply to insert the average cross section (averaged over the Maxwell distribution) into (527). Assuming that the first resonance energy E_r is large compared to both the level width Γ and the thermal energy $E\approx kT$, we find

$$B_{\mathrm{th}}=\surd 3q'L(w/M)\tfrac{1}{4}\sigma_0\Gamma^2E_r^{-\frac{3}{2}}(E^{-\frac{1}{2}})_{\mathrm{Av}}, \tag{532}$$

where $(E^{-\frac{1}{2}})_{\mathrm{Av}}$ is to be calculated as the average over the thermal neutrons striking the absorber. Since the number of such neutrons in the energy interval dE is proportional to the number of Maxwell neutrons of energy E, $M(E)dE$, times the velocity $v\sim E^{\frac{1}{2}}$, we have

$$(E^{-\frac{1}{2}})_{\mathrm{Av}}=\frac{\int M(E)dE}{\int M(E)E^{\frac{1}{2}}dE}=\frac{1}{(E^{\frac{1}{2}})_a}, \tag{532a}$$

where $(E^{\frac{1}{2}})_a$ is the average of $E^{\frac{1}{2}}$ over the Maxwell distribution,

$$(E^{\frac{1}{2}})_a=(4/\pi)^{\frac{1}{2}}(kT)^{\frac{1}{2}}. \tag{532b}$$

Thus (532) becomes

$$B_{\mathrm{th}}=\tfrac{1}{8}(3\pi)^{\frac{1}{2}}q'L(w/M)\sigma_0\Gamma^2E_r^{-\frac{3}{2}}(kT)^{-\frac{1}{2}}. \qquad (533)$$

As in Section B, we integrate the activities (531) (533) over all the paraffin and obtain for the ratio

$$Y=\frac{\int B_{\mathrm{res}}d\Omega}{\int B_{\mathrm{th}}d\Omega}=4\left(\frac{\pi}{3}\right)^{\frac{1}{2}}\frac{l_0}{L}\frac{(E_rkT)^{\frac{1}{2}}}{\Gamma}. \qquad (534)$$

With the values $l_0=0.9$ cm, $L=2.1$ cm, this gives

$$\frac{\Gamma}{(E_rkT)^{\frac{1}{2}}}=\frac{1.75}{Y}. \qquad (534a)$$

In contrast to (528), Γ is the *natural width* of the line. The denominator on the left-hand side is proportional to the Doppler width; therefore we have (cf. (513a), (514b))

$$\xi=\frac{\Gamma}{\Delta}=2\left(\frac{\pi}{3}\right)^{\frac{1}{2}}\left(\frac{M}{m}\right)^{\frac{1}{2}}\frac{l_0}{L}\frac{1}{Y}=\frac{0.88A^{\frac{1}{2}}}{Y}. \qquad (535)$$

For the three elements listed in Table XXIII, the results are given in Table XXIV.

The result is fairly satisfactory only for Rh. This substance is just the one for which the one-level formula may be expected to hold best because it has a resonance level at very low energy which in addition is fairly broad and can therefore be expected to give a large contribution to the cross section at thermal energies.

For Ag, there is no agreement at all. From the one-level formula we should conclude a natural width much larger than the Doppler width while actually we found that the natural width is negligible compared to the Doppler width. This means that the cross section for thermal neutrons is much larger (Y much smaller) than would be expected from the one-level formula. This again is understandable because the Ag level is extremely narrow and thus cannot be expected to give a large contribution to the thermal cross section. Probably the "B" level (cf. Table XXII) gives a much larger contribution.

For I, the data on the thermal activity which are available are rather scarce, the value of Y being estimated. The situation is here the opposite of that with Ag, showing that the Γ obtained in Table XXIII is probably grossly overestimated. We suggested already in view of the very high value of Γ obtained in Table XXIII that the "resonance" activity in iodine is presumably due to a number (n, say) of levels. Assuming this, (528) will give n times the average value of $\Gamma_{\mathrm{eff}}/E_r$. Furthermore, if it is assumed that the interference terms between the contributions of various levels to the thermal cross section average out, B_{th} will be given by a formula of the type (532) with n times the average of $\sigma_0\Gamma^2E_r^{-\frac{3}{2}}$ over all the levels replacing the simple $\sigma_0\Gamma^2E_r^{-\frac{3}{2}}$. Therefore, with this assumption of negligible interference, formula (534) should give the average value of ξ for the resonance levels contributing, while from Table XXIII we should get approximately n times this average value. This would mean that about $40/3.3=12$ levels contribute to the iodine activity which seems plausible. Furthermore, it seems reasonable to assume that the ξ derived from (534) may still be sufficiently in error to make the actual average of ξ of order unity or even smaller, corresponding to natural widths of the individual iodine levels of the same order as for Rh.

D. Widths from the one-level formula

From the considerations of Section C, we may conclude that the use of the one-level formula is justified for substances which have a not too narrow resonance level at low energies. For these substances, the one-level formula may be used to calculate the width from the *absorption* coefficients for resonance and thermal neutrons, in the absence of measurements of the total activity.

For this purpose, we compare the cross section for self-indication (520) with the average cross section for thermal neutrons which we calculate from the one-level formula (530). We have from (520) (520b) (521a)

$$\sigma_s=\tfrac{1}{2}\pi\sigma_0\Gamma/\Gamma_{\mathrm{eff}} \qquad (536)$$

and from (530) (532a) (532b)

$$\sigma_{\mathrm{th}}=\tfrac{1}{8}\pi^{\frac{1}{2}}\sigma_0\Gamma^2/E_r^{\frac{3}{2}}(kT)^{\frac{1}{2}}. \qquad (537)$$

The ratio of the absorption coefficients is thus

$$\frac{K_r}{K_{\mathrm{th}}}=\frac{\sigma_s}{\sigma_{\mathrm{th}}}=4\pi^{\frac{1}{2}}\frac{E_r^{\frac{3}{2}}(kT)^{\frac{1}{2}}}{\Gamma\Gamma_{\mathrm{eff}}}, \qquad (537a)$$

a formula which could have been obtained also from (528) and (534). If the Doppler width is small compared to the natural width (which can

TABLE XXV. *Widths from one-level formula.*

SUBSTANCE	Rh 44″	In 54′	Ir	Au	Ag 22″ (Group A)	Ag 22″ (B)
K_r (cm²/g)	19	3.8	1.0	4	20	0.3
K_{th} (cm²/g)	0.7	0.6	1.0	0.25	0.2	0.2
E_r (cf. Table XXII)	1.3	1.6	1.9	3.0	3.0	5.5
Γ (from 538, volts)	0.13	0.34	1.0	0.34	0.14	1.7
$\xi=\Gamma/\Delta$	4	9	30	9	2.7	23
Calculated Y (cf. (535))	2.2	1.0	0.4	1.3	3.4	0.4
Observed Y	1.16	—	—	—	0.72	∼0.3

be checked from the result) we have, according to (522):

$$\Gamma=2\pi^{-\frac{1}{4}}E_r^{\frac{3}{4}}(kT)^{\frac{1}{4}}(K_{\text{th}}/K_r)^{\frac{1}{2}}. \tag{538}$$

The condition $\Gamma\gg\Delta$ is equivalent to (cf. (513a))

$$\frac{K_r}{K_{\text{th}}}\ll\frac{M}{\pi^{\frac{1}{2}}m}\left(\frac{E_r}{kT}\right)^{\frac{1}{2}}. \tag{538a}$$

This is fulfilled for all cases except for the A group of Ag.

The results for various elements are given in Table XXV. The absorption coefficients are due to Amaldi and Fermi.[40c] The absorption of thermal neutrons must be corrected if more than one activity is produced in the same substance, as in Ag and In. In Ag, the ratio of the long period to the short period activity (initial activity after infinite irradiation) is about 1 : 3 (private communication of Professor Goudsmit) so that $\frac{3}{4}$ of the absorption coefficient is due to the short period. In Indium, the ratio of the activities is $\text{In}_{54'}:\text{In}_{16''}=3:2$ (cf. A11) so that 60 percent of the absorption of thermal neutrons is due to the long period.[41] These figures have been used in Table XXV. All other substances have only one strong activity. Where possible, the absorption coefficients measured with detector of the same substance as the absorber were taken. kT is taken as 0.026 volt.

As can be seen from the Table XXV, the natural width Γ turns out to be much larger than the Doppler width in all cases except for the A group of Ag. Thus the widths derived ought to be fairly reliable for all cases except the Ag A group. A check is provided by calculating Y from the Γ obtained in Table XXV, and comparing it to the observed ratio Y of resonance to thermal activity wherever the latter is known. The calculated values of Y seem to be of the right order of magnitude (about unity) for all cases except Ag A, the observations giving approximately equal activity from resonance and thermal neutrons for most substances (A11), (G17). A reasonable agreement (as good as for Rh) is also obtained for the B group of Ag, supporting the hypothesis that the B level gives the main contribution to the cross section at thermal energies.

The widths found vary from 0.13 to 1.6 volts which seems reasonable. Broad levels are generally connected with small absorption coefficients for the resonance neutrons. One of the broader levels is that of Ir. The width given for it may, however, not be quite reliable because the absorption coefficient was measured by Amaldi and Fermi with Rh and In as detectors, and the resonance energy of Ir seems to be somewhat higher than for these two substances. This makes it likely that K_r has been underestimated, and therefore Γ overestimated, in Table XXV.

E. Overlapping of levels

In some cases, energy levels of various substances are known to overlap. This is true for Rh, In and Ir ("D group") and for Au and the A level of Ag (A group). In these cases, an estimate of the difference of the resonance energies of the substances in question may be obtained by measuring the mutual absorption coefficients of the resonance radiation and comparing them with the absorption coefficients of each substance for its own radiation. The width must be known for at least one of the substances.

If the resonance energies are E_1 and E_2, the widths Γ_1 and Γ_2, the absorption coefficient of substance 1 for the activity of substance 2 is given by

$$K_{12}=K_1(0)\int\frac{dE}{[1+(E-E_1/\frac{1}{2}\Gamma_1)^2][1+(E-E_2/\frac{1}{2}\Gamma_2)^2]}\bigg/\int\frac{dE}{1+(E-E_2/\frac{1}{2}\Gamma_2)^2}, \tag{539a}$$

[40c] Except for Rh, which is due to v. Halban and Preiswerk and to Manley, Goldsmith and Schwinger.

[41] Since we are here only interested in the ratio of the absorption coefficients, the abundance of the isotope responsible for the absorption is immaterial.

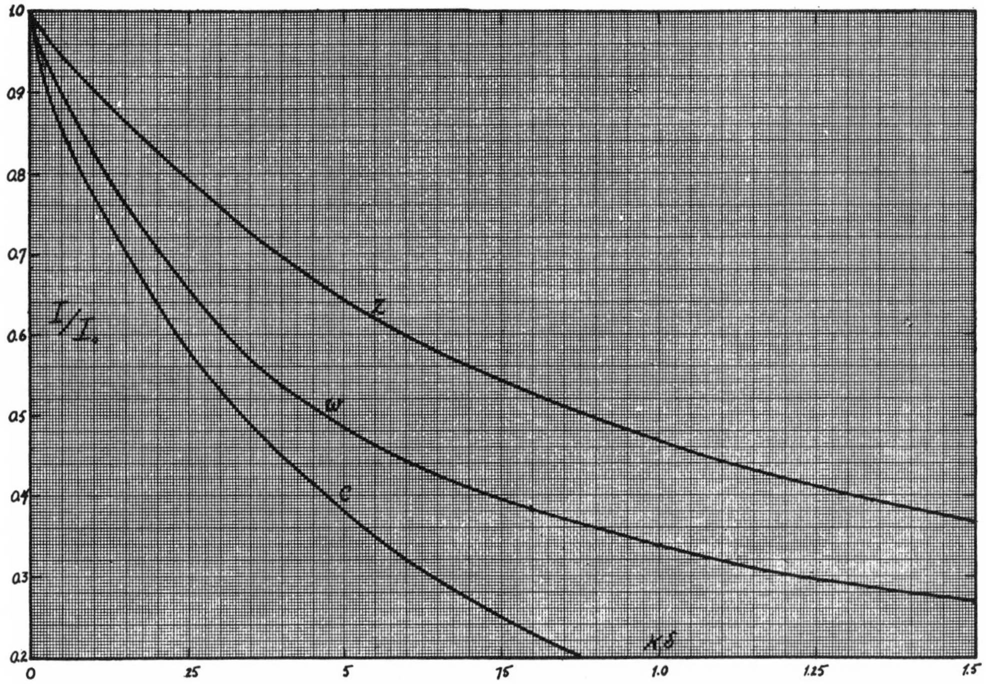

FIG. 16a. Absorption of resonance radiation. Abscissa: $K_r\delta$ where K_r is the average absorption coefficient of the resonance radiation for self-indication and small absorber thickness δ. (The figures on the abscissa should read 0.25, 0.5, 0.75, 1.0 etc.) Ordinate: Transmitted intensity. Curve z corresponds to a collimated incident beam of neutrons, curve w to an angular distribution $\cos\vartheta+\sqrt{3}\cos^2\vartheta$. For comparison, curve c gives the absorption for the same angular distribution with *constant* absorption coefficient, according to Amaldi and Fermi (A11, Fig. 2).

where $K_1(0)$ is the absorption coefficient of substance 1 at its resonance. If the natural width is large compared to the Doppler width, $K_1(0)$ is twice as large as the absorption coefficient for self-indication, K_{11}. Then we obtain by evaluating the integrals

$$\frac{K_{12}}{K_{11}}=\frac{2\Gamma_1(\Gamma_1+\Gamma_2)}{4(E_1-E_2)^2+(\Gamma_1+\Gamma_2)^2}. \tag{539}$$

For $E_1=E_2$ and $\Gamma_1=\Gamma_2$, the right-hand side reduces to unity.

If all four absorption coefficients K_{11} K_{12} K_{21} K_{22} are known, the ratio of the widths can be determined from (539)

$$\Gamma_1/\Gamma_2=K_{12}K_{22}/K_{11}K_{21}. \tag{540}$$

The right-hand side is the product of the two absorption coefficients with the second substance as indicator, divided by the product with the first as indicator. This determination of the ratio of the widths does not make use of the one-level formula except in the immediate neighborhood of the resonance where it is well justified. The difference of the resonance energies becomes, according to (539)

$$|E_1-E_2|=\tfrac{1}{2}(\Gamma_1+\Gamma_2)\times\left[\frac{2K_{11}K_{22}}{K_{12}K_{22}+K_{11}K_{21}}-1\right]^{\frac{1}{2}}. \tag{541}$$

The only case in which all necessary data are available,[41a] is that of Rh (44″) and In (54′). According to Amaldi and Fermi,

$$K_{\mathrm{Rh\,Rh}}=2.0,\qquad K_{\mathrm{Rh\,In}}=1.6,$$
$$K_{\mathrm{In\,Rh}}=3.0,\qquad K_{\mathrm{In\,In}}=3.8, \tag{541a}$$

showing that for a given absorber the absorption is noticeably stronger when the detector is of the same material as the absorber. The ratio of the widths becomes according to (540)

$$\Gamma_{\mathrm{In}}/\Gamma_{\mathrm{Rh}}=2.0\cdot3.0/1.6\cdot3.8=1.0 \tag{540a}$$

in good agreement with the result found from

[41a] *Note added in proof:* With the large change in $K_{\mathrm{Rh\,Rh}}$ (from 2 to 19) necessary according to Manley, Goldsmith and Schwinger (M2a), the data given in (541a) and the conclusions drawn below seem very unreliable.

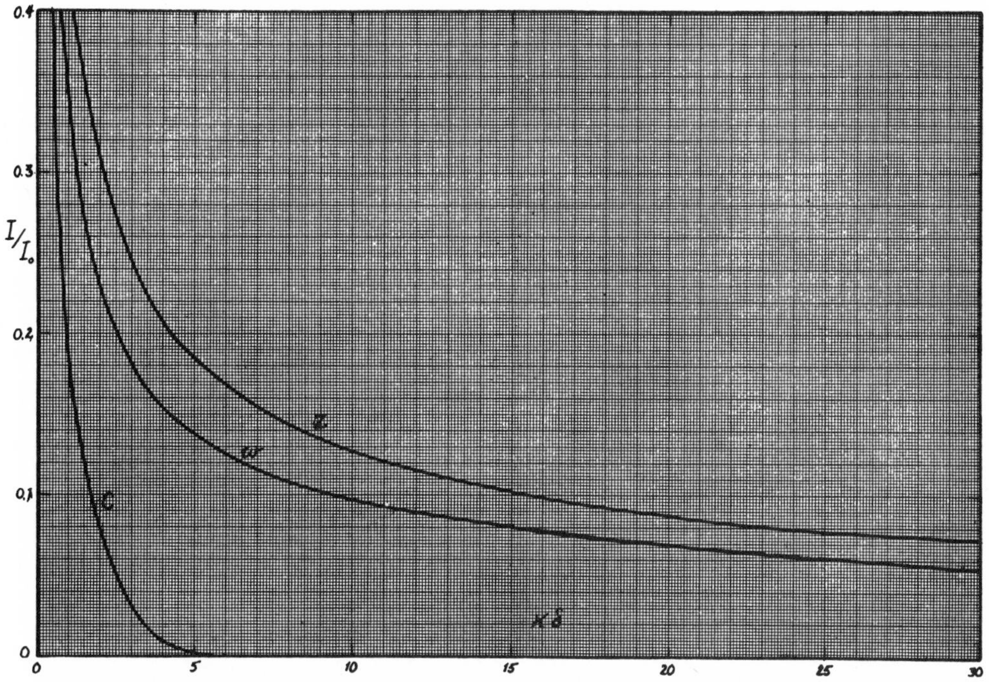

FIG. 16b.

the one-level formula with the use of $K_{\mathrm{Rh\,Rh}} = 2.0$, *viz.*

$$\Gamma_{\mathrm{In}}/\Gamma_{\mathrm{Rh}} = 0.34/0.40 = 0.8. \qquad (540b)$$

The difference in the resonance energies is (cf. (531))

$$|E_{\mathrm{In}} - E_{\mathrm{Rh}}| = \tfrac{1}{2}(\Gamma_1 + \Gamma_2)\left[\frac{2\cdot 2.0\cdot 3.8}{2.0\cdot 3.0 + 1.6\cdot 3.8} - 1\right]^{\frac{1}{2}}$$

$$= \tfrac{1}{2}(\Gamma_1 + \Gamma_2)\cdot 0.50. \qquad (541b)$$

With $\Gamma_{\mathrm{In}} = 0.34$ ev (Table XXV) and $\Gamma_{\mathrm{In}} = \Gamma_{\mathrm{Rh}}$, we have 0.17 volt for the energy difference. The sign cannot be determined from these experiments, but it can be found from the absorption coefficients of other substances, measured with a Rh and In detector. According to Amaldi and Fermi, the absorption coefficient of Hg is larger (0.07 cm^2/g) with Rh than with In (0.04 cm^2/g) as detector, while the ratio is inverted for Ag (0.06 cm^2/g with Rh, 0.09 cm^2/g with In detector) and Au (0.03 and 0.04). Since Hg has a resonance level at very low energies (cf. end of §60, §61G), these measurements suggest that the In level lies higher, in agreement with the conclusions from the boron absorption measurement (Table XXII).

Measurements of the overlapping of levels would be useful (1) to check the level widths obtained by other methods, (2) to check the positions of the levels obtained from the boron method, (3) to test whether the activity of a given substance is due to a single level or to several—the latter conclusion being suggested in any case where there are serious discrepancies in the tests (1) and (2). The "mutual absorption" method may be one of the most sensitive for indicating more than one level.

F. Absorption of resonance radiation

Thus far, we have only considered the absorption coefficient for very thin absorbers. In the case of thick absorbers, the transmitted intensity is given by

$$I(\delta) = \frac{I_0}{\pi}\int \frac{dx}{1+x^2} e^{-2K_r\delta/(1+x^2)}, \qquad (542)$$

where δ is the thickness of the absorber in g/cm^2, K_r the absorption coefficient for thin absorbers (=one-half the absorption for exact resonance, cf. (520)) and $x = (E - E_r)/\tfrac{1}{2}\Gamma$. It has been assumed that the Doppler width is negligible compared to the natural width which is true in every case investigated thus far, except for the A group of Ag.

The transmission coefficient $z(K_r\delta)=I/I_0$ is given in Fig. 16, a and b, as a function of $y=K_r\delta$. We have

$$z(y)=e^{-y}J_0(iy), \tag{542a}$$

where J_0 is the Bessel function of zero order. For small and large thickness, this gives:

$$z=1-y+\tfrac{3}{4}y^2-(5/2)y^3+\cdots \qquad \text{if } y=K_r\delta\ll 1, \tag{542b}$$

$$z=1/(2\pi y)^{\frac{1}{2}} \qquad \text{if } y=K_r\delta\gg 1. \tag{542c}$$

According to (542c), the transmitted intensity decreases very slowly with thickness for thick absorbers.

If the incident neutrons are distributed according to the law $\cos\vartheta+\sqrt{3}\cos^2\vartheta$ (cf. (497)), the transmitted intensity is

$$w(y)=\int_0^{\pi/2}\sin\vartheta d\vartheta(1+3^{\frac{1}{2}}\cos\vartheta) \times[1+\tfrac{1}{2}3^{\frac{1}{2}}]^{-1}z(y/\cos\vartheta), \tag{543}$$

where the detector is again assumed to be thin. w is also given in Fig. 16. For small and large argument we have

$$w=1-0.31\,y(\log y+2.3) +0.40\,y^2(\log y+3.75)+\cdots \quad (y\ll 1) \tag{543a}$$

$$w=0.29\,y^{-\frac{1}{2}} \quad (y\gg 1). \tag{543b}$$

Further corrections must be applied for thick detectors.

Investigations of the whole absorption curve of a given detector for its own resonance radiation might show deviations from the laws (542a) to (543b). Such deviations, in particular a still slower decrease of the intensity at large thicknesses than that indicated by (543b), would indicate that the substance possesses more than one resonance level.

G. Very low resonance levels

As already mentioned at the end of §60, several elements have very low resonance levels, near the thermal region, whose energy cannot be measured directly. A method for determining the position and width of these levels is to measure the deviation from the $1/v$ law in two different energy regions.

The possible experiments are:

(1) In the region of very low energies: Determination of the dependence of the capture cross section on temperature.

(2) Near kT: Wheel experiment of Rasetti and collaborators (R4, R6) described in §58B.

(3) Measurement of the "absorption limit" of the substance in question by determining the boron absorption coefficient of the neutrons penetrating through the substance, using boron or lithium indication (L32).

(4) Determination of the absorption coefficient of the substance for D group neutrons (Rh detector) and comparison with the absorption coefficient for thermal neutrons (A11).

The first two methods have the advantage that they make use of the "one-level dispersion formula" only over a very limited energy region (0 to kT). On the other hand, the experiments are not very accurate; in case (2) because of the smallness of the effect, and in case (1) because it is very difficult to cool neutrons efficiently to a temperature below room temperature (F19) so that it is not clear what to insert for the effective temperature of the neutrons. Method (4) depends on the validity of the one-level formula in the substance investigated up to 1 volt (resonance level of Rh); moreover it is experimentally rather difficult to measure the very small absorption of the substance for the D group. However, such measurements have been carried out by Amaldi and Fermi for Cd and Hg; the absorption coefficients for D neutrons are 0.05 and 0.07 cm²/g, respectively, as compared to 14 and 1 cm²/g for thermal neutrons. Method (3) is, in principle capable of higher accuracy than (1) and (4), but thus far only preliminary data on Cd are available (F33; Hoffman, Livingston and Bethe, H34, L32).

The experiments are conveniently discussed in terms of the capture *probability* p, i.e., the cross section times the neutron velocity, or the absorption coefficient times $E^{\frac{1}{2}}$. We have

$$p(E)\sim\sigma E^{\frac{1}{2}}\sim[(E-E_r)^2+(\tfrac{1}{2}\Gamma)^2]^{-1}. \tag{544}$$

The measurements (1) (3) and (4) give ratios of the capture probabilities at different energies. Besides the absorption coefficient of Cd at thermal energy, $p_{th}=15$ cm²/g, we shall use the "absorption limit" of Cd. This absorption limit has been measured (L32) by determining the boron absorption coefficient of the neutrons just able to penetrate $\frac{1}{2}$ mm of Cd. We may estimate that these neutrons have in Cd an absorption coefficient of about

$$p_L \approx 2 \text{ cm}^2/\text{g}.$$

The boron absorption coefficient was found to be 9.5 cm²/g as compared to 36 cm²/g for thermal neutrons. This means that the energy of the absorption limit is

$$E_L = 0.026 \cdot (36/9.5)^2 = 0.37 \text{ volt.} \tag{544a}$$

According to (544), we have

$$\frac{p_{th}}{p_L} = \frac{15}{2}\left(\frac{kT}{E_L}\right)^{\frac{1}{2}} = \frac{(E_L - E_r)^2 + \frac{1}{4}\Gamma^2}{(E_r - kT)^2 + \frac{1}{4}\Gamma^2}. \tag{544b}$$

Besides, it follows from (544a) that Cd absorbs not only thermal neutrons but also neutrons of considerably higher energy.

Rasetti's wheel experiment gives the logarithmic derivative of the capture probability with respect to the energy, at thermal energy. This derivative can conveniently be expressed by the quantity

$$\epsilon = (dE/d \log p)_{E=kT}, \tag{545}$$

which has the dimension of an energy. If $\Delta p/p$ is the relative change in the absorption coefficient when the direction of rotation of the wheel is reversed, then

$$\epsilon = 2muv_a \sin \vartheta \; p/\Delta p, \tag{545a}$$

where $v_a = (2kT/m)^{\frac{1}{2}}$ is the Maxwell velocity of the neutrons (2200 meter/sec. for room temperature), u the velocity of the edge of the wheel (140 m/sec. in the experiments of Rasetti and collaborators) and ϑ the angle between neutron beam and the normal to the wheel (65° in experiments). The observed value of $\Delta p/p$ was 6.3 percent. Therefore, for Cd,

$$\epsilon = 0.09_5 \text{ volt.} \tag{545b}$$

Theoretically, according to (544) (545)

$$\epsilon = \frac{(E_r - kT)^2 + (\frac{1}{2}\Gamma)^2}{2(E_r - kT)}. \tag{546}$$

ϵ is positive (and therefore $E_r > kT$) if the capture probability increases with increasing neutron energy, i.e., if the absorption coefficient is larger when the wheel moves opposite to the neutron beam.

From (544b) and (546) we have

$$E_r = \frac{E_L^2 + 2\lambda kT - (kT)^2}{2(E_L + \lambda - kT)}, \tag{546a}$$

where

$$\lambda = \epsilon(p_{th} - p_L)/p_L. \tag{546b}$$

We find for Cd

$$E_r = 0.16 \text{ volt,} \tag{547}$$

$$\Gamma = 0.17 \text{ volt.} \tag{547a}$$

This result is, unfortunately, very sensitive to small errors in $\Delta p/p$ (cf. (545a)) and in the "absorption limit" E_L of Cd. E.g., if ϵ is changed to 0.11 volt, the result is $E_r = 0.15_7$ and $\Gamma = 0.21_6$ volt. Therefore, all that can be said at the moment is that E_r and Γ are of the same order of magnitude and are both about 0.15 volt. Similar conclusions are reached by using the absorption coefficient for D neutrons (B15).

The cross section as a function of the energy is given in Fig. 11 (§58), for $\Gamma/E_r = \sqrt{2}$. The curve shows a very broad region of approximately constant cross section which, for Cd, will extend from about thermal energy (0.026 volt) to about 0.2 volt.

Other elements with very low resonance levels are Sm and Hg. The main evidence comes, in both cases, from the temperature dependence of the cross section. Sm, like Cd, shows hardly any change of the absorption coefficient upon cooling; its resonance level should therefore also lie slightly above thermal energies. For Hg, the increase of the absorption at low temperatures is slightly over normal which suggests a resonance level at *negative* energy. The position of this level may be estimated from the ratio of the absorption coefficients for D neutrons and thermal neutrons which is about 1 : 15 (A11) as compared to 1 : 7 for boron and other substances obeying the $1/v$ law. A value $E_r \approx -0.3$ volt would fit the experimental data.

§62. The Neutron Width and the Absolute Cross Section (B15, A11, D23)

The cross section at exact resonance is, according to (430b),

$$\sigma_0=\frac{1.30_5\cdot 10^{-18}}{E_r}\left(1\pm\frac{1}{2i+1}\right)\frac{\Gamma_N}{\Gamma},\qquad (548)$$

where Γ_N is the "neutron width" of the resonance level, Γ the total (γ-ray) width, E_r the resonance energy, i the spin of the capturing nucleus. The $\pm$ sign stands according to whether the angular momentum of the compound state (resonance level) is $i+\frac{1}{2}$ or $i-\frac{1}{2}$; since we have no way of knowing this angular momentum, we neglect the term $1/(2i+1)$. σ_0 is measured in cm², E_r in volts. The absorption coefficient, in cm²/g, at exact resonance becomes then

$$K_0=\frac{\sigma_0 La}{A}=7.87\cdot 10^5\frac{a}{A}\frac{\Gamma_N}{\Gamma E_r},\qquad (548a)$$

where A is the atomic weight of the capturing element, a the abundance of the capturing isotope in the element in question, and L Avogadro's number.

The *observed* resonance absorption coefficient with self-indication is connected to K_0 by [cf. (520) (521a)]

$$K_r=(\pi\Gamma/2\Gamma_{\rm eff})K_0,\qquad (548b)$$

so that

$$\Gamma_N=\frac{E_rAK_r\Gamma_{\rm eff}}{1.23\cdot 10^6 a}.\qquad (549)$$

If the total activity is known, $\Gamma_{\rm eff}$ can be expressed by (528c) so that:

$$\Gamma_N=\frac{E_r{}^2A}{1.23\cdot 10^6 a}\frac{3^{1/2}L}{l_0}K_{\rm th}Y.\qquad (549a)$$

In this formula, only directly observable quantities occur. However, in many cases the resonance absorption coefficient is known when the total activity is unknown. In this case, (549) may be used directly, and $\Gamma_{\rm eff}$ may be replaced by $\pi\Gamma$ if it is known that the natural width is large compared to the Doppler width (cf. §61A). The results are given in Table XXVI. For Cd, the absorption coefficient at resonance is about 1.7 times that at thermal energies, if the data (547, 547a) are used. The isotope responsible for the large cross section is probably one of odd mass (cf. B13, p. 340). Both the odd isotopes (111, 113) have about the same abundance (10 and 12 percent) so that $A/a\sim 1000$.

The neutron widths observed vary mostly between 0.5 and 2 millivolt, except for iodine for which the data are uncertain because of the superposition of several levels (§60A, 61C). The neutron widths are very fundamental for nuclear theory because they are the only existing data on the particle width for heavy nuclei. They form the basis of the calculation of the nuclear radius from the α-decay (§67) and also the basis for the estimate of the "sticking probability" (§54) of nuclear particles for which they give a value of 0.01 to 1.

According to the theory (§58), the neutron widths should *cet. par.* increase as the velocity of the neutrons so that $\Gamma_N'=\Gamma_N E_r{}^{-1/2}$ should, in the average, be independent of E_r. In fact, when the very uncertain value for Cd is excluded no trend of $\Gamma_N E_r{}^{-1/2}$ with the energy can be observed in the data in Table XXVI, all values lying in the very small region between 0.4 and $1.2\cdot 10^{-3}$ volt$^{1/2}$.

Inserting $\Gamma_N'E_r{}^{1/2}$ for the neutron width Γ_N, we find for the cross section at exact resonance (548)

$$\sigma_{0r}\sim\Gamma_N'/\Gamma E_r{}^{1/2}\qquad (550)$$

Table XXVI. *Neutron widths.*

Substance	Rh 44″	In 54′	Ir	Au	Ag(A)	Ag(B)	I	Cd
E_r	1.3	1.6	1.9	3.0	3.0	5.5	140	0.16
A/a	100	120	190	200	220	220	130	∼1000
K_r (cm²/g)	19	3.8	1.0	4	20	0.3	0.38	∼25
$\Gamma_{\rm eff}$* (volts)	0.22*	1.1	3.1	1.1	0.11*	5.3	∼1†	0.55
Γ_N (millivolts)	0.44	0.65	0.9	2.1	1.2	1.6	6	1.8
$1000\Gamma_N E_r{}^{-1/2}$ (volt$^{1/2}$)	0.4	0.5	0.6	1.2	0.7	0.7	0.5	4.5
$1000\Gamma_N/\Gamma$	6	1.8	0.9	6	>30	0.9	?	12

* When $\Gamma_{\rm eff}$ was obtained from the activity, the value is marked by an asterisk. The other values of $\Gamma_{\rm eff}$ are from the one-level formula (§61D)
† $(2\pi)^{1/2}\Delta$ (cf. Table XXIII). It is assumed that the width of the iodine levels is primarily determined by Doppler broadening.

and for the total activity (531)

$$B_r \sim \Gamma_N' E_r^{-\frac{1}{2}}. \qquad (550a)$$

Therefore, if we assume Γ_N' and Γ to be in the average independent of the number r of the resonance level, the absorption coefficient for neutrons at exact resonance becomes inversely proportional to the square root of the resonance energy E_r. This "$1/v$ law" holds also for the *average* absorption coefficient, averaged over an energy region large compared to the spacing D of the resonance levels which is (cf. §56)

$$\sigma_a = (\sigma_{0r}\pi\Gamma/2D)_{\text{Av}} \sim \Gamma_N'/DE^{\frac{1}{2}}. \qquad (550b)$$

From (550a) it follows that the contribution of a resonance level E_r to the total activity is, under otherwise equal conditions, proportional to $E_r^{-\frac{1}{2}}$, a fact which we used in interpreting Table XXII.

In the last row, we have tabulated the ratio of the neutron width to the γ-ray width. This ratio determines the ratio of resonance scattering to capture (§58, 63). The ratio is seen to be extremely small, of the order of one-thousandth, except for Ag (A group). Only for this latter element may we thus expect to find observable resonance scattering, and even there the predicted magnitude of the scattering is very low ($\sigma_{\text{sc}}/\sigma_{\text{capt}} > 1/30$).

The values given for the neutron width should be very reliable for Ag (A group) and Rh 44″ (D group) because in these cases the total activity has been determined. The other neutron widths (In, Ir, Au and Ag B) depend on the applicability of the one-level dispersion formula to thermal neutrons (§61D), which, however, is probably justified in these cases. The data for iodine and cadmium are rough estimates.

The cross sections observed at resonance are very large, ranging up to about 10^{-20} cm². They are usually much larger than at thermal energies so that many resonance groups may be recognized by the fact that the activity transmitted by cadmium is more strongly absorbed in the substance of the detector itself than the part of the activity which is absorbed in cadmium. Table XXVII gives the known cross sections for thermal neutrons, resonance neutrons and "fast" neutrons (one to several MV). The measurements of thermal and fast neutrons are mostly due to Dunning, Pegram, Fink and Mitchell (D23) supplemented by data of Amaldi and Fermi (A11), Hevesy and Levi (H31) and of Powers, Fink and Pegram (P14). The resonance data are due to Amaldi and Fermi. The figures refer to the cross section at *exact* resonance which is twice the observable cross section for thin absorbers.

TABLE XXVII. *Neutron cross sections (in 10^{-24} cm²).*

Element	Main Process	Cross Section: Therm.	Res.	Fast
1 H	E	40*	14*	1.7
1 D	E	4.0		1.7
3 Li^6	α	900†		1.8
4 Be	E	5.3		1.5
5 B^{10}	α	3000†		1.6
6 C	E	4.1		1.3
7 N	$E+p$	11.3		1.8
8 O	E	3.3		
9 F		2.5		
11 Na		4.2		
12 Mg		3.5		
13 Al		1.5		2.4
14 Si		2.5		
15 P		14.7		
16 S		1.4		2.6
17 Cl		39		
19 K		8.2		
20 Ca		11.0		
21 Sc		25?		
22 Ti		11.9		
23 V		10		
24 Cr		4.9		
25 Mn		14.3		
26 Fe	$E+C$	12.0		3.0
27 Co		35		
28 Ni	$E+C$	15.4		
29 Cu		7.5		3.2
30 Zn		4.7		3.3
32 Ge		~75		
33 As		9		
34 Se		19		
35 Br		12		
38 Sr		~9		
39 Y		30		
40 Zr		17		
41 Cb		~14		
42 Mo		7		
44 Ru		12		
45 Rh	44″	125	6000	
46 Pd		10		
47 Ag	22″A†	} 80	14000	
	22″B†		200	
	2.3′†	30		
	total	55		
48 Cd	C	2600	(~80000†)	
49 In	54′†	$85/a_1$	$1500/a_1$	
	16″†	$55/a_2$		
	total	140		
50 Sn		4.0		4.3
51 Sb		8		
52 Te		8		
53 I		9.4	150	4.6
56 Ba		140		
57 La		80		
58 Ce		~25		
59 Pr		25		
60 Nd		220		
62 Sm	C	4300		
63 Eu		2500		
64 Gd		30000		
65 Tb		<1000		
66 Dy		700		
67 Ho		150		
68 Er		120		
69 Tu		500		
70 Yb		90		
71 Lu		~400		
73 Ta		27		
74 W		23		5.3
75 Re		90		
76 Os		27		
77 Ir	C	280	600	
78 Pt		25		
79 Au	C	90	2500	
80 Hg	C	440		5.8
81 Tl		11		
82 Pb		9		5.7
83 Bi		8		
90 Th		32		
92 U		43		

* Cf. §59C. The figure for "resonance" refers to slow neutrons above the thermal region.

† Cross section of the isotope responsible for the absorption. In the case of indium, a_1 and a_2 give the abundance of the two isotopes; it is not known which isotope is responsible for each of the two activities. For Ag, the abundance was taken as 50 percent for each of the two isotopes. For Cd, 12 percent was assumed.

C = capture,
E = elastic scattering,
α, p = α and proton emission.

Remarks on Table XXVII:

The *main process* responsible for the observed absorption *of thermal neutrons* is indicated wherever there was direct experimental evidence about it. Thus hydrogen, deuterium, carbon and oxygen are known to capture neutrons only very slightly, showing that the cross section observed is mainly due to elastic scattering (E). For cadmium, it has been observed that the scattering is extremely small compared to the capture (C) (cf. §63). On the other hand, iron and nickel are known from scattering experiments to have large elastic scattering besides capture (§63). Mainly capture has also been assumed in all cases where the cross section is known to change rapidly with energy and to show marked resonance effects such as Rh, Ag, In, Sm. Generally, capture may be assumed whenever the cross section is large (more than, say, $10\cdot 10^{-24}$ cm^2) for not too light nuclei ($Z>10$, say); but in these cases no remark about the main process has been made in the table. Small cross sections are usually due to scattering. α-disintegration is produced by slow neutrons in Li and B, proton disintegration in nitrogen.—For fast neutrons, scattering, α- and proton disintegration occur.

The data about the rare earths are in general not very reliable because of the difficulties in purification. The cross sections given for some of the rare earths may be spurious due to impurities of Gd, Sm or Eu.

§63. Scattering of Slow Neutrons (D23, M17–20, B5)

A

The elastic cross section for neutrons ($s=\frac{1}{2}$), including potential scattering, is in the case when only one resonance level is important:

$$\sigma^N{}_N=\frac{\pi}{2(2i+1)}\left[(2J+1)\left|2R+\lambda\!\!\!^{-}_r\frac{\Gamma_N}{E-E_r+\frac{1}{2}i\Gamma}\right|^2+(4i+1-2J)4R^2\right] \quad (551)$$

$$=4\pi R^2+\frac{\pi}{2}\left(1\pm\frac{1}{2i+1}\right)\lambda\!\!\!^{-}_r\Gamma_N\frac{4R(E-E_r)+\lambda\!\!\!^{-}_r\Gamma_N}{(E-E_r)^2+\frac{1}{4}\Gamma^2}.$$

Here R is the (effective) nuclear radius, $\lambda\!\!\!^{-}_r$ the wave-length at exact resonance, E_r the resonance energy, Γ_N the neutron width and Γ the total width of the resonance level. In the first line, the first term in the square bracket gives the effect of resonance and potential scattering for angular momentum J, the second term is the potential scattering for the other value of the angular momentum possible by selection rules, *viz.* $2i-J$, for which there is no resonance scattering. In the last line of (551), the first term represents the potential scattering, the second the resonance scattering and the interference term between the two.

The behavior of the cross section (551) near the resonance level E_r will be governed by the ratio

$$\rho=\lambda\!\!\!^{-}_r\Gamma_N/R\Gamma. \quad (552)$$

This ratio determines the ratio of the cross section at exact resonance to that of the potential scattering alone,

$$\sigma_r/\sigma_p=1+\tfrac{1}{2}(1\pm 1/(2i+1))\rho^2. \quad (552a)$$

According to Table XXVI, the ratio Γ_N/Γ in (552) is of the order 10^{-3} for most resonance levels observed, at resonance energies of a few volts, corresponding to $\lambda\!\!\!^{-}_r\sim 3\cdot 10^{-10}$ cm. With $R\sim 10^{-12}$ cm, this gives values of ρ smaller than unity. This means that the resonance scattering will, for the resonance levels listed in Table XXVI, not differ markedly from the potential scattering. Only for cases like Ag A (cf. Table XXVI) where Γ_N/Γ is about 30 times larger, may we get an appreciable increase of the scattering cross section at resonance, to 100 or more times the potential scattering.

Introducing the abbreviation ρ, and writing

$$x=(E-E_r)/\tfrac{1}{2}\Gamma, \quad \sigma_p=4\pi R^2, \quad (553a)$$

we have

$$\frac{\sigma^N{}_N}{\sigma_p}=1+\tfrac{1}{2}\left(1\pm\frac{1}{2i+1}\right)\rho\frac{\rho+2x}{1+x^2}. \quad (553)$$

This cross section has a minimum and a maximum near the resonance energy E_r. These extrema occur at

$$x_{\min}{}^{\max}=-\tfrac{1}{2}\rho\pm(\tfrac{1}{4}\rho^2+1)^{\frac{1}{2}} \quad (553b)$$

and have the values

$$\sigma_{\min}{}^{\max}/\sigma_p=1+\tfrac{1}{2}(1\pm 1/(2i+1))\rho[\pm(\tfrac{1}{4}\rho^2+1)^{\frac{1}{2}}+\tfrac{1}{2}\rho]. \quad (554)$$

If $\rho\ll 1$, the resonance will have no appreciable effect on the scattering. We have a minimum at $E=E_r-\frac{1}{2}\Gamma$ and a maximum at $E_r+\frac{1}{2}\Gamma$, but they will differ only by the relative amount $\rho(1\pm 1/(2i+1))$. Essentially we have simply the potential scattering $\sigma_p=4\pi R^2$. Even for $\rho\gg 1$, this potential scattering will be the main term if the energy E is far from resonance. In almost all elements, the scattering observed for thermal neutrons will be largely potential scattering.

If ρ is large, the minimum of the cross section occurs at

$$x_{\min}=-\rho; \quad E_{\min}=E_r-\Gamma_N\lambda\!\!\!^{-}_r/2R \quad (555a)$$

and has the value

$$\sigma_{\min}=\tfrac{1}{2}\sigma_p(1\mp 1/(2i+1)). \quad (555)$$

The maximum occurs practically exactly at E_r, its value is given by (552a).

However, as already mentioned, the resonance scattering will in general be negligible compared to the potential scattering, except in such extraordinary cases as the A level of Ag (22″) where the total width is unusually small and therefore the neutron width a large percentage of the total.

Thus, as a rule, we can consider the scattering cross section as *independent* of the neutron energy.

B

The measurement of the scattering offers some experimental difficulties, mainly because it must be made sure that the observed scattering is actually due to the scatterer investigated. Ordinarily, when scattering is measured, the detector is shielded against the direct neutron beam by absorbing materials. Even then, there will be some activity induced in the detector even without the scatterer in position, because there is always some "stray" scattering material around. It is, in this case, not always sufficient simply to subtract the activity without scatterer from that measured with the scatterer in place, because of the possibility of double scattering from the stray material and the scatterer investigated. All substances containing hydrogen (paraffin, water, wood, etc.) are, of course, particularly dangerous in this respect.

The dependence of the scattering on energy may be measured[42] by using suitable detectors, such as Ag and Rh, and absorbers, such as Cd or the detector material itself. In such experiments, the change of neutron energy in the scattering process must be considered; but this represents a negligible correction unless the scattering cross section depends sensitively on energy in the region investigated, i.e., in all cases except the resonance scattering.

For the resonance scattering, we have to find out whether the change of the neutron energy in the scattering process will throw the neutrons out of the resonance region so that they can no longer be detected by a detector of the same material as the scatterer. For backward scattering by an atom of weight A, the decrease in the netron energy is $4E_r/A$ where E_r is the original neutron energy. The energy loss in the scattering will be immaterial if it is smaller than the effective width of the neutron level:

$$\Gamma_{\text{eff}} \gg 4E_r/A. \qquad (556)$$

This is fulfilled for all neutron levels listed in Table XXVI, except for the A level of Ag, for which the two quantities are of the same order. But only in this last case is the investigation of the resonance scattering of interest, because in all other cases the resonance scattering is expected not to differ appreciably from the ordinary potential scattering (cf. above). It is generally true that the resonance scattering is large only if Γ is small (cf. (552)), but that small Γ means, at the same time, that the condition (556) will in general not be fulfilled. Thus, whenever the resonance scattering is large, the neutrons will be thrown off resonance by the scattering process itself, so that the scattered intensity can no longer be measured by a detector of the same material as the scatterer.

[42] In these experiments, even greater care has to be taken to avoid secondary scattering, especially by substances containing hydrogen. Firstly, the scattering of hydrogen is known to depend sensitively on energy (chemical bond, cf. §59C), and secondly, the scattering by hydrogen means a considerable energy loss. Thus neutrons of high energy may be scattered by the primary scatterer, then scattered again by the secondary scatterer and slowed down to the resonance energy of the detector. Then the detector would measure the scattering cross section of the scatterer for a much higher energy than the detector's resonance region.

C

We want to discuss the actual evaluation of scattering experiments for the case of backward scattering, assuming there is no secondary scattering material. For simplicity, we shall treat the problem in one dimension, and shall later discuss the three dimensional problem in some special cases.

Let $f_+(x)$ be the number of neutrons moving to the right (away from the source) at a depth x from the entrance plane of the scatterer, and f_- the number moving to the left. Then of the neutrons moving to the right, a fraction a will be absorbed (captured) per unit path, and a fraction $\frac{1}{2}s$ will be scattered back where a is the absorption and s the scattering coefficient. ($\frac{1}{2}s$ enters because, with isotopic scattering, half of the scattered neutrons will still move forward.) In this way, we find the differential equations

$$\begin{aligned} df_+/dx &= -(a+\tfrac{1}{2}s)f_+ + \tfrac{1}{2}sf_-, \\ -df_-/dx &= -(a+\tfrac{1}{2}s)f_- + \tfrac{1}{2}sf_+. \end{aligned} \qquad (557)$$

The solution is

$$\begin{aligned} f_+ &= \alpha s e^{-\lambda x} + \beta s e^{\lambda x}, \\ f_- &= (2a+s-2\lambda)\alpha e^{-\lambda x} + (2a+s+2\lambda)\beta e^{\lambda x} \end{aligned} \qquad (557a)$$

with

$$\lambda = (s+a)^{\frac{1}{2}} a^{\frac{1}{2}}. \qquad (557b)$$

α and β are two integration constants. If the thickness of the absorber is d, we must have

$$\beta = -\alpha e^{-2\lambda d}(2a+s-2\lambda)/(2a+s+2\lambda). \qquad (557c)$$

The intensity of the back scattering (reflection coefficient) is

$$R = \frac{f_-(0)}{f_+(0)} = \frac{s(1-e^{-2\lambda d})}{2a+s+2\lambda-(2a+s-2\lambda)e^{-2\lambda d}}. \qquad (558)$$

For small thickness, this reduces to the well-known expression

$$R = \tfrac{1}{2} sd. \tag{558a}$$

For large thicknesses, we have

$$R = \frac{s}{2a+s+2\lambda} = \frac{2a+s}{s} - 2\left(\frac{a}{s}\frac{a+s}{s}\right)^{\frac{1}{2}}, \tag{559}$$

a formula first derived by Bayley and Goudsmit (B5). Thick scatterers are, in general, more convenient especially for strong neutron absorbers, of which extremely thin layers would have to be taken in order to make (558a) valid. According to (559), the back scattering from thick scatterers gives directly the ratio of scattering to absorption (capture) cross section $\sigma_s/\sigma_a = s/a$. If $\sigma_s \gg \sigma_a$ (good scatterers), (559) reduces to[43]

$$R = 1 - 2(\sigma_a/\sigma_s)^{\frac{1}{2}} \qquad \sigma_a \ll \sigma_s, \tag{559a}$$

whereas for small scattering cross section

$$R = \sigma_s/4\sigma_a \qquad \sigma_a \gg \sigma_s. \tag{559b}$$

These last two formulae may easily be checked by a three-dimensional calculation. The case of small absorption was treated in §59F for paraffin. For the reflection coefficient (albedo) we found (cf. (503)) exactly the formula (559a) [N in (503) is defined (cf. 484) as σ_s/σ_a]. For strong absorption, the observed back scattering will depend on the angular distribution of the incident neutrons. We may treat the cases

(a) of a collimated beam striking the scatterer perpendicularly

(b) of an angular distribution $f(\vartheta) \sim \cos\vartheta + \tfrac{3}{2}c\cos^2\vartheta$ of the incident neutrons (c a constant). In case a, the number of neutrons arriving at a depth x, is e^{-ax}. Of these, a fraction $\tfrac{1}{2}s \sin\vartheta d\vartheta$ is scattered into the direction ϑ. The probability that the scattered neutrons can escape from the scatterer, is $e^{-ax/\cos\vartheta}$. Thus we have

$$R = \tfrac{1}{2}s\int_0^{\pi/2} \sin\vartheta d\vartheta \int_0^\infty e^{-ax(1+1/\cos\vartheta)}dx,$$
$$= (s/2a)(1-\log 2) = 0.153 s/a. \tag{560}$$

In case (b), a somewhat longer calculation gives

$$R = \frac{\tfrac{2}{3}(1-\log 2)+3c/16}{1+c}\frac{s}{a} = \frac{0.205+0.187c}{1+c}\frac{s}{a}, \tag{560a}$$

so that R is approximately equal to $\sigma_s/5\sigma_a$, in close but not exact agreement with the "one dimensional" formula (559b).

D

Actual *experimental data* on the scattering are very scarce. Measurements have been made by Dunning, Pegram, Fink and D. P. Mitchell (D23) and by A. C. G. Mitchell, Murphy, Langer and Whitaker (M17–20). Dunning and collaborators measured the scattering from cadmium and found that the cross section must be less than 1 percent of the capture cross section. This was of great importance in disproving the one-body theory of neutron phenomena (cf. §57), but is in agreement with the present many-body theory. According to Table XXVI, we expect the resonance scattering at thermal energies to be about 0.45 percent of the capture cross section (with a considerable uncertainty), while the cross section due to potential scattering may be estimated as 6 to $8\cdot 10^{-24}$ cm² which is 0.2–0.3 percent of the capture cross section.

Mitchell and his collaborators showed that the scattering of thermal neutrons is very small from Ag and Cd but large from C, Mg, Al, S, Fe, Ni, Cu, Zn, Sn, Pb and Bi, with Cr, Mn and Hg giving medium values. This is to be expected from the magnitude of the total absorption cross sections listed in Table XXVII. Assuming a scattering cross section of a few times 10^{-24} in each case, and attributing the rest of the observed absorption to capture, we find that the cross sections of absorption and scattering will be comparable for all the substances for which "large scattering" was found in the experiments but that the scattering should be negligible compared to the capture in Ag and Cd. A quantitative interpretation of the experiments of Mitchell and Murphy seems not possible because the scatterer was placed on top of a paraffin block so that the scattered neutrons could again be scattered by the paraffin, then again by the substance etc., thus increasing greatly the observed scattering coefficient (B5).

[43] In this case, measurements of the back scattering will serve to determine the *capture* rather than the scattering cross section, the scattering cross section being practically equal to the total cross section measured in absorption experiments.

§64. Disintegration by Slow Neutrons with Emission of Charged Particles (A8, C12)

Disintegrations by slow neutrons with emission of charged particles have been found in three cases:

$$Li^6 + n^1 = H^3 + He^4, \tag{I}$$

$$B^{10} + n^1 = Li^7 + He^4, \tag{II}$$

$$N^{14} + n^1 = C^{14} + H^1. \tag{III}$$

Probably the process

$$B^{10}+n^1=Be^{10}+H^1 \qquad \text{IV}$$

is also possible. All other processes giving protons are energetically impossible with slow neutrons (§57D), although they are caused with great probability by fast neutrons. Other reactions giving α-particles also seem energetically impossible, on the basis of the atomic masses, for all elements up to phosphorus except for the rare isotopes Ne^{21} and Si^{29} (Table LXI). Even for these, the energy of the emitted α-particle would be so small (1.0 and 1.7 MV, respectively) that it would have a very small probability of escaping through the potential barrier.

The processes are observed through the ionization produced by the charged particle emitted, *viz.* H^3, He^4 and H^1 in the three processes mentioned. They are used as sensitive detectors for slow neutrons, their absorption coefficient following the $1/v$ law (§58B). The absorption in boron is used as an energy gauge for slow neutrons (§60A).

In the reactions I and II above, the particles produced are rather fast, *viz.* 4.6 MV and 3.0 MV, respectively (total kinetic energy of both particles). They will therefore go over the top of the Coulomb potential barrier.[44] Therefore it is reasonable to assume that the "widths" Γ_Q corresponding to the emission of these charged particles are of the same order as for neutrons of the same energy. With this assumption and assuming the width to be proportional to the velocity, we have then for the neutron width

$$\gamma_N(E)=c(E/E_Q)^{\frac{1}{2}}\Gamma_Q, \tag{561}$$

where E is the neutron energy, E_Q the particle energy and c a constant of order unity. Using the one-level formula (262), we have therefore:

$$\sigma^N{}_Q=\tfrac{1}{2}\pi\lambda^2\left(1\pm\frac{1}{2i+1}\right)\frac{\gamma_N\Gamma_Q}{(E-E_r)^2+\frac{1}{4}\Gamma^2}, \tag{561a}$$

$$\approx 2\pi c\lambda\lambda'\frac{\Gamma^2}{4(E-E_r)^2+\Gamma^2}, \tag{561b}$$

where λ' is the wave-length of a neutron of energy E_Q. Γ has been put equal to Γ_Q, and $1/(2i+1)$ has been neglected. With $E_Q=3.0$ MV and a neutron energy of $kT=0.026$ volt, we have $\lambda'=2.6\cdot 10^{-13}$ cm and $\lambda=2.8\cdot 10^{-9}$ cm. Therefore, neglecting $E=kT$ compared to E_r, we obtain

$$\sigma^N{}_Q=5\cdot 10^{-21}c/[1+4E_r^2/\Gamma^2]. \tag{562}$$

The observed cross section of B^{10} for thermal neutrons is

$$\sigma^N{}_Q=3.0\cdot 10^{-21}\ \text{cm}^2 \tag{562a}$$

(cf. Table XXVII). From a comparison of (562) and (562a) we conclude that the resonance level E_r must be of the same order as Γ, which may, by analogy to other levels of light nuclei, be assumed to be of the order of 100,000 volts. For the compound nucleus B^{11} this means an excited level with an excitation energy approximately equal to the difference in mass between $B^{10}+n^1$ and B^{11}, i.e. (cf. Table LXXIII)

$$B^{10}+n^1-B^{11}=11.5\ \text{MV}. \tag{562b}$$

The cross section of Li^6 for thermal neutrons is smaller, *viz.*

$$0.9\cdot 10^{-21}\ \text{cm}^2. \tag{562c}$$

Therefore the lithium resonance level should be somewhat farther away from zero neutron energy than the boron level. The corresponding resonance level of the Li^7 compound nucleus lies at an excitation energy around 7 MV.

In the case III above, the protons emitted are very slow, having an energy of about 0.62 MV (including C^{14} recoil, cf. §102B). The potential barrier of C^{14} for protons is probably (599) about 1.7 MV, i.e. much higher than the proton energy. The penetrability of the potential barrier for protons of 0.62 MV is about (cf. 600, Fig. 18)

$$P=e^{-3\cdot 3}=1/27. \tag{563a}$$

The proton width will then be

$$\Gamma_Q=G_QP, \tag{563}$$

where P is the penetrability of the barrier and G_Q the proton width without barrier. If $E_r \ll \Gamma$ which is certainly true in our case, the cross section is proportional to Γ_Q (cf. (561a)). The observed total cross section is $11\cdot 10^{-24}$ cm² (Table XXVII), including the elastic scattering which may be $3\cdot 10^{-24}$ cm². This leaves $8\cdot 10^{-24}$ cm² for the cross section of process III, which is just one percent of the cross section of Li^6. This is about what would be expected from the penetrability of the proton of 1/30, and from the fact that even

[44] From the general formula (599), the barriers would be 1.4 and 2.2 MV in the two cases.

G_Q must be expected to be smaller than the Γ_Q for the Li^6 reaction because the width without barrier is known to increase with the energy of the particle. Thus the observed cross section of N^{14} for slow neutrons may be interpreted by assuming a resonance level at about the same neutron energy as in Li^6, probably a few hundred kilovolts.

All the numerical values given in this section should be regarded as estimates only; actually at least the cross sections at two different neutron energies (e.g. zero and a few hundred kilovolts) are necessary to determine width and position of the resonance levels.

§65. Fast Neutrons (B32, B33, W7, D2, E1, A7, K7, K9, L19, F13)

A. Classification of processes

Fast neutrons, of energies of several MV, may interact with nuclei in a great variety of ways, *viz.:*

(*a*) Elastic scattering.

(*b*) Inelastic scattering, the initial nucleus being left in an excited state.

(*c*) Simple capture, leading to the formation of an isotope one unit higher in atomic weight.

(*d*) Disintegration with emission of α-particles.

(*e*) Disintegration with emission of protons.

The processes *c*, *d*, *e* may be observed very easily if the nucleus formed in the transmutation is radioactive. Many instances of such transmutations have been found, particularly by Fermi and his collaborators (F13, cf. §102 of this report). If the resultant nucleus is stable, the processes *d*, *e* may be ascertained by observing the particles emitted in the process itself (α-particles or protons). This is usually done in a cloud chamber (cf. §94).

Process *b* (inelastic scattering) is certainly very probable for all nuclei of medium and high atomic weight. This was first shown by Danysz, Rotblat, Wertenstein and Zyw (D2) and then confirmed by Amaldi, d'Agostino, Fermi, Pontecorvo, Rasetti and Segrè (A7) and by Ehrenberg (E1). These authors investigated the change of the radioactivity induced in various substances due to interposing various "scatterers" between a source of fast neutrons and the detector. Scatterers made of C, SiO_2, Ag, Au and Pb produced a marked increase in the radioactivity induced in Ag and Rh indicators but a decrease in the radioactivity in Al and Si. The two latter detectors are made radioactive by processes *d* and *e* mentioned above (α- and proton emission after the capture of the neutron) which, on energetic grounds, are only possible with fast neutrons. On the other hand, the radioactivities in Ag and Rh are produced by radiative neutron capture which is much more probable with slow neutrons (cf. below). Thus the experiments can simply be interpreted as showing that *all the scatterers investigated are very effective in slowing down the neutrons.*

This would be quite impossible if all collisions between the neutron and the scattering nuclei were elastic, because then the maximum possible decrease in energy in one collision would be only $4/A$ times the initial neutron energy, for a scatterer of atomic weight A. Such a decrease would not be noticeable at all for heavy substances such as Ag, Au and Pb. It might be objected that a large number of elastic collisions, even with a heavy substance, would slow down the neutrons: but with the thicknesses of material used (1–3 cm) and the scattering coefficient for fast neutrons known from other measurements (0.2–0.4 cm^{-1}), ordinarily only one collision can take place. This was also proved directly by Ehrenberg, who found a linear increase of the activity of a Ag detector with the thickness of the Ag scatterer interposed.

The slowing down must thus be ascribed to *inelastic scattering* of the neutrons.[45] The nucleus is excited to some excited state while the neutron loses the corresponding amount of kinetic energy. After the neutron has left, the nucleus will lose its excitation energy by emitting one or several γ-rays. These γ-rays from the "noncapture" excitation of nuclei by fast neutrons have actually been found by Lea (L19) and by Kikuchi, Aoki and Husimi (K7, K9) and their intensity has been measured.

No direct evidence is available for the elastic scattering (process *a*), but it may safely be assumed to occur by analogy with the inelastic scattering.

---

[45] Inelastic scattering seems much more likely than "radiative scattering," i.e., a process in which an incident fast neutron produces a slower neutron plus a γ-ray in the field of the nucleus which itself is not excited in the process.

B. Theory

The theoretical treatment of fast neutron effects must, of course, again be based on the general theory developed in §52. If the spacing of the nuclear levels is comparable to the neutron energies used, resonance effects may be observed. It was pointed out by Gamow (G5) that some of the resonance levels to be expected could be predicted from the known resonance levels in α-particle reactions. E.g., in the bombardment of Si^{28} by neutrons, the compound nucleus Si^{29} is formed whose resonance levels are known from the reaction $Mg^{25}+He^{4}\rightarrow Si^{29*}\rightarrow Al^{28}+H^{1}$ (observable through the radioactivity of Al^{28}). A great number of similar examples could be given. However, it is not quite certain whether the neutron resonances thus predicted will actually be observable: The α-particles producing the resonances have energies less than or comparable to the potential barrier, and therefore can only penetrate into the nucleus when they have small orbital momentum (§78) while the neutrons have high energy and no potential barrier and therefore can have high orbital momentum. This means that many more resonance levels will interact with neutrons than with α-particles, which may smooth out the resonances in the neutron case.

For heavier nuclei, the nuclear energy levels will become very closely spaced and it will no longer be possible to define the energy of the incident neutron accurately enough to observe any resonances. All we can do in this case is to observe an *average* value of the cross section, averaged over an energy region large compared to the spacing between the levels as discussed in §56.

Applying formula (407) of §56 to our case, and summing over all states q of the final nucleus, we obtain

$$\sigma^{N0}{}_{Q}=\pi R^{2}\xi\Gamma_{Q}/\Gamma \quad (564)$$

for the total probability of the production of particle Q by a fast neutron. If we further sum over all possible particles Q which may be produced we find

$$\sigma_{N0}=\pi R^{2}\xi. \quad (565)$$

This expression does not contain the potential scattering σ_{pot}. Since the mixed terms due to interference of potential scattering and resonance elastic scattering (cf., e.g., the term $4R(E-E_0)$ in (551)) vanish upon integration over a large energy interval, we have for the total cross section

$$\sigma_{N0}'=\pi R^{2}\xi+\sigma_{\text{pot}}. \quad (566)$$

This cross section will determine the total "absorption" of fast neutrons as measured by the decrease in the intensity of a collimated beam of fast neutrons when passing through matter. On the other hand, the elastic scattering may be practically eliminated from the measured "absorption" by surrounding, e.g., the source of fast neutrons with a sphere of the material to be investigated and using a detector which responds only to fast neutrons. Now the elastic scattering will be practically equal to the potential scattering.[46] Therefore, if we eliminate experimentally the elastic scattering, we shall measure the cross section (565), which gives immediately the "sticking probability" ξ.

Experimental data

Quantitative data on fast neutrons are not easy to interpret. The reasons are firstly, that the neutrons obtained from most sources are not monochromatic so that it is not known to which neutron energy the results refer. Secondly, the methods for detecting fast neutrons vary enormously in efficiency with changes in the neutron energy: E.g., the number of recoil protons formed by a neutron in a thick layer of a hydrogenous substance is about proportional to the neutron energy (§94). Endoergic reactions such as $Si^{28}+n^{1}=Al^{28}+H^{1}$ are only possible above a certain neutron energy, and their yield will depend on the energy even at higher energies. For most of these reactions, the dependence of the yield on the neutron energy is unknown so that no exact interpretation is possible. The experimental data can thus give only qualitative ideas about the probability of various neutron processes.

C. The cross section for inelastic scattering

The cross section for inelastic scattering may be deduced either from measurements of the γ-rays emitted by the nuclei after excitation by

[46] The resonance part of the elastic scattering is certainly for fast neutrons quite small compared to the inelastic scattering, because there is only one possible final state, the ground state, in the elastic resonance scattering but a great number of final states are possible for the inelastic scattering.

fast neutrons (L19, K7, K9) or from measurements of the total absorption of fast neutrons after elimination of the elastic scattering.

The *production of γ-rays* by fast neutrons was first measured by Lea (L19), using neutrons from a Po+Be source which have in the average energies of the order of 5 MV. He deduced from his experiments a cross section of about 2 or $3 \cdot 10^{-24}$ cm^2 for Fe and Pb, and much less for C. He could also show that the γ-rays were in all probability due to inelastic scattering rather than capture of the neutrons. This is confirmed by the experiments on the slowing down of neutrons by heavy nuclei (see below). The cross section may be considered an upper limit because the excited nucleus will, in returning to its ground state, often emit more than one quantum (§90).

Kikuchi, Aoki and Husimi (K7, K9) measured the γ-rays produced by the fast neutrons of 2 MV from a D+D source. They investigated a great number of elements all over the periodic table. The cross section was about $1 \cdot 10^{-24}$ cm^2 for Cu and Fe, i.e., slightly less than found by Lea with his faster neutrons. For heavier elements, larger cross sections were found,[47] for light elements, smaller ones. No detectable γ-rays (less than 5 percent of those from Cu) were emitted by any element up to oxygen. This is easily explained by assuming that these light nuclei do not possess any excited states below 2 MV, and can therefore not be excited by 2 MV neutrons. This is in agreement with Lea's result that carbon has a small but definite excitation probability with his faster neutrons. The increase of the excitation probability with increasing weight of the nucleus should partly be due to the fact that there will be more and more energy levels below 2 MV, partly simply to the increase in the nuclear radius (cf. (565)), partly perhaps to an increase in the "sticking probability" ξ of the neutron.

Experiments on the *effective absorption of fast neutrons* will give the cross section for inelastic scattering only if all other processes, such as capture, α- and proton disintegration, are rare compared to the inelastic scattering. This condition is probably fulfilled for not too light nuclei (cf. Section D). As a measure of the intensity of fast neutrons, the activity induced in a silicon detector may be used (D2, E1). The activity is due to the process $Si^{28}+n^1=Al^{28}+H^1$ which requires neutrons of at least 3 MV energy (cf. Table LXIII). Ehrenberg found that this activity was reduced by 23 percent by a Ag cylinder of 15 mm. thickness surrounding the source of fast neutrons (Be+Po). This would correspond to a cross section of about $3 \cdot 10^{-24}$ cm^2 for Ag, which is compatible with the γ-ray evidence of Lea and Kikuchi.

The *amount of energy lost* by a fast neutron in exciting a nucleus in the Ag scatterer may be estimated experimentally from the increase in the activity induced in a detector, which responds preferably to slower neutrons (e.g. Ag or Rh). According to the experiments (D2, E1) the radioactivity induced in Ag increases about twice as much as that in Si decreases. Thus, if we assume that the decrease in the Si activity measures the number of inelastically scattered neutrons, the scattered neutrons must be captured three times as easily by Ag as the original ones are. Hence we conclude that the energy loss of a fast neutron in an inelastic collision with a nucleus in the Ag scatterer must be fairly large.

This is in agreement with theoretical expectations. The nucleus possesses much more levels at higher excitation energies and each level has, in the average, the same probability of being the final state in the inelastic collision (§54D). Using the theoretical expressions for the level density of the final nucleus (§53), we find that the average kinetic energy of the scattered neutron will be only of the order of the "temperature" of the residual nucleus (evaporation model, §54E). This temperature is only of the order of 1 MV for heavy nuclei and 10 MV excitation energy (§53, Table XXI) so that fast neutrons lose in the average perhaps 90 percent of their kinetic energy in an inelastic scattering process. Weisskopf (W7) has investigated this problem in detail.

In conclusion, it may be pointed out that the observed total cross sections for fast neutrons (cf. Table XXVII) are of the same order as the cross sections for inelastic scattering. This shows that certainly a very large fraction of the "total cross section" observed is due to inelastic scattering. Elastic (potential) scattering can, accordingly, at best be of the same order as the inelastic scattering.

[47] The exceptionally high value found for Cd (more than 4 times the copper value) is probably due to an admixture of slow neutrons in the beam, giving capture γ-rays.

If we take a value of $3 \cdot 10^{-24}$ for the total cross section for inelastic scattering, and put $R=10^{-12}$ cm (medium sized nuclei), formula (565) gives a "sticking probability" near unity (cf. §54D, E).

D. Capture of fast neutrons

A quantitative estimate of the capture cross section for fast neutrons is even more difficult to obtain than for the inelastic scattering. Since the capture cross section is known to be extremely large for slow neutrons, a small admixture of slow neutrons in a fast neutron beam may produce larger effects than the fast neutrons themselves. Doubts of this kind seem justified in view of the fact that Fermi, Amaldi, d'Agostino, Rasetti and Segrè (F13) found, in their pioneer work with supposedly fast neutrons, large activities in all those elements which were later found to be strongly activated by slow neutrons, and in no other cases of capture reactions. Now it is inconceivable that the capture of fast neutrons should have anything to do with that of slow ones. Large capture of slow neutrons is, as we know, due to strong low resonance levels which have no influence on fast neutrons.

The *activity* produced by fast neutrons should be of the same order of magnitude for *all* heavier nuclei which become radioactive by neutron capture, which must be true of all nuclei of odd charge, for simple stability reasons[47a] (cf. §10). Thus a measure of the capture probability for fast neutrons may be found in the smaller activities observed by Fermi and collaborators with odd elements. In this way, one may estimate cross sections of the order of 10^{-25} cm^2 or less for the capture of fast neutrons.

Such a figure seems compatible with theoretical considerations. If only scattering and capture are possible, the cross-section for capture is

$$\sigma^N{}_\gamma = \pi R^2 \xi \Gamma_\gamma/(\Gamma_N+\Gamma_\gamma), \tag{567}$$

where Γ_γ is the total radiation width and Γ_N the total neutron width. The radiation width will presumably not change very much with the neutron energy (cf. §87); its value is, according to the slow neutron experiments, about 0.1 to 1 volt which agrees very well with the theoretical value obtained in §87 ($\sim\frac{1}{2}$ volt). The neutron width Γ_N may be written as

$$\Gamma_N = \sum_{U_q<E} \Gamma_{Nq}, \tag{568}$$

where the sum goes over all levels of the final nucleus (which is, in this case, identical with the initial nucleus) which can be excited by the incident neutron, i.e., whose excitation energy U_q is less than the kinetic energy E of the incident neutron. Each partial width Γ_{Nq} is, at not too high energies, proportional to the velocity of the neutron which is emitted when the nucleus is left in state q, i.e.,

$$\Gamma_{Nq} = \Gamma_{Nq}'(E-U_q)^{\frac{1}{2}}. \tag{568a}$$

The constant Γ_{Nq}' can be estimated from Table XXVI and is about $\frac{1}{2}$ to $1 \cdot 10^{-3}$ volt$^{\frac{1}{2}}$. Thus Γ_{Nq} is of the same order as Γ_γ if the neutron energy $E-U_q$ is of the order of 1 MV. Therefore, even if no excitation of the nucleus (i.e., no inelastic scattering) is possible, the neutron width will be larger than the radiation width for neutron energies larger than about 1 MV. Then we may write approximately

$$\Gamma_N/\Gamma_\gamma \approx N(E)E^{\frac{1}{2}}, \tag{569}$$

where $N(E)$ is the number of states of the scattering nucleus with an excitation energy less than E, and E is measured in MV.

Since the number of states of a nucleus increases extremely rapidly with increasing excitation energy, the capture cross section (567) becomes negligibly small as soon as the neutron energy is sufficient to excite many levels of the scattering nucleus. Thus really fast neutrons (of several MV energy) should have extremely small capture probability, and only medium fast neutrons (E of the order of a few hundred kilovolts) will be appreciably captured. It should be possible to confirm this point by experiments with neutrons of definite energy such as those from the H^2+H^2 reaction.

We may thus distinguish altogether four energy regions for neutron capture:

(1) Thermal region.

(2) Resonance region ($\frac{1}{2}$ to, perhaps, 1000 or 10,000 volts).

[47a] Only if the radioactive element produced has a very long life can the apparent activity be smaller than usual.

(3) Medium fast neutrons (1 or 10 to about 500 kv).

(4) Fast neutrons ($>\frac{1}{2}$ MV).

In the region of medium fast neutrons, it will scarcely be possible to investigate the individual resonance levels. Only the general trend of the capture cross-section with energy can be found. This is obtained by averaging the dispersion formula over the resonance levels which gives (cf. (405), $s=\frac{1}{2}$, $J=i$)

$$(\sigma^N{}_\gamma)_a=\pi^2\frac{\hbar^2}{2ME}\frac{\Gamma_{N0a}\Gamma_{\gamma a}}{\Gamma_a D}. \tag{570}$$

The subscript a denotes that an average over the resonance levels near the neutron energy E should be taken, D is the average spacing of these levels. Since the region of medium energy is defined by the fact that the radiation width is larger than the neutron width, we have $\Gamma_a=\Gamma_{\gamma a}$. Furthermore, we put

$$\Gamma_{N0a}=\Gamma_{N0}'E^{\frac{1}{2}}. \tag{570a}$$

Then

$$(\sigma^N{}_\gamma)_a=\pi^2\frac{\hbar^2}{2MD}\frac{\Gamma_{N0}'}{E^{\frac{1}{2}}}=2.03\cdot 10^{-18}\frac{\Gamma_{N0}'}{DE^{\frac{1}{2}}}\ \mathrm{cm}^2, \tag{571}$$

where D and E are to be measured in volts, Γ_{N0}' in volt$^{\frac{1}{2}}$. If we put for medium heavy nuclei (cf. Table XXVI, and after Table XXII)

$$D=10\ \text{volts},\ \Gamma_{N0}=10^{-3}\ \text{volts}^{\frac{1}{2}}, \tag{571a}$$

we have

$$(\sigma^N{}_\gamma)_a=2\cdot 10^{-22}E^{-\frac{1}{2}}\ \mathrm{cm}^2. \tag{572}$$

For $E\sim 100$ kv, this is of the order of 10^{-24} cm^2.

For *high energies*, the ratio of radiation width to neutron width may be expressed approximately in terms of the entropy, with the help of the formula (359) for the partial neutron width, (347) for the number of levels of the initial nucleus below the excitation energy E, and (729a) for the radiation width. The result is

$$\Gamma_N/\Gamma_\gamma\sim 1\cdot 10^6 EA^{\frac{1}{3}}\xi e^{S(E)-S(E+Q)}. \tag{573}$$

Here E is the neutron energy in MV, ξ the sticking probability, A the atomic weight of the scattering nucleus, $S(E)$ its entropy corresponding to an excitation energy E, and $S\ (E+Q)$ the entropy of the compound nucleus whose excitation energy is the kinetic energy of the neutron plus the neutron dissociation energy Q. Table XXVIII gives Γ_N/Γ_γ, i.e., the ratio of inelastic scattering to capture, for various nuclei and neutron energies, for $\xi=1$. Table XXVIII shows the extreme smallness of the capture probability for high neutron energy.

E. α-particle and proton emission

Reactions in which an incident fast neutron causes the emission of an α-particle or a proton are known in great number among the lighter nuclei. The cross sections seem to be of the order 10^{-24} cm^2, as should be expected from the inelastic scattering of neutrons (Section C). As long as there is no potential barrier to prevent the charged particles from escaping from the nucleus, the "proton width" and "α-particle width" should be of the same order of magnitude as the fast neutron width.

In many cases, the α-particles and protons produced have been observed in the cloud chamber (for references cf. §102) and their energies measured. As far as these measurements are reliable, they show that the residual nucleus is left in an excited state as often as not.[48] This is to be expected from our general theory, because the

TABLE XXVIII. *Ratio of the probabilities of inelastic scattering and radiative capture for fast neutrons.*

Neutron Energy	$A=20$	50	100	200
2 MV	$1.6\cdot 10^4$	$6\cdot 10^3$	$4\cdot 10^3$	$4\cdot 10^3$
5 MV	10^5	$5\cdot 10^4$	$4\cdot 10^4$	$4\cdot 10^4$
10 MV	$3.5\cdot 10^5$	$2\cdot 10^5$	$1.8\cdot 10^5$	$2\cdot 10^5$

[48] With a monochromatic group of incident neutrons, the produced particles should fall into several groups according to the state q of the residual nucleus, the energy of the emitted particle being

$$E_Q=E_P+W_A-W_B(q),$$

where W_A is the energy of the initial nucleus in the ground state, $W_B(q)$ that of the final nucleus B in the excited state q. Unfortunately, the neutron sources available do not give monochromatic neutrons, so that the neutron energy itself has to be determined from the resultant momentum of particle Q and recoil nucleus B. Such determinations are extremely uncertain, giving apparent neutron energies of all orders of magnitude, including many which are much higher than the known maximum energy of the neutrons in the beam (B41). Such experiments can therefore easily lead to spurious excitation levels, or even to results entirely irreconcilable with reasonable expectations (K31).

"partial widths" Γ_{Qq} corresponding to the various states q of the residual nucleus, will be of the same order whether q is an excited or the ground state.

For elements heavier than Zn ($Z=30$), the potential barriers become too high to allow the escaping of charged particles from the nucleus to any appreciable extent (B12). Capture of fast neutrons plus subsequent emission of charged particles has not been observed with elements of higher nuclear charge than 30 (F13) except for thorium and uranium (H7, M15). For this element, as for other natural radioactive elements, the large energy set free by spontaneous α-disintegration, plus the kinetic energy of the neutron, are sufficient to bring the produced α-particle over the potential barrier.

XI. α-Radioactivity

§66. Theory of α-Radioactivity According to the One-Body Model (G6, G7, G8, G10, C32, L4, B42, S10, S11)

The theory of the emission of α-particles by radioactive nuclei was the first successful application of quantum theory to nuclear phenomena. As is well known, the theory was given simultaneously by Condon and Gurney (C32) and by Gamow (G7). Subsequently, a great number of authors (G8, G10, G6, L4, B42, S10, S11) have given alternative, and partly more rigorous, mathematical methods for arriving at the same result. All these methods are based on the one-body model, the α-particle being considered as moving in a certain potential created by the nucleus. They must, accordingly, be modified to take into account the principles of the many-body problem (§67).

The starting point of the one-body theory is to assume a suitable potential between an α-particle and a nucleus of charge[49] Z. If the α-particle is far away from the nucleus, the two particles will repel each other according to the Coulomb law, the potential energy being

$$V=zZe^2/r, \tag{574}$$

where $z=2$ is the charge of the α-particle and r the distance between the α-particle and the center of the nucleus. When the α-particle is inside the nucleus, the Coulomb potential (574) breaks down and is to be replaced by a much lower potential energy. This potential energy may be assumed to be constant over the interior of the nucleus, so that

$$V=V_0 \quad (r<R), \tag{574a}$$

where R is the radius of the nucleus. The exact value of V_0 is of no great importance. It is usually positive but in any case smaller than the kinetic energy E of the α-particles which may be emitted from the radioactive nucleus.

The potential as a function of the distance r is shown in Fig. 17. Its maximum occurs at $r=R$ and has the value

$$B=zZe^2/R. \tag{575}$$

B is often called the top of the potential barrier. It is, for the natural α-emitters, much *larger* than the energy E of the emitted α-particles.

If we consider α-particles of a given energy E, we may divide the whole space into three regions, *viz.:*

(1) The interior of the nucleus, $r<R$: Here, the potential energy V_0 is *less* than the energy E of the α-particle.

(2) The region of the potential barrier, i.e., r between R and

$$r_E=zZe^2/E. \tag{575a}$$

In this region the potential energy is *greater* than the energy of the α-particle.

(3) The outside region, $r>r_E$, in which again $V<E$.

According to classical mechanics, the α-particle could only move in regions 1 or 3. Once it was confined in one of these regions, e.g. inside the nucleus, it would be compelled to stay there forever. In wave mechanics, we have the well-known possibility of penetration through the "forbidden region" 2. This enables a particle originally in the nucleus, to "leak out" and to appear, sooner or later, as a free α-particle outside the nucleus (in region 3). Wave mechanics

[49] Z is, of course, the charge of the nucleus which remains after the emission of the α-particle. The radioactive nucleus itself has therefore the charge $Z+2$.

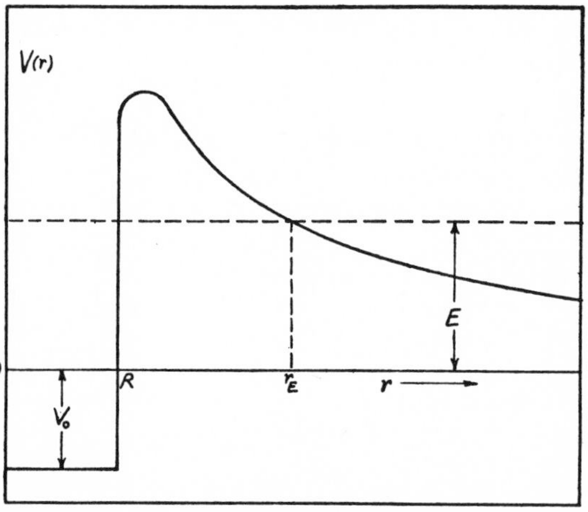

FIG. 17. Potential energy of an α-particle in the field of a nucleus according to Gamow.

allows us to calculate the probability for this to happen in a given time, and thus to compute the lifetime of radioactive nuclei in a perfectly straightforward way.

Hardly any other problem in quantum theory has been treated by so many authors in so many different ways as the radioactive decay. All the proposed methods are, of course, equivalent (insofar as they are correct) but they differ in rigor and complication. We shall give a method which seems about the simplest of the correct ones.

First of all, we shall assume in this section that the wave function of the α-particle is spherically symmetrical, corresponding to an "s state" of the α-particle in the nucleus. Then the wave function of the α-particle can be written

$$\psi=u/r, \tag{576}$$

where u satisfies the equation

$$d^2u/dr^2+2M\hbar^{-2}(E-V)u=0, \tag{576a}$$

V being given by (574, 574a) and E being the experimental α-particle energy. At large distances from the nucleus, V is small (cf. 574) so that the solution of (576a) becomes simply a sine or cosine function, or more generally

$$u=Ae^{ikr}+Be^{-ikr} \tag{577}$$

with

$$k=(2ME)^{\frac12}/\hbar \tag{577a}$$

and A and B some constants. The first term represents an outgoing wave, the second a wave converging towards the nucleus, as can be seen by multiplying the time factor $e^{-iEt/\hbar}$ with (577). In the physical problem, an α-particle may leave the nucleus but none comes towards it from outside, and we are therefore interested only in the first term, *viz.*

$$u=Ae^{ikr}. \tag{577b}$$

The constant A determines the number of α-particles leaving the nucleus per unit time:

$$I=v4\pi r^2|\psi|^2=4\pi|A|^2v, \tag{577c}$$

while the *form* of (577b) determines also the form of the wave function at smaller distances from the nucleus.

To obtain the wave functions for smaller values of r, and especially inside the nucleus, it is simplest to use the Wentzel-Kramers-Brillouin (WKB) approximation (W10, K24, B61) in the form due to Kramers (K24). According to this method, the solution of (576a) is approximately

$$u_1(r)=\Phi^{-\frac14}(r)\cos\left(\int_{r_E}^{r}\Phi^{\frac12}(\rho)d\rho+\tfrac14\pi\right) \tag{578}$$

or

$$u_2(r)=\Phi^{-\frac14}(r)\cos\left(\int_{r_E}^{r}\Phi^{\frac12}(\rho)d\rho-\tfrac14\pi\right) \tag{578a}$$

in region (3) ("outside" region). Here

$$\Phi(r)=2M(E-V)/\hbar^2 \tag{578b}$$

is proportional to the kinetic energy $E-V(r)$ which the α-particle has when at the point r. r_E is the classical distance of closest approach of an α-particle of energy E falling on the nucleus from outside as defined in (575a). The most general solution of the Schrödinger equation "outside" is

$$u=B_1u_1+B_2u_2. \tag{579}$$

In order to make u have the form (577b) we must choose

$$B_2=iB_1. \tag{579a}$$

Since for large distances r, the potential energy may be neglected, we have in this case

$$\Phi=2ME/\hbar^2=k^2 \tag{580}$$

and, therefore, comparing (577b) and (578):

$$|A|=|B_1|k^{-\frac12}. \tag{580a}$$

For r between R and r_E, Φ (cf. 578b) would be negative. The solution u in this region is of an exponential rather than a wave type. The continuations of the functions u_1u_2 (cf. 578, 578a) in the barrier region are:

$$u_1(r)=|\Phi(r)|^{-\frac14}\exp\left[+\int_r^{r_E}|\Phi(\rho)|^{\frac12}d\rho\right], \tag{581}$$

$$u_2(r)=\tfrac12|\Phi(r)|^{-\frac14}\exp\left[-\int_r^{r_E}|\Phi(\rho)|^{\frac12}d\rho\right], \tag{581a}$$

u_1 decreases from the nucleus outwards, u_2 increases.

Inside the nucleus, we have assumed constant potential energy $V=V_0$ so that the wave function becomes

$$u=c\sin\kappa r, \tag{582}$$

where

$$\kappa=(2M)^{\frac12}(E-V_0)^{\frac12}/\hbar. \tag{582a}$$

At the boundary of the nucleus ($r=R$), the solution (582) must be identical with the outside solution [(579) to (581a)] as to the value of the wave function and its first derivative. In all practical cases, $u_1(R)$ will be much larger than $u_2(R)$, because u_1 contains an exponential with a large positive exponent, u_2 one with large negative exponent. We may therefore put

$$u(R)=c\sin\kappa R=B_1u_1(R)=B_1|\Phi(R)|^{-\frac14}\exp\left[\int_R^{r_E}|\Phi(\rho)|^{\frac12}d\rho\right], \tag{583}$$

$$(du/dr)_R=\kappa c\cos\kappa R=-B_1|\Phi(R)|^{\frac14}\exp\left[\int_R^{r_E}|\Phi(\rho)|^{\frac12}d\rho\right]. \tag{583a}$$

Dividing the lower equation by the upper, we have

$$\kappa \cot \kappa R = -|\Phi(R)|^{\frac{1}{2}}. \qquad (583b)$$

Since the Coulomb potential V is very large at the top of the barrier, $\Phi(R)$ is large, and usually much larger than κ. Therefore practically[50]

$$\cot \kappa R = -\infty, \qquad (584)$$

$$\kappa R = \pi, \qquad (584a)$$

$$\cos \kappa R = -1. \qquad (584b)$$

This gives, according to (583a),

$$B_1 = \kappa c |\Phi(R)|^{-\frac{1}{2}} e^{-C} \qquad (585)$$

with

$$C = \int_R^{r_E} |\Phi(\rho)|^{\frac{1}{2}} d\rho. \qquad (585a)$$

The constant c may be obtained by normalizing the internal wave function (582) to unity:

$$4\pi \int u^2 dr = 1, \qquad (586)$$

which gives

$$c = (2\pi R)^{-\frac{1}{2}}. \qquad (586a)$$

From (577c) (580a) (584a) (585) (586a) we obtain for the number of α-particles emitted per second

$$\lambda' = \frac{2\pi^2 v}{kR^3 |\Phi(R)|^{\frac{1}{2}}} e^{-2C} = \sqrt{2} \frac{\pi^2 \hbar^2 e^{-2C}}{M^{\frac{1}{2}} R^3 (zZe^2 R^{-1} - E)^{\frac{1}{2}}}. \qquad (587)$$

The half-life is given by $\tau = (\log 2)/\lambda'$. If we insert[51] for R the value $0.9 \cdot 10^{-12}$ cm derived with this model from experiment in §68, for Z the average nuclear charge of radioactive elements, i.e., about 86, and for E some average energy, let us say, 6 MV, we find

$$\tau = 3.3 \cdot 10^{-21} e^{2C}. \qquad (588)$$

The most important factor in this formula is e^{2C}, where C is given by (585a). Inserting Φ from (578b) and V from (574), we have

$$C = (2M)^{\frac{1}{2}} \hbar^{-1} \int_R^{r_E} (zZe^2 r^{-1} - E)^{\frac{1}{2}} dr. \qquad (588a)$$

The integration is straightforward and yields

$$C = \frac{(2M)^{\frac{1}{2}}}{\hbar} \frac{zZe^2}{E^{\frac{1}{2}}} \left[\arccos \left(\frac{ER}{zZe^2} \right)^{\frac{1}{2}} - \left(\frac{ER}{zZe^2} \right)^{\frac{1}{2}} \left(1 - \frac{ER}{zZe^2} \right)^{\frac{1}{2}} \right] \qquad (589)$$

$$= (2zZe^2/\hbar v)[\arccos x^{\frac{1}{2}} - x^{\frac{1}{2}}(1-x)^{\frac{1}{2}}]$$

with (cf. (575)) $x = ER/zZe^2 = E/B$. (589a)

For a rough approximation, we may neglect E compared to $B = zZe^2/R$, and replace arc cos $x^{\frac{1}{2}}$ by $\frac{1}{2}\pi - x^{\frac{1}{2}}$; then

$$C = \pi zZe^2/\hbar v - (2e/\hbar)(2zZMR)^{\frac{1}{2}}. \qquad (589b)$$

The error in C committed when using this approximate formula, is about 1.5 for $E = 6$ MV. Correspondingly, the penetrability e^{-2C} would come out about 20 times too large.

According to (589, 589b) the "Gamow exponent" C is the larger, the higher the nuclear charge Z, the slower the α-particle, and the smaller the nuclear radius R. This is very plausible, since increasing Z increases the height, decreasing v increases the breadth, and decreasing R increases both height and breadth of the potential barrier, and the penetration of the barrier will of course be the more difficult the higher and broader the barrier.

The formula (587) for the lifetime may be written as follows:

$$\tau = \tau_0 / P \qquad (590)$$

where P is the transmission coefficient of the potential barrier,

$$P = e^{-2C} \qquad (591)$$

with C given in (589), and

$$\tau_0 = 3.3 \cdot 10^{-21} \text{ sec.} \qquad (591a)$$

would be the lifetime without potential barrier which is of the same order as the "period of vibration" of the α-particles in the potential of the nucleus.

For the comparison with experiment, cf. §68.

§67. Many-Body Theory of the α-Decay (B14)

According to the many body concept (cf. Chapter IX), the α-particle must not be considered as moving freely in a potential created by the residual nucleus. Only when the α-particle has left the nucleus and is sufficiently far away from it, will it be justified to consider α-particle and residual nucleus as two separate entities. When the α-particle is "inside" the nucleus, its four constituent particles take part in the complicated motion of the compound (initial)

[50] Other solutions, *viz.* $\kappa R = 3\pi$, 5π etc. would correspond to excited α-particle states. Since α-particles obey Bose statistics, there is no exclusion principle for them. In the ground state of the nucleus, all α-particles will therefore be in the lowest level.

[51] Since λ' is not sensitive to the factor multiplying e^{-2C}, we may replace that factor by an approximate value.

nucleus and are in no way distinguishable from the other particles in that nucleus. The α-particle can thus not be considered to "exist" in the radioactive nucleus before its emission but only as formed at the moment of its emission. Accordingly, the probability of α-decay is the product of two factors:

(1) The probability of formation of the α-particle.

(2) The probability of penetration through the potential barrier.

The first named probability will presumably be quite small, making the decay constant of α-radioactive nuclei small even in the absence of a potential barrier. The lifetime of radioactive nuclei would, perhaps, be 10^{-15} sec. (cf. (593a)) if they had no potential barrier. This time, it is true, is much shorter than the actual lifetimes of α-radioactive nuclei, but it is very much longer than the "period of vibration" $\tau_0 \approx 10^{-21}$ sec. (cf. 591a) which would give the lifetime without barrier in the *one*-body model.

Since the lifetime without barrier is much longer in the many-body than in the one-body theory, it is obvious that much lower and narrower barriers are required in order to explain the observed lifetimes of the α-radioactive nuclei. Indeed, the nuclear radii deduced from the many-body concept are about 40 percent larger than those derived from the one-body model (cf. Table XXIX), corresponding to a potential barrier of only seventy percent of the height required in the one-body theory.

In order to derive the nuclear radius from the experimental data, it is, of course, necessary to separate the probability of penetration through the barrier ("2" above) from the probability of formation of the α-particle ("1" above). In order to estimate the latter probability we assume that the *emission of* an *α-particle* by a nucleus would, in the absence of the potential barrier, be just *as probable as that of a neutron of the same energy.* The latter can be obtained from the "neutron widths" measured for *slow* neutrons (§60), if we admit that the neutron width is proportional to the neutron velocity (§54, 56).

According to Table XXVI we have in the average

$$\Gamma_N E_0^{-\frac{1}{2}} \approx 4 \cdot 10^{-4} \text{ volts}^{\frac{1}{2}}, \quad (592)$$

where Γ_N is the neutron width of a resonance level at the neutron energy E_0. Taking $E_0 = 6$ MV which is about the average energy of α-rays, we find

$$\Gamma_N \approx 1 \text{ volt.} \quad (592a)$$

We assume for the α-radioactive nuclei that the "α width in the absence of a potential barrier" would also be

$$G_\alpha = 1 \text{ volt.} \quad (593)$$

The decay constant in the absence of a barrier would then be

$$G_\alpha/\hbar \approx 10^{15} \text{ sec.}^{-1} \quad (593a)$$

and in the presence of the barrier

$$\lambda = \Gamma_\alpha/\hbar = G_\alpha P/\hbar, \quad (594)$$

where P is the penetrability given in (591). Numerically we have;

$$\log_{10} \lambda = \log_{10} P + 15.2 = 15.2 - 0.869C, \quad (594a)$$

where C is given by (589).

§68. Comparison with Experiment (G9, G10)

The relation (588), (589) between the lifetime and the disintegration energy of α-radioactive nuclei was discovered by Geiger and Nuttall (G13) as early as 1911, on a purely empirical basis. With the help of the approximate formula (589b), the relation can be written in the form

$$\log \tau = 2\pi z Z e^2/\hbar v - K, \quad (595)$$

where K is a constant involving, according to (589b), the radius of the radioactive nucleus (or, more accurately, of the nucleus *produced* in the α-disintegration). Assuming the radius to be about the same for all radioactive nuclei, we find *a linear relation between the logarithm of the lifetime and the reciprocal velocity of the α-particle.* Therefore, relatively small changes of the α-particle energy correspond to very large differences in the lifetime: The slowest α-particle, that from thorium, has an energy of 4.3 MV, the fastest (from ThC′) one of 8.9 MV, i.e., only a little more than twice as much. The corresponding lifetimes are $2 \cdot 10^{10}$ years and $2 \cdot 10^{-8}$ sec., differing by a factor 10^{25}. This huge variation of the lifetime is correctly represented by (595), with almost the same K throughout, i.e., almost the same nuclear radius.

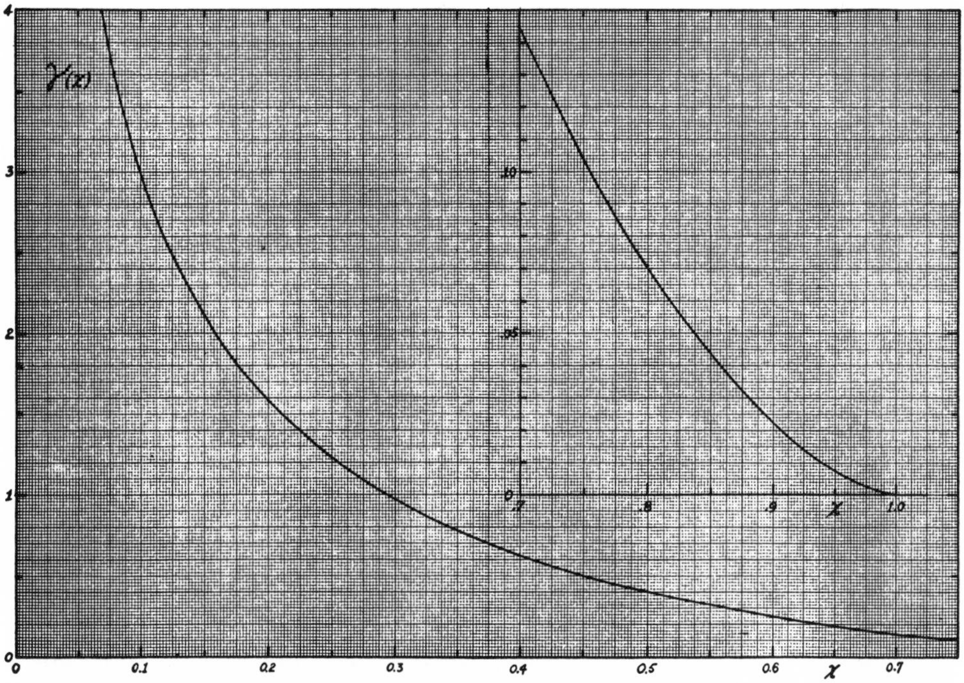

FIG. 18. The function $\gamma(x)$ determining the penetrability of the potential barrier (cf. (600)). x is the ratio of particle energy to height of barrier.

Gamow (G9, G10) has compared all experimental data with the theoretical formula (595). It is most convenient to use the experimental data for the decay constant and the energy of the α-particle, and to compute the nuclear radius from these data, with the help of the theoretical formula. If the formula is correct, the radius R must come out about the same for all radioactive nuclei. This is actually the case for both the one and the many-body theory, as can be seen from Table XXIX. With the one-body method, all radii are between 8.2 and $9.8\cdot 10^{-13}$ cm, with the many-body model, they vary between 11.3 and $13.2\cdot 10^{-13}$ cm. Exceptions are, in both cases, the radii for the C and C′ products.

Our values for the nuclear radii from the one-body model are, in the average, about $0.5\cdot 10^{-13}$ cm larger than Gamow's. The main reason for this is that we have used the exact formula (589) rather than the approximate one (589b), for calculating C.

Much larger are the differences between the radii derived from the one- and the many-body concept. In part A of this report, the radius from the one-body concept was used (cf. §8, 26, 30, etc.). The effect of the introduction of the new, larger radius on calculations concerning the stability of nuclei is considerable. In the present part B the radius derived from the many-body problem has been used throughout.

To judge the accuracy of the nuclear radius obtained from the many-body problem, we investigate the influence of an error in the estimate of G_α (cf. (593)) on the calculated radius. According to (594), a given error in G_α means an equally large error in P, only in the opposite direction. Now, according to (589b), small variations of R and P are connected by the formula

$$\frac{\delta P}{P}=\frac{\delta R}{R}\frac{2e}{\hbar}(2MzZR)^{\frac{1}{2}}. \qquad (596)$$

Inserting the numerical values ($R=12.3\cdot 10^{-13}$ cm), this gives

$$\delta P/P\approx 50\delta R/R. \qquad (596a)$$

Now our estimate for G_α will probably be correct within a factor 100. This means an uncertainty of ± 10 percent in the nuclear radius. Thus we may write our result for the average radius of the α-radioactive nuclei

$$R = 12.3 \cdot 10^{-13}\ \text{cm} \pm 10\ \text{percent}. \qquad (597)$$

From this radius, we may obtain the radii of other nuclei if we assume proportionality between nuclear volume and number of particles in the nucleus:

$$R = r_0 A^{1/3}, \qquad (598a)$$

where A is the atomic weight. Taking the average of all the radioactive nuclei with the exception of the C products, we find

$$r_0 = 12.3_5 \cdot 10^{-13} \cdot 219.6^{-1/3} = 2.05 \cdot 10^{-13}\ \text{cm}. \qquad (598)$$

Accordingly, the height of the potential barriers becomes

$$B = zZe^2/R = 0.70 zZA^{-1/3}\ \text{MV}. \qquad (599)$$

In the particular case of light nuclei we may put $A = 2Z$ so that

$$B = 0.55 zZ^{2/3}. \qquad (599a)$$

With the values (598) (599) for nuclear radius and height of the potential barrier, the formula for the penetrability of the potential barrier of a given nucleus of charge Z and mass number A by particles of charge z and mass number a takes the simple form (cf. (591a) (589))

$$P = \exp\left[-2g\gamma(E'/B)\right], \qquad (600)$$

where

$$g = (2MzZe^2R)^{1/2}/\hbar$$
$$= 0.37_5 (ZzAa/A+a)^{1/2} A^{1/6}, \qquad (600a)$$

$$\gamma(x) = x^{-1/2} \arccos x^{1/2} - (1-x)^{1/2}, \qquad (600b)$$

For M the reduced mass has been inserted. E' is the total kinetic energy (of particle *plus* recoil nucleus) in the system of reference in which the center of gravity of nucleus and particle is at rest. The function $\gamma(x)$ is plotted in Fig. 18 as a function of x. If we neglect in (600a) a compared to A and put then $A = 2Z$ we have

$$g = 0.42 (az)^{1/2} Z^{2/3}. \qquad (600c)$$

Values of g for various nuclei are found in Table XXXIII, §70.

In computing Table XXIX, it is of course, necessary to take for v in formula (589) the *relative* velocity of α-particle and recoil nucleus which is $(A+4)/A$ times the velocity of the α-particle itself if A is the atomic weight of the product nucleus. Similarly, the energy E includes the kinetic energy of the recoil nucleus.

Table XXIX includes samarium which is known (L35) to give α-particles of 1.28 cm range, i.e., 2.46 MV energy (the figure in the table includes again the recoil energy). The number of α-particles emitted is about 4 per cm² of Sm_2O_3 per min. (R17). From the range of the α-particles and the stopping power of Sm and O, we may estimate that about $1.5 \cdot 10^{19}$ Sm atoms per cm² are close enough to the surface so that their α-particles may be detected. Considering the (uniform) angular distribution of the alphas, one in four emitted particles will actually be detected. Assuming that the radioactive isotope of Sm has an abundance of 10 percent, we find for the decay constant $\lambda = 4 \cdot 4/60 \cdot 1.5 \cdot 10^{18} = 1.7 \cdot 10^{-19}\ \text{sec.}^{-1}$. The radius turns out to be $9.4 \cdot 10^{-13}$ cm, i.e., about what would be expected from (598a).

It is of interest to note that no α-particles of less than 4 MV have been observed from nuclei of the radioactive families. Indeed, an energy somewhat below 4 MV would make the penetration of the potential barrier so rare that no radioactivity could be detected. Assuming that one α-particle per hour per cm² constitutes the limit of observability, we may easily calculate the minimum energy of

TABLE XXIX. *α-disintegration.*

Nucleus: Disintegrating	Nucleus: Product	Z	Energy E MV	Decay constant λ in sec.$^{-1}$	Nuclear radius R in 10^{-13} cm: Old	Nuclear radius R in 10^{-13} cm: New
Th	MThI	88	4.34	$1.2 \cdot 10^{-18}$	8.7	11.3
RdTh	ThX	88	5.52	$1.15 \cdot 10^{-8}$	8.8	12.3
ThX	Thn	86	5.79	$2.20 \cdot 10^{-6}$	9.0	12.5
Thn	ThA	84	6.40	$1.27 \cdot 10^{-2}$	9.1	12.3
ThA	ThB	82	6.90	5.0	8.9	12.7
ThC	ThC″	81	6.20	$6.7 \cdot 10^{-5}$	7.0	10.6
ThC′	ThD	82	8.95	$[4 \cdot 10^{7}]$*	—	(13.9)
UI	UXI	90	4.15	$5.0 \cdot 10^{-18}$	9.8	13.2
UII	Io	90	4.76	$(7 \cdot 10^{-14})$	9.6	13.0
Io	Ra	88	4.67	$2.9 \cdot 10^{-13}$	9.5	13.1
Ra	Rn	86	4.88	$1.42 \cdot 10^{-11}$	9.0	12.5
Rn	RaA	84	5.59	$2.10 \cdot 10^{-6}$	9.0	12.6
RaA	RaB	82	6.11	$3.8 \cdot 10^{-3}$	9.0	12.5
RaC	RaC″	81	5.61	$2.3 \cdot 10^{-7}$	7.2	10.9
RaC′	RaD	82	7.83	$7 \cdot 10^{+4}$	9.4	13.9
RaF	RaG	82	5.40	$5.7 \cdot 10^{-8}$	8.2	11.5
Pa	Ac	89	5.16	$6.9 \cdot 10^{-13}$	8.6	11.6
RdAc	AcX	88	6.16	$4.25 \cdot 10^{-7}$	8.4	11.4
AcX	An	86	5.82	$7.2 \cdot 10^{-7}$	8.8	12.1
An	AcA	84	6.95	$1.77 \cdot 10^{-1}$	8.5	12.0
AcA	AcB	82	7.51	$3.5 \cdot 10^{+2}$	8.8	12.8
AcC	AcC″	81	6.74	$5.3 \cdot 10^{-3}$	7.3	10.6
AcC′	AcD	82	7.58	$[1 \cdot 10^{+4}]$*	—	(13.9)
Sm	Nd	60	2.55	$1.7 \cdot 10^{-19}$	6.7	9.4

* Calculated from radius by means of the theoretical relation. The radius was assumed equal to that of RaC′.

the α-particle necessary to give an observable radioactivity. If we assume a range of the α-particles of 1.5 cm air, and make observations with a pure element, there are about $2\cdot 10^{19}$ atoms per cm² whose α-particles may be detected. Then an activity of one particle per hour per cm² is equivalent to a decay constant $\lambda = 1/3600\cdot 2\cdot 10^{19}$. According to (594a), this means about $C=44$. Inserting this into the approximate formula (589b), and assuming the nuclear radius R to be proportional to $Z^{1/3}$, we find for the limit of α-activity

$$E_{\text{min.}} = 3.9Z^2/(44+2.41Z^{1/3})^2. \tag{601}$$

As a function of the nuclear charge, this gives:

Z	10	20	30	40	50	60	70	80	90
$E_{\text{min.}}$	0.13	0.41	0.77	1.2	1.65	2.15	2.65	3.15	3.7 MV

Thus α-particles of less than 3.7 MV energy would give no observable radioactivity for a nuclear charge of 90, and even for Z as low as 40, an energy of more than 1.2 MV would be required. This fact is important for considerations about the stability of nuclei against α-decay (§8).

Of particular interest among α-emitters is Be^8. This nucleus is known to be formed in some transmutation processes (Chapter XVII) and seems to have a mass just larger than that of two α-particles, the difference being of the order of a few hundred thousand volts. Be^8 should therefore break up spontaneously into two α-particles (This process has not yet been observed.) The lifetime will depend on the "size" of the α-particle, i.e., on the maximum distance R up to which two α-particles exert attractive forces upon each other. For the values $R=2.5\cdot 10^{-13}$ and $5\cdot 10^{-13}$, which are probably too low and too high respectively, and assuming $G_\alpha = 1$ MV for a nucleus as light as Be^8, we find the values for the lifetime given in Table XXX. The lifetime should therefore be exceedingly short, even if the energy of Be^8 is only 50,000 volts higher than that of two α-particles.

TABLE XXX. *Estimated lifetime of Be^8 as a function of the energy excess E over two α-particles (in seconds).*

E (IN MV)	0.05	0.10	0.20	0.30	0.40
τ for $R=2.5\cdot 10^{-13}$ cm	$4\cdot 10^{-13}$	$3\cdot 10^{-16}$	$2\cdot 10^{-18}$	$2\cdot 10^{-19}$	$5\cdot 10^{-20}$
τ for $R=5\cdot 10^{-13}$ cm	$7\cdot 10^{-14}$	$4\cdot 10^{-17}$	$3\cdot 10^{-19}$	$3\cdot 10^{-20}$	$1\cdot 10^{-20}$

The derivation of nuclear radii given here is open to criticism on three points: Firstly, it is questionable whether the "width without barrier" is the same for α-particles as for neutrons; secondly, the width may be larger for the ground state than for a highly excited state; and thirdly the radius obtained is probably the sum of the radii of nucleus and α-particle rather than that of the nucleus alone.

On the first point, we have very little information. It is not known experimentally whether the partial width for the emission of α-particles is in the average larger or smaller than that for the emission of neutrons or protons (leaving out the penetrability of the potential barrier in each case). Some experiments with light nuclei indicate a larger probability for α-emission (e.g., $F^{19}+n^1 \rightarrow N^{16}+He^4$ or $O^{19}+H^1$, cf. §102), others a larger probability for emission of protons (e.g., $Na^{23}+H^2 = Na^{24}+H^1$ or $Ne^{21}+He^4$, cf. §101). The rather scarce evidence on heavy nuclei is compatible with about equal *a priori* probabilities for α and proton emission. Theoretically, it may be argued that the neutron width should be larger because the neutron is an elementary particle, but also that the α-width should be larger because there might be some slight tendency for the preformation of α-particles in nuclei due to their great stability. We are rather inclined to think that, if anything, the neutron width would be larger.

As regards the second point, some dependence of the width on the excitation of the nucleus seems plausible. If nuclear states are described in terms of "configurations" of the individual particles (Hartree approximation), the ground state may be a mixture of a smaller number of such configurations (perhaps a hundred) than an excited state (containing perhaps a million configurations). If this is true the partial width of the ground state corresponding to disintegration into a definite state of the residual nucleus, plus an α-particle, might be considerably larger than that of an excited state; this would give some intermediate value for the nuclear radii between the one-body and the many-body result. Perhaps a better estimate of the α-particle width (without barrier) of the ground state will ultimately come from the matrix element of β-disintegrations (Nordheim and Yost, in course of publication).

The third question is: what exactly is meant by "top of potential barrier" and "nuclear radius"? It must be admitted that the α-particle itself has a finite radius, and also that the nuclear forces have a finite range. Therefore it will not be necessary for a breakdown of the Coulomb law that the center of the α-particle enters the nucleus and not even that the α-particle "touches" the nucleus. The effective radius, measured in the radioactive decay, will be approximately the sum of the radii of the nucleus itself and the α-particle, *plus* a quantity of the order of the range of the nuclear forces. Therefore the radius of the nucleus itself may be appreciably less than the value derived from the radioactive decay constant. Thus, presumably, a smaller radius should be used in calculations of the nuclear properties such as the contribution of the Coulomb energy to the nuclear forces (§9), the semi-empirical formula for nuclear binding energies (§30) and the density of nuclear levels (§53). We have used the larger radius derived from (594a) in all our calculations but mainly because there is at the moment no way of obtaining the radius of the nucleus itself.

For calculating the probability of nuclear disintegrations and especially the transmission coefficients of nuclear potential barriers, it seems to be far preferable to use directly the nuclear radius derived from the many-body picture of radioactive decay. In all transmutation problems, the effective radius will be determined again by the sum of the radii of the initial nucleus and the incident particle, plus a quantity of the order of the nuclear forces. If the particle is an α-particle, the situation will therefore be exactly the same as in the natural α-decay and the nuclear radius may be taken over immediately. For protons and neutrons, the size of the particle itself is zero but the extension of the forces will probably be larger than for the saturated α-particle so that in effect the α-particle radius may again be a good approximation. For deuterons, a larger effective radius may be taken, perhaps $2\cdot 10^{-13}$ may be a fair estimate of the additional term in the effective radius (i.e., of the difference between the radii of α-particle and deuteron).

While these problems about the exact meaning of the nuclear radius remain to be settled, there can be no doubt that in principle the many-body picture must be applied to the natural α-decay just as much as to other nuclear processes. This can most easily be seen by going to the limit of extremely large nuclei (which do not exist in nature), for which all the correction terms mentioned would be negligible.

§69. Complex α-Spectra. Levels of Radioactive Nuclei (R5a, R12, R14, R14a, R22, R23, L23, G6)

In many cases, the α-rays emitted from a radioactive nucleus are not homogeneous but consist of groups differing in energy. These groups have been studied by Rosenblum (R12, R14, R14a) and by Rutherford, Wynn-Williams, Lewis and Bowden (R22, R23, L23) using a magnetic spectrograph. The main results are given in Table XXXI. The elements not quoted, *viz.* UI, UII, Io, Rn, RaA, Po, Pa, AcA, AcC′, Th, Thn, and ThA, emit homogeneous α-rays. The disintegration energy is equal to the kinetic energy of the α-particle times $A/(A-4)$ where A is the atomic weight of the emitting nucleus. The relative numbers of particles in the various groups are given in percent for the spectra denoted as "type I," in particles per million for "type II."

The complex α-spectra fall naturally into two types, I and II. The spectra of type I consist of rather closely spaced lines, the spacing being of the order of 100 kv, of *about equal* intensity, the groups of *lower energy* being usually *less intense.* This type comprises most of the α-spectra. Spectra of type II contain one very strong group and several very *weak groups of* much *higher energy*, the energy differences being of the order of 1 MV and the *ratio of intensities between 10^4 and 10^6*. This type is only found for the very short-lived α-emitters ThC′ and RaC′. The interpretation is as follows:

Type I spectra correspond to transitions from the ground state of the initial nucleus to various excited states of the final nucleus. This hypothesis, which was first suggested by Gamow (G6) is confirmed by the internal conversion of the γ-rays emitted by the final nucleus when left in an excited state (§88A). (The energies of the conversion electrons correspond to the electronic energy levels of the atom *produced* in the α-disintegration rather than of the atom emitting the α-ray.) The longest range α-group observed (group 0) must be attributed to a transition to the ground state. The difference in disintegration energy between any group n and group 0 gives the excitation energy of the level in which the final nucleus is left after emission of group n.

The excitation energies derived in this way are listed in the fourth column of Table XXXI. The spacing between adjacent levels is smallest (about 40 kv) for the nucleus produced by α-emission from radioactinium, *viz.* actinium X; in the average over all radioactive elements, the spacing is about 100 kv.

Transitions to excited levels are in general less probable because of the smaller penetrability of the potential barrier for the slow α-particles corresponding to such transitions. From the general formula (589b) we find that the penetrability decreases approximately by a factor

$$P(E)/P(E-\delta E)=\exp\,(170\delta E/E^{3/2}) \quad (602)$$

for a decrease in disintegration energy by δE, where δE and the disintegration energy E itself are measured in MV. For $E=6$ MV, this corresponds to a factor of about 3 for each 100 kv decrease in disintegration energy. This means e.g. that the disintegration probability *without barrier* must be about the same for the groups 1 and 4 in the disintegration of thorium C, since the observed numbers of particles for the two groups are about in the ratio of the expected penetrabilities. On the other hand, groups 7 and 9 of radioactinium have evidently an enormously larger disintegration probability without barrier than group 0 because they contain about as many particles as the latter group although the penetrability is about 30 times smaller for them. The decrease of the penetrability is obviously the explanation for the absence of α-groups of very low disintegration energy, i.e., very high excitation energy of the final nucleus.

The nuclear energy levels obtained from α-groups can be checked by the γ-rays emitted from the product nucleus after the α-disintegration (§88). The most extensive comparison has been made for ThC″, i.e., the nucleus formed by α-decay of ThC. Table XXXII gives the possible combinations between the levels of ThC″ as derived from the α-groups ("calculated" $h\nu$, column 2) and the observed γ-rays (column 3).

TABLE XXXI. *Complex α-spectra.*

Element*	Group No.	Disint. Energy MV	Excit. Energy kv	Relat. Number of Part.
		Type I		
Ra	0	4.879	0	—
	1	4.695	184	—
RaC	0	5.612	0	45
	1	5.550	62	55
RaTh	0	5.517	0	85
	1	5.431	86	15
ThC	0	6.2007	0	27.2
	1	6.1607	40.0	69.8
	2	5.8729	327.8	1.80
	3	5.7283	472.4	0.16
	4	5.7089	491.8	1.10
RaAc†	0	6.159	0	24
	1	6.127	32	2
	2	6.097	62	19
	3	6.075	84	1
	4	6.030	129	3
	5	5.975	184	4
	6	5.921	238	3
	7	5.869	290	22
	8	5.847	312	1
	9	5.822	337	18
	10	5.776	383	3

Element*	Group No.	Disint. Energy MV	Excit. Energy kv	Relat. Number of Part.
AcX	0	5.823	0	40
	1	5.709	114	35
	2	5.634	189	18
	3	5.543	280	7
An	0	6.953	0	70
	1	6.683	270	15
	2	6.556	397	11
	3	6.343	610	4
AcC	0	6.739	0	84
	1	6.383	356	16
		Type II		
RaC′	0	7.829	0	10^6
	1	8.437	608	0.43
	2	9.112	1283	0.45
	3	9.242	1412	22
	4	9.493	1664	0.38
	5	9.673	1844	1.35
	6	9.844	2015	0.35
	7	9.968	2139	1.06
	8	10.097	2268	0.36
	9	10.269	2440	1.67
	10	10.342	2513	0.38
	11	10.526	2697	1.12
	12	10.709	2880	0.23
ThC′	0	8.948	0	10^6
	1	9.674	726	34
	2	10.745	1797	190

* Element *emitting* the α-particles.
† Rosenblum, Guillot and Perey (R14a) report 18 instead of 11 α-groups for RaAc and about 10 for AcX. According to the authors, most of these new groups require confirmation as to their exact position and some as to their existence.

TABLE XXXII. *Possible combinations between nuclear levels of ThC″ and observed γ-rays.*

Combination	$h\nu$ (kv) calc.	$h\nu$ (kv) obs.	Designation	Combination	$h\nu$ (kv) calc.	$h\nu$ (kv) obs.	Designation
4→3	19.4	—	—	3→1	432.4	431.7	γ_2
4→2	164.0	—	—	3→0	472.4	470.9	γ_3
4→1	451.8	451.1	γ_1	2→1	287.8	286.9	γ_4
4→0	491.8	—	—	2→0	327.8	326.7	γ_5
3→2	144.6	—	—	1→0	40.0	39.9	γ_6

The agreement is perfect within the experimental accuracy. Of the 10 possible combinations, 6 are "allowed" by optical selection rules. From observations of the internal conversion of the γ-rays it can be shown (§88B) that all the observed γ-rays correspond to quadrupole radiation.

Type II α-spectra are interpreted as being due to various states of the *initial nucleus emitting the α-particle*. Such an emission of α-particles *from* an excited state is only possible if the emission probability is comparable with the probability of the emission of a γ-ray by which the nucleus would go over into a lower quantum state. Since the lifetime of nuclear states against γ-emission is only about 10^{-13} sec. (see below, and §87, 88), α-particle emission from excited states is only possible if the respective α-particles have extremely high energies. Accordingly, the emission of α-rays of longer range than the normal group has only been observed for the nuclei RaC′ and ThC′ for which the normal α-particles are already very energetic.

Neither RaC′ nor ThC′ seems to have any α-spectrum of type I, i.e., any groups of α-particles having energy lower than and intensity comparable to the main group (0). Thus it seems that the final nuclei formed in the disintegrations, *viz.* RaD and ThD, respectively, have no excited levels of importance. The observed long range α-groups should thus give directly the excited levels of the initial nuclei RaC′ and ThC′. The level schemes obtained from the α-groups can again be checked, to a certain extent, by the γ-rays accompanying the α-disintegration. These γ-rays are, in this case, *alternative* to the emission of long range α-particles and not following it.

The number of α-rays in a given long range group is proportional to

$$N_\alpha = F\Gamma_\alpha/(\Gamma_\gamma+\Gamma_\alpha). \quad (603)$$

Here F is the probability of formation of the excited state from which the group originates, in the preceding disintegration, i.e., in the β-disintegration of RaC and ThC, respectively. Γ_α and Γ_γ are the probabilities of emission of α-rays and γ-rays from the given excited level. The number of γ-rays emitted from the same excited level would be

$$N_\gamma = F\Gamma_\gamma/(\Gamma_\gamma+\Gamma_\alpha). \quad (603a)$$

By comparing the number of γ-rays and α-rays, we can therefore determine the ratio $\Gamma_\gamma/\Gamma_\alpha$. The results are approximately (cf. §88) $\Gamma_\gamma/\Gamma_\alpha \approx 10^6$, 90, 4000 and 100 for the levels 1 and 3 of RaC′ and the levels 1 and 2 of ThC′, respectively. This means that γ-emission is more probable in all cases which agrees with reasonable theoretical expectations (§88).

The most important application of the observed ratio $\Gamma_\gamma/\Gamma_\alpha$ is to estimate Γ_γ. The probability of emission of α-rays may be estimated by the usual formula $\Gamma_\alpha = G_\alpha e^{-2C}$ where we put the "α-ray width without barrier" G_α again equal to one volt, and calculate the penetrability exponent C from the general formula (589), with the nuclear radius $R = 13.9\cdot 10^{-13}$ cm as observed for the ground state (group 0) of RaC′. Thus we obtain $\hbar/\Gamma_\alpha = 2.7\cdot 10^{-7}$, $9.5\cdot 10^{-9}$, $1.6\cdot 10^{-9}$ and $6\cdot 10^{-11}$ sec., and therefore $\hbar/\Gamma_\gamma = 3$, 800, 4 and $6\cdot 10^{-13}$ sec. for the four levels mentioned. This is the only way available for determining the lifetimes of excited states of radioactive nuclei for γ-radiation. For a discussion of this lifetime, cf. §87, 88C.

Since Γ_γ is always much larger than Γ_α, the number of nuclei *formed* in the excited state is practically equal to the number of emitted γ-quanta. For all the four levels of RaC′ and ThC′ mentioned, the probability of formation is much less than the probability of formation of the ground states of the respective nuclei, *viz.* about 0.4 and 0.0025 for the excited levels no. 1 and 3 of RaC′, and 0.14 and 0.02 for the two excited levels of ThC′. The reason for this small probability of formation is obviously the fact that the probability of β-decay increases with increasing β-energy (§41) and is therefore greater when the β-decay leads to the ground state unless this transition is forbidden.

Complete level schemes using both α- and γ-ray data have been given by Ellis (E2, E3, E4,

E7, E8, E9) for the radioelements ThB, C, C′, C″ and D. From these level schemes, the first evidence for the neutrino hypothesis was derived (§39), since it could be shown that the total energy evolution in the disintegration ThC—C′—D was the same as in the alternative disintegration ThC—C″—D when the maximum energy of the β-rays was considered as the disintegration energy. Ellis (E8, E9) has also given a scheme of levels for RaC′.—In all cases, the relative excitation probabilities of the various levels have been worked out. From these, the average excitation energy can be calculated which may be compared to the total energy evolution in the disintegration as measured calorimetrically. The agreement is fairly good (E2).—Attempts have been made to assign quantum numbers (angular momentum) to the various states, using the results about the character (dipole or quadrupole radiation) of the various γ-rays emitted and the selection rules indicated by the absence of certain γ lines. The results are still questionable.

XII. Scattering of Charged Particles by Nuclei

§70. General Remarks

The scattering of α-particles by nuclei is one of the oldest methods for exploring nuclear fields. More recently, the scattering of protons by protons (§18) gave the most convincing proof for the existence of forces between like particles. The scattering of protons and deuterons by deuterons and α-particles will probably provide an excellent test of our theoretical ideas about the nuclear forces and of the approximate wave functions used for these light nuclei (cf. Chapter III). The scattering by slightly heavier nuclei, from Li to Si or P, has given and will give evidence about nuclear resonance levels. It appears that the scattering of charged particles is very sensitive to small deviations from the Coulomb field and is therefore well suited to investigations of nuclear forces and resonance levels. For this reason, it is regrettable that so little attention has been paid to this field of nuclear physics in more recent years. Of course, it must be admitted that scattering experiments are useful only for the lighter nuclei; for the heavier ones, the Coulomb scattering is too large and masks all other effects even if particles of sufficient energy are available to penetrate through the Coulomb field into the nucleus.

The theoretical treatment of the scattering is, of course, most conveniently carried out in a coordinate system in which the center of mass of scattered particle and scattering nucleus is at rest (C system, §14). If M_1 is the mass of the scattered particle, M_2 that of the nucleus, and ϑ the deflection of the scattered particle in the C system, then the *observed* deflection Θ of the particle in the laboratory coordinate system (R system) is given by

$$\tan\Theta = \frac{M_2 \sin\vartheta}{M_1 + M_2\cos\vartheta}, \tag{604}$$

provided the scattering nucleus is initially at rest. Conversely, ϑ may be deduced from the observed deflection Θ:

$$\sin(\vartheta - \Theta) = (M_1/M_2)\sin\Theta \tag{604a}$$

or, approximately, for $M_2 \gg M_1$:

$$\vartheta = \Theta + (M_1/M_2)\sin\Theta. \tag{604b}$$

The scattering nucleus itself experiences a recoil; the angle between its direction of motion and that of the incident particle is simply

$$\varphi = \tfrac{1}{2}(\pi - \vartheta). \tag{604c}$$

The number of particles scattered through an angle between Θ and $\Theta + d\Theta$ in the ordinary coordinate system, is

$$N(\Theta)\sin\Theta d\Theta = N(\vartheta)\sin\vartheta d\vartheta$$
$$= \frac{N(\vartheta)}{M_2}\frac{[M_1\cos\Theta + (M_2^2 - M_1^2\sin^2\Theta)^{\frac{1}{2}}]^2}{(M_2^2 - M_1^2\sin^2\Theta)^{\frac{1}{2}}}\sin\Theta d\Theta, \tag{605}$$

where $N(\vartheta)$ is the number scattered per unit solid angle in the C system. For $M_1 \ll M_2$ this reduces to:

$$N(\vartheta)[1 + 2(M_1/M_2)\cos\Theta]\sin\Theta. \tag{605a}$$

The mass to be inserted into the Schrödinger equation is the *reduced mass*

$$M = M_1M_2/(M_1 + M_2). \tag{606}$$

The kinetic energy associated with the relative motion of scattered particle and scattering nucleus is

$$E = M_2E_0/(M_1 + M_2) = \tfrac{1}{2}Mv^2, \tag{606a}$$

where v is the velocity of the incident particle in the laboratory system and $E_0 = \frac{1}{2}M_1v^2$ its kinetic energy measured in the ordinary way. The energy transferred to the scattering nucleus in the collision is

$$E' = 4E_0M_1M_2(M_1 + M_2)^{-2}\sin^2\tfrac{1}{2}\vartheta. \tag{607}$$

The maximum fraction of the energy of the incident particle which can be transferred is thus

$$E'_{\max}/E_0 = 4M_1M_2/(M_1 + M_2)^2; \tag{607a}$$

the scattered particle goes, in this case, back into the

direction from which it came ($\Theta=\vartheta=180°$) with a velocity equal to $(M_2-M_1)/(M_2+M_1)$ times its original velocity. All these relations follow from elementary application of conservation of energy and momentum.

If only the Coulomb force acts between scatterer and scattered particle, the cross section[52] for scattering through an angle between ϑ and $\vartheta+d\vartheta$, is given by the well-known Rutherford law

$$\sigma(\vartheta)\sin\vartheta d\vartheta=2\pi(zZe^2/2Mv^2)^2 \times(\sin\tfrac{1}{2}\vartheta)^{-4}\sin\vartheta d\vartheta, \quad (608)$$

where Ze is the charge of the scattering nucleus, ze that of the scattered particle, M the reduced mass (606) and v the velocity of the incident particle.

The Rutherford law (608) has proved correct for the scattering of α-particles and protons by all heavy nuclei. As is well known, the scattering of α-particles by heavy nuclei was the original basis of Rutherford's proposal of the nuclear atom (R26). The accuracy of the Rutherford law for heavy nuclei is so great that the nuclear charge can be determined from the scattering of α-particles (C11).

We are interested in *deviations* from the Rutherford law. Such deviations will occur (1) if the incident particle has an energy great enough to overcome the potential barrier of the nucleus, and (2) if there is resonance, i.e., if the kinetic energy of the incident particle plus the internal energy of the initial nucleus is equal to the energy of one of the states of the compound nucleus.

Experiments on the scattering of particles by nuclei can therefore be used to find resonance levels and also to find the height of the potential barrier. If there is a strong deviation from the Rutherford law which is restricted to a small energy interval, we shall attribute it to resonance. On the other hand, a deviation which sets in rather gradually with increasing energy, and does not disappear upon further increase, is in general interpreted as showing that the particle can go over the top of the potential barrier. (For the exact explanation, cf. §71.)

It must be admitted that the heights of potential barriers obtained from scattering are not very accurate. The reasons for this are (1) that the penetration through the barrier is already fairly probable when the particle energy is still well below the height of the barrier, (2) that the penetration of particles of higher angular momentum keeps increasing even for energies higher than the barrier (§78), (3) that resonances may occur even for energies above the barrier height (§§78, 82).

As far as the evidence goes (§§74, 75), the radii deduced from scattering experiments seem compatible with the assumption that the nuclear volume is proportional to the number of particles in the nucleus. With this assumption, we have found (cf. (598), (598a))

$$R=2.05\cdot10^{-13}A^{\frac{1}{3}}\text{ cm} \quad (608a)$$

and the expressions (599), (599a) for the height of the potential barrier, B. In applying these formulae, it must be kept in mind that only the *relative* kinetic energy is available for overcoming the potential barrier. With the notations used in the beginning of this section, the relative kinetic energy is $M_2/(M_1+M_2)$ times the total kinetic energy of the incident particle, provided the bombarded nucleus is at rest. Therefore, if $a=M_1/M_0$ (M_0=proton mass) is the atomic weight of the incident particle, the kinetic energy must be greater than

$$B=0.70\,zZ(A+a)A^{-4/3}\text{ MV} \quad (609)$$

in order to overcome the potential barrier. Table XXXIII gives the effective heights of the potential barriers of various nuclei for various incident particles according to (609).

The fastest α-particles available are those from ThC′ which have an energy of about 9 MV. Thus

TABLE XXXIII. *Effective heights B of potential barriers in MV. "Characteristic orbital momenta" g(Z).*

Z		2	4	10	20	30	50	70	92
A		4	9	20	40	66	112	174	238
NUCLEUS		He	Be	Ne	Ca	Zn	Sn	Yb	U
Height of barrier for	Protons	1.1	1.5	2.7	4.2	5.3	7.3	8.8	10.4
	Deuterons	1.3	1.6	2.8·	4.3	5.4	7.4	8.9	10.5
	α-particles	3.5	3.9	6.2	9.0	11.0	15.0	18.0	20.9
Critical orbital momentum	Protons	0.6	1.0	1.9	3.1	4.1	5.9	7.5	9.1
	Deuterons	0.8	1.4	2.6	4.3	5.8	8.3	10.6	12.8
	α-particles	1.3	2.5	5.0	8.4	11.5	16.4	21.0	25.5

[52] The cross section is so defined that the *number* of particles scattered by a substance containing N scattering nuclei per cm^3, is

$$I=NI_0\,\sigma(\vartheta)\sin\vartheta d\vartheta,$$

where I_0 is the number of incident particles per cm^2 per sec., and I the number of particles scattered per sec. through an angle between ϑ and $\vartheta+d\vartheta$.

the scattering of α-particles may give evidence about the nuclear field for nuclei up to about $Z=20$ (calcium). Actual experiments have been carried out up to $Z=13$ (Al, cf. §75).

From Table XXXIII we see furthermore that about 10 MV is required for protons and deuterons to go over the top of the potential barrier of uranium. Such energies will probably soon be available from cyclotrons.

An important point for the scattering, especially for the angular distribution of the scattered particles, is the orbital momentum of the incident particle. If the kinetic energy of the particle is large compared to the potential barrier, particles of orbital momentum l will in general approach the nucleus to a minimum distance of $l\lambda\!\!\!^{-}$ (cf., e.g. (54b, c, d)). The particles strongly affected by the nuclear field will therefore be those for which $l\lambda\!\!\!^{-}<R$, i.e., whose orbital momentum is less than

$$l_0=R/\lambda\!\!\!^{-}. \qquad (610)$$

At low energies, when the Coulomb field is important, the problem is much more complicated (§72, end). We may, however, formally apply (610) to an energy just sufficient to overcome the potential barrier: Then the wave-length is

$$\lambda\!\!\!^{-}_B=\hbar R^{\frac{1}{2}}/(2Me^2zZ)^{\frac{1}{2}} \qquad (610a)$$

and (610) becomes (cf. (600a))

$$l_0(B)=g(Z). \qquad (611)$$

§71. General Theory of Scattering of Charged Particles

The general theory of scattering is contained in §55 in which the dispersion formula for nuclear processes was derived. It is only necessary to make a few slight changes in order to represent correctly the asymptotic behavior of the wave functions of a particle in a Coulomb field. This is particularly essential because in the Coulomb scattering the contributions of high orbital momenta is very important, in contrast to the scattering in all other types of fields.

In a pure Coulomb field, the wave function of the scattered particle normalized per unit current is (cf. M32, p. 35, and this article (100))

$$\psi_0=v^{-\frac{1}{2}}e^{ikz+i\alpha\log k(r-z)}+(zZe^2/2Mv^2r\sin^2\tfrac{1}{2}\vartheta)\times v^{-\frac{1}{2}}e^{ikr-i\alpha\log 2kr-i\alpha\log\sin^2\frac{1}{2}\vartheta+i\pi+2i\eta_0}, \qquad (612)$$

where M is the reduced mass (606),

$$\alpha=zZe^2/\hbar v, \qquad (612a)$$

$$k=Mv/\hbar=(2ME)^{\frac{1}{2}}/\hbar, \qquad (612b)$$

$$e^{i\eta_0}=\Gamma(1+i\alpha)/|\Gamma(1+i\alpha)|. \qquad (612c)$$

The first term in (612) represents the incident, the second the scattered wave. The square of the absolute value of the second term, multiplied by $r^2d\omega$, gives just the Rutherford scattering cross section for the element of solid angle

$$d\omega=2\pi\sin\vartheta d\vartheta.$$

Alternatively, we may write the Coulomb wave function as a sum over the partial waves of different l, *viz.:*

$$\psi_C=\sum_l A_l f_l(r)Y_{l0}(\vartheta)/r. \qquad (613)$$

The radial function f_l behaves asymptotically as (cf. M32, p. 39)

$$f_l=\sin(kr-\alpha\log 2kr-\tfrac{1}{2}l\pi+\eta_l), \qquad (613a)$$

where

$$e^{i\eta_l}=\frac{\Gamma(l+1+i\alpha)}{|\Gamma(l+1+i\alpha)|}=e^{i\eta_0}\frac{(1+i\alpha)\cdots(l+i\alpha)}{(1+\alpha^2)^{\frac{1}{2}}\cdots(l^2+\alpha^2)^{\frac{1}{2}}} \qquad (613b)$$

and

$$A_l=\frac{(4\pi)^{\frac{1}{2}}(2l+1)^{\frac{1}{2}}i^l}{v^{\frac{1}{2}}k}e^{i\eta_l}. \qquad (613c)$$

The wave scattered by the nucleus has, according to (373a) (383b), the general form

$$\psi_N=\sum_{lm}\beta_{Plm}(if_l+g_l)Y_{lm}(\vartheta\varphi)/r, \qquad (614)$$

where f_l is the regular solution (613a) in the Coulomb potential[53] and g_l the solution which is

[53] This means that we identify the particle potential U_P (cf. §55, (367)) with the Coulomb potential. Actually, it would probably be a better approximation to take, inside the nucleus, a constant repulsive potential. We would then have the Coulomb scattering, potential scattering from the repulsive auxiliary potential between nucleus and particle, and resonance scattering. We have not included the nuclear potential scattering in order not to make our formulae too complicated; also, not much useful information can be gained by including this potential scattering.

irregular at $r=0$. We have (cf. (376), (376b)) asymptotically

$$g_l+if_l=e^{i(kr-\alpha\log 2kr-\frac{1}{2}l\pi+\eta_l)}. \quad (614a)$$

The coefficients β can be expressed by the A's (cf. (613c)); we have (cf. 380, 391, 392a)

$$\beta_{Plm}=-\pi\sum_{rl'}\frac{A_{l'}H^{C_r}{}_{Pl'0}H^{C_r}{}_{Plm}{}^*}{E-E_r+\frac{1}{2}i\gamma_r}. \quad (615)$$

A change of the magnetic quantum number of the orbital motion of the particle from 0 to m is, of course, only possible if at the same time the spin of the scattering nucleus or of the scattered particle or both change their direction. We therefore write (615) more fully:

$$\beta_{Plm}{}^{\mu_i'\mu_s'}{}_{\mu_i\mu_s}=$$

$$-\pi\sum_{rl'M}\frac{A_{l'}H^{rM}{}_{Pl'0\mu_i'\mu_s'}H^{rM}{}_{Plm\mu_i\mu_s}{}^*}{E-E_r+\frac{1}{2}i\gamma_r}, \quad (615a)$$

where $\mu_i'\mu_s'$ are the components of the spins of scattering nucleus and scattered particle before the collision, $\mu_i\mu_s$ the same quantities after collision, and M the magnetic quantum number of the compound state. We have the selection rule

$$m+\mu_i+\mu_s=M=\mu_i'+\mu_s'. \quad (615b)$$

We insert (613c, 614a, 615a) into (614) and add the result to (612). Then we obtain for a given pair of spin quantum numbers $\mu_i\mu_s$ of nucleus and particle,

$$\psi^{\mu_i'\mu_s'}{}_{\mu_i\mu_s}=v^{-\frac{1}{2}}e^{ikz+i\alpha\log k(r-z)}\delta_{\mu_i\mu_i'}\delta_{\mu_s\mu_s'}$$

$$-v^{-\frac{1}{2}}e^{ikr-i\alpha\log 2kr+2i\eta_0}r^{-1}f^{\mu_i'\mu_s'}{}_{\mu_i\mu_s} \quad (616a)$$

with

$$f^{\mu_i'\mu_s'}{}_{\mu_i\mu_s}=\frac{zZe^2}{2Mv^2\sin^2\frac{1}{2}\vartheta}e^{-i\alpha\log\sin^2\frac{1}{2}\vartheta}\delta_{\mu_i\mu_i'}\delta_{\mu_s\mu_s'}$$

$$+2\pi^{\frac{3}{2}}\lambda\sum_{lm}Y_{lm}(\vartheta\varphi)\sum_{l'rM}i^{l'-l}(2l'+1)^{\frac{1}{2}}e^{i(\eta_l+\eta_{l'}-2\eta_0)}$$

$$\times H^{CrM}{}_{Pl'0\mu_i'\mu_s'}H^{CrM}{}_{Plm\mu_i\mu_s}{}^*/(E-E_r+\tfrac{1}{2}i\gamma_r). \quad (616)$$

The cross section for elastic scattering through the angle ϑ is then:

$$\sigma(\vartheta)d\omega=\frac{d\omega}{(2i+1)(2s+1)}\sum_{\mu_i'\mu_s'\mu_i\mu_s}|f^{\mu_i'\mu_s'}{}_{\mu_i\mu_s}(\vartheta\varphi)|^2 \quad (616b)$$

In (616b), the average has been taken over all possible directions of the spins of nucleus A and particle P in the initial state and the sum over all directions in the final state. As a result, the cross section (616b) is independent of the azimuth φ.

The expression (616), (616b) for the cross section is rather complicated and no simplified form of it has yet been worked out in the general case. However, in some special cases it is not difficult to reduce the expression to a simpler one.

A. Scattering nucleus and scattered particle having zero spin

In this case, the sum over $\mu_i'\mu_s'\mu_i\mu_s$ in (616b) reduces to a single term $\mu_i'=\mu_s'=\mu_i=\mu_s=0$. The selection rules (615b) require that the compound state r have zero magnetic quantum number M. Moreover, its angular momentum J must be equal to l'. Similarly, $H^{C_r}{}_{Plm}$ can only be different from zero if $l=J=l'$ and $m=0$. Furthermore, we have

$$2\pi|H^{CrJ}{}_{Pl'0}|^2=\gamma^{rJ}{}_{P},\quad (617)$$

since the given state r of the compound nucleus can only emit particles of angular momentum l' and magnetic quantum number zero, again owing to the selection rules. Inserting $e^{2i(\eta_l-\eta_0)}$ from (613b), and remembering that

$$Y_{l0}=(4\pi)^{-\frac{1}{2}}(2l+1)^{\frac{1}{2}}P_l(\vartheta),$$

we have from (616):

$$\sigma(\vartheta)=\Bigg|\frac{e^2zZ}{2Mv^2\sin^2\frac{1}{2}\vartheta}e^{-i\alpha\log\sin^2\frac{1}{2}\vartheta}$$

$$+\tfrac{1}{2}\lambda\sum_J(2J+1)P_J(\vartheta)$$

$$\times\frac{(1+i\alpha)^2\cdots(J+i\alpha)^2}{(1+\alpha^2)\cdots(J^2+\alpha^2)}\sum_r\frac{\gamma^{rJ}{}_{P}}{E-E_r+\frac{1}{2}i\gamma_{rJ}}\Bigg|^2. \quad (618)$$

Of special interest is the case of resonance, in which the sum over r and J reduces to a single term. Replacing the "effective widths" γ by the "true widths" Γ (cf. §52), and introducing the abbreviations

$$(2J+1)\frac{\lambda\!\!\!^{-}\Gamma^{rJ}{}_{Pp}}{\Gamma_{rJ}}\frac{2Mv^2\sin^2\frac{1}{2}\vartheta}{e^2zZ}P_J(\vartheta)=2(2J+1)$$

$$\times(\hbar v/e^2zZ)(\Gamma^r{}_{Pp}/\Gamma_r)\sin^2\tfrac{1}{2}\vartheta P_J(\vartheta)=\rho, \quad (618a)$$

$$e^{i\alpha\log\sin^2\frac{1}{2}\vartheta}\frac{(1+i\alpha)^2\cdots(J+i\alpha)^2}{(1+\alpha^2)\cdots(J^2+\alpha^2)}=e^{i\zeta}, \quad (618b)$$

$$2(E-E_r)/\Gamma_r=x, \quad (618c)$$

$$(e^2zZ/2Mv^2\sin^2\tfrac{1}{2}\vartheta)^2=\sigma_0, \quad (618d)$$

we obtain for the cross section $\sigma(\vartheta)$:

$$\frac{\sigma(\vartheta)}{\sigma_0}=1+\frac{\rho^2+2\rho\sin\zeta+2\rho x\cos\zeta}{1+x^2}, \quad (619)$$

σ_0 is the Rutherford scattering cross section.

The ratio of resonance scattering to Coulomb scattering is primarily determined by ρ. This quantity is (cf. 618a) in general greater for large scattering angles than for small ones. Thus the backward scattering ($\vartheta=\pi$) will show the most pronounced deviations from the Rutherford formula (cf. §74, 75). ρ increases with increasing energy as $v\sim E^{\frac{1}{2}}$ (see, however, remark 1 below). For a given energy, it is larger for light than for heavy particles ($v\sim M^{-\frac{1}{2}}$) and larger for a small charge z of the scattered particle than for a large charge. From this it would follow that protons will show more pronounced resonance scattering than deuterons, and deuterons more than α-particles. However, this factor is probably more than offset by the considerations given below, cf. remark 3. Resonances with levels of high angular momentum J are stronger than for low J, provided they can occur at all (cf. end of §72).

The factor of greatest importance is $\Gamma^r{}_{Pp}/\Gamma_r$. It represents the ratio of the partial width of the resonance level (corresponding to emission of the incident particle P with the scattering nucleus being left in the ground state p) to the total level width. In order that this ratio be large, it is necessary that no other processes but elastic scattering can occur with great probability. The conditions for this are:

(1) The scattered particle must not have too high energy, because otherwise inelastic scattering will occur. Particularly if the energy E_P is high enough so that many levels of the scattering nucleus can be excited, the total width corresponding to all the excited states of the scattering nucleus will in general be much greater than the partial width corresponding to the ground state. This fact will in many cases more than offset the factor v in (618a), and will lead to a *decrease in the strength of resonances with increasing energy.*

(2) The scattering nucleus must not be too heavy because otherwise it will have many excited states of low energy so that again inelastic scattering will be very probable.

(3) The compound nucleus C must not disintegrate with emission of particles other than P with any great probability. This condition will be most easily fulfilled if P is an α-particle. Because of the large mass defect of the α-particle, most nuclear reactions produced by α-particles are endoergic, and unless the kinetic energy of the α-particle is very high, no nuclear reaction can in general occur, especially if the scattering nucleus itself has a very high mass defect. This is the case for most nuclei of zero spin, which are, strictly speaking, the only ones for which our formulae are valid (see above).

Deuterons, and to a lesser extent protons, can almost with every nucleus cause a variety of nuclear processes, owing to their large internal energy. In these cases the total level width Γ_r will in general be much larger than the partial width referring to the incident particle, $\Gamma^r{}_{Pp}$. It is therefore likely that the strongest resonance effects are found in the scattering of α-particles, less strong effects with protons and very weak resonances in the deuteron scattering. This will be true in spite of the factor $1/z$ in (618a) which would tend to make the proton and deuteron resonances stronger than those for α-particles.

To obtain an estimate of the absolute magnitude of ρ, we may consider the α-particles of polonium (energy 5.3 MV). Let us assume that the partial width $\Gamma^r{}_{Pp}$ is equal to the total Γ_r, and that the scattering angle is 180°. Then approximately

$$\rho=7.3(2J+1)/Z. \quad (619a)$$

Thus for elements up to $N(Z=7)$, ρ would be larger than unity even for $J=0$, i.e. for S levels of the compound nucleus. This means, according to (619), that the resonance scattering backward will, at its maximum, be more than twice the Coulomb scattering whatever J. For elements heavier than nitrogen, the resonance scattering by an S level of the compound nucleus will increase the Coulomb scattering by less than a factor two: With the limited accuracy of the scattering experiments, it will then be difficult to observe resonance scattering of Po α-particles by S resonance-levels. P resonances ($J=1$), on the other hand, will correspond to an increase in the cross section by a factor of about 10 for nitrogen and will therefore be easily observable up to rather high atomic number, probably as long as the nuclear potential barrier is at all penetrable for Po α-particles. This is *a fortiori* true for D resonances.

We shall now discuss the behavior of the scattering cross section (619) near resonance. As in all cases, the resonance scattering is restricted to an energy region of width Γ_r about the resonance energy E_r. An investigation of the scattering cross section as a function of the energy of the incident particle will therefore indicate the width of the resonance level. The cross section has a minimum and a maximum at

$$x_{\min}{}^{\max} = \pm(1+c^2)^{\frac{1}{2}} - c, \tag{620}$$

where

$$c = (\rho/2\cos\zeta) + \tan\zeta. \tag{620a}$$

The maximum and minimum cross section are given by

$$\sigma_{\min}{}^{\max}/\sigma_0 = 1 + \rho[\pm(\tfrac{1}{4}\rho^2 + \rho\sin\zeta + 1)^{\frac{1}{2}} + \tfrac{1}{2}\rho + \sin\zeta]. \tag{621}$$

If ρ is large, i.e., if the resonance scattering is large compared to the Coulomb scattering, the minimum of the cross section occurs at

$$x_{\min} = -\rho/\cos\zeta, \tag{622}$$

i.e. (cf. 618c),

$$E = E_r - \tfrac{1}{2}\Gamma_r\rho/\cos\zeta. \tag{622a}$$

The maximum occurs at

$$x_{\max} = \cos\zeta/\rho, \tag{622b}$$

which, for large ρ, is practically at the resonance energy. The minimum cross section is, for large ρ:

$$\sigma_{\min} = \sigma_0 \sin^2\zeta \tag{623}$$

and the maximum cross section

$$\sigma_{\max} = \sigma_0\rho(\rho + 2\sin\zeta) \approx \sigma_0\rho^2. \tag{623a}$$

Inserting the values of σ_0 and ρ from (618a, d), this becomes

$$\sigma_{\max} = (2J+1)^2 \lambda\!\!\!^-{}^2 (\Gamma^r{}_{Pp}/\Gamma_r)^2 P_J{}^2(\vartheta). \tag{624}$$

This is the same cross section as without the Coulomb field. It should, however, be kept in mind that the condition $\rho \gg 1$ will never be fulfilled for small scattering angles ϑ, so that for small angles the Coulomb scattering is always predominant.

B. Any spin of scattering nucleus and scattered particle, but only zero orbital momentum important

This condition will be fulfilled for slow particles. The sum over l and m in (616) reduces to a single term $l=m=0$, the sum over l' to a single term $l'=0$. The magnetic quantum number of the compound state M is equal to the sum of the magnetic quantum numbers of the spins of scattering nucleus and scattered particle $\mu_i' + \mu_s' = \mu_i + \mu_s$. A fairly simple calculation involving the spacial symmetry properties of the matrix elements H in (616) (similar to the calculation in the appendix of B15) gives for the scattering cross section

$$\frac{\sigma(\vartheta)}{\sigma_0} = 1 + \frac{2J+1}{(2i+1)(2s+1)} \frac{\rho^2 + 2\rho\sin\zeta + 2\rho x\cos\zeta}{1+x^2} \tag{625}$$

similar to (619). Here i and s are the spins of scattering nucleus and scattered particle, σ_0 is the Rutherford scattering cross section (618d), x the distance from resonance divided by half the width of the resonance level (cf. 618c), while ζ and ρ have somewhat simpler forms than in (618a, b) because the orbital momentum of the incident particle is now zero rather than J. We have

$$\zeta = \alpha \log \sin^2 \tfrac{1}{2}\vartheta, \tag{625a}$$

$$\rho = (2\hbar v/e^2 zZ)(\Gamma^r{}_{Pp}/\Gamma_r)\sin^2\tfrac{1}{2}\vartheta. \tag{625b}$$

The "shape of the resonance line" (625) is the same as for zero spin of nucleus and particle (cf. 619). If we compare the scattering of particles of orbital momentum zero in both cases, we see that the intensity of the resonance scattering is *reduced* by a factor $(2J+1)/(2i+1)(2s+1)$ by the existence of the spins of particle and nucleus.[54] On the other hand, there will be *more* resonances: if the spins s and i are zero, and the orbital momentum $l=0$, only S-states ($J=0$) of the compound nucleus can give rise to resonance scattering, whereas, if s and i are different from zero, any compound state with J between $|s-i|$ and $s+i$ will give resonance scattering.[54]

[54] It may be mentioned that there will be no reduction in the intensity, and no increase in the number of resonances if only *one* of the two spins, s or i, is different from zero. However, this is only true for orbital momentum zero; for arbitrary l, *one* spin different from zero is sufficient to decrease the intensity and increase the number of effective levels.

C. Very high density of levels (high energy)

If we average over an energy interval large compared to the spacing of levels (cf. §56, §65), the interference term between Coulomb scattering and resonance scattering will disappear. Then the scattering cross section will be equal to the Coulomb scattering, plus the average resonance scattering, the latter having the same form as for neutrons. The most probable process will in general be a disintegration or inelastic scattering. The elastic scattering will be primarily Coulomb scattering at small angles and nuclear potential scattering at large angles.

§72. Penetration of the Potential Barrier and Angular Momentum

The scattering is, according to (616, 618a), primarily determined by the matrix elements H, or by the partial width $\Gamma^r{}_{Pp}$ of the nuclear resonance level, the latter being proportional to the square of the former. The matrix elements involve the wave function of the incident particle in the nucleus (or at its surface, if the nuclear potential suggested in §54A is accepted). This wave function introduces the well-known penetrability of the potential barrier as a factor into the matrix elements, and therefore into the partial width of the levels and into the scattering cross section.

Accurate tables of the wave functions of a particle in a Coulomb field have been given by Yost, Wheeler and Breit (Y2). They must be used whenever the fundamental assumptions of the theory are sufficiently justified to warrant exact calculations, as, e.g., in the scattering of protons by protons (B53). Similarly, when only one resonance level is of importance, and all the angular momenta concerned are well known, the use of exact wave functions is desirable. Such a case seems to be, e.g., the resonance scattering of protons of 440 kv by Li^7 nuclei (cf. §75).

For estimates, especially when the properties of the nuclei concerned are not well known, it will be sufficient to use the WKB method for calculating wave functions. If we assume the Coulomb potential to hold everywhere (down to $r=0$), the wave function of the particle must *decrease* exponentially as we approach the nucleus. In the notation of §66, we have thus to take the function u_2 (except for the inside part). Therefore, if the wave function behaves asymptotically as (cf. 578a)

$$\begin{aligned}\psi(r) &= v^{-\frac{1}{2}} \cos\left(\int_{r_E}^{r} \Phi^{\frac{1}{2}}(\rho)d\rho - \tfrac{1}{4}\pi\right) \\ &= v^{-\frac{1}{2}} \sin(kr - \alpha \log 2kr + \eta_l),\end{aligned} \tag{626}$$

its value at the surface of the nucleus ($r=R$) will be

$$\psi(R) = \tfrac{1}{2}(m/\hbar)^{\frac{1}{2}} |\Phi(R)|^{-\frac{1}{4}} \times \exp\left(-\int_{R}^{r_E} |\Phi(\rho)|^{\frac{1}{2}} d\rho\right), \tag{626a}$$

where $$|\Phi(R)| = 2m\hbar^{-2}(zZe^2/R - E), \tag{626b}$$

E is the particle energy, and

$$r_E = zZe^2/E. \tag{626c}$$

The exponent in (626a) is *exactly* the quantity C (cf. (585a)); therefore the square of the exponential is simply the penetrability of the potential barrier.

We may now change our assumptions and suppose that there is a constant repulsive potential V_0 acting on the particle when it is inside the nucleus (cf. §54A). We assume that the transition from the Coulomb potential to this nuclear potential is gradual so that the WKB solution holds throughout. Then we have inside the nucleus

$$\psi(r) = \tfrac{1}{2}(\tfrac{1}{2}m)^{\frac{1}{4}}(V_0 - E)^{-\frac{1}{4}} e^{-C} e^{-\kappa(R-r)} \tag{627}$$

with

$$\kappa = (V_0 - E)^{\frac{1}{2}}(2m)^{\frac{1}{2}}/\hbar \tag{627a}$$

as in (335). If there were no potential barrier, we should have instead of (627) (cf. 335)

$$\psi_0(r) = (\tfrac{1}{2}m)^{\frac{1}{4}} E^{\frac{1}{4}} V_0^{-\frac{1}{2}} e^{-\kappa(R-r)}. \tag{627b}$$

The ratio is

$$\psi/\psi_0 = \tfrac{1}{2}[V_0^2/E(V_0 - E)]^{\frac{1}{4}} e^{-C}. \tag{628}$$

The most important factor here is, of course, the penetrability e^{-C}. However, there is, in addition, a factor $\tfrac{1}{2}[V_0^2/E(V_0-E)]^{\frac{1}{4}}$. This factor arises from the fact that the potential must be assumed to change *suddenly* from zero to V_0 when there is *no potential* barrier, "suddenly" being understood in comparison with the wave-length of the particle outside the nucleus. On the other hand, the potential changes *gradually* when there is a *Coulomb field*. The sudden change of potential causes a reflection of particles at the surface of the nucleus, especially if the particles are very slow ($E \ll V_0$), tending to make the wave function of slow particles inside the nucleus relatively smaller in the case of free particles than in the case of a Coulomb field. The factor $v^{\frac{1}{2}} \sim E^{\frac{1}{4}}$ in the matrix element, and correspondingly $v \sim E^{\frac{1}{2}}$ in the width of the levels, does not occur in the Coulomb field but only in the case of free particles.

Practically, this point is of little consequence, since very slow charged particles will never be able to penetrate into the nucleus on account of the potential barrier. Therefore we shall, for our purposes, write the "partial width" for charged particles in the form

$$\Gamma^r{}_{Pp} = G^r{}_{Pp} e^{-2C} = G^r{}_{Pp} P, \qquad (629)$$

where $G^r{}_{Pp}$ is the width of the level without potential barrier, and P the penetrability of the barrier. A convenient formula for P was given in (600).

The influence of the orbital momentum of the particle on the penetrability may be calculated by the WKB method (cf. §66). The only difference as compared to the case of $l=0$ is that we have instead of (578b)

$$\Phi_l(r) = 2m\hbar^{-2}(E-V) - l(l+1)/r^2. \qquad (630)$$

Defining C_l in analogy with (585a), *viz.*

$$C_l = \int_R^{r_l} |\Phi_l(\rho)|^{\frac{1}{2}} d\rho, \qquad (630a)$$

where r_l is defined as the value of r for which Φ_l vanishes, we obtain

$$\frac{C_l}{g} = \tfrac{1}{2}x^{-\frac{1}{2}}\left(\tfrac{1}{2}\pi + \arcsin \frac{1-2x}{(1+4xy)^{\frac{1}{2}}}\right) - (y+1-x)^{\frac{1}{2}}$$
$$+ y^{\frac{1}{2}} \log \frac{1+2y^{\frac{1}{2}}[y^{\frac{1}{2}}+(y+1-x)^{\frac{1}{2}}]}{(1+4xy)^{\frac{1}{2}}}. \qquad (631)$$

Here g is the abbreviation introduced in (600a), $x=E/B$ is the ratio of particle energy to barrier height and

$$y = l(l+1)/g^2. \qquad (631a)$$

For values of l small compared to the "critical orbital momentum" g, i.e., for $y \ll 1$, and for $x<1$, this may be approximated by

$$C_l = C_0 + \tfrac{2}{3}g[(y+1-x)^{\frac{3}{2}} - (1-x)^{\frac{3}{2}}]. \qquad (632)$$

For energies not too near the top of the barrier, and below the top, we may further simplify (632) and obtain for the penetrability for the orbital momentum l

$$P_l = e^{-2C_l} = P_0\, e^{-2gy(1-x)^{\frac{1}{2}}}. \qquad (632a)$$

This expression falls to $1/e$ times its value for

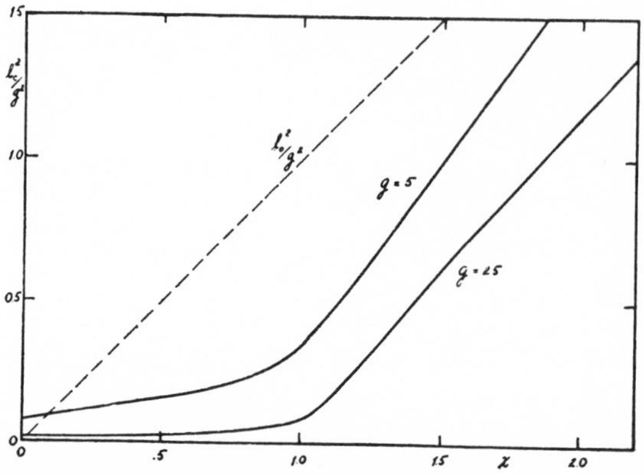

FIG. 19. Critical orbital momentum l_c of charged particles as a function of energy. Abscissa: Ratio of particle energy to barrier height. Ordinate: l_c^2/g^2 where g is the characteristic orbital momentum for the given nucleus. $g=5$ corresponds approximately to deuterons on zinc, $g=25$ to α-particles on uranium. The broken curve corresponds to neutral particles. The curves show clearly the sharp break when the particle energy becomes equal to the barrier height, and the rapid rise at energies above the barrier. They also show that at very low energies the critical angular momentum is larger for charged than for neutral particles.

$l=0$ when

$$y_c = \tfrac{1}{2}g^{-1}(1-x)^{-\frac{1}{2}}, \qquad (632b)$$

which corresponds to

$$l_c + \tfrac{1}{2} = g y_c^{\frac{1}{2}} = (\tfrac{1}{2}g)^{\frac{1}{2}}(1-x)^{-\frac{1}{4}}. \qquad (633)$$

This critical orbital momentum is, for low energy ($x \ll 1$), approximately the square root of one-half the critical momentum g given in Table XXXIII. Thus, e.g., slow α-particles ($E \ll 20$ MV) will be about equally effective in disintegrating uranium when they have orbital momentum 3 as for $l=0$, but much less effective when $l \gg 3$. Therefore the angular distribution of the particles produced in the disintegration of a heavy nucleus by a charged particle will not be spherically symmetrical, even if the wave-length of the incident particle is large compared to the nuclear radius (Teller, unpublished; cf. §78).

With increasing energy, the critical orbital momentum l_c increases slowly. For $x \approx 1$, i.e., for particles which can just go over the top of the barrier, (633) is no longer valid, but (632) reduces to ($C_0=0$)

$$C_l = \tfrac{2}{3}g y^{\frac{3}{2}}. \qquad (634)$$

The critical orbital momentum is then

$$l_c(B) = (\tfrac{3}{4})^{\frac{1}{3}} g^{\frac{2}{3}} = 0.91 g^{\frac{2}{3}}. \qquad (634a)$$

For α-particles on uranium, this gives $l_c(B)=8$. Fig. 19 gives l_c as a function of x (see §78).

TABLE XXXIV. *Penetrabilities of potential barriers for various orbital momenta.*

	Beryllium (B=3.9 MV)				Calcium (Z=20, B=9.0 MV)			
		Penetrability P_l				P_l		
x	E(MV)	$l=0$	$l=1$	$l=2$	E	$l=0$	4	8
1.6	6.2			1	14.4			0.51
1.4	5.5			0.82	12.6		1	0.165
1.2	4.7		1	0.55	10.8		0.95	0.035
1.0	3.9	1	0.74	0.29	9.0	1	0.60	$4\cdot10^{-3}$
0.8	3.1	0.63	0.35	0.105	7.2	0.28	0.039	$1.7\cdot10^{-3}$
0.7	2.7	0.40	0.20	0.051	6.3	0.073	$7.5\cdot10^{-3}$	$2\cdot10^{-5}$
0.6	2.3	0.20	0.092	0.021	5.4	0.011	$7.5\cdot10^{-4}$	$1.3\cdot10^{-6}$
0.5	1.9	0.077	0.028	$6.5\cdot10^{-3}$	4.5	$6.5\cdot10^{-4}$	$3.5\cdot10^{-5}$	$3.5\cdot10^{-8}$
0.4	1.6	0.020	$8.5\cdot10^{-3}$	$1.2\cdot10^{-3}$	3.6	$1.2\cdot10^{-5}$	$7.5\cdot10^{-7}$	$2\cdot10^{-10}$
0.3	1.2	$2.0\cdot10^{-3}$	$8\cdot10^{-4}$	$1.4\cdot10^{-4}$	2.7	$2.5\cdot10^{-8}$	$6\cdot10^{-10}$	$2\cdot10^{-13}$
0.2	0.8	$5\cdot10^{-5}$	$1.4\cdot10^{-5}$	$2\cdot10^{-6}$	1.8	$4\cdot10^{-13}$	$8\cdot10^{-15}$	10^{-18}
0.1	0.4	$6\cdot10^{-9}$	$2\cdot10^{-9}$	$2\cdot10^{-10}$	0.9	10^{-24}	$4\cdot10^{-26}$	10^{-29}

For energies well above the potential barrier, the penetrability is no longer important, and particles of angular momentum up to $l_0=R/\lambda\!\!\!^-$ will be effective. As already mentioned at the end of §70, this means that, above the barrier, higher angular momenta become rapidly more important. Therefore there will often be a further increase of the nuclear scattering above the top of the barrier, making the barrier appear higher than it actually is. Indications of this effect can be seen in the data given in §75 where the height of the potential barriers as deduced from scattering is sometimes higher than that following from our general interpolation formula (609).

As illustrations, we give in Table XXXIV the penetrabilities for α-particles of the barriers of two nuclei, a very light one (Be) and a medium heavy one (Ca), as functions of the energy of the particle for various orbital momenta l. The critical orbital momentum g is about 2 for Be and 8 for Ca while l_c (cf.(633)) is, for small energy, about 1 and 2, respectively. It is seen that, for the same ratio of energy to barrier height, the penetrability is much greater for Be than for Ca. Also it is shown by the table that the penetrability is decidedly less for the higher angular momenta than for the lower, in accord with the considerations above.

§73. The One-Body Theory of Scattering (M32)

For light nuclei, it may be a fair approximation to treat the scattering problem as a one-body problem, assuming a suitable potential to act between the scattering nucleus and the scattered particle. This method has been used widely in the past for the analysis and interpretation of experimental results; mainly for this reason, we discuss this method here although we are aware of its limitations.[55]

If we denote by ψ_0 the wave function in the pure Coulomb field (cf. (612)), the wave function in the field of the nucleus may be written (cf. Mott and Massey, p. 24, Eq. (17))

$$\psi=\psi_0+\sum_l\left[u_l-A_lf_l(r)\right]Y_{l0}(\vartheta)/r, \tag{635}$$

where A_l is given in (613c), f_l is the radial wave function in the Coulomb field whose asymptotic behavior is indicated in (613a), and u_l is the radial wave function in the actual field, i.e. the Coulomb plus the specifically nuclear field. The asymptotic behavior of u_l will be similar to the Coulomb wave function f_l, except for a phase shift δ_l, *viz.:*

$$u_l=B_l\sin\left(kr-\alpha\log 2kr-\tfrac{1}{2}l\pi+\eta_l+\delta_l\right). \tag{635a}$$

δ_l depends on the nuclear potential and can be calculated only if this potential is known. B_l must be determined so that $u_l-A_lf_l$ is purely an outgoing wave. This leads to

$$B_l=A_le^{i\delta_l}. \tag{635b}$$

Inserting the values of A_l, B_l, f_l, u_l and ψ_0 in (635), we find for the scattering cross section per unit solid angle

$$\sigma(\vartheta)=\left|\frac{e^2zZ}{2Mv^2\sin^2\frac{1}{2}\vartheta}e^{-i\alpha\log\sin^2\frac{1}{2}\vartheta}+\tfrac{1}{2}i\lambda\!\!\!^-\times\sum_l(2l+1)(e^{2i\delta_l}-1)\frac{(1+i\alpha)^2\cdots(l+i\alpha)^2}{(1+\alpha^2)\cdots(l^2+\alpha^2)}P_l(\cos\vartheta)\right|^2. \tag{636}$$

This formula is similar to that obtained with the many-body model and zero spin of nucleus and particle (618). The main difference is that the sum over the levels r in (618) has been replaced by $i(e^{2i\delta_l}-1)$. The behavior near resonances is also quite similar to that discussed in §71A, only the factor $(\Gamma^r{}_{Pp}/\Gamma_r)^2$ in the maximum cross section (624) is replaced by unity because no processes other than elastic scattering are being considered.

The similarity between (636) and the many-body formula for particles with zero spin is not accidental. If the spins of both scattering nucleus and scattered particle are zero, the scattered wave contains only partial waves of zero magnetic quantum number. Therefore, for large distances from the nucleus, we can certainly write the wave function in the form

$$\psi=\psi_0+\sum_l v_lY_{l0}(\vartheta)/r, \tag{636a}$$

where v_l is an outgoing spherical wave. Furthermore, we know that the orbital momentum l of a particle cannot change in the scattering process; therefore if we analyze the total wave function ψ in spherical harmonics and split the radial function, multiplying each spherical harmonic, into an incident and an outgoing part, the amplitudes of these two parts must have the same absolute value. This leads uniquely to the *form* (636) for the scattering cross section. However, the phases δ_l can in general *not* be found

[55] The method will give incorrect results (1) if the spins of nucleus or particle are different from zero, (2) if disintegrations have a probability comparable to scattering, (3) if inelastic scattering has a probability comparable to the elastic process.

as the phase differences between the solutions of the radial Schrödinger equations with and without the nuclear potential, but they must be regarded as arbitrary constants of much more complicated significance. Thus it will in general be possible to determine "empirical" phases δ_l suitable to represent the observed scattering, provided scattering nucleus and scattered particle have zero spin. But it will usually not be possible to find a suitable nuclear potential which gives the "observed" δ_l's (cf. §74).

§74. Scattering by Very Light Nuclei (up to He^4) (C2, R25, C3, W22, T6, T7, W13)

The scattering of particles by very light nuclei claims particular interest. In some cases it has been possible to draw conclusions about the fundamental forces between elementary nuclear particles from the scattering of very light nuclei (§14, 15, 18, and below). Since the scattering of elementary particles (protons and neutrons) by other elementary particles has already been treated (§18, 14, 15), we shall here be concerned with the scattering of protons, neutrons, deuterons and α-particles by deuterons and α-particles.

In all scattering experiments carried out thus far with protons and deuterons, the effective de Broglie wave-length $\lambdabar$ was larger than the presumable range of the nuclear forces. Under these conditions, it is known (§14) that only the particles of angular momentum zero are affected strongly by the nuclear forces. This fact has been made use of in theoretical investigations (T6, T7, M8). When protons of several MV, fast neutrons and deuterons are used for scattering experiments, the situation will be different.

To facilitate estimates of whether angular momenta higher than 0 will play any role in any particular scattering process, we have listed in Table XXXV the effective de Broglie wave-lengths *viz.*

$$\lambdabar = 4.52 \cdot 10^{-13}(A+a)/A(aE_0)^{\frac{1}{2}}, \qquad (637)$$

Table XXXV. *Scattering of light nuclei.*

Scattering Nucleus	Wave-lengths $\lambda/2\pi=\lambdabar$ (in 10^{-13} cm for $E_0=1$ MV) Scattered Particle			Range of the Forces (in 10^{-13} cm) Scattered Particle		
	H, n	D	He^4	H, n	D	He^4
H, n	9.06	9.60	11.30	2	5*	4.5
D	6.78	6.38	6.78	5*	7*	6*
H^3, He^3	6.04	5.28	5.27	4.5*	6*	5*
He^4	5.65	4.79	4.52	4.5	6*	4

* Estimates (other figures from experiments, see below).

for an energy $E_0=1$ MV. (a and A are the atomic weights of scattered particle and scattering nucleus, E_0 the kinetic energy of the incident particle in MV in the laboratory coordinate system.) In addition, we have tabulated the estimated effective radius of interaction for the particular pair of interacting nuclei (see below, under the discussion of the separate processes). According to the discussion in §14, strong scattering of angular momentum l can be expected only if $\lambdabar l$ is amaller than the range of the forces. E.g., for the scattering of protons by α-particles, $\lambdabar = 5.65E_0^{-\frac{1}{2}} \cdot 10^{-13}$ cm while the range of the forces is about $4 \cdot 10^{-13}$ cm. Therefore for proton energies above 2 MV an influence of the angular momentum $l=1$ is to be expected.

We turn now to the discussion of the various processes.

A. Scattering of neutrons by deuterons

The range of interaction between neutron and deuteron is determined to a large extent by the size of the deuteron. The average distance between proton and neutron in the deuteron is (cf. §12) $4.5 \cdot 10^{-13}$ cm, therefore the average distance of either proton or neutron from the center of the deuteron is $2.2 \cdot 10^{-13}$ cm. With a range of the force between two elementary particles of about 2 to $2.5 \cdot 10^{-13}$ cm (§21), we may estimate that a neutron will begin to be influenced by a deuteron at a distance of about $5 \cdot 10^{-13}$ cm.

The wave-length characteristic for the process is (Table XXXV) $6.78 \cdot 10^{-13}E_0^{-\frac{1}{2}}$ where E_0 is the kinetic energy of the neutron in MV. Thus the scattering may be expected to be approximately spherically symmetrical for neutron energies up to about 2 MV. For faster neutrons, the partial waves $l=1$ etc. will be affected so that the scattering will no longer show spherical symmetry. The "spherical symmetry" refers, of course, to the C system (cf. §70); this means (cf. (604)) a distribution of the recoil deuterons according to a law $\sin 2\varphi d\varphi$, φ being the angle between the motions of neutron and recoil deuteron.

If the scattering is spherically symmetrical ($E_0 \ll 2$ MV), the cross section is determined by the phase δ_0 of the partial wave function $l=0$ (cf. (635a)). The phase shift δ_0 has been calculated by Massey and Mohr (M8) with the assumption that the interaction between deuteron and neutron can be represented by a potential (cf. below). For a potential hole of radius $4.5 \cdot 10^{-13}$ cm which should about correspond to reality, they find a pronounced "Ramsauer effect" (Fig. 1 of their paper), i.e., the scattering is low at high energies, rises to a maximum at about 0.8 MV neutron energy and falls off again for smaller energies. The cross section at the maximum is $7 \cdot 10^{-24}$ cm^2, at zero energy only $2.5 \cdot 10^{-24}$ cm^2.

Experimentally, the scattering of neutrons by deuterons has been investigated by Dunning, Pegram, Fink and Mitchell (D23). They found the scattering cross section to be $4 \cdot 10^{-24}$ cm^2 for very slow neutrons (absorbable by Cd, energy probably less than 1 volt, cf. §61G) and $1.7 \cdot 10^{-24}$ for "fast" neutrons, being a mixture of all energies from 1 volt to about 10 MV.

No better agreement between theoretical and experimental value at low energy is to be expected since the influence of the deuteron on the neutron cannot be appropriately represented by a potential. The deuteron should be regarded as a dynamic system, and the proper wave function of all *three* particles, the two neutrons and the proton, be calculated with the actual interaction between them. This proper treatment of the problem as a three-particle problem is of course mathematically very complicated. If the problem is treated as a two-particle one (deuteron+neutron), then at least the exchange of the incident neutron with the one contained in the deuteron and the Majorana type of the forces should be considered. This refinement of the theory should be more necessary for the scattering by the deuteron than by any other nucleus, owing to the small binding energy and the small number of particles in the deuteron.

B. Scattering of protons by deuterons

If the nuclear forces are symmetrical in neutrons and protons (§6, 22), the force between proton and deuteron should be exactly the same as between neutron and deuteron, except for the Coulomb force. The phase shifts δ_l should therefore be identical for the two cases, only the scattering should be calculated from (636) for protons, from (53) for neutrons. A comparison of the scattering of neutrons and protons *of the same energy* by deuterons would therefore provide a very accurate test of the assumption that nuclear forces are symmetrical in neutrons and protons. Of course, it would be necessary to find the angular distributions in both cases. Experiments on the angular distribution and the absolute cross section for the scattering of protons by deuterons have been made by Tuve, Heydenburg and Hafstad (unpublished).

C. Deuteron-deuteron

From the diameter of the deuteron ($4.5 \cdot 10^{-13}$ cm) we estimate that the mutual influence of two deuterons will be appreciable up to distances of about $7 \cdot 10^{-13}$ cm. Accordingly, marked deviations from the Rutherford scattering should set in at very low energies, of the order of a few hundred kv. The partial wave $l=0$ only should be strongly influenced for energies up to about 1 MV. Experiments have been made by Tuve, Heydenburg and Hafstad (unpublished).

D. α-particle—proton

The size (radius) of the α-particle is only about $2 \cdot 10^{-13}$ cm (§21). Therefore the range of the forces between α-particle and protons may be estimated to be about $4 \cdot 10^{-13}$ cm. Consequently, if α-particles are scattered in hydrogen, only the partial wave $l=0$ should be materially affected as long as the α-particle energy is less than $(11.3/4)^2 \approx 8$ MV (cf. Table XXXV).

Very careful experiments on the scattering of α-particles of different velocities in hydrogen have been carried out by Chadwick and Bieler (C2) as early as 1921: The experiments consisted in observing the number and angular distribution of the protons projected by the α-particles from a hydrogen-containing substance such as paraffin. Because of their longer range (§95) the protons are much easier to observe than the scattered α-particles. The angle φ between the directions of proton and incident α is connected to the deflection ϑ of the α-particle in the center-of-mass system by the simple relation $\varphi = \frac{1}{2}(\pi - \vartheta)$ (cf. 604c). The protons emitted in a forward direction correspond thus to maximum deflection ($\vartheta = 180°$) of the α's and therefore their number should show the largest relative deviation from the Rutherford formula (cf. (618a) and text following (619)).

The results of Chadwick and Bieler show that the scattering probability is many times as large as the Rutherford law would indicate. The scattering ratio (ratio of actual to Rutherford scattering) increases with increasing energy of the α-particles and surpasses 40 for α-particles of the highest energy ($E_0=7.5$ MV). For small energies, below about 3 MV, the scattering becomes essentially equal to the Rutherford scattering. This behavior is to be expected. If we assume a range of the nuclear interaction of about $R=4.5\cdot 10^{-13}$ cm (see below) the height of the potential barrier is $2e^2/R=0.6$ MV. But the kinetic energy in the relative coordinate system in only one-fifth of the kinetic energy of the α-particle (cf. 606a) so that the α-particles are able to go over the top of the potential barrier if their energy is more than 3 MV.

The data have been analyzed theoretically by Taylor (T6). He finds that the angular distributions can be explained very satisfactorily on the one-body model by assuming a single phase shift δ_0 for the partial wave of zero angular momentum. This is to be expected for α-particles of less than 8 MV (see above). The "experimental" phases δ_0 can be interpreted theoretically by assuming a potential energy of the rectangular-hole type between proton and α-particle. The depth of the hole which gives best agreement with the experimental data, is 6 MV, the width $4.5\cdot 10^{-13}$ cm.

In such a potential hole, there should be a stable energy level at -0.8 MV, corresponding to a stable nucleus Li^5 of 0.8 MV binding energy. Such a nucleus is not known and probably does not exist. Its energy must be higher than the energy of He^5 by the amount of the Coulomb repulsion between the proton and the α-particle. But even He^5 probably does not exist (W21c) and is unstable by 0.9 MV against disintegration into an α-particle and a neutron. Thus the potential assumed by Taylor for the scattering, does not correctly represent the interaction between proton and α-particle in the stationary state. However, it can hardly be expected that this interaction can be described by a potential. Moreover, it seems that only a slight change of the potential would be required to make Li^5 unstable, and such a change seems quite possible in view of the crude approximation given by the one-body model.

Experimental results apparently different from those of Chadwick and Bieler have been obtained by Pose and Diebner (P13). They give a scattering probability equal to more than 100 times the Rutherford scattering for α-particles of only 5 MV, for which Chadwick and Bieler found only 10 times the Rutherford scattering. Horsley has pointed out (H37) that such a large scattering ratio could not be explained if only the angular momentum $l=0$ were affected by the nuclear field, and therefore postulates an influence of the nuclear field on the p-scattering ($l=1$), contrary to our estimates above. It is true that p-scattering might occur in a limited energy interval due to a p-resonance level. However the discrepancy between Pose-Diebner's and Chadwick-Bieler's results seemed to us far too large to be explained by experimental uncertainties. Indeed, there is obviously an error in the evaluation of the data of Pose and Diebner: They observe all protons emitted in all directions making angles less than 5° with the incident alphas, corresponding to a solid angle $2\pi\ (1-\cos 5°)=2.4\cdot 10^{-2}$. The value of the solid angle given in their paper, and obviously used in their evaluation, is $2.5\cdot 10^{-3}$. All their results should therefore be divided by 10, which makes them agree with Chadwick and Bieler's values. This change invalidates the conclusions of Horsley about the p-resonance scattering.

E. Neutron—α-particle

The interaction here should be the same as between proton and α-particle, except for the Coulomb force. Therefore the scattering should be spherically symmetrical for neutron energies up to 2 MV. Using Taylor's potential hole for the proton-alpha-interaction, the scattering cross section should be of the order $6\cdot 10^{-24}$ cm² for slow neutrons (below about 1 MV). Experimental investigations of the angular distribution and the scattering cross section are not available, in spite of the fact that recoil helium nuclei have been used as a means of detecting neutrons (B40).

F. α-particle—α-particle

More extensive work has been done on this scattering process, both theoretically and experimentally, than on any of the processes previously mentioned. The theoretical treatment is slightly different from that of the other processes because the symmetry of the wave function in the two α-particles must be considered (M31, T6, T7, and §18 of this article). α-particles obey Bose statistics (§§4, 48) therefore the wave function must be *symmetrical* in the two particles. The particles possess no spin, so that the symmetry requirement applies to the positional wave function directly. This symmetry condition has the immediate consequence that *no odd angular momenta l occur* in an expansion of the wave function in spherical harmonics; for we have shown in §12 that a wave function containing a spherical harmonic of odd order l, changes sign when the two particles are interchanged, while an even harmonic remains unchanged. This fact affects, of course, the condition for the appearance of phase shifts other than δ_0 in the scattering: The first partial wave besides $l=0$ is $l=2$ rather than $l=1$; but $l=2$ will only be affected by the nuclear field if the range of the forces is more than twice the effective wave-length $\lambda\!\!\!^{-}$. From Table XXXV we find that $\lambda\!\!\!^{-}=4.5\,E_0^{-\frac{1}{2}}\cdot 10^{-13}$ cm for the scattering of α-particles by α-particles. Estimating the range of the forces to be about $4\cdot 10^{-13}$ cm, we would expect that the second spherical harmonic will contribute to the scattering above about 5 MV. This seems to be roughly correct (see below). The higher spherical harmonics $l=4$ etc. should, according to this reasoning, not come in at energies below about 20 MV. However, there seems to be a resonance level for $l=4$ at about 10 or 12 MV α-energy which increases the influence of the partial wave $l=4$.

The scattering is most conveniently expressed in terms of its ratio to the classical scattering in a Coulomb field,

$$\sigma_{\text{cl}} \sin\Theta d\Theta = 8\pi\left(\frac{4e^2}{M_\alpha v^2}\right)^2\left(\frac{1}{\sin^4\Theta}+\frac{1}{\cos^4\Theta}\right) \times \sin\Theta\cos\Theta d\Theta, \quad (638)$$

where M_α is the mass of the α-particle and $\Theta=\frac{1}{2}\vartheta$ its deflection in the *ordinary* coordinate system. The first term in the bracket, $(\sin\Theta)^{-4}$, gives the number of α-particles deflected through Θ; the second term, $(\cos\Theta)^{-4}$, gives the number of α's deflected through $\frac{1}{2}\pi-\Theta$ each of which is accompanied by a recoil α-particle in the direction Θ (cf. beginning of §14, and (604c)). The ratio of the actual scattering to the Rutherford scattering (638) is (cf. (636); §18; M32, Chapter V, or T6, T7)

$$P=\frac{\left|\dfrac{e^{-i\alpha\log\sin^2\Theta}}{\sin^2\Theta}+\dfrac{e^{-i\alpha\log\cos^2\Theta}}{\cos^2\Theta}+\dfrac{2i}{\alpha}\sum_{l\text{ even}}(2l+1)(e^{2i\delta_l}-1)\dfrac{(1+i\alpha)^2\cdots(l+i\alpha)^2}{(1+\alpha^2)\cdots(l^2+\alpha^2)}P_l(\cos 2\Theta)\right|^2}{(\sin^{-4}\Theta+\cos^{-4}\Theta)}, \quad (639)$$

where

$$\alpha=4e^2/hv. \quad (639\text{a})$$

The first two terms arise from the symmetrization of the wave function in incident and scattering α-particle. These terms alone give deviations from the classical scattering which were first pointed out by Mott (M31). In particular, for $\Theta=45°$, we have $\sin^2\Theta=\cos^2\Theta=\frac{1}{2}$ which makes the two terms equal and gives $P=2$. Thus at 45° the scattering of α-particles is twice as large as it would be in classical mechanics, even if there is no action of the nuclear force at all. This has actually been proved experimentally for slow α-particles by Chadwick (C3). At angles other than 45°, the deviations from classical scattering change periodically with the velocity and the angle Θ, giving sometimes larger, sometimes smaller scattering than classical. This effect has been checked for various angles and velocities by Blackett and Champion (B27).

The sum over l in (639) represents the influence of the nuclear field. The factor 2 in front of it arises again from the symmetrization of the wave function. The sum extends only over the even values of l, again because of the symmetry requirement. In all other respects, the expression is analogous to (636).

Experiments were carried out by Rutherford and Chadwick (R25), by Chadwick (C3), by Wright (W22) and by Mohr and Pringle (unpublished). A theoretical analysis was made by Taylor (T6, T7), Wright (W22) and recently by Wheeler (to be published in Phys. Rev.).[56] Wheeler, whose analysis is the most complete, represented the observations in terms of the three phase shifts $\delta_0\delta_2\delta_4$ for the three lowest values of l possible. For each energy, the phase shifts are determined provisionally from the scattering at three angles, preferably 19.6°, 27.4° and 35.0° because for these angles either $P_2(2\Theta)$ or $P_4(2\Theta)$ vanishes. In general, 8 different sets of values for $\delta_0\delta_2\delta_4$ satisfy the data for the three angles. The experimental scattering at a fourth angle, e.g. 45°, is then used to decide between these 8 sets, the decision being in general unique.

[56] We are indebted to Professor Wheeler for communication of his results before publication.

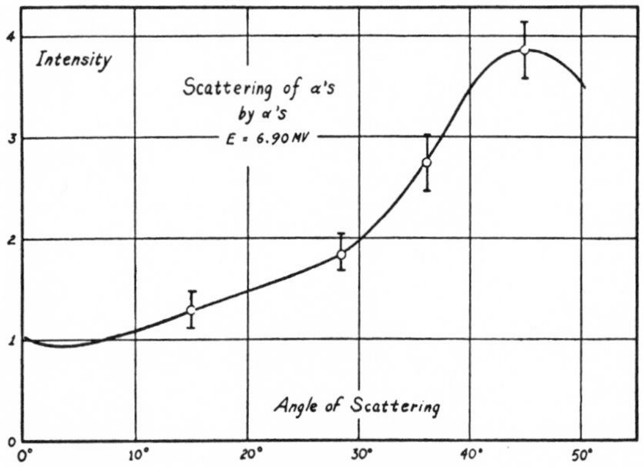

FIG. 20. Scattering of α-particles of 6.9 MV by He nuclei. Solid curve calculated by Wheeler to pass through the four experimental points indicated. (We are indebted to Professor Wheeler for the permission to publish this and the following figure.)

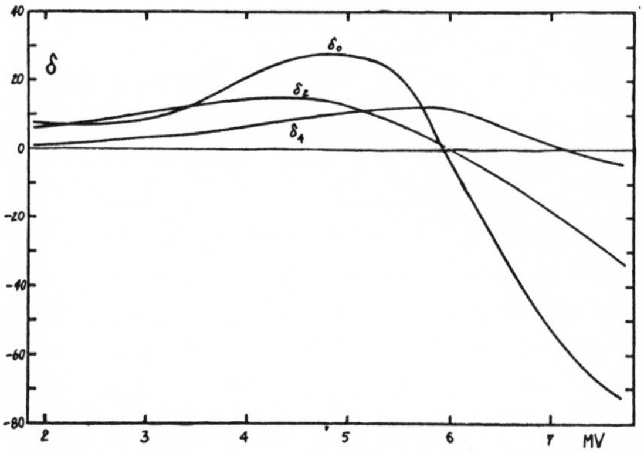

FIG. 21. Phase shifts (in degrees) of the partial waves $l=0$, 2 and 4 in the scattering of α-particles by α-particles according to Wheeler.

The agreement obtained with experiment is very satisfactory, as can be seen from Fig. 20 which shows the scattering of α-particles of 6.9 MV, the solid curve being obtained from (639) with the suitably chosen values of the δ's.

Fig. 21 gives the phase shifts obtained in this way, as functions of the α-particle energy. The δ's are quite small and vary regularly up to about 5.5 MV. Above this energy, a rather sudden change occurs which is probably connected to the level of Be^8 at 3 MV excitation energy which is known from the nuclear reaction $B^{11}+H^1$ (§88). (The energy in the center of gravity system is half the kinetic energy of the incident α-particle!) —Wheeler has also tried to calculate the scattering on purely theoretical grounds, using only the known forces between neutrons and protons. These calculations, being very laborious, are not yet completed.

§75. Other Experimental Results on Scattering (R8, R9, W11)

Most experimental results on scattering have been obtained with α-particles as the scattered particles. The most comprehensive experimental paper is by Riezler (R8). In addition, older experiments by Rutherford and Chadwick (R25) and more recent supplementary experiments by Riezler (R9) must be mentioned.

Wenzel (W11) has analyzed Riezler's results theoretically.[57] His analysis is based on the one-body model (§73) which, as we have mentioned, is justified (to the extent that it is used) if both the scattered particle and the scattering nucleus have spin zero. For the scattered particle (α-particle) this is fulfilled, for the scattering nucleus it is true in the case of C, O and Ne (among the nuclei investigated), but not for Be, B and Al. In fact, Wenzel's analysis gives very satisfactory results for carbon, and much less satisfactory ones for Be and B (the other cases were not analyzed).

If the spins of both nuclei are zero, the orbital momentum l of the particle is identical with the angular momentum J of the compound nucleus. Formula (618) (or (636)) is valid for the scattering, and it is easy to estimate J from the resonance maximum of the scattering (cf. the discussion of carbon, below).

All the results enumerated in the following refer to the scattering of α-particles by the respective nuclei.

Lithium

No experimental data.

Beryllium

Riezler (R8) finds for α-particles of 7.5 MV a scattered intensity exceeding greatly the Rutherford scattering. The ratio (observed scatt./Coulomb scatt.) goes up to 43 for a deflection of 165° (in the center-of-gravity system). Such a large scattering could not be produced if only the wave of orbital momentum zero were affected by

[57] Wenzel's formulae contain an error. In his formula (13) etc. σ_l should be replaced by $2\sigma_l$. The error caused by this is not very great for $l=0$ and 1, but very considerable for $l=2$.

the nuclear field. p-scattering ($l=1$) accounts fairly satisfactorily for the observations if the one-body model is used (Wenzel). In the many-body model a higher orbital momentum is ordinarily required to give the same scattering (§71B). From the size of the Be nucleus and the wave-length of the α-particles it should be expected that just the s-, p- and d-scattering are seriously affected by the nuclear field, which is compatible with our reasoning above. Since the top of the potential barrier lies at about 4 MV for the scattering of α-particles by beryllium, Riezler's α-particles go well over the top. No experiments on the variation of the scattering with the α-particle energy are available.[58]

Boron

Riezler observed the scattering as a function of the angle for 7.5 MV energy, finding a high maximum at 180° as should be expected from (618), (636). He also observed the scattering at 160° as a function of the energy, finding the expected rise with increasing energy. The highest ratio observed is 25 (i.e., observed scattering = 25 times Rutherford scattering) at 167° and 7.5 MV. The increase of the scattering ratio occurs for α-particle energies between 4 and 7 MV. The potential barrier of B for α-particles is expected to be about 5 MV high. Therefore the anomalous scattering is obviously connected with the overcoming of the barrier. The height of the barrier as determined from the experiments turns out in accord with our expectations, thus proving approximate proportionality of nuclear volume and atomic weight (cf. §3, and §25 to §30). From the nuclear radius and the wave-length of the α-particles, we find again that s-, p- and d-particles should be influenced by the nuclear field. In fact, the observed scattering ratio at 7.5 MV is too high to be explained by s scattering alone while p scattering would be sufficient in the one-body model (W11) (cf. the remarks on Be). No resonance maximum of the scattering was observed.

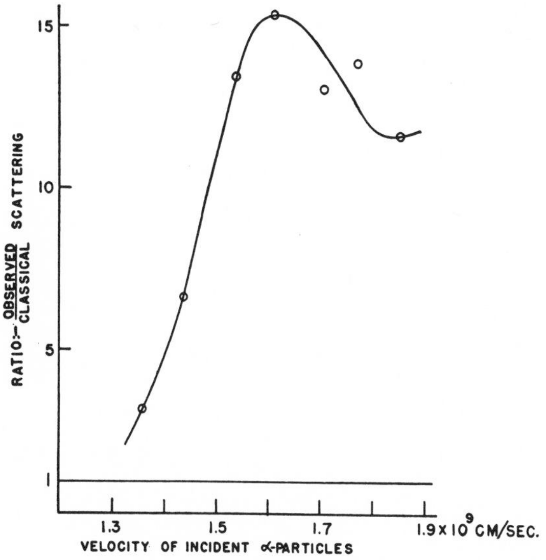

FIG. 22. The scattering of α-particles by carbon, according to experiments of Riezler.

Carbon

In contrast to the scattering by Be and B, the scattering ratio for carbon does not increase continuously with α-particle energy, but shows a marked maximum at about 5 MV (cf. Fig. 22). This is about equal to the height of the potential barrier for C (cf. 609). The observed maximum should be ascribed to resonance, as Wenzel has pointed out. The scattering ratio at the maximum is 15 for 166° deflection. A perturbation of the s scattering alone by the nuclear field could, according to (624), not give a scattering ratio greater than about 3. Therefore the maximum must be ascribed to a resonance level with J different from zero. Indeed, the observed maximum of the scattering ratio agrees closely with the theoretical maximum for p scattering (W11). The width of the resonance peak is of the order of 1 MV, which seems reasonable.

Nitrogen

No observations.

Oxygen

According to the second paper of Riezler (R9), the scattering through about 90° begins to deviate from Rutherford scattering at about 4 MV, *decreases* continuously with the energy, and reaches 0.5 times the Rutherford value at

[58] α-particles scattered backwards by Be have an energy of only $(9-4)^2/(9+4)^2=0.15$ times their original energy. If the initial energy is low, the range of the scattered alphas is too short for observation.

5.3 MV. The scattering can be explained by assuming a height of the potential barrier of about 6 MV, as compared to about 5.5 from (609).

Neon

Up to energies of 5.3 MV, no deviation from Rutherford scattering was observed by Riezler (R8). This is compatible with the expected height of the barrier (6 MV).

Aluminum

For about 7 MV energy, Riezler found the scattering to be considerably smaller than Rutherford scattering at angles below 158°, and larger for the largest scattering angles. From the magnitude of the deviations, it seems that the energy used (7 MV) is approximately equal to the height of the potential barrier, which agrees with our general formula (609).

Thus the scattering of α-particles by medium light nuclei confirms, at least qualitatively, the proportionality of the nuclear volume with the mass number.

A case of considerable interest, although no experiments are yet available, is the scattering of protons by Li^7. The compound nucleus, Be^8, is known to have a resonance level at 440 kv kinetic energy of the protons. From the general formula (625) it follows that the resonance scattering should increase the Coulomb scattering in the backwards direction by a factor of about 3 (B15, p. 478). In calculating this figure, the assumption has been made that the width of the level is almost entirely due to the disintegration $Be^8 \rightarrow Li^7 + H^1$ which is very plausible because no other process seems possible (cf. §81) except the emission of γ-rays which certainly gives a very small contribution to the width of the level.

XIII. Distintegrations Produced by Charged Particles

§76. Classification of Processes

The general scheme of a nuclear disintegration is, as we know,

$$A + P \rightarrow C \rightarrow B + Q. \qquad (640)$$

The incident particle P falls on the initial nucleus A and combines with it to form the compound nucleus C. The latter splits into an outgoing particle Q and a residual nucleus B. In this chapter, we shall be concerned with processes produced by charged particles P, of which the most important are protons, deuterons and α-particles.

A. The main nuclear process

The processes may first be classified into two types according to the character of the outgoing particle Q. Q may be either a light quantum or a material particle. In the first case, we speak of the simple capture of particle P, in the second case of a disintegration in the strict sense, or a "particle disintegration."

In any nuclear process, charge, mass number, energy and momentum must be conserved. The conservation of energy and momentum will be discussed in detail in §96. It determines the kinetic energy of the outgoing particle in terms of the kinetic energy of the incident particle, and of the difference in internal energy between the initial nuclei $A+P$ and the final nuclei $B+Q$. In the case of simple capture, practically all the momentum will be taken up by the nucleus B, and practically all the energy by the light quantum Q. In the case of particle-disintegrations, energy and momentum will be shared between the two resultant nuclei B and Q (see §96).

A special case of a particle disintegration is scattering (elastic or inelastic). In this case, the outgoing particle is identical with the incident one. Apart from the potential scattering by the Coulomb field, there is no essential difference between scattering and disintegration, and the probabilities of the two processes will, in general, be of the same order. Experimental data on scattering are scarce except for α-particles (§74, 75). Data on inelastic scattering, especially, are almost unavailable and the existence of the process can only be inferred on theoretical grounds and from analogy with neutron experiments (§65C).

Simple capture is ordinarily less probable than particle disintegration, including scattering, because the "radiation width" of nuclear levels is usually smaller than the particle width (§81). Simple capture is therefore probable only when particle disintegrations are impossible or very improbable for some special reason, e.g.,

(1) Because sufficient energy is not available to make up the masses of the possible produced particles, or (2) because the potential barrier is too high for particles which, on energetic grounds, might be emitted, or (3) because only enough energy is available for slow particles to be emitted, so that the particle width is made small by the factor v contained in it (cf. §52B), or (4) because the emission of particles is forbidden by selection rules.

The last case is realized in the simple capture of protons by Li^7 and F^{19} (cf. §81), the first in the capture by C^{12}. The conditions 2 and 3 are fulfilled for the simple capture of slow neutrons by heavy nuclei. The emission of charged particles is then impossible because of the high barriers; the reemission of neutrons (elastic scattering) is improbable because of the small neutron velocity. In some cases, it is possible to establish the existence of simple capture experimentally, in spite of its small probability. This will be the case when the product nucleus B could not have been produced in any other way and can be detected, e.g., by β-radioactivity.

In any nuclear process, the residual nucleus B may be left either in the ground state or in an excited state. All the excited states which are energetically possible and not improbable because of selection rules (§83) or because of considerations of the penetration through the potential barrier (§78), will in general be formed. To each excited level of nucleus B there corresponds a "group" of outgoing particles Q with a certain definite energy. The group of highest kinetic energy will correspond to the nucleus B being left in the ground state, and the other groups, of decreasing kinetic energy, correspond to excited states of nucleus B of increasing excitation energy. This will be true for simple capture as well as for particle disintegrations. In the case of simple capture, the $h\nu$ of the γ-ray replaces the kinetic energy of the outgoing particle. By measuring the kinetic energies of the various groups of outgoing particles, or the spectrum of the γ-rays from the simple capture process, the excited energy levels of the final nucleus B can be deduced. This has been done in a number of cases in which the outgoing particles are protons, neutrons or α-particles (Chapter XVII, and §109).

The probability is, in general, of the same order of magnitude for an excited final state of the residual nucleus B as for the ground state (cf. §54). Thus, if many excited states are energetically possible, only a small fraction of all disintegration (or simple capture) processes will lead directly to the ground state of the nucleus. Therefore, it may sometimes be difficult to find the group of particles, or the line in the capture γ-ray spectrum, which corresponds to the ground state. On the other hand, the *total* disintegration probability will be greatly enhanced if many final states are possible (§79).

B. Secondary processes due to the residual nucleus

If the residual nucleus is left in an excited state (which, according to the foregoing, is the rule rather than the exception), we may again have two cases. Either the excited state is below the dissociation limit or above it.

1. γ-ray spectrum.—In the first case, the only "particles" which can be emitted by the nucleus B, are light quanta. Therefore the residual nucleus will emit one or more γ-rays until it finally arrives in its ground state. These γ-rays from the residual nucleus must not be confused with the γ-rays emitted in capture processes. The latter are part of the main nuclear process (640) itself, while the former are emitted *after* the main process is finished. γ-rays from the residual nucleus have a discrete spectrum depending only on the levels of nucleus B, while the frequency of the capture γ-rays depends on the kinetic energy of the incident particle so that their spectrum is continuous if the incident particles are not "monochromatic," i.e., if they do not all have the same energy.

The spectrum of the γ-rays from the residual nucleus has been observed in several cases (§89, 99–102). It provides a very important check on the scheme of energy levels of nucleus B derived from the groups of particles emitted in the nuclear process itself. In cases for which these groups have not been measured, the γ-rays from the residual nucleus may also be used directly to obtain a scheme of energy levels; however, the analysis of the spectrum is often difficult, especially if there are many lines.

It need hardly be mentioned that γ-rays from the residual nucleus may be emitted after capture processes as well as after particle disintegrations. In the former case, it is extremely difficult to separate the two kinds of γ-rays.

2. *Three-particle disintegration.*—If the residual nucleus B is, in the main nuclear process, left in an excited state above its dissociation energy, it may break up further, emitting material particles according to the scheme

$$B^* \rightarrow D+S. \qquad (641)$$

The asterisk denotes an excited state, D is the second residual nucleus, and S the second emitted particle. Experimentally, such a process will appear as a *three-particle disintegration*, according to the scheme

$$A+P \rightarrow C \rightarrow B^*+Q \rightarrow D+S+Q. \qquad (641a)$$

Such three-particle disintegrations have been observed in a number of cases (§85, 79, 80, 101D, E, 102D) and it could be shown that the mechanism is actually as indicated by the scheme (641a), at least in the case

$$B^{11}+H^1 \rightarrow C^{12} \rightarrow Be^{8*}+He^4 \rightarrow 3He^4. \qquad (641b)$$

It seems that a breaking up into three particles in a single process is very improbable.

The emission of a material particle S by the excited residual nucleus B^*, when energetically possible, is in general more probable than the emission of γ-rays, just as, in the "main" nuclear process, particle disintegrations are more likely than simple capture. The reasons and the exceptions are the same as discussed above, in A.

A particular example of a three-particle disintegration is the case in which one or both of the emitted particles S and Q are of the same kind as the incident particle P, so that we have in effect the reaction

$$A+P \rightarrow D+P+Q. \qquad (642)$$

Then it appears from the final result as if particle P had not taken part at all in the reaction, except by giving part of its energy to the other particles. The reaction appears as a noncapture disintegration according to the scheme

$$A \rightarrow D+Q. \qquad (642a)$$

However, it should be remembered that actually the mechanism of the reaction is more complicated. The incident particle is first absorbed by the initial nucleus A, to form the compound nucleus C. It is then reemitted, A being left in an excited state. This excited state then disintegrates finally into $D+Q$. Or else, the compound nucleus may first emit particle Q, leaving the residual nucleus B in an excited state, whereupon the reemission of particle P follows as the last stage. The two schemes of "noncapture disintegrations" are thus:

$$A+P \rightarrow C \rightarrow A^*+P, \quad A^* \rightarrow D+Q \qquad (642b)$$

or

$$A+P \rightarrow C \rightarrow B^*+Q, \quad B^* \rightarrow D+P. \qquad (642c)$$

Both are, in principle, equally probable.

"Noncapture" disintegrations seem to occur in several cases with fast neutrons, the neutron losing part of its energy which is used to knock another neutron out of the nucleus (§85, 102D).

The same statements could be made about nucleus D, as about B. D may again be formed in an excited state and may again emit γ-rays, or, if excited above the dissociation energy, may disintegrate with emission of another particle, etc.

C. β-radioactivity of final nucleus

The nucleus which is finally formed, after all particles, γ-rays, etc. have been emitted, may be β-radioactive. It seems preferable not to include this β-decay in the nuclear reaction at all, not even in the secondary reactions of the type (641), because the time required for the β-decay is of an entirely different order of magnitude than the time of nuclear reactions. Even the slowest nuclear reactions are completed in about 10^{-13} sec., this figure corresponding to a width of the compound nuclear level of about 0.01 volt which is smaller than all widths found experimentally thus far (cf. §61, Table XXVI and §81). The same figure will represent something of an upper limit to the lifetime of the excited levels of the residual nucleus B against γ-emission, except possibly for low metastable levels (§87D). On the other hand, the shortest lifetime ever observed for a β-radioactive nucleus is 1/50 sec., with most lifetimes ranging from a few seconds to a few years. This is 10^{11}, or more, times longer than the duration of a nuclear reaction.

Thus it seems justified to draw a definite line between the nuclear process itself and the β-emission which may follow it. From this standpoint we have to say that electrons (positive or negative) will never be produced in a nuclear reaction, but can only be given off later, by the nucleus produced in the reaction. The interaction of nuclei with the Fermi field of electrons is so much weaker than all other interactions that we can neglect it entirely in nuclear dynamics. We consider the interaction of the compound nucleus with the "fields" of heavy particles of all kinds, and with the electromagnetic radiation field, but we leave out the extremely small interaction with the β-ray field. For this reason, β-emission was described in part A of this report, which dealt with stationary states of nuclei rather than with dynamic processes.

The chief function of β-radioactivity in the study of nuclear reactions is to provide a convenient indicator of the production of certain product nuclei. In fact, most of the capture processes of neutrons (Chapter X) and many of the reactions produced by charged particles (this chapter) have been discovered through detecting the radioactivity of the product nucleus.

β-emission may also lead to an excited state of the final nucleus which will therefore emit γ-rays. These γ-rays can easily be separated from the γ-rays produced in the original nuclear reaction (capture γ-rays) or emitted by the residual nucleus of the original reaction, because the γ-rays emitted after β-emission appear with a time lag equal to that of the β-emission itself.

§77. General Theory

The general expression for the disintegration cross section is (cf. (260))

$$\sigma^{Pp}{}_{Qq}=\frac{\pi\lambda^2}{(2s+1)(2i+1)}\sum_J(2J+1)\times\left|\sum_r\frac{u^{rJ}{}_{Pp}\,u^{rJ}{}_{Qq}}{E-E_r+\frac{1}{2}i\gamma_r}\right|^2. \quad (643)$$

The notations are the same as in §52. The matrix element $u^{rJ}{}_{Pp}$ contains the wave function of the incident particle P, and therefore (cf. §72) the factor $P_{Pp}{}^{\frac{1}{2}}$ where P_{Pp} is the transmission probability (penetrability) of the potential barrier of the initial nucleus A for the incident particle P. We put

$$u^r{}_{Pp}=w^r{}_{Pp}\,P_{Pp}{}^{\frac{1}{2}}, \quad (644)$$

where w is the matrix element in the absence of the potential barrier and P the penetrability. Similarly,

$$u^r{}_{Qq}=w^r{}_{Qq}\,P_{Qq}{}^{\frac{1}{2}}, \quad (644a)$$

where P_{Qq} is the penetrability for the outgoing particle. It will depend on the kind Q of the outgoing particle as well as on its energy which, for given energy E of the incident particle, is determined by the quantum state q of the residual nucleus. The "matrix elements without barrier" w are connected to the level widths without barrier, G, introduced in (594), by

$$G^r{}_{Qq}=(w^r{}_{Qq})^2. \quad (644b)$$

For a given energy E of the incident particle, the penetration factors P are the same for all levels of the compound nucleus (§52). Therefore we may write

$$\sigma^{Pp}{}_{Qq}=P_{Pp}P_{Qq}\,s^{Pp}{}_{Qq}, \quad (645)$$

where

$$s^{Pp}{}_{Qq}=\frac{\pi\lambda^2}{(2s+1)(2i+1)}\sum_J(2J+1)\times\left|\sum_r\frac{w^r{}_{Pp}\,w^r{}_{Qq}}{E-E_r+\frac{1}{2}i\gamma_r}\right|^2 \quad (645a)$$

The cross section is thus a product of three factors, *viz.*

(1) The penetrability of the potential barrier for the incident particle P_{Pp}
(2) The penetrability for the outgoing particle P_{Qq}
(3) The "internal disintegration probability" $s^{Pp}{}_{Qq}$.

If either the incident or the outgoing particle, or both, have sufficient energy to go over the top of the barrier, the corresponding penetrability factors must be replaced by unity.

The discussion of formula (645, 645a) is conveniently carried out separately for light and for heavy nuclei.

A. Light nuclei

The spacing between the resonance levels is known to be wide, of the order of 1 MV (cf. §81, 82). The resonances will therefore in general be observable if particles of suitable energy are available, and if the width of the resonance levels is small compared to the spacing so that there is actually a marked increase of the scattering cross section at resonance. This latter condition will be fulfilled if the energy of the resonance level is not too high. For very high excitation energy, the ratio of the total width to the spacing is known to increase rapidly (§54). Therefore it is improbable that resonances will be observed with deuterons as the incident particle, since the high internal energy of the deuteron makes the energy of any compound nucleus which is formed by adding a deuteron to an initial nucleus, very high also.

1. In the neighborhood of a fairly narrow resonance level, the penetrabilities P_{Pp} and P_{Qq} may be regarded as constant and be replaced by their values at exact resonance. (645a) reduces to a single term, and (643) to the well-known one level formula

$$\sigma^{Pp}{}_{Qq}=\frac{\pi\lambda\!\!\!^{-2}(2J+1)}{(2s+1)(2i+1)}\frac{\Gamma^r{}_{Pp}\Gamma^r{}_{Qq}}{(E-E_r)^2+\frac{1}{4}\Gamma_r{}^2} \quad (646)$$

with

$$\Gamma^r{}_{Pp}=G^r{}_{Pp}P_{Pp}, \quad \Gamma^r{}_{Qq}=G^r{}_{Qq}P_{Qq}. \quad (646a)$$

The G's are the partial widths without barrier, the Γ's the widths with barrier, and the P's the penetrabilities of the barriers at resonance. The total width Γ_r may or may not contain other partial widths besides those corresponding to the emission of Pp and Qq. The experimental evidence on resonance disintegration, and the determination of the partial and total widths from the experimental results, will be discussed in §81, 82.

2. For very low energies there will certainly be a region in which the penetrability factor P_{Pp} in (645) changes much more rapidly than the internal disintegration probability s. This region corresponds to the $1/v$ region in neutron disintegrations (§58B). The extension of the "penetrability region" is determined in a way similar to that of the $1/v$ region for neutrons. The penetrability factor will govern the disintegration function as long as the kinetic energy of the incident particle is small compared to the energy or to the width of the first resonance level, whichever is larger.

A narrow resonance level with small partial widths $\Gamma^r{}_{Pp}\Gamma^r{}_{Qq}$ can be disregarded for the purpose of determining the extension of the "penetrability region." Such a level will only interrupt the general trend of the disintegration function for a small energy interval.

In the "penetrability region," the cross section may be written

$$\sigma=\text{const}\cdot P_{Pp}P_{Qq}/E, \quad (647)$$

since the factor $\lambda\!\!\!^{-2}$ in (645a) is proportional to $1/E$. This formula was first suggested by Gamow and is well confirmed for small energies of the incident particle (§78).

For higher energies, the fluctuations of the internal disintegration cross section (645a) with energy will greatly affect the disintegration function. It will then no longer be possible to deduce the penetrability of the barrier directly from the disintegration function, as has been done in the past. Some of the important factors besides the penetrability are discussed in §79.

B. Heavy nuclei

The spacing between the levels of heavy nuclei is known to be of the order of only a few volts, from neutron experiments (§60). No available source gives fast charged particles homogeneous enough in energy to observe such closely spaced levels. Thus we cannot hope to observe resonance levels in heavy nuclei with charged particles. All we can observe is an *average* cross section, averaged over an energy interval large compared to the spacing of levels. This average has been calculated in §56 (405).

The rather intricate influence of the angular momentum is treated in §78B. As will be shown there, we may write

$$\Gamma^J{}_{Pp}=G_{Pp}P_{Pp}\Pi^J{}_{Pp}. \quad (648)$$

Here G_{Pp} is the width without potential barrier which is, in the average, independent of J, and is connected to the "sticking probability" ξ (§§54, 56) approximately by (cf. 360):

$$G_{Pp}=(1/2\pi)(2i+1)(2s+1)D_J\xi_{Pp}. \qquad (649)$$

P_{Pp} is the penetrability of the potential barrier for particles of zero orbital momentum, and Π a factor which decreases with increasing J and takes into account the dependence of the penetrability on the angular momentum. A formula similar to (649) holds also for particle Q.

(405) may now be rewritten

$$(\sigma^{Pp}{}_{Qq})_{\mathrm{Av}}=\pi\lambda\!\!\!^{-2}\xi_{Pp}P_{Pp}\,l_c{}^2(\Gamma_{Qq}/\Gamma)_{\mathrm{Av}}, \qquad (650)$$

where the factor $l_c{}^2$ comes from the summation over the angular momenta J (cf. §78) and increases slowly with the energy, while Γ_{Qq} is a suitable average of the $\Gamma^J{}_{Qq}$ over J. Two cases should now be distinguished:

(a) The total level width Γ contains mainly contributions which are not very sensitive to the energy, i.e., contributions from the emission of neutrons or of charged particles of energy high enough to go over the top of the potential barrier. Then it is convenient to replace Γ_{Qq} by $G_{Qq}P_{Qq}$ and to write

$$(\sigma^{Pp}{}_{Qq})_{\mathrm{Av}}=s^{Pp}{}_{Qq}P_{Pp}P_{Qq}, \qquad (651)$$

where the "internal disintegration probability" s is defined by

$$s^{Pp}{}_{Qq}=\pi\lambda\!\!\!^{-2}l_c{}^2\xi_{Pp}G_{Qq}/\Gamma \qquad (652)$$

and changes slowly with the energy as compared to the penetrabilities P_{Pp} and P_{Qq}.

(b) The total width contains mainly contributions from the emission of slow charged particles which must penetrate the potential barrier. Then the influence of the potential barrier on σ is expressed by

$$\sigma\sim P_{Pp}P_{Qq}/P_{\mathrm{Av}}, \qquad (651a)$$

where P_{Av} is an average of the penetrability over all charged particles which can be emitted.

§78. Penetration Through Potential Barrier

As we mentioned in §77, case A2, there will always be a region at low energies in which the disintegration function is governed by the probability of penetration through the potential barrier, *viz.*

$$\sigma=\mathrm{const}\cdot P_{Pp}P_{Qq}/E. \qquad (653)$$

P_{Pp} is the penetrability for the incident, P_{Qq} that for the outgoing particle and E the energy of the incident particle. The penetrability may be written in the form (600):

$$P=e^{-2g\gamma(E/B)}, \qquad (600)$$

where g is given in (600a), γ in (600b) and B is the height of the potential barrier (609).

In Table XXXVI (see also Fig. 23), we have compared the "Gamow disintegration function" (653) to the experimental data in a few cases. The agreement is fairly satisfactory for the reactions $Li^6+H^2\rightarrow 2He^4$, $Li^7+H^1\rightarrow 2He^4$ in which only incident particles of energy well below the potential barrier were used. In other cases, the agreement is not so satisfactory especially when the particle energies extend over a wide range. This is to be expected, according to the general theory given in §77, because the internal disin-

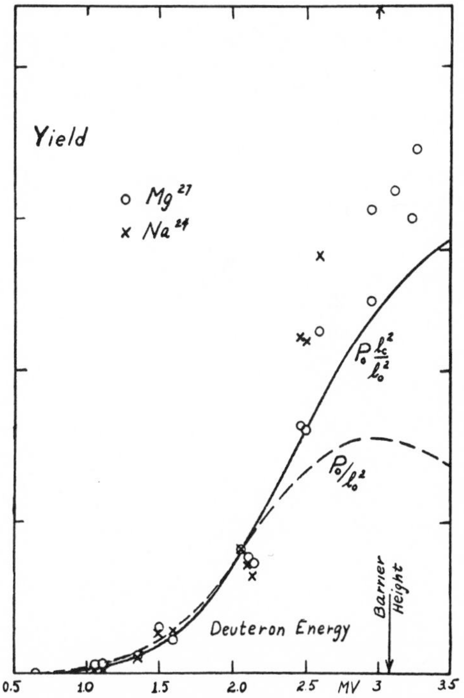

FIG. 23. Yield of the disintegrations of Mg^{26} by deuterons. Solid curve theoretical penetration function, broken curve same without consideration of angular momentum. Circles and crosses observed points for the formation of Mg^{27} and Na^{24} respectively. It is seen that the experimental disintegration functions rise more steeply than the "theoretical" curves. The agreement is slightly better when the angular momentum is taken into account. The remaining discrepancy should be attributed to the increasing number of possible states of the final nucleus (§79).

TABLE XXXVI. *Excitation functions of nuclear reactions. Energies in kv, excitation functions in arbitrary units. All excitation functions are reduced to thin targets.*

Energy kv	Excit. Funct. Theor.	Excit. Funct. Exp.	Energy kv	Excitation Function Theor. $l=0$	Excitation Function Theor. all l	Excitation Function Exp.
$Li^6+H^2=2He^4$ Oliphant, Kinsey and Rutherford (O3)			$Be^9+H^2=B^{10}+n^1$ Crane, Lauritsen and Soltan* (C38)			
			320	1.0	0.95	0.3
120	0.33	0.19	400	2.2	2.1	1.5
130	0.50	0.43	440	2.9	2.8	2.5
147	0.92	0.92	480	3.8	3.8	3.7
158	1.26	1.16	520	4.8	4.8	4.8
173	1.52	1.82	560	5.7	5.8	6.6
182	2.13	2.42	600	6.8	7.0	8.8
			640	7.6	7.9	10.7
$Li^7+H^2=2He^4+n^1$ Oliphant, Kinsey and Rutherford (O3)			$F^{19}+H^1=O^{16}+He^4$ Henderson, Lawrence and Livingston† (H22)			
104	0.34	0.10	730	0.74	0.65	0.14
115	0.51	0.29	1050	2.5	2.35	1.1
130	0.94	1.01	1180	3.4	3.3	1.45
151	1.61	1.88	1280	4.0	4.0	4.0
175	3.2	2.8	1420	5.0	5.2	5.7
			1770	7.1	8.3	12.9

* The energies given by the authors have been reduced by 20 percent (see §92C).

† The energies have been deduced from the *ranges* given. They are in the average 8 percent higher than the *energies* given by the authors, which were also deduced from ranges but with the help of an incorrect range-energy relation.

tegration probability s will change appreciably over energy regions of the order of a MV or more.[59]

A. Influence of the angular momentum, theory

Except for the very lightest nuclei, particles with various orbital momenta will be effective for the disintegration. Attention must then be paid to the dependence of the disintegration probability on the orbital momentum l of the incident particle. We shall treat this problem with the following simplifying assumptions which will correspond to reality for sufficiently heavy nuclei:

(1) The density of levels is supposed to be so great that only the average cross section will be measured (cf. §56, §77B).

(2) The contribution of each orbital momentum l is proportional to the corresponding penetrability of the potential barrier P_l (cf. (631) etc.).

Furthermore we assume

(3) The spins of scattering nucleus and scattered particle are zero, so that the orbital momentum l of the incident particle is identical with the angular momentum of the compound state. This restriction can easily be removed.

The averaged cross section is given by (405), *viz.* $(i=s=0)$

$$(\sigma^{Pp}{}_{Qq})_{\mathrm{Av}}=2\pi^2\lambda^2\sum_J(2J+1)\left(\frac{\Gamma^J{}_{Pp}\Gamma^J{}_{Qq}}{D_J\Gamma_J}\right)_{\mathrm{Av}}. \quad (654)$$

We insert

$$\Gamma^J{}_{Pp}=G_{Pp}P_J, \quad (655)$$

where G_{Pp} is the average partial width without barrier which is supposed to be the same for all values of J, and P_J is the penetrability for particles of orbital momentum J. (655) corresponds to assumption 2 above. For the outgoing particle we assume an energy sufficient to carry it over the top of the barrier, and to give $\Gamma^J{}_{Qq}$ practically the same value for all values of J for which the penetrability P_J for the incident particle is appreciable. Then (654) may be replaced by

$$\sigma^{Pp}{}_{Qq}=2\pi^2\lambda^2(G_{Pp}\Gamma_{Qq}/D\Gamma)_{\mathrm{Av}}\sum_J(2J+1)P_J. \quad (655a)$$

The sum over J may, for energies below the top of the barrier, be extended from 0 to ∞ because there is no restriction on J but that imposed by the decrease of the penetrability P_J with increasing J. (655a) may also be expressed in terms of the sticking probability for the incident particle, using (649a):

$$\sigma^{Pp}{}_{Qq}=\pi\lambda^2\xi_{Pp}(\Gamma_{Qq}/\Gamma)\sum_J(2J+1)P_J. \quad (656)$$

The penetrability as a function of $l=J$ was calculated at the end of §72. From the formulae given there, we may compute the "effective orbital momentum" l_c which we define by

$$l_c{}^2=\sum_l(2l+1)P_l/P_0. \quad (657)$$

[59] Henderson, Lawrence and Livingston (H22) to whom the experiments on the disintegration function of the reaction $F^{19}+H^1$ are due, found what seemed to be good agreement with the "Gamow formula" (653). However, they approximated the function γ in the penetrability formula (600) by

$$\gamma'=\tfrac{1}{2}\pi x^{-\frac{1}{2}}-2, \quad (a)$$

which gives for the penetrability the "simplified Gamow formula" (589b), i.e.

$$P=\text{const}\cdot\exp\,(-2\pi e^2zZ/\hbar v). \quad (b)$$

The use of this formula is unjustified if the energy of the particle is comparable with the height of the barrier.

The details of the straightforward calculation will be given in a future paper by Bethe and Konopinski. Fig. 19 gives l_c^2/g^2 as a function of x, where x is the ratio of particle energy to barrier height and g the characteristic angular momentum given in Table XXXIII. One curve ($g=5$) corresponds approximately to the disintegration of Zn by deuterons, the other ($g=25$) to the disintegration of uranium by α-particles. For comparison, the figure contains also $l_0^2/g^2=x$ where l_0 is the "critical orbital momentum" of neutral particles $l_0=R/\lambdabar$.

The effective orbital momentum l_c becomes equal to l_0 at high energies because then the Coulomb potential may be neglected. At somewhat lower energies but still above the barrier, we have $l_c<l_0$ because the combined action of Coulomb repulsion and centrifugal force produces an "effective potential barrier" for some of the orbital momenta just below l_0. As the energy decreases, this effect becomes more pronounced. For an energy just equal to the potential barrier, (cf. (634a))

$$l_c(B)\approx g^{\frac{2}{3}}, \tag{657a}$$

while $l_0(B)=g$ (by definition, cf. 611). The ratio is here $l_c/l_0\sim g^{-\frac{1}{3}}$ which is about 1/3 for uranium $+\alpha$-particles. For energies below the potential barrier, l_c does not decrease much further while l_0 keeps on decreasing as $E^{\frac{1}{2}}$. At very low energies ($<\frac{1}{2}B$), l_c reaches a *finite limit* (cf. (633))

$$l_c(0)=(\tfrac{1}{2}g)^{\frac{1}{2}}, \tag{658}$$

whereas l_0 goes to zero. This effect, which was already discussed in §72, means that, at low energies, particles of higher angular momentum penetrate the nucleus *relatively* more easily when they are charged than when they are neutral (relatively compared to particles of $l=0$; the absolute penetration probability is of course very much less for charged particles). The Coulomb potential, so to speak, carries the particles smoothly into the nucleus.

With our definition of l_c (cf. (657)), we obtain for the cross section

$$\sigma\sim\lambdabar^2P_0l_c^2, \tag{659}$$

if we neglect (cf. 655a) the variation of Γ_{Qq}/Γ and of G_{Pp} with energy. (659) represents the corrected penetration function of the potential barrier which replaces the simple Gamow function $\lambdabar^2P_0$. The formula, as well as the considerations leading to it, are valid if we assume that the critical orbital momentum l_c' for the outgoing particle is much greater than l_c. If the reverse is true, the critical orbital momentum will be determined by the outgoing rather than by the incident particle. If incident and outgoing particle have about the same l_c, a rule-of-thumb formula for the cross section is

$$\sigma\sim\lambdabar^2P_0l_c^2l_c'^2/(l_c^2+l_c'^2), \tag{659a}$$

where the primed quantities refer to the outgoing, the unprimed ones to the incident particle.

B. Physical consequences of the angular momentum

The behavior of l_c as a function of the energy has consequences for the cross section, the angular distribution of disintegration particles and the occurrence of resonances.

(1) *The cross section* is given by (659).

(a) At very low energies, l_c^2 becomes constant and different from zero. In this region, the cross section depends on the energy as in the Gamow theory, i.e., as $E^{-1}P_0$. However, the magnitude of the cross section is considerably (by a factor l_c^2) larger than would be expected if only particles of zero orbital momentum could enter the nucleus, and also larger than the cross section for neutral particles multiplied by the penetrability P_0. The factor l_c^2 goes, according to (658), up to 13 for α-particles on uranium. The effect of the angular momentum is thus quite appreciable although very small compared to the effect of the penetrability P_0.

(b) At higher energies (about $\frac{1}{3}B$ to $2B$), the cross section is *smaller* than that for neutral particles times the penetrability P_0. The ratio is given by l_c^2/l_0^2 which reaches a minimum of about 0.05 for α-particles on uranium.

(c) The cross section for charged particles *increases considerably for energies higher than the potential barrier*. E.g., for α-particles on uranium, the expression (659) increases by a factor 11 from an energy just equal to the height of the barrier, to twice that energy. Since $P_0=1$ in this region, the increase is entirely due to the increased importance of higher angular momenta. This effect is probably the most important effect of the angular momentum on the cross section; it means that the disintegration function will, in general, keep increasing beyond the top of the potential barrier so that barrier heights deduced from disintegration experiments are apt to be too high.

(2) The angular distribution of the disintegration particles will not be spherically symmetrical, even at zero energy of the incident particle, because even then orbital momenta of the incident particle up to l_c are effective. This was first pointed out by Teller (unpublished). It is remarkable that, in this respect, charged incident particles behave quite differently from neutral ones which always give, at small energy, spherical distribution of the disintegration products.

(3) The rapid increase of the number of effective angular momenta above the top of the potential barrier accounts for the fact that in α-particle disintegrations, resonance maxima can only be observed for α-energies up to approximately the height of the barrier (§82).

Table XXXVI and Fig. 23 give examples of the influence of the angular momentum on the disintegration function. In Table XXXVI we have examples of reactions on which measurements were made only well below the top of the barrier ($B=1.6$ MV for Be^9+H^2 and $B=2.5$ MV for $F^{19}+H^1$). The effect in this energy region is not very great, but the agreement with experiment is slightly improved.

The same is true for the reactions $Mg^{26}+H^2=Mg^{27}+H^1$ and $Mg^{26}+H^2=Na^{24}+He^4$ whose experimental disintegration functions are shown in Fig. 23 (data from Henderson, H23). The reaction giving α-particles has, experimentally, a steeper disintegration function. From the standpoint of the theory presented in this section, the disintegration functions should be approximately the same because in both cases the produced particle can go over the top of the potential barrier. (In §79, the reason for the difference will be explained.) The figure contains the theoretical curves with and without the angular momentum factor l_c^2; the former curve, which rises more steeply, agrees much better with experiment but is still not quite steep enough. As will be explained in the next section, this difference is also understandable.

C. Energies required for given yields

It is often useful to know in advance what energies are required in order to obtain observable results in disintegration experiments. The number of disintegrations per incident particle is approximately

$$p=N\,\delta R\,\pi\lambda\!\!\!^{-2}l_c^2P_P\xi_P\Gamma_Q/\Gamma, \tag{660}$$

where δR is the part of the range of the incident particles in which they are effective and N the number of atoms per cm³ in the disintegrated substance. We want to compute only the order of magnitude and we are especially interested in heavy nuclei because for light ones the observation of disintegrations offers no difficulty. Therefore, we use the following rough approximations: We put $\Gamma_Q=\Gamma$, $\xi_P=1$, $l_c^2=7$, $\lambda\!\!\!^{-}=1.5\cdot10^{-13}$ cm (deuterons of 5 MV). Furthermore we replace $N\delta R$ for the substance by one-third of that quantity for air, corresponding to a stopping power of 3 per atom (cf. §95); we have $N_{\text{air}}=5\cdot10^{19}$ and $\delta R_{\text{air}}\approx5$ cm considering that the particles will only be effective in the first part of their range. With these constants,

$$p\approx10^{-4}P_P. \tag{660a}$$

With a current of incident particles of 10 μa (good average for cyclotrons) the number of disintegrations per sec. is then

$$p\approx10^{10}P_P, \tag{660b}$$

so that, with good detecting apparatus, a penetrability of 10^{-10} should be just sufficient for establishing a process. Table XXXVII gives the energies of protons and α-particles required to give penetrabilities of 10^{-2}, 10^{-5} and 10^{-10}. For deuterons the energies required are about the same as for protons for the larger penetrabilities (10^{-2}) and somewhat higher than for protons for the smaller penetrabilities (10^{-10}).

TABLE XXXVII. *Penetration probabilities for various particles. (The table gives the energy in MV required for given penetration probabilities.)*

Penetration Prob. P	Particle	$Z=3$	7	11	19	29	50	92
10^{-2}	Proton	0.12	0.5	0.8	1.4	2.2	3.7	6.3
	α-particle	0.94	2.2	3.1	4.8	6.7	10.0	16.0
10^{-5}	Proton	0.04	0.15	0.3	0.6	1.1	2.1	4.0
	α-particle	0.34	1.1	1.7	3.0	4.6	7.5	13.0
10^{-10}	Proton	0.02	0.06	0.12	0.3	0.5	1.2	2.5
	α-particle	0.08	0.6	0.9	1.8	2.9	5.2	10.0

§79. The Competition Between the Nuclear Processes

A. General formulae

The cross section for a nuclear process, averaged over an energy region large compared to the spacing between resonance levels, may be written

$$\sigma^{Pp}{}_{Qq}=\pi\lambda\!\!\!^{-2}\xi_{Pp}P_{Pp}l_c^2\Gamma_{Qq}/\Gamma, \qquad (661)$$

where

$$\Gamma_{Qq}=G_{Qq}P_{Qq} \qquad (661a)$$

(cf. (656) to (659)). All the quantities such as Γ and G denote averages over a large number of energy levels in the energy region in question. P_{Qq} is the penetrability for the outgoing particle; it has been assumed that this penetrability does not depend sensitively on the orbital momentum of the particle (otherwise cf. (659a)). P_{Pp} is the penetrability of the incident particle for angular momentum zero, ξ its sticking probability and l_c the critical angular momentum (cf. §78).

(661) may be written as the product of the probability of formation of the compound nucleus and the probability of the particular mode Qq of its disintegration. The probability of formation,

$$\sigma_{Pp}=\pi\xi_{Pp}\lambda\!\!\!^{-2}P_{Pp}l_c^2 \qquad (662)$$

contains the penetration factor $\lambda\!\!\!^{-2}P_{Pp}l_c^2$ which was discussed in detail in the last section, and the "sticking probability" ξ which is expected (§54) to increase slowly with increasing energy. For large energies, $P=1$, $l_c=l_0=R/\lambda\!\!\!^{-}$ (cf. §78) and therefore

$$\sigma_{Pp}=\pi R^2\xi_{Pp}, \quad (E\gg B) \qquad (662a)$$

(cf. (408)). σ_{Pp} is the total cross section of all processes which may be produced by particle P; it depends, at low energies, mainly on the penetrability, at high energies mainly on the sticking probability of particle P.

The fraction of processes which lead to the emission of particle Q and to the state q of the final nucleus, is

$$\Gamma_{Qq}/\Gamma \qquad (663)$$

and the total relative probability for an emission of particle Q by the compound nucleus:

$$\Gamma_Q/\Gamma. \qquad (663a)$$

The probability of emission of a given kind of particles Q is therefore determined by a *competition of this process with other possible processes.*

B. Relative probability of emission of various kinds of particles

From (661a):

$$\Gamma_Q=\sum_q G_{Qq}P_{Qq}. \qquad (664)$$

According to the considerations above and in §54, an individual G_{Qq} will not depend very much on the energy. Therefore the width corresponding to a given kind Q of particles will depend

(1) on the number of possible states q of the final nucleus

(2) on the penetrabilities P_{Qq}.

Both these factors will depend on the energy available in the reaction. Let

$$Q_0=(M_A+M_P-M_B-M_Q)c^2 \qquad (665)$$

be the energy evolved in the reaction when nucleus B is formed in the ground state; the M's denote the masses of the respective nuclei. Then E_P+Q_0 will be the total energy available. It will, in general, be shared between excitation energy of nucleus B and kinetic energy of particle Q. We may now distinguish between two cases:

(a) The *available energy* E_P+Q_0 is *larger than* the *height of* the *barrier* B_Q for the outgoing particle. Then there will be some states of the final nucleus B for which the penetrabilities P_{Qq} are unity. These states will, then, in general give the largest contribution to the particle width Γ_Q (cf. also Fig. 10, §54). Therefore, if $N(U)$ denotes the number of quantum states of the *final* nucleus B below the excitation energy U, and G_{Qa} the average of G_{Qq} over the various states q of nucleus B, we may write

$$\Gamma_Q=N(E_P+Q_0-B_Q)G_{Qa}. \qquad (666)$$

Thus the particle width increases as the number of available levels of the final nucleus. There will be $N(E_P+Q_0-B_Q)$ different groups of outgoing particles with different kinetic energies. Most of the outgoing particles will have energies between B_Q and $B_Q+\tau_B$ where τ_B is the "temperature" of the residual nucleus B corresponding to the excitation energy $E_P+Q_0-B_Q$ (cf. §54).

(b) The available energy E_P+Q_0 is *smaller than* the *height of* the *barrier* B_Q. Then the particle Q will be compelled to penetrate the barrier, whatever the state q of the final nucleus. In general, the penetrability P_{Qq} will be very much less for highly excited states of B than for the ground state. Therefore the nucleus B will, in this case, in general be formed in a low state. Accordingly, there will be only comparatively few strong groups of outgoing particles. The partial width is of the same order as for the ground state alone, *viz.*

$$\Gamma_Q = G_{Q0} P_{Q0}. \qquad (666a)$$

It is obvious that the width Γ_Q in case (a) will ordinarily be much greater than in case (b). Therefore the total width Γ will be determined by those outgoing particles for which the net available energy $E_P+Q_0-B_Q$ is largest. Since E_P is the same for all outgoing particles, the conditions for large contribution to the width are

(1) Large energy evolution Q_0, i.e., the final nucleus and outgoing particle must be as stable as possible.

(2) Low barrier B_Q, i.e., neutrons are preferred compared to protons and deuterons, and these compared to α-particles.

(3) In addition to these factors, of course, the probability of formation G_Q plays a part. This factor is probably not very different for neutrons and protons, but presumably smaller for deuterons and α-particles and very much smaller for γ-rays.

From these conditions we conclude:

(1) The formation of stable nuclei is in general more probable than that of radioactive ones if both can be formed from the same compound nucleus.

(2) The emission of neutrons is, for heavy nuclei, by far the most probable nuclear process whatever the incident particle.

(3) The emission of γ-rays usually gives a negligible contribution to the total width, unless there are special reasons which make the other contributions exceptionally small. Such special reasons seem to exist only in the case of slow neutrons (cf. especially §61, 62) where the γ-ray width is 100 and more times larger than the neutron width. The reason for this is that only *one* final state is possible for the nucleus after neutron emission, *viz.* the ground state of the initial nucleus, but very many states are possible after γ-emission, *viz.* all states of the final nucleus with excitation energies below about 8 MV (binding energy of the neutron).

(4) The emission of α-particles will in general have a probability of the same order of magnitude as that of protons in spite of the higher potential barrier for alphas, because this will be compensated by the greater energy evolution. If $\Delta(A)$ is the mass excess (difference between exact mass and mass number) in energy units, as a function of the mass number A, and U_C is the excitation energy of the compound nucleus, the available energy will be for proton emission,

$$U_H = U_C + \Delta(A) - \Delta(H) - \Delta(A-1), \qquad (667)$$

for α-emission,

$$U_\alpha = U_C + \Delta(A) - \Delta(\alpha) - \Delta(A-4), \qquad (667a)$$

and therefore the difference is:

$$U_\alpha - U_H = \Delta(H) + \Delta(A-1) - \Delta(\alpha) - \Delta(A-4). \qquad (667b)$$

If we assume Δ to be a regular function of the atomic number, and insert the values $\Delta(H)=7.5$ MV, $\Delta(\alpha)=3.7$ MV, we have

$$U_\alpha - U_H = 3.8 \text{ MV} + 3d\Delta/dA. \qquad (667c)$$

Using the semi-empirical formula for nuclear masses (cf. §30) up to $A=150$, and the average energy of natural α-particles for $A=200$ and 240, we find the values given in Table XXXVIII. In this table, we have also listed the difference in the heights B of the potential barrier for α-particles and protons, according to (609).[60] It can be seen from the table that the difference in energy evolution is, in most cases, almost as great as or even greater than the difference in height of the potential barrier. Apart from the factor G_Q which cannot be estimated easily, the other factors in

TABLE XXXVIII. *Average energy evolution for α-particle and proton emission in MV.*

A	50	100	150	200	240
$d\Delta/dA$	-0.9	0	0.9	$\sim$1.9	$\sim$ 2.5
$U_\alpha - U_H$	1.1	3.8	6.5	9.5	11.3
$B_\alpha - B_H$	4.3	6.5	8.0	9.0	10.0

[60] The energy evolved in the reaction, U_α and U_H resp., is available in the center-of-gravity system. Therefore it must be compared to the barrier height in the same system

the partial width Γ_Q will be about as favorable for α-particle as for proton emission.

(5) The emission of deuterons is usually energetically unfavorable because of the high internal energy of deuterons.

We therefore come to the conclusion that in most cases the level width Γ will be determined primarily by the neutron emission. If the neutron emission leads to a radioactive nucleus and the nuclear charge is not too high, proton emission may give the main contribution to Γ. For a number of light compound nuclei which are multiples of the α-particle, *viz.* Be^8 C^{12} O^{16} Ne^{20} and, perhaps, Mg^{24} and Si^{28}, the most important mode of disintegration will be the emission of an α-particle. In no known instance is deuteron emission the most probable process. γ-ray emission is the predominant process only for low excited levels of the compound nucleus, *viz.* when no other process is energetically possible or when the available energy for other modes of disintegration is very small.

The ratios of the probabilities of various processes may be of the order ten or a hundred for medium light nuclei. For heavy nuclei, however, these ratios will reach very large values. Consider, e.g., the disintegration of heavy nuclei ($A \approx 200$) by deuterons of about 5 MV. Then the available energy U_B of the residual nucleus will be about 8.5 MV when neutrons are emitted, 9 MV for proton emission and 18 MV for α-emission. Thus protons and α-particles will just be able to go over the top of the barrier when the final nucleus is left in the ground state, and at the best only about 10 levels of the final nucleus will be possible. For neutrons, on the other hand, all the final levels below 8.5 MV can be reached which are perhaps (cf. §53) one million in number. Thus the neutron emission may be about 100,000 times as probable as the emission of a proton or an α-particle.

The probability of the predominant process will have about the same properties as the over-all probability of all possible processes (cf. 661). The probability of all other processes will behave approximately as

$$\sigma^{Pp}{}_Q \approx \sigma_{Pp}\Gamma_Q/\Gamma', \tag{668}$$

where Γ' is the partial width due to the predominant process. Now the predominant process is usually characterized by a large energy evolution, larger than in the other possible processes. But the partial width Γ_Q increases in general at the greatest rate when the energy evolution is small, whether the increase is due to increased penetrability or to an increase in the number of possible states of the final nucleus. Therefore it is likely that Γ_Q in (668) increases much faster with energy than Γ'. Consequently, *the less probable processes have in general a greater rate of increase with increasing energy.* (An exception is the emission of γ-rays. This process is intrinsically improbable in spite of the fact that the energy evolution is greatest in this case. It therefore becomes even more improbable with increasing energy, cf., e.g., Table XXVIII, §65.)

More quantitatively, we may say that the probability of a less probable process will be approximately proportional to the ratio of the number of possible final levels for this process to the number for the predominant process, if in both cases the produced particle can go over the top of the barrier. If only the particles in the predominant process can do so, the probability of the less probable process will be approximately proportional to the penetrability for the particles produced in this process, divided by the number of final levels possible in the predominant process, the latter factor being presumably less important than the former. Finally, if even the particles of the predominant process have to go through a barrier, the probability of the other process will be proportional to the *ratio* of the transmission coefficients of the barrier for the particles in question and the predominant ones. This is entirely different from the *elementary* theory (one-body theory) in which, of course, a penetrability never appears in the denominator of a cross section formula.

C. Example: The disintegration of Mg by deuterons

An illustration of our considerations is the disintegration of Mg^{26} by H^2 for which the disintegration functions are shown in Fig. 23. The most probable process in this case is no doubt

$$Mg^{26}+H^2 \rightarrow Al^{28} \rightarrow Al^{27}+n^1. \tag{669}$$

Less probable is

$$Mg^{26}+H^2 \rightarrow Al^{28} \rightarrow Mg^{27}+H^1, \tag{669a}$$

and still less probable

$$Mg^{26}+H^2\rightarrow Al^{28}\rightarrow Na^{24}+He^4. \quad (669b)$$

The energy evolution in the first case is probably about 5.5 MV (cf. Table LIX), in the second 4.2 MV (Table LVIII), while in the last it is 3.0 MV (Table LVI). The heights of the barriers are 0, 2.8, and 5.35 MV respectively. Therefore we have for the usable energy evolution, i.e. the difference between energy evolution and potential barrier:

$$\begin{aligned} Q-B=&5.5 \text{ MV for neutron emission} \\ &1.4 \text{ MV for proton emission} \\ &-2.4 \text{ MV for } \alpha\text{-particle emission.} \end{aligned} \quad (669c)$$

In the experiments of Henderson (H22), a deuteron energy up to 3.5 MV was used. Therefore, with neutron emission, all levels of Al^{27} up to an excitation energy of 9.0 MV are possible as final states. The number of these levels will probably amount to about a hundred (cf. Table XXI). With proton emission, Mg^{27} may be left in excited states up to about 4.9 MV excitation energy which may, perhaps, correspond to about 10 possible levels. The nucleus Na^{24} which is left after α-emission, on the other hand, can only have up to 1 MV excitation energy—otherwise the α-particle could not go over the top of the potential barrier. There will hardly be more than one or two excited levels in this region, besides the ground state of Na^{24}. Thus we expect that the neutron emission is much more probable than the proton emission, and the latter in turn more probable than α-emission.

On the other hand, an increase in energy will multiply the number of final levels for Na^{24} at a much faster rate than the number of possible levels for Mg^{27} or Al^{27}. This is due to the fact that in the formula for the density of levels (cf. §53),

$$\log \rho(U)=a+bU^n. \quad (670)$$

(a, b, n constants) the exponent n is smaller than unity ($\frac{1}{2}$ for free particles, 4/7 for the liquid drop model). Therefore the relative increase in the number N of levels below a certain energy U becomes approximately

$$\frac{1}{N}\frac{dN(U)}{dU}\approx\frac{1}{\rho}\frac{d\rho}{dU}=bnU^{n-1}, \quad (670a)$$

which *decreases* with increasing excitation energy U.

These predictions agree with the experimental results of Henderson for the disintegration functions of the reactions (669a, b) which were measured by means of the radioactivity of the produced nuclei. He showed that the disintegration function for the α-reaction (669b) was smaller but increased more steeply (cf. Fig. 23) with energy than that for the proton reaction (669a). The difference in slope cannot be attributed to the penetration of the α-particles through the barrier because the α-particles produced by deuterons of more than 2.5 MV energy are able to go over the top of the barrier. The effect of the angular momentum (§78) can be shown to be small owing to the large mass of the α-particle. Therefore the difference in slope should be attributed to the increase in the number of possible states of the Na^{24} nucleus.

D. Energy distribution of disintegration products

Of great interest is also the distribution of the particles of a given kind among the groups, at a fixed energy of the incident particle. For charged outgoing particles, the intensity in any group is proportional to the penetrability P_{Qq} of the potential barrier for that particular group; for outgoing neutrons, P_{Qq} should be replaced by unity. P_{Qq} (or unity) should be multiplied by $l_c^2(Qq)$ (or $l_0^2(Qq)$) if the critical angular momentum l_c (or l_0) is less for the outgoing than for the incident particle (see (659a) for a more accurate expression). If we assume that the factors l_0^2 and l_c^2 must be taken into account, the number of emitted particles with an energy between E_Q and E_Q+dE_Q is for neutrons

$$n_N(E_Q)dE_Q\sim E_Q\,\rho(E_P+Q_0-E_Q)dE_Q, \quad (671)$$

for charged particles

$$n_C(E_Q)dE_Q\sim l_c^2P_Q(E_Q) \times\rho(E_P+Q_0-B_Q-E_Q)dE_Q. \quad (672)$$

Here $\rho(U)$ is the density of levels of the final nucleus at an excitation energy U, Q_0 the energy evolution in the reaction, B_Q the height of the potential barrier, $P_Q(E_Q)$ the penetrability of the barrier for particles of energy E_Q.

The density of levels ρ increases with increasing

excitation energy. Therefore neutrons will preferably be emitted with small kinetic energies (cf. §54, 65) of the order of the "temperature" of the residual nucleus B.

For charged particles, the energy distribution (672) is plotted in Fig. 10 (§54). It shows a sharp maximum at a kinetic energy E_Q very near the height B_Q of the potential barrier and falls off rapidly on the low energy side because of the penetration factor, on the high energy side because of the Boltzmann factor. The "width" of the distribution curve is of the order of the "temperature" τ which, for heavy nuclei, is small compared to the barrier height B. The use of the energy distribution of charged particles emitted from heavy nuclei for a determination of the height of the potential barrier seems very promising. Such a determination would probably be more accurate and more direct than that from the excitation function (§78) or the natural α-decay (§68).

E. Secondary processes for very heavy nuclei

As we have shown in B, by far the most probable process for heavy nuclei is the emission of a neutron. According to Section D, the energy E_Q of this neutron is in general of the order of the temperature τ, i.e., about one MV. Therefore the excitation energy of the remaining nucleus is approximately

$$U_B \approx \Delta(P) - \Delta(Q) - (a-1)d\Delta/dA + E_P - E_Q, \quad (673)$$

where a is the mass number and $\Delta(P)$ the mass excess of the incident particle, and the other symbols have the same meaning as above. If the incident particle is a deuteron, we have $\Delta(P) = 13.7$ MV, $\Delta(Q) = 8.3$ MV so that (cf. Table XXXVIII) for $A = 200$:

$$U_B \approx 3.5 + E_P - E_Q. \quad (673a)$$

With deuterons of $E_P = 5$ MV, and a neutron energy of 1 MV, we get

$$U_B \approx 7.5 \text{ MV}. \quad (673b)$$

Now a heavy nucleus is always energetically unstable against α-disintegration, even in its ground state. The energy available for α-decay is approximately, for the ground state and $A = 200$,

$$4d\Delta/dA - 3.7 \approx 4 \text{ MV}. \quad (673c)$$

Therefore a nucleus of 7.5 MV excitation energy may emit α-particles of 11.5 MV. This is more than the energy of the fastest natural α-particles (§69) and, in addition, the nuclear charge is somewhat less than for natural α-emitters. Therefore the emission of α-particles by the residual nucleus will be even more probable than that of the long range α-groups from RaC′ and ThC′. These latter groups have an emission probability somewhat smaller than the probability of γ-emission; consequently, in our case the emission of α-particles will be at least equally probable and perhaps more probable than that of γ-rays. Therefore the primary reaction will, in many cases, be followed by the emission of an α-particle from the residual nucleus. The complete reaction scheme would thus be

$$Z^A + H^2 \rightarrow (Z+1)^{A+2} \rightarrow (Z+1)^{A+1} + n^1$$
$$\rightarrow (Z-1)^{A-3} + n^1 + He^4. \quad (674)$$

Some evidence for such a double process was obtained by Cork and Lawrence (C35) in the bombardment of platinum by deuterons. According to our consideration in B, the most probable process should be the capture of the deuteron followed by emission of a neutron. This would lead to the formation of a gold nucleus, and from the most abundant Pt isotopes, $Pt^{194,\,195,\,196}$, the gold isotopes $Au^{195,\,196,\,197}$ should be obtained of which the first two should be positron radioactive. Actually, *no gold activity* was observed. This seems very surprising, although not too much emphasis should be placed on negative evidence because the isotopes in question may have very short or very long lives and thus escape detection. Instead of a gold activity, two radioactive isotopes of iridium were found (besides Pt isotopes which are formed by the Oppenheimer-Phillips mechanism, cf. §80). The formation of an Ir isotope would require either a reaction of the type

$$Pt^A + H^2 \rightarrow Ir^{A-2} + He^4 \quad (674a)$$

or a double process of the type (674). The emission of an α-particle in the primary reaction, according to (666), must be (cf. Section B) several thousand times less probable than neutron emission. Thus, if we restrict ourselves to primary processes only, it would be difficult to understand why radioactive iridium is formed and not radioactive gold. However, the results become immediately understandable when we accept the double process (674). We must only assume that practically all the Au nuclei formed in the primary (dn)-process will disintegrate with α-emission so that as a result Ir remains rather than Au.

According to the scheme (674), the Ir isotopes formed from the most abundant Pt isotopes 194, 195, 196 would be $Ir^{191, 192, 193}$, of which $Ir^{191, 193}$ are the known stable isotopes so that only Ir^{192} can be radioactive. Observed are two active substances, both emitting negative electrons. The second of these must therefore be formed from the somewhat rarer Pt isotope, Pt^{198} (abundance 7.2 percent), and must thus be Ir^{195}. In addition, the formation of Ir^{189} from Pt^{192} might be expected; this iridium isotope should be positron active but should only be formed in very small amounts because the abundance of Pt^{192} is only 0.8 percent.

The emission of an α-particle by the residual nucleus should become even more probable with faster incident particles. It may also occur as a secondary process after the inelastic scattering of very fast neutrons (§65). It is probably restricted to rather heavy nuclei because the energetic instability of the nucleus in the ground state against α-decay is essential. However, very high kinetic energy of the incident deuteron will always produce the effect.

An emission of neutrons by the residual nucleus will occur at still higher deuteron energy. Such a process would be of the type

$$Z^A+H^2=(Z+1)^A+2n, \qquad (674b)$$

i.e., it would lead to an isobar of the original nucleus. If this isobar emits positrons of 1 MV energy, it must be 2 MV heavier than the original nucleus. The two neutrons are 3 MV heavier than the deuteron; in addition, they require about 1 MV kinetic energy each. Thus process (674b) will become probable for deuterons of about 7 MV energy: it will then, by competition, make the process (674) improbable.

F. Validity of assumptions

The considerations given here apply primarily to heavier nuclei and high excitation energies. These two conditions are needed to ensure a sufficient density of levels of the compound nucleus in the energy region in question. The excitation energy of the compound nucleus is

$$U_C=E_P+\Delta(P)+\Delta(A-A_P)-\Delta(A), \qquad (675)$$

where A is the mass number of the compound nucleus and A_P that of the incident particle. If we again assume the mass excess Δ to be a regular function of the mass number—which, of course, is a very crude assumption for all nuclei except the very heavy ones—we have

$$U_C=E_P+\Delta(P)-A_P d\Delta/dA. \qquad (675a)$$

If we take, e.g., nuclei of mass number around 20, for which these considerations probably just begin to become valid, we have $d\Delta/dA\approx-1.3$ MV. Then we find for the most important projectiles:

$$U_C=E_P+13.7+2.6$$
$$=E_P+16 \text{ MV for deuterons} \qquad (675b)$$

$$E_P+\ 8.3+1.3$$
$$=E_P+10 \text{ MV for neutrons} \qquad (675c)$$

$$E_P+\ 7.5+1.3$$
$$=E_P+\ 9 \text{ MV for protons} \qquad (675d)$$

$$E_P+\ 3.9+5.2$$
$$=E_P+\ 9 \text{ MV for } \alpha\text{-particles.} \qquad (675e)$$

Thus the excitation energy of the compound nucleus will always be very high when deuterons are used as projectiles: With 3 MV kinetic energy of the deuterons, the total excitation energy would be about 19 MV (in the average) and the spacing of the energy levels, according to Table XXI only of the order of about 100 volts for atomic weights around 20. Thus we believe that the theory developed here will be applicable to reactions produced by deuterons for atomic weights as low as 20, or even lower. With neutrons, protons and α-particles as projectiles, the excitation energies will be much lower, unless the kinetic energies are extremely high, and the spacing between levels will, for $A=20$, be at least several kilovolts. In fact, in many cases resonance levels have been observed for α-particle disintegrations which have spacings of the order of hundreds of kilovolts. This large spacing is probably partly due to selection rules (§82), partly, perhaps, to incomplete resolution in the experiments. Quite generally, the properties of the individual levels of the compound nucleus will not average out as completely for neutrons, protons and α-particles as projectiles as for deuterons. However, for atomic weights of 40 to 50 or higher, the "averaged" theory given in this section will probably be true regardless of the incident particle used, if only the energy of the incident particles has an inhomogeneity of several kilovolts or more.

For lighter nuclei the individual properties of the particular nucleus in question will be more important. However, the qualitative results of this section, concerning the relative probability of various processes, the importance of the number of groups of outgoing particles etc. will be valid even for light nuclei, as is shown to a surprising extent by the observed energy distributions of the emitted particles (cf. above, and (366)).

§80. Disintegration by Deuterons (Oppenheimer-Phillips Theory) (O11)

Oppenheimer and Phillips (O11) have suggested that nuclear reactions of the type

$$Z^A + D^2 \rightarrow Z^{A+1} + H^1 \qquad (676)$$

follow a different mechanism from other reactions which makes them more probable and less dependent on energy than would be expected from the ordinary theory. The applicability of the Oppenheimer-Phillips (O-P) theory is restricted to heavy nuclei for which it will, indeed, increase the disintegration probability considerably. For light nuclei (up to about $Z=30$), the difference between the O-P theory and the ordinary Gamow-Condon-Gurney (G-C-G) theory is unobservably small, as will be shown below. The disintegration functions for nuclei such as Na, Al, etc. by deuterons can therefore not be used as evidence for the O-P theory.

The mechanism proposed by Oppenheimer and Phillips may be described as "partial entry" of the deuteron into the nucleus. Since the proton in the deuteron is reemitted in the nuclear process, it is actually not necessary for it to enter the nucleus but it is only necessary for the *neutron* of the deuteron to come inside the nucleus Z^A. This partial entry has the great advantage that the neutron does not need to overcome the potential barrier of the nucleus because the electrostatic repulsion acts only on the proton in the deuteron. The partial entry is possible by virtue of the small dissociation energy of the deuteron (2.2 MV). Due to this fact, the deuteron has a very large radius even in the free state, and the separation between neutron and proton will be increased considerably when the deuteron is subjected to a large electric field near a heavy nucleus which will repel the proton and not act on the neutron.

A. Disproof for light nuclei

The first condition for the applicability of the Oppenheimer-Phillips theory is that it give a greater probability for the entry of a neutron than the ordinary theory of Gamow and Condon-Gurney gives for the entry of the deuteron as a whole. Now the neutron has to overcome the dissociation energy I of the deuteron in order to enter the nucleus alone, whereas the deuteron has to overcome the Coulomb potential barrier B in order to enter as a whole. Therefore the partial entry will be easier than the total entry roughly if

$$B > I. \qquad (676a)$$

Now I is 2.2 MV. The height of the barrier for deuterons is just of the same order of magnitude for nuclear charge around 10. Therefore the Oppenheimer-Phillips theory will give approximately the same result for the entrance proba-

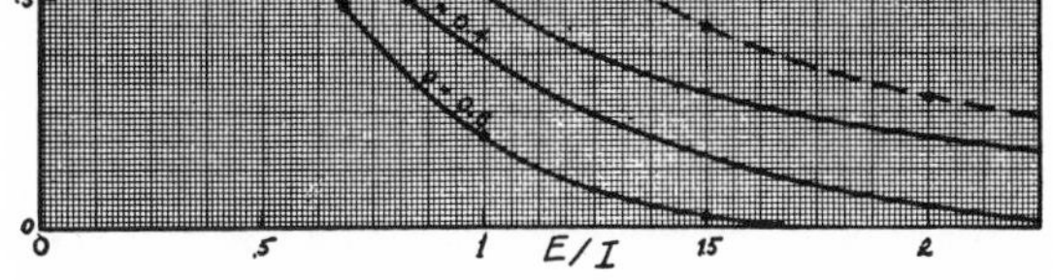

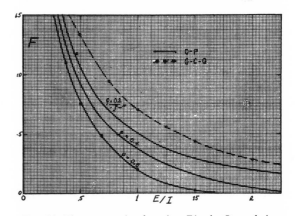

Fig. 24. The penetration function F in the Oppenheimer-Phillips theory. Curves are given for various ratios $\rho = IR/Ze^2$ of the binding energy I of the deuteron to the height of the potential barrier. $\rho=0.6$ corresponds approximately to $Z=17$, 0.4 to $Z=35$, 0.2 to $Z=100$. For $\rho=0.2$ the penetration function following from the Gamow theory is given by the broken line. For $\rho=0.6$, the Gamow points fall on the O-P curve.

bility as the Gamow-Condon-Gurney theory for nuclei with charges around 10.

This qualitative result is confirmed by a quantitative calculation which will be given in more detail in an article to appear shortly in the *Physical Review* by the author (B17). Fig. 24 gives the entrance probability according to the Oppenheimer-Phillips theory as a function of the energy. The entrance probability has been written in the form

$$P=\exp\left[-\frac{2Ze^2}{\hbar}\left(\frac{2M}{I}\right)^{\frac{1}{2}}F(E/I)\right], \quad (677)$$

where M is the mass of the deuteron. F is plotted in Fig. 24 as a function of the ratio of the kinetic energy of the deuteron E to the dissociation energy I of the deuteron, for both the Oppenheimer-Phillips and the Gamow-Condon-Gurney theory. It depends on the ratio of the dissociation energy of the deuteron to the height of the potential barrier, *viz.*

$$\rho=IR/Ze^2. \quad (678)$$

With the expression (609) for the height of the potential barrier, we have

$$\rho=4.0Z^{-\frac{2}{3}}. \quad (678a)$$

The value $\rho=0.6$, which gives the lowest curve in Fig. 24, will therefore correspond to $Z=17$. For this value of ρ, the Gamow-Condon-Gurney theory and the Oppenheimer-Phillips theory differ by an almost unnoticeable amount. The difference in F is 0.04 unit at an energy as low as $\frac{1}{4}I=0.55$ MV; at higher energies, the difference is much less. Now

$$(2Ze^2/\hbar)(2M/I)^{\frac{1}{2}}=0.60Z. \quad (677a)$$

Therefore a difference of 0.04 unit in F means, for $Z=17$, a factor of about 1.5 in the entrance probability (cf. (677)). This means that for $Z=17$ and $E=0.55$ MV, the entry of the neutron alone is 60 percent more probable than that of the deuteron as a whole. The difference is smaller for higher deuteron energy, because higher kinetic energy helps the deuteron to overcome the Coulomb potential barrier while it does not help the dissociation of the deuteron. E.g., for 1.1 MV energy the difference may be about 20 percent, and for deuteron energies above 2.2 MV, it becomes unnoticeable. Such small differences are quite unobservable and are entirely insignificant compared to the changes of the internal disintegration probability (§77) which must be expected over such large energy intervals.

These considerations hold *a fortiori* for elements lighter than $Z=17$, such as Na, Mg, Al, etc. This invalidates the interpretation given for the excitation functions observed in (dp) disintegrations[61] of these elements by Lawrence, McMillan and Thornton (L17) and by Henderson (H23). These authors found that their experimental results did not agree with the excitation function predicted by the simple Gamow-Condon-Gurney (G-C-G) theory (entry of the deuteron as a whole) but agreed well with the Oppenheimer-Phillips (O-P) theory. This is in contrast to our result that the two theories give identical results for the light nuclei in question. The reason for this difference is that Oppenheimer and Phillips calculated the excitation functions only for zero nuclear radius, i.e., for $\rho=0$ (cf. (678)). In this case there is, of course, a very great difference between the probability for partial entry and that for total entry of the deuteron, because the potential barrier for the deuteron would become enormously high if the Coulomb potential continued down to $r=0$. This explains the large difference between the "theoretical" curves for O-P and G-C-G theory obtained in the papers mentioned. Moreover, it happens that the excitation functions for different values of the nuclear radius differ only by an almost constant factor in the O-P theory; at small distances from the center of the nucleus the splitting of the deuteron into neutron plus proton would obviously be more favorable, and the penetration of the neutron into the nucleus depends in a first approximation only on the dissociation energy of the deuteron and not on the Coulomb barrier. On the other hand, the G-C-G excitation function becomes steeper for the smaller nuclear radii. Consequently, we should expect that of the disintegration functions calculated with zero nuclear radius the O-P function will depend on the energy in about the correct

[61] By a (dp) disintegration we understand a process in which the bombarding particle is a deuteron and the emitted one a proton.

way whereas the G-C-G function will not. This is the reason why agreement was found between experiment and the simplified O-P theory.[62]

B. Discussion for heavy nuclei

The difference between O-P and G-C-G theory will be much larger for heavy nuclei. For uranium, ρ (cf. (678a)) will be about 0.2. Then the difference between the values of F for the two theories is about 0.09 for 4.4 MV deuteron energy, which corresponds, according to (677a), to a factor $e^5=150$ in the penetration probability, in favor of partial entry. According to this, the O-P theory should give markedly different results from the G-C-G theory for heavy nuclei, and should therefore be applied in this case.

Before it is possible to say anything definite about the applicability of the O-P as against the G-C-G theory to heavy nuclei, it is necessary to investigate both in the light of the Bohr model. The O-P process follows the scheme

$$Z^A+H^2 \rightarrow Z^A+H^1+n^1 \rightarrow Z^{A+1}+H^1, \quad (679)$$

while the ordinary G-C-G process may be written as

$$Z^A+H^2 \rightarrow (Z+1)^{A+2} \rightarrow Z^{A+1}+H^1. \quad (679a)$$

Thus the O-P process is quite unusual inasmuch as the compound nucleus is identical with the final nucleus. The process is a pure absorption rather than a dispersion phenomenon. It requires, therefore, a treatment different (essentially simpler) from that of the usual processes. The result for the cross section for a given final state q is approximately (cf. B17)

$$(\sigma^{Ap}{}_{Bq})_{\text{O-P}}=2\pi^2\lambda^2\kappa\Gamma^q{}_{Np}/E_H. \quad (680)$$

Here $\Gamma^q{}_{Np}$ is the partial width of the compound (=final) state q corresponding to the emission of neutrons and to the state p (=ground state) of the initial nucleus, E_H is the energy of the outgoing proton and κ a constant of the order unity.

[62] As a special argument for the O-P theory of (dp) disintegrations, it was pointed out by Henderson (H23) that the reaction $Mg^{26}+H^2=Na^{24}+He^4$ has a much steeper disintegration function than $Mg^{26}+H^2=Mg^{27}+H^1$ although the initial particles are the same. This was interpreted as showing that in the first case the whole deuteron must enter the nucleus, in the second only the neutron. The correct explanation is that the number of final levels of the final nucleus increases with the energy of the deuteron more rapidly for the $(d\alpha)$ than for the (dp) reaction (§79).

We may express Γ in terms of the penetrability of the barrier and of the sticking probability (cf. 649), *viz.*

$$\Gamma^Q{}_{Np}=D_{Bq}\,\xi_N(U_q)\,P_{\text{O-P}}/2\pi, \quad (680a)$$

where D_{Bq} is the spacing of the levels of the final nucleus near the state q, $P_{\text{O-P}}$ the Oppenheimer-Phillips penetrability of the barrier and ξ_N the sticking probability of the neutron which is a function of the excitation energy U_q of nucleus B. Summing over all states q, we have

$$\sum_q \Gamma^q{}_{Np}=(P_{\text{O-P}}\xi_{N\,\text{Av}}/2\pi)\sum_0^{U'} D_{BQ}$$
$$=P_{\text{O-P}}\xi_{N\,\text{Av}}U'/2\pi, \quad (680b)$$

where U' is the maximum possible excitation energy of the final nucleus (see below). Since U' is of the order of E_H, we find finally

$$(\sigma^{Ap}{}_{Bq})_{\text{O-P}}=\pi\lambda^2\kappa\;\xi_{N\,\text{Av}}(U_q)\;P_{\text{O-P}}. \quad (681)$$

This should be compared with the cross section for total entry of the deuteron, *viz.*

$$(\sigma^{Ap}{}_{Bq})_{\text{G-C-G}}=\pi\lambda^2\xi_D(U_r)P_{\text{G-C-G}}\Gamma_H/\Gamma, \quad (681a)$$

where $P_{\text{G-C-G}}$ is the Gamow-Condon-Gurney penetrability, ξ_D the sticking probability for the deuteron, Γ_H the proton width and Γ the total width of the compound level (compound nucleus $(Z+1)^{A+2}$). The ratio is therefore

$$\frac{\sigma_{\text{O-P}}}{\sigma_{\text{G-C-G}}}=\frac{P_{\text{O-P}}}{P_{\text{G C-G}}}\cdot\frac{\xi_N(U_q)}{\xi_D(U_r)}\cdot\frac{\Gamma}{\Gamma_H}\kappa. \quad (682)$$

Of the factors occurring in (682), the first was found to be about 100 in favor of the Oppenheimer-Phillips theory. As regards the second factor, we have shown in §54 that the sticking probability will presumably increase with increasing excitation energy U. The excitation energy U_r in the case of the ordinary G-C-G processes is (cf. (675b)) extremely high, *viz.* of the order of 20 MV. In the case of the O-P process, we have

$$U_q=\Delta(H^2)-\Delta(H^1)-d\Delta/dA+E_D-E_P, \quad (683)$$

where Δ is the mass excess (cf. (667)), E_D the kinetic energy of the incident deuteron and E_P that of the outgoing proton. It can be shown

(B17) that the proton energy is, in the average, roughly equal to deuteron binding energy plus half the kinetic energy of the incident deuteron, i.e.,

$$E_H = 2.2 \text{ MV} + \tfrac{1}{2}E_D. \qquad (683a)$$

With $\Delta(H^2) = 13.7$ MV, $\Delta(H^1) = 7.5$ MV and $d\Delta/dA = 1.9$ MV ($A \approx 200$), we find

$$U_q \approx 2 \text{ MV} + \tfrac{1}{2}E_D. \qquad (683b)$$

For $E_D = 5$ MV, this is about two MV less than the excitation energy of a nucleus formed by capture of a slow neutron. Therefore the sticking probability ξ_N will be somewhat less than for slow neutrons, perhaps about 1/100 (cf. 357b). On the other hand ξ_D can be expected to be of the order unity. Therefore the second factor (sticking probability) will approximately cancel the first (penetration probability).

Therefore it will depend on Γ_H/Γ which of the two mechanisms will give the greater cross section. If the proton width Γ_H gives the largest contribution to the total width, the two kinds of processes will be about equally probable. However, for a heavy nucleus by far the largest contribution to the width will come from the emission of neutrons (cf. §79) because the number of possible levels of the residual nucleus is much smaller for proton than for neutron emission. Therefore, $\Gamma_H \ll \Gamma$. This makes the Oppenheimer-Phillips mechanism much more probable for heavy nuclei than the Gamow-Condon-Gurney mechanism.

Our considerations show that the reason for the applicability of the O-P theory is quite different from that naïvely expected. What matters is not so much the facilitated entry of the deuteron as the facilitated escape of the proton. This, again, is not directly affected by the potential barrier but rather indirectly. The fastest protons produced by deuterons of about 5 MV will, in practically all cases, be able to go over the top of the barrier. But the barrier reduces the number of proton groups able to go over the top, and this is the reason which makes the probability of a (dp) reaction small in the ordinary scheme.

At the same time, we see that the probability of emission of a proton according to the Oppenheimer-Phillips mechanism should be about the same as that of a neutron in the ordinary mechanism. This fact may be useful for interpretation of experiments.

A possible experimental test of the O-P theory is afforded by the energy distribution of the emitted protons. According to the ordinary theory, most of the protons should have energies about equal to the height of the potential barrier. According to the O-P theory, the protons do not come as near to the nucleus and therefore do not need to have such high energy in order to escape easily. The most probable energy of the protons will be approximately equal to the deuteron binding energy plus half the kinetic energy of the incident deuterons. Thus, according to the O-P theory, the protons should be slower and have an energy depending on that of the deuterons.

Since the outgoing protons have comparatively small energy, the residual nucleus will be left rather highly excited, the excitation energy being, in the average, about $U_q = 2 \text{ MV} + \tfrac{1}{2}E_D$ (cf. (683a)). This excitation energy will, for heavy nuclei and very fast deuterons, be sufficient to make the emission of an α-particle by the residual nucleus about as probable as that of a γ-ray (cf. §79E). The available energy for α-emission is, at $A = 200$, about $6 \text{ MV} + \tfrac{1}{2}E_D$ (cf. (673c)). The energy required to make α-emission as probable as γ-emission is about 10 MV (energy of the fastest natural alphas). This means that deuterons of 8 MV will in general produce the double process

$$Z^A + H^2 = (Z-2)^{A-3} + H^1 + He^4 \qquad (684)$$

but deuterons of 5 MV will in general not do so to any appreciable extent; they will cause the simple O-P process (676) instead. The process (684) has therefore a higher threshold than the analogous process (674) in which a neutron and an α-particle are emitted. The reason is, of course, that the proton emitted in the O-P process has in general a higher energy than the neutron emitted in a dn process.

A process of the type (684) has actually been observed by Cork and Thornton (C36) who bombarded gold with 7 MV deuterons and observed the formation of a radioactive iridium isotope. Cork and Thornton suggested the reaction

$$Au^{197}+H^2 \rightarrow Ir^{194}+He^4+H^1, \quad (684a)$$

which seems to be well justified theoretically by our considerations.

Cork and Thornton observed also the simple O-P reaction

$$Au^{197}+H^2 \rightarrow Au^{198}+H^1. \quad (684b)$$

This reaction is presumably caused by the deuterons already slowed down (to 5 MV, perhaps) in the target. It would be of great interest to observe the excitation functions of these two reactions. Up to a certain critical energy E_c, the simple process (684b) should be the most probable and (684a) should be much less probable. At E_c, the relative probabilities will change rather suddenly, owing to the very rapid increase of the α-particle penetrability with increasing energy, so that above E_c the double process (684a) will predominate. Finally, for deuteron energies of, perhaps, 10 MV, the nucleus formed in the primary O-P process will retain enough energy to emit a neutron, so that the complete process will be

$$Z^A+H^2 \rightarrow Z^{A+1}+H^1 \rightarrow Z^A+H^1+n^1, \quad (684c)$$

which is, of course, unobservable except by the neutron emission itself. For these energies, the double process (684) will therefore cease to be probable, because the emission of α-particles will be less probable (barrier!) than that of neutrons as soon as the latter can occur on energetic grounds.

§81. Resonance Phenomena in the Simple Capture of Protons (H4, H6)

Resonance effects with charged incident particles have been observed for two types of processes, *viz.*

(1) processes produced by α-particles (emission of neutrons or protons).

(2) simple capture of protons.

The simple capture of protons has been studied extensively in the following three cases:

$$Li^7+H^1 \rightarrow Be^{8*} \rightarrow Be^8+\gamma, \quad (685)$$

$$C^{12}+H^1 \rightarrow N^{13*} \rightarrow N^{13}+\gamma, \quad (685a)$$

$$F^{19}+H^1 \rightarrow Ne^{20*} \rightarrow Ne^{20}+\gamma. \quad (685b)$$

In all three cases, the dependence of the yield on the proton energy was studied (H6, H4) and resonance maxima were found. The γ-spectrum has also been investigated (§90, D13). The yield in the reactions (685, 685b) is studied by measuring the γ-rays, whereas in case (685a) the radioactivity of the product nucleus N^{13} is used. Other simple capture phenomena which are known to occur are

$$B^{10}+H^1 \rightarrow C^{11}+\gamma, \quad (685c)$$

$$B^{11}+H^1 \rightarrow C^{12}+\gamma, \quad (685d)$$

$$O^{16}+H^1 = F^{17}+\gamma. \quad (685e)$$

The first and third of these are detected through the radioactivity of C^{11} (C39) and F^{17} (D28), respectively, the second is identified by means of the very energetic γ-rays it produces (C50). The first process has not been investigated for resonance. The second seems to have a resonance maximum at 180 kv proton energy (G13a) with a width of about 15 kv. (The same resonance level appears in the reaction $B^{11}+H^1=Be^8+He^4$, cf. §88 and W21a.) The reaction (685e) does not seem to show resonances (DuBridge, private communication).

The resonances observed by Hafstad, Heydenburg and Tuve in the capture of protons by Li^7 and F^{19} are extremely sharp. The resonance energies and the widths of the levels are given in Table XXXIX. The width in the case of Li is about 11 kv, at a proton energy of 440 kv. For the width of the F levels, only upper limits could be given because the observed width is not larger than the inhomogeneity of the incident proton beam. The data for C are taken from earlier experiments of Hafstad and Tuve in which the voltage definition of the incident particles was

Table XXXIX. *Resonance levels in the simple capture of protons.*

Capturing Nucleus / Compound Nucleus	Li^7 / Be^8	C^{12} / N^{13}	F^{19} / Ne^{20}		
Resonance energy E_0 (kv)	440	420	328	892	942
Width Γ (kv)	11	<40	<4	<12	<15
Angular momentum J (probably)	1	½?	odd (or even)		
Parity (probably)	odd	?	even (or odd)		
Yield at resonance (cm^2)	$\sim 10^{-27}$	—	—	—	—
Integrated yield (10^{-24} volt cm^2)	17	0.9	0.9	20	9
γ-ray width (volts)	4	0.08	0.6*	18*	8*
Reciprocal proton penetrability	3.5	50	6000	13	10.5
Proton width without barrier (kv)	40	<2000	<24000	<160	<160

* For $J=0$. For $J=1$, one-third of these figures should be taken.

less sharp; therefore the width given in the table is probably much larger than the actual width of the resonance level.

The small widths observed make it likely that, in all five cases listed, there is no "probable" process by which the compound nucleus may disintegrate. E.g., the compound state of Be^8 concerned in the capture of protons by Li^7 presumably is not able to disintegrate into two α-particles because such a disintegration would give a very large width of the resonance level. Similarly, the three states of the compound nucleus Ne^{20} probably cannot disintegrate into an O^{16} nucleus in the normal state and an α-particle. *We shall assume that in all cases* listed in Table XXXIX, *the compound nucleus can only disintegrate with the emission of protons or* γ*-rays.*

There may be two different reasons for such a situation: (a) It may be that the emission of protons and γ-rays are the only processes which are energetically possible. This seems to be the case for the N^{13} state. Disintegrations with emission of neutrons, deuterons and α-particles would lead to the highly unstable or even nonexistent nuclei N^{12}, C^{11} and B^9, respectively; therefore these processes will be energetically impossible.

(b) The emission of other particles may be forbidden by selection rules. It is very easy to see how this may occur for an excited state of Be^8. This nucleus may break up into two α-particles. As is well known, the spin of an α-particle is zero and its internal wave function has even parity.[63] Moreover, two α-particles obey Bose statistics, therefore the wave function describing the relative motion of their centers of gravity must also have even parity and must contain only even orbital momenta (cf. §74F). Thus the complete wave function of a "final state" containing two α-particles must have even parity and even total angular momentum. Therefore a state of the compound nucleus Be^8 which has odd parity and/or odd angular momentum, cannot disintegrate into two α-particles. We therefore ascribe odd J or odd parity or both to the level of Be^8 responsible for the capture of 440 kv protons by Li^7.

In the case of Ne^{20}, we seek a selection rule forbidding the disintegration into $O^{16}+He^4$. These two nuclei have again zero spin and even parity. However, there is no such strict symmetry requirement on the wave function describing the relative motion of the two nuclei with respect to each other as for Be^8. The only requirement is that the wave function should have even parity for even orbital momentum and odd parity for odd l. Thus the disintegration of an excited state of Ne^{20} will be forbidden, if it has even angular momentum combined with odd parity, or *vice versa*. This conclusion is strongly confirmed by the absence of a line corresponding to a transition to the ground state in the γ-ray spectrum (§90).

The disintegration of the compound states of Be^8 and Ne^{20} with emission of neutrons and deuterons is again energetically impossible as can easily be seen from the nuclear masses. There remain therefore only the disintegrations with proton and γ-ray emission. Of these processes, the γ-ray emission certainly does not give a large contribution to the level width, because we know the interaction between matter and radiation to be small. The main contribution to the observed width of the resonance levels must therefore be attributed to the protons, i.e., to the incident particles themselves.[64]

The proton width may be split into a penetrability factor P and a factor G giving the width without barrier. These factors are listed in Table XXXIX. For Be^8, a value of 40 kv is obtained for G, giving

$$G_P' = G_P E_P^{-\frac{1}{2}} = 40{,}000 \cdot 340{,}000^{-\frac{1}{2}} = 70 \text{ volts} \quad (686)$$

for the "width at one volt energy"[65] (cf. (550)). This is to be compared to values of the order of a millivolt for the same quantity deduced from experiments on the capture of slow neutrons by heavy nuclei. The great difference (factor 100,000) is due to the much smaller number of levels in the light nucleus Be^8 as compared to heavy nuclei. The average spacing between the energy levels of Be^8 may be estimated to be of the order of 1 MV, either from theoretical calculations such as those of Wigner and Feenberg (F10), or from the empirical fact that just one resonance level has been

[63] This means that ψ does not change sign upon "inversion," i.e., upon change of the sign of the coordinates of all elementary particles in the α-particle.

[64] For a discussion of other possibilities, cf. H6.

[65] 340 kv is the energy of the proton with respect to the center of gravity; $340 = 440\ (7/8)^2$.

observed in the energy region below 1 MV which was investigated experimentally. This spacing is about 100,000 times larger than the average spacing between neutron resonance levels in heavy nuclei (§60). Therefore the reduced particle width G' seems to be approximately proportional to the spacing of energy levels, corresponding to a sticking probability (§54) independent of the nuclear mass.

For the other resonance levels, only upper limits for the proton width without barrier can be deduced from the experimental data. These upper limits are rather high, and if the proton width is really the only appreciable contribution to the total level width, it is likely that the actual widths of the resonance levels are much smaller than the upper limits given in Table XXXIX.

The γ-ray width Γ_γ may be deduced from the cross section at resonance which, with $s=\frac{1}{2}$ for the proton spin, has the value (cf. (262))

$$\sigma^P{}_\gamma = 2\pi\lambda\!\!\!^{-2}(2J+1/2i+1)\Gamma_\gamma/\Gamma, \qquad (686a)$$

where $\Gamma=\Gamma_P$ is the total width and $\lambda\!\!\!^-$ the proton wave-length. The resonance cross section was only determined for the capture of protons by Li^7; it is about 10^{-27} cm^2 in this case. The spin of the compound state is probably $J=1$, because this state has a strong optical transition to the ground state (§90) which is almost certainly a ^1S state. The spin of the capturing nucleus Li^7 is known to be $i=\frac{3}{2}$. The wave-length of the protons is $7.8\cdot 10^{-13}$ cm. With these data, we find $\Gamma_\gamma=4$ volts. This seems plausible in comparison with the γ-ray widths of a few tenths of a volt found for heavy nuclei (cf. §61, 90).

The γ-ray width can also be inferred from the integrated (thick-target) cross section (cf. (517))

$$\int \sigma^P{}_\gamma dE = \pi^2\lambda\!\!\!^{-2}(2J+1/2i+1)\Gamma_\gamma. \qquad (687)$$

If the energy loss of the bombarding particle per cm of the bombarded material is written in the form (cf. §95)

$$-dE/dx = N\epsilon \qquad (687a)$$

where N is the number of disintegrable nuclei per cm^3 and ϵ a quantity of the dimension energy times area, then the probability of capture for each incident particle is

$$p^P{}_\gamma = \int \sigma^P{}_\gamma\, dE/\epsilon. \qquad (688)$$

This method may be used to determine the γ-ray widths of the carbon and the fluorine levels.

For carbon, Hafstad and Tuve (H4) found, from the intensity of the produced radioactivity, a capture probability of about 1 in 10^{10}. This value was confirmed by Allison (A5). For protons of 420 kv, we have (§95) about $\epsilon=9\cdot 10^{-15}$ volt cm^2. Therefore $\int\sigma dE=9\cdot 10^{-25}$ volt cm^2. With $i=0$, $J=\frac{1}{2}$ (most probable because of selection rules), $\lambda\!\!\!^-=7.6\cdot 10^{-13}$, we find then $\Gamma_\gamma=0.08$ volt. The much smaller value found here as compared to the Li case, is probably due primarily to the smaller energy of the γ-rays (2.3 as compared to 17 MV).

The width of the fluorine levels can only be estimated very roughly by comparing the yields reported by Hafstad and Tuve (H4) for Li and F. They found that the γ-rays from a thick LiOH target produced an ionization of 0.26 divisions/min./μa in an ionization chamber 30 cm from the target, while the γ-rays from CaF_2 which corresponded to the 330 kv resonance capture caused an ionization of 0.21 div./min./μa at a distance of 12 cm. This means a ratio of the ionizations of $0.21\cdot 12^2/0.26\cdot 30^2=0.13$. We may assume the ionization power of γ-rays to be proportional to the number of Compton electrons produced times their average energy (or range). This gives about equal ionization power for the fluorine and lithium gammas. From the cross section σ_0 of the Li capture at resonance (10^{-27} cm^2) we find then:

Integrated cross section for Li: $\frac{1}{2}\pi\Gamma\sigma_0=\frac{1}{2}\pi\cdot 11000\cdot 10^{-27}$ $=1.7\cdot 10^{-23}$ cm^2 volt.

Energy loss constant (ϵ) for 440 kv protons in air: $1.03\cdot 10^{-14}$ cm^2 volt.

Stopping power of LiOH: 2.0 times air per atom of Li, 2.3 times air per atom of Li^7 (considering abundance).

Thus, number of quanta per proton: $1.7\cdot 10^{-23}/2.3\cdot 1.03\cdot 10^{-14}$ $=7\cdot 10^{-10}$.

Number of quanta per proton in fluorine: $0.13\cdot 7\cdot 10^{-10}$ $=9.10^{-11}$.

Energy loss constant for 330 kv protons in CaF_2: $2.3\cdot 10^{-14}$ cm^2 volt per F atom.

Integrated cross section for 330 kv level of F: $9\cdot 2.3\cdot 10^{-25}$ $=2\cdot 10^{-24}$ cm^2 volt.

The intensities (integrated cross sections) for the other two levels of F are, according to Hafstad, Heydenburg and Tuve (H4), about 22 and 10 times those of the 330 kv level, respectively. This corresponds to integrated cross sections of 4.5 and $2\cdot 10^{-23}$ cm^2 volt. Inserting the nuclear spin $i=\frac{1}{2}$ of the fluorine nucleus, and the proton wave-lengths of 8.4, 5.1 and $4.9\cdot 10^{-13}$ cm for the three resonance levels, we obtain the figures listed in Table XXXIX for $J=0$. It is seen that the γ-ray widths of the two higher levels would appear rather large if $J=0$, especially considering the much lower quantum energy of the fluorine as compared to the lithium radiation.

It may be mentioned that besides the five resonance levels listed in Table XXXIX, some indications of other levels have been obtained for the proton capture by Li^7, C^{12} and F^{19} (H6, H4). There seems to be a very broad level of Be^8 giving rise to capture of protons of 800 kv and more by Li^7, a weak narrow level and a weak broad level of Ne^{20} corresponding to capture of 400–700 kv protons by F^{19}, and a multiplet structure of the level of N^{13}. This shows that, at least for Ne^{20}, the density of nuclear energy levels is already rather large, the spacing being of the order of 100 kv at an excitation energy of about 14 MV. This is, however, a somewhat larger spacing than was calculated theoretically in §53 from the liquid drop model, showing that this model is probably not adequate for such a light nucleus.

Some explanation may be necessary for the fact that no appreciable capture is observed outside the rather narrow resonance regions, while other processes, leading to particle emission, are observable for all energies of the incident particle. Presumably, the latter result is due to the effect of very broad resonance levels whose width is comparable to their distance apart (§84). Now the cross section for γ-ray emission near resonance is inversely proportional to the total width of the level (cf. (686a)) and will therefore be much smaller for a broad level than for a narrow one. On the other hand, it must be admitted that the integrated cross section (687) is independent of the width of the level and depends only on the partial width for γ-emission. Therefore, in thick targets, there should be about as many capture processes due to narrow resonance levels as due to broad ones. The present experimental evidence does not extend over sufficiently many cases to say much about this point; but DuBridge and his collaborators have shown that at least the capture of protons by O^{16} does not arise from narrow resonance levels.

§82. Resonances in α-Particle Disintegrations. Resonance and Barrier Height

A great number of resonance levels have been observed in disintegrations produced by α-particles. A description of the experiments is found in §99. The experimental results are given in Table XXXX. Most reactions have been investigated by several authors, in these cases we have given in the table the results which we consider most reliable; usually, these are the most recent ones.

A. Barrier height and resonances

We have included in Table XXXX the height of the potential barrier for α-particles according to two methods, *viz.* (a) as derived from the interpolation formula (609) and (b) the "experimental" values given by Chadwick and Feather (C10). These authors pointed out that disintegrations should occur at any α-energy higher than the top of the barrier, while for lower α-energies observable intensities can only be obtained in the resonance regions.

This consideration while correct in the old one-body picture of nuclear disintegrations, might at first sight seem somewhat doubtful in the many-body picture. According to this picture, the width of resonance levels is determined by the *most probable* mode of disintegration of the compound nucleus. Now in all cases listed in Table XXXX, the particle *produced* in the reaction

TABLE XXXX. *Barrier heights and resonance levels from α-particle disintegrations.*

Compound Nucl.	Reaction	Barrier in MV		Resonance Levels		Width MV	Ref.
		Theor.	Exp.	α-energy in MV	Excitation En. of Comp. Nucl. (MV)		
C^{13}	$Be^9+\alpha=C^{12}+n$	3.9	3.5	3.4; 4.8?	12.8; 13.8?	0.3; ~0.3	B7
N^{14}	$B^{10}+\alpha=C^{13}+H$	4.6	3.6	4.2	14.8	0.5	M16
N^{15}	$B^{11}+\alpha=N^{14}+n$	4.3	3.7	3.2?	13.4?	~0.4	C7
F^{18}	$N^{14}+\alpha=O^{17}+H$	5.2	4.1	3.6	8.2	~0.15?	P2
Na^{23}	$F^{19}+\alpha=Ne^{22}+H$	5.7	5.0	3.7; 4.1	14.5; 14.8	0.10; 0.13	C13
Si^{28}	$Mg^{24}+\alpha=Al^{27}+H$	6.8	6.5	5.7; 6.3	13.7; 14.2	0.12; 0.13	C13
P^{31}	$Al^{27}+\alpha=Si^{30}+H$	6.9	6.8	4.0; 4.5; 4.9; 5.3; 5.8; 6.6	12.0; 12.4; 12.8; 13.1; 13.6; 14.3	~0.10 0.07; 0.13 ~0.12	C10 D19

(neutron or proton) can go over the top of the barrier even when the energy of the α-particle is low. Therefore the width of the levels will be determined by the proton or neutron emission, especially if the penetrability of the barrier for α-particles is small. From this point of view, there should be little difference in the width of levels below and above the top of the α-particle barrier.

However, the selection rules for the angular momentum may cause a difference in the excitation function below and above the top of the barrier. Below the barrier, we know that only α-particles of small orbital momentum can enter the nucleus (§78) while sufficiently far above the barrier, much higher orbital momenta are possible. Therefore below the barrier there will be resonance only with compound levels of small angular momentum J, while above also compound levels of large J will be important. This will make the spacing between "important" resonance levels much smaller above the barrier than below it, and since the width of the levels will stay about constant, it is to be expected that the resonances will overlap each other at energies above the height of the barrier, and the maxima and minima will thus become unobservable.

This consideration justifies the reasoning of the older theory and makes it possible to deduce approximate barrier heights from the cessation of pronounced resonance phenomena. This has been done by Chadwick and Feather; their results agree surprisingly well with the "theoretical" values obtained from the rule that the nuclear volume is proportional to the number of particles. This can be considered as a fairly good experimental confirmation of this rule.

In accord with our considerations, most of the resonances lie below the calculated top of the barrier. The only exception is a very weak and somewhat doubtful resonance observed by Bernardini for the compound nucleus C^{13} (reaction $Be^9+He^4=C^{12}+n^1$). We must expect that such resonances above the barrier occur occasionally, especially for very light nuclei.—Several resonance levels are found very near the top of the barrier, e.g., the highest levels in the compound nuclei N^{14}, Si^{28} and P^{31}.

B. Spacing of the resonances

In judging the figures in Table XXXX, we must keep in mind that α-particle energies of less than 3–4 MV have not been investigated because for these energies the penetrability of the α-particle would be so small that no observable disintegration effects could be obtained, even at resonance. It is to be hoped that the greater intensity of artificial sources will make the investigation of the region of low α-particle energies possible.

The general impression from the figures in Table XXXX is that the resonance levels become more numerous and therefore more closely spaced for heavier nuclei. This is entirely in accord with out theoretical expectations. The average spacing is apparently slightly more than 1 MV for the lightest, and about $\frac{1}{2}$ MV for the heaviest nuclei listed.

The excitation energies U of the compound nuclei corresponding to resonance are listed in the sixth column of Table XXXX. If M_A, M_α and M_C are the masses of initial nucleus, α-particle and compound nucleus (ground state), respectively, we have

$$U=(M_A+M_\alpha-M_C)c^2+(M_A/M_C)E_\alpha. \quad (689)$$

Most of the excitation energies listed are about 12–15 MV.

C. Width

The width of the resonance levels may be obtained in various ways (cf. C13, C4).

1. If there are few resonances of comparatively large width the simplest procedure is to measure the total yield of the reaction from a thin target, as a function of the α-particle energy. This has been done for the reactions $Be^9+\alpha=C^{12}+n$, $B^{10}+\alpha=C^{13}+H$, $B^{11}+\alpha=N^{14}+n$. The method is apt to give too large widths because of the finite thickness of the target and the straggling of the α-particles. Closely spaced narrow resonances may disappear entirely.

2. With a thick target, the intensity of a given *group* of produced particles (usually protons) may be investigated as a function of the energy of the α-particles entering the target. From the α-particle energy at which the group first appears, and that at which it attains its full intensity, the

width of the level may be inferred. The method is subject to similar criticism as (1). It has been the one most widely used for determining widths (C4).

3. The inhomogeneity in energy of the *protons* of a given group will also give the width of the level, due account being taken of the momentum relations (recoil of the disintegrated nucleus). This method is free from corrections due to the finite thickness of the target and to the straggling of the α-particles but is influenced by the straggling of the protons. Also, the proton energy depends on the direction of emission of the protons which introduces a further inhomogeneity. Thus the widths deduced from this method will again be too large. We have used this method for estimating the widths in the reaction $F^{19}+He^4=Ne^{22}+H^1$.

4. The *maximum* range of the protons of a group may be determined for the α-energy at which the group first appears, and for the α-energy at which it attains full intensity. This method is the most satisfactory in eliminating straggling. Unfortunately, it is probably the most difficult one experimentally. It has been used by Chadwick and Constable to deduce the width of the 4.9 MV level in the reaction $Al^{27}+\alpha=Si^{30}+H$. It was found that the maximum range of the protons increased by about 2 cm (from 28 to 30) when the α-particle range increased from 3.25 (first appearance of group) to 3.55 cm. This corresponds to an increase in proton energy of about 0.18 MV. Considering that the protons were observed *in* the direction of the incident α-particles, we obtain from the momentum relations (§96) that this corresponds to an increase of the α-energy of about 0.17 MV. For the excitation energy only the kinetic energy in the center of gravity system is available; therefore we must reduce the width by a factor 27/31 which gives 0.15 MV. Half this value has been taken in Table XXXX (cf. below).

In accord with our custom, we have given in the table the *width at half-maximum*. Where the shape of the yield-energy curve was not obtained experimentally, it was assumed that the disintegration becomes unobservable when the cross section is about one-fifth of its maximum value. According to the resonance formula, this means that the disintegration will be observable over an energy region equal to *twice* the width Γ of the level. Therefore the values given in the table are smaller than those given, e.g., by Chadwick and Feather, by about a factor of 2.

The width of the resonance levels must probably be attributed mainly to the *produced* particle, i.e., proton or neutron. The partial width due to the α-particle will be smaller because of the potential barrier. In almost all cases, several groups are emitted from each resonance level, corresponding to several states of the final nucleus. E.g., 5 proton groups have been observed for the reaction $B^{10}+\alpha=C^{13}+H$, 4 proton groups for each resonance level of $Al^{27}+\alpha=Si^{30}+H$, and at least three neutron groups for $Be^9+\alpha=C^{12}+n$. The width corresponding to a *given* final state is therefore only about 0.1, 0.06 and 0.02 MV for the compound nuclei C^{13}, N^{14} and P^{31}, respectively. The neutrons and protons emitted have energies of the order of 2–10 MV in each case. Taking an average of 5 MV, we find for the reduced width (width at one volt energy) $\Gamma'=\Gamma E^{-\frac{1}{2}}$ (cf. 685) about 25–50 volts for the lighter nuclei C^{13} and N^{14}, and about 10 volts for the heavier compound nucleus P^{31}. These values fit in very well with the value obtained from the capture of protons by Li^7 (cf. §80) *viz.* 70 volts. They also show the expected tendency to decrease with increasing mass of the nucleus.

D. α-particle width

The α-particle width can be deduced from the integrated disintegration probability if it is admitted

(1) That the α-particle width is smaller than the width for the produced particle, and

(2) That no other process of comparable probability can occur.

Both conditions are probably fulfilled, at least approximately, in the cases listed in Table XXXX.

Chadwick and Constable have given the total yield of protons from α-disintegration of Al and F. For Al, each resonance level gives about $3.5\cdot 10^{-7}$ protons per α-particle. The "energy loss constant ϵ" (cf. (687a), §95) is, for α-particles of 5 MV in Al, about $2.3\cdot 10^{-14}$ volt cm². According to (688), this corresponds to an integrated cross section of $0.8\cdot 10^{-20}$ cm² volts. Since for α-particles

$s=0$, the formula for the integrated cross section is (cf. (687))

$$\int \sigma dE = 2\pi^2 \lambda\!\!\!^{-}{}_\alpha^2 \Gamma_\alpha (2J+1)/(2i+1). \quad (690)$$

The spin of Al^{27} is $i=\frac{1}{2}$, the wave-length of 5 MV α-particles (with reduced mass!) $1.17\cdot 10^{-13}$ cm. Therefore we find

$$(2J+1)\Gamma_\alpha = 60 \text{ kv}. \quad (690a)$$

Since the total width is only 90 kv, Γ_α must be considerably smaller than this amount. Therefore we must assume that J is not zero. For the given ratio of α-particle energy to barrier height, orbital momenta up to about 3 or 4 will be important. For $J=2$, the α-ray width would come out to be 12 kv. The penetrability of the barrier is 1/14, giving for the α-width without barrier about 170 kv, i.e., somewhat more than the proton width per final level (20 kv).

In a CaF_2 target, the yield is 9.2 and $7.2\cdot 10^{-7}$ protons per α-particle for the two resonance levels of fluorine observed. In pure fluorine, the yield would be 1.77 times as large (C4) since a calcium atom has a stopping power for α-particles about 1.54 times as large as a F atom (§95). This would give in the average for the two levels $14.5\cdot 10^{-7}$ protons per α-particle. The energy loss constant in F is $2.35\cdot 10^{-14}$ volt cm^2 for an α-energy of 3.9 MV (corresponding to the average of the two levels), therefore the integrated cross section is $3.4\cdot 10^{-20}$ volt cm^2. The spin of F^{19} is again $i=\frac{1}{2}$ (Table XVII), the wave-length $1.40\cdot 10^{-13}$ cm. This gives $(2J+1)\Gamma_\alpha=175$ kv. Since the total widths of the levels are about 100–130 kv, the angular momenta of the levels are obviously again greater than zero.

§83. Selection Rules (G15)

We have repeatedly drawn attention to the importance of selection rules. Selection rules forbidding otherwise probable disintegrations will reduce the width of many nuclear levels and will thus give rise to the very sharp resonance levels which we observe in the simple capture of protons (§81). Selection rules influence greatly the γ-ray spectrum of nuclei (§90). Selection rules combined with the small penetrability of the potential barrier for particles of high orbital momentum will prevent all compound levels of high angular momentum from being effective in disintegrations caused by α-particles as long as the α-particle cannot go over the top of the potential barrier (§82A), thus making the resonances observable for energies below the height of the barrier while they are not for higher energies.

In this section we shall show that selection rules also regulate to a considerable extent the probability of disintegrations which show no marked resonance effects, in particular disintegrations produced by protons and deuterons in light nuclei. This was first pointed out by Goldhaber (G15) who concluded, from such considerations, that the ground states of Li^6 and Li^7 should be a ^3S and a $^2P_{3/2}$ state respectively, and that B^{10} should have a triplet, B^{11} a doublet as its ground state. The prediction for Li^6 which was at the time against evidence from hyperfine structure (G25, and ref S2 of part A), has since then been proved correct by the atomic beams method (M2). Moreover, all the predictions mentioned agree with our present theoretical ideas about nuclear energy levels (F10).

The selection rules may be divided into two classes: Those which should hold generally, for any "coupling scheme" in the nucleus, and those which hold only for Russell-Saunders coupling. The first class concerns only the total angular momentum and the parity of initial nucleus, incident particle, final nucleus and outgoing particle. The second class will give information on the behavior of orbital momentum and spin separately.

The selection rules for total angular momentum and for orbital momentum will involve the orbital momentum of the relative motion of the incident and outgoing particle. The existence of useful selection rules depends therefore on the existence of restrictions on this orbital momentum. Such restrictions will exist (a) for very slow particles for which only the orbital momentum zero gives a high probability of entering the nucleus, and (b) when the two particles (*viz.* incident particle and initial nucleus, or produced particle and residual nucleus) are identical. In case (b) the symmetry requirements for the wave function will forbid certain values of the orbital momentum. The most notable case is that of two α-particles: Because of the Bose statistics and the zero value of the spin, only even orbital momenta are possible (§73F).

A. Selection rules between initial and final nuclei which hold for any coupling scheme

The vector sum of the total angular momenta of initial nucleus and incident particle must be equal to the vector sum of the total angular momenta of final nucleus and outgoing particle. Let ss' be the spins of incident and outgoing particle, i and i' the (total) intrinsic angular momenta of initial and final nucleus. If, then, both incident and outgoing particle are slow we have

$$\begin{aligned} |i-s| &\leq i'+s' \\ i+s &\geq |i'-s'|. \end{aligned} \tag{691}$$

This selection rule very seldom gives any useful information because it leaves too many possibilities.

The more useful one of the general selection rules is that relating to the parity. Any nuclear energy level has a certain parity, e.g., the ground states of all nuclei up to He^4 have even parity, those from He^4 to O^{16} are probably even and odd for even and odd mass number, respectively. The wave function of the relative motion of two particles has even or odd parity according to whether the orbital momentum is even or odd. The selection rule requires that the parity of the system as a whole, i.e., the product of the intrinsic parity and that of the center-of-gravity motion, remains unchanged in the process. If both incident and outgoing particle are slow, and none of these particles is heavier than He^4, this means that the parities of initial and final nucleus must be the same. According to the above rule, this would forbid all processes in which a nucleus of even mass between 4 and 16 is transmuted into one of odd mass in the same region in the ground state or *vice versa*, i.e., all dp, dn, αp and αn reactions (cf. Chapter XVII). This very restrictive selection rule holds, however, only if both incident and outgoing particle are slow (wave-length larger than nuclear radius), and in no practical case are *both* particles slow at the same time.

However, there is one reaction which is definitely "forbidden" according to this parity rule, if only the *incident* particle is slow. This is the oldest of all reactions produced by artificially accelerated particles, *viz.*

$$Li^7 + H^1 = 2He^4. \tag{692}$$

The ground state of Li^7 has almost certainly odd parity (see above) while the intrinsic parity of all other particles is even. The wave function describing the relative motion of the two α-particles must be even (Bose statistics). Therefore the process can only occur at all if the wave function of the motion of the proton relative to the Li^7 nucleus is odd. This requires at least $l=1$, which makes the process improbable at low proton energy. In fact, its probability is 30 times smaller than that of the "probable" reaction

$$Li^6 + H^2 = 2He^4. \tag{692a}$$

We may, conversely, conclude from the observed relative probabilities of reactions (692), (692a) that the parity of Li^6 is even, that of Li^7 odd in agreement with theoretical views.

As Breit (O15) has pointed out, the energy dependence of the disintegration cross section is not appreciably affected by the orbital momentum for slow charged particles, in contrast to slow neutral particles. The ratio of the penetrabilities of the potential barrier for various orbital momenta is practically independent of the energy of the particle if this is small compared to the barrier height (§72, 78). The reason for this is that the particle energy is negligible compared to the Coulomb potential and the centrifugal force.

B. Selection rules between initial and final nucleus holding only for Russell-Saunders coupling

It is reasonable to assume that in light nuclei the resultant spin σ and the resultant orbital momentum λ of all particles contained in the nucleus are good quantum numbers (F10, I1, R10). This will be true if the magnetic spin-orbit interaction is small compared to the average spacing between levels. It is probable (R10) that the spin-orbit interaction is of the order $\frac{1}{2}$ MV. Therefore Russell-Saunders coupling will break down when the spacing between the energy levels of the *compound nucleus* (not the initial or final nucleus!) becomes of the order $\frac{1}{2}$ MV or less. This will probably occur for atomic weights around 15 or 20, so that the Russell-Saunders coupling is restricted to the lightest nuclei.

The first selection rule to be added to the two general ones mentioned in A, requires that the resultant spin of the system does not change, i.e.,

$$|s-\sigma| \leq s'+\sigma', \quad |s'-\sigma'| \leq s+\sigma, \quad (693)$$

where the primed quantities refer to the final, the unprimed ones to the initial nuclei and s refers to the particle, σ to the nucleus. It was from this selection rule that Goldhaber derived the spins of Li^6 Li^7 B^{10} and B^{11}. E.g., from the fact that (692a) is a probable reaction, and that the α-particles have zero and the deuteron unit spin, it follows that Li^6 must also have unit spin. The same follows from the probable reaction $B^{10}+H^2=3He^4$ for B^{10}. Similarly, the reaction $B^{11}+H^1=3He^4$ is probable, therefore B^{11} presumably has spin $\frac{1}{2}$. For Li^7 the situation is a little more complicated. From the reactions $Li^6+H^2=Li^7+H^1$, $Li^7+H^2=2He^4+n^1$ which are both "probable" we can only conclude that the spin of Li^7 is either $\frac{1}{2}$ or $\frac{3}{2}$. On the other hand, the reaction $Li^7+H^1=2He^4$ is improbable (cf. above (692a)). But we had already found a reason for this, *viz.* the change of parity provided the orbital momentum of the incident proton is zero. It may be argued that the reaction is not improbable enough for a reaction violating *two* independent selection rules so that the total spin does not change. This would lead to $\sigma=\frac{1}{2}$ for Li^7, in agreement with present theories.

The selection rule for l is not very useful because there are in general no restrictions on the orbital momentum of the outgoing particle.[66] Where there are restrictions; e.g. in the reactions (692, 692a), this rule gives the same result as the parity rule.

C. Selection rules in resonance disintegrations

As already pointed out in the beginning of this section, more stringent selection rules hold for resonance disintegrations than for nonresonance processes. Angular momentum as well as parity have to be conserved. This means quite generally that either only even or only odd values of the orbital momentum are possible when a compound nucleus in a given state breaks up into two given final nuclei. For the parity is determined for the compound state C as well as for the final nuclei B and Q. Therefore the parity of the motion of the nuclei B and A with respect to each other is also given, and this parity determines whether the orbital momentum is even or odd. Explicitly, we have:

Only even l if all the nuclei C, B and Q have even parity or if two of them have odd parity and one even.

Only odd l if all parities are odd, or two even and one odd.

An application of this parity rule is the excited state of Ne^{20} formed in the proton capture by F^{19} (§81). Other applications are connected with the γ-ray emission (§90). A further application may be the capture of slow neutrons by B^{10}, according to the scheme

$$B^{10}+n^1=B^{11*}=Li^7+He^4 \quad (693a)$$

for which the experimental determinations (cf. §102) indicate that Li^7 is always formed in its excited state of 0.44 MV excitation energy (cf. Table LXXIV).

Another important consequence of the even-odd rule for the orbital momentum is, of course, that the angular distribution of the disintegration products must be symmetrical with respect to the plane perpendicular to the direction of the incident beam. This was actually found by Kempton, Browne and Maasdorp (K2) for the disintegration

$$H^2+H^2=H^3+H^1,$$

which may be considered as evidence that even this disintegration has resonance character.

§84. Absolute Probability of Disintegration Processes; Angular Distribution of Disintegration Products

The absolute probability of processes showing pronounced resonance effects was already discussed in §81, 82 and was used to deduce the partial widths for γ-rays and α-particles of the levels concerned. The absolute probability of processes showing no marked resonances is a much less clean-cut problem.

[66] This seems to be the case, e.g., in the reactions $2H^2=H^3+H^1$ and $2H^2=He^3+n$ where the outgoing particles are fast enough for orbital momentum 1 or even 2. We therefore do not believe that Goldhaber's conclusion of zero intrinsic orbital momentum for H^3 and He^3 from these reactions is legitimate. From purely theoretical reasons (Chapter IV) zero orbital momentum seems, of course, very probable.

A. Light nuclei, total cross section

It is very probable that these processes should also be described in the general scheme of the formation and disintegration of a compound nucleus. From this point of view, even the processes showing no marked resonance are in reality resonance effects; but the resonance levels are so broad compared to their spacing that the maxima and minima are leveled out. Indications of a very weak resonance effect are found in the reaction most closely investigated, *viz.*

$$Li^7+H^1=2He^4. \tag{694}$$

The disintegration function for this reaction was observed experimentally by various authors (D16, H6, H27, H20, O6) for energies from 8 to 1400 kv, with thin and thick targets. Breit and his collaborators (O14, 15, 16) worked out the theoretical disintegration function with a special model, *viz.* assuming that the proton moves in the "potential" created by the Li^7 nucleus, and that this potential is a simple rectangular hole. They found that for certain values of the depth of this hole, agreement could be obtained with experiment. Furthermore they found that the relatively high yield of the reaction at low proton energies is irreconcilable with an orbital momentum 1 of the incident proton unless one assumes a resonance level fairly near zero proton energy. The value 1 for the proton orbital momentum is required by the selection rules for the parity (cf. §83A) which should be strictly obeyed. The conclusion concerning the necessity of a resonance for slow protons is independent of the special rectangular hole potential assumed but is simply a consequence of the penetration of p protons ($l=1$) through the Coulomb potential barrier.

Thus the calculations and experiments may be regarded as evidence for a resonance level of Be^8 at about 17 MV excitation energy, having even parity and probably angular momentum zero[67] (in contrast to the level responsible for the radiative capture of protons by Li^7, §81) and governing the probability of the reaction (694). This level probably has a width of $\frac{1}{2}$ MV or more, which is presumably mainly due to the disintegration into two α-particles. There can be little doubt that other processes, if investigated as carefully as $Li^7+H^1=2He^4$, will exhibit similar weak indications of resonance phenomena.

TABLE XXXXI. *Absolute cross sections.*

REACTION TARGET	$D+D=He^3+n^1$ D_2O	$Li^7+D=2He^4+n^1$ Li	$Be^9+D=B^{10}+n^1$ Be	$Li^7+H=2He^4$ Li
$10^7 p$ (800 kv)	8.0	17.1	9.5	
$10^7 p$ (700 kv)	6.2	8.6	5.9	
ϵ(750 kv)(in 10^{-15} cm² volt)	12	10	11	
Cross sect. σ(in 10^{-26} cm²)	2.2	8.5	4	0.33
$\lambdabar$ (in 10^{-13} cm)	7.4	4.8	4.5	6.0
Penetrability P	1	0.5	0.4	0.7
Sticking prob. ξ	0.013	0.25	0.16	0.004

Independent of the description of the processes as resonance effects, the cross section for probable processes must be of the order of

$$\sigma^P{}_Q=\pi\lambdabar_P{}^2P_P\xi_P, \tag{695}$$

where $\lambdabar_P$, ξ_P and P_P are wave-length, sticking probability and penetrability for the incident particle. The partial width Γ_Q for the outgoing particle has been assumed equal to Γ which is approximately true for "probable" processes. The quantity usually measured is the total number p of disintegrations per particle of a given energy E; from this σ can be deduced with the help of the energy loss cross section ϵ (cf. (688) and §95), *viz.*,

$$\sigma=\epsilon dp/dE. \tag{696}$$

Table XXXXI gives the absolute yield for four reactions for which the yield has been investigated with special care by Amaldi, Hafstad, Heydenburg and Tuve (H6, A11a). The first and second row give the total yield p (disintegrations per incident particle) at two energies (700 and 800 kv) of the incident particle, the third line gives the energy loss constant ϵ for the target substance, the fourth the cross section (at 750 kv) derived from these data, the next two lines give wave-length and penetrability for the incident particle and the last the sticking probability (cf. (695)).

The "sticking probabilities" obtained are of the order unity for the processes $Li^7+H^2=2He^4+n^1$ and $Be^9+H^2=B^{10}+n^1$, as should be expected. This justifies the assumption that these

[67] J must be even because disintegration into two α-particles is possible (§81). It cannot be greater than $l+i+s$ where $l=1$ is the orbital momentum, $i=3/2$ the spin of Li^7 and $s=1/2$ that of the proton. This reduces the possible values for J to 0 and 2. The angular distribution of the α-particles is spherically symmetrical which makes $J=0$ probable.

processes are "probable" processes. For the reaction $Li^7+H=2He^4$, the sticking probability comes out to be very small, about 1/50 of that for the two first named processes, in accord with the characterization of this process as, in first approximation, "forbidden" (cf. §83). The probability of the disintegration of deuterons by deuterons also comes out to be small. In this case, it must of course be considered that there is the alternative mode of disintegration into H^3 and a proton, which is experimentally (and theoretically for symmetry reasons) about as probable as the disintegration into He^3+n^1. Even so, the sticking probability would be only 1/10 of that for Li^7+D, and only 0.025 in absolute value. The reasons for this small internal disintegration probability are unknown; calculations using special models (D15) gave even greater theoretical probabilities (and therefore a greater discrepancy) than we would obtain for $\xi=1$. For a possible explanation cf. B.

B. Light nuclei, angular distribution

No theoretical calculations on the angular distribution of disintegration particles have yet been made. Except in the resonance case, such calculations would require the use of a special model.

Experimentally, the angular distribution has been investigated for the α-particles produced in the reaction

$$Li^7+H^1=2He^4 \tag{697}$$

(G14a), for the protons and neutrons from the processes

$$H^2+H^2=H^3+H^1, \tag{697a}$$

$$H^2+H^2=He^3+n^1, \tag{697b}$$

(K3, N4a) and for the α-particles from

$$B^{11}+H^1=Be^8+He^4 \tag{697c}$$

(N4a; homogeneous group of alphas of 4 cm range). In the first case, spherical symmetry was found with protons of 200 kv, in the other three cases, marked maxima in the forward and backward direction (about 60 percent more than at right angles) were observed at 100–200 kv deuteron energies. In all cases, the distribution is symmetrical about the "equatorial plane" as is required for resonance disintegrations (§83C, end).

In the Li case, we assume (§83) that the incident protons have orbital momentum one and might therefore expect an anisotropic distribution of the emitted α-particles; we conclude that the compound state of Be^8 must have $J=0$ (cf. A). In the deuteron case, at low energies primarily deuterons of zero orbital momentum should be effective, giving a spherically symmetrical distribution, unless $l=0$ is forbidden by selection rules. The most tempting hypothesis is to assume that the compound state of He^4 involved in reactions (697a, b) has odd parity. Since deuterons have even intrinsic parity, this means that only odd orbital momenta would be allowed for their relative motion, so that $l=1$ gives the main contribution. At the same time, this would explain the smallness of the absolute cross section (cf. A). Moreover, an odd state (^1P) of He^4 is expected to lie near the He^4 dissociation energy according to calculations of Feenberg (F9a). It only seems difficult to understand why even compound states should give no contribution at all.

The nonspherical distribution of the alphas in (697c) shows again that $l=0$ is probably forbidden by selection rules. The simplest assumption is that the state of C^{12} involved has even parity; then, since B^{11} has odd intrinsic parity, only odd l will be allowed for the incident proton. The outgoing α must then have even l', and since both Be^8 and He^4 have spin zero, J must be even for the compound state. Since $J=0$ is excluded on account of the nonspherical symmetry, and $J\leqslant l+i+s=2$, we must have $J=2$.

C. Heavy nuclei

In §77, we have derived a general formula for the probability of disintegration processes produced in heavy nuclei which was supplemented in §80 by a discussion of the special case of reactions of the dp type. Thus far, quantitative studies of the yield are only available for the disintegration of Pt by deuterons of 4–5 MV. This energy, though large, is decidedly lower than the potential barrier of Pt ($\sim$9 MV). Thus we may apply formulae (650), (633), *viz.*

$$\sigma^{P_P}{}_Q=\pi\lambda\!\!\!^{-2}\frac{g}{(1-x)^{\frac{1}{2}}}P_P\xi_P\frac{\Gamma_Q}{\Gamma}. \tag{698}$$

Here the wave-length is about $1.5\cdot10^{-13}$ cm (for

4.5 MV deuterons), $g=11.6$ for deuterons on Pt, x = deuteron energy/height of potential barrier $=\frac{1}{2}$. ξ_P is the sticking probability for the incident particles, P_P the penetrability of the barrier for them. Γ_Q is the width corresponding to the produced particles, Γ the total width. According to §79, the main contribution to the latter will probably come from neutron emission, a smaller part from emission of protons, a still smaller part from α-particles and a negligible fraction from deuteron emission. We shall in the following assume neutron emission; as was shown in §80, the emission of protons, according to the Oppenheimer-Phillips mechanism, is about as probable as that of neutrons according to the ordinary mechanism. Then we may put $\Gamma_Q=\Gamma$.

With these assumptions,

$$\sigma^{Pp}{}_Q=1.0\cdot 10^{-24}P_P\xi_P. \qquad (698a)$$

We assume further that $\xi=1$ because of the high excitation energy of the compound nucleus formed (cf. §54). The penetrability P_P depends on the nuclear radius. With $R=12\cdot 10^{-13}$ cm we obtain

$$P=5\cdot 10^{-5},\quad \sigma=5\cdot 10^{-29}. \qquad (698b)$$

If we take account of the rather large size of the deuteron by assuming an effective radius of $15\cdot 10^{-13}$ cm, we have instead

$$P=1.2\cdot 10^{-3},\quad \sigma=1.2\cdot 10^{-27}. \qquad (698c)$$

This last figure agrees as to order of magnitude with the observed cross sections for the production of radioactivity in Pt which are of the order 10^{-28} to 10^{-27} cm^2 (C35).

In all reactions produced in Pt by deuterons, fluctuations of the yield with energy have been observed which resemble resonances. It seems practically certain that they must not be interpreted as such. The excitation energy of the compound nucleus formed by adding a deuteron to a heavy nucleus, was estimated in §78 to be about 12 MV plus the kinetic energy of the deuteron, i.e. more than 16 MV in our case. The spacing of levels at such an energy for a nucleus as heavy as Pt is probably of the order of a millivolt or less (§53). The observed fluctuations of the yield have maxima and minima spaced by about $\frac{1}{2}$ MV. They can therefore not be due to resonance. A possible explanation might be secular fluctuations in the matrix elements over large energy regions, or competition of various modes of disintegration.

§85. Three-Particle Disintegrations

As was already mentioned in the general discussion in §76, the residual nucleus may be left in the primary nuclear reaction in a state above the dissociation energy. Then it will in general disintegrate further, emitting a second heavy particle. A "cascade reaction" of this type will appear experimentally as a disintegration in which three nuclei, including the final nucleus, are produced.

The best investigated reaction of this type is

$$B^{11}+H^1\rightarrow C^{12*}\rightarrow Be^{8*}+He^4\rightarrow 3He^4. \qquad (699)$$

The fact that the mechanism of this reaction is actually as indicated in (699), was first established by Dee and Gilbert (D9) from cloud chamber experiments. Similar to (699) is the reaction

$$B^{10}+H^2\rightarrow C^{12*}\rightarrow Be^{8*}+He^4\rightarrow 3He^4. \qquad (699a)$$

Other three-body reactions with light nuclei are:

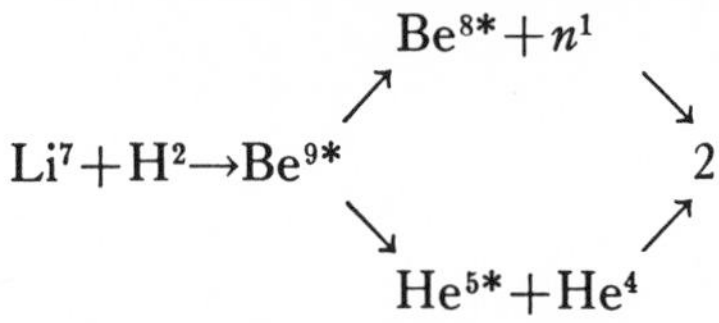

(699b)

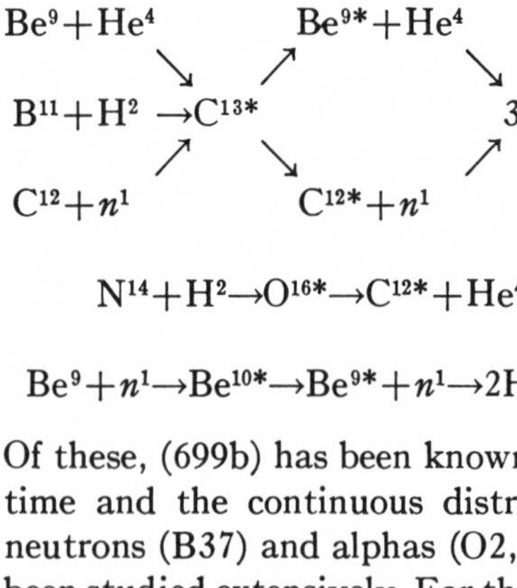

(699c)

$$N^{14}+H^2\rightarrow O^{16*}\rightarrow C^{12*}+He^4\rightarrow 4He^4, \qquad (699d)$$

$$Be^9+n^1\rightarrow Be^{10*}\rightarrow Be^{9*}+n^1\rightarrow 2He^4+2n^1. \qquad (699e)$$

Of these, (699b) has been known for a fairly long time and the continuous distributions of both neutrons (B37) and alphas (O2, K2, W21c) have been studied extensively. For the other reactions, similar but less complete evidence has been given (cf. Chapter XVII for discussion and references).

With heavy nuclei, evidence has been presented by Heyn (H33) for processes in which an incident fast neutron knocks out another neutron from a nucleus, thus decreasing the latter's atomic weight by one unit. An example of this type of reaction is

$$Cu^{63}+n^1 \rightarrow Cu^{64*} \rightarrow Cu^{63*}+n^1 \rightarrow Cu^{62}+2n^1 \quad (700)$$

detected by the positron activity of Cu^{62}. Pool, Cork and Thornton (P11b) have confirmed reaction (700) and found similar reactions with Cu^{65}, Ag^{107}, Ag^{109}, N^{14} and O^{16} giving Cu^{64}, Ag^{106}, Ag^{108}, N^{13} and O^{15}, respectively. Other examples are probably found in uranium and other very heavy nuclei (H7, M15).

Another three particle disintegration has been reported by Cork and Thornton (C35, 36) who found that deuteron bombardment of Au produced a radioactive isotope of Ir. This means the emission of three units of charge from the compound nucleus so that they suggested the process

$$Au^{197}+H^2=Au^{198}+H^1=Ir^{194}+H^1+He^4. \quad (700a)$$

Such a process is very probable with the Oppenheimer-Phillips disintegration as the first step (cf. §80). We have also given arguments (§79E) for assuming that a similar process with emission of neutrons and alpha-particles is very probable in the reaction of deuterons with heavy nuclei, e.g.,

$$Pt^{195}+H^2=Au^{196}+n^1=Ir^{192}+n^1+He^4. \quad (700b)$$

According to the general theory, three-particle reactions will have probabilities of the same order as two-particle disintegrations provided the necessary energy is available and no potential barriers prevent the escaping of the emissible particles. With large surplus energy, it may even happen that three-particle disintegrations become more important than two-particle ones, because of the frequently mentioned tendency of the residual nucleus to retain a large fraction of the available energy (evaporation model, §§54, 65, 79).

With the now accepted mechanism of three body reactions, it is, of course, not possible to deduce the reaction energy from the upper limit of the energy spectrum of the disintegration products as has formerly been done (O2, O3). The maximum energy of the product particles depends on the particular energy levels of the nucleus formed in the intermediate stages of the decay of the compound nucleus.

A. The disintegration of B^{11} by protons

After considerable controversy (K15, O2, O9, L5) it has been shown (D9) that the α-particles emitted after proton bombardment of boron fall into three groups, *viz.* (for 0.2 MV proton energy):

(1) A group of homogeneous energy of about 5.7 MV;

(2) A fairly homogeneous group at about 3.85 MV;

(3) A continuous distribution extending from very low energies to about 5 MV.

Group 3 contains roughly twice as many α-particles as 2, while group 1 contains, at low proton energies ($\sim$200 kv) only about 1 percent of all the particles produced.

Group 1 is to be attributed to the reaction

$$B^{11}+H^1=Be^8+He^4 \quad (701)$$

as was first pointed out by Kirchner and Neuert (K15). The Be^8 nucleus is, in this case, formed in the ground state. Its mass is almost identical with that of two α-particles (§108). The energy evolution in the reaction (701) is 8.5 MV, of which the α-particle receives $\frac{2}{3}$, the Be^8 nucleus $\frac{1}{3}$, according to the law of conservation of momentum.

Group 2 should be attributed to the first part of reaction (699), i.e.,

$$B^{11}+H^1=Be^{8*}+He^4. \quad (701a)$$

Since the average energy of the α-particles in this group is 3.85 MV, and the Be^8 recoil energy is one-half of this amount, the energy evolution turns out to be 5.77 MV. Several corrections to this figure have to be applied (§96, 97) but they happen just to cancel. The total reaction energy is about 8.57 MV (from the distribution of group 3, see below). This gives for the excitation energy of Be^{8*} a value $U=2.80$ MV.

The particles of group 2 do not all have the same energy, but their energy distribution corresponds about to the dispersion formula, with a width at half-maximum[68] of 0.51 MV. Since the

[68] The most accurate way of determining the width from the experimental data of Oliphant, Kempton and Rutherford seems to be the following: The total number of particles counted by these authors (O9, Fig. 7) is 2245. Of these, 30 belong to group 1. 25 particles should be subtracted because O. K. R. overestimated the number of very slow particles, assuming the distribution to be homogeneous down to zero energy, while this is not the case according to the theoretical distribution (Fig. 25). Of the remaining 2190 particles, two-thirds = 1460 belong to group 3. Since this group extends almost uniformly from 0 to 4.7 MV energy, the number of group 3 particles

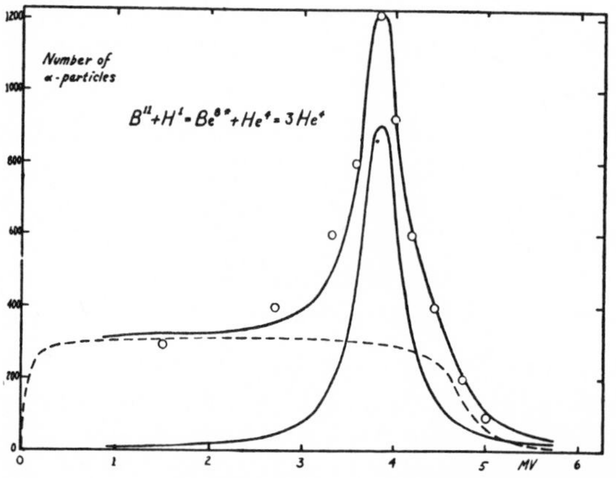

FIG. 25. Energy distribution of the α-particles emitted in the proton bombardment of B^{11} (without the discrete group corresponding to the reaction $B^{11}+H^1=Be^8+He^4$). Lower solid curve: theoretical distribution of the α-particles emitted in the primary process, *viz.* $B^{11}+H^1=Be^{8*}+He^4$. Broken curve: α-particles emitted in the break-up of the excited Be^8 (secondary reaction). Upper solid curve: total distribution, theoretical. Circles: experimental points (Oliphant, Kempton and Rutherford, O9).

kinetic energy of the recoiling Be^8 nucleus must always be one-half of the energy of the α-particle, the width of the Be^8 level turns out to be

$$\Gamma=\tfrac{3}{2}\cdot 0.51=0.77\ \text{MV}. \qquad (702)$$

This means a lifetime of the excited Be^8 nucleus of

$$\tau=\hbar/\Gamma=1.04\cdot 10^{-27}/1.60\cdot 10^{-6}\cdot 0.77$$
$$=0.85\cdot 10^{-21}\ \text{sec}. \qquad (702a)$$

Group 3 is due to the disintegration of the excited Be^8 nuclei:

$$Be^8\rightarrow 2He^4+U. \qquad (703)$$

If Q (=8.57 MV) is the total energy evolution in the whole reaction

$$B^{11}+H^1=3He^4+Q \qquad (703a)$$

and U that in the disintegration of the excited Be^8, we have for the velocity of the center of gravity of the Be^{8*} nucleus

$$v_8=(Q-U/3M_\alpha)^{\frac{1}{2}} \qquad (704)$$

and for the velocity of each α-particle relative to that center of gravity

$$v_4=(U/M_\alpha)^{\frac{1}{2}}. \qquad (704a)$$

If we assume that there is no correlation between the directions of the two velocities, we find that the energies of the α-particles of group 3 will be uniformly distributed between the limits.

$$E_{\min}{}^{\max}=\tfrac{1}{6}[(Q-U)^{\frac{1}{2}}\pm(3U)^{\frac{1}{2}}]^2. \qquad (705)$$

With the numerical values given above,

$$E_{\min}=0.04\ \text{MV} \quad \text{and} \quad E_{\max}=4.68\ \text{MV}.$$

This distribution, when modified by the width of the level of Be^8, is well confirmed by the experimental data of Oliphant, Kempton and Rutherford (O9, cf. D9).

The theoretical distributions for groups 2 and 3 and the total distribution are given in Fig. 25. The agreement with experiment is excellent on the high energy side of the main maximum, but not so good on the low energy side. Part of this may be due to uncertainties in the range-energy relation.

The same order of magnitude (∼3 MV) for the energy of the first excited level of Be^8 was obtained from theoretical calculations by Wigner and Feenberg (F10). The level is, theoretically, a ^1D level which is compatible with the fact that it can disintegrate into two α-particles.

B. Other three-particle disintegrations of light nuclei

The reaction

$$B^{10}+H^2=Be^{8*}+He^4=3He^4 \qquad (699a)$$

follows undoubtedly the same mechanism as (699). The total reaction energy is considerably higher, *viz.* 18.0 MV. If Be^8 is left in the 3 MV state, the first α-particle (corresponding to group 2 in the disintegration of B^{11} by protons) should receive an energy of $\tfrac{2}{3}\cdot 15=10$ MV, corresponding to a range of 10.5 cm. An almost homogeneous group of this range seems to exist (C28) but its width seems to be larger than in reaction (709). It is very likely that, owing to the larger energy available, a number of higher excited levels of Be^8 can also be formed in this reaction which makes the observable energy

---

per MV energy interval will be 1460/4.7=310. At the maximum, the number of particles observed is 1220 per MV. Thus 910 of these have to be attributed to group 2, and generally the number of particles of energy E in group 2 will be $910/[1+4(E-E_r)^2/\gamma^2]$ (dispersion formula, γ = width at half-maximum). On the other hand, the total number in group 2 must be $\tfrac{1}{3}\cdot 2190=730$. Therefore we have

$$910\cdot\tfrac{1}{2}\pi\gamma=730$$

from which $\gamma=0.51$ MV.

distribution of the α-particles more uniform and therefore the interpretation more difficult.

The reaction

$$Li^7+H^2=Be^{8*}+n^1=2He^4+n^1 \qquad (706)$$

or

$$Li^7+H^2=He^{5*}+He^4=2He^4+n^1 \qquad (706a)$$

has been studied by Bonner and Brubaker (B37, neutron distribution) and by Kempton, Browne and Maasdorp (K2, α-particles). The neutrons consist of a homogeneous group of about 13.2 MV energy, corresponding to the reaction

$$Li^7+H^2=Be^8+n^1 \qquad (706b)$$

and a practically continuous distribution extending at least up to 12.4 MV. If we wanted to explain the occurrence of neutrons of 12.4 MV on the basis of the mechanism (706a), we should have to assume an excited state of Be^8 at $9/8\ (13.2-12.4)=0.9$ MV. This is extremely improbable in view of the fact that no trace of such a state was observed in the process $B^{11}+H^1=3He^4$, and also from the theoretical calculations of Feenberg and Wigner. Moreover, in order to explain the continuous distribution of neutrons on this basis, it would be necessary to assume a continuous spectrum for Be^8. This is again quite contrary to the result from the $B^{11}+H^1$ reaction, and also contrary to the fundamental assumptions of Bohr's theory of nuclear processes. This shows that the mechanism indicated in (706a), with He^5+He^4 as intermediate products, besides the mechanism (706), must certainly play a part in the reaction.[69]

The reaction

$$Be^9+He^4=3He^4+n^1 \qquad (707)$$

was suggested by Bohr (B33) in order to explain the large number of very slow neutrons among the neutrons produced by α-bombardment of Be^9. According to the evaporation model, the neutrons formed in the reaction

$$Be^9+He^4=C^{12}+n^1 \qquad (707a)$$

should have, in the average, energies of the order of the "temperature" of the C^{12} nucleus formed, i.e., about 2–3 MV. In reality, most of the neutrons have less than 1 MV energy (§99, and D21). The mechanism of reaction (707) is one of the following:

$$Be^9+He^4 \rightarrow C^{13*} \rightarrow$$

$$Be^{9*}+He^4 \nearrow Be^{8*}+He^4+n^1 \searrow 3He^4+n^1, \qquad (707b)$$

$$Be^{9*}+He^4 \searrow He^{5*}+2He^4 \nearrow 3He^4+n^1$$

so that the neutrons are emitted either in the break-up of an excited Be^9 or an excited He^5 nucleus.

The reaction (707) is confirmed by the fact that the compound nucleus C^{13} when formed from other initial particles, is known to break up into 3 α-particles and one neutron. E.g., the reaction

$$B^{11}+H^2=3He^4+n^1 \qquad (707c)$$

has been established both by observing large numbers of very slow neutrons (B40) and large numbers of slow α-particles (C28), in both cases forming a continuous energy distribution. The reaction

$$C^{12}+n^1=3He^4+n^1 \qquad (707d)$$

has also been observed, with fast neutrons in a cloud chamber (C8).

The process

$$Be^9+n^1=Be^9+2n^1 \qquad (708)$$

is established by observing an increase in the number of neutrons when neutrons are allowed to fall on beryllium (R20). The evidence for

$$N^{14}+H^2=4He^4 \qquad (709)$$

is the observation of very many slow α-particles, forming a continuous distribution in energy (C29).

[69] This is in contrast to the interpretation of Kempton, Browne and Maasdorp. *Note added in proof:* Our view is strongly confirmed by a recent paper of Williams, Shepherd and Haxby (W21c). These authors have investigated the distribution of the α-particles more accurately and find, superposed upon the continuous distribution, a fairly homogeneous group of energy 7.8 MV which they attribute to reaction (706a) with He^5 being left in the ground state. This gives for He^5 the very reasonable mass value of 5.0137, unstable by 0.8 MV against disintegration into an α-particle and a neutron. The width of the α-group is ~ 0.3 MV; therefore the width of the He^5 ground state $(9/5)\cdot 0.3=0.5_5$ MV, corresponding to a lifetime of $1.2\cdot 10^{-21}$ sec. which again appears very plausible. It is possible that He^5 can also be formed in excited states.

XIV. γ-Rays

§86. Classification of γ-Rays

The γ-rays emitted in nuclear processes fall into two main classes:

1. γ-rays emitted during the nuclear process itself.
2. γ-rays emitted in a secondary process following the proper (primary) nuclear process.

The most important case[70] for class 1 are the γ-rays emitted in simple capture processes. Such γ-rays have been observed from the capture of slow neutrons by heavy nuclei and of protons by light nuclei, and their energy has been measured in a number of cases.

The γ-rays of class 2 are emitted whenever a primary nuclear process of any kind leads to an excited state of the resultant nucleus. The primary nuclear process may be:

(a) Any nuclear transmutation process, produced by any kind of incident particle and with emission of any kind of outgoing particle (proton, neutron, deuteron, α-particle, γ-ray, etc.). Incident and outgoing particle may be of the same kind (inelastic scattering).

(b) A β-transformation.

(c) A natural α-decay.

Since it is the rule rather than the exception that nuclear disintegrations lead to excited states of the resultant nucleus, γ-rays of class 2 are an extremely frequent phenomenon and accompany practically every nuclear transmutation in which any nucleus of atomic weight greater than 6 is formed.

The probability of the emission of γ-rays of class 2 is given by the probability of the primary process which usually is connected with emission of material particles. Since we know that ordinarily the emission of material particles in the primary process is more probable than that of γ-rays, it follows that in general the γ-rays of class 2 are much more intense than those of class 1. Therefore the capture γ-rays can only be observed:

(a) When, contrary to the rule, simple capture is the only possible process. This case is realized for slow neutrons with heavy nuclei (§57, 60–62), for the capture of protons by C^{12} (§81) and probably by a few other light nuclei.

(b) When a primary process with particle emission is possible but never leads to an excited state of the residual nucleus. This is true for Li^7+H^1 since the α-particles emitted apparently can not be formed in excited states.

(c) When the capture is for certain energies enhanced by resonance while the class 2 γ-rays are not. This is true for $F^{19}+H^1$, Li^7+H^1, $C^{12}+H^1$ and perhaps other nuclei.

(d) When the energy of the capture γ-rays is much higher than the maximum possible energy of class 2 γ-rays.

In all other cases, the capture γ-rays will be swamped by the much more intense (by a factor 10,000 for light nuclei) γ-rays from the residual nucleus ("class 2").

Even in obvious capture processes such as the capture of slow neutrons by heavy nuclei, not all the γ-radiation observed is actually capture radiation, i.e., emitted in the capture process itself. For we know that any primary nuclear process will in general lead to an *excited* state of the residual nucleus especially if a great number of excited states are available as is the case for heavy nuclei. This is, of course, also true for the simple capture (§90). Therefore the capture process proper will be followed by secondary processes in which the residual nucleus loses its excitation energy by emission of one or more further γ-quanta of class 2. Thus the observed "capture radiation" is actually a mixture of γ-rays of classes 1 and 2.

§87. The γ-Ray Width

It has long been known from the study of the internal conversion of "natural" γ-rays (§88) that nuclear γ-radiation has about the same intensity whether it is dipole or quadrupole radiation. This is very surprising indeed since we should expect that the ratio of the two kinds of radiation is approximately as $(R/\lambda\!\!\!^{-})^2$ where $2\pi\lambda\!\!\!^{-}$ is the wave-length of the emitted

[70] Apart from capture processes, γ-rays may be emitted in a transition of the compound nucleus from its initial state to another state still above the dissociation energy. Then the later state may afterwards disintegrate with emission of a particle. These "noncapture" γ-rays emitted in the nuclear process itself have not yet been observed and are presumably very rare.

γ-ray and R the radius of the nucleus. Now for γ-rays of 1 MV energy, we have $\lambdabar = 2\cdot 10^{-11}$ cm whereas, even for the heaviest nuclei, $R=1.3\cdot 10^{-12}$ cm. This gives $(R/\lambdabar)^2=1/250$. Thus the quadrupole radiation should be about 250 times weaker than the dipole radiation, in striking contrast to experiment.

A. Elementary calculation of width

This discrepancy is emphasized by considerations of the absolute probability of emission of γ-rays. This absolute probability has been measured for γ-rays from natural radioactive nuclei (§88), for γ-rays from neutron capture (§61, 90) and from proton capture (§81, 90). In all cases, much smaller probabilities are found experimentally than would be expected from a simple theoretical consideration.

Generally, the probability of a radiative transition from a state m to a state n is per unit time

$$\frac{\Gamma_\gamma}{\hbar}=\frac{4}{3}\frac{\omega^3}{\hbar c^3}D_{mn}{}^2, \tag{710}$$

where D_{mn} is the matrix element of the electric dipole moment corresponding to the transition in question, and ω the frequency of the emitted γ-ray

$$\omega=(E_n-E_m)/\hbar. \tag{710a}$$

The dipole moment may be expressed by the oscillator strength which we define by

$$f_{mn}=2M\omega\hbar^{-1}(D_{mn}/e)^2, \tag{710b}$$

where M and e are mass and charge of the proton. Then

$$\Gamma_\gamma/\hbar=(2e^2\omega^2/3Mc^3)f. \tag{711}$$

If we express the quantum energy $\hbar\omega$ in MV, we obtain

$$\Gamma_\gamma/\hbar=8.0\cdot 10^{15}(\hbar\omega)^2 f \text{ sec.}^{-1}, \tag{711a}$$

$$\Gamma_\gamma=5.24(\hbar\omega)^2 f \text{ volts}. \tag{711b}$$

For strong radiative transitions, we might expect f to be of the order unity, as in atoms. Or more generally, we may expect the sum of f over all the lines starting from a given nuclear level to have a value of the order unity.[71]

[71] A still larger value for Γ_γ is obtained by assuming the dipole moment D_{mn} in (710) to be of the order of a single elementary charge times the nuclear radius.

Actually, the observed lifetimes of excited states of natural radioactive nuclei are of the order 10^{-12} sec. (§69, 88), i.e. 10,000 times longer than would follow from (711a) with $\hbar\omega=1$ and $f=1$. Similarly, the γ-ray width of nuclear levels Γ_γ which is simply $\hbar$ times the reciprocal lifetime, is much smaller than the values obtained from (711b). One of the largest widths measured thus far is that of the excited state of Be^8 formed in the capture of protons by Li^7 (§81): it is 4 volts at a quantum energy of 17 MV, as compared to 1500 volts from (711b) with $f=1$. Again, the widths observed in the capture of slow neutrons (§61) and of protons by F^{19} (§81) are of the order of 0.1 to a few volts, for quantum energies of about 4 to 6 MV (§90) for which we should expect about 100 volts from (711b).

In all these cases, the probability of γ-ray emission is "too small" by a factor of about 100 for dipole radiation and therefore would be about right for quadrupole radiation if we use a similarly simple estimate for the intensity of the latter as in (711b). This fact, in combination with the observation mentioned above that quadrupole radiation is of the same order of intensity as dipole radiation, confronts us with the problem of explaining why dipole radiation is relatively so weak in nuclei.

B. Dipole and quadrupole radiation

A clue to the solution may be found in the fact that a system of particles *all having the same specific charge*, will not emit any dipole radiation but will emit quadrupole and higher multipole radiation. The general expression for the "effective electric moment" corresponding to a radiative transition $m\rightarrow n$ is (cf. B16, Eq. (38.10))

$$\mathbf{D}_{mn}=\hbar\omega^{-1}\int\Psi_n{}^*\sum_j \exp(i\mathbf{k}\cdot\mathbf{r}_j)(e_j/m_j)\nabla_j\Psi_m d\tau, \tag{712}$$

where e_j and m_j are charge and mass of particle number j, $\mathbf{r}_j$ its position, ∇_j the gradient operator with respect to r_j, $\mathbf{k}$ is the wave vector of the light quantum, ω its frequency, and $\Psi_m\Psi_n$ initial and final wave function of the radiating system. For dipole radiation, the retardation factor $\exp(i\mathbf{k}\cdot\mathbf{r}_j)$ should be replaced by unity:

$$\mathbf{D}_{mn}=\hbar\omega^{-1}\int\Psi_n{}^*\sum_j(e_j/m_j)\nabla_j\Psi_m d\tau. \quad (712a)$$

Quadrupole radiation is obtained by taking the second term in the expansion of the exponential. With $k=\omega/c$, we have then[72]

$$\mathbf{D}_{mn}=i\hbar c^{-1}\int\Psi_n{}^*\sum_j(e_j/m_j)(\mathbf{r}_j)_k\,\nabla_j\Psi_m d\tau. \quad (712b)$$

$(\mathbf{r}_j)_k$ is the component of $\mathbf{r}_j$ in the direction $\mathbf{k}$ of propagation of the light wave. Since the direction of the dipole moment $\mathbf{D}$ is identical with the direction of polarization of the light, $\mathbf{k}$ must always be taken perpendicular to $\mathbf{D}$ (i.e., to the direction of the gradient ∇_j).

We must now consider that the wave functions $\Psi_m\Psi_n$ will, of course, depend only on the relative coordinates of the particles with respect to the center of gravity, and not on the absolute coordinate of the latter. If the $\mathbf{r}_j$ denote *absolute* coordinates, the center of gravity is

$$\mathbf{R}=\sum_j m_j\mathbf{r}_j/M, \quad M=\sum_j m_j \quad (713)$$

and the relative coordinates

$$\boldsymbol{\varrho}_j=\mathbf{r}_j-\mathbf{R}. \quad (713a)$$

Let x_j, X and ξ_j be the x components of $\mathbf{r}_j$, $\mathbf{R}$ and $\boldsymbol{\varrho}_j$, respectively. Then

$$\frac{\partial\Psi_m}{\partial x_j}=\sum_i\frac{\partial\Psi_m}{\partial\xi_i}\frac{\partial\xi_i}{\partial x_j}=\frac{\partial\Psi_m}{\partial\xi_j}-\frac{m_j}{M}\sum\frac{\partial\Psi_m}{\partial\xi_i} \quad (713b)$$

and

$$\sum_i\frac{e_j}{m_j}\frac{\partial\Psi_m}{\partial x_j}=\sum_i\left(\frac{e_j}{m_j}-\frac{\sum e_i}{M}\right)\frac{\partial\Psi_m}{\partial\xi_j}. \quad (714)$$

Thus the transformation to relative coordinates makes the apparent charge of particle j different from e_j, *viz.*

$$e_j{}'=e_j-m_j\epsilon/M, \quad (715)$$

where
$$\epsilon=\sum_i e_i \quad (715a)$$

is the total charge of the system. In the particular case of nuclei, all the particles (neutrons and protons) have the same mass m_i, and $M=m_iA$ where A is the number of particles. Thus

$$e_j{}'=e_j-eZ/A, \quad (716)$$

which gives for protons ($e_j=e$)

$$e_j{}'=e(1-Z/A) \quad (716a)$$

and for neutrons

$$e_j{}'=-eZ/A. \quad (716b)$$

For the problem of emission of dipole radiation, neutrons should therefore be considered as having a negative effective charge equal to about half an elementary charge whereas protons have a positive effective charge which is only about half their true charge.

For a system whose particles have *all the same specific charge* (*i.e., ratio of charge to mass*), we obtain from (715)

$$e_j{}'\equiv 0. \quad (717)$$

This means that for such a system *the dipole moment would vanish identically whatever the wave functions* Ψ_m and Ψ_n. Thus we may understand the smallness of the dipole radiation if we can show that the nucleus, though actually composed of particles of different charge (neutrons and protons), acts like a system of particles all having the same charge. E.g., if the nucleus contains as many neutrons as protons ($A=2Z$), this will be true if we assume the nuclear particles to be combined in α-particles as subunits. Quite generally, the dipole moment will be zero if the centers of gravity of the neutrons alone and of the protons alone coincide so that the matrix element of their difference vanishes for all transitions. Owing to the large forces between neutrons and protons, this does not seem unlikely.

An alternative explanation due to Wigner would attribute the absence of dipole radiation in the spectra of natural radioactive elements to the fact that all low energy levels of a nucleus have the same "partition" (W17). This explanation would indeed seem much more attractive than the vague statement that the neutrons and protons will probably be distributed in a similar way. However, it does not explain why the γ-ray widths of the *high* excited states observed in neutron and proton capture are so much smaller than would be expected from the elementary calculation in *A*.

[72] D as defined in (712b) is k times the "quadrupole moment" corresponding to the transition, cf. Section C.

Whatever the explanation for the smallness of the dipole radiation may be, there will be no factors reducing the quadrupole and higher multipole radiation below the value expected from elementary considerations. E.g., if we repeat the calculations leading to (714) for quadrupole radiation propagated in the z direction, we find

$$ik\sum_i\left(\frac{e_j}{m_j}\zeta_j-\frac{\sum e_i\zeta_i}{M}\right)\frac{\partial\Psi_m}{\partial\xi_j}. \qquad (718)$$

The expression in the parenthesis does *not* vanish when all particles have the same specific charge because it contains the coordinates of the particles.

This shows, at the same time, that multipole radiation higher than quadrupole will be improbable compared to quadrupole radiation by about the factors expected from elementary consideration, i.e., by a factor $(R/\lambda\!\!\!^{-})^2$ for each successive multipole. Higher multipole radiation will therefore, in general, not occur except from metastable states.

C. The radiation width from the liquid drop model

Bohr has suggested the use of the liquid drop model (§53C) to calculate the probability of emission of radiation. Since in this model only the density of nuclear matter is considered and not the density of neutrons and protons separately, the model can obviously only give quadrupole radiation (cf. B). Moreover, at low "temperature" of the nucleus, only surface vibrations are excited and therefore only the emission of quadrupole radiation by these modes of vibration is important.

The intensity of radiation I from a system of currents periodic in time

$$\mathbf{j}(\mathbf{r},t)=\mathbf{j}_0(\mathbf{r})\cos\omega t \qquad (719)$$

is, according to classical electrodynamics,

$$I=\frac{1}{8\pi}\frac{\omega^2}{c^3}\int d\sigma|\mathbf{n}\times\mathbf{j}|^2, \qquad (720)$$

where $$\mathbf{j}=\int\mathbf{j}_0(\mathbf{r})\exp(i\omega\mathbf{r}\cdot\mathbf{n}/c)d\tau \qquad (720a)$$

and the integration $\int d\sigma$ extends over all directions of the vector $\mathbf{n}$ (direction of propagation of the light). Quadrupole radiation corresponds to the second term in the Taylor expansion of the exponential in (720a) so that

$$|\mathbf{n}\times\mathbf{j}|=(\omega/c)\left|\int\mathbf{r}\cdot\mathbf{n}\ \mathbf{n}\times\mathbf{j}_0\,d\tau\right|. \qquad (721)$$

Here the current density $\mathbf{j}_0$ of the radiating system may be expressed in terms of the charge density ρ; we have from (719) and the continuity equation:

$$\rho=\rho_0\sin\omega t,\quad \rho_0=\operatorname{div}\mathbf{j}_0/\omega. \qquad (721a)$$

Neglecting the magnetic dipole radiation, i.e., assuming

$$\int(j_{0x}y-j_{0y}x)d\tau=0, \qquad (721b)$$

we find for (721)

$$|\mathbf{n}\times\mathbf{j}|=(\omega^2/2c)\left|\int\rho_0\,\mathbf{r}\cdot\mathbf{n}\ \mathbf{r}\times\mathbf{n}\,d\tau\right|. \qquad (722)$$

Taking the direction of emission $\mathbf{n}$ parallel to x, we have

$$|\mathbf{n}\times\mathbf{j}|=(e\omega^2/2c)([xy]^2+[xz]^2)^{\frac{1}{2}}, \qquad (722a)$$

where $$e[xy]=\int\rho_0\,xy\,d\tau \qquad (722b)$$

is the quadrupole moment of the emitting system. If we assume that all quadrupole moments, *viz.* $[xy]$, $[xz]$, $[yz]$ etc., are equal in size (statistical equilibrium), the total radiation intensity (720) becomes

$$I=e^2(\omega^6/4c^5)[xy]^2. \qquad (723)$$

The γ-ray width is $\hbar$ times the number of quanta emitted:

$$\Gamma_\gamma=\hbar I/\hbar\omega=\tfrac{1}{4}(\omega/c)^5e^2[xy]^2. \qquad (723a)$$

For surface vibrations, the density of the nuclear liquid is constant (charge density $=Ze/\Omega$ where Ω is the total volume of the nucleus), and only the shape changes. If R is the normal radius of the nucleus, and

$$\zeta(\vartheta,\varphi)=\zeta_0(\vartheta,\varphi)\sin\omega t, \qquad (724)$$

the displacement of the surface at a given point ϑ, φ, then

$$e[xy]=(Ze/\Omega)\int \sin\vartheta d\vartheta d\varphi \times\int_0^{R+\zeta_0} r^2\cos\vartheta\sin\vartheta\cos\varphi\, r^2dr, \quad (724a)$$

assuming the polar axis $\vartheta=0$ to be in the x direction. Integration over r gives

$$[xy]=(3Z/4\pi R^3)\int \sin^2\vartheta\cos\vartheta\cos\varphi d\vartheta d\varphi \times[\tfrac{1}{5}R^5+R^4\zeta_0+0(\zeta_0^2)]. \quad (724b)$$

The term R^5 vanishes upon integration over the angles, the terms containing higher powers of ζ_0 may be neglected so that

$$[xy]=(3ZR/4\pi)\int \zeta_0\sin^2\vartheta\cos\vartheta \times\cos\varphi d\vartheta d\varphi. \quad (725)$$

For each normal mode of surface vibration of a sphere, the displacement ζ depends on the angles ϑ and φ as a spherical harmonic Y_{lm}. Higher l corresponds to higher frequencies. (725) shows that of all the normal modes only one will give a contribution to the quadrupole moment, *viz.* that for which ζ depends on the angles as

$$\zeta_0=b\cos\vartheta\sin\vartheta\cos\varphi. \quad (725a)$$

This vibration corresponds to the deformation of the sphere into an ellipsoid; it is the vibration of lowest frequency possible. The calculation of the frequency is similar to that given in §53C; the index l of the spherical harmonic corresponds, for high l, to kR in the old calculation. For any l, (311) must be replaced by

$$\omega^2=(G/\rho S)(l+2)l(l-1)R^{-3}, \quad (726)$$

so that, with (313) to (314), we have for $l=2$

$$\omega^2=8\Gamma/(3Mr_0^2A), \quad (726a)$$

$$\hbar\omega=(8/3)^{\frac{1}{2}}(\Gamma P)^{\frac{1}{2}}A^{-\frac{1}{3}}. \quad (726b)$$

The potential energy is given by an expression similar to (307a), *viz.*

$$V=(G/2S)\int \sin\vartheta d\vartheta d\varphi \times\left[\left(\frac{d\zeta}{\partial\vartheta}\right)^2+\frac{1}{\sin^2\vartheta}\left(\frac{\partial\zeta}{\partial\varphi}\right)^2-2\zeta^2\right]. \quad (727)$$

For a given l, the integrand reduces to $(l+2)\times(l-1)\zeta^2$; and if ζ is given by (724, 725a), the total energy in the normal mode, which is twice the time average of V, becomes

$$E=2b^2G/15R^2=2b^2\Gamma/15r_0^2 \quad (727a)$$

(cf. (314)). Since the frequency of the vibration is rather low, its excitation energy will in general be equal to the nuclear temperature[73] τ, so that

$$b=(15/2)^{\frac{1}{2}}r_0(\tau/\Gamma)^{\frac{1}{2}}. \quad (727b)$$

Inserting (725a) (727b) into (725), we have

$$[xy]=(1/5)ZRb=(3/10)^{\frac{1}{2}}Zr_0^2\Gamma^{-\frac{1}{2}}A^{\frac{1}{3}}\tau^{\frac{1}{2}} \quad (728)$$

and, with (723a) (726b)

$$\Gamma_\gamma=\frac{2^{5/2}}{3^{3/2}\cdot 5}\frac{e^2}{\hbar c}\frac{\Gamma^{3/2}P^{1/2}}{(Mc^2)^2}\frac{(2Z)^2}{A^{11/6}}\tau. \quad (729)$$

Inserting the numerical values $\Gamma=P=10$ MV, $Mc^2=930$ MV, we have

$$\Gamma_\gamma=2\cdot 10^{-7}\tau(2Z)^2A^{-11/6}. \quad (729a)$$

The factor $(2Z)^2A^{-11/6}$ practically does not change with the nuclear mass; it increases only from 1 at low atomic weight to 1.5 at high A. The nuclear temperature τ is, for excitation energies of the order of 10 MV, about 1 to 2 MV. Therefore generally

$$\Gamma_\gamma=0.2 \text{ to } 0.5 \text{ volt}. \quad (729b)$$

This agrees surprisingly well with the average radiation width observed in the capture of slow neutrons by heavy nuclei (§61). For proton capture by light nuclei, larger widths up to 5 volts have been observed, probably due to dipole transitions which are not so rare in this case.

The radiation widths of low states of radioactive nuclei (§88) are much smaller, *viz.* of the order of millivolts. In our model, this may be understood by assuming the excitation energy

[73] This will not be true for low excitation energy of the whole nucleus, e.g., for the γ-rays emitted by natural radioactive nuclei. In this case, the average excitation energy per normal mode will be lower.

per normal vibration (727a) to be much smaller than at higher temperatures. A more elementary and more satisfactory treatment seems to be to deduce the quadrupole moment $[xy]$ from the observed width by means of (723a). If we measure ω in MV and the quadrupole moment $[xy]$ in 10^{-24} cm², we have

$$\Gamma_\gamma = 0.013[xy]^2\omega^5 \text{ volts.} \qquad (730)$$

The quadrupole moments of the transitions listed in Table XXXXIV are of the order 10^{-24} cm² which seems plausible.

For all but these very low states, the γ-ray width (729a) depends only very slightly on the excitation energy of the nucleus and on its size. This fact is in agreement with the observations (see above) and in striking contrast to the particle width which, for any kind of particle, increases enormously with increasing excitation energy (cf., e.g., §65, 78, 79) and decreases very much with increasing atomic weight (§54, 81). This means that the emission of radiation (simple capture) can in general only compete with that of particles (disintegration) if the nucleus concerned is heavy and the excitation energy low. The best known example is the capture of slow neutrons which is only probable for heavy nuclei (§57ff.). The capture of fast neutrons (§65) is very improbable compared to inelastic scattering, the capture of particles by light nuclei (§64, 81) improbable compared to particle disintegrations.

The γ-ray spectrum would, according to our liquid drop model, consist of a single line with a frequency given by (726b), i.e., about 1 MV for heavy, 2–3 MV for lighter nuclei. Of course, this consequence of the model should not be taken seriously; but perhaps the spectrum has actually an intensity maximum near the frequency (726b).

The formulae given (especially 729 to 729b) refer to the total γ-ray width. The width corresponding to a given γ-ray transition is smaller by a factor of the order of the number of possible final states. An estimate of the partial γ-ray width will be given in §90.

D. Metastable states

Weizsäcker has pointed out (W9) that some nuclei may have low excited states whose angular momentum is very different from that of the ground state, and that such states would be "metastable" and may have a very long lifetime against γ-emission. As we shall show below, lifetimes of several seconds, or even of years, may occur. This would explain the "isomerism" observed with various β-emitting nuclei. The best established case of this kind is a radioactive nucleus formed in Br by capture of slow neutrons. Bromine possesses two stable isotopes, Br^{79} and Br^{81} so that two radioactive nuclei (Br^{80} and Br^{82}) would be expected from slow neutron capture. Actually, three different radioactive nuclei with three different periods (18 min., 4.2 hr., 36 hr.) have been observed (K34), all of them chemically indentical with Br. Since no other process can occur but neutron capture, and since no stable isotope of bromine has been found despite very careful search (B28), we are forced to conclude that either Br^{80} or Br^{82} exists in two modifications with different half-lives.

A similar case may be indium which also has two stable isotopes (In^{113} and In^{115}) and gives three different periods under neutron bombardment (S28). However, it is possible (L26) that one of these activities ($3\frac{1}{2}$ hr.) is produced by fast neutrons through a *np* reaction and is in reality an isotope of Cd. Another case of isomerism may be the natural β-emitters UX_2 and UZ (G4).

The existence of "isomeric" β-emitting nuclei violates the rule that β-emission is in general slower than any other nuclear process, especially γ-radiation. As a rule, a nucleus formed in any way, e.g., by neutron capture, natural α-decay etc., will first emit γ-radiation until it arrives in its ground state, and the time required for that is usually of the order of 10^{-12} sec. (cf. Table XXXXIV). The β-emission which takes several seconds or more, can ordinarily not at all compete with γ-emission so that, in general, β-particles cannot be emitted from excited states of nuclei. From this consideration, any nucleus should have *one* characteristic lifetime with respect to β-decay, and isomeric nuclei could never exist.

This will be different only if there are selection rules making the emission of γ-radiation from a certain (in general the first) excited state much less probable than it is ordinarily. If this is the case, the metastable state will have its own decay period. Its decay may occur in two different ways, *viz.:*

(1) The metastable state may go over into the ground state with emission of γ-radiation, whereupon the ground state will decay further with β-emission. The half-life of the activity due to the metastable state is then determined by the probability of the forbidden γ-transition. It can only be observed as a separate period if it is longer than the natural β-period of the ground state. The energy spectrum of the β-rays is exactly the same for the two periods.

(2) The metastable state may emit a β-particle directly. This will be the case if the probability of γ-emission is smaller than that of β-emission for the given metastable state. The lifetime of the metastable state may in this case be longer or shorter than that of the ground state. The β-emissions from the metastable and from the ground state will in general lead to different states of the final nucleus because the angular momenta are very different (cf. below). Thus at least one of the β-transformations will be followed by a γ-ray from the residual nucleus, and the upper limits of the two β-spectra will not be the same. In the case of Br, the 18 min. period is also produced by photoelectric disintegration of Br (§91, 103) and should therefore belong to Br^{80}. Neither of the other two periods has been found as yet in the photoelectric process, thus we may provisionally ascribe them to Br^{82} which would then be the nucleus possessing the metastable level. One of the periods (35 hr.) gives β-rays of 0.8 MV (A1) and gammas (S16a), the other (4.2 hr.) gives betas of 2.05 MV and no gammas. If our interpretation is correct, the γ-rays should have an energy of about 1.25 MV. An exact measurement of the γ-ray energy should decide whether the 35 hr. or the 4.2 hr. activity belongs to the metastable level of Br^{82}.

The probability of emission of multipole radiation of order l is obtained by taking in (720a) the lth term in the Taylor expansion of the exponential. With the direction of observation $\mathbf{n}$ parallel to the x axis, we have

$$|\mathbf{n}\times\mathbf{j}| = \frac{(\omega/c)^{l-1}}{(l-1)!}\left|\int x^{l-1}\mathbf{n}\times\mathbf{j}_0 d\tau\right|. \quad (731)$$

It seems rather difficult to replace the current distribution $\mathbf{j}_0$ by the charge distribution ρ_0 in as rigorous a way as that used for the quadrupole radiation in (722a). However, it seems to be a fair approximation to replace (731) by

$$|\mathbf{n}\times\mathbf{j}| = (\omega^l/l!c^{l-1})\int x^{l-1}\mathbf{n}\times\mathbf{r}\,\rho_0 d\tau \quad (731a)$$

$$= [e\omega^l/l!c^{l-1}]([x^{l-1}y]^2+[x^{l-1}z]^2)^{\frac{1}{2}}. \quad (731b)$$

Roughly, we may estimate

$$[x^{l-1}y]\sim\kappa R^l, \quad (731c)$$

where κ is of the order unity. Then the probability of γ-emission per second becomes, in analogy to (723a),

$$\Gamma_\gamma/\hbar = (\omega/c)^{2l+1}(e^2/\hbar)\kappa^2R^{2l}/l!^2. \quad (732)$$

Inserting the numerical values $\kappa=1$, $R=10^{-12}$ cm etc., we find for the lifetime against γ-emission

$$\tau=\hbar/\Gamma_\gamma=5\cdot10^{-21}l!^2(20/\hbar\omega)^{2l+1}\text{ sec.}, \quad (733)$$

where $\hbar\omega$ is measured in MV.

Table XXXXII gives the lifetimes calculated from (733) for various l's and excitation energies. 200 kv is about the average excitation energy of the first excited state in a heavy nucleus; 50 kv occurs frequently, e.g., in 3 of the naturally radioactive nuclei; and 10 kv may occur occasionally. Differences of the angular momentum of as much as 4 units should not be very rare considering that several nuclei are known with an angular momentum of 9/2 in their ground state. Thus it seems possible that some nuclei may possess metastable states with lifetimes of a year or more against γ-emission which, of course, is ample for the explanation of two "lifetimes" of the same radioactive isotope. If the lifetime against γ-emission is as long as this, the excited state of a β-active nucleus will in general decay by direct emission of a β-particle (mode 2 above).

From our considerations, it is obvious that sufficiently metastable states will in general not exist for light nuclei, firstly because of the small

TABLE XXXXII. *Lifetime of metastable states.*

Excitation Energy (kv)	Change of Angular Momentum in Transition to Ground State			
	$l=2$ (Quadrupole)	$l=3$ (Octopole)	$l=4$	$l=5$
10	$6\cdot10^{-4}$ sec.	7 hrs.	$5\cdot10^{4}$ yrs.	$5\cdot10^{12}$ yrs.
50	$2\cdot10^{-7}$ sec.	0.3 sec.	10 days	10^{5} yrs.
200	$2\cdot10^{-10}$ sec.	$2\cdot10^{-5}$ sec.	3 sec.	10 days

density of levels for such nuclei and secondly because of the relatively small angular momenta. For heavy nuclei, we may expect metastable states roughly to be as frequent as "stable" neighboring isobars (§43), of which there are perhaps 10 examples altogether. Metastable states will, of course, not be restricted to β-radioactive nuclei but will also exist in stable nuclei: In this case, their existence is difficult to observe because the transition to the ground state will be accompanied by emission of very soft γ-rays only. Finally, there is a (very small) chance that some stable nucleus may possess a metastable state with a lifetime of the order of 10^{10} years; such a nucleus might then appear under normal conditions in two modifications, distinguishable by their spins.

§88. γ-Rays from Natural Radioactive Substances

A. Frequency of natural γ-rays and internal conversion

More accurate data are available on natural γ-rays than on those emitted during or after transmutations. This is partly due to the fact that the natural γ-rays have been studied over a longer period of time. But the main reason is that there is a method of study available for natural γ-rays which is not applicable to the γ-rays from transmutations of the light elements, *viz.* the internal conversion.

Crudely speaking, the internal conversion may be regarded as a photoelectric effect which the γ-ray produces in the same atom from whose nucleus it was emitted. This internal photoelectric effect is very probable because of the high intensity of the γ-radiation in the emitting atom itself. The process will lead to the emission of one of the external electrons[74] of the atom instead of the γ-ray. If the conversion electron is ejected from the nth electron shell whose ionization potential is E_n, its kinetic energy will be

$$E = h\nu - E_n. \tag{734}$$

By magnetic deflection of the conversion electron its kinetic energy may be measured quite accurately. If the origin of the electron is known the quantum energy $h\nu$ can be inferred from the kinetic energy by means of (734). The determination of $h\nu$ will be quite unambiguous when conversion electrons are ejected from several shells by the same γ-ray. This can be recognized by the fact that the difference between the energies of two groups of conversion electrons is equal to the known energy difference between two electronic levels of the atom. Then the assignment of the level of origin to each electron group is straightforward.

In one case (RaB), 8 groups of conversion electrons have been observed all originating from the same γ-ray (E8) by internal conversion in the electron shells[75] L I, L II, L III, M I, M II, M III, N I and O. From each of the groups, the same energy for the γ-quantum (52.91 kv) can be inferred, with a mean deviation of the individual determinations of about 0.1 percent. In most other cases, the internal conversion in outer electron shells is too weak to be observed because these electrons are too far from the source of the γ-radiation (nucleus). However, for a great number of γ-rays at least 3 groups of conversion electrons have been observed, *viz.* in the K, L I and M I shell. Only for very weak γ-ray lines, the number of conversion electron groups is reduced to one. In this case, there is of course no check on the γ-ray energy as in the case of several groups, but it is safe to assume that a single group of conversion electrons always originates from the K shell[76] because the probability of internal conversion decreases approximately as the inverse third power of the principal quantum number n of the ejected electron (for s electrons).

In this way, about 50 γ-ray lines originating from 12 different nuclei have been observed. A list of the results for the quantum energies $h\nu$, and of the conversion groups observed in each case, is found in Rasetti's book on *Nuclear Physics*, p. 124, etc.

[74] From the historical development of the subject (cf., e.g., Rasetti, *Nuclear Physics*, p. 121) it is still customary to call these secondary electrons "β-particles" and to speak of them as "discrete β-spectrum" in distinction from the continuous, true β-spectrum. In the light of our present knowledge, this must be regarded as a misnomer. The terms α-, β-, γ-particles should be reserved for particles coming from the nucleus itself. The internal conversion electrons originate from the electron shells of the atom and should be named in a way indicating their origin, e.g., conversion electrons.

[75] There is no conversion in the K shell because the energy of the γ-ray is not sufficient.

[76] This conclusion is quite safe if the energy of the conversion electron is greater than the difference of the ionization potentials of the K and L shell. If this is not the case, the electron may have originated from the L I shell.

The measurements of the electronic energies are accurate enough to determine the atomic number of the element from which the conversion electrons, and therefore the γ-rays, originate. The γ-rays observed in a radioactive transformation $A \rightarrow B$ must in almost all cases be ascribed to the product nucleus B, i.e., an emission *after* the α- or β-particle characteristic of the main transformation. This is in accord with theoretical expectation (cf. §69).

The γ-rays observed have been compared with the groups of α-particles found in α-disintegrations (§69) and gratifying agreement has been obtained between the $h\nu$ of the γ-rays and the differences between the energies of various α-groups (Table XXXII). Complete level schemes have been worked out for various nuclei from α- and γ-ray data (cf. §69, end).

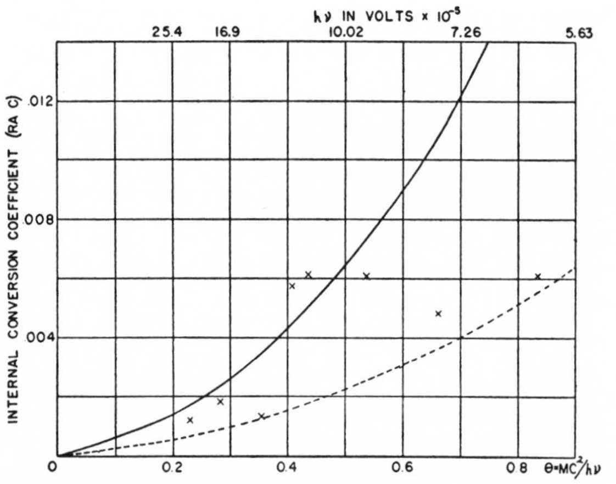

FIG. 26. Internal conversion coefficient for γ-rays in heavy nuclei ($Z=84$), according to Taylor and Mott. Solid curve: quadrupole radiation, broken curve: dipole radiation. Crosses represent experimental values determined by Ellis and Aston.

B. Intensity of γ-rays. Dipole and quadrupole radiation

From the observed number of internal conversion electrons, the number of emitted γ-rays may be deduced if the coefficient of internal conversion, α, is known. This coefficient is defined as the ratio of the number of conversion electrons to the number of quanta emitted. The theory of internal conversion has been given by Taylor, Mott, Hulme and F. Oppenheimer (H39, T4, T5). It is found that the internal conversion coefficient decreases rapidly with increasing quantum energy and with increasing principal and azimuthal quantum number of the electron shell from which the conversion electron originates. The first fact is mainly due to the rapid increase of the probability of radiation with increasing energy. The dependence on the quantum numbers is explained by the smaller probability of coming near the nucleus for electrons with higher n and l. The conversion coefficient depends also on the nature of the radiation, being about 3 times larger for quadrupole than for dipole radiation. The theoretical dependence of the conversion coefficient in the K shell, α_K, on the quantum energy $h\nu$ for dipole and quadrupole radiation is shown in Fig. 26.

The expressions for the internal conversion coefficient are rather complicated owing to the complicated mathematical form of the Dirac wave functions. An approximate expression may be obtained (R5a, p. 139) by neglecting relativity effects and at the same time assuming the frequency of the γ-ray to be large compared to the K absorption limit of the element in question. The result is, if $\alpha_K \ll 1$:

$$\alpha_K = \pi\left(4\frac{\hbar c}{e^2}\right)^3\left(\frac{Ry}{\hbar\omega}\right)^4 Z^4 \frac{e^{-4n \operatorname{arc\,cot} n}}{e^{2\pi n}-1}, \qquad (735)$$

where Ry is the Rydberg energy ($=13.54$ volts), $\hbar\omega$ the quantum energy of the γ-ray, Z the nuclear charge and

$$n^2 = Z^2Ry/(\hbar\omega - Z^2Ry). \qquad (735a)$$

The values of the conversion coefficient obtained from (735) are, for natural radioactive nuclei, about 5 times too small, owing to the unjustified approximations made.[77]

Experimentally, the internal conversion coefficient has been measured for a few γ-ray lines by comparing the number of internal conversion electrons with the number of photoelectrons produced in some material (usually lead) exposed to the γ-rays. Since the coefficient of photoelectric absorption is approximately known, both theoretically and experimentally, the absolute intensity of the γ-rays can be deduced from the number of photoelectrons observed. These experi-

[77] However, the formula would be correct for small nuclear charge Z. It shows that in this case the internal conversion coefficient would be exceedingly small so that there seems to be no chance of observing internal conversion of γ-rays in light atoms.

TABLE XXXXIII. *Intensities of γ-ray lines from ThC″.*

Line	$10^4 p\alpha_K$	Theoretical $\alpha_K(\%)$		Calculated $p(\%)$		Excitation from α-groups
		Dipole	Quadrupole	Dipole	Quadrupole	
γ_1	2.2	0.95	2.9	2.3	0.76	1.1
γ_2	2.2	1.02	3.2	2.2	0.69	0.16 (γ_2, γ_3)
γ_3	0.9	0.88	2.7	1.0	0.33	
γ_4	28	1.76	11.3	16	2.5	1.8 (γ_4, γ_5)
γ_5	6.1	1.49	7.5	4.1	0.81	

ments are not very accurate and give values for the internal conversion coefficient which are slightly (about 20–30 percent) higher than the theoretical values. The agreement is, however, sufficient to decide in most cases whether the γ-radiation investigated is dipole or quadrupole radiation (cf. Fig. 26).

Accepting the theory as correct, the internal conversion may be used for finding out the nature of the radiative transition when the correspondence between α-ray groups (§69) and γ-rays has been established and the intensities of the α-groups are known. This is, e.g., possible for ThC″ (cf. Table XXXII, §69). We denote by $p_1 \cdots p_6$ the intensities of the six γ-lines $\gamma_1 \cdots \gamma_6$ observed, in number of quanta per α-disintegration, and by $q_1 \cdots q_4$ the "excitations" of the four excited states of ThC″, i.e., the number of α-particles from the parent nucleus ThC corresponding to transitions to the four excited levels of ThC″. Then we have (cf. Table XXXI and Table XXXII)

$$\begin{aligned} q_4 &= p_1 &&= 1.10\%, \\ q_3 &= p_2 + p_3 &&= 0.16\%, \\ q_2 &= p_4 + p_5 &&= 1.80\%, \\ q_1 &= -p_1 - p_2 - p_4 + p_6 &&= 69.8\ \%. \end{aligned}$$

The observed number of conversion electrons from the K shell, $p\alpha_K$, is listed in the second column of Table XXXXIII for the five[78] γ-rays $\gamma_1 \cdots \gamma_5$. In the next two columns, the theoretical conversion coefficients for the known frequencies of the γ-rays are listed, for dipole and quadrupole radiation. Dividing the observed $p\alpha$ by these figures, the intensities p of the γ-rays are obtained (next two columns). It is seen that for $p_4 + p_5$ reasonable agreement with the value derived above (1.80 percent) is obtained by assuming both the γ-lines γ_4 and γ_5 to be quadrupole radiation while the assumption of dipole radiation would give about 10 times too much γ-radiation. For level 3 ($\gamma_2 + \gamma_3$), no agreement is obtained either way but the assumption of quadrupole radiation gives a less serious discrepancy. For the line γ_1, either dipole or quadrupole radiation may be assumed. Thus the evidence shows that either all the five γ-lines from ThC″ or at least four of them are due to quadrupole transitions.

In one case, *viz.* the third excited level of RaC′ (cf. Table XXXI) (excitation energy 1412 kv), no actual γ-ray has been observed but only an internal conversion electron. This is interpreted by assuming that the radiative transition from level 3 to the ground state is completely forbidden, e.g., because the two states both have angular momentum zero. Then the internal conversion is the only way in which the nucleus can go from state 3 to the ground state.

C. Absolute probability of γ-emission

The lifetime of nuclear states against γ-emission may be estimated by comparing the intensity of the long range α-particle groups from the excited levels of ThC′ and RaC′ with that of the γ-rays from the same levels. Table XXXXIV gives, for two levels each of RaC′ and ThC′, the observed number N_γ of γ-rays (per normal

[78] γ_6 does not have sufficient quantum energy to eject a K electron.

TABLE XXXXIV. *Lifetimes of excited states of RaC′ and ThC′.*

Nucl.	Level	$\hbar\omega$ (MV)	N_γ	N_α	E_d(MV)	τ_α(sec.)	τ_γ(sec.)	Γ_γ (millivolts)	f_0	$[xy]$ (10^{-24} cm²)
RaC′	1	0.61	0.4	$0.43\cdot10^{-6}$	8.437	$2.7\cdot10^{-7}$	$3\cdot10^{-13}$	2	$1\cdot10^{-3}$	1.8
	3*	1.41	0.0025	$22\cdot10^{-6}$	9.242	$9.5\cdot10^{-9}$	$8\cdot10^{-11}$	0.01	—	—
ThC′	1	0.73	0.14	$34\cdot10^{-6}$	9.674	$1.6\cdot10^{-9}$	$4\cdot10^{-13}$	1.5	$6\cdot10^{-4}$	0.6
	2	1.80	0.02	$190\cdot10^{-6}$	10.745	$6\cdot10^{-11}$	$6\cdot10^{-13}$	1	$6\cdot10^{-5}$	0.004

* Emits only conversion electrons.

α-disintegration), the observed number N_α of long range α-particles, the disintegration energy E_d for these α-particles, the lifetime against α-decay τ_α calculated from the energy by formulae (594, 600) with $G_\alpha=1$ volt, the lifetime against γ-emission

$$\tau_\gamma=(N_\alpha/N_\gamma)\tau_\alpha, \tag{736}$$

the γ-ray width $\Gamma_\gamma=\hbar/\tau_\gamma$, the oscillator strength f_0 calculated from (711b), and the quadrupole moment $[xy]$ associated with the transition according to (730).

The numbers of the excited levels are the same as in Table XXXI. N_γ is as estimated by Ellis and Aston (E8, E9). The estimates are based primarily on observations of the number of photoelectrons ejected from a Pt foil by the γ-rays (E9) supplemented by the intensities deduced from internal conversion using the theoretical value of the internal conversion coefficient. For the level 3 of RaC′ the figure given is the number of internal conversion electrons itself because for this transition no actual γ-ray is emitted (end of Section B). The calculated lifetimes are all of the same order, *viz.* a few times 10^{-13} sec., except for the highly forbidden transition 3 of RaC′ for which it is, of course, much larger. The γ-ray widths of the excited levels are of the order of a millivolt; they are, of course, smaller than the widths of neutron capture levels (§61) because from the levels considered here only *one* radiative transition can occur while transitions to many different final levels are possible from compound states responsible for neutron capture. The quadrupole moments are of the order 10^{-24} cm^2 as should be expected.

§89. γ-Rays from Residual Nuclei Produced by Transmutations

Not much experimental material is yet available on γ-rays from nuclear transmutations. Accounts of the experimental methods and results are given in §99–102. The most important points in the study of γ-rays are the following:

1. Nuclear energy levels. The levels of the residual nucleus can be deduced from the groups of outgoing particles emitted in the primary nuclear reaction (§§99–101, 109). The differences between the energy levels so deduced must agree with the energies of the γ-rays emitted by the residual nucleus. Thus the γ-rays provide an important check on the energy levels derived from particle groups. With light nuclei, it seems also possible to compare the experimental energy levels with theoretical expectation (F10) and to assign quantum numbers to them.

Checks between the data from particle groups and the γ-rays observed have been obtained for the nuclei B^{10} (produced in the reaction $Be^9+D=B^{10}+n^1$), B^{11} (reaction $B^{10}+D=B^{11}+H$), N^{15} (from $N^{14}+D=N^{15}+H$) and especially C^{12} (reactions $B^{11}+D=C^{12}+n^1$, $Be^9+He^4=C^{12}+n^1$ and $N^{14}+D=C^{12}+\alpha$). The particles produced in the reactions are neutrons, protons, and α-particles. The particle energies can in most cases be determined more accurately than the energy of the γ-rays, but the agreement obtained in the cases mentioned is satisfactory (cf. §99, 101).

2. Transition probabilities. A determination of the *absolute* radiative transition probabilities between the discrete levels of light nuclei does not seem feasible because, in contrast to natural radioactive nuclei (§88C), there is no alternative process of known probability (α-decay!) with which the γ-ray intensity could be compared.

Measurements of the relative intensities of the γ-ray lines starting from the same excited level of the nucleus will, however, give the relative transition probabilities. It seems feasible for light nuclei to calculate these transition probabilities theoretically from nuclear wave functions such as those given by Wigner and Feenberg (F10), using the Hartree model.

The sum of the intensities of all γ-ray lines starting from a given level of the residual nucleus, minus the intensities of the γ-ray lines ending at that level, must be equal to the intensity of the particle group in the primary reaction which corresponds to the formation of the nuclear level in question. Such comparisons of the intensity would serve in the first place to confirm the assignment of γ-rays to given transitions between nuclear energy levels. In fact, such assignments should not be called definite before a comparison of the intensities has been made. Moreover, intensity comparisons of the kind described might give more information about the efficiency of the apparatus used for detecting heavy particles as well as γ-rays.

A special kind of "γ-rays from the residual

nucleus" are the γ-rays emitted after an inelastic collision. Inelastic collisions have thus far only[79] been observed with fast neutrons (§65). The γ-rays emitted after the collision have been observed by Lea (L19) and by Kikuchi and collaborators (K7, K9, K5) for a great number of substances throughout the periodic table. The cross section for the production of the γ-rays was found to correspond approximately to the known scattering cross section for fast neutrons as should be expected. Measurements of the frequency are not yet available, but in view of the many levels of heavy nuclei there can be no doubt that the γ-spectra must be highly complex. Just as in capture processes (§90), the γ-emission will ordinarily take place in steps. It would be of interest to determine the average number of quanta emitted per inelastically scattered neutron.

Strictly speaking, all the γ-rays from capture processes but the one emitted first fall under the heading of γ-rays from the residual nucleus. This distinction is, however, not of much practical use.

§90. Capture γ-Rays

A. Light nuclei

The γ-rays from the capture of protons have been observed for the processes:

$$Li^7+H^1=Be^8+\gamma, \quad (I)$$
$$B^{10}+H^1=C^{11}+\gamma, \quad (II)$$
$$B^{11}+H^1=C^{12}+\gamma, \quad (III)$$
$$F^{19}+H^1=Ne^{20}+\gamma. \quad (IV)$$

I. The γ-rays from the proton bombardment of lithium were previously reported as very complex (C51). More recent experiments (D13) have shown that there is only one strong line at 17.1 MV and probably one or more weak ones between 10 and 17 MV. The first line corresponds to the transition to the ground state of Be^8. From the masses given in Table LXXIII (§108) the mass difference $Li^7+H^1-Be^8$ is found to be 17.0 MV. To this we must add $\frac{7}{8}$ (relative motion!) of the energy of the incident proton of 440 kv, giving a total of 17.4 MV in good agreement with the observed 17.1 MV. A line at 14 MV may be expected from a transition to the excited state of Be^8 of 2.8 MV excitation energy which is known from the study of the disintegration $B^{11}+H^1=Be^{8*}+He^4=3He^4$ (§85). If this is true, the line should be rather broad since the state at 2.8 MV has a width of about 0.8 MV (§85).

The occurrence of radiative transitions to excited states of Be^8 has been confirmed by Lauritsen and his co-workers (private communication) by the observation of groups of short range α-particles when Li^7 was bombarded by protons of 440 kv (resonance energy). The observation of these α-particles would give the most direct evidence on the position and width of the excited states of Be^8.

From the intensity of the γ-rays and simple theoretical considerations, we can obtain rather definite information on the angular momentum and parity of the resonance level of Be^8 responsible for the capture process. It seems certain theoretically that the ground state of Be^8 is a ^1S state of even parity. Since the γ-ray width of the resonance level is fairly large and the transition to the ground state the most intense, this transition is almost certainly "allowed"; it may still be either dipole or quadrupole. If the transition is a dipole transition, the resonance level must have odd parity and $J=1$. For a quadrupole transition, we must have even parity and $J=2$.[80] However, this is impossible because then the resonance level could disintegrate into two α-particles (§81). Therefore the resonance level must have $J=1$ and odd parity, and the γ-ray emitted must be dipole radiation. The first excited state of Be^8 is presumably an even ^1D state (F10); therefore the γ-ray transition to this state is also dipole radiation.

II and III. The γ-rays from these two capture processes are both observed when the mixed element boron is bombarded by protons (C50). The data do not yet allow definite conclusions but it seems certain that there is one γ-ray line corresponding to a transition into the ground state of C^{12} (energy about 15 MV) and several lines corresponding to transitions to excited states of that nucleus (cf. §101B).

IV. The γ-radiation from the capture of protons by fluorine is especially interesting since it is an excellent example of the working of selection rules. There seems to be a single γ-ray line having

[79] Except for the isolated example $Li^7+He^4=Li^{7*}+He^4$, cf. §99.

[80] While in general J may change by one unit in quadrupole radiation, the transition $1\rightarrow0$ is forbidden.

a quantum energy of 6.0±0.2 MV (D14). The energy evolution from nuclear masses is 12.9 MV, to which the kinetic energy of the incident protons (0.33 MV resonance energy) has to be added. It seems therefore that the transition to the ground state is forbidden. This conclusion is even more stringent because the method used for investigating the γ-rays, *viz.* the observation of electron pairs, is most sensitive to high energy γ-rays.

The fact that the transition to the ground state must be forbidden, may be deduced directly from the known properties of the resonance level responsible for the capture of the protons. We know that this resonance level must have either even J and odd parity, or *vice versa*, because otherwise it would disintegrate into $O^{16}+He^4$ which would be irreconcilable with its small width (§81). Now the ground state of Ne^{20} is, again, almost certainly a ^1S state of even parity. Such a state combines optically only with states of $J=1$ and odd parity (dipole transitions) or of $J=2$ and even parity (quadrupole transition). Both these possibilities are excluded for the resonance level; therefore there can be no strong γ-ray corresponding to the direct transition into the ground state, in agreement with observation.[81]

The observed γ-radiation of 6.0 MV leads to a state of Ne^{20} with an excitation energy of 7.2 MV. This is 2.5 MV more than the dissociation energy of Ne^{20} into an α-particle and an O^{16} nucleus. Unless there are again selection rules forbidding it, this state of Ne^{20} will probably disintegrate into $O^{16}+He^4$ rather than emit a further γ-ray, because the potential barrier of O^{16} for α-particles is not very high (4.5 MV). It would be interesting to find these slow α-particles (2.0 MV)

The absolute probability of radiative transitions can be found from the γ-ray width of resonance levels responsible for proton capture (§81). The results are given in Table XXXXV. Besides the radiation width Γ_γ observed, and the lifetime of the excited state against γ-radiation, $\hbar/\Gamma_\gamma$, we have given the effective oscillator strength f of the γ-ray line emitted in the capture process as calculated from (711b). The values obtained ($3\cdot 10^{-3}$ to 0.08) are larger than for radioactive nuclei (§88C). This is presumably due to the fact that the number of γ-lines starting and ending at a given level is very much greater in a heavy (naturally radioactive) than in a light nucleus, and therefore the f value for any single line smaller.

[81] The difference between the Ne^{20} and the Be^8 case (cf. I) is that the resonance level of Be^8 is allowed to be an odd level with $J=1$ while that of Ne^{20} is not. The reason for this is that Be^8 may only disintegrate into two like particles (α-particles) which are not capable of existing in a state of odd parity while Ne^{20} disintegrates into unlike particles ($O^{16}+He^4$) which may have a wave function of odd parity if only the angular momentum is also odd.

B. Heavy nuclei

The γ-ray spectrum emitted in the capture of neutrons will no doubt be enormously complex because of the extremely large number of energy levels between the ground state and the compound state responsible for the capture of the neutron (resonance level). Because of this complexity, it will, of course, not be possible to deduce the binding energy of the captured neutrons from the frequency of the emitted γ-rays. However, a study of the frequency distribution will give very valuable information on the distribution of energy levels over the region between the ground state and the dissociation energy of the nucleus.

We assume that the dipole moment D_{rs} will, in the average, be of the same order for transitions between any two states r and s. More accurately, D_{rs} shall not show any general trend with the excitation energies U_r and U_s of the two states or with their difference but shall only vary irregularly from level to level. Then the partial width of a level r due to a radiative transition to level s, will be in the average (cf. (710))

$$\gamma_{rs} \approx c\omega^3 = c'(U_r - U_s)^3, \qquad (737)$$

where c and c' are constants.

With this assumption, we can easily calculate the primary γ-ray spectrum emitted in the capture of a slow neutron. In this case, the excitation energy U_r of the initial state is equal to the dissociation energy Q of the nucleus. Let $D(U)$ be the spacing of levels at the excitation energy U. Then the number of quanta of energy between $\hbar\omega$ and $\hbar(\omega+d\omega)$ emitted is given by

TABLE XXXXV. *Absolute probabilities of capture of protons.*

	Process	Radiation Width (Volts)	Lifetime sec.	$h\nu$ MV	$10^3 f$
(I)	$Li^7+H^1=Be^8+\gamma$	4	$1.6\cdot 10^{-16}$	17	3
(V)	$C^{12}+H^1=N^{13}+\gamma$	0.08	$8\cdot 10^{-15}$	2.4	3
(IV)	$F^{19}+H^1=Ne^{20}+\gamma$	0.6	$1\cdot 10^{-15}$	6.0	3
		18	$3.5\cdot 10^{-17}$	6.5?	80
		8	$8\cdot 10^{-17}$	6.6?	40

$$I(\omega)d\omega = \gamma_{rs}\hbar d\omega/D(Q-\hbar\omega) \sim \omega^3 d\omega/D(Q-\hbar\omega). \quad (738)$$

We assume for D a formula similar to that derived in §53, *viz.*

$$D(U) = Q\cdot\exp(-kU^n). \quad (738a)$$

($n=\frac{1}{2}$ for free particle model, 4/7 to $\frac{3}{4}$ for liquid drop model.) Then we have

$$I(\omega) \sim (\hbar\omega)^3 \exp[k(Q-\hbar\omega)^n]. \quad (739)$$

This intensity has a maximum at $\hbar\omega = Qx_0$ where x_0 is determined by

$$(1-x_0)^{1-n}x_0^{-1} = \tfrac{1}{3}nkQ^n = \tfrac{1}{3}n\log(Q/D(Q)). \quad (739a)$$

The intensity falls to half its maximum value for $\hbar\omega = Q(x_0 \pm \delta x)$ where (approximately)

$$\delta x \approx x\left[\frac{1-x}{1-xn}\right]^{\frac{1}{2}}\left(\frac{\log 2}{3}\right)^{\frac{1}{2}}. \quad (739b)$$

The position of the intensity maximum and the half-width are given in Table XXXXVI for $n=\frac{1}{2}$ and $\frac{3}{4}$ and for various values of the spacing of levels D at dissociation energy, i.e., for various atomic weights (cf. table XXI). It is seen that the maximum should occur at a quantum energy of between $\frac{1}{4}$ and $\frac{1}{2}$ of the dissociation energy while the half-width should be somewhat less than one-half of the quantum energy at the intensity maximum. These theoretical figures compare fairly well with the experimental determinations of the quantum energy by Rasetti (R3) and by Fleischmann (F25) who found approximately 4–5 MV, i.e., about half the dissociation energy, by a method which favors the harder components of the γ-radiation.

The intensity maximum should lie at a relatively higher quantum energy if the spacing of levels is large, i.e., for lighter nuclei. This tendency is supported by the increase of the dissociation energy itself with decreasing atomic weight. Evidence for this trend of the quantum energy with mass number was also found by Fleischmann. The apparent quantum energy of the neutron capture radiation was found to be 7.7, 4.1, and 4.2 for Fe, Cd and Pb, respectively.

TABLE XXXXVI. *Position of intensity maximum x_0 and half-width δx of capture spectrum from theoretical formula (approximate).*

Spacing of Levels at Dissoc. Energy (Volts)	$n=0.5$ x_0	δx	$n=0.75$ x_0	δx
1	0.31	0.13	0.23	0.10
10	0.35	0.15	0.27	0.12
100	0.41	0.17	0.32	0.14
1000	0.48	0.20	0.39	0.17

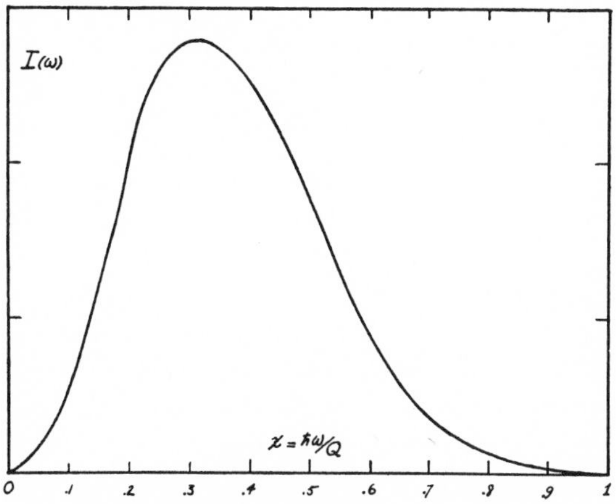

FIG. 27. Theoretical frequency distribution of the γ-rays emitted in the capture of a neutron by a heavy nucleus. The curve refers to the radiation emitted in the capture process itself, not including the γ-rays emitted afterwards by the residual nucleus. It has been assumed that the optical dipole moment is the same for the transition to each state of the final nucleus.

The expected γ-ray spectrum for $n=0.6$ and $D(Q)=10$ volts is shown in Fig. 27. The very small intensity near the maximum possible quantum energy ($\hbar\omega = Q$) should be noticed.

As the γ-ray emitted in the capture process itself (primary γ-ray) uses up only one-quarter to one-half of the energy available, several γ-rays will in general be emitted after this first one. This secondary γ-radiation will ordinarily be softer than the primary radiation. The number of γ-rays emitted in each capture process will in the average be between three and ten. Experimental evidence for this fact was obtained by Griffiths and Szilard (G23).

The total radiation width of neutron resonance levels has been determined from neutron experiments (§61) and theoretically (§87C) to be of the order of 0.1 to 1 volt. From this value, we may determine the average dipole moment $r_0^2 = (\mathbf{r}_{rs}^2)_{Av}$ corresponding to transition to *one* of the final states. With the same assumptions as above, we have for the total radiation width

$$\Gamma^r_\gamma = \frac{4}{3}\sum_s \frac{e^2\omega_{rs}^3}{c^3} r_0^2 = \frac{4}{3}\frac{e^2 r_0^2}{\hbar^3 c^3}\int_0^Q \frac{(Q-U)^3}{D(U)}dU \quad (740)$$

$$= (\Gamma^r_{\gamma 0})_{Av} N_{eff}(Q), \quad (740a)$$

where

$$(\Gamma^r_{\gamma 0})_{Av} = \tfrac{4}{3}e^2 r_0^2 Q^3/\hbar^3 c^3 \quad (740b)$$

is the partial γ-ray width for the transition to the ground state *if* the dipole moment for this transition has the "normal" value r_0, and

$$N_{eff}(Q) = \int_0^Q \left(\frac{Q-U}{Q}\right)^3 \frac{dU}{D(U)} = \int_0^1 x^3 e^{kQ^n(1-x)^n}dx \quad (741)$$

may be called the "effective" number of quantum states of the nucleus below dissociation energy, in distinction from the true number of levels

$$N(Q)=\int dU/D(U)=\int_0^1 \exp\left[kQ^n(1-x)^n\right]dx. \qquad (741a)$$

The quantities mentioned are listed in Table XXXXVII, for $n=0.6$ (approximately corresponding to the liquid drop model, §53) and various values of the spacing of levels at dissociation energy, $D(Q)$.

The "effective" number of levels is seen to be much smaller than the true number because most of the levels lie very near the dissociation energy and are therefore unimportant as final states for γ-ray emission. The partial width due to radiative transitions to the ground state alone turns out to be of the order of a few millivolt for spacings of the order 10 to 100 volts as they are found for most elements for which the capture of neutrons has been investigated. This is just the same order of magnitude as for the low excited states of radioactive nuclei (§88C). The corresponding oscillator strength (cf. (711)) comes out between 10^{-5} and 10^{-4} (for natural radioactive nuclei 10^{-4} to 10^{-3}), and the average dipole moment for each transition is of the order of one thousandth of the nuclear radius.

§91. Nuclear Processes Produced by γ-Rays

Two kinds of processes are possible when γ-rays fall on nuclei: (a) The γ-ray may be absorbed and a material particle emitted from the nucleus (photodissociation). (b) The γ-ray may be scattered, either with or without giving part of its energy to the nucleus (Raman and Rayleigh scattering).

Since the emission of material particles is in general more probable (§65, 76), photodissociation will ordinarily be the more frequent process if it can occur energetically. Exceptions will be found (1) when the emission of the material particle is improbable because of the potential barrier and (2) when the emitted particle is very slow so that the factor v in the particle width (§52) makes the emission probability small. Scattering will be the only process occurring if the energy of the γ-ray is not sufficient to dissociate the nucleus.

TABLE XXXXVII. *Data on the capture γ-ray spectrum for heavy nuclei (Total γ-ray width $\Gamma_\gamma=0.5$ volt, dissociation energy $Q=8$ MV, $n=0.6$).*

Spacing at dissociation energy, $D(Q)$ (volts)	1	10	100	1000
Total number of levels below Q, $N(Q)$	$8\cdot10^5$	10^5	$1.2\cdot10^4$	$1.5\cdot10^3$
"Effective" number of levels, $N_{\text{eff}}(Q)$	3000	500	90	20
Partial width for ground st. transition $(\Gamma_{\gamma0})_{\text{Av}}$ (millivolts)	0.15	1	5	25
Oscillator strength for trans. to ground st. f_0	$5\cdot10^{-6}$	$3\cdot10^{-5}$	$1.5\cdot10^{-4}$	$7\cdot10^{-4}$
Average dipole moment r_0 (cm)	$4\cdot10^{-16}$	$9\cdot10^{-16}$	$2\cdot10^{-15}$	$4\cdot10^{-15}$

The photodissociation has been observed for the deuteron (§16), for Be^9, and, more recently, for a great number of heavier nuclei (B47b, c, d). In all cases, one of the particles into which the nucleus dissociates is a neutron. There can be no doubt that, in principle, all nuclei can be disintegrated by γ-rays of sufficiently high quantum energy. The required dissociation energies are listed in Table LXVII (§103); they vary from 1.5 to over 10 MV for light nuclei. It must also be expected that particles other than neutrons can be split off a nucleus by γ-rays, e.g. protons, α-particles or deuterons. The relative probability of these various types of photodissociation will be determined by the respective "partial widths" of the levels of the compound nucleus, i.e. primarily by the penetrability of the potential barrier for the particle to be emitted; the probability will therefore be small for heavier nuclei.[81a]

If the emitted particle is a slow neutron, the corresponding partial width will be proportional to the neutron velocity (cf. §52, 265a). Therefore the cross section for a photodissociation into a slow neutron and a residual nucleus will be

$$\sigma=\text{const}\cdot E^{\frac{1}{2}}, \qquad (742)$$

where

$$E=h\nu-Q \qquad (742a)$$

is the kinetic energy of the neutron and Q the dissociation energy of the initial nucleus. The validity of (742) is, of course, restricted in a similar way as the $1/v$ law for neutron capture, *viz.* to neutron energies small compared to the energy of the first resonance level (§58b). Moreover, (742) holds only if the emitted neutron can have orbital momentum zero. This means that the angular momenta of initial and final nucleus must not differ by more than 3/2, i.e., the sum of the neutron spin (1/2) and the angular momentum of a γ-quantum (1 for dipole radiation).[82]

[81a] Photodissociation with emission of an α-particle will be probable when the energy of the γ-ray is insufficient for neutron-dissociation but sufficient to make the α-width larger than the γ-width (cf. §§79E, 80B).

[82] A stricter selection rule holds if Russell-Saunders coupling is valid in the initial and final nucleus. Then the internal *orbital* momenta of the two nuclei may differ by 0 or 1 but must not both be zero while the "multi-

For light nuclei, resonance effects will be observable in the photodissociation. E.g., if the γ-rays emitted in the capture of protons by B^{11} fall on a C^{12} nucleus, this nucleus will be dissociated into $B^{11}+H^1$. The cross section for this dissociation σ_{ph} is connected to the cross section for the emission of γ-rays in the proton bombardment of B^{11} by

$$\sigma_{ph}=\frac{(2i+1)(2s+1)}{(2i'+1)(2s'+1)}\frac{\lambda_\gamma^2}{\lambda_p^2}\sigma_{c0}. \qquad (743)$$

Here i' is the angular momentum of the dissociated nucleus (C^{12}, $i'=0$); i and s those of the dissociation products, *viz.* B^{11} and H^1 so that $i=\frac{1}{2}$ (probably, cf. R10a) and $s=\frac{1}{2}$; $2s'+1$ is the "statistical weight" for radiation, i.e., 2 (two directions of polarization); λ_γ and λ_p are the wave-lengths of γ-ray and proton, and σ_{c0} is that part of the capture cross section which corresponds to the formation of C^{12} in the ground state. Let us, e.g., assume that there is a resonance level for proton capture at 0.5 MV proton energy, corresponding to $\lambda_p=6.5\cdot10^{-13}$ cm, and that the corresponding cross section is 10^{-27} cm^2, as for the capture of protons by Li^7. The quantum energy of the γ-rays emitted in our process is 15 MV according to the masses of the nuclei concerned, so that $\lambda_\gamma=1.3\cdot10^{-12}$ cm. With these data, σ_{ph} would be $8\cdot10^{-27}$ cm^2 which should be observable without too great difficulty.

For heavy nuclei, it will, of course, be impossible to observe resonance maxima in the cross section for photodissociation. All we can hope to observe is the average cross section, averaged over energy regions large compared to the spacing between energy levels. Unlike that of material particles, the wave-length of γ-rays will be larger than the nuclear radius for quantum energies below about 20 MV. Therefore only dipole and quadrupole radiation will ever be of any importance. For dipole radiation, only compound states of angular momentum $J=i-1$, i and $i+1$ will be important where i is the angular momentum of the ground state of the nucleus to be dissociated. The average cross section for photodissociation becomes thus (cf. (405))

$$\sigma_{ph}=\frac{\pi^2\lambda_\gamma^2}{(2i+1)}\sum_J(2J+1)\frac{\Gamma^J{}_{\gamma0}\Gamma^J{}_Q}{D_J\Gamma_J}, \qquad (744)$$

where λ_γ is the "wave-length" of the incident γ-ray, $\Gamma^J{}_{\gamma0}$ the partial width corresponding to emission of a γ-ray which leaves the residual nucleus in the ground state, averaged over all levels in the energy region in question, $\Gamma^J{}_Q$ the partial width corresponding to emission of particale Q (photodissociation), Γ_J the total width and D_J the average spacing of the levels of angular momentum J. Assuming these quantities to be independent of J and summing over the three possible values of J ($=i-1$, i and $i+1$), we find

$$\sigma_{ph}=3\pi^2\lambda_\gamma^2\frac{\Gamma_{\gamma0}\Gamma_Q}{D\Gamma}. \qquad (744a)$$

If the particle width Γ_Q is large compared to the γ-ray width Γ_γ which will be true for large kinetic energies of the emitted particle (cf., e.g., §65), we may put $\Gamma_Q=\Gamma$. Moreover, the partial width $\Gamma_{\gamma0}$ is equal to the total γ-ray width, Γ_γ, divided by the "effective number of quantum states $N_{eff}(h\nu)$" of the compound nucleus below the excitation energy $h\nu$, N_{eff} being defined as in the preceding section. We may write

$$N_{eff}(U)=U/D(U)\kappa(U), \qquad (745)$$

where κ is a function which varies slowly with the excitation energy U (slowly compared to $D(U)$). According to the values given in Table XXXXVII, κ will be of the order of one to a few thousand. Inserting (745) into (744a), we obtain

$$\sigma_{ph}=3\pi^2\kappa\lambda_\gamma^2\Gamma_\gamma/h\nu. \qquad (745a)$$

From neutron experiments we know that the total γ-ray width of the levels of compound nuclei of atomic weight about 100, is, in the average, of the order $\frac{1}{2}$ volt. $h\nu$ may be estimated to be about 10 MV, corresponding to $\lambda_\gamma=1.9\cdot10^{-12}$

plicities" must differ just by one. According to this rule, the photodissociation of the deuteron is forbidden, because the internal orbital momentum of the deuteron is zero, and that of the "residual nucleus," *viz.* the proton, is, of course, also zero. Indeed, the photodissociation of the deuteron does not lead to a state of the dissociated system of zero orbital momentum (s state) but to a p state ($l=1$), as was shown in §16. Correspondingly, the cross section is, in this case, proportional to $E^{\frac{3}{2}}$ rather than $E^{\frac{1}{2}}$ (cf. (265a)). This will no longer be true if magnetic dipole radiation is taken into account (§17).

cm. With $\kappa=2000$, we obtain

$$\sigma_{ph}=10^{-26}\ \text{cm}^2. \qquad (746)$$

The cross section observed by Bothe and Gentner in their experiments on the photodissociation of heavy nuclei (B47b, c, d) is of the order 10^{-27} cm^2. The agreement is satisfactory, in view of the crudeness of both the theoretical and the experimental estimate.

If the energy of the γ-ray is only just sufficient to dissociate the nucleus, the cross section for photodissociation will be small because of the factor $E^{\frac{1}{2}}$ in the particle width. This decrease of the cross section will set in when the particle width becomes smaller than the γ-ray width (cf. §65) which will occur for a kinetic energy of the particle of the order of a few hundred thousand volts.

When the γ-ray energy is insufficient for dissociation, the only process which may occur is the scattering of the γ-rays by the nucleus. Since, in this case, Γ_γ is identical with the total width Γ, (745a) now holds for the *scattering* cross section. The cross section for the nuclear scattering of γ-rays is thus of the order of[83] 10^{-26} cm^2. This is very much less than the ordinary Klein-Nishina scattering which amounts to $5\cdot10^{-23}$ cm^2 for atomic number $Z=50$ and a γ-ray energy of 8 MV. Thus the nuclear scattering of γ-rays will probably be unobservable.

[83] Above the dissociation energy, the scattering cross section will be smaller by a factor Γ_γ/Γ.

References

The following list of references is not intended to be a complete list. It does contain all papers referred to explicitly in the text, both in Part B and in Part C. The references are arranged alphabetically and are denoted by a letter indicating the initial of the primary author and a number representing the position in the alphabetical list. The paragraphs in which the references appear are indicated.

A1. Alichanian, Alichanow and Dzelepow, Physik. Zeits. Sowjetunion **10**, 78 (1936). (§87, 99,101 105)

A2. Alichanow, Alichanian and Dzelepow, Nature **133**, 871 (1934). (§99, 105)

A2a. Alichanow, Alichanian and Dzelepow, Nature **133**, 950 (1934). (§99)

A3. Alichanow, Alichanian and Dzelepow, Nature **135**, 393 (1935). (§102, 105)

A4. Allen, Phys. Rev. **51**, 182 (1937). (§100)

A5. Allison, Proc. Camb. Phil. Soc. **32**, 179 (1936). (§81)

A6. Alvarez, McMillan and Snell, Phys. Rev. **51**, 148 (1937). (§92)

A7. Amaldi, D'Agostino, Fermi, Pontecorvo, Rasetti and Segrè, Proc. Roy. Soc. **149**, 522 (1935). (§57, 59, 65, 101, 102, 105)

A8. Amaldi, Nuovo Cimento **12**, 223 (1935). (§57, 102, 105)

A9. Amaldi and Fermi, Ricerca Scient. **1**, 56 (1936). (§57)

A10. Amaldi and Fermi, Ricerca Scient. **1**, 310 (1936). (§57)

A11. Amaldi and Fermi, Phys. Rev. **50**, 899 (1936). (§58, 59, 60, 61, 62, 92, 101, 102, 105)

A11a. Amaldi, Hafstad and Tuve, Phys. Rev., in press. (§102)

A12. Anderson and Neddermeyer, Phys. Rev. **50**, 263 (1936). (§104)

A13. Andersen, Nature **137**, 457 (1936). (§102, 105)

A14. Andersen, Zeits. f. physik. Chemie **32**, 237 (1936). (§102, 105)

A15. Andersen, Nature **138**, 76 (1936). (§102, 105)

A16. Arsenjewa-Heil, Heil and Westcott, Nature **138**, 462 (1936). (§60)

A17. Aston, Nature **135**, 541 (1935). (§107)

A18. Aston, *Mass Spectra and Isotopes* (E. Arnold, 1933). (§107)

A19. Aston, Nature **137**, 357 (1936). (§107)

A20. Aston, Nature **137**, 613 (1936). (§107)

A21. Aston, Nature **138**, 1094 (1936). (§107)

A22. Aoki, Nature **139**, 372 (1937). (§65)

B1. Bainbridge and Jordan, Phys. Rev. **49**, 883 (1936). (§107)

B2. Bainbridge and Jordan, Phys. Rev. **51**, 384 (1937). (§101, 107)

B2a. Bainbridge and Jordan, Phys. Rev. **50**, 282 (1936). (§107)

B3. Banks, Chalmers and Hopwood, Nature **135**, 99 (1935). (§103)

B4. Bardeen, Phys. Rev. **51**, 799 (1937). (§53)

B4a. Barnes, DuBridge, Wiig, Buck, and Strain, Phys. Rev. **51**, 775 (1937). (§100, 105)

B5. Bayley, Curtis, Gaerttner and Goudsmit, Phys. Rev. **50**, 461 (1936). (§63)

B5a. Bayley and Crane, Phys. Rev. (Abst. 73, Washington Meeting, 1937). (§101, 105)

B6. Beams and Trotter, Phys. Rev. **45**, 849 (1934). (§92)

B6a. Bennet, Proc. Roy. Soc. **155**, 419 (1936). (§95)

B7. Bernardini, Zeits. f. Physik **85**, 555 (1933). (§82, 99)

B8. Bethe, Ann. d. Physik **5**, 325 (1930). (§95)

B9. Bethe and Heitler, Proc. Roy. Soc. **146**, 83 (1934). (§93)

B10. Bethe and Peierls, Int. Conf. Phys., London (1934). (§103)

B11. Bethe, Phys. Rev. **47**, 633 (1935). (§99, 101, 107)

B12. Bethe, Phys. Rev. **47**, 747 (1935). (§57, 65, 102)
B13. Bethe, Phys. Rev. **50**, 332 (1936). (§52)
B14. Bethe, Phys. Rev. **50**, 977 (1936). (§101)
B15. Bethe and Placzek, Phys. Rev. **51**, 450 (1937). (§52, 54, 61, 71, 91, 102)
B16. Bethe, *Handbuch der Physik*, Vol. 24, 273 (1933). (§51, 95)
B17. Bethe, Phys. Rev. to be published shortly. (§80)
B18. Bethe, Phys. Rev. (Abstract 32, Washington Meeting, 1937). (§79, 80)
B19. Bethe, Zeits. f. Physik **76**, 293 (1932). (§95)
B20. Birge, Phys. Rev. **37**, 1669 (1931). (§101)
B21. Bjerge and Westcott, Nature **134**, 286 (1934). (§102, 105)
B22. Bjerge and Westcott, Nature **134**, 177 (1934). (§102)
B23. Bjerge, Nature **137**, 865 (1936). (§102)
B24. Bjerge and Broström, Nature **138**, 400 (1936). (§102, 105)
B25. Blackett and Lees, Proc. Roy. Soc. **134**, 658 (1932). (§95)
B26. Blackett, Proc. Roy. Soc. **135**, 132 (1932). (§95)
B27. Blackett and Champion, Proc. Roy. Soc. **130**, 380 (1931). (§74)
B28. Blewett, Phys. Rev. **49**, 900 (1936). (§87, 105)
B29. Bloch, Ann. d. Physik **16**, 285 (1933). (§95)
B30. Bloch and Gamow, Phys. Rev. **50**, 260 (1936). (§87)
B31. Bloch, Zeits. f. Physik **81**, 363 (1933). (§95)
B32. Bohr, Nature **137**, 344 (1936). (§51, 57, 100, 102)
B33. Bohr and Kalckar, Kgl. Dansk Acad. (1939). (§51–56, 77, 79, 81–90)
B33a. Bohr, Phil. Mag. **25**, 10 (1913). (§95)
B34. Bohr, Phil. Mag. **30**, 581 (1915). (§97)
B35. Bonner and Mott-Smith, Phys. Rev. **46**, 258 (1934). (§99)
B36. Bonner and Brubaker, Phys. Rev. **48**, 469 (1935). (§102)
B37. Bonner and Brubaker, Phys. Rev. **48**, 742 (1935). (§85, 97, 101)
B38. Bonner and Brubaker, Phys. Rev. **49**, 19 (1936). (§97, 101)
B39. Bonner and Brubaker, Phys. Rev. **49**, 778 (1936). (§102, 105)
B40. Bonner and Brubaker, Phys. Rev. **50**, 308 (1936). (§74, 85, 97, 99, 101, 108)
B41. Bonner and Brubaker, Phys. Rev. **50**, 781 (1936). (§65, 95, 102)
B42. Born, Zeits. f. Physik **58**, 306 (1929). (§66)
B43. Bothe and Becker, Zeits. f. Physik **66**, 289 (1930). (§99)
B44. Bothe and Frãnz, Zeits. f. Physik **49**, 1 (1928). (§95)
B45. Bothe and Baeyer, Gött. Nachr. **1**, 195 (1935). (§94)
B46. Bothe, Physik. Zeits. **36**, 776 (1935). (§99)
B47. Bothe, Zeits. f. Physik **100**, 273 (1936). (§99, 101)
B47a. Bothe and Maier-Leibnitz, Naturwiss. **25**, 25 (1937). (§110)
B47b. Bothe and Gentner, Naturwiss. **25**, 90 (1937). (§91, 103)
B47c. Bothe and Gentner, Naturwiss. **25**, 126 (1937). (§91, 103)
B47d. Bothe and Gentner, Naturwiss. **25**, 191 (1937). (§91, 103)
B48. Brasch, Naturwiss. **21**, 82 (1933). (§92)
B49. Brasch, Lange, Waly, Banks, Chalmers, Szilard and Hopwood, Nature **134**, 880 (1934). (§92)
B50. Brasefield and Pollard, Phys. Rev. **50**, 296 (1936). (§99)
B51. Breit and Wigner, Phys. Rev. **49**, 519 (1936). (§52, 57, 86, 100, 102)
B52. Breit and Condon, Phys. Rev. **49**, 904 (1936). (§103)
B53. Breit, Condon and Present, Phys. Rev. **50**, 825 (1936). (§72)
B53a. Breit, R. S. I. **8**, 95 (1937). (§99)
B54. Breit, Phys. Rev. **51**, 248, 778 (1937). (§84)
B54a. Breit and Wigner, Phys. Rev. **51**, 593 (1937). (§105)
B55. Briggs, Proc. Roy. Soc. **114**, 313 (1927). (§97)
B56. Briggs, Proc. Roy. Soc. **114**, 341 (1927). (§96)
B57. Briggs, Proc. Roy. Soc. **118**, 549 (1928). (§95)
B58. Briggs, Proc. Roy. Soc. **139**, 638 (1933). (§95)
B59. Briggs, Proc. Roy. Soc. **143**, 604 (1934). (§95)
B60. Briggs, Proc. Roy. Soc. **157**, 183 (1936). (§92)
B61. Brillouin, Comptes rendus **183**, 24 (1926). (§66)
B62. Brown and Mitchell, Phys. Rev. **50**, 593 (1936). (§105)
B63. Burcham and Goldhaber, Proc. Camb. Phil. Soc. **32**, 632 (1936). (§102)
B64. Blau, J. de phys. et rad. **61** (1934). (§99)
B65. Bruche and Scherzer, *Geometrische Elektronenoptik* (J. Springer, 1934). (§106)
B66. Brubaker and Pollard, Phys. Rev. (Abst. 82, Washington Meeting, 1937). (§99)
B67. Buck, Strain, and Valley, Phys. Rev. (Abst. 75, Washington Meeting). (§100, 105)
B68. Burhop, Proc. Camb. Phil. Soc. **32**, 643–7 (1936). (§92, 100, 101)
C1. Carlson and Oppenheimer, Phys. Rev. **51**, 220 (1937). (§104)
C2. Chadwick and Bieler, Phil. Mag. **42**, 923 (1921). (§76)
C3. Chadwick, Proc. Roy. Soc. **128**, 114 (1930). (§74)
C4. Chadwick and Constable, Proc. Roy. Soc. **135**, 48 (1931). (§82, 99)
C5. Chadwick, Proc. Roy. Soc. **136**, 692 (1932). (§99)
C6. Chadwick, Nature **129**, 312 (1932). (§99)
C7. Chadwick, Proc. Roy. Soc. **142**, 1 (1933). (§82, 99)
C8. Chadwick, Feather and Davies, Proc. Camb. Phil. Soc. **30**, 357 (1934). (§85, 112)
C9. Chadwick and Goldhaber, Nature **134**, 237 (1934). (§91, 92)
C10. Chadwick and Feather, Int. Conf. Phys., London (1934). (§82, 99)
C11. Chadwick, Phil. Mag. **40**, 734 (1920). (§70)
C12. Chadwick and Goldhaber, Nature **135**, 65 (1935). (§102)
C13. Chadwick, Constable and Pollard, Proc. Roy. Soc. **130**, 463 (1931). (§82)

C14. Chadwick and Goldhaber, Proc. Roy. Soc. **151**, 479 (1935). (§93, 103)

C15. Chadwick, Phil. Mag. **2**, 1056 (1926). (§99)

C16. Chadwick and Goldhaber, Proc. Camb. Phil. Soc. **31**, 612 (1935). (§57, 102)

C17. Coates, Phys. Rev. **46**, 542 (1934). (§104)

C18. Cockcroft and Walton, Proc. Roy. Soc. **129**, 477 (1930). (§92)

C19. Cockcroft and Walton, Proc. Roy. Soc. **137**, 229 (1932). (§92, 100)

C20. Cockcroft and Walton, Nature **129**, 242 (1932). (§95)

C21. Cockcroft and Walton, Nature **129**, 649 (1932). (§100)

C22. Cockcroft and Walton, Proc. Roy. Soc. **136**, 619 (1932). (§92)

C23. Cockcroft and Walton, Nature **131**, 23 (1933). (§100)

C24. Cockcroft, Gilbert and Walton, Nature **133**, 328 (1934). (§100, 105)

C25. Cockcroft and Walton, Proc. Roy. Soc. **144**, 704 (1934). (§92, 96, 101)

C26. Cockcroft, Int. Conf. Phys., London (1934). (§96, 100, 101)

C27. Cockcroft, Gilbert and Walton, Proc. Roy. Soc. **148**, 225 (1935). (§92, 101)

C28. Cockcroft and Lewis, Proc. Roy. Soc. **154**, 246 (1936). (§85, 96, 101)

C29. Cockcroft and Lewis, Proc. Roy. Soc. **154**, 261 (1936). (§85, 96, 101, 107)

C30. Collie, Nature **137**, 614 (1936). (§93, 102)

C31. Condon and Breit, Phys. Rev. **49**, 229 (1936). (§59)

C32. Condon and Gurney, Phys. Rev. **33**, 127 (1929). (§66, 92)

C33. Cooksey and Lawrence, Phys. Rev. **49**, 866 (1936). (§101)

C34. Cork, Richardson and Kurie, Phys. Rev. **49**, 208 (1936). (§101, 105)

C35. Cork and Lawrence, Phys. Rev. **49**, 788 (1936). (§79, 84, 85, 101, 105)

C36. Cork and Thornton, Phys. Rev. **51**, 59 (1937). (§79, 85, 92, 101, 105)

C36a. Cork and Thornton, Phys. Rev. **51**, 608 (1937). (§101, 105)

C37. Crane, Lauritsen and Soltan, Phys. Rev. **44**, 514 (1933). (§101)

C38. Crane, Lauritsen and Soltan, Phys. Rev. **44**, 692 (1933). (§78, 101)

C39. Crane and Lauritsen, Phys. Rev. **45**, 497 (1934). (§81, 100, 105)

C40. Crane and Lauritsen, Phys. Rev. **45**, 430 (1934). (§101, 105)

C41. Crane, Lauritsen and Soltan, Phys. Rev. **45**, 507 (1934). (§101)

C42. Crane, Delsasso, Fowler and Lauritsen, Phys. Rev. **46**, 531 (1934). (§100)

C43. Crane and Lauritsen, Int. Conf. Phys., London (1934). (§99)

C44. Crane, Delsasso, Fowler and Lauritsen, Phys. Rev. **46**, 1109 (1934). (§99, 101)

C45. Crane, Delsasso, Fowler and Lauritsen, Phys. Rev. **47**, 782 (1935). (§100, 101)

C46. Crane, Delsasso, Fowler and Lauritsen, Phys. Rev. **47**, 887 (1935). (§101, 105)

C47. Crane, Delsasso, Fowler and Lauritsen, Phys. Rev. **48**, 484 (1935). (§100, 101)

C48. Crane, Delsasso, Fowler and Lauritsen, Phys. Rev. **47**, 971 (1935). (§101, 105)

C49. Crane, Delsasso, Fowler and Lauritsen, Phys. Rev. **48**, 100 (1935). (§101)

C50. Crane, Delsasso, Fowler and Lauritsen, Phys. Rev. **48**, 102 (1935). (§81, 90, 100)

C51. Crane, Delsasso, Fowler and Lauritsen, Phys. Rev. **48**, 125 (1935). (§90, 94, 100)

C52. Crane, Phys. Rev. **51**, 58 (1937). (§92)

C53. Curie and Joliot, J. de phys. et rad. **5**, 153 (1934). (§94)

C54. Curie and Joliot, Comptes rendus **198**, 254 (1934). (§99, 105)

C55. Curie and Joliot, Comptes rendus **197**, 237 (1933). (§99)

C56. Curie and Joliot, J. de phys. et rad. **4**, 21 (1933). (§99)

C57. Curie and Joliot, J. de phys. et rad. **4**, 494 (1933). (§105)

C58. Curie and Joliot, Comptes rendus **198**, 559 (1934). (§105)

C59. Curie and Joliot, J. de phys. et rad. **4**, 278 (1933). (§99)

C60. Curie, Halban and Preiswerk, J. de phys. et rad. **7**, 361 (1935). (§102, 105)

C61. Curie and Preiswerk, Comptes rendus **203**, 787 (1936). (§102)

D1. Dahl, Hafstad and Tuve, R. S. I. **4**, 373 (1933). (§94)

D2. Danysz, Rotblat, Wertenstein and Zyw, Nature **134**, 970 (1934). (§65)

D3. Danysz and Zyw, Acta Phys. Polonica **3**, 485 (1934). (§99)

D3a. Darling, Curtis and Cork, Phys. Rev. (Abst. 67, Washington Meet., 1937). (§101, 105)

D4. Darrow, Bell S. Tech. J. **10**, 628 (1931). (§99, 105)

D5. Darrow, R. S. I. **5**, 66 (1934). (§99)

D6. Dee and Walton, Proc. Roy. Soc. **141**, 733(1933). (§100, 101)

D7. Dee, Nature **133**, 564 (1934). (§101)

D8. Dee and Gilbert, Proc. Roy. Soc. **149**, 200 (1935). (§101)

D9. Dee and Gilbert, Proc. Roy. Soc. **154**, 279 (1936). (§85, 100, 101)

D10. Delbruck and Gamow, Zeits. f. Physik **72**, 492 (1931). (§87)

D11. Delsasso, Fowler and Lauritsen, Phys. Rev. **48**, 848 (1935). (§101)

D12. Delsasso, Fowler and Lauritsen, Phys. Rev. **50**, 389 (1936). (§100)

D13. Delsasso, Fowler and Lauritsen, Phys. Rev. **51**, 391 (1937). (§81, 90, 100)

D14. Delsasso, Fowler and Lauritsen, Phys. Rev. **51**, 527 (1937). (§90, 100)
D15. Dolch, Zeits. f. Physik **100**, 401 (1936). (§84)
D16. Doolittle, Phys. Rev. **49**, 779 (1936). (§84, 100)
D17. Döpel, Zeits. f. Physik **91**, 796 (1934). (§100)
D18. Duncanson, Proc. Camb. Phil. Soc. **30**, 102 (1934). (§95)
D19. Duncanson and Miller, Proc. Roy. Soc. **146**, 396 (1934). (§82, 99)
D20. Dunning and Pegram, Phys. Rev. **45**, 295 (1934). (§92)
D21. Dunning, Phys. Rev. **45**, 586 (1934). (§59, 85, 99)
D22. Dunning, R. S. I. **5**, 387 (1934). (§94)
D23. Dunning, Pegram, Fink and Mitchell, Phys. Rev. **48**, 265 (1935). (§57, 58, 60, 62, 63, 74)
D24. Dunning, Pegram, Fink and Mitchell, Phys. Rev. **47**, 970 (1935). (§102)
D25. Dunning, Pegram, Fink and Mitchell, Phys. Rve. **48**, 704 (1935). (§60)
D26. Dunning, Fink, Pegram and Segrè, Phys. Rev. **49**, 199 (1936). (§60)
D27. Dempster, Phys. Rev. **11**, 316 (1918). (§106)
D28. DuBridge, Barnes and Buck, Phys. Rev. (Abst. 74, Washington Meet., 1937). (§100, 105)
D29. Dyer, *The Long Death* (Chas. Scribner's Sons, 1937). (§92)
E1. Ehrenberg, Nature **136**, 870 (1935). (§65, 102)
E2. Ellis, Int. Conf. Phys., London (1934). (§69)
E3. Ellis and Mott, Proc. Roy. Soc. **139**, 369 (1933). (§69)
E4. Ellis, Proc. Roy. Soc. **136**, 396 (1932). (§69)
E5. Ellis and Henderson, Nature **135**, 429 (1935). (§99, 101, 105)
E6. Ellis and Henderson, Nature **136**, 755 (1935). (§105)
E7. Ellis, Proc. Roy. Soc. **138**, 318 (1932). (§88)
E8. Ellis, Proc. Roy. Soc. **143**, 350 (1934). (§69, 103)
E9. Ellis and Aston, Proc. Roy. Soc. **129**, 180 (1930). (§88)
E10. Ellis and Henderson, Proc. Roy. Soc. **156**, 358 (1936). (§99, 105)
E11. Evans and Livingston, Rev. Mod. Phys. **7**, 229 (1935). (§98)
F1. Fahlenbrach, Zeits. f. Physik **96**, 503 (1935). (§99, 105)
F2. Farkas and Farkas, *Ortho-hydrogen, Para-hydrogen and Heavy hydrogen*. (§101)
F3. Fay, Phys. Rev. **50**, 560 (1936). (§65)
F4. Fea, Nuovo Cimento **12**, 368 (1935). (§101)
F5. Feather, Proc. Roy. Soc. **136**, 709 (1932). (§102)
F6. Feather, Nature **130**, 257 (1932). (§102)
F7. Feather, Proc. Roy. Soc. **141**, 194 (1933). (§95)
F8. Feather, Proc. Roy. Soc. **142**, 689 (1933). (§93, 94, 102)
F9. Feather, Nature **136**, 468 (1935). (§103)
F9a. Feenberg, Phys. Rev. **49**, 328 (1936) (§84)
F10. Feenberg and Wigner, Phys. Rev. **51**, 95 (1937). (§81, 83, 89, 99, 101, 110)
F11. Feenberg and Phillips, Phys. Rev. **51**, 597 (1937). (§110)
F12. Fermi, Ric. Sci. **2**, 12 (1933). (§102, 105)
F13. Fermi, Amaldi, D'Agostino, Rasetti and Segrè, Proc. Roy. Soc. **146**, 483 (1934). (§57, 65, 94, 102, 105)
F14. Fermi, Nature **133**, 757 (1934). (§102)
F15. Fermi, Pontecorvo and Rasetti, Ricerca Scient. **2** (1934). (§105)
F16. Fermi and Rasetti, Nuovo Cimento **12**, 201 (1935). (§105)
F17. Fermi, Ric. Sci. **7**, 13 (1936). (§59, 102)
F18. Fermi, *Zeeman Jubilee* (1935), p. 128. (§59)
F19. Fink, Dunning, Pegram and Mitchell, Phys. Rev. **49**, 103 (1936). (§60, 61)
F20. Fink, Phys. Rev. **50**, 738 (1936). (§59)
F21. Fischer-Colbrie, Ak. W. Wien, **145**, 283 (1936). (§99)
F22. Fisk, Schiff and Shockley, Phys. Rev. **50**, **1090** (1936). (§59)
F23. Fisk and Morse, Phys. Rev. **51**, 54 (1937). (§59)
F24. Fleischmann and Gentner, Zeits. f. Physik **100**, 440 (1936). (§102, 103)
F25. Fleischmann, Zeits. f. Physik **103**, 113 (1936). (§57, 90, 102)
F26. Flügge, and Krebs, Physik. Zeits. **36**, 466 (1935). (§101)
F26a. Flügge and Krebs, Physik. Zeits. **38**, 13 (1937). (§101)
F27. Fomin and Houtermans, Physik. Zeits. Sowjetunion **9**, 273 (1936). (§102, 105)
F28. Fowler, Delsasso and Lauritsen, Phys. Rev. **49**, 561 (1936). (§101, 105)
F29. Fox and Rabi, Phys. Rev. **48**, 746 (1935). (§101)
F30. Frenkel, Physik. Zeits. Sowjetunion **9**, 533 (1936). (§53)
F31. Frisch, Nature **133**, 721 (1934). (§99, 101, 105)
F32. Frisch, Nature **136**, 220 (1935). (§99, 105)
F33. Frisch and Placzek, Nature **137**, 357 (1936). (§58, 61, 102)
F34. Furry, Phys. Rev. **51**, 592 (1937). (§59)
G1. Gaerttner, Turin and Crane, Phys. Rev. **49**, 793 (1936). (§105)
G2. Gaerttner and Crane, Phys. Rev. **51**, 58 (1937). (§100)
G3. Gaerttner and Crane, Phys. Rev. **51**, 49 (1937). (§100)
G4. Gamow, Nature **133**, 833 (1934). (§87, 105)
G5. Gamow, Phys. Rev. **49**, 946 (1936). (§65)
G6. Gamow, Physik. Zeits. **32**, 651 (1931). (§69)
G7. Gamow, Zeits. f. Physik **52**, 510 (1929). (§66, 92)
G8. Gamow, Nature **122**, 805 (1929). (§66)
G9. Gamow, *Atomic Nuclei and Radioactivity* (Cambridge, 1937). (§68)
G10. Gamow and Houtermans, Zeits. f. Physik **52**, 496 (1929). (§66, 68)
G11. Geiger, Proc. Roy. Soc. **83**, 492 and 505 (1910). (§95)
G12. Geiger, *Handbuch der Physik*, Vol. 24 (1927), 137. (§95)
G13. Geiger and Nuttall, Phil. Mag. **22**, 613 (1911). (§68)

G13a. Gentner, Naturwiss. **25**, 12 (1937)
G14. Gerthsen and Reusse, Physik. Zeits. **34**, 478 (1933). (§95)
G14a. Giarratana and Brennecke, Phys. Rev. **49**, 35 (1936). (§84)
G15. Goldhaber, Proc. Camb. Phil. Soc. **30**, 561 (1934). (§83, 101)
G16. Goldsmith and Cohen, Phys. Rev. **45**, 850 (1934). (§102)
G17. Goldsmith and Rasetti, Phys. Rev. **50**, 328 (1936). (§58, 60, 61)
G18. Goldsmith and Manley, Phys. Rev. **51**, 382 (1937). (§58, 60)
G18a. Goldsmith and Manley, Phys. Rev. (Abst. 126, Washington Meeting, 1937). (§60)
G18b. Goloborodko and Rosenkewitsch, Physik. Zeits. Sowjetunion **11**, 78 (1937). (§103)
G19. Goudsmit, Phys. Rev. **49**, 406 (1936). (§59)
G20. Goudsmit, Phys. Rev. **51**, 64 (1937). (§53)
G21. Gray, Proc. Camb. Phil. Soc. **27**, 103 (1931). (§93)
G22. Greinacher, Zeits. f. Physik **36**, 364 (1926). (§94)
G23. Griffiths and Szilard, Nature **139**, 323 (1937). (§90)
G24. Gurney, Nature **123**, 565 (1929). (§99)
G25. Güttinger and Pauli, Zeits. f. Physik **67**, 743 (1931). (§83)
G26. Gentner, Naturwiss. **25**, 12 (1937). (§100)
G27. Grahame, Seaborg and Gibson, Phys. Rev. **51**, 590 (1937). (§65)
G28. Gibson, Seaborg and Grahame, Phys. Rev. **51**, 370 (1937). (§65)
H1. Hafstad, Tuve and Brown, Phys. Rev. **45**, 746 (1934). (§101)
H2. Hafstad and Tuve, Phys. Rev. **47**, 506 (1935). (§92, 100)
H3. Hafstad and Tuve, Phys. Rev. **47**, 507 (1935). (§100)
H4. Hafstad and Tuve, Phys. Rev. **48**, 306 (1935). (§81, 92, 100, 105)
H5. Hafstad, Heydenburg and Tuve, Phys. Rev. **49**, 866 (1936). (§92, 100)
H6. Hafstad, Heydenburg and Tuve, Phys. Rev. **50**, 504 (1936). (§56, 81, 84)
H7. Hahn and Meitner, Naturwiss. **23**, 320 (1935). (§65, 85)
H8. von Halban and Preiswerk, Nature **137**, 905 (1936). (§58, 60)
H9. von Halban and Preiswerk, Helv. Phys. Acta **9**, 318 (1936). (§58, 60)
H10. von Halban and Preiswerk, J. de phys. et rad. **8**, 29 (1937). (§60)
H11. Hall, Phys. Rev. **49**, 401 (1936). (§103)
H11a. Halpern, Lueneburg and Clark, Phys. Rev., in press. (§59)
H12. Harkins, Gans, Newson, Phys. Rev. **44**, 529 (1933). (§102)
H13. Harkins and Gans, Phys. Rev. **46**, 397 (1934). (§102)
H14. Harkins, Gans and Newson, Phys. Rev. **47**, 52 (1935). (§102)
H15. Harkins, Kamen, Newson and Gans, Phys. Rev. **50**, 980 (1936). (§59)
H16. Hartree, Proc. Roy. Soc. **139**, 311 (1933). (§95)
H17. Haxel, Zeits. f. Physik **83**, 323 (1933). (§99)
H18. Haxel, Zeits. f. Physik **93**, 400 (1935). (§99)
H19. Haxel, Physik. Zeits. **36**, 804 (1935). (§99)
H19a. Haxel, Zeits. f. Physik **104**, 540 (1937). (§102)
H20. Henderson, Phys. Rev. **43**, 98 (1935). (§84)
H21. Henderson, Livingston and Lawrence, Phys. Rev. **45**, 428 (1934). (§100, 101, 105)
H22. Henderson, Livingston and Lawrence, Phys. Rev. **46**, 38 (1934). (§78, 100)
H23. Henderson, Phys. Rev. **48**, 480 (1935). (§78, 80)
H24. Henderson, Phys. Rev. **48**, 855 (1935). (§101, 105)
H25. Henneberg, Zeits. f. Physik **83**, 555 (1933). (§51)
H26. Henneberg, Zeits. f. Physik **86**, 592 (1933). (§95)
H27. Herb, Parkinson and Kerst, Phys. Rev. **48**, 118 (1935). (§84, 100)
H28. Herb, Parkinson and Kerst, Phys. Rev. **51**, 75 (1937). (§92)
H29. Hevesy and Levi, Nature **135**, 580 (1935). (§102, 105)
H30. Hevesy and Levi, Nature **136**, 103 (1935). (§102, 105)
H31. Hevesy and Levi, Nature **137**, 185 (1936). (§60, 62, 102, 105)
H32. Hevesy, Nature **135**, 1051 (1935) (§102, 105)
H33. Heyn, Nature **138**, 723 (1936). (§85, 102, 105)
H33a. Heyn, Physica **4**, 160 (1937). (§65, 102)
H34. Hoffman and Bethe, Phys. Rev. (Abstract 125, Washington Meeting, 1937). (§61)
H35. Hönl, Zeits. f. Physik **84**, 1 (1933). (§95)
H36. Hopwood and Chalmers, Nature **135**, 341 (1935). (§93)
H37. Horsley, Phys. Rev. **48**, 1 (1935). (§74)
H38. Horvay, Phys. Rev. **50**, 897 (1936). (§59)
H39. Hulme, Mott, Oppenheimer and Taylor, Proc. Roy. Soc. **155**, 315 (1936). (§88)
I1. Inglis, Phys. Rev. **50**, 783 (1936). (§83)
J1. Jaeckel, Zeits. f. Physik **96**, 151 (1935). (§102)
J2. Johnson and Hamblin, Nature **138**, 504 (1936). (§102, 105)
J3. Johnson and Johnson, Phys. Rev. **50**, 170 (1936). (§94)
J4. Jordan and Bainbridge, Phys. Rev. **49**, 883 (1936). (§107)
J5. Jordan and Bainbridge, Phys. Rev. **50**, 98 (1936). (§107)
J6. Jordan and Bainbridge, Phys. Rev. **51**, 385 (1937). (§107)
K1. Kapitza, Proc. Roy. Soc. **106**, 602 (1924). (§95)
K2. Kempton, Browne and Maasdorp, Proc. Roy. Soc. **157**, 372 (1936). (§83, 85, 101)
K3. Kempton, Browne and Maasdorp, Proc. Roy. Soc. **157**, 386 (1936). (§84, 85, 101)
K4. Kikuchi, Aoki and Husimi, Proc. Phys. Math. Soc. Japan **17**, 369 (1935). (§57)
K5. Kikuchi, Husimi and Aoki, Proc. Phys. Math. Soc. Japan **18**, 35 (1936). (§57, 89)

K6. Kikuchi, Aoki and Husimi, Nature **137**, 186 (1936). (§102)
K7. Kikuchi, Aoki and Husimi, Nature **137**, 398 (1936). (§65)
K8. Kikuchi, Aoki and Husimi, Nature **137**, 745 (1936). (§102)
K9. Kikuchi, Aoki and Husimi, Proc. Phys. Math. Soc. Japan **18**, 115 (1936). (§57, 65)
K10. Kikuchi, Husimi and Aoki, Nature **137**, 992 (1936). (§57)
K11. Kikuchi, Husimi and Aoki, Proc. Phys. Math. Soc. Japan **18**, 188 (1936). (§57, 102)
K12. Kikuchi, Aoki and Husimi, Nature **138**, 84 (1936). (§102)
K12a. Kikuchi, Aoki and Husimi, Proc. Phys. Math. Soc. Japan **18**, 727 (1936). (§65)
K13. Kinsey, Phys. Rev. **50**, 386 (1936). (§92, 104)
K14. Kirchner, Ergeb. d. exact. Naturwiss. **13**, 57 (1934). (§101)
K15. Kirchner and Neuert, Physik. Zeits. **35**, 292 (1934). (§85, 100)
K16. Kirchner and Neuert, Physik. Zeits. **36**, 54 (1935). (§95, 100)
K17. Klarmann, Zeits. f. Physik **95**, 221 (1935). (§102)
K18. Klarmann, Zeits. f. Physik **87**, 411 (1933). (§94)
K19. Klein and Nishina, Zeits. f. Physik **52**, 853 (1928). (§93)
K20. Knol and Veldkamp, Physica **3**, 145 (1936). (§102, 105)
K21. Konig, Zeits. f. Physik **90**, 197 (1934). (§99)
K22. Konopinski and Uhlenbeck, Phys. Rev. **48**, 7 (1935). (§105)
K23. Konopinski and Bethe, Phys. Rev. **52** (1937). (§78)
K23a. Konopinski and Bethe, Phys. Rev. (Abst. 33, Washington Meeting, 1937). (§54)
K24. Kramers, Zeits. f. Physik **39**, 828 (1926). (§66)
K25. Kraus and Cork, Phys. Rev. **51**, 383 (1937). (§101, 105)
K26. Kruger and Green, Phys. Rev. **51**, 57 (1937). (§92)
K26a. Kruger, Shoupp and Stallmann, Phys. Rev., to be published shortly (§59)
K27. Kurie, R. S. I. **3**, 655 (1932). (§94)
K28. Kurie, Phys. Rev. **43**, 771 (1933). (§102)
K29. Kurie, Phys. Rev. **43**, 1056 (1933). (§59)
K30. Kurie, Phys. Rev. **45**, 904 (1934). (§102)
K31. Kurie, Phys. Rev. **47**, 97 (1935). (§65, 102)
K32. Kurie, Richardson and Paxton, Phys. Rev. **48**, 167 (1935). (§105)
K33. Kurie, Richardson and Paxton, Phys. Rev. **49**, 368 (1936). (§99, 101, 105)
K34. Kurtchatov, Kurtchatov, Myssowsky and Roussinow, Comptes rendus **200**, 1201 (1935). (§87, 102, 105)
K35. Kurtschatow, Nemenow and Selinow, Comptes rendus **200**, 2162 (1935). (§102, 105)
K36. Kurtschatow, Latyschew, Nemenow and Selinow, Physik. Zeits. Sowjetunion **8**, 589 (1935). (§102, 105)
K37. Kronig, Physica **4**, 171 (1937). (§85)
L1. Ladenburg, Roberts and Sampson, Phys. Rev. **48**, 467 (1935). (§92)
L1a. Ladenburg and Kanner, Phys. Rev. (Abst. 129, Washington Meeting, 1937). (§105)
L1b. Laporte, Phys. Rev., in press (§61)
L2. Laslett, Phys. Rev. **50**, 388 (1936). (§101, 105)
L3. Latimer, Hull and Libby, J. Am. Chem. Soc. **57**, 781 (1935). (§102, 105)
L4. Laue, Zeits. f. Physik **52**, 726 (1929). (§66)
L5. Lauritsen and Crane, Phys. Rev. **45**, 493 (1934). (§85, 101)
L6. Lauritsen and Crane, Phys. Rev. **45**, 63 (1934). (§100)
L7. Lauritsen and Crane, Phys. Rev. **45**, 345 (1934). (§94, 101)
L8. Lauritsen and Crane, Phys. Rev. **45**, 550 (1934). (§101)
L9. Lauritsen and Crane, Phys. Rev. **32**, 850 (1928). (§92)
L10. Lawrence, Phys. Rev. **46**, 746 (1934). (§101)
L11. Lawrence, Livingston and White, Phys. Rev. **42**, 150 (1932). (§92)
L12. Lawrence, Livingston and Lewis, Phys. Rev. **44**, 56 (1933). (§101)
L13. Lawrence and Livingston, Phys. Rev. **45**, 220 (1934). (§101)
L14. Lawrence and Livingston, Phys. Rev. **45**, 608 (1934). (§92)
L15. Lawrence, Phys. Rev. **47**, 17 (1935). (§101, 105)
L16. Lawrence, McMillan and Henderson, Phys. Rev. **47**, 273 (1935). (§96, 101)
L17. Lawrence, McMillan and Thornton, Phys. Rev. **48**, 493 (1935). (§80, 101)
L18. Lawrence and Cooksey, Phys. Rev. **50**, 1131 (1936). (§92)
L19. Lea, Proc. Roy. Soc. **150**, 637 (1935). (§65, 83, 102)
L20. Lewis, G. N., and Macdonald, J. Chem. Phys. **1**, 341 (1933). (§101)
L21. Lewis, G. N., Livingston and Lawrence, Phys. Rev. **44**, 55 (1933). (§101)
L22. Lewis, G. N., Livingston and Lawrence, Phys. Rev. **44**, 317 (1933). (§101)
L23. Lewis, W. B. and Wynn-Williams, Proc. Roy. Soc. **136**, 349 (1932). (§69, 92)
L24. Lewis, W. B., Burcham and Chang, Nature **139**, 24 (1937). (§101, 105)
L25. Livingood and Snell, Phys. Rev. **48**, 851 (1935). (§92, 104)
L26. Livingood, Phys. Rev. **50**, 425 (1936). (§101, 105)
L27. Livingood and Seaborg, Phys. Rev. **50**, 435 (1936). (§101, 105)
L28. Livingston, Henderson and Lawrence, Phys. Rev. **44**, 316 (1933). (§100)
L29. Livingston, Henderson, and Lawrence, Phys. Rev. **46**, 325 (1934). (§102)
L30. Livingston and McMillan, Phys. Rev. **46**, 437 (1934). (§101)
L31. Livingston, R. S. I. **7**, 55 (1936). (§92)
L32. Livingston and Hoffman, Phys. Rev. **50**, 401 (1936). (§102)

L33. Livingston and Hoffman, Phys. Rev. (Abst. 124, Washington Meeting, 1937). (§61)
L34. Livingston, Genevese and Konopinski, Phys. Rev. **51**, 835 (1937). (§95)
L35. Lyford and Bearden, Phys. Rev. **45**, 743 (1934). (§68)
L36. Lyman, Phys. Rev. **51**, 1 (1937). (§105)
L37. Lukirsky and Careva, Comptes rendus (U.S.S.R.) **3**, 411 (1936). (§102)
L38. Li, Proc. Roy. Soc. **158**, 571 (1937). (§88)
M1. Madsen, Nature **138**, 722 (1936). (§102, 105)
M2. Manley and Millman, Phys. Rev. **51**, 19 (1937). (§83)
M2a. Manley, Goldsmith and Schwinger, Phys. Rev. (Abst. 127, Washington Meeting, 1937). (§61)
M3. Mano, Comptes rendus **197**, 47 (1933). (§95)
M4. Mano, Ann. d. Physik **1**, 407 (1934). (§95)
M5. Mano, Comptes rendus **194**, 1235 (1932). (§95)
M6. Mano, J. de phys. et rad. **5**, 628 (1934). (§95)
M7. Marsh and Sugden, Nature **136**, 102 (1935). (§102, 105)
M8. Massey and Mohr, Proc. Roy. Soc. **148**, 206 (1935). (§74)
M9. Massey and Burhop, Phys. Rev. **48**, 468 (1935). (§51)
M10. Mattauch, Phys. Rev. **50**, 617, 1089 (1936). (§107)
M10a. Mattauch, Physik. Zeits. **35**, 567 (1934). (§107)
M10b. Mattauch, Naturwiss. **25**, 156 (1937). (§107)
M10c. Mattauch, Naturwiss. **25**, 170 (1937). (§107)
M11. May and Vaidyanathan, Proc. Roy. Soc. **155**, 519 (1936). (§99)
M12. Meitner and Philipp, Naturwiss. **20**, 929 (1932). (§102)
M13. Meitner and Philipp, Zeits. f. Physik **87**, 484 (1934). (§102)
M14. Meitner, Naturwiss. **22**, 420 (1934). (§99, 105)
M15. Meitner and Hahn, Naturwiss. **24**, 158 (1936). (§65, 85, 105)
M16. Miller, Duncanson and May, Proc. Camb. Phil. Soc. **30**, 549 (1934). (§82, 99, 105)
M17. Mitchell, A. C. G., and Murphy, Phys. Rev. **47**, 881 (1935). (§58, 63)
M18. Mitchell, A. C. G., and Murphy, Phys. Rev. **48**, 653 (1935). (§58, 63)
M19. Mitchell, A. C. G., Murphy and Langer, Phys. Rev. **49**, 400 (1936). (§58, 63)
M20. Mitchell, A. C. G., Murphy and Whitaker, Phys. Rev. **50**, 133 (1936). (§58, 63)
M21 Mitchell, D. P., Phys. Rev. **49**, 453 (1936). (§93, 102)
M22. Mitchell, D. P., Rasetti, Fink and Pegram, Phys. Rev. **50**, 189 (1936). (§91, 103)
M23. Mitchell, D. P., and Powers, Phys. Rev. **50**, 486 (1936). (§60)
M24. Mohr and Pringle, Nature **137**, 865 (1936). (§74)
M25. Moller, Ann. d. Physik **14**, 531 (1932). (§95)
M26. Moon and Tillman, Proc. Roy. Soc. **153**, 476 (1936). (§57, 102)
M28. Morse, Fisk and Schiff, Phys. Rev. **50**, 748 (1936). (§59)
M29. Morse, Fisk and Schiff, Phys. Rev. **51**, 706 (1937). (§59)
M30. Marsden and Taylor, Proc. Roy. Soc. **88**, 443 (1913). (§95)
M31. Mott, Proc. Roy. Soc. **126**, 259 (1929). (§74)
M32. Mott and Massey, *Theory of Atomic Collisions* (Oxford, 1933). (§51, 59, 70)
M33. Mott, Proc. Camb. Phil. Soc. **27**, 553 (1931). (§95)
M34. Mott-Smith and Bonner, Phys. Rev. **45**, 554 (1934). (§92)
Mc1. McLennan, Grimmett and Read, Nature **135**, 147 (1935). (§102, 105)
Mc2. McLennan, Grimmett and Read, Nature **135**, 505 (1935). (§102)
Mc3. McLennan and Rann, Nature **136**, 831 (1935). (§102, 105)
Mc4. McMillan, Phys. Rev. **46**, 868 (1934). (§93, 100, 101)
Mc5. McMillan and Lawrence, Phys. Rev. **47**, 343 (1935). (§101)
Mc6. McMillan and Livingston, Phys. Rev. **47**, 452 (1935). (§94, 101, 105)
Mc7. McMillan, Phys. Rev. **49**, 875 (1936). (§101, 102, 105)
N1. Nahmias and Walen, Comptes rendus **203**, 71 (1936). (§102, 105)
N2. Naidu and Siday, Proc. Phys. Soc. London **48**, 330 (1936). (§105)
N3. Naidu, Nature **137**, 578 (1936). (§102, 105)
N4. Neuert, Physik. Zeits. **36**, 629 (1935). (§95, 100)
N4a. Neuert, Physik. Zeits. **38**, 122 (1937). (§100, 101)
N5. Newson, Phys. Rev. **48**, 482 (1935). (§101, 105)
N6. Newson, Phys. Rev. **48**, 790 (1935). (§96, 99, 105)
N7. Newson, Phys. Rev. **49**, 208 (1936). (§101)
N7a. Newson, Phys. Rev. **51**, 620 (1937). (§101)
N7b. Newson, Phys. Rev. **51**, 624 (1937). (§101, 105)
O1. Oliphant, Shire and Crowther, Proc. Roy. Soc. **146**, 922 (1934). (§100, 101)
O2. Oliphant and Rutherford, Proc. Roy. Soc. **141**, 259 (1933). (§85, 92, 100)
O3. Oliphant, Kinsey and Rutherford, Proc. Roy. Soc. **141**, 722 (1933). (§78, 85, 100, 101)
O4. Oliphant, Shire and Crowther, Nature **133**, 377 (1934). (§101)
O5. Oliphant, Harteck and Rutherford, Nature **133**, 413 (1934). (§101)
O6. Oliphant, Int. Conf. Phys., London (1934). (§84, 100, 101)
O7. Oliphant, Harteck and Rutherford, Proc. Roy. Soc. **144**, 692 (1934). (§92, 101)
O8. Oliphant, Kempton and Rutherford, Proc. Roy. Soc. **149**, 406 (1935). (§96, 100, 101)
O9. Oliphant, Kempton and Rutherford, Proc. Roy. Soc. **150**, 241 (1935). (§85, 96, 100, 101)
O10. Oppenheimer, Phys. Rev. **43**, 380 (1933). (§101)
O11. Oppenheimer and Phillips, Phys. Rev. **48**, 500 (1935). (§80, 101)
O12. Oppenheimer and Serber, Phys. Rev. **50**, 391 (1936). (§53)
O13. Ortner and Stetter, Zeits. f. Physik **54**, 449 (1929). (§94)

O14. Ostrofsky, Breit and Johnson, Phys. Rev. **49**, 196 (1936). (§84)

O15. Ostrofsky, Breit and Johnson, Phys. Rev. **49**, 22 (1936). (§84, 100)

O16. Ostrofsky, Bleick and Breit, Phys. Rev. **49**, 352 (1936). (§84)

P1. Paton, Phys. Rev. **46**, 229 (1934). (§99)

P1a. Paneth and Lollett, Nature **136**, 950 (1935). (§99)

P2. Paton, Zeits. f. Physik **90**, 586 (1934). (§82)

P3. Paton, Phys. Rev. **47**, 197 (1935). (§99)

P4. Paxton, Phys. Rev. **49**, 206 (1936). (§101)

P5. Peierls, Proc. Roy. Soc. **149**, 467 (1935). (§105)

P6. Perrin and Elsasser, J. de phys. et rad. **6**, 194 (1935). (§57)

P6a. Polanyi and Wigner, Zeits. f. physik. Chemie **139**, 439 (1928). (§51)

P7. Pollard, Proc. Leeds Phil. Lit. Soc. **2**, 324 (1932). (§99)

P8. Pollard and Brasefield, Phys. Rev. **50**, 890 (1936). (§99)

P9. Pollard and Brasefield, Phys. Rev. **51**, 8 (1937). (§99)

P10. Pollard, Schultz and Brubaker, Phys. Rev. **51**, 140 (1937). (§99)

P11. Pool and Cork, Phys. Rev. **51**, 383 (1937). (§101, 105)

P11a. Pool and Cork, Phys. Rev. (Abst. 66, Washington Meeting, 1937). (§101, 105)

P11b. Pool, Cork and Thornton, Phys. Rev. **51**, 890 (1937).

P12. Pose, Physik. Zeits. **30**, 780 (1929). (§99)

P13. Pose and Diebner, Zeits. f. Physik **90**, 773 (1934). (§74)

P14. Powers, Fink and Pegram, Phys. Rev. **49**, 650 (1936). (§61, 62)

P15. Preiswerk and Halban, Comptes rendus **201**, 722 (1935). (§102, 105)

R1. Rabi, Phys. Rev. **43**, 838 (1933). (§65)

R2. Rasetti, Zeits. f. Physik **78**, 165 (1932). (§99)

R3. Rasetti, Zeits. f. Physik **97**, 64 (1935). (§57, 90, 93, 94, 102)

R4. Rasetti, Segrè, Fink, Dunning and Pegram, Phys. Rev. **49**, 104 (1936). (§58, 61, 102)

R5. Rasetti, Mitchell, Fink and Pegram, Phys. Rev. **49**, 777 (1936). (§58, 60)

R5a. Rasetti, *Elements of Nuclear Physics* (Prentice Hall, 1936). (§69, 88)

R6. Richardson, Phys. Rev. **49**, 203 (1936). (§105)

R6a. Richardson and Emo, Phys. Rev. (Abst. 86, Washington Meeting, 1937). (§103)

R7. Richardson and Kurie, Phys. Rev. **50**, 999 (1936). (§101, 105)

R7a. Ridenour, Henderson, Henderson and White, Phys. Rev. (Abst. 78, Washington Meeting, 1937). (§99)

R8. Riezler, Proc. Roy. Soc. **134**, 154 (1932). (§75)

R9. Riezler, Ann. d. Physik **23**, 198 (1935). (§75)

R9a. Risser, Phys. Rev. (Abst. 79, Washington Meeting, 1937). (§102, 105)

R10. Rose and Bethe, Phys. Rev. **51**, 205 (1937). (§83)

R10a. Rose and Bethe (Errata), Phys. Rev., in press. (§91)

R11. Rose, Phys. Rev. (Abstract 135, Washington Meeting, 1937). (§90)

R12. Rosenblum and Chamie, Comptes rendus **194**, 1154 (1932). (§69)

R13. Rosenblum, Ann. d. Physik **10**, 408 (1928). (§95)

R14. Rosenblum, Comptes rendus **202**, 1274 (1936). (§69)

R14a. Rosenblum, Guillot and Perey, Comptes rendus **204**, 175 (1937). (§69)

R15. Rotblat, Nature **136**, 515 (1935). (§102, 105)

R16. Rotblat, Nature **138**, 202 (1936). (§102)

R17. Ruark and Fussler, Phys. Rev. **48**, 151 (1935). (§68)

R17a. Ruben and Libby Phys. Rev. **51**, 776 (1937). (§60)

R18. Rüchardt, Ann. d. Physik **71**, 377 (1923). (§95)

R19. Rumbaugh and Hafstad, Phys. Rev. **50**, 681 (1936). (§101)

R20. Rusinow, Physik. Zeits. Sowjetunion **10**, 219 (1936). (§65, 85, 102)

R21. Rutherford, Chadwick and Ellis, *Radiation from Radioactive Substances* (1930). (§94, 99)

R21a. Rutherford, Phil. Mag. **47**, 277 (1924). (§95)

R22. Rutherford, Lewis, Bowden, Proc. Roy. Soc. **142**, 347 (1933). (§69, 95)

R23. Rutherford, Wynn-Williams, Lewis and Bowden, Proc. Roy. Soc. **139**, 617 (1933). (§69, 92, 95)

R24. Rutherford and Chadwick, Phil. Mag. **42**, 809 (1921). (§99)

R25. Rutherford and Chadwick, Phil. Mag. **4**, 605 (1927). (§74)

R26. Rutherford, Phil. Mag. **21**, 669 (1911). (§70)

S1. Sagane, Phys. Rev. **50**, 1141 (1936). (§101, 105)

S2. Sampson, Ridenour and Bleakney, Phys. Rev. **50**, 382 (1936). (§105)

S3. Sampson and Bleakney, Phys. Rev. **50**, 732 (1936). (§105)

S4. Savel, Comptes rendus **198**, 368 (1934). (§99)

S5. Savel, Ann. d. Physik **4**, 88 (1935). (§99)

S6. Savel, Comptes rendus **198**, 1404 (1934). (§99)

S7. Schnetzler, Zeits. f. Physik **95**, 302 (1935). (§99)

S8. Schüler, Zeits. f. Physik **66**, 431 (1930). (§83)

S8a. Schultz, Phys. Rev. (Abst. 130, Washington Meeting). (§99)

S9. Segrè, Nuovo Cimento **12**, 232 (1935). (§94)

S10. Sexl, Zeits. f. Physik **81**, 163 (1933). (§66)

S11. Sexl, Zeits. f. Physik **56**, 62 and 72: **59**, 579 (1929). (§66)

S12. Sizoo and Koene, Physica **3**, 1053 (1936). (§105)

S13. Sloan and Coates, Phys. Rev. **46**, 539 (1934). (§92, 104)

S14. Sloan, Phys. Rev. **47**, 62 (1935). (§92)

S15. Snell, Phys. Rev. **49**, 555 (1936). (§101, 105)

S16. Snell, Phys. Rev. **51**, 142 (1937). (§101, 105)

S16a. Snell, Phys. Rev. (Abst. 69, Washington Meeting, 1937). (§99, 101, 105)

S17. Soden, Ann. d. Physik **19**, 409 (1934). (§51)

S18. Sosnowski, Comptes rendus **200**, 391 (1935). (§101)

S19. Sosnowski, Comptes rendus **200**, 1027 (1935). (§105)

S20. Sosnowski, Comptes rendus **200**, 922 (1935). (§102, 105)

S21. Stegmann, Zeits. f. Physik **95**, 72 (1935). (§99)

S22. Stetter, Zeits. f. Physik **100**, 652 (1936). (§99)
S23. Steudel, Zeits. f. Physik **77**, 139 (1932). (§99)
S24. Stone, Livingston, Sloan and Chaffee, Radiology **24**, 153 (1935). (§92)
S25. Sugden, Nature **135**, 469 (1935). (§102, 105)
S26. Szilard and Chalmers, Nature **134**, 494 (1934). (§92, 103)
S27. Szilard and Chalmers, Nature **134**, 462 (1934). (§94)
S28. Szilard and Chalmers, Nature **135**, 98 (1935). (§87, 105)
S29. Szilard, Nature **136**, 950 (1935). (§57, 102)
T1. Taylor and Goldhaber, Nature **135**, 341 (1935). (§94)
T2. Taylor, Proc. Phys. Soc. London **47**, 873 (1935). (§102)
T3. Taylor and Dabholkar, Proc. Phys. Soc. London **48**, 285 (1936). (§102)
T4. Taylor and Mott, Proc. Roy. Soc. **142**, 215 (1933). (§88)
T5. Taylor and Mott, Proc. Roy. Soc. **138**, 665 (1932). (§88)
T6. Taylor, Proc. Roy. Soc. **136**, 605 (1932). (§73, 74)
T7. Taylor, Proc. Roy. Soc. **134**, 103 (1931). (§74)
T8. Teller, Phys. Rev. to be published shortly? (§78)
T9. Thomson and Saxton, Phil. Mag. **23**, 241 (1937). (§104)
T10. Thornton, Phys. Rev. **49**, 207 (1936). (§101, 105)
T11. Thornton and Cork, Phys. Rev. **51**, 383 (1937). (§101, 105)
T12. Traubenberg, Eckardt and Gebauer, Naturwiss. **21**, 26 (1933). (§92, 100)
T13. Traubenberg, Eckardt and Gebauer, Zeits. f. Physik **80**, 557 (1933). (§100)
T14. Tuve and Hafstad, Phys. Rev. **45**, 651 (1934). (§101)
T15. Tuve and Hafstad, Phys. Rev. **48**, 106 (1935). (§101)
T16. Tuve, Hafstad and Dahl, Phys. Rev. **48**, 315 (1935). (§92)
U1. Urey, Brickwedde and Murphy, Phys. Rev. **40**, 1 (1932). (§101)
U2. Urey and Teal, Rev. Mod. Phys. **7**, 34 (1935). (§101)
V1. Van Atta, Northrup, Van Atta and Van de Graaff, Phys. Rev. **49**, 761 (1936). (§92)
V2. Van Atta, Van de Graaff and Barton, Phys. Rev. **43**, 158 (1933). (§92)
V3. Van de Graaff, Compton and Van Atta, Phys. Rev. **43**, 149 (1933). (§92)
V4. Van de Graaff, Phys. Rev. **38**, 1919 (1931). (§92)
V5. Van Vleck, Phys. Rev. **48**, 367 (1935). (§57)
V6. Van Voorhis, Phys. Rev. **49**, 889 (1936). (§101, 105)
V7. Van Voorhis, Phys. Rev. **50**, 895 (1936). (§101, 105)
V8. Veldkamp and Knol, Physica **4**, 166 (1937). (§102, 105)
W1. Walke, Phys. Rev. **51**, 143 (1937). (§99, 101, 102, 105)
W1a. Walke, Phys. Rev. (Abst. 70, Washington Meeting, 1937). (§99, 101, 102, 105)
W2. Ward, Wynn-Williams and Cave, Proc. Roy. Soc. **125**, 715 (1929). (§94)
W3. Waring and Chang, Proc. Roy. Soc. **157**, 652 (1936). (§99)
W4. Webster, Proc. Roy. Soc. **136**, 428 (1932). (§99)
W5. Weekes, Livingston and Bethe, Phys. Rev. **49**, 471 (1936). (§58, 102)
W6. Weisskopf and Wigner, Zeits. f. Physik **63**, 54 (1930). (§52)
W7. Weisskopf, Phys. Rev., in press. (§54, 56, 65, 79)
W8. Weizsäcker, Zeits. f. Physik **96**, 431 (1935). (§110)
W9. Weizsäcker, Naturwiss. **24**, 813 (1936). (§87)
W10. Wentzel, Zeits. f. Physik **38**, 518 (1926). (§66)
W11. Wenzel, Zeits. f. physik. Chemie **90**, 754 (1934). (§75)
W12. Wertenstein, Nature **133**, 564 (1934). (§99, 105)
W13. Wheeler, Phys. Rev. to be published shortly, (§73)
W13a. White, Henderson, Henderson and Ridenour, Phys. Rev. (Abst. 77, Washington Meeting). (§99)
W14. Whitmer and Pool, Phys. Rev. **47**, 795 (1935). (§104)
W15. Wick, Phys. Rev. **49**, 192 (1936). (§59)
W16. Wigner and Breit, Phys. Rev. **50**, 1191 (1936). (§101)
W17. Wigner, Phys. Rev. **51**, 106 (1937). (§87, 110)
W18. Wigner, Phys. Rev. **51**, 947 (1937). (§105, 110)
W19. Wilson, Phil. Trans. Roy. Soc. **193**, 289 (1913). (§94)
W20. Williams, E. J., Proc. Roy. Soc. **135**, 108 (1932). (§51, 97)
W21. Williams, J. H., and Wells, Phys. Rev. **50**, 187 (1936). (§100)
W21a. Williams, J. H., Wells, Tate and Hill, Phys. Rev. **51**, 434 (1937). (§100)
W21b. Williams, J. H., Shepherd and Haxby, Phys. Rev. ·bst. 68, Washington Meeting, 1937). (§101)
W21c. Williams, J. H., Shepherd and Haxby, Phys. Rev. **51**, 888 (1937)
W22. Wright, Proc. Roy. Soc. **137**, 677 (1932). (§72)
W23. Wynn-Williams, Proc. Roy. Soc. **132**, 295 (1931). (§94)
Y1. Yost, D. M., Ridenour and Shinohara, J. Chem. Phys. **3**, 133 (1935). (§101)
Y2. Yost, F. L., Wheeler and Breit, Phys. Rev. **49**, 174 (1936). (§72)
Z1. Zeleny, Brasefield, Bock and Pollard, Phys. Rev. **46**, 318 (1934). (§92, 104)
Z1a. Zahn, Phys. Rev., in press. (§61)
Z2. Zinn and Seeley, Phys. Rev. **50**, 1101 (1936). (§92, 101)
Z3. Zyw, Nature **134**, 64 (1934). (§99, 105)

NUCLEAR PHYSICS

Part C

Nuclear Dynamics, Experimental

REVIEWS OF MODERN PHYSICS

VOLUME 9 — JULY, 1937 — NUMBER 3

Nuclear Physics

C. Nuclear Dynamics, Experimental*

M. STANLEY LIVINGSTON AND H. A. BETHE†

Cornell University, Ithaca, New York

TABLE OF CONTENTS

XV. Experimental Methods

IN this discussion of experimental methods the field will be divided into three sections, dealing respectively with the sources of nuclear projectiles, the properties of nuclear radiations and the detecting and recording instruments and methods.**

§92. Sources of Nuclear Projectiles

Artificial disintegration of elements was first accomplished by Rutherford in 1919 using the high speed alpha-particles from Ra C′ as projectiles, and more than 10 years passed before any other projectiles were considered. By 1930 several laboratories were working on the development of high voltage apparatus to produce high speed ions artificially. It was expected that energies sufficient to penetrate the force fields of nuclei would be required, i.e., several million volts, and the proton was considered the most likely possibility as a projectile. At this stage Gamow (G7) and Condon and Gurney (C32) introduced the quantum mechanical interpretation of natural α-decay according to which there exists a certain probability of penetration of a potential barrier by particles with considerably less energy than required to go over the top of the barrier. Cockcroft and Walton (C18) realized that this might make disintegrations possible with relatively low energy protons and performed the first successful experiment resulting in the disintegration of Li (C19). Lawrence, Livingston and White (L11) immediately checked these observations with the 1 MV protons then available with the magnetic resonance accelerator, and others soon followed, some using much lower proton energies and proving the validity of the hypothesis.

At the present time the projectiles that have been successfully used for nuclear disintegration are: the *alpha-particle*, artificially accelerated *protons* and *deuterons*, *neutrons* occurring as products of other nuclear reactions, *gamma-rays* and accelerated ions of *He* and *Li*.

A. Natural radioactive sources

Alpha-particles are obtained from high intensity prepared sources of many radioactive materials. Of considerable importance has been the use of polonium (Ra F) since this material can be electrochemically separated from parent materials which are beta- and gamma-active, has no radioactive product and very weak gamma-radiation, and so supplies a pure source of alphas. They have a mean range of 3.805 cm at 15°C and 76 cm Hg pressure and a corresponding energy of 5.303 MV (R23). Higher energy alphas, such as those from Ra C′ with a mean range of 6.87 cm and an energy of 7.6802 MV (B60) have been used for much of the work on disintegration. Lower energies can be obtained by the use of absorbing material, either in the form of thin foils or as gases. The range-energy relations for alpha-particles are well known, both experimentally and theoretically (§95), and also the straggling after absorption (§97), so that the homogeneity of the alpha-beam is known for all energies. The ranges have been measured accurately with ionization chambers (L23) and the energies by magnetic analysis (B60, R23). In addition, alpha-particles have been used in experiments on nuclear scattering (Chap. XII), the results of which can be interpreted to show the extent and something of the character of the nuclear force fields.

The advantages due to the simplicity of naturally radioactive sources are unmistakable but the technique has always been handicapped by the low intensities available. Sources have seldom been used which yield more than 1.8×10^9 alphas/second (equivalent to 500 millicuries of radon), and observations must be made on small numbers of disintegration particles over extended times.

The *gamma-radiation* from naturally radioactive materials has been, up to a very recent date, the only source of photons of energies

* Part A of this report ("Stationary States of Nuclei," by H. A. Bethe and R. F. Bacher) appeared in these Reviews in the issue of April, 1936; part B ("Nuclear Dynamics, Theoretical") in the April, 1937 issue. The present part contains also the references to part B while part A contains its own reference section.

† The authors are greatly indebted to Drs. E. J. Konopinski and M. E. Rose for help in computing tables, drawing figures and especially for a critical revision of the manuscript.

** The descriptive material of this chapter is of necessity limited to include only a few of the many experimental devices and techniques now in use in nuclear physics laboratories. The discussion deals with those considered to be the most representative and those which have been most productive of experimental data.

greater than 1 MV. The commonly known radium source has a radiation consisting of many lines with an average energy of 1.5 MV and so is considerably less useful than Th C″ which has a single strong line of 2.62 MV and relatively small intensities of other energies. Other radioactive materials have their characteristic rays, although in general with less intensity due to the small amounts of radioactive material concentrated in units.

Natural gamma-ray energies have been accurately measured through the β-ray lines due to internal conversion (cf. §88). Unfortunately this method cannot be applied to the gammas from disintegration processes in light elements and other techniques are required. In §93 these methods are discussed in some detail.

As disintegrating agents natural gamma-rays have been successful for only two targets, H^2 (C9) and Be^9 (S28), requiring energies of greater than 2.20 and 1.74 MV in the two cases. Similar disintegrations of other elements require higher energies than are available from natural sources (see §103). Certain disintegration processes produced with high speed ions in apparatus to be described later yield gamma-rays with energies up to 17.5 MV, usually either monochromatic or having a definite line structure. Another promising source is from certain of the induced radioactive materials, such as Na^{24}, in which the ejected electron is accompanied by gamma-radiation of about 3 MV, and already produced with intensities equal to weak radioactive sources. X-rays of about 1.5 to 2.0 MV energy have been used to produce disintegration of Be (B49), but higher energies are not easily obtained.

Neutrons are emitted as disintegration products of certain light elements such as Be^9, Li^7 and B^{11} on alpha-particle bombardment. They are also ejected from H^2 and Be^9 by gamma-radiation. In fact the discovery of the neutron occurred as a result of the studies of the highly penetrating radiations from Be on alpha-bombardment. The neutrons from this reaction have a wide spread of energies ranging up to 13.0 MV for Rn alphas and with an intensity maximum at about 4.8 MV (D20) (see §99). A standard type of radioactive neutron source is a sealed capsule containing finely divided Be and Rn gas, which is alpha-active with a half-life of 3.8 days. With 1 millicurie of Rn about 27,000 neutrons per second are obtained (A11) and this yield varies roughly with the 3/2 power of the energy of the alpha if obtained from other sources than Rn. Gamma-rays from the Rn decay products accompany the neutrons from this source and it must be properly shielded in experiments where such gamma-rays would bias the interpretation of the results. Po is sometimes used where such radiation is disadvantageous. For many experiments the small size of such a source makes observations exceedingly simple and effective due to the large solid angle available for targets.

Alphas on B produce neutrons of a different energy distribution, in this case with a maximum intensity at <1 MV and an upper limit of 4.2 MV (M34). Sources of neutrons of different energies are valuable for analyzing certain processes and especially in determining reaction cross sections as a function of energy. The inhomogeneity of neutron energies from the reactions noted is of considerable disadvantage, and in order to avoid this use has been made of neutrons from other processes in which energies are homogeneous. Chief of these is the reaction of deuterons on deuterium, to be discussed in detail later.

B. Condenser-rectifier voltage multiplier

Of the several types of high voltage ion sources in use for nuclear research at present the first to produce disintegrations and explore the field of proton-produced reactions was that of Cockcroft and Walton (C22) in the Cavendish Laboratories, where earlier Rutherford had opened the door to such research with his alpha-produced disintegrations. It has undergone many improvements since 1932 but still embodies the voltage-multiplier circuit originally used. In this design the "voltage doubling" principle of charging condensers in parallel and discharging in series across a load has been extended to triple the voltage output from a transformer of standard design supplying about 250 kv of 60 cycle alternating current. The use of rectifier tubes as switches provides the method with a means of charging the condensers 60 times a second and supplies an essentially constant direct potential of about 800 kv across the final bank of 3 condensers. The condensers are mounted on suitable insulating supports. The rectifier tubes are of special design and are continuously pumped, with the filaments heated by insulated transformers. In operation the transformer charges a feeding condenser through a pair of rectifiers which are conducting during one half-cycle. In the succeeding half-cycle these rectifiers are nonconducting and the feeding condenser divides its charge with another through a second set of rectifiers, and so on until all condensers are fully charged.

The high voltage discharge tube in which the ions are accelerated is in two sections for voltage distribution. (Single sections have never been operated satisfactorily with potentials above 400 kv.) The tube was originally built of large glass cylinders but these have been replaced with porcelain (C25); joints are sealed with wax and the chamber is continuously pumped. Canal-ray sources of hydrogen ions are placed at the high potential end of the tube and the ions are accelerated through large tubular electrodes and strike a target placed at the grounded end. The original source produced up to 10 microamperes of ions at 800 kv but recent improvements (C25) have increased this yield to 100 microamperes. Voltages are determined with sphere-gap sparking measurements and although calibration details are not published they are probably accurate to better than 5 percent. A magnetic deflection chamber has been added (C27) to separate the H^1 and H^2 focal spots of the ion beam and these measurements are used to check the sphere gap voltage calibration. Fluctuations in the supply voltage are less than 20 kv, introducing an error of 2.5 percent. The focal spot on the target is 1 cm in diameter and with the magnetic deflecting field the ion energies can be considered as essentially homogeneous.

C. Cascade transformer

The development of high voltage transformers for transmission line testing and as power sources for deep therapy x-ray tubes led to the "cascade transformer" potential source used by Lauritsen and Crane and their collaborators (L9) at the Kellogg Radiation Laboratory of the California Institute of Technology in Pasadena. Three transformers are used, each producing about 350 kv. They are arranged in "series"; the primaries of the second and third transformers are activated from low voltage tertiaries at the high potential ends of the high voltage windings of the first and second transformers respectively. The second and third transformers are mounted on platforms of porcelain insulators so they may be operated at 350 and 700 kv above ground potential. This yields a 60-cycle voltage of about 1.0 MV root mean square between ground and the high end of the third transformer. The alternating character of the voltage gives potentials both positive and negative with respect to ground and so can be used to accelerate either electrons or positive ions, depending on the type of source used. The source supplies the rectification and when used with positive ions the stray electrons accelerated in the odd half-cycles give a small intensity of x-rays from the target.

The discharge tube is in two sections, using two standard conical transformer bushings of porcelain of the type used on commercial 220 kv transformers and having a 750 kv sparkover limit. The porcelains are sealed end on end with wax to metal plates which also support the interior accelerating tubes, and are evacuated with high speed pumps. The accelerating electrodes are tubular and re-entrant so that the porcelain wall is protected from discharge originating in the ion beam. The intermediate electrode is connected to the mid-point of the voltage supply for distribution of potential. Ions are supplied by a hydrogen discharge tube at the high potential end of the accelerator and are focused by the accelerating voltage to a spot of less than 1.2 cm diameter on the target. Steady currents of as much as 20 microamperes of hydrogen ions are obtained up to 900 kv, and 100 microamperes have been obtained for short time intervals synchronized with a cloud chamber for certain observations. It is to be expected that different types of ions, such as H^1, H_2 and H^2 will be accelerated simultaneously and that a spread of energies will result due to the sinusoidal character of the voltage supply. Magnetic focusing is not used but tests indicate that there is a considerable predominance of high energy ions, and that with proper care in purity of gases entering the source the light and heavy hydrogen beams can be kept relatively pure.

Maximum ion energies are determined by spark gap measurements of voltage, based on the Westinghouse high voltage calibrations and are subject to some criticism. Corona from terminals and wires at this high potential (easily visible) introduces fluctuations estimated to be about 5 percent at 900 kv but considerably less for lower voltages. Data of Hafstad and Tuve (H2) on the resonance values of proton energies in the "simple capture" by C do not check with similar values given by Lauritsen, showing a discrepancy between the voltage calibrations of the two laboratories by as much as 30 percent. Many experiments, however, do not require an accurate voltage calibration, and such is the case for much of the published data from this laboratory. Most voltage excitation functions of nuclear processes are steeply-rising curves with increasing voltage and the experimenters take advantage of this fact by subtracting out the small effects due to ions of less energy or of complex character.

The target is ideally located at the end of a long tube through which the focused ion beam travels, and can readily be shielded with lead or paraffin. Due to the method of locating the target this type of apparatus has been used more successfully than other types for measurement of emitted gamma-radiation.

A recent installation by Crane (C52) yields 1 MV in 5 sections and with an ion intensity of 250 microamperes.

D. Electrostatic generator

A simplified method for the production of high electrostatic voltages involving a large sphere charged by a rapidly moving belt was developed by Van de Graaff from a small apparatus capable of delivering a few microamperes at 80,000 volts (V4) to the present Round Hill installation (V1) yielding several milliamperes at 5 MV. During this time a paralleling development of the application of this method to discharge tubes has been carried on by Tuve, Hafstad and Dahl (T16) in the Bureau of Terrestrial Magnetism of the Carnegie Institution at Washington. They have placed the emphasis on the development of a discharge tube and other operational details with the result that positive ion currents of 20 microamperes at energies up to 1.2 MV have been available for experiments. With this technique a large sphere, or rather a cylinder with spherical caps over the ends, of 2 meters diameter, is charged by means of a belt on which charge is sprayed by a high voltage rectifier unit. Insulation for the high potentials is achieved by mounting the sphere on Textolite supports. The multiple-section discharge tube contains about 14 hollow cylindrical electrodes supported at equal intervals along the length of the glass tube wall and each equipped with an external doughnut-shaped corona shield. These shields are varied in curvature and points are adjusted between them where needed to distribute the sphere voltage uniformly between the electrodes. This design protects the discharge tube and the voltage limit is set by the corona from the surface of the sphere or by sparks down the belt or across the gaps of minimum clearance.

When used to accelerate positive ions the source is placed in the sphere which is maintained at a high positive potential by spraying positive charge on the belt. Since it is possible to utilize the interior as well as the exterior curvature of the sphere for field distribution a smaller sphere is mounted concentric with and inside the 2 meter one and is maintained at a positive potential with respect to it. Two accelerating sections are placed between the two shells and by means of corona points which are variable by means of long insulating strings the potentials between these two sets of electrodes can be varied to adjust the focusing of the ion beam. The hydrogen discharge tube used as a source of ions is inside the central sphere where gas flow and potentials are also regulated by strings. A generator to supply the power for operation of the discharge is driven by an auxiliary belt.

The focusing of the ions is determined chiefly by the first two accelerating stages mentioned above, and is maintained down the long column of electrodes by the accelerating fields between them. This produces a focal spot at the target of 3 to 4 mm in diameter. A magnetic field deflects the beam, analyzes it into the component ions and maintains the energy to within a maximum spread of 20,000 volts. Both H^1_2 and H^2 ions reach the target when the mass 2 spot is in focus and so a subtraction technique is required to analyze results for mixed gases. Due to irregularities in the corona discharge and other conditions the high voltage itself has a fluctuation of about 20,000 volts, and this varies greatly with changing atmospheric conditions.

Proton energies were at first determined by direct range measurements on the primary beam for energies above 500 kv. The ions passed through a thin foil and out into air where the range was measured visually or electrically. In the first instance the blue ionization glow was found to have a relatively sharp terminus, enabling estimates of range to within 1 or 2 mm. The voltage fluctuations of the source resulted in a flicker of intensity in the last few mm of range. A shallow (1 mm deep) ionization chamber and electrometer were used to measure the range electrically, but due to the straggling of the ions and the extreme sensitivity of the electrical method a range was observed of some 2 mm greater than the visual one. Due to doubts as to the reliability of the then existing range-energy relations an assumed (H4) 10 mm range for 500 kv and 36 mm for 1200 kv protons was chosen in stating ion energies up to August, 1935. More exact range-energy relations are given in §95 and where used later in this discussion the data from this laboratory are corrected by the proper factors. A fair estimate of the probable error in the measurement of proton ranges is 2 mm, equivalent to 50 kv at 1 MV and 80 kv at 500 kv. Below 500 kv sphere gap readings were used to calibrate voltages and possible errors in estimating the potential between inner and outer spheres result in an error which may be as high as 50 kv. A recent advance (H5) (June, 1936) in the accuracy of voltage calibration has resulted from current measurements through standard high resistances. The present calibrations, which check closely with the range-energy curves of §95 are 23 mm for 1.0 MV and 29 mm for 1.2 MV protons.

A modification by Herb, Parkinson and Kerst (H28) consists of the use of higher pressures to increase the voltage limits of the electrostatic generator. The cylindrical generator is mounted inside a tank 20 ft. long and $5\frac{1}{2}$ ft. in diameter which contains air at 7 atmospheres pressure. Ions have been obtained of up to 2.16 MV energy through a discharge tube extending longitudally along the cylinder. The success of this method suggests that it will be of importance for future high voltage experiments.

E. Low potential-high intensity discharge tubes

Certain disintegration reactions in light elements such as the disintegration of Li by protons can be studied with relatively low energy ions if sufficient intensity is available. Several laboratories have developed low voltage techniques for this purpose, of which that of Oliphant (O2) and his collaborators is an example. An intense discharge tube source yields 100 microamperes of protons or deuterons which are accelerated to 200 kv in a single stage tube supplied by a voltage doubler circuit and a 100 kv transformer. A 90° magnetic analyzer separates ions of different masses and energies. The voltage standard is a calibrated spark-gap, checked by the magnetic deflection of the ions, and is accurate to better than 1 percent. The voltage output was later raised to 400 kv by changes in the apparatus (O7) and the intensity increased to 200 microamperes.

The value of this type of apparatus lies in the accuracy with which the energies of the bombarding projectiles can be determined. A 1 percent error at 100,000 volts represents only 0.000001 mass units, negligible in comparison with other errors. So reaction energies can be obtained to within the accuracy of the measurements of the energies of the ejected particles. An added virtue is the freedom from scattered primary ions due to their very short range, which means that very short range groups of product particles can be detected. The method is of considerable importance in accurate studies of reaction energies. It is also valuable as a means of obtaining the neutrons of homogeneous energy resulting from the reaction $H^2+H^2 \rightarrow He^3+n^1$. An apparatus developed by Ladenburg, *et al.* (L1), produces as many neutrons as a 260 mC Rn-Be source, at 200 kv and 100 microamps. Zinn and Seeley (Z2) use 4 milliamperes of ions at 60 kv to get neutrons equivalent to 500 mC of Rn and Be.

Extremely low voltage disintegration of lithium has been realized by Traubenberg, *et al.* (T12) who observed disintegrations by protons of as low as 13 kv, using ion currents of 1 milliampere. Burhop (B68) has disintegrated Li and H^2 with 8 kv ions.

Zeleny (Z1) and his associates at Yale University have accelerated Li ions to such energies (240 kv) that they yield alphas when bombarding hydrogen. This is the same reaction as is observed when protons bombard a lithium target, and yields can be calculated from the relative velocities of the two ions.

F. The magnetic resonance accelerator

The highest energies obtained by artificial acceleration of ions have been achieved with the magnetic resonance accelerator or "cyclotron" of Lawrence and Livingston (L14) of the Radiation Laboratory of the University of California. In this method ions are brought to high energies not by high voltages but by many successive small accelerations in a low voltage, high frequency field with which they are in resonance. The ions revolve in a series of increasing semi-circles in a large and uniform magnetic field and receive an increment of energy each time they cross a diametral gap between two semi-circular hollow electrodes on which the high frequency is impressed. The angular velocity of ions moving in a uniform magnetic field is constant for all energies, which means that the time for the traversal of a circular path is constant, regardless of the radius of the path. If the magnetic field is such that this time interval is the period of the imposed high frequency oscillations the ions remain in "resonance" until they achieve an energy determined only by the strength of magnetic field and the radius of the final path. Ions are produced near the center of the chamber in hydrogen or deuterium gas by an electron beam from a filament. The magnetic field collimates the electron beam so that it is concentrated on the region between the electrodes where the source is desired. The ions receive about 100 to 200 accelerations of 30 to 60 kv each to reach the final energy, which is about 6 MV for H^2 atomic ions (deuterons). The normal intensities are 10 to 20 microamperes (L18), but recent changes involving increased electrode aperture have resulted in as much as 50 microamperes at somewhat lower energy (4 to 4.5 MV). A magnetic analysis of mass and velocity of the ions is inherent in the resonance principle, so only ions having the same e/m can be accelerated. H_2 and H^2 ions have the same e/m but widely different mean free paths, and if the gas pressure in the chamber is high enough all the molecular ions will be absorbed and only deuterons will reach the collector, so purity of gases is not essential. The energy spread is equal to twice the voltage between the electrodes and can be made as small as desired at the expense of beam intensity. In practice this spread is usually less than 2 percent of the ion energy. This condition is obtainable only with small sources and uniform magnetic fields; without these features much wider energy distributions are sometimes observed and must be corrected by "shimming" (see below).

Energies are calculable from the resonance value of magnetic field and the radius of the final path and are accurate to the limits stated above only if the ion paths are concentric about the center of the chamber. Iron "shims" near the periphery are used to increase the region of uniformity of the magnetic field. They also help to correct for any asymmetrical properties of the magnetic field which would result in a migration of the ion paths and so a lower final energy. Due to this feature energies are usually determined by the range of the ion beam in air, which involves a possible error of about 2 mm. In the original arrangement the focal spot was about 4 by 6 mm and targets were located opposite a large observation port just inside the edge of the magnetic field. For experiments on induced radioactivity the re-entrant observation port was placed deep enough to allow the ions to pass inside through a foil mounted on the side of the port; targets were activated at atmospheric pressure and removed for observation.

Recent improvements in the accelerating chamber (L18) allow the ion beam to be brought out tangentially by means of electrostatic deflecting fields. The beam passes through a thin window and emerges into the air where it can be observed visually as a blue ionization glow having a range of some 26 cm (6 MV). Still further development (A6) has resulted in bringing a fraction of this beam through an evacuated tube to a point 6 feet from the cyclotron.

Several other installations are now completed

or in progress. Small compact models yielding 1 to 2 MV are in operation at Cornell University (L31) and the University of Illinois (K26). A record high voltage of 6.7 MV has been obtained at the University of Michigan and results of several experiments have been reported (C36). Preliminary experiments using cyclotrons have recently been reported from the University of Rochester (B4a) and Princeton University (W13a, D29). Several other cyclotrons in this country and abroad are nearing completion.

Very intense yields of neutrons are obtained from high voltage apparatus by bombarding Li, Be, B, etc., by high speed deuterons. These processes have probabilities proportional to the intensity of the deuteron beam but varying approximately with the 3/2 power of the energy above a characteristic threshold value in each case, so the most intense neutron sources have been obtained with the magnetic accelerator. An estimate made by Lawrence of the total number of neutrons is 10^9 to 10^{10} per second.[1]

G. Other sources—general

Several other types of apparatus for accelerating ions have been successful in disintegration experiments to some extent, and others are at present under development. The "linear accelerator" of Sloan and Coates (S13) produced 2.8 MV Hg^+ ions by a radiofrequency resonance method similar to that of the magnetic accelerator, but no disintegrations were observed. An application of this method to Li ions by Kinsey (K13) has resulted in the observation of the $Li^7+H^1\rightarrow He^4+He^4$ reaction. Van Atta, *et al.* (V2) are at present engaged in developing a discharge tube for the large Van de Graaff generator (V3). A surge generator supplying a low vacuum discharge tube developed by Brasch, *et al.* (B48) yields great intensities of a variety of ions but has not been brought under control except for disintegrations by x-rays as previously indicated. Beams and Trotter (B6) have accelerated protons to over a million volts energy in an apparatus utilizing transmission lines to supply the accelerating electrodes, but intensities are still too small for experiments. The "resonance transformer" x-ray tube, developed by Sloan (S14, S24) has been applied by Livingood and Snell (L25) to the search for radioactivity produced by 0.8 MV electrons with no positive results.

§93. Properties of Nuclear Radiations

A. Alpha-particles

The methods whereby alpha-particles are used as projectiles for disintegration experiments have been discussed in §92. When occurring as products of nuclear reactions they are detected through the dense ionization they produce. Several instruments such as the cloud chamber and the pulse amplifier utilize this property and are sufficiently sensitive to detect individual particles (§94).

The specific ionization varies with velocity (the Bragg curve); about 2500 ion pairs/mm being formed by alphas having a residual range of 5 cm and 7000 ion pairs/mm at 3 mm from the end of the range. Alphas have been observed having ranges from a few mm to 14 cm in air at standard conditions. The range-energy relation is known to great accuracy (see Figs. 29(a)(b), §95), especially in the region calibrated by alphas of known energy from natural sources. Observations are usually made on the extrapolated range and are reduced to mean range for energy determinations (§97).

B. Protons

Protons produce less ionization along their tracks than alphas, about 32 percent of the specific ionization of the alpha at the maxima of the Bragg curves and 25 percent for residual ranges of more than 2 cm. This results in the cloud chamber in a fainter track, and is observed in the ionization chamber as a smaller current pulse for the same depth of penetration, making it possible to distinguish the two particles.

[1] It seems important to point out to prospective experimenters in the field that these intensities of neutrons are sufficient to produce harmful biological effects. The high concentration of hydrogen makes the body an unusually efficient absorber of neutrons and the ionization produced by recoil atoms has already been shown to have biological effects similar to those from gamma-radiation. Quantum-for-quantum the neutrons are more effective than gammas. The neutrons obtained from the magnetic accelerator are found to have the biological effectiveness of 100 g of radium. Thick baricades of hydrogenous materials are required to reduce neutron intensities materially (6 cm of paraffin absorbs about $\frac{1}{2}$ the fast neutrons), and the experimenter should arrange to stay at considerable distances from the apparatus.

Groups of protons are separated by observing only the stronger pulses in the amplifier, equivalent to counting only those protons which have the maximum of the Bragg curve within the ionization chamber. Protons have much longer ranges than alphas, those observed ranging from a few cm to as much as 200 cm in air, which last is equivalent to 13.65 MV. The range is again usually observed as the extrapolated value; see §97 for the relation between mean and extrapolated ranges. The range energy relation, as inferred from the alpha-particle curve, is given in Figs. 30(a)(b)(c) of §95. Accurate experimental determinations are scarce, but the evidence that does exist indicates that the relation is essentially correct.

C. Deuterons

Deuterons are of chief interest as projectiles, being accelerated to high speeds by several types of apparatus, as discussed in §92. In the one reaction yielding deuterons as products ($Be^9+H^1\rightarrow Be^8+H^2$) they could not be separated from alphas in deflection experiments due to their having the same e/m. Identification came through the observation of their lower specific ionization. The specific ionization is about 4/3 that for protons and about 1/3 that for alpha-particles for the same residual range. The range of a deuteron of energy E is twice that for a proton of half the energy. Using this relation deuteron energies can be determined from the proton range-energy curves of §95. (Figs. 30(a) (b) (c).)

D. Neutrons

Neutrons, having no charge, produce no effects in electrical counting instruments. They are detected either through recoil with atoms to form charged ions or through the disintegrations they produce.

Recoil protons from hydrogen have energies ranging up to the full neutron energy, depending upon the angle of projection. Only those projected in the direction of the incident neutrons are satisfactory for measurements, since neutrons may be scattered through large angles by adjacent heavy materials and yet retain a large share of their energy. Such neutrons, on recoil, will give high energy protons which, if interpreted as coming from nonscattered neutrons, will indicate excessively high energies. This feature is no doubt responsible for many erroneous measurements indicating high neutron energies. When care is taken on this point the recoil distribution-in-range curves show definite group structure and reasonably sharp end-points, but with somewhat less resolution than equivalent curves for a proton beam. In hydrogen or methane-filled cloud chambers the direction of the individual protons may be ascertained and this condition fulfilled.

With a paraffin layer in front of a shallow ionization chamber about 3.7 protons are projected from the surface per 1000 neutrons of 5 MV energy. The efficiency is approximately proportional to the energy; 0.35 proton results from 1000 neutrons of 0.5 MV energy. The paraffin should be equal in thickness to the equivalent range of the protons produced, for highest intensity; for best resolution of groups the paraffin should be thin compared to the proton range.

An alternative method of measuring neutron energies is through recoil with other atoms, such as C, O or N (F8). This is seriously handicapped by the uncertain range-energy relations for such particles, and data will probably serve best as calibrations of such relations when neutron energies are better known.

The other general property through which neutrons are observed and measured is the disintegration of certain target elements to yield observable radiations. Three types of disintegration serve this purpose; the n-α reaction is known to occur with slow neutrons only in Li and B, the n-p type reaction is endoergic except for N, while the simple capture reaction is always exoergic and often results in induced radioactivity.

The cross section for the $Li^6+n^1\rightarrow He^4+H^3$ reaction is about 900×10^{-24} cm^2, that for the $B^{10}+n^1\rightarrow Li^7+He^4$ process is 3000×10^{-24} cm^2. The range of the H^3 from Li is about 6 cm, and the He^4 from B only 0.8 cm, so the number of observable disintegrations from a solid target are about the same in the two cases. These substances are used to line the inside of ionization chambers in which the disintegration products are recorded (C14). With layers essentially equal to the range of the product particles, the number of particles entering the chamber far enough to be recorded is about 7/1000 neutrons for B and 10/1000 neutrons for Li. A still more efficient technique is the use of large ionization chambers filled with a B gas (C30). With BF_3 at atmospheric pressure in a chamber about 20 cm deep 200 counts/1000 incident neutrons are recorded if the neutrons are assumed to travel parallel to the axis of the chamber; the efficiency varies with the geometry of source and ionization chamber.

The simple capture reaction induces electron activity in many substances and may be used as a detector of neutrons through observation of these activities. Average cross sections are in general smaller than those mentioned for the particle processes but the simplicity of the technique and the recording devices (counters, electroscopes, etc.) have made this a useful alternative. With pressure ionization chambers (in which a large share of the electron ionization is observed) and sensitive electrometers Fermi and his group at Rome have been able to get sensitivities nearly as good as those obtained from counters, and better accuracies due to the more constant background.

Neutrons may be classified as fast or slow, on the basis of their widely different cross sections for the capture process. Slow neutrons may be further subdivided into those of thermal energies and of energies in the "resonance" region (see §60). Fast neutrons are slowed down by elastic and inelastic impacts with the nuclei in their paths. In hydrogenous materials they have, on the average, an energy $1/e^n$ of their initial energy after n collisions. Relatively few impacts are required to reduce neutron energies to the thermal range (about 20), and in the process they pass through the region of resonance energies (from 0.1 to 1000 volts). In paraffin the mean free path for a 5 MV neutron is about 5 cm; this rapidly decreases as the energy becomes lower until it reaches the value of 3 mm for those of thermal energy. So a paraffin sphere of about 6 cm radius results in a maximum density of slow neutrons near its surface. Thicker paraffin absorbs the fast neutrons in an exponential manner, resulting in a reduced number of slow neutrons. Fink (M21) has designed a slow neutron "howitzer" which, by taking advantage of the back scattering of neutrons starting away from the target and by reducing the absorption of those neutrons directed toward the target, produces slow neutron intensities of 3 to 5 times the intensity obtainable from a sphere of paraffin.

Certain simple capture reactions resulting in stable rather than radioactive products have very large cross sections, observed as absorption of slow neutrons. Cd, for example, has a cross section of 3000×10^{-24} cm^2, while Sm has the largest known value, of 9000×10^{-24} cm^2. Such substances are valuable as absorbers of slow neutrons; a layer of 0.5 mm of Cd is sufficient to remove all thermal neutrons from a beam. This makes it possible to study the disintegrations due to neutrons of "resonance" energies (see §60).

Thermal neutrons have an undirected motion resulting in diffusion through the paraffin. They pass and repass many times through a target immersed in the paraffin, which feature assists in producing the large efficiencies of slow neutron processes. This diffusion has been utilized in certain experiments in which the neutrons are "piped" through paraffin to a more distant point or around corners (H36). It makes it possible to shield targets or detecting instruments from the gamma-radiation of a Rn source with lead blocks, in which case the slow neutrons diffuse around the lead to the target (R3). The slow neutrons are finally absorbed, after covering a region of influence of about 3.5 cm radius, with the emission of photons following the reaction:

$$H^1 + n^1 \rightarrow H^2 + h\nu.$$

E. Gamma-radiation

The absorption of gamma-radiation by matter is a complex phenomenon, involving three processes of absorption. The photoelectric absorption effect, prominent for low energy x-rays, still persists but is relatively unimportant for energies above 1 MV (in lead). The absorption is known to follow an exponential law: $I/I_0 = ce^{-\tau x}$, where x is the thickness of material and τ is the photoelectric absorption coefficient, for a monochromatic beam of gamma-rays. Hulme, McDougall, Buckingham and Fowler (H38a) have calculated the magnitude of this coefficient and the variation with energy. The result is in accord with an empirical relation of Gray (G21). The second process, predominant for energies between 0.6 and 2.5 MV (in lead) is the "Compton" absorption, in which recoil electrons and scattered quanta are produced. The absorption coefficient, σ, is given by the equation of Klein and Nishina (K19). Both coefficients decrease with increasing energy, τ much more rapidly than σ. The third process is the absorption due to the formation of electron pairs and involves the transformation of part of the gamma-ray energy into the mass of the electrons. It does not occur for energies less than 2 mc^2 (electron mass), which corresponds to about 1.0 MV, and rises with increasing energy in a manner predicted by the theoretical work of Bethe and Heitler (B9) and others. In lead the absorption coefficient for this process, κ, is essentially equal to σ at 2.5 MV, and in aluminum at 10–12 MV. This leads to a minimum in the total absorption coefficient near these energies and means that an observed coefficient may indicate one of two energies, one greater and one less than the value at the minimum. Even though the measurement of the absorption in different materials gives a unique value of the energy (a technique used by certain experimenters (Mc4)) the variation of the coefficient with energy near the minimum is so slow that accurate results are not possible.

The three processes mentioned above involve different types of reduction of the energy, τ resulting in complete absorption of the quantum, σ producing only partial transformation into beta-rays and yielding scattered quanta of degraded energies while κ results in the production of positrons which, on annihilation, produce photons of 0.5 MV. Equilibrium of these secondary radiations with the primary gammas will be achieved only after considerable thickness of absorber, so thickness is an additional complicating factor, and the absorption is not truly exponential.

The most accurate method of measuring gamma-ray energies in the disintegration of light elements is through the observation of the high energy limit of the secondary electrons, or through measuring the total energy of pairs. Energies are measured through the curvatures of the electron paths in magnetic fields. The cloud chamber, equipped with Helmholtz coils, has been used most successfully for the observations, but magnetic spectrographs with coincidence counters have also been satisfactory. From the energy and momentum relations of the Compton effect it is found that the electron can have a maximum energy just 0.25 MV less that the energy of the gamma-quantum. Photoelectrons acquire essentially the full energy of the gamma-ray (less the K ionization limit of the element in which they are absorbed). The theory of pair production indicates that the gamma-quantum is completely absorbed, so the sum of electron and positron energies plus their mass (1.02 MV) is a measure of the initial energy. The thickness of absorber in which the secondaries are produced determines the resolution of the technique; gamma-ray energies are determined from the extrapolated upper limits of electron energy distributions.

F. Electrons and positrons

Nuclear electrons (β-particles) and positrons are observed with maximum energies ranging from 0.3 to 12 MV. An electron-active substance emits electrons having a distribution in energy from zero to a poorly defined maximum energy in each case—the characteristic continuous β-ray spectrum. Even monokinetic electrons do not have a well-defined range as do the heavier particles, but are strongly scattered in passing through an absorber, so that there is no distinct extrapolated range. Absorption of the electrons of a β-spectrum results in an almost exponential absorption curve, having no unique characteristic. Measurements of the half-value thickness (thickness of absorber to reduce intensities to half) have been used by certain experimenters as a measure of the electron energy. This can only be approximate since the distribution curves are not identical.

Magnetic analysis, as used for measurements of secondary electrons from gamma-rays, is at present the best technique for energy measurements. Beta-ray spectrographs and cloud chambers are used for this purpose, and only recently have there been reports of satisfactory measurements. The shape of the β-spectrum involves the theoretical problem of explaining the apparent loss of energy and has resulted in the hypothesis of the neutrino. Only the upper limit of the spectrum has any significance in the energy measurements, and attempts to explain the shape of the distribution and predict this end-point have met with only partial success (§105).

§94. Instruments for Detection and Observation

The first recorded physical evidence of nuclear radiations was the photographic effect of certain uranium salts, which led Becquerel, in 1896, to the discovery of natural radioactivity in this mineral. The electrical method of observation with an ionization chamber was initiated by Mme. Curie in 1898, and with electroscope and electrometer recording such instruments are still in use. In 1903 Sir William Crookes and also Elster and Geitel discovered that alpha-rays are capable of producing brilliant luminosity of a fluorescent screen, and led to the use of the scintillation technique for individual particles. Rutherford and Geiger, in 1908, developed the forerunner of the Geiger-Müller tube counter, now in wide use. With the development of the Wilson cloud expansion chamber starting from 1912 the tracks of individual ions have been observable. A most valuable development has followed from the use of vacuum tube amplifiers by Wynn-Williams to record the ionization of single charged particles.

A. The scintillation method

Although it has nearly outlived its usefulness in view of later developments of apparatus for recording individual high speed ions the scintillation screen was for years the most sensitive instrument for making such observations. Its use must necessarily be dependent on an observer and so introduces human fallibility and fatigue to some degree. It was successful, at least in the hands of the Cavendish Laboratory group, in yielding all the essential information about alpha-particle disintegration and scattering. Only high speed, heavy, charged ions are observable, such as the alpha-particle and the proton. In its usual form the screen consists of a thin layer of zinc sulphide or some other fluorescent salt mounted on a glass plate and observed on a dark field by a low power microscope. Observers must have their eyes dark-adapted and attempt observations only for relatively short periods to avoid errors due to fatigue. The scintillations are distinct and unmistakable with the eye in proper focus; the optimum size of field and frequency of counts are determined for the individual observer. As the end of the alpha-article range is approached the intensity of the scintillations fades gradually. Rutherford states that alphas of velocity $V=0.15V_0$ (V_0 is the initial velocity of the Ra C′ alpha) are not detectable, so that ranges obtained with this method are subject to variations between observers and are in general short. A thorough discussion of the technique and applications is contained in the book by Rutherford, Chadwick and Ellis (R21).

B. The cloud expansion chamber

Since its inception by Wilson (W19) the "cloud chamber" has undergone many modifications and at present exists in a variety of forms, each adapted to some special purpose. In general, a volume of gas containing a saturated vapor is adiabatically expanded to give a short interval of supersaturation. During this interval the ions produced along the paths of alpha-particles, protons or electrons traversing the chamber act as condensation nuclei for the vapor and under suitable illumination sharp tracks of droplets of liquid are observable.

The improvement of illumination and photographic technique has resulted in essentially perfect recording of these tracks. Stereoscopic photography, introduced by Wilson, makes it possible to calculate ranges and directions of the tracks, or they can be directly observed and measured by reprojection through the two films onto a suitably oriented translucent screen. A modification by Kurie (K27) replaces the two cameras with a single camera and lens system equipped with mirrors and a prism to take the two stereoscopic views on motion picture film. As a source of illumination the carbon arc is usually used, with an arrangement whereby a resistance in the power line can be momentarily shorted out to give an instant of intense illumination. Capillary mercury arcs (D1) and exploded wires are also satisfactory. For dense tracks (alphas or protons) "photoflood" or overvoltage lamps are successful. Recent improvements in technique have made this method satisfactory for electron tracks as well.

Mechanically the expansion is achieved in a variety of ways. A reciprocating piston to compress the gas slowly and expand it rapidly can be operated by a cam, by an electromagnet or by changing pressure in the chamber under the piston. Sylphon bellows may take the place of the piston and have their length altered suddenly by valving air under pressure. Rubber diaphragms with similar pressure regulation provide another satisfactory method. Excessive turbulence is avoided by maintaining a geometrical simplicity in the design, and with fine-mesh screens just below the active portion of the chamber. This region is usually several cm deep and, depending on the use, from a few inches to 20 inches in diameter. Most installations are operated cyclically by clockwork or pendulum and arranged to repeat the cycle at regular intervals, so spaced that temperature equilibrium is maintained.

With high pressures and large diameters the particle range observable in the chamber can be increased to include the whole of even the fastest proton tracks. The use of low pressures (down to a few cm of Hg) of gas inverts the results so that particles of small energy such as heavy atom recoils from disintegration or scattering processes can be observed. Ranges come directly from the known gas density and constitution and can be measured to a reasonable accuracy, depending upon the density of track, etc. The expansion ratio is determined from the known constants of the gas and vapor in the chamber which produce supersaturation. It is usually determined experimentally to give the sharpest definition of tracks. For water vapor in air this ratio is in the region 1.25 to 1.38, but is much lower for other vapors such as propyl alcohol, or in other gases such as He or A.

The cloud chamber has been used with great success in cosmic-ray studies by Anderson and by Blackett, and much of the technique developed for this work is directly applicable to disintegration studies. The use of the chamber to observe and measure the charged particle products of disintegration is obvious. When the gas in the chamber contains the target substance and the bombarding projectile enters the chamber the complete disintegration reaction is recorded. Neutron produced reactions are distinguishable as sharp angle tracks, often with definite differences in the density to indicate the particle and the recoil atom, and neutron energies and directions are calculable from the momenta of the particle tracks. There is still, however, insufficient knowledge of the range-energy relations for the recoil nuclei, and this handicaps the accuracy of the observations. Blackett and Lees (B25) and Feather (F8) have made attempts to evaluate these but more experimental work is required.

For the observation of the secondary electrons from gamma-rays a magnetic field is superimposed by the use of a pair of Helmholtz coils, and the curvatures of the electron tracks in this field can be used to calculate gamma-energies. This technique has been highly developed by Lauritsen and his collaborators (C51). The same principle is applied to the measurement of the radioactive betas and positrons from induced radioactive materials. The chief difficulty in these experiments is the excessive number of photographs required to give sufficient data for an exact statistical value of the electron energy distribution. These workers have also initiated a method for measuring gamma-radiation by observing the energies of both the positive and negative electrons of pairs and adding to this sum the 1 MV equivalent of the mass of the electrons. The pairs are formed in the walls of the chamber or laminae inserted in the chamber and so may have been partially absorbed, but if sufficiently thin laminae are used a relatively few pairs are sufficient to determine the gamma-ray energy.

C. Ionization chamber and pulse amplifier

The most satisfactory method of measuring the ranges of individual charged particles is in the shallow ionization chamber and with the use of the linear pulse amplifier, originally developed by Greinacher (G22), as a recording instrument. The amplifier has been perfected by Ward, Wynn-Williams and Cave (W2) and by Dunning (D22) and others with the addition of control features and the reduction of the background or noise level. With the instruments in use at present a chamber 2 mm deep suffices to give alpha-particle pulses large enough in comparison with the background to make accurate measurements possible. The voltage pulse due to the individual alpha-particle in the chamber is amplified by 3 or 4 stages of resistance-capacity coupled vacuum tube amplifiers, and is made to operate an oscillograph or electrical counter. The theory of the amplifying circuits and the resolving time of the instrument has been worked out by Ortner and Stetter (O13) and by Johnson and Johnson (J3). The resolving time is usually smaller than that of the mechanical counter used to record the counts. The chamber is usually constructed of two parallel electrodes, one a wire grid or thin foil through which the particles enter and upon which a suitable saturation voltage is applied; the other a plate connected to the grid of the first amplifier tube (near ground potential). The microphonic sensitivity of the instrument requires vibrationless supports for the chamber and at least the first tube of the amplifier. This also limits the use of thin metal foils as defining walls for the chamber, since such a foil records sound vibrations in the air as a condenser microphone.

In such shallow chambers electrons do not yield sufficient ionization to record and so protons and alpha-particles are observable even when accompanied by rather intense beta- or gamma-radiation. Protons and other light nuclei recoiling from fast neutrons are readily detected. If the chamber walls are coated with lithium or boron energetic alpha-particles are emitted in neutron disintegration processes and will be recorded by the amplifier. These reactions are greatly increased in probability with slow neutrons, and so this method constitutes one of the best methods of observing them.

With a properly constructed instrument the natural contamination alpha-particle background is less than one count per minute. The background due to the minimum "shot effect" of the individual thermionic electrons in the amplifier tubes and the "Johnson effect" of electron currents in the resistances in the circuits define the ultimate amplification possible with such an instrument. Careful shielding of the amplifier in iron and copper boxes has made it possible to reduce the extraneous effects nearly to this limit, and the background to approximately 1/50 of the amplitude of the alpha-particle pulse in a 2 mm chamber. The amplitude of this background fluctuation determines the minimum depth of the chamber, and a "bias" voltage must be applied to the output to prevent such fluctuations from operating the recording instruments. The "thyratron" or grid controlled relay tube is often used for this purpose. Here the grid is biased to any desired value and only those pulses of greater amplitude than this bias will operate the tube, which in turn operates a mechanical counter. The thyratron "scale of 8" counter (W23) is often used to increase the rate of recording. With such an instrument operating a high speed counter as many as 5000 counts/min. statistically distributed in time can be recorded with small error.

Due to the different specific ionizations of protons and alpha-particles these two ions will give pulses of different amplitudes and can be separated in the counting with the use of the bias voltage mentioned above. Also some success has been achieved in the use of the "differential ionization chamber," essentially two equivalent chambers with opposite applied potentials through which the ions pass. This makes it possible to obtain no counts from particles which cross both chambers, but only from those stopping after crossing one chamber. By this means short ranges of particles can be counted, even though longer range particles are traversing the chamber. A somewhat similar effect is obtained with the single chamber if sufficiently high bias is applied to allow the recording of only those ions which have their maximum specific ionization in the chamber. Due to the variation of the specific ionization with velocity (the "Bragg" curve) this means that only those ions near the end of their range will be counted.

D. Integrating ionization chambers

Electroscopes and integrating ionization chambers in general have been the basic instruments in the experimental development of natural radioactivity. They are the simplest of all sensitive electrical instruments and in their present form still supply the best values of ionization intensities. Background effects due to electrical leakage, cosmic rays and contamination of radioactive materials give a steady rate of discharge which can be subtracted from observations to give very accurate results. Where the nature of the radiation is known and the chamber is designed for proper absorption of that radiation the results are unique and nearly as sensitive to small intensities as individual counting methods. For the observation of the induced beta or positron radioactivity in various elements the standard techniques are immediately adaptable since decay periods can be measured directly.

A rather large chamber with several atmospheres pressure of CO_2 which has a thin aluminum window is used by Fermi and his group (S9) for the measurement of neutron induced activities. A sensitive electrometer is the observing instrument. It is accurately calibrated for sensitivity and background effects and is peculiarly adapted to the study of relatively long-lived materials.

A modification of the electroscope, designed by Lauritsen (L7) uses a small metalized quartz fiber whose deflections are observed by a small microscope focused on the tip. Fibers of about 5 microns diameter and about 6 mm long are stable for any orientation of the instrument. The low capacity of this small fiber gives a high charge sensitivity. It is charged and deflected electrically and the rate of discharge is observed in the microscope. It has the advantage of small size and can be used to measure any radiation observable above the background. Its chief value is for studies of induced radioactivity and as a qualitative instrument for the detection of γ's and n's. It has been used in certain experiments in which interchangeable lead and Cellophane walls make it differentially sensitive to neutron and gamma-radiations. The ionization is largely due to absorption of radiation in the walls of the chamber, the volume of air being small, and this absorption is a function of the wall materials.

E. Tube and point counters

The Geiger-Müller tube counter and its contemporary the Geiger point counter have been used for the detection of all types of radiations. In fact, this universal sensitivity constitutes its chief disadvantage since all kinds of radiations are identically recorded. It responds to cosmic rays and radioactive contaminations so that a natural background is always present. Many workers have studied the physical and electrical conditions influencing the action of the counters and in the hands of a relatively few, proper conditions of stability have been achieved. In such hands it has been a valuable and consistent instrument, and due to its extreme sensitivity it has been successful in certain exploratory observations where other instruments would have failed. It consists of a system of electrodes between which a steady potential is applied which is just under that for which a continuous discharge occurs. The tube counter has a cylindrical cathode with a central thin wire anode. The materials should be of low photoelectric sensitivity and free from contaminations. A tungsten or iron wire of 1- to 10-thousandths of an inch with a copper or nickel cylinder of 1 to 5 cm diameter would represent a typical instrument, although different workers recommend widely different specifications. The potential range in which the counter is sensitive is a rather narrow one, requiring a carefully regulated supply of from 1000 to 1800 volts determined by the construction.

The point counter consists of a cylindrical anode with a central cathode terminating in a point located at a distance from the end of the anode about equal to the radius of the cylinder. The end is usually closed with a thin metal foil to define the active region. The point counter can be operated at atmospheric pressure (which makes it valuable for certain experiments) while the tube counter has its greatest sensitivity for a few cm of Hg pressure. The type of gas in the counter is also a function of the experimenter, involving variations with purity of gas which are not easily explainable. The tube counter has a wider range of voltage for the sensitive condition and is not so readily disturbed by discharge.

A relatively few ions formed in the chamber by an ionizing radiation produce what is essentially a cumulative ionization resulting in a discharge. Proper adjustment of external circuit constants extinguishes the discharge within a very short time interval and reestablishes the sensitive condition. A simple one- or two-stage tube amplifier is sufficient to amplify the voltage pulse for recording.

For gamma-rays thick walled chambers may be used, although the practice is to keep the amount of material at a minimum in order to reduce the contamination background. Secondary electrons released chiefly from the metal tube walls produce the discharge. When intended for the detection of beta-particles or positrons either thin windows or thin chamber walls are used to admit the radiation. Constructed in this way the chamber is sensitive to gamma-radiation and is shielded with lead when necessary. Heavy particles such as alphas or protons are most readily observed with the point counter which may have thinner chamber walls due to operation at atmospheric pressure. With somewhat lower operating voltages the instrument will respond only to heavy particles—the "proportional counter." When filled with hydrogen the chamber records neutrons, but not efficiently enough to make it valuable for this purpose.

Coincidence methods with two or more counters, such as are used for cosmic-ray experiments, are also available. Such techniques have been used by Rasetti (R3) to measure the secondary electrons from the gamma-rays emitted by the absorption of slow neutrons in various materials, following earlier experiments by Bothe. The method also serves to determine the association of different radiations emitted from a reaction simultaneously but in different directions and recorded as coincidences. Bothe and Baeyer (B45) used a central counter with 8 others surrounding it and from the duplicate coincidences (requiring the central counter and two others to discharge simultaneously) concluded that one gamma-ray is emitted with each beta-particle from Ra E.

F. Photographic effects

Microscopic examination of suitably sensitized photographic plates shows a linear arrangement of developed grains in the emulsion along the track of an alpha-particle or proton. Such tracks are not as dense as those in a cloud chamber and due to the questionable estimates of density and nature of the emulsion, range measurements, obtained with a micrometer microscope, are not as accurate as from other methods. Neutrons give recoil protons which can be observed, but electron tracks are not dense enough to measure. The chief value is in the spatial orientation of tracks, as with the cloud chamber. Taylor and Goldhaber (T1) have been successful in showing that in the disintegration of boron by slow neutrons two particles (straight tracks in the emulsion) resulted and so the reaction was identified. The one peculiar advantage of the technique is that the records are permanent and cumulative so that long exposures with weak sources will give observable effects.

G. Chemical separations

As in the development of experimental results on natural radioactivity the studies of the new induced radioactivities have been materially assisted by the use of certain standard chemical separation techniques. A good guess as to the nature of a disintegration reaction makes it possible to separate the possible products and by observing in which of the separated residues the activity lies the chemical nature of the element can be determined. It is usually accomplished by using a radioactively inert element of the kind to be separated, as a carrier, in order to have a sufficient quantity to work with. It was introduced for alpha induced radioactivity by Curie and Joliot (C53) and for neutron processes by Fermi (F13). Positron radioactivity obtained by deuteron bombardment has been studied in like manner (Mc6).

A valuable modification was initiated by Szilard and Chalmers (S27). They realized that chemical bonds would be broken by the recoil of the activated atom and so the free element would be released from a chemical compound. Radioactive iodine can thus be separated from an iodine compound by reduction of the free iodine. Precipitation of the radioactive material from a large sample used as a target concentrates the activity into a small sample which can be brought close to the recording instrument, and so increases the observable intensity.

XVI. Auxiliary Data for the Evaluation of Experiments

§95. The Range-Energy Relation

Practically in all experiments on nuclear transmutations, the energy of the emitted particle is deduced from its range. Thus the relation between range and energy is of paramount importance for the determination of nuclear reaction energies. By "range" we understand in this section the mean range in air of 15°C temperature and 760 mm pressure. The relation between mean range and extrapolated range will be discussed in §97.

A. Experimental determinations

The ranges and energies of most of the natural α-ray groups emitted from radioactive sources have been measured with extremely great accuracy, by Rutherford, Wynn-Williams, Lewis and Bowden (R23, L23), by Briggs (B58–60) and by Rosenblum (R12, R14, R14a). The energy was measured by magnetic deflection of the particles, the error was in some cases as small as 1 in 10,000 for energy as well as range. The energies of the α-particles for which such precision measurements are available, extend from 5.3 MV (polonium) to 10.5 MV (long range α-particles from Ra C′ and Th C′). In this energy region we can therefore regard the range-energy relation for α-particles as absolutely certain.

For lower energies, careful measurements have been carried out by Mano (M4, M5). Mano measured the magnetic deflection of α-particles from Th C′ and Th C after these particles had traversed measured thicknesses of air. The accuracy claimed is about 1 in 1000 which is sufficient to determine α-particle energies from their range to about 5 kv. The main difficulty as compared to the measurements mentioned before is the straggling of the energy of the α-particles (cf. §97). While the α-particles emerging from the source all have exactly the same energy (for a given radioactive substance and given α-group!), the energies of α-particles which have traversed stopping material will be distributed over a more or less broad energy band. Mano chose consistently the most probable energy as significant.

Mano's measurements extend from 8.8 MV down to about 2.1 MV energy. They agree well with the data obtained from natural α-rays in the region from 5 to 8 MV. For energies below 5 MV, Mano's measurements seem internally consistent, following a smooth curve with the possible exception of the two lowest energies (2.1 and 2.5 MV) for which the ranges obtained seem to be too large as compared to the other measurements.

Earlier measurements by Briggs (B56), also based on magnetic deflection of partially stopped α-particles, gave ranges considerably higher than those found by Mano. Moreover, Briggs found a range of 3.93 cm for α-particles of 5.30 MV energy, which is the energy of natural α-particles from polonium. The observed range of Po α-particles is only 3.805 cm (R23). This discrepancy is found in spite of the fact that Briggs' ranges have been multiplied by 6.87/6.90 in order to bring his range of natural α-particles from Ra C′ (6.90 cm) into agreement with the now accepted range (6.870 cm). There can thus be no doubt that Briggs' values are too high throughout. For energies below 2 MV, the energies obtained by Briggs for a given range, should be increased by about 9 percent in order to obtain the energy values accepted in Section C.

A point at lower energies may be obtained from disintegration data, using the reaction

$$Li^6+H^1=He^4+He^3 \qquad (746a)$$

which was studied carefully by Neuert (N4). The He^3 particles were found to have a (most probable) range of 1.19 cm, the He^4 particles of 0.82 cm. These figures should be corrected for the penetration of the protons into the Li target. The protons used had an energy of 0.16 MV. Following the procedure of §97, we find that the most probable range corresponds to a penetration of the protons of about 0.03 cm. Thus the true ranges of the particles would be 1.22 and 0.85 cm, respectively. Now a He^3 particle of energy E has a range 3/4 times as great as that of an α-particle of energy 4/3 E (cf. Section B). Interpolating between Mano's experimental points, we find that the energy of an α-particle of $4/3\cdot 1.22 = 1.62_7$ cm range would be 2.90 MV. The energy of the He^3 is therefore $3/4\cdot 2.90=2.17_5$ MV. From the conservation of momentum, it follows that the He^4 produced in the reaction (746a) receives an energy equal to 3/4 times the energy of the He^3, i.e., 1.63 MV. A correction for the momentum of the incident proton lowers this figure to 1.62 MV. Thus we conclude that an α-particle of range 0.85 cm has an energy of 1.62 MV. This value was used in the accepted range-energy relation.

The region of *very low energies* was investigated by Blackett and Lees (B25). The production of such slow α-particles by slowing down fast (natural) alphas is very inconvenient because of the large straggling which prevents accurate measurements. Blackett and Lees therefore used *collisions* to reduce the energy of the α-particles: The "close" collisions of α-particles near the end of their range with helium nuclei were photographed in a cloud chamber. From the angle of deflection, the ratio of the energies after and before the collision can be obtained (§70). The energy before the collision can be found from the fraction of the range between α-particle source and collision; this procedure is safe because the incident α-particle has a fairly high energy so that the straggling of ranges is relatively not so important. The range of the scattered α-particle (from the point of collision) can be measured directly.

Unfortunately, Blackett and Lees had to use Briggs' range-energy relation for determining the energy of the incident α-particle, no other determination being available at the time. Assuming that in most cases the energy of the incident α-particle was less than 2 MV, the Briggs energies may be expected to be 9 percent low, so that we may expect the same error in Blackett and Lees' values. For ranges up to 0.4 cm (energies up to 0.66 MV) we have used the experimental energy values of Blackett and Lees, plus nine percent, for constructing our final range-energy curve (Fig. 29).

Blackett and Lees have also measured the ranges of *slow protons* produced by collisions of α-particles with hydrogen nuclei. The ranges were measured in a cloud chamber filled with a mixture of hydrogen and air, and the energies calculated from that of the incident α-particle in the same way as before. The proton ranges were reduced to those in pure air using approximate values for the stopping power of hydrogen relative to air.

It was found that slow protons have much smaller ranges than α-particles of the same velocity. On the other hand, for velocities greater than about $6\cdot 10^8$ cm sec.$^{-1}$, the range of α-particles and protons increases by the same amount for a given increase in velocity, both theoretically (Section B) and experimentally (B26). According to the experiments of Blackett and Lees, we have

$$R_H(v) = R_\alpha(v) - 0.2 \text{ cm} \tag{747}$$

for $v > 6\cdot 10^8$ cm/sec. This relation (or the more accurate one, (760), (760a)) is used for calculating the ranges of faster protons (Section C). The difference of 0.2 cm is purely empirical.

B. Theory

The energy loss of charged particles when passing through matter is mainly due to ionization and excitation of the atoms of the substance traversed. The probability of these inelastic collisions with the atoms depends on velocity and charge of the particle. At not too low energies (>1 MV for α-particles, >0.1 MV for protons) it is safe to assume that α-particles are doubly, protons singly charged. When the particles have been slowed down sufficiently, they will capture electrons: Thus, e.g., an α-particle of 0.8 MV is found to be singly ionized about as often as doubly ionized (R21a). The probability of single ionization increases with decreasing energy, until finally neutral atoms appear. This reduction of the average charge of a particle was studied by Kapitza (K1, cf. B) for α-particles in H_2 and by various authors for canal rays of H and He (R18). The reduction of charge will tend to decrease the energy loss of slow particles. On the other hand, the process of capture and loss of electrons itself will use up some energy so that the stopping is not reduced as much as would be expected from the decrease of the average charge.

The theory of the energy loss by inelastic collisions has been given by Bethe (B8), on the basis of Born's approximate treatment of collision processes. Bethe's formula is valid only if the velocity of the incident particle is large compared to the velocities of the electrons in the atom, in other words if

$$E \gg (M/m)E_{el}, \tag{748}$$

where E is the energy of the incident particle,

E_{el} the ionization potential of the electrons, and M and m the masses of incident particle and electron. Under these conditions, the energy loss per cm path is

$$-\frac{dE}{dX}=\frac{4\pi e^4 z^2}{mv^2}NB \qquad (749)$$

with $$B=Z\log(2mv^2/I). \qquad (749a)$$

Here v is the velocity and ez the charge of the incident particle, N the number of atoms per cm^3 of the material, Z the nuclear charge and I the average excitation potential of the atom (cf. Section C). B is a convenient dimensionless quantity proportional to the stopping power, we call it the "stopping number."

A relativistic treatment by Bethe (B19) and Møller (M25) showed that a term

$$-\log(1-\beta^2)-\beta^2 \qquad (750)$$

had to be added to the log in (749a). For small $\beta=v/c$, (750) is proportional to β^4 and is therefore negligible for the velocities occurring in nuclear processes (less than $5\cdot 10^9$ cm $sec.^{-1}$).

On the basis of classical mechanics, Bohr (B33a) derived a formula for the stopping power which contained essentially an additional factor $\hbar v/e^2$ in the argument of the logarithm in (749a). Bloch (B29) has investigated the reasons for this discrepancy between classical and quantum theory. By taking into account approximately the perturbation of the wave functions of the atomic electrons by the incident particle, he arrived at a formula which contains the quantum theoretical formula (749a) and the classical formula of Bohr's as limiting cases for high and low velocities, respectively. Bloch's correction is negligible when

$$v\gg e^2z/\hbar. \qquad (751)$$

The right-hand side is the velocity of an electron moving around the incident particle in a 1s-orbit. This velocity is, in all practical cases, *much smaller* than the velocity of a K electron in the atoms of the stopping material—except, of course, if the stopping material has smaller nuclear charge than the incident particle. Now the calculations of Bloch, as well as those of Bethe, were made with the assumption that the particle velocity is greater than the velocity of all electrons in the atoms constituting the stopping material. Therefore the Bloch correction is always negligible in the domain of validity of Bloch's formula. It should therefore *not* be included in calculations of the stopping power, contrary to the procedure adopted by Mano (M4) and by Bloch himself (B31) for calculating range-energy relations.

The condition (748) is quite a serious restriction of the validity of the stopping formula (749). For oxygen, e.g., the ionization potential of the K shell is about 540 volts, and therefore the right-hand side of (748) is almost 4 MV for α-particles. This means that the condition (748) is never very well fulfilled for α-particles, even with a stopping substance as light as oxygen. While we may expect the stopping due to L electrons of oxygen (valence electrons!) to be correctly represented by (749) down to quite low energies (less than 1 MV for α-particles), this will not be the case for the stopping by the K electrons. If the theoretical formula for the stopping is to be used for obtaining a range-energy relation of any precision, the stopping by the K electrons must be taken into account more accurately.

Fortunately, it can be shown that *the general principles of Born's approximation method can be applied to the collisions of heavy particles with atoms down to velocities much lower than that of the atomic electron*, and that only some minor approximations made in deriving (749) are no longer justified. The validity of Born's approximation was first proved by Mott (M33). He points out first that the motion of the incident heavy particle (proton, α-particle etc.) may be treated by classical mechanics since its wavelength is very small compared to atomic dimensions. The incident particle will then be equivalent to a perturbing potential whose center moves with constant velocity. Mott then shows that the transitions caused by such a potential are exactly the same as those found by the Born method provided the incident particle (if considered stationary at a given place inside the atom) does not distort greatly the wave functions of the atomic electrons. This condition will be fulfilled when the charge of the incident particle is small compared to that of the atomic nucleus, i.e., in all cases when the velocity of the K electrons of the atom is high. Thus we get the somewhat paradoxical result that the Born method as such is the better justified the less condition (748) is fulfilled.

The same result was obtained by Henneberg (H26) using a more analytical argument. Henneberg also calculated the probability of ionization of atoms in the K shell by slow protons and α-particles and found good agreement with experimental data of Gerthsen and Reusse (G14) and of Bothe and Fränz (B44). More recent experiments of Livingston, Genevese and Konopinski (L34), using higher intensities, also agree well with Henneberg's theory.

According to the Born method, the contribution to the "stopping number" B (cf. (749a)) due to excitation of the two K electrons is (cf. B16).

$$B_K = \int_{\vartheta}^{\infty} \epsilon d\epsilon \int_{Q_0(\epsilon)}^{\infty} \frac{dQ}{Q} \varphi(\epsilon, Q). \qquad (752)$$

Here ϵ is the energy given to the atomic electron divided by $Z^2_{\text{eff}}Ry$ where Ry is the ionization potential of the hydrogen atom and $Z_{\text{eff}} \approx Z - 0.3$ the effective nuclear charge in the K shell. The kinetic energy of the ejected K electron is thus

$$E_{\text{kin}} = \epsilon Z^2_{\text{eff}}Ry - E_K = (\epsilon - \vartheta)Z^2_{\text{eff}}Ry, \qquad (753)$$

where E_K is the (observed) ionization potential of the K shell

and
$$\vartheta = E_K / Z^2_{\text{eff}}\,\text{Ry} \qquad (753a)$$

is the ratio of E_K to the "ideal ionization potential" in the absence of "outer screening," Z^2_{eff} Ry (cf. B16, p. 478). Q is defined by

$$Q = (\mathbf{p} - \mathbf{p}')^2 / (2mZ^2_{\text{eff}}Ry), \qquad (754)$$

where $\mathbf{p}$ and $\mathbf{p}'$ are the momenta of the incident particle before and after collision. $Q_0(\epsilon)$ is the smallest value of Q possible for a given energy transfer ϵ; from energy and momentum considerations one finds easily

$$Q_0(\epsilon) = \epsilon^2 / 4\eta \qquad (754a)$$

with $\eta = mv^2/2Z^2_{\text{eff}}Ry = Em/MZ^2_{\text{eff}}Ry. \qquad (754b)$

φ is the transition probability from the K shell to a state of energy (753), *viz.*

$$\varphi(\epsilon, Q) = Q^{-1} \left| \int \psi_K(\mathbf{r}) \psi_\epsilon(\mathbf{r}) \exp\left(i(\mathbf{p} - \mathbf{p}') \cdot \mathbf{r}/\hbar\right) d\tau \right|^2, \qquad (755)$$

where ψ_K and ψ_ϵ are the wave functions of a K electron and an electron of energy (753), respectively. For not too low nuclear charge (a condition which we have always assumed to hold), Coulomb wave functions may be taken for ψ_K and ψ_ϵ and we find

$$\varphi(\epsilon, Q) = \frac{2^7(Q + \frac{1}{3}\epsilon)}{[(q+k)^2+1]^3[(q-k)^2+1]^3} \times \frac{\exp\left(-(2/k)\arctan 2k/(q^2 - k^2 + 1)\right)}{1 - e^{-2\pi/k}} \qquad (755a)$$

with
$$q = Q^{\frac{1}{2}}, \quad k = (\epsilon - 1)^{\frac{1}{2}}. \qquad (755b)$$

For a given value of the energy of the incident particle (or of η, (754b)), the energy loss (752) can be calculated by numerical integration. The result is given in Fig. 28, for $\vartheta = 0.7$ (cf. (753a)) which is very nearly correct for all elements from about carbon to aluminum. For high energies ($\eta \gg 1$), it is convenient to write

$$B_K(0.7, \eta) = 1.81 \log 3.63\eta - C_K(\eta), \qquad (756)$$

where C_K approaches zero as η increases. 1.81 is the "effective number of K electrons," or, more accurately, the total oscillator strength of all optical transitions from the K shell into the continuous spectrum (and to the unoccupied discrete levels). This value was checked by calculating the oscillator strength corresponding to the transition to the $2p$ shell using the Hartree wave functions for oxygen (H16). An oscillator strength of 0.15_5 was found for the transition of *one* K electron to the $2p$ shell. Since 4 of the 6 substates of the $2p$ shell are occupied by electrons, this means an oscillator strength of $(4/6) \cdot 2 \cdot 0.15_5 = 0.21$ for the transitions to the occupied states, leaving $2 - 0.21 = 1.79$ for the transitions to unoccupied discrete and continuous levels, in close agreement with the value of 1.81 obtained by integrating (752).

The argument of the log in (756) is

$$3.63\eta = \frac{2mv^2}{1.103Z^2_{\text{eff}}Ry}. \qquad (756a)$$

Therefore, for high energies ($C_K = 0$), (756) takes the same form as (749a), with the average excitation potential of the K shell being given by

$$I_K = 1.103Z^2_{\text{eff}}Ry. \qquad (757)$$

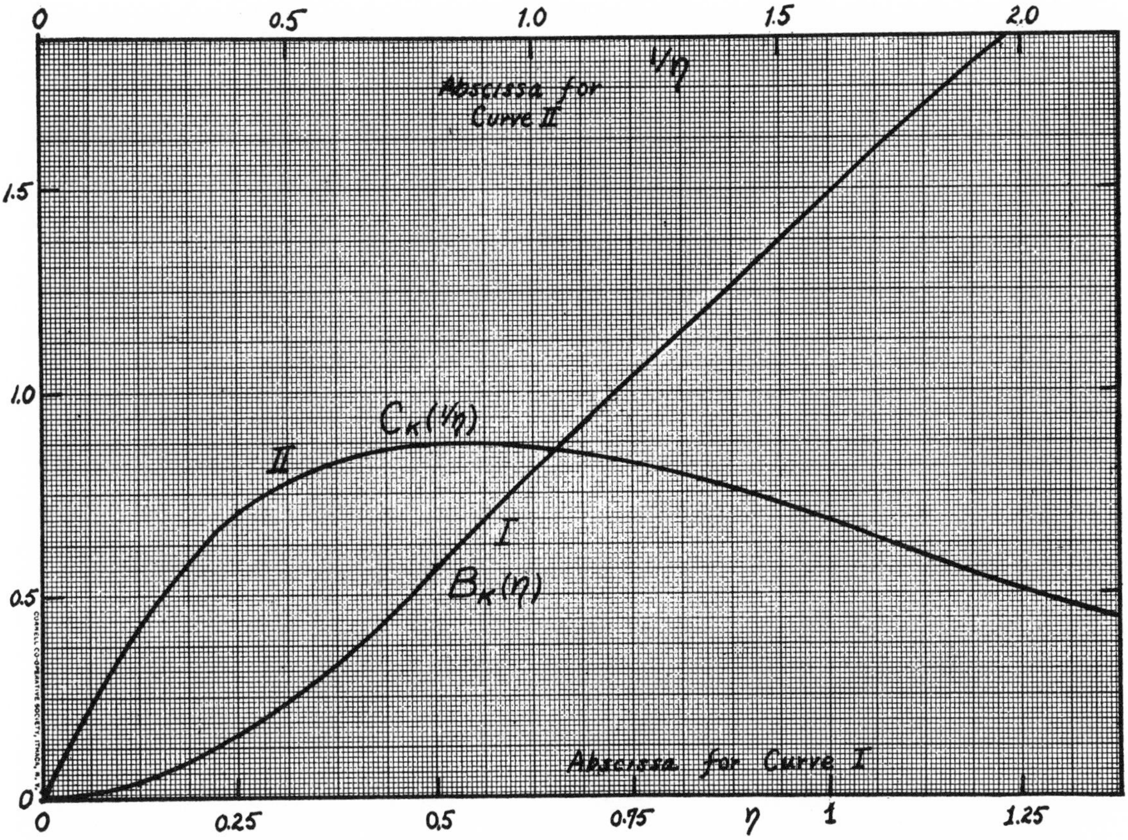

FIG. 28. Curve I: Contribution B_K of the K shell to the "stopping number" B (cf. (749a)). Abscissa: η = (velocity of the incident particle/velocity of a K electron)^2 = (electron mass×energy of the incident particle)/(particle mass× "ideal" ionization potential Z^2 Ry of the K shell). Curve II: Correction C_K to the high energy stopping number for K electrons against $1/\eta$. From C_K, the stopping number of K electrons can be obtained using (756), or the stopping number of the complete atom from (758).

If we denote the average excitation potential of the electrons outside the K shell by I', we can write the total stopping number

$$B=(Z-1.81)\log(2mv^2/I')+B_K, \quad (757a)$$

where $Z-1.81$ is the "effective" number of electrons outside the K shell. Using (756), we find

$$B=Z\log(2mv^2/I)-C_K, \quad (758)$$

where the average ionization potential of the whole atom is given by

$$\log I=(1-1.81/Z)\log I' + (1.81/Z)\log I_K. \quad (758a)$$

As can be seen from Fig. 28, C_K is positive for high energies η. This means that the K electrons are less effective in stopping than would be expected from the simple formula (749), (749a). The deficiency in stopping power C_K increases as the energy decreases down to about 1.5 times the "critical energy" $(M/m)Z^2_{\text{eff}}Ry$, then remains almost constant down to 0.8 times the critical energy and decreases from then on. For very low energies ($\eta \ll 1$), the stopping B_K due to K electrons becomes negligibly small. It will, of course, never become negative. This point has not been taken into account in previous calculations. For $\eta \ll 1$, the stopping number B is simply given by the first term in (757a), i.e., the effective number of stopping electrons as well as the average excitation potential are smaller than for high energies.

C. Determination of constants in stopping power formula

It is not easy to determine the average excitation potential I purely theoretically with any accuracy. An early attempt of Bethe (B8) in this

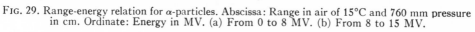

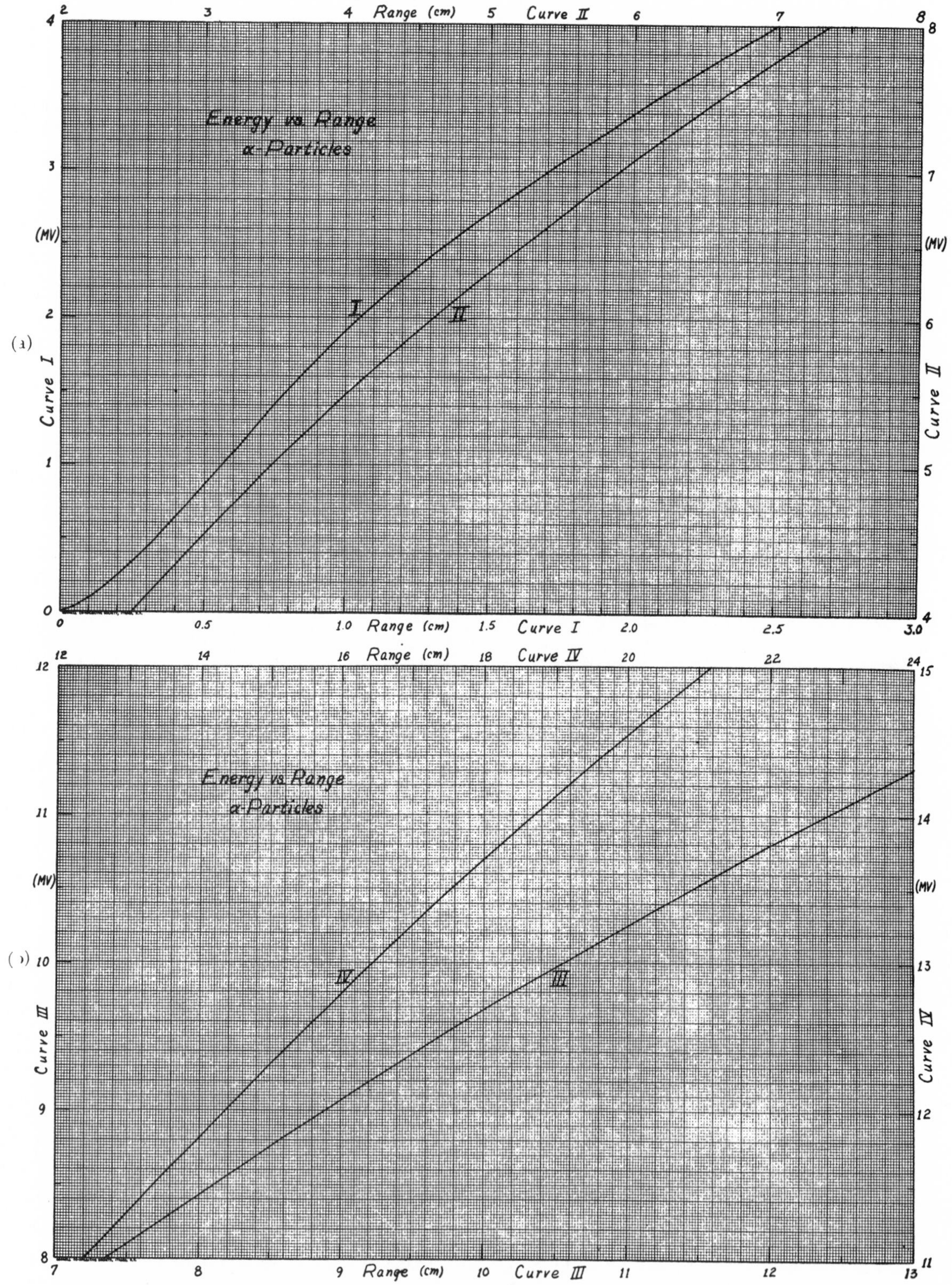

FIG. 29. Range-energy relation for α-particles. Abscissa: Range in air of 15°C and 760 mm pressure in cm. Ordinate: Energy in MV. (a) From 0 to 8 MV. (b) From 8 to 15 MV.

direction gave much too low values for I. Therefore I must be determined from the experimental data on stopping. This procedure was first suggested by Blackett (B26) and later also adopted by Duncanson (D18) and Mano (M4). Blackett and Duncanson used the simple Bethe formula (749), (749a) and regarded Z as well as I as an adjustable parameter. Mano used the formula of Bloch mentioned above. We have redetermined the constant I, using the more accurate formula (758).

The value of I for air was determined so that the difference between the ranges of a Th C′ long range α-particle and a Po α-particle agrees with experiment. This yielded $I=80.5$ volts for air. From (758a) we obtain then $I'=40.3$ volts, inserting for Z the value[2] 7.22. This is a reasonable value for the average excitation potential of the L shell of nitrogen and oxygen, since it must be expected that most optical transitions from states in the L shell lead to states of rather high energy in the continuous spectrum (Bethe, unpublished).

When I has been determined, the stopping power is known for all energies. In order to obtain the range for a given energy, (749) has to be integrated. The integration constant is fixed by making the range at a given energy agree with experiment. In our calculations, we chose the integration constant so that agreement was obtained *in the average* over the region in which measurements of natural α-ranges are available (5.3 to 11.5 MV). In order to achieve this, we had to make the "theoretical" range of Ra C′ α-particles equal to 6.865 cm instead of the observed value of 6.870 cm.

The agreement obtained with the observed ranges of natural α-particles was quite satisfactory, the maximum difference being 0.012 cm for Po and Th C′ (long) α-particles, in both cases the theoretical ranges are too high. In the final range-energy relation (Fig. 29), we adopted the experimental values which can at present not be equalled in accuracy by any theoretical calculation.

[2] $2Z$ was taken equal to the average number of effective electrons per molecule. For nitrogen and oxygen, $Z=7$ and 8, respectively. For argon, the two K electrons are not effective ($\eta \ll 1$, see Section B), leaving 16 effective electrons, i.e., $Z=8$. With 78 percent nitrogen and 1 percent argon, the average of Z is 7.22.

Between 2.5 and 5 MV, the agreement with Mano's observations is fairly satisfactory. However, as the differences seemed just outside the experimental error and since the theoretical calculations are less accurate at low energies, we adopted an empirical correction of the theoretical energies by -10, -25, -40, -60 and -65 kv, at 4, 3.4, 2.9, 2.3 and 1.5 MV respectively. With this correction our curve passes about midway between Mano's experimental points, being slightly higher than his points at 2.1 and 2.6 MV, slightly lower than his result at 3.0 MV. Our curve is also made to give a range of 0.85 cm for 1.62 MV as is required by the observations on the disintegration (746a).[3]

No attempt was made to continue the theoretical calculations to energies below 1 MV, because then capture and loss of electrons will become important and no adequate theory is yet available for dealing with this case. Instead, Blackett and Lees' experimental values, increased by 9 percent (cf. Section A), were adopted up to 0.7 MV. These values join to the theoretical curve above 1 MV very smoothly, which was taken as an indication of the reliability of the adopted range-energy relation even in the intermediate region.

The extension to velocities greater than those of natural α-particles is straightforward, using (749, 749a). The theoretical range-energy relation may be considered particularly reliable in this region because no corrections like that for diminished stopping due to K electrons or capture and loss of electrons need to be applied. It must only be kept in mind that $\frac{1}{2}Mv^2$ can no longer be identified with the energy of the particle, the relativistic correction amounting up to 2 percent for the velocities considered here. This was not taken into account by Duncanson (D18) who was the first to give a theoretical range-energy relation for very fast particles (protons). The relativistic correction is the main reason why our range-energy relation differs appreciably from Duncanson's at high energies (239.3 cm range against Duncanson's 230.6 cm for protons of 15.03 MV energy).

[3] It might be thought that the theoretical relation is useless if it has to be corrected so extensively. However, the theoretical relation is still useful in order to provide a simple and accurate way of interpolation between the observed points.

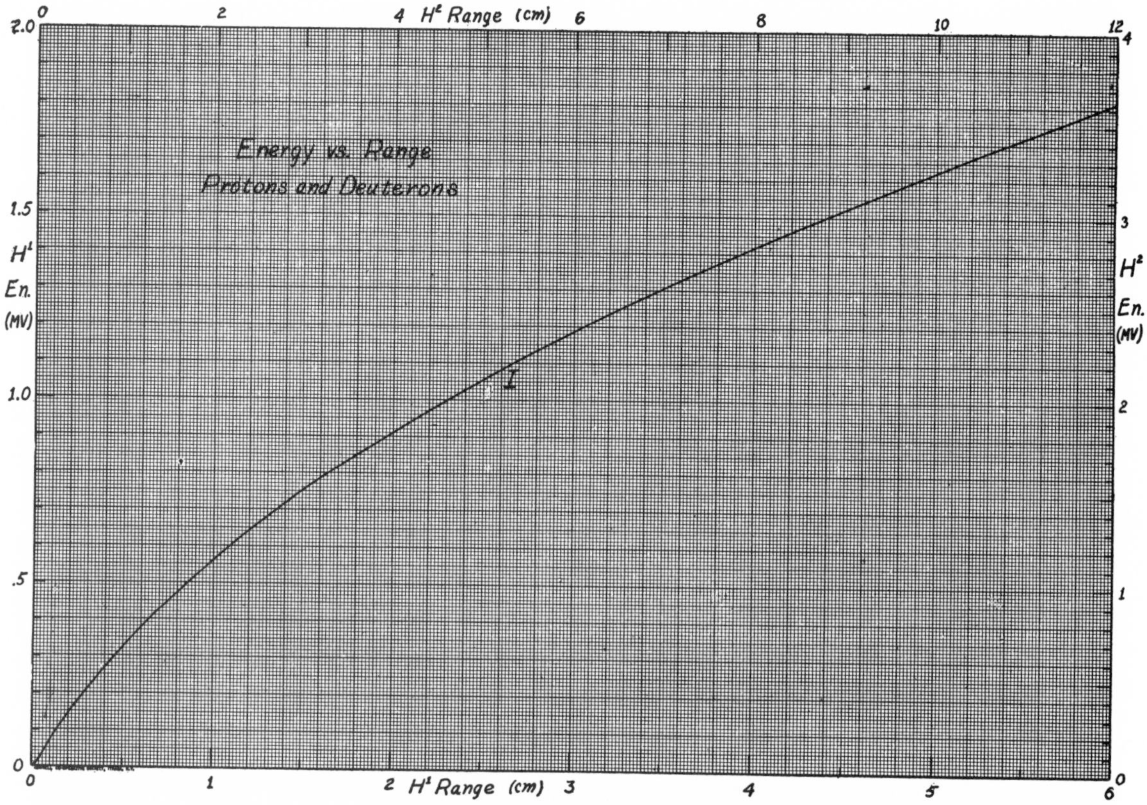
Energy vs. Range
Protons and Deuterons
H^2 Range (cm)
0
2
4
6
8
10
12
H^1 En. (MV)
2.0
1.5
1.0
.5
0
H^2 En. (MV)
4
3
2
1
0
I
H^1 Range (cm)
0
1
2
3
4
5
6

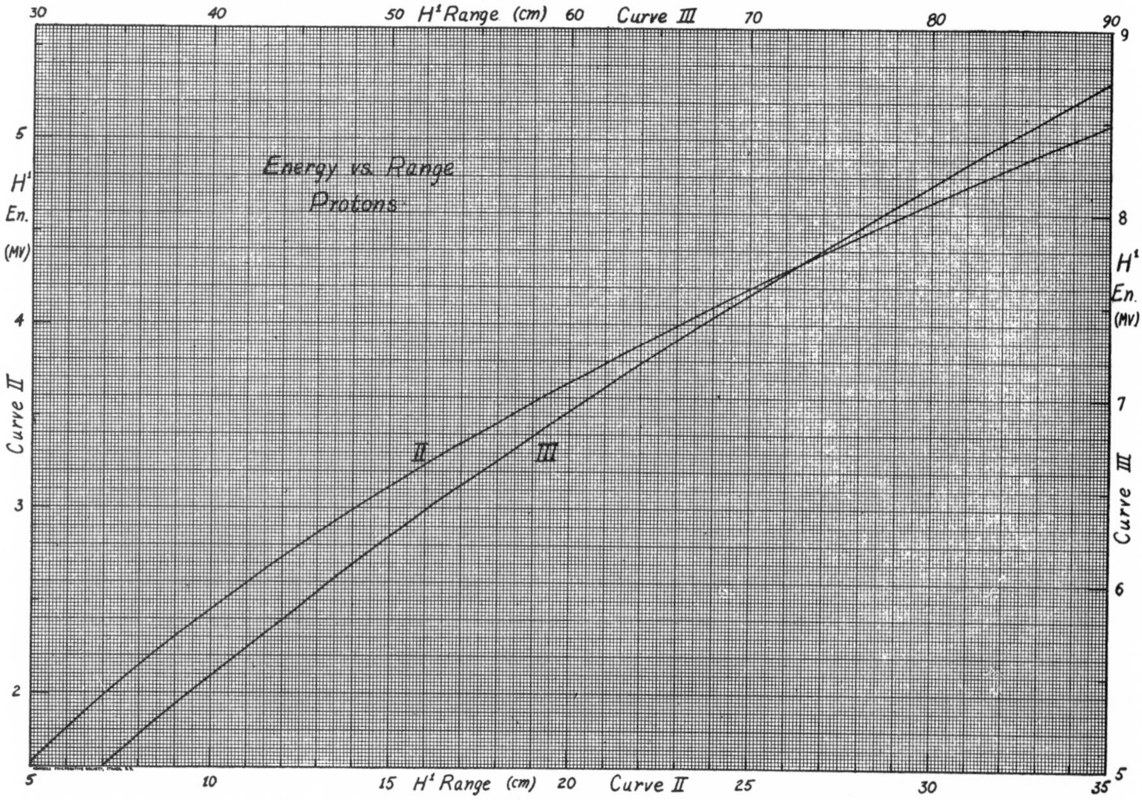
Energy vs. Range
Protons
H^1 Range (cm) Curve III
30
40
50
60
70
80
90
H^1 En. (MV)
Curve II
5
4
3
2
H^1 En. (MV)
Curve III
9
8
7
6
5
II
III
H^1 Range (cm) Curve II
5
10
15
20
25
30
35

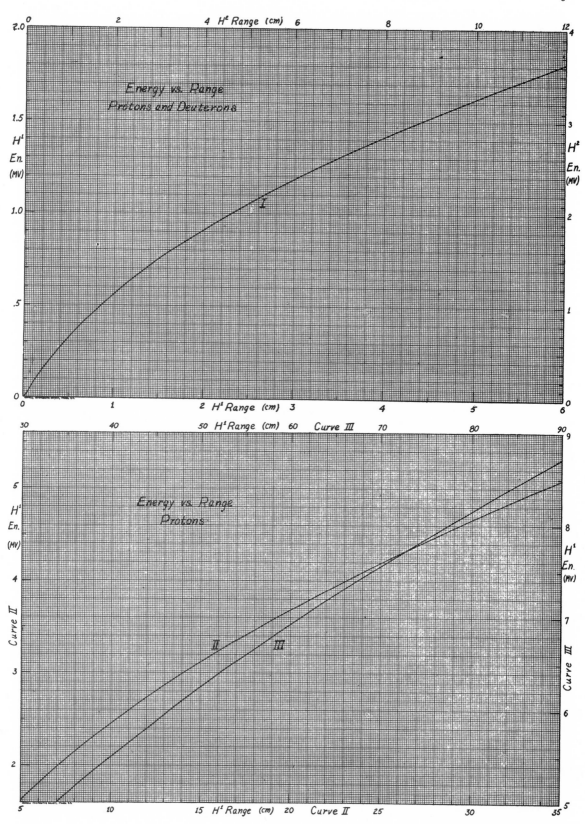

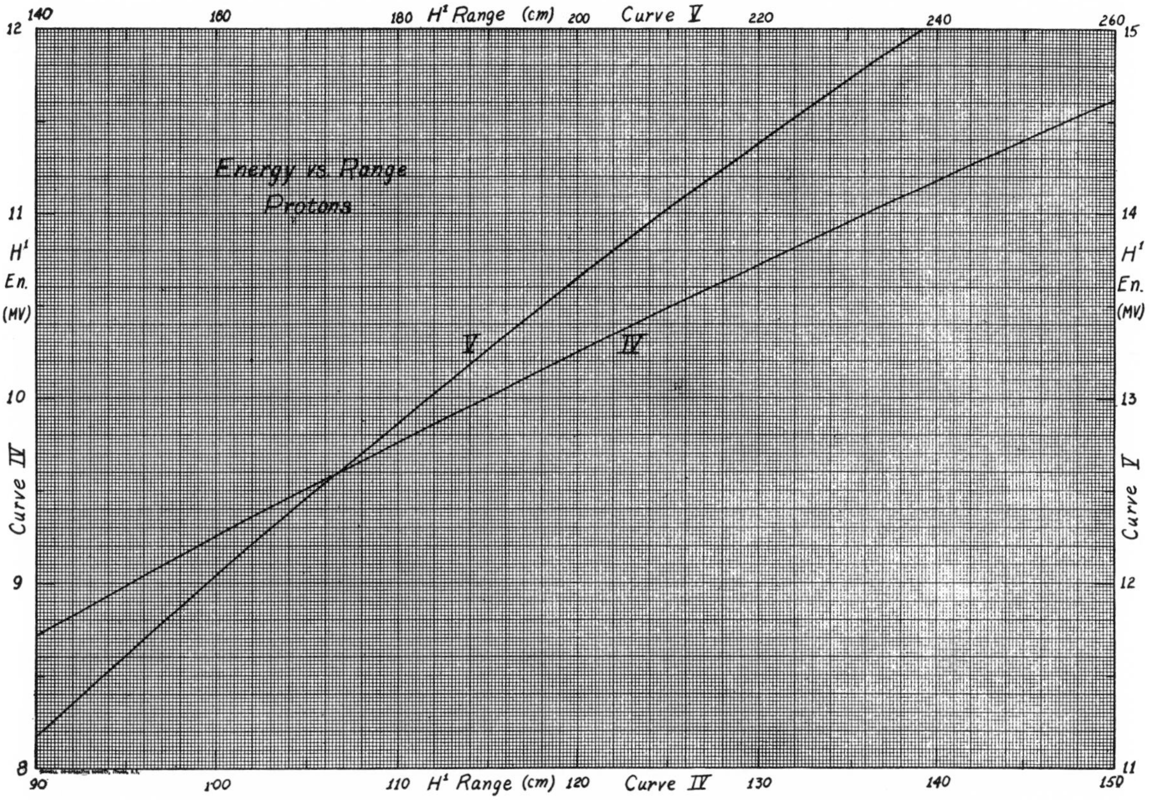

FIG. 30. (a) (b) (page 268) (c) (above) Range-energy relation for protons. (For deuterons, the ranges and the energies should be multiplied by 2.)

In the actual computation, the careful work of Duncanson was used as a starting point. The "stopping number" B was computed according to (758) with relativity corrections, and from Duncanson's formula. The percentage difference was calculated and Duncanson's ranges corrected accordingly. This saved the laborious computation of the integral exponential $Ei(x)$ which occurs in the integrated expression for the range (D18). Only at low energies (below 2.5 MV) Duncanson's calculation was found to be too much in error so that a completely new calculation was made.

For *protons* of high energy (i.e., high enough so that capture and loss of electrons is irrelevant) the energy loss per cm should be just one-quarter of that of an α-particle of the same velocity (cf. (749)), the energy loss being proportional to the square of the charge. Since E is, for a given velocity, proportional to the mass M of the particle, we should have generally

$$R(v, M, z) - R(v', M, z) = Mz^{-2}[f(v) - f(v')], \quad (759)$$

where v and v' are two velocities both large enough to minimize capture and loss of electrons, and f is a function of the velocity independent of mass and charge of the particle. From (759) it follows that

$$R_H(v) = (M_H/M_\alpha)(z_\alpha/z_H)^2 R_\alpha(v) - c, \quad (759a)$$

where c is a constant. With the known masses of hydrogen and helium, and with the constant c as determined experimentally by Blackett (Section A), we have

$$R_H(v) = 1.0072 R_\alpha(v) - 0.20 \text{ cm} \quad (760)$$

or

$$R_H(E) = 1.0072 R_\alpha(3.971E) - 0.20 \text{ cm}. \quad (760a)$$

The constant takes account of the capture and loss of electrons at low energies which affects α-particles more than protons. Its value (0.20

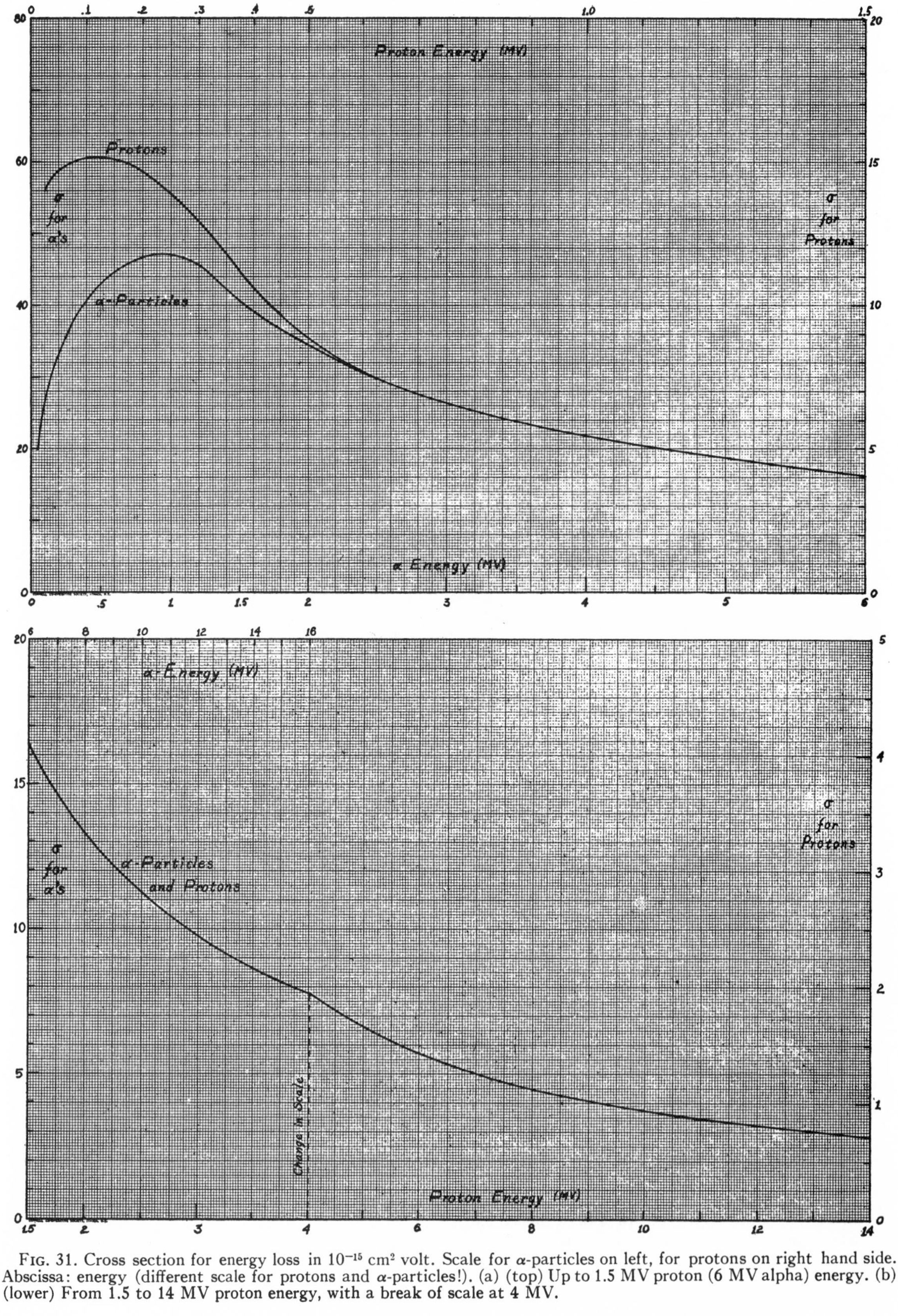

FIG. 31. Cross section for energy loss in 10^{-15} cm^2 volt. Scale for α-particles on left, for protons on right hand side. Abscissa: energy (different scale for protons and α-particles!). (a) (top) Up to 1.5 MV proton (6 MV alpha) energy. (b) (lower) From 1.5 to 14 MV proton energy, with a break of scale at 4 MV.

cm) is probably not very accurate. However, it is satisfactory that the range-energy relation for protons as obtained from (760a) passes very near two experimental points obtained by Cockcroft and Walton (C20). The lowest part of the curve (up to 0.3 MV) was again obtained from Blackett and Lees' experiments, with the energies increased by 9 percent. The rest of the curve is theoretical, with the same empirical corrections applied as for α-particles.

The range-energy relations for the ions of the other hydrogen and helium isotopes (H^2, H^3, He^3) may be obtained immediately from those of proton and α-particle, respectively. Capture and loss of electrons being the same for all hydrogen isotopes (or all helium isotopes) we have *exactly* (i.e., to a better approximation than (759a))

$$R_{z,\,M}(E)=(M/M_0)R_{z,\,M_0}(EM_0/M), \quad (761)$$

where $R_{z,\,M}(E)$ is the range of a particle of charge z, mass M and energy E.

The finally adopted range-energy relations for protons and α-particles are given in Figs. 29, 30. Experimental determinations are not indicated. We believe that our relation is accurate to 10 kv for α-particles between 5 and 10 MV, to about 30 kv between 2 and 5 MV and above 10 MV, to 50 kv between 0.5 and 2 MV and to 10 percent below $\frac{1}{2}$ MV. For protons, we estimate the accuracy to be about 10 percent below 0.2 MV and 20 or 30 kv above. However, more accurate experimental data are very much needed, particularly at low energies where the theoretical relation is least reliable.

D. Stopping cross section and range exponent

For some purposes, particularly for estimating disintegration cross sections from thick target experiments (cf., e.g., §81, 82) the "stopping cross section" is useful. This quantity is defined as the energy loss per cm, divided by the number of atoms per cm³, or, according to (749):

$$\sigma=\frac{4\pi e^4 z^2}{mv^2}B. \quad (762)$$

The dimension of σ is energy times area. The stopping cross section for protons and α-particles in air (per *atom*, not molecule) is given in Fig. 31 as a function of the energy. It is of the order of a few times 10^{-15} volt cm². The stopping cross section in other materials is of the same order of magnitude (section E).

For a number of corrections which must be applied to the observed range of particles (§96, 97) the logarithmic derivative of the range with respect to the energy is important. We define

$$n=2\ d\log R/d\log E. \quad (763)$$

For nonrelativistic energies,

$$n=d\log R/d\log v. \quad (763a)$$

If n were constant, R would be proportional to v^n. A relation of this type, with $n=3$, was first proposed by Geiger (G11) and is not very far wrong for natural α-particles. Actually, n varies considerably with energy, increasing from about 1.4 for slow α-particles and protons to over 3.6 for fast protons. For still higher energies, the "range exponent" n approaches 4.

E. Stopping power of substances other than air

According to (749a), the stopping number B (and therefore the stopping power per atom) will in general increase with increasing number Z of electrons in the atom. However, the stopping power increases more slowly than Z itself because the average excitation energy I which occurs in the log is also larger for high atomic number Z Moreover, it follows immediately that the relative stopping power must be a function of the velocity of the particle. The stopping power of heavy atoms with high excitation potential will increase relatively with increasing particle energy.

The quantity customarily given is the stopping power relative to air,

$$s=B/B_{\text{air}}, \quad (764)$$

where B is the "stopping number" of the material as defined in (749a). The experimental data for s are rather conflicting, the more recent determinations giving consistently higher values for s than the older ones, at least for solid materials. This may be due to nonuniformity of the sheets used for the stopping in the older experiments. As it is most convenient to determine the reduction of the *extrapolated* range of the α-particles by passing through the sheet, the stopping power of the thinnest spot in the absorbing foil will be measured rather than an average. In the more recent experiments which are mainly due to Rosenblum

(R13), the *average* reduction in velocity was measured magnetically as a function of the thickness of the material.

Table XLVIII compares the older and the newer data for the relative stopping power. Part A of this table gives the thickness of various materials (in mg/cm²) equivalent to 1 cm air (1.22 mg/cm²) for the stopping of α-particles of initial range 6 cm, according to older measurements of Marsden and Richardson (quoted by Geiger, G12) and to the more recent ones of Rosenblum (R13). The latter are consistently about 6 percent lower. The old value for mica was obtained by Briggs, the new one calculated (see below). Part B of Table XLVIII gives the atomic stopping powers of various substances for "α-particles of medium velocity" according to Geiger's article in the *Handbuch der Physik* (old measurements) and according to Mano's tables which are based on newer measurements. Mano's values were taken for a velocity of $1.75\cdot10^9$ cm/sec. (energy about 6 MV); they are consistently about 10 percent higher for the heavier elements than Geiger's values, with the exception of the rare gases for which no error due to non-uniformity is possible.

TABLE XLVIII. *Relative stopping power of various materials for α-particles of about 6 MV.*

A. Thickness (in mg/cm²) equivalent to 1 cm of air

Substance	Marsden and Richardson	Rosenblum
Al	1.62	1.51
Cu	2.26	2.09
Ag	3.86	2.71
Au	3.96	3.74
Mica	1.45*	1.43**

B. Atomic stopping power relative to air

Subst.	Geiger	Mano	Subst.	Geiger	Mano
$\frac{1}{2}H_2$	0.22	0.20	Cu	2.29	2.57
He	0.42	0.35	Kr	2.89	2.92
Li	0.53	0.50	Mo	2.75	3.20
$\frac{1}{2}N_2$	0.98	0.99	Ag	3.04	3.36
$\frac{1}{2}O_2$	1.10	1.07	Sn	3.19	3.59
Ne	1.24	1.23	Xe	3.94	3.76
Al	1.40	1.50	Au	4.02	4.50
A	1.92	1.94	Pb	4.25	4.43

* Briggs.
** Bennett.

In computing theoretical range-energy relations, it must be kept in mind that the innermost electrons can practically not be excited at all by the incident particle. Excitation is very improbable if $\frac{1}{2}mv^2=mE/M$ is small compared to the ionization potential of the electron concerned. For α-particles of 8 MV energy, we have $mE/M=1000$ volts, so that all electrons of more than 1000 volts (75 Rydberg) excitation potential will be ineffective in stopping. If we take, e.g., gold ($Z=79$), all the electrons in the K, L and M shells are to be excluded on this ground, i.e. 28 electrons altogether. However, this does not mean that the number of effective electrons is reduced to $79-28=51$, because the stopping effect of an electron is given (B8) by the oscillator strength of the optical transitions into the continuous spectrum. This oscillator strength is smaller than unity for inner, larger for outer electrons. It has been calculated by Hönl (H35) for the K and L shell; for Au the result is about 0.57 and 0.60 for each K and L electron, respectively. Assuming 0.65 for M electrons, the number of electrons effective for stopping would be $79-2\cdot0.57-8\cdot0.60-18\cdot0.65=61.4$ in gold. In a similar way, this number may be determined for other substances.

Figure 32 shows the range-velocity relation for gold with $Z=61.4$ and $I=520$ volts. The agreement with the experimental points of Rosenblum (R13) is satisfactory, except for the last three points. This is understandable because at these low velocities the ionization potential of the N shell becomes comparable to $\frac{1}{2}mv^2$ so that a correction C_N, similar to C_K in (756) and whose exact form has not yet been calculated, should be applied. The curve marked "3.77 air" is obtained if it is assumed that 3.77 mg/cm² of gold are equivalent to 1 cm of air, i.e., that the stopping power of one gold atom is 4.42 times that of an air atom irrespective of the velocity of the particle. The deviations are considerable.

As an illustration of the dependence of the atomic stopping power (764) on the energy, we give in Table XLIX the stopping powers of some

TABLE XLIX. *Atomic stopping power for various velocities (semi-empirical, air = 1).*

v(10⁹ cm/sec.)	1.0	1.5	2.0	2.5	3	4	5
E_α (MV)	2.07	4.66	8.3	12.95	18.6	33.2	51.9
E_H (MV)	0.52	1.17	2.09	3.26	4.70	8.36	13.06
$\frac{1}{2}H_2$	0.26	0.224	0.209	0.200	0.194	0.186	0.181
C	0.94	0.932	0.921	0.914	0.908	0.899	0.892
Al	1.45	1.51	1.53	1.54	1.55	1.57	1.59
Cu	(1.92)	2.41	2.62	2.73	2.80	2.89	2.95
Ag	(2.25)	3.08	3.43	3.64	3.76	3.93	4.04
Au	(2.42)	3.96	4.64	5.00	5.25	5.57	5.79

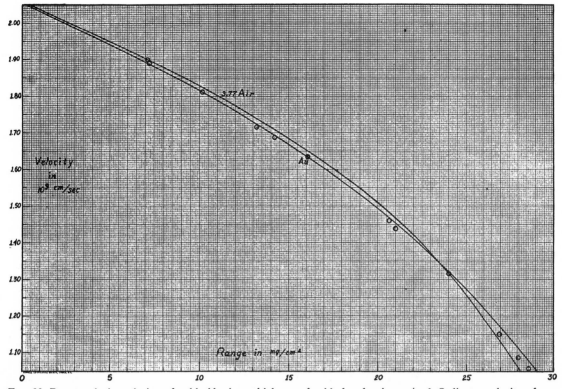

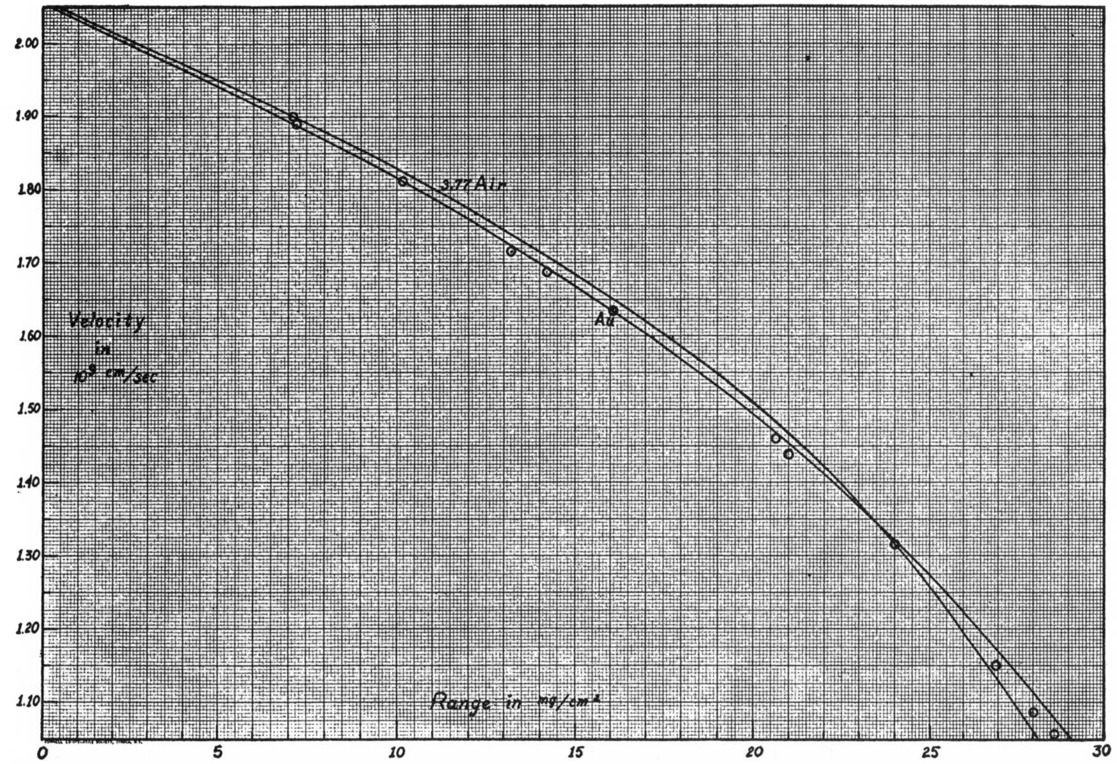

FIG. 32. Range-velocity relation of gold. Abscissa: thickness of gold absorber in mg/cm². Ordinate: velocity of α-particles in 10^9 cm/sec. The curve marked "Au" gives the theoretical range-velocity relation for Au. The curve marked "3.77 air" is computed assuming that 3.77 mg/cm² of Au are equivalent to 1 cm of air. The circles represent experimental points of Rosenblum. They fall closely on the Au curve except at the end where the N electrons of the Au will cease to follow the elementary theory of stopping. The figure shows the variation with velocity of the stopping power of gold relative to air.

elements as functions of the velocity (or energy) of the particle. The effective number of electrons used for Cu and Ag was 27.4 and 39.6, respectively. The average excitation potentials of Al, Cu, Ag and Au were determined so as to make the stopping power near $2\cdot 10^9$ cm/sec. agree with Rosenblum's experiments. Since no correction was made for the electrons whose ionization potential is of the same order as $\frac{1}{2}mv^2$ (L, M and N electrons for Cu, Ag and Au, respectively), the value obtained for the lowest velocity ($1.0\cdot 10^9$ cm/sec.) is probably considerably in error for Cu, Ag and Au; the correct value should be higher. For Al and C the correction C_K (Section B) was applied. The average ionization potential of carbon was estimated from that of N_2, O_2 and Ne (cf. Mano) by extrapolation, that of hydrogen was taken from Mano's experiments.

Because of the large variation of the relative stopping power with energy, it is not legitimate to use foils of heavy materials such as Au or Ag for the stopping of particles and to assume them to be equivalent to a given thickness of air. As the range-energy relation in these heavy materials is not yet accurately known either theoretically or experimentally, they should be avoided in range measurements.

For light elements the range-energy relation is sufficiently established to correct for the variation of stopping power with energy. In Fig. 33, we have given the stopping power per atom of hydrogen and carbon relative to air. Hydrogen and methane are frequently used as stopping gases in determinations of neutron energies (§94) by means of the proton recoils. The evaluation of a hypothetical experiment would go as follows: Proton range determined in methane of normal temperature (15°C) and pressure (760 mm) 64.3 cm. Estimated range in air (need not be accurate) 55 cm. For 55 cm, Fig. 33 gives for the stopping powers of C and H the values 0.903_2 and 0.189_1, respectively. Thus stopping power of half a mole-

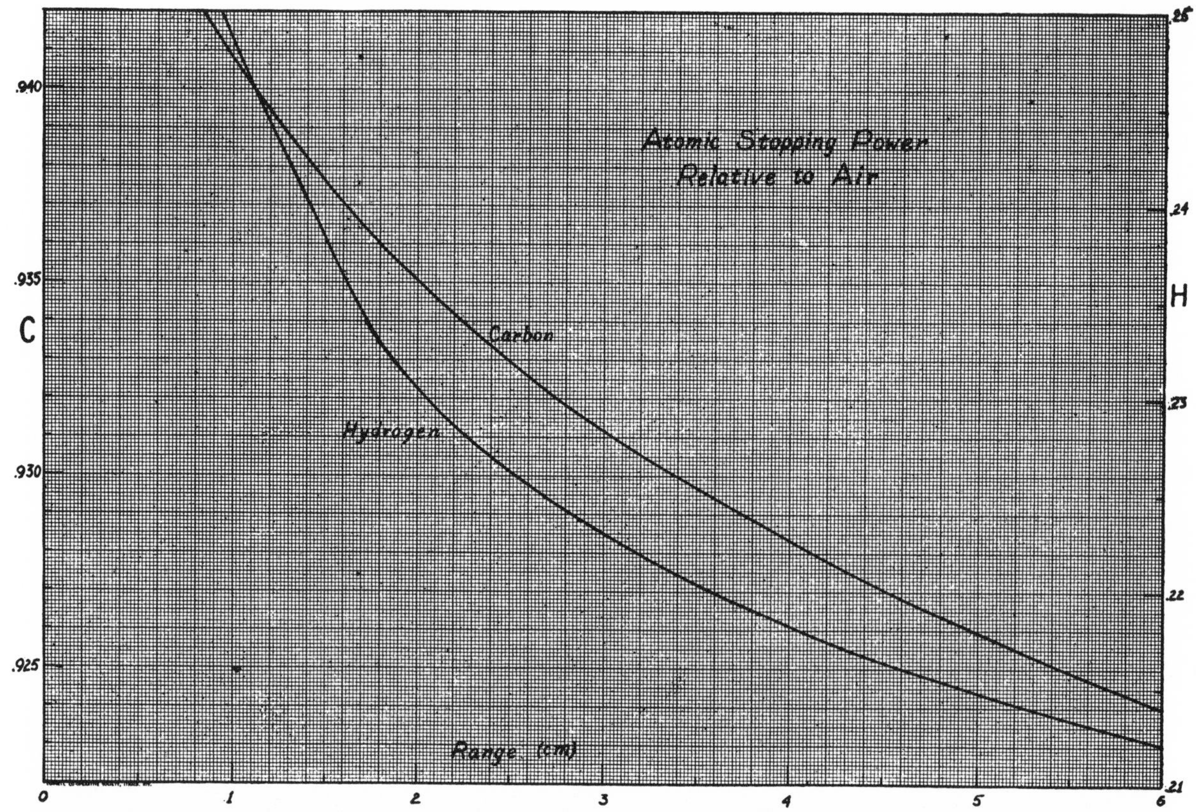
Atomic Stopping Power
Relative to Air
Carbon
Hydrogen
Range (cm)
C
H
.940
.935
.930
.925
.25
.24
.23
.22
.21
0
1
2
3
4
5
6

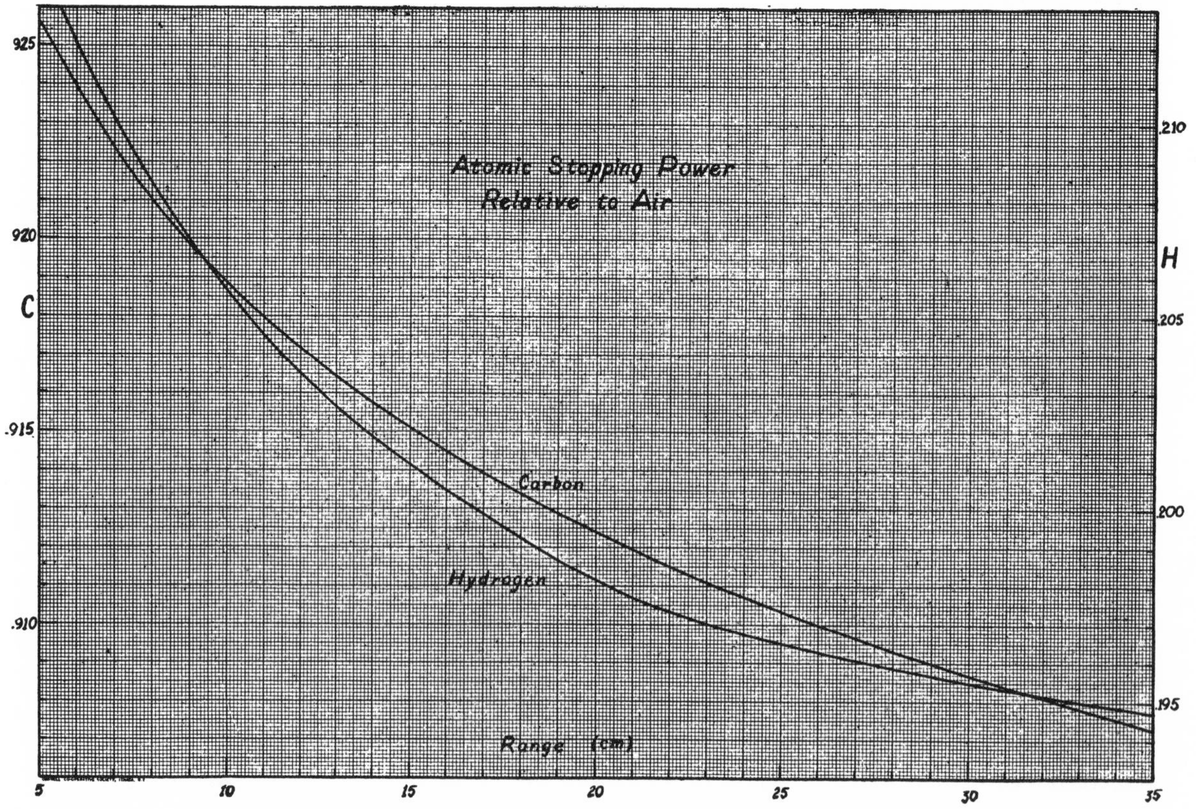
Atomic Stopping Power
Relative to Air
Carbon
Hydrogen
Range (cm)
C
H
.925
.920
.915
.910
.210
.205
.200
.195
5
10
15
20
25
30
35

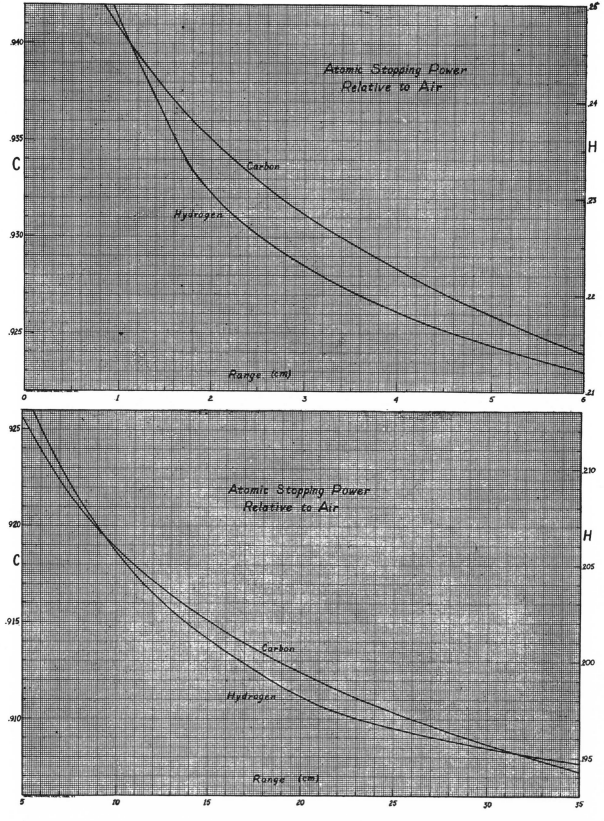

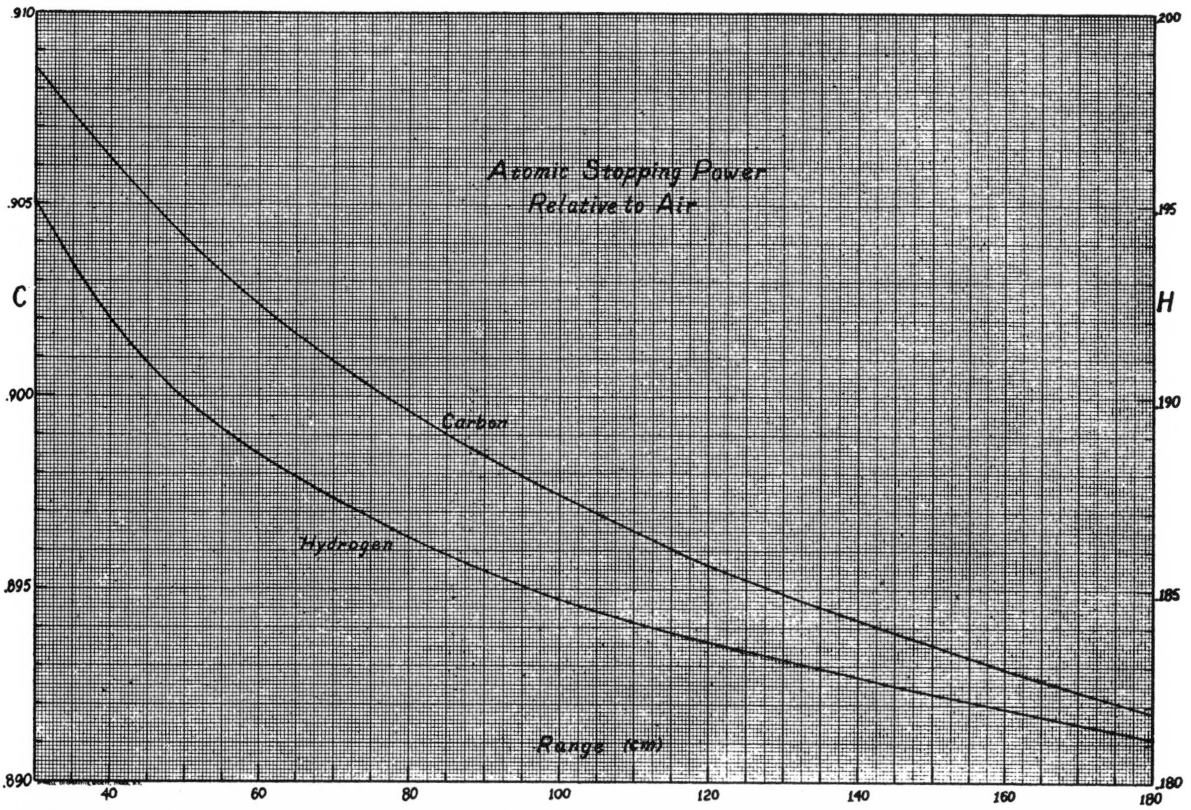

FIG. 33. (a) (b) (page 274) (c) (above) Stopping power of carbon and hydrogen. Abscissa: Range of particles in air. Ordinate: stopping power per atom relative to air. The curve is intended for the evaluation of experiments with hydrogen or methane filled cloud chambers. It is valid for α-particles and protons.

cule of $CH_4=\frac{1}{2}\cdot 0.903_2+2\cdot 0.189_2=0.830$. Therefore range in air $64.3\cdot 0.830=53.3_7$ cm. (For this value of the range, the exact stopping power of methane would be 0.830_8, corresponding to a range in air of 53.4_2 cm. The difference is insignificant.)

For Al we have given the correction in a different form. As a first approximation, the stopping power of an Al atom may be put equal to 1.500. This means that 1.51_8 mg/cm^2 of Al are equivalent to one cm of air. With this reduction factor, the thickness of Al may be expressed in equivalent centimeters of air. To the "equivalent range" obtained in this way, a correction must be added which is given in Fig. 34 as a function of the equivalent range itself. The result is the range in air from which the energy is obtained in the usual way from Figs. 29, 30.[4]

In re-evaluating experimental data (Chapter XVII and XVIII) we have, therefore, not made any corrections for the absolute stopping power of mica. We have, however, corrected for the change of the stopping power with velocity by adding half the aluminum correction given in Fig. 34, corresponding to the fact that about half the stopping power of mica is due to the heavy atoms Al, Si and K, and half to oxygen and hydrogen.

F. Range of particles heavier than helium

Nuclei of Li and heavier elements are ordinarily only partially ionized at the velocities with which they are commonly produced in nuclear collisions. Since the dependence of the effective charge on the energy is unknown, no theoretical treatment of the range-energy relation is possible.

[4] Frequently, aluminum is only used for the first part of the stopping, the rest of the path being in air. In this case, the correction may also be read off Fig. 34. Suppose, e.g., the total range of an α-particle consists of 5.93 equivalent cm in Al and then 3.54 cm air, together 9.47 cm. The correction given in Fig. 34 for 9.47 cm is 0.041 cm, for 3.54 cm we find −0.026 cm. The total correction is the difference of these two figures, i.e. 0.067 cm, therefore the range in air 9.54 cm.

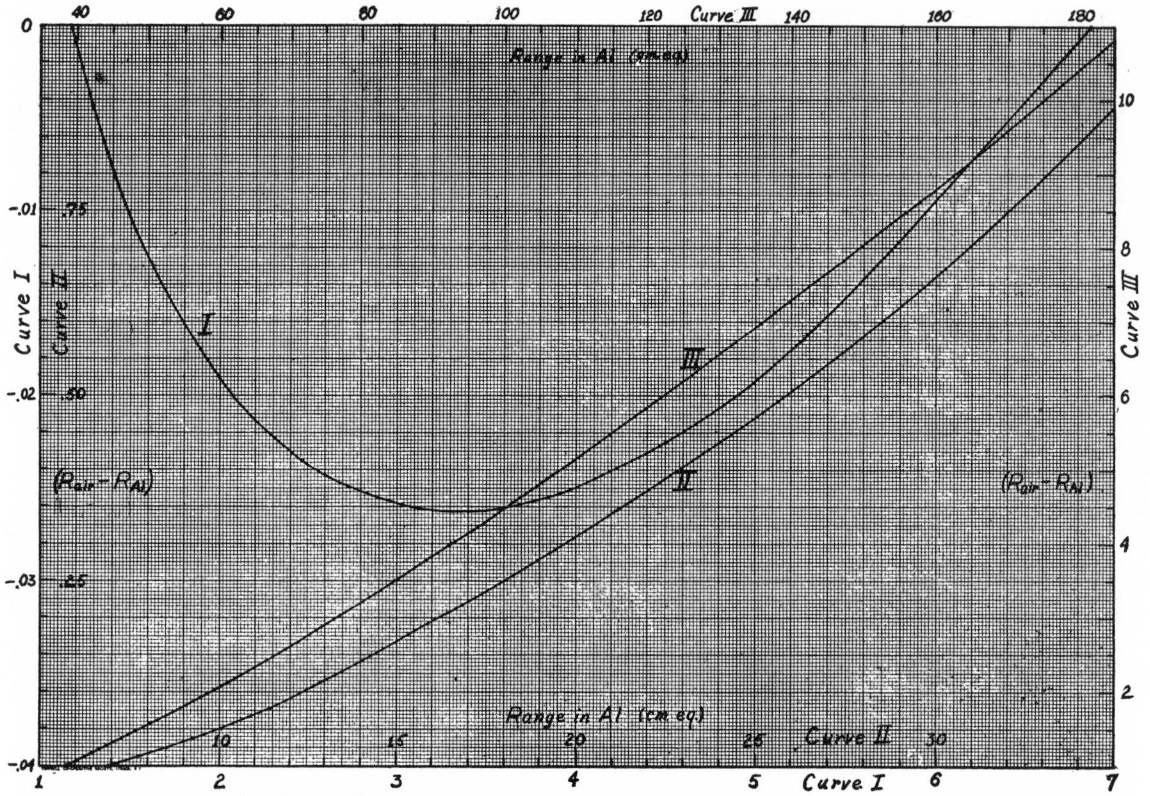

FIG. 34. Range correction for Al. The range of the particles is supposed to be determined by absorption in aluminum foils whose air equivalent has been measured beforehand using natural α-particles for calibration or using a suitable conversion factor (e.g., 1.52 mg/cm² Al = 1 cm air). For any value of the equivalent range in air thus obtained (abscissa), the curve gives the correction to be applied for the variation of the stopping power of Al with velocity (ordinate). For mica, one-half the correction should be applied. Curve valid for protons and α-particles.

Experimental data have been obtained by Blackett and Lees (B25) for N^{14}, O^{16}, O^{17} and A^{40} nuclei, by Feather (F7) for C^{12} and F^{19}. The method used was in all cases to observe, in a cloud chamber, the range of recoil atoms produced in collisions by α-particles of known energy. The range-energy relations obtained are rather irregular and the accuracy is quite small owing to the small number of recoil tracks measured. Therefore any experimental results which are based on the range-energy relations for heavy nuclei are very unreliable. This is true, e.g., for all measurements of disintegration energies in disintegrations produced by fast neutrons (§102) as has been pointed out by Feather (F8) and by Bonner and Brubaker (B41).

§96. MOMENTUM RELATIONS AND RECOIL ENERGY

In the most common type of nuclear reactions, we have two nuclei (initial nucleus and incident particle) in the beginning, which are transformed in the reaction into two other nuclei. Ordinarily, one of these resultant nuclei is light (mass 4 or less) the other heavier. The lighter one has ordinarily greater velocity and therefore longer range; it is therefore easier to observe. We call the particle which is actually observed in the experiments the "emitted particle" and the other nucleus produced in the reaction the "residual" or "recoil" nucleus. Our problem in this section is to determine the total energy Q evolved in the reaction when the range of the emitted particle is known. This is done by calculating the recoil energy of the residual nucleus from the law of conservation of momentum.

Let the subscript 0 refer to the initial nucleus, 1 to the incident particle, 2 to the produced particle and 3 to the residual nucleus. Let $M_0 \cdots M_3$ be the masses of the four nuclei, and $M = M_0 + M_1 = M_2 + M_3$ the total mass (mass of the compound nucleus). The kinetic energies of the particles shall be denoted by $E_0 \cdots E_3$. The initial nucleus will in all practical cases be at rest so that $E_0=0$. The total energy evolution is

$$Q=E_2+E_3-E_1. \quad (765)$$

This quantity is the difference between the internal energies (masses) of the initial and final products of the reaction.

The law of conservation of momentum is

$$\mathbf{p}_3=\mathbf{p}_1-\mathbf{p}_2, \quad (766)$$

where $\mathbf{p}_i$ denotes the momentum of particle i, and $\mathbf{p}_0$ has been assumed to be zero (see above). We denote the angle between the directions of motion of incident and produced particle by ϑ. Using the relation[5]

$$p_i^2=2M_iE_i \quad (766a)$$

we obtain then

$$M_3E_3=M_1E_1+M_2E_2 -2(M_1M_2E_1E_2)^{\frac{1}{2}}\cos\vartheta. \quad (767)$$

Relation (767) gives the recoil energy E_3 if the angle ϑ is known. In most experiments, the geometrical arrangement is such that the produced particles leave the target perpendicularly to the incident beam ($\vartheta=90°$). Then (767) reduces to

$$E_3=(M_1E_1+M_2E_2)/M_3. \quad (767a)$$

Inserting this into (765), we obtain

$$Q=(M_2+M_3/M_3)E_2-(M_3-M_1/M_3)E_1. \quad (768)$$

Another arrangement often used in α-particle experiments is observation of the particles emitted *in* the direction of the incident beam ($\vartheta=0$). Then

$$E_3=(M_1^{\frac{1}{2}}E_1^{\frac{1}{2}}-M_2^{\frac{1}{2}}E_2^{\frac{1}{2}})^2/M_3. \quad (769)$$

In all cases, relative masses (atomic weights) may be used; and in almost all cases the exact atomic weights may be replaced by the mass numbers.

The formulae given have been widely used for calculating nuclear reaction energies. However, it must be pointed out that no experimental arrangement yet devised is geometrically so perfect that it is justified to put the angle ϑ exactly equal to 90° (or exactly equal to zero). In all practical cases, there will be a certain finite solid angle about the average of, say, 90° in which particles are observed. Since the recoil energy is very sensitive to the exact value of ϑ, corrections are necessary.

We assume an arrangement of the kind most commonly used (e.g. O8, C28) where the most direct path from target to detector is perpendicular to the incident beam but particles emitted in a rather wide solid angle may also reach the detector. According to (767), the recoil energy will be smaller if particle 2 is emitted in a more forward direction, and therefore the energy of particle 2 itself will be larger in this case. E_2 may be calculated as a function of ϑ if we express E_3 in (767) in terms of E_1E_2 and Q by (765). Solving for E_2, we obtain

$$ME_2^{\frac{1}{2}}=(M_1M_2)^{\frac{1}{2}}E_1^{\frac{1}{2}}\cos\vartheta +(MM_3Q+M_0M_3E_1-M_1M_2E_1\sin^2\vartheta)^{\frac{1}{2}}. \quad (770)$$

For small values of $\cos\vartheta$, this reduces to

$$E_2=E_2^0+2(M_1M_2E_1E_2^0)^{\frac{1}{2}}\cos\vartheta/M, \quad (770a)$$

$$\text{where } E_2^0=(M_3/M)Q+(M_3-M_1/M)E_1 \quad (770b)$$

is the energy of the emitted particle for emission at right angles to the incident beam.

The particles emitted in a more forward direction have, according to (770a), more energy and therefore longer range than those emitted at 90°. Since in most experiments the longest range is measured, the experimental determinations do not refer to emission at 90°. On the other hand, emission in the most forward direction possible geometrically is often not the most favorable case either, because then the particles have to

[5] Relativistic corrections need practically never be considered. The relativistically correct relation would be

$$p_i^2=2M_iE_i+E_i^2/c^2. \quad (A)$$

In the most important case $\vartheta=90°$ this gives a correction to the recoil energy

$$\delta E_3=(E_1^2+E_2^2-E_3'^2)/2M_3c^2, \quad (B)$$

where E_1 and E_2 are the measured energies of incident and emitted particle and E_3' the recoil energy calculated from the nonrelativistic formula (767). If the incident particle is slow compared to the emitted one, (B) reduces to

$$\delta E_3=\frac{E_2^2}{2M_3c^2}\frac{M_3^2-M_2^2}{M_3^2}. \quad (C)$$

As a typical example, we may choose the reaction $B^{10}+H^2 \rightarrow Be^8+He^4$. The energy E_2 of the emitted α-particles is about 12 MV, for small deuteron energy. Thus

$$\delta E_3=\frac{12^2}{2\cdot 8\cdot 930}\cdot\frac{3}{4}=0.007 \text{ MV}. \quad (D)$$

In most other cases, the relativistic correction is still smaller.

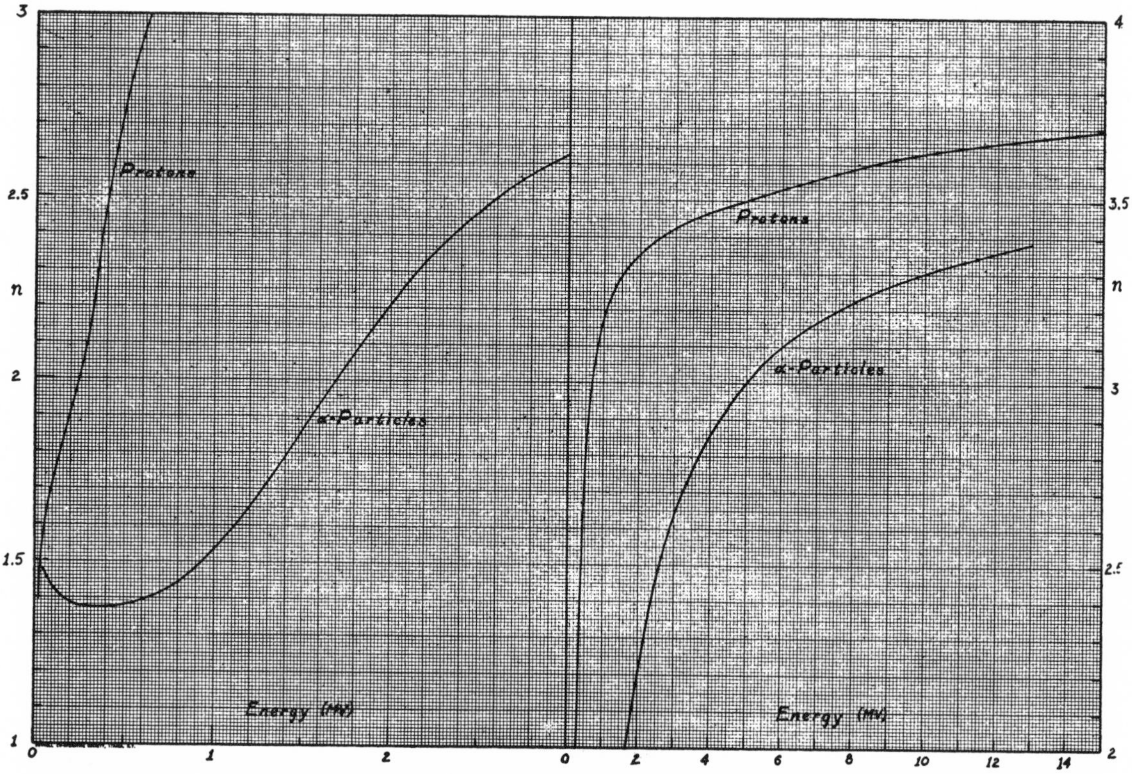

FIG. 35. Range exponent n for protons and α-particles. Over small regions, the range is proportional to the nth power of the velocity. Analytical definition: $n=2\ d \log R/d \log E$. The large deviations from Geiger's law ($n=3$) are apparent. The range exponent is important for thick target correction, angular straggling etc. (§§ 96, 97.)

traverse the material (air or absorbing foils) between target and detector obliquely. In general, there will be an optimum angle ϑ_0 for which the component x of the particle range R in the direction from target to detector is largest. We have

$$x=R \sin \vartheta \approx R(1-\tfrac{1}{2} \cos^2 \vartheta). \qquad (771)$$

Furthermore, if R_0 is the range corresponding to emission at right angles ($\vartheta=90°$), and $n=2\ d \log R/d \log E$ (cf. §95 D and Fig. 35), we have

$$R=R_0[1+n(M_1M_2)^{\frac{1}{2}}M^{-1}(E_1/E_2{}^0)^{\frac{1}{2}} \cos \vartheta]. \qquad (771a)$$

According to (771), (771a), x has a maximum at ϑ_0 where

$$\cos \vartheta_0=n\frac{(M_1M_2)^{\frac{1}{2}}}{M}\left(\frac{E_1}{E_2{}^0}\right)^{\frac{1}{2}}. \qquad (772)$$

We may now distinguish two cases: "Good geometry," i.e., small aperture of the observing apparatus, and "poor geometry," i.e. large aperture. The condition for "poor geometry" is that particles emitted at an angle ϑ_0 may pass freely from target to detector. "Good geometry" means that all particles emitted in the direction ϑ_0 are prevented from reaching the detector. Both these cases are easy to treat; only the intermediate case would offer difficulties.[6] Fortunately, all of the more important experiments were made either with "good" or with "poor" geometry. We have, e.g., "good geometry" in the experiments of Oliphant, Rutherford and collaborators (O8, O9) and in those of Cockcroft and Lewis (C28, C29) "poor geometry" in the older experiments of Cockcroft and Walton (C23, C25) and in the older experiments of the California group (L16). The values of ϑ_0 vary widely: e.g., for α-particles from the reaction $Li^7+H^1=2He^4$, we have about $\cos \vartheta_0=0.13$ for $E_1=100$ kv, and $\cos \vartheta_0=0.4$ for $E_1=1$ MV. For protons from $Na^{23}+H^2=Na^{24}+H^1$, we would have $\cos \vartheta_0=0.08$ only for

[6] Therefore an improvement of the geometry is not always an advantage, unless the aperture can be reduced to decidedly less than $\cos \vartheta_0$.

$E_1=1$ MV. Thus it is fairly difficult to obtain "good geometry" for reactions with heavy nuclei, but in this case the recoil energy is not particularly important.

We treat now the two possible cases separately:

a. "*Poor geometry*" (aperture$\gg\cos\vartheta_0$). The most penetrating particles are those emitted in the direction ϑ_0. They penetrate through a thickness of material

$$x_0=R_0(1+\tfrac{1}{2}\cos^2\vartheta_0). \tag{772a}$$

If their energy is calculated from their "range" x_0, it is found to be

$$E_2{}^m=E_2{}^0\left(1+\frac{1}{n}\cos^2\vartheta_0\right) \tag{772b}$$

or, inserting (772)

$$E_2{}^m=E_2{}^0+n\frac{M_1M_2}{M^2}E_1. \tag{772c}$$

The reaction energy Q may now be expressed in terms of the "observed" energy $E_2{}^m$ of the emitted particle by inserting the value of $E_2{}^0$ from (772c) into (768):

$$Q=\frac{M}{M_3}E_2{}^m+E_1\frac{M_1}{M_3}\left(1-n\frac{M_2}{M}\right)-E_1. \tag{773}$$

The term nM_2/M represents the correction for the forward emission of the particles. It often cancels or even over-compensates the term E_1M_1/M_3.

In the actual evaluation of experiments, it must be remembered that x_0 is only the effective range of the most penetrating particles but that we do not have a homogeneous group of particles of range x_0. If the direction of actual emission forms an angle χ with the "most favorable direction" (either because ϑ is larger or smaller than ϑ_0, or because of a sideways deviation from the plane defined by incident beam, target and detector, or both), the effective range will be $x=x_0\cos\chi$. Within certain limits given by the geometry, the number of particles emitted in the direction χ is proportional to $\sin\chi d\chi$. Therefore the number of particles of observable range between x and $x+dx$ is proportional to dx for all values of x below x_0 and above a certain lower limit x_1 given by the geometry. This will be of importance for determining the mean range from the extrapolated range (§97).

b. "*Good geometry*" (aperture$\ll\cos\vartheta_0$). The variation of the particle range because of the variation of ϑ will be slight. It may most conveniently be considered as a straggling of the ranges of the emitted particles which is added to the ordinary straggling arising from the process of stopping (§97A). We calculate this angular straggling assuming that the beam of emitted particles is defined by two circular openings of radius a at a distance b from each other ($b\gg a$). Such an arrangement has been used by Oliphant, Kempton and Rutherford (O8, O9) by Cockcroft and Lewis (C28, C29) and others.

A beam of particles which passes the first opening at an angle of χ with the normal, will be displaced by the amount

$$c=b\tan\chi \tag{774}$$

when it arrives at the second opening. The fraction of particles passing through this second circle is therefore proportional to the common area of two circles of radius a whose centers are a distance c apart. This area is, apart from the trivial factor a^2:

$$f(c)=2\varphi-\sin 2\varphi, \tag{774a}$$

where $$\cos\varphi=c/2a. \tag{774b}$$

The function $f(c)$, although not a Gaussian function, may be approximated by one. This is convenient for comparison with the ordinary straggling due to stopping. As the best Gaussian function approximating $f(c)$, we may take that which gives the same value for the average of c^2. From (774a) we have

$$\frac{(c^2)_{\text{Av}}}{a^2}=\frac{4\int_0^{\pi/2}\cos^2\varphi(2\varphi-\sin 2\varphi)d\varphi}{\int_0^{\pi/2}(2\varphi-\sin 2\varphi)d\varphi}$$
$$=2\frac{\pi^2-8}{\pi^2-4}=0.637, \tag{774c}$$

so that the "best" Gaussian function is

$$g(c)=\exp(-c^2/1.274a^2). \tag{775}$$

The energy of the particles depends on the angle ϑ with the incident beam. We assume again that the direction of observation is approximately $\vartheta=90°$. Denoting by ψ the angle between the direction of emission of a particle and the plane formed by the incident beam and the axis of the detecting system, we have since χ is small:

$$\tan^2 \chi = \chi^2 = \psi^2 + (\cos \vartheta)^2. \quad (775a)$$

The number of particles with a given ϑ detected is now proportional to

$$\int_{-\infty}^{\infty} g(c)d\psi \sim \exp\,(-b^2 \cos^2 \vartheta / 1.274a^2). \quad (775b)$$

The extension of the integration over ψ from $-\infty$ to $+\infty$ is allowable because of the rapid decrease of $g(c)$. Expressing $\cos \vartheta$ in terms of the range R of the particle by (771a), we find that the number of particles of range R observed will be proportional to[7]

$$g(R) = \exp\,(-\pi(R-R_0)^2/4\gamma^2), \quad (776)$$

where
$$\gamma^2 = \pi \frac{\pi^2-8}{\pi^2-4} \frac{a^2}{b^2} n^2 R_0^2 \frac{M_1 M_2}{M^2} \frac{E_1}{E_2^0}. \quad (776a)$$

The numerical factor is almost exactly unity, so that

$$\frac{\gamma}{R_0} = \frac{a}{b} n \frac{(M_1 M_2)^{\frac{1}{2}}}{M} \left(\frac{E_1}{E_2^0}\right)^{\frac{1}{2}}. \quad (777)$$

Since the geometry is "good," the ratio a/b is small. Also, in general the masses of incident and emitted particle, M_1 and M_2, are small compared to the total mass M, and the energy of the incident particle is small compared to that of the outgoing. Therefore the "angular straggling" γ is small compared to the range R. However, it is usually of the same order as the natural straggling due to stopping.

In this discussion of the case of good geometry, we have neglected entirely the *geometrical* effect of the obliquity of the particles on the apparent range which played a rather important rôle in the case of poor geometry. The apparent range of a particle moving at an angle χ with the axis of the detecting system is reduced by a factor

$$\cos \chi \approx 1 - \tfrac{1}{2}\chi^2. \quad (778)$$

With the distribution function (775), this gives in the average

$$(\cos \chi)_{\text{Av}} = 1 - a^2/\pi b^2. \quad (778a)$$

[7] This form of the exponential is convenient for calculating the extrapolated range, §97,

In the experiments of Oliphant and collaborators (O8, O9), we have $a/b = 1/12$, in those of Cockcroft and Lewis (C28, C29), $a/b = 0.1125$. Thus the correction term in (778a) amounts to 0.22 and 0.40 percent, respectively, which is just noticeable when the reaction energies are given to 0.01 MV as is customary (cf. Tables LII ff).

However, the actual correction is even less than this. In the experiments of Oliphant, the ranges of the disintegration particles were compared to those of Th C″ α-particles. The latter are, of course, subject to the same obliquity correction as the former. Therefore, the obliquity correction should, in these experiments, only be applied to the *difference* between the particle range and the range of the Th C″ alphas. In a case like the α-particles from $Li^7 + H^1 = 2He^4$, which have almost the same range as Th C″ alphas, the correction is therefore entirely negligible, and even for the long range alphas from $Li^6 + H^2 = 2He^4$, it amounts only to 0.02 cm (=0.02 MV in Q). Therefore, we have applied this correction only to the best investigated reactions.

If the range is determined in mica or another substance whose stopping power must be measured before the actual experiment, no obliquity correction should be applied at all because the percentage correction is the same for the particles used for the calibration as for the disintegration particles themselves.

Besides the change of the average range, there is also a very small contribution to the angle straggling (776a) from the obliquity.

Three-particle disintegrations

Disintegrations leading to three product nuclei follow, according to the general theory (§85) the scheme

$$A + P \to C \to B^* + R; \qquad B^* \to D + S. \quad (779)$$

The indices 0, 1, . . . 5 shall refer to the particles A, P, R, B, S and D, respectively. M_i denotes the mass, E_i the energy of particle i, $M = M_0 + M_1$ is the mass of the compound nucleus C, Q_1 the energy evolution in the first part of the reaction, Q_2 the energy set free in the break-up of nucleus B^*. The velocity of the center of gravity is

$$v_0 = (2M_1E_1)^{\frac{1}{2}}/M. \quad (779a)$$

In the center of gravity system the energies of the particles are:

$$E_2' = [Q_1 + (M_0/M)E_1]M_3/M, \quad (780)$$

$$E_3' = [Q_1 + (M_0/M)E_1]M_2/M, \quad (780a)$$

$$[(E_3'M_4)^{\frac{1}{2}}-(Q_2M_5)^{\frac{1}{2}}]^2/M_3<E_4'<[(E_3'M_4)^{\frac{1}{2}}+(Q_2M_5)^{\frac{1}{2}}]^2/M_3 \quad (780b)$$

and E_5' correspondingly. The distribution of the values of E_4' between the limits indicated in (780b) is uniform if the direction of emission of the particles D and S is independent of the direction of motion of nucleus B.

The energy of any particle i emitted at an angle ϑ with the incident beam is in the *ordinary* coordinate system

$$E_i=E_i^0\left(1+2\frac{(M_iM_1)^{\frac{1}{2}}}{M}\left(\frac{E_1}{E_i^0}\right)^{\frac{1}{2}}\cos\vartheta\right), \quad (781)$$

where E_i^0 is the energy of a particle emitted at right angles to the incident beam:

$$E_i^0=E_i'-M_iM_1M^{-2}E_1, \quad (781a)$$

E_i' being the energy in the center of gravity system. Formulae (781, 781a) are valid for *any* type of disintegration and any mechanism (cf. (770a)). They show that the corrections for the deviation of the direction of emission from 90° are the same for three as for two particle disintegrations.

The energies of the particles emitted at right angles in the three-particle disintegration are, according to (780) to (781a):

Particle R: $E_2^0=Q_1M_3/M+E_1(M_3-M_1)/M$. (781b)

Particle S: for $E_1\ll Q_1$ and Q_2, maximum and minimum energies:

$$(E_4)_{\min}^{\max}=\left[\left(Q_1\frac{M_2M_4}{MM_3}\right)^{\frac{1}{2}}\pm\left(Q_2\frac{M_5}{M_3}\right)^{\frac{1}{2}}\right]^2+\frac{M_4}{M^2}E_1\left[\frac{M_0M_2}{M_3}\left|1\pm\left(\frac{Q_2M_5M}{Q_1M_4M_2}\right)^{\frac{1}{2}}\right|-M_1\right]. \quad (781c)$$

With the help of (781b) and (781c), the reaction energies Q_1 and Q_2 can be deduced from the observations. This has been done, for the reaction $B^{11}+H^1\rightarrow 3He^4$, in §85.

Determination of reaction energies from the angular distribution of recoil nuclei

It has been suggested by Newson (N6) that the energy evolved in nuclear reactions can be determined from the angular distribution of the recoil nuclei. This can be conveniently done if the residual nucleus is radioactive.

Let χ denote the angle between the incident beam and the direction of motion of the recoil nucleus in the center of gravity system, φ the same angle in the ordinary coordinate system, v_0 the velocity of the center of gravity and v that of the recoil nucleus in the center of gravity system. Then in the ordinary coordinate system the velocity component parallel to the incident beam is $v_0+v\cos\chi$, the perpendicular component $v\sin\chi$, and therefore

$$\cot\varphi=\frac{v_0}{v\sin\chi}+\cot\chi. \quad (782)$$

We must distinguish two cases:

(a) $v_0<v$: Then, as χ goes from 0 to π, φ goes through the same interval.

(b) $v_0>v$: In this case, $\cot\varphi$ has a minimum for $\cos\chi=-v/v_0$.

At the minimum, we have

$$\cot\varphi_0=+(v_0^2/v^2-1)^{\frac{1}{2}}, \quad (782a)$$

$$\sin\varphi_0=v/v_0. \quad (782b)$$

This means that the recoil nuclei can only be emitted into a cone around the direction of the incident beam whose half-aperture φ_0 is less than 90°. By measuring the maximum angle at which recoil nuclei appear, v/v_0 can be found. This in turn determines Q; we have

$$v_0=(2M_1E_1)^{\frac{1}{2}}/M, \quad (783)$$

$$v=\left[\frac{2M_2}{MM_3}\left(Q+\frac{M_0}{M}E_1\right)\right]^{\frac{1}{2}}, \quad (783a)$$

$$Q=\frac{E_1}{M}\left[\frac{M_1M_3}{M_2}\left(\frac{v}{v_0}\right)^2-M_0\right]. \quad (783b)$$

The method is well applicable only in case (b), i.e., if $v_0>v$. Assuming that initial and resultant nucleus are heavy compared to incident and outgoing particle, this condition is equivalent to

$$M_1E_1>M_2(Q+E_1), \quad (783c)$$

$$Q<(M_1-M_2)E_1/M_2. \quad (783d)$$

If the incident particle is *lighter* than the outgoing, this requires a highly endoergic process. Therefore the method is practically never applicable to processes in which the emitted particle is an α-particle and the incident one a lighter particle (proton, neutron, deuteron). If the incident particle is heavier than the outgoing one, the condition requires that the process be not too exoergic. E.g., if the incident particle is a deuteron, the outgoing a proton or neutron, the condition is $Q<E_1$. In this case, the method is applicable when the incident particle is fast enough.

The angular distribution itself can be calculated if we assume that in the center of gravity system any direction of emission is equally probable. Then the number of particles per unit solid angle in the ordinary coordinate system is

$$N(\varphi)=\frac{\sin\chi d\chi}{\sin\varphi d\varphi}. \quad (784)$$

In case (a), this expression has the value

$$N(\varphi)\propto(1-\gamma^2\sin^2\varphi)^{-\frac{1}{2}}+2\gamma\cos\varphi, \quad (784a)$$

while in case (b):

$$N(\varphi)\propto(1-\gamma^2\sin^2\varphi)^{-\frac{1}{2}};\ \gamma=v_0/v. \quad (784b)$$

In this latter case, $N(\varphi)$ becomes infinite near the limiting angle φ_0 which should facilitate observation of this angle.

§97. Extrapolated and Mean Range, Corrections for Thick Target, etc.

A. Straggling of energy loss

Since the loss of energy by charged particles occurs in discrete amounts, the energy lost after a given length of path will show statistical fluctuations. It can be shown that the mean square fluctuation of the energy is given by

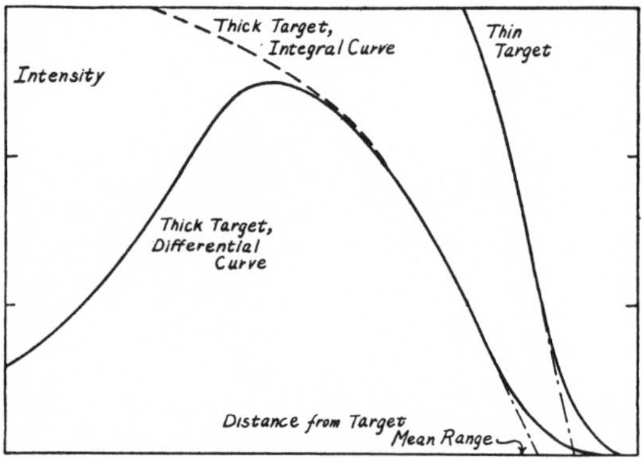

FIG. 36. Theoretical range distribution of disintegration products from a thin and thick target. For the thick target, the distribution is given (a) for a detector counting *all* particles ("integral" curve), (b) for a "differential detector" counting only particles near the end of their range. All curves are "extrapolated" along the steepest tangent. The extrapolated range for the thin target is seen to be much longer than for the thick target while for the latter it is practically independent of the type of detector used ("integral" or "differential"). The mean range is also indicated; the difference between extrapolated and mean range is much larger for the thin than for the thick target.

$$[(E^2)_{\text{Av}}-(E_{\text{Av}})^2]_X = N\int_0^X dx \sum_n E_n{}^2\sigma(E_n). \quad (785)$$

Here X is the amount of stopping material traversed, N the number of atoms per cm³ in the material, E_n the excitation energy of the nth excited level of these atoms, and $\sigma(E_n)$ the cross section for the excitation of this level by the incident particle. $\sigma(E_n)$ is given by

$$\sigma(E_n)=\frac{2\pi e^4z^2}{mv^2}\int\frac{dQ}{Q^2} \times\left|\int \Psi_0\Psi_n{}^*\sum_j \exp\left(i(\mathbf{p}-\mathbf{p}')\cdot\mathbf{r}_j/\hbar\right)d\tau\right|^2, \quad (785a)$$

where $\mathbf{p}$ and $\mathbf{p}'$ are the momenta of the incident particle before and after collision, $\mathbf{r}_j$ the position of the electron j in the atom, Ψ_0 and Ψ_n the atomic wave functions of ground state and excited state, and

$$Q=(\mathbf{p}-\mathbf{p}')^2/2m, \quad (785b)$$

differing by a constant factor from the definition given in (754). σ is, of course, a function of the energy of the particle and therefore of the distance X traveled from the source.

Equation (785) may be evaluated by inverting the order of summation over n and integration over Q and using the completeness relation (cf. B8). In this way, an expression is obtained which involves only the wave function Ψ_0 of the ground state. We may then assume that Ψ_0 can be written as an antisymmetrical product (determinant) of wave functions of the individual electrons in the atom (Hartree-Fock approximation). Then we find

$$\frac{d}{dX}[(E^2)_{\text{Av}}-(E_{\text{Av}})^2]=\frac{2\pi e^4z^2N}{mv^2}\sum_k S_k, \quad (786)$$

where S_k is the contribution of the kth electron in the atom to the straggling (except for a constant factor). We have

$$S_k=\int_{Q_{\min}}^{Q_{\max}} dQ\left[1+\frac{4}{3}\frac{E_k-V_{ka}}{Q}-\frac{1}{Q^2}\sum_{l(\text{occ})}(E_l-E_k)^2 \times\left|\int \exp\left(i(\mathbf{p}-\mathbf{p}')\cdot\mathbf{r}/\hbar\right)\psi_k\psi_l d\tau\right|^2\right]. \quad (786a)$$

Here E_k is the energy of the kth electron, V_{ka} its average potential energy in the atomic field so that E_k-V_{ka} is the average kinetic energy. The sum over l extends over all electronic states occupied by electrons in the atom in question; it takes account of the Pauli principle. ψ_k and ψ_l are the wave functions of the electronic states k and l. $Q_{\max}$ is the largest value which Q can take when momentum is conserved between the incident particle and the electron ejected from the atom,[8] *viz.*

$$Q_{\max}=2mv^2=(4m/M)E. \quad (786b)$$

$Q_{\min}$ is a suitable average of the quantity Q_0 defined in (754a); in sufficient approximation we may put

$$Q_{\min}=MI_k{}^2/4mE=I_k{}^2/2mv^2, \quad (786c)$$

where I_k is the average excitation potential of the kth electron defined similarly as in (749a). The contribution of small Q's to (786a) is small so that the exact value of $Q_{\min}$ does not matter.

For small values of $Q(<I_k)$, the second term in the square bracket in (786a) is partly compensated by the last term (sum over l), while the first term is comparatively small. For larger Q, the

[8] It is easy to show (B8) that for larger values of Q the expression (785*a*) becomes very small for any state n. (786*b*) corresponds to a velocity $2v$ of the ejected electron.

sum is negligible. Therefore we may write

$$S_k = Q_{\max} + \kappa'(E_k - V_{ka}) \log (Q_{\max}/Q_{\min}), \quad (787)$$

where κ' is between about 2/3 and 4/3. (For hydrogen, it is exactly 4/3.) The average potential energy V_{ka} would be equal to $2E_k$ for a pure Coulomb field (hydrogen) but is very much larger ($10E_k$ and more) for the outer electrons of heavy atoms. Thus $E_k - V_{ka}$ will in general be larger than the average excitation potential I_k. We may thus replace $\kappa'(E_k - V_{ka})$ by κI_k where κ is larger than κ', maybe about 4/3 or larger. Inserting (786b, c), this gives

$$S_k = 2mv^2 + 2\kappa I_k \log (2mv^2/I_k) \quad (787a)$$

and (786) becomes

$$\frac{d}{dX}[(E^2)_{\text{Av}} - (E_{\text{Av}})^2] = 4\pi e^4 z^2 N\left(Z' + \sum_n \kappa_n \frac{I_n Z_n}{mv^2} \log \frac{2mv^2}{I_n}\right). \quad (788)$$

Here Z' is the total number of effective electrons as defined in §95E, Z_n the number of electrons in the nth shell, I_n their average excitation energy and κ_n the respective value of the constant κ.

For high energies (788) reduces to

$$d[(E^2)_{\text{Av}} - (E_{\text{Av}})^2]/dX = 4\pi e^4 z^2 NZ'. \quad (788a)$$

This formula (only with Z' replaced by the actual number of electrons in the atom, Z) had been given in 1915 by Bohr on the basis of classical mechanics. For lower energies of the particle, the sum over n in (788) should be added. This sum may be fairly large, e.g., for α-particles of 5 MV in air the K electrons ($I_K \sim 700$ volts, $mv^2 \sim 1300$ volts, $\kappa_K \sim 4/3$, $Z \sim 1.8$) contribute about 1.7, the L electrons ($I_L \sim 40$, $\kappa_L \sim 4/3$, $Z_L = 5.4$) about 0.9, so that the second term in (788) is about 40 percent of the first ($Z' = Z = 7.2$). This shows that the straggling may, at lower energies of the particle, be considerably larger than Bohr's classical value (788a), in agreement with observation. However, the deviation from Bohr's value can certainly not be as large (100 percent) as earlier measurements by Briggs (B55) would indicate. The most recent measurements by Bennet (B6a) with 8 MV alphas in mica give 20 percent deviation, in satisfactory agreement with (788).

B. The extrapolated range for a homogeneous group of particles

Formula (788) gives the fluctuation of the energy of particles which have traversed a certain thickness of material. We are primarily interested in the reverse, *viz.* the fluctuation of the distance traveled by particles which have lost the same amount of energy. We have (cf. B34)

$$\frac{d}{dE}[(X^2)_{\text{Av}} - (X_{\text{Av}})^2] = \frac{d}{dX}[(E^2)_{\text{Av}} - (E_{\text{Av}})^2]\left(\frac{dE}{dX}\right)^{-3}. \quad (789)$$

Inserting (788) and integrating from the initial energy E_0 of the particle to zero, we obtain for the mean square fluctuation of the particle range:

$$(R - R_0)^2_{\text{Av}} = \int_0^{E_0} \frac{4\pi e^4 z^2 NZ'}{(dE/dX)^3} \times \left(1 + \sum_n \frac{\kappa_n I_n Z_n}{mv^2 Z'} \log \frac{2mv^2}{I_n}\right) dE. \quad (790)$$

From this relation the fluctuation in range may be calculated if the κ_k are known. For the actual calculations, we have put $\kappa = 4/3$ for both K and L shell of air. The dependence of all other quantities in (790) on the energy is known.

To a sufficient approximation, the distribution of the ranges of a homogeneous group of particles is given by a Gaussian distribution:

$$p(R)dR = \pi^{-\frac{1}{2}} \alpha e^{-\alpha^2 (R - R_0)^2} dR. \quad (791)$$

The range fluctuation is then

$$(R - R_0)^2_{\text{Av}} = \int p(R)(R - R_0)^2 dr = 1/2\alpha^2. \quad (791a)$$

From this equation and (790) α is determined.

Most of the important detecting devices (except the cloud chamber) measure the number of particles reaching a certain distance r from the source, in other words the particles whose range is greater than r, *viz.*

$$P(r) = \int_r^\infty p(R)dR = \tfrac{1}{2}[1 - \Phi(\alpha(r - R_0))]. \quad (792)$$

$P(r)$ is given in Fig. 36 (curve marked "thin target"). The mean range R_0 is that range which is reached by just one-half of the particles. This would seem to allow an easy determination of R_0. However, this method is restricted to a beam of particles perfectly homogeneous in the beginning. The material emitting the particles must be infinitely thin in order to provide no stopping of its own and its surface must be perfectly smooth. Even so, the beam will be homogeneous in energy only if emitted by a natural radioactive substance. The energy of the particles from artificial transmutations always depends on the angle of their emission (§96) which prevents complete homogeneity of the beam.

For an inhomogeneous beam the point at which the intensity is reduced to one-half has no particular significance. If, in particular, the inhomogeneity arises from a finite thickness of the source, the particles of larger range will be more significant as they come from the top layers of the source. For this reason it is in general preferable to measure the "extrapolated" range rather than the mean range. The extrapolated range is obtained by drawing the steepest tangent to the number-range curve (i.e., the experimental curve giving the number of particles detected as a function of the distance r from the source). The intersection of this tangent with the axis of abscissae gives the extrapolated range (cf. Fig. 36).

The range distribution function $P(r)$ for homogeneous particles has its steepest slope at $r=R_0$. We have

$$P(R_0)=\tfrac{1}{2};\quad P'(R_0)=-\alpha/\sqrt{\pi}. \qquad (792a)$$

Therefore the extrapolated range is

$$R_{\text{extr}}=R_0-P(R_0)/P'(R_0)=R_0+\tfrac{1}{2}\sqrt{\pi}/\alpha. \qquad (792b)$$

We denote by s the difference between extrapolated and mean range and obtain

$$s^2=\tfrac{1}{2}\pi(R-R_0)^2_{\text{Av}}, \qquad (793)$$

$$\alpha=\tfrac{1}{2}\sqrt{\pi}/s, \qquad (793a)$$

$$p(R)dR=\tfrac{1}{2}e^{-\pi(R-R_0)^2/4s^2}dR/s. \qquad (794)$$

The difference s between mean and extrapolated range has been measured carefully for the α-particles from Th C′ (long), Ra C′ and Po, with the results 0.111, 0.075 and 0.043 cm, respectively. On the other hand, s may be calculated from (793) (790). The result of such a calculation for Po would be rather unreliable because the theory of stopping is known to be inaccurate at low energies. However, the difference between the values of s^2 for Po and the two other groups of α-particles should be comparable with the theoretical value. We have

	$s^2-s_{\text{Po}}^2$ (in 10^{-4} cm²) experimental	theoretical
Ra C′ α's	38	36
Th C′	105	121

The agreement is satisfactory, especially considering that s^2 is rather sensitive to small experimental errors in s. We are therefore justified in using the theoretical relation to calculate the "straggling" s for particles faster than Th C′ α-particles for which the theory should be even more accurate.

The result is given in Fig. 37 in which the straggling is given as a function of the range of the particles. The straggling of α-particles is about 1.2 percent of their range for energies around 4 MV, about 1.0 percent for 16 MV and would be about 0.85 percent for 50 MV energy.

The dependence of the straggling on the mass and charge of the particle can be found immediately from (790). The bracket $1+\cdots$ in (790) is a function of the velocity only. The energy loss dE/dx is (§95) proportional to z^2 times a function of the velocity. The energy E itself is $\tfrac{1}{2}Mv^2$. Therefore

$$s=M^{\frac{1}{2}}z^{-2}g(v), \qquad (795)$$

g being another function of v. The straggling of particles of mass (atomic weight) M, charge z and range R can thus be expressed in terms of the straggling of protons of range z^2R/M, *viz.*:

$$(s/R)_{M,\,z,\,R}=M^{-\frac{1}{2}}(s/R)_{1,\,1,\,z^2M^{-1}R}. \qquad (795a)$$

In this way, curve 37 which is given for protons may be used for any kind of particle.

C. Thick target

For reasons of intensity, most nuclear transmutation experiments are made in "thick" targets, i.e., targets thick compared to the range of the incident particles in the target material.[9]

[9] The use of thick targets is no disadvantage, as it can be easily corrected for. However, "medium thin" targets,

The particles produced in the reaction are therefore partly stopped in the target itself, so that their observable range is reduced. Only the particles produced in the surface layer of the material will have the full range. In order to determine this full range, *it is not sufficient to "extrapolate" the number-range curve along its steepest tangent* as described in the preceding section and to consider the extrapolated range so obtained as the extrapolated range for the particles produced in the top layer of the target. The reason why this leads to fallacious results is that the shape of the number-range curve is altered by the presence of the particles coming from greater depths of the target. We shall, in this section, derive a method for the determination of the mean range of the particles coming from the top layer of the target, from the measured number-range curve.

We assume that the incident beam strikes the target at an angle of 45°, and that the particles produced in the reaction are observed again at an angle of 45° with respect to the target, and at right angles with the incident beam. This arrangement is the most symmetrical and most commonly used; it means that the incident particles travel the same distance in the target as the produced ones. Generalization of the formulae to other arrangements is obvious.

Consider particles produced at a depth X in the target, X being measured along the path of the particle in cm air equivalent. Let $R(X)$ be the residual range of these particles after leaving the target. Then $R(X)$ will decrease with increasing depth X for two reasons: (1) because the part X of the particle range itself lies in the target, and (2) because the incident particles which have penetrated to the depth X have lost some energy, and therefore impart less energy to the produced particles. If the emitted particle is observed exactly at right angles to the incident beam, we have from (770b) for not too large X:

$$R(X)=R(0)-X\left[1+\frac{M_3-M_1}{M}\frac{(dE/dX)_1}{(dE/dX)_2}\right]. \quad (796)$$

The notations are the same as in §96, the indices 1 and 2 referring to incident and emitted particle, respectively. The stopping power dE/dX is proportional to the stopping cross section σ shown in Fig. 31. Alternatively we may write

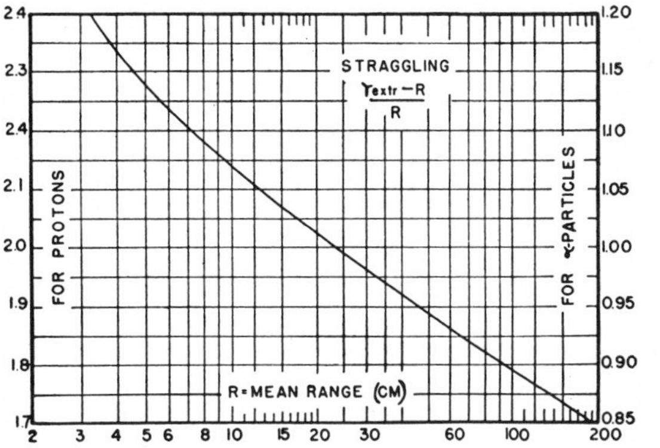

FIG. 37. Straggling of protons and α-particles. The difference between extrapolated and mean range, in percent of the latter, is given as a function of the mean range.

$$\frac{(dE/dX)_1}{(dE/dX)_2}=\frac{\sigma_1}{\sigma_2}=\frac{E_1}{E_2}\frac{R_2n_2}{R_1n_1}, \quad (796a)$$

where R is the range and n the range exponent given in Fig. 35. The quantities E_1, R_1 etc., should be taken about at the depth $\frac{1}{2}X$. The first term in the square bracket in (796) is more important if the produced particle is heavy and slow, in the other cases the second term is usually more important.

If the incident particle is charged (for neutrons cf. below), its efficiency in producing the nuclear reaction decreases approximately exponentially with X. Since we are only calculating a fairly small correction, it is sufficient to take the simple Gamow formula for the reaction probability (cf. §78, footnote 59), *viz.*

$$\varphi(v)\sim\exp\left(-2\pi\frac{zZe^2}{\hbar v}\right)$$
$$=\exp\left(-0.99\frac{zZM_1^{\frac{1}{2}}}{E^{\frac{1}{2}}}\right), \quad (797)$$

where M_1 is the atomic weight of the incident particle and E its energy in MV at any point in the target. The number of particles produced at a depth X is thus

$$\varphi(X)dX\sim\exp\left[-\frac{1}{2}0.99\frac{zZM_1^{\frac{1}{2}}}{E_1^{3/2}}\left(\frac{dE}{dX}\right)_1X\right]dX, \quad (797a)$$

where E_1 is a suitable average of the energy of the

of a thickness comparable to the range of the particles, are decidedly undesirable.

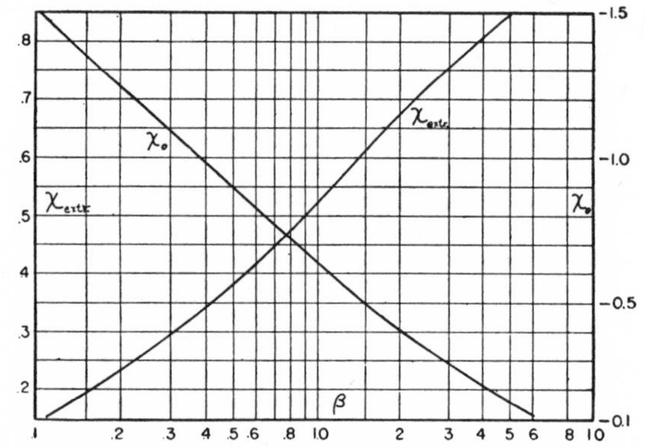

FIG. 38. Thick target correction. x_{extr} and x_0 as functions of the constant β which describes the depth of penetration into the target (cf. (798a)). To obtain the mean range from the most probable range, sx_0 should be added where s is the straggling (including angle straggling, see text). From the *extrapolated* range, the mean range is obtained by subtracting sx_{extr}.

incident particle in the target. The probability that the produced particle has a mean residual range r may be written

$$N(r)dr = (\beta/s)e^{-\beta(R_0-r)/s}dr, \tag{798}$$

where s is the straggling of a homogeneous group of particles as calculated in section B, R_0 is the mean range of the particles produced in the top layer of the target, and (cf. (796)–(798))

$$\beta = 0.99zZ\left(\frac{M_1}{E_1}\right)^{\frac{1}{2}}\frac{s_2}{R_2}\left[\frac{R_1}{R_2}n_1 + \frac{E_1}{E_2}\frac{M_3-M_1}{M}n_2\right]^{-1} \tag{798a}$$

In this formula suitable average values should be inserted for energy and range of the incident particle. It is more convenient to use the approximate formula[10]

$$\beta = \left[zZ\left(\frac{M_1}{E_1}\right)^{\frac{1}{2}} + 4\right]\frac{s_2}{R_2}\Big/\left[\frac{R_1}{R_2}n_1 + \frac{E_1}{E_2}\frac{M_3-M_1}{M}n_2\right], \tag{799}$$

in which now E_1 and R_1 refer to the *initial* energy and range of the incident particle.

Formula (798) gives the number of particles with a *mean* range r which leave the target. For a given mean range, the number of particles having an actual range R is given by (794). Thus the number of particles with actual range R coming from a thick target is given by

$$t(R)dR = dR(\beta/2s^2)\int_0^{R_0} dr \exp\left(-\pi(R-r)^2/4s^2 - \beta(R_0-r)/s\right). \tag{800}$$

We introduce the abbreviations

$$\frac{R-R_0}{s} = x; \qquad \frac{R_0-r}{s} = y \tag{800a}$$

and find

$$t(x)dx = \tfrac{1}{2}\beta e^{\beta(x+\beta/\pi)} \times \{1-\Phi(\tfrac{1}{2}\pi^{\frac{1}{2}}x + \pi^{-\frac{1}{2}}\beta)\}dx, \tag{801}$$

where Φ is the error integral.

The maximum of $t(x)$ lies always at a negative x, i.e., at a range smaller than the mean range of the particles emitted from the surface of the target. For large β (rapid decrease of the disintegration probability with the depth) the most probable range corresponds to

$$x_0 = -1/\beta \tag{801a}$$

for small β (deep penetration) we have approximately

$$x_0 = -2\pi^{-\frac{1}{2}}|\log 2\beta|^{-\frac{1}{2}}. \tag{801b}$$

For intermediate values of β, the most probable x_0 is given in Fig. 38. This quantity is important for the evaluation of cloud chamber measurements: The mean range of the particles emitted from the surface of a thin target is equal to the most probable range determined from experiment, plus $s|x_0|$.

The total number of particles reaching a given point x is given by

$$T(x) = \int_x^\infty t(\xi)d\xi = \tfrac{1}{2}[1-\Phi(\tfrac{1}{2}\pi^{\frac{1}{2}}x)] - \tfrac{1}{2}e^{\beta x+\beta^2/\pi} \cdot [1-\Phi(\tfrac{1}{2}\pi^{\frac{1}{2}}x + \pi^{-\frac{1}{2}}\beta)]. \tag{802}$$

A typical number-range curve of this type is shown in Fig. 36 (thick target, integral curve), together with the corresponding thin target

[10] This formula amounts to the assumption that the number of disintegrations produced by incident particles with energies between E and $E+dE$ is proportional to $\exp\left[-\frac{E_1-E}{E_1}\left(\frac{1}{2}zZ\left(\frac{M_1}{E_1}\right)^{\frac{1}{2}}+2\right)\right]$ where E_1 is the initial energy of the incident particle. For large $\frac{1}{2}zZ(M_1/E_1)^{\frac{1}{2}}$, this reduces to (797); in the opposite case, it gives for the *average* value of the energy of the incident particle just $\frac{1}{2}E_1$ as it should be.

curve. The extrapolated range is given by

$$x_{\text{extr}} = x_0 + T(x_0)/t(x_0), \tag{802a}$$

where x_0 (cf. above) is the maximum of the function t, i.e., the point of steepest slope of the number-range curve. In the limiting cases we have

$$x_{\text{extr}} = 1 - \tfrac{1}{2}\beta \qquad \beta \gg 1, \tag{802b}$$

$$x_{\text{extr}} = \beta(x_0^2 + \tfrac{1}{2}) \quad \beta \ll 1. \tag{802c}$$

For intermediate values of β, x_{extr} is given in Fig. 38. For small β (deep penetration into the target), x_{extr} goes to zero, i.e., the *extrapolated range obtained* in the usual way *from* the *experimental* number-range curve *gives* directly the *mean range* of the *particles* coming *from* the *top layer* of the target. In the general case, the amount sx_{extr} should be subtracted from the experimental extrapolated range in order to obtain the mean range mentioned.

Many experiments have been carried out with "differentiating" detectors, especially with ionization chamber plus linear amplifier at high bias. In this case, only particles near the end of their range will be detected. The evaluation is the same as described above, provided the detection extends over a portion of the range large compared to the straggling s. An example is given in Fig. 36 (thick target, differential curve). It is seen that this curve coincides completely with that obtained from a nondifferentiating detector "thick target, integral curve" in the region important for the determination of the extrapolated range.

If the incident particle is a *neutron*, the number of produced particles emitted at a depth x is in first approximation independent of x. In this case we have to put $\beta = 0$ in the above theory. The number of particles of "range" x is then (cf. (801), divide first by β!)

$$t(x)dx = \tfrac{1}{2}[1 - \Phi(\tfrac{1}{2}\pi^{\frac{1}{2}}x)]dx, \tag{803}$$

i.e., it reaches no maximum but remains constant for small ranges. Similarly, $T(x)$ increases linearly with decreasing range, *viz.*

$$T = (1/\pi)e^{-(\pi/4)x^2} - \tfrac{1}{2}x[1 - \Phi(\tfrac{1}{2}\pi^{\frac{1}{2}}x)]. \tag{803a}$$

The determination of the extrapolated range causes no difficulty; it is, as already mentioned, equal to the mean range of the particles emitted from the surface of the target.

D. Evaluation of experiments with "good" geometry

As we have shown in §96, the spread in angle of emission is, in the case of "good" geometry equivalent to additional straggling. If s is the straggling due to the process of stopping as discussed in section B (range straggling), the total straggling s' is given by

$$s'^2 = s^2 + \gamma^2 = s_2^2 + n_2^2\left(\frac{a}{b}\right)^2 R_2^2 \frac{M_1 M_2}{M^2} \frac{E_1}{E_2^0}. \tag{804}$$

Here γ is given in (777); the indices 1 and 2 refer again to incident and outgoing particle, a and b are radius and length of the cylindrical channel defining the beam; R is the range and n its logarithmic derivative with respect to the velocity. s' has to be used instead of s in the formula (799) for β.

We give an example of the determination of the reaction energy Q in the case of "good geometry." We choose the reaction

$$N^{14} + H^2 = C^{12} + He^4 + Q, \tag{805}$$

which was studied by Cockcroft and Lewis (C28, C29). The beam of α-particles was defined by a channel of $a = 0.45$ cm radius and $b = 4$ cm length at right angles to the incident deuteron beam. The energy of the incident deuterons was $E_1 = 0.575$ MV; the observed extrapolated range of the α-particles 11.37 ± 0.1 cm. Without correction, this would correspond (Fig. 29) to $E_2 = 10.45$ MV. The criterion for "good geometry" is that $\cos\vartheta_0$ (cf. (772)) is greater than $a/b = 0.1125$. We have from (772):

$$\cos\vartheta_0 = 3.3 \cdot \frac{(2\cdot 4)^{\frac{1}{2}}}{16} \cdot \left(\frac{0.575}{10.45}\right)^{\frac{1}{2}} = 0.14.$$

Thus the condition of good geometry is just fulfilled. We have then:

(1) Correction because the range was measured in mica instead of air: Fig. 34 gives for Al and 11.4 cm range the correction 0.08 cm; the mica correction is one-half of this amount, i.e., +0.04 cm.

(2) Straggling: (a) Range straggling (according

to Fig. 37) $s=1.06$ percent (of the range) for α-particles of 11.4 cm range.

(*b*) Angle straggling (cf. (777). For α-particles of about 10.5 MV energy, Fig. 32 gives $n=3.32$. Therefore

$$\frac{\gamma}{R}=\frac{0.45}{4.0}\cdot 3.32\frac{(2\cdot 4)^{\frac{1}{2}}}{16}\cdot\left(\frac{0.575}{10.45}\right)^{\frac{1}{2}}=0.0154,$$

i.e. 1.54 percent.

(*c*) Total straggling $s'=(s^2+\gamma^2)^{\frac{1}{2}}=1.87$ percent of range.

(3) Thick target correction. For the incident deuterons ($E_1=0.575$ MV) Fig. 30 gives $R_1=0.86$ cm and Fig. 35 gives $n_1=2.1$. Therefore (799) gives

$$\beta=\frac{[1\cdot 7(2/0.575)^{\frac{1}{2}}+4]\cdot 0.0187}{2.1\cdot(0.86/11.37)+3.32(0.575/10.5)(10/16)}$$

$$=\frac{17\cdot 0.0187}{0.160+0.11_4}=1.18.$$

From Fig. 38 we find that for $\beta=1.18$

$$x_{\text{extr}}=0.57.$$

Therefore the difference between extrapolated and mean range is

$$0.57s'=0.57\cdot 1.87\cdot 11.37/100=0.12.$$

Thus: Mean range $=11.37+0.04-0.12=11.29$ cm ± 0.15 cm. (Estimated error: 0.10 cm in measurement itself, 0.08 each in mica correction and straggling correction, total $(0.0100+2\cdot 0.0064)^{\frac{1}{2}}=0.15$.)

Energy of α-particle when emitted from top layer of target at 90° from Fig. 29 (corresponding to mean range of 11.29 ± 0.15 cm)

$$E_2=10.41\pm 0.08 \text{ MV}.$$

From (768) we find then

$$Q=\frac{16}{12}\cdot 10.41-\frac{10}{12}\cdot 0.575=13.88-0.48,$$

$$Q=13.40\pm 0.11 \text{ MV}$$

$$=14.39\pm 0.12 \text{ milli-mass-units}.$$

E. Experiments with "poor" geometry

Let us suppose that particles from the target can enter the detector at all angles χ up to χ_0, where χ is the angle between the particle beam and the normal to the detector. If the target is thin, and if the straggling as well as the dependence of the particle energy on χ is neglected, the particles will all have the same true range, say, R_0. The apparent range (i.e., the range component normal to the detector) is then $X=R_0\cos\chi$. Since the number of particles emitted at an angle between χ and $\chi+d\chi$, is proportional to $\sin\chi d\chi$, the number of particles with an apparent range between X and $X+dX$, will be a constant times dX for X between $R_0\cos\chi_0$ and R_0. The total number of particles detectable at the distance X from the target, will then be

$$T(X)=\begin{cases}0 & \text{for } X>R_0\\ R_0-X & \text{for } R_0\cos\chi_0<X<R_0\\ R_0(1-\cos\chi_0) & \text{for } X<R_0\cos\chi_0,\end{cases}\qquad(806)$$

leaving out a constant factor.

The straggling will modify this distribution near the upper end $X\approx R_0$. If the straggling s is small compared to the region over which the apparent ranges are distributed, i.e., if[11]

$$s\ll R_0(1-\cos\chi_0)\qquad(806\text{a})$$

the distribution function (806) is replaced by

$$T(X)=\tfrac{1}{2}\left(\frac{R_0-X}{s}\right)\{1-\Phi[\tfrac{1}{2}\pi^{\frac{1}{2}}(X-R_0)/s]\}$$

$$+(1/\pi)\exp\left(-\frac{\pi}{4}(R_0-X)^2/s^2\right),\qquad(806\text{b})$$

which goes over into the straight line $T(X)=(R_0-X)/s$ for $R_0-X\gg s$. The extrapolation of this straight line gives exactly the mean range R_0. This means that with a thin target and sufficiently poor geometry, the extrapolated range is equal to the mean range, just as for an infinitely thick target and good geometry (end of Section C).

The dependence of the energy on direction does not appreciably affect the range distribution. It merely changes the most favorable angle (i.e., the angle giving the longest extrapolated range) from $\chi=0$ to $\chi=\frac{1}{2}\pi-\vartheta_0$ (cf. 772). The extrapolated range R_0 becomes equal to the mean range of the particles emitted at the most favorable angle ϑ_0, times $\sin\vartheta_0$. If $E_2{}^m$ is the energy cor-

[11] $s<\frac{1}{2}R_0\,(1-\cos\chi_0)$ is in general sufficient.

responding to the measured extrapolated range, Q follows from (773).

With a thick target and poor geometry, the number of particles increases faster than linearly with decreasing range X, since a linear decrease is already obtained from a thin target. It is therefore *impossible to find an extrapolated range by linear extrapolation*. The most suitable procedure seems to be to plot the square root of the number of counts as a function of the thickness of stopping material between detector and target, and to extrapolate *this* curve along the steepest tangent.

Since the experiments with poor geometry are never very accurate, it was not felt worth while to calculate the exact relation between the "penetration constant" β (cf. (799)) and the extrapolated range obtained from the extrapolation of the square root curve as described above. The relation was only determined in the limiting case of large and small β. We use again the abbreviation

$$x_{\text{extr}} = (R_{\text{extr}} - R_0)/s, \qquad (807)$$

where R_{extr} is the experimental extrapolated range and R_0 the required mean range for the most favorable direction ϑ_0. Then we have

$$(a) \text{ for large } \beta: \ x_{\text{extr}} = 1.21 - 1/\beta, \qquad (807a)$$

$$(b) \text{ for small } \beta: \ \frac{x_{\text{extr}}}{1+2/(\pi x^2_{\text{extr}})} = \frac{6}{\pi}\beta. \qquad (807b)$$

Examples for "poor" geometry are very rare in the more recent experimental literature. In order to explain the principles of evaluation, we discuss the old experiments of Cockcroft and Walton (C25) on the reaction

$$C^{12} + H^2 = C^{13} + H^1, \qquad (808)$$

which have since been superseded by the more recent experiments of Cockcroft and Lewis (C29) with "good" geometry. The energy of the incident deuterons was 0.50 MV, the approximate range of the protons 13.7 cm, corresponding to an energy of 2.95 MV and to $n_2 = 3.4$ (Fig. 35). Thus (cf. (772))

$$\cos \vartheta_0 = 3.4 \cdot \frac{(1\cdot 2)^{\frac{1}{2}}}{14} \cdot \left(\frac{0.50}{2.95}\right)^{\frac{1}{2}} = 0.14.$$

The aperture of the detecting apparatus is not stated, but was apparently wide, so that particles with $\cos \vartheta = 0.14$ could enter freely. This is sufficient for "poor" geometry.

When a plot of the square root of the number of particles observed is plotted against the apparent range, the extrapolated range turns out to be about 13.72 ± 0.20 cm. The straggling of protons of this range (Fig. 37) is $s = 2.1$ percent. With $R_1 = 0.73$ cm and $n_1 = 2.0$ for the incident deuterons, (799) gives

$$\beta = \frac{[6\cdot(2/0.5)^{\frac{1}{2}} + 4]\cdot 0.021}{\frac{0.73}{13.7}\cdot 2.0 + \frac{0.50}{2.95}\cdot\frac{11}{14}\cdot 3.4}$$

$$= 16\cdot 0.021/(0.106 + 0.454) = 0.60.$$

Therefore we have approximately $x_{\text{extr}} = 0.42$ (cf. curve 38, which, however, is not directly applicable to "poor" geometry). Thus straggling $\approx 0.42 \cdot 2.1\% = 0.88\%$ of the range $= 0.12$ cm, giving for the mean range

$$R_0 = 13.60 \pm 0.2 \text{ cm}.$$

The corresponding energy is $E_2^m = 2.94 \pm 0.04$ MV. From (773) we obtain then

$$Q = \frac{14}{13}\cdot 2.94 + \frac{2}{13}\cdot 0.50\left(1 - \frac{3.4\cdot 1}{14}\right) - 0.50$$

$$= 2.72 \pm 0.05 \text{ MV}.$$

The newer experiments of Cockcroft and Lewis give 2.71 ± 0.05 MV.

F. Measurement of neutron energies by proton recoil

The most exact measurements of neutron energies (B37, B38, B40) are based on measurements of the range of recoil protons produced by the neutrons in hydrogen gas or hydrogenic substances. The energy of the neutrons depends on their direction of emission ϑ, and on the depth in the target at which they are produced, just as with charged particles. The only difference is that the "range" of the neutrons is essentially infinite so that the first term in the denominator of (799) is absent. For s_2/R_2 in (799), the values for the recoil protons should be inserted.

If only the recoil protons emitted exactly

forward (i.e., in the direction of the neutron motion) are counted, the evaluation is the same as for charged particles. The quantity a/b in (777) should be replaced by one-half the angle subtended by the cloud chamber (or other detecting apparatus for neutrons) at the middle of the target. s' is then found from (804), s being the natural range straggling of the recoil protons. β is calculated from (799) as described above, x_{extr} or x_0 found from Fig. 38, and the mean range of the recoil protons found from their extrapolated or from the most probable range as usual. The corresponding proton energy gives immediately the energy of neutrons emitted at right angles to the incident beam.

If recoil protons within an angle χ_0 of the neutrons are taken into consideration, s' should be replaced by

$$s''=\left(s'^2+\frac{n^2}{48}R_2^2\chi_0^4\right)^{\frac{1}{2}}, \tag{809}$$

where R_2 is again the proton range and n the "range exponent" for the proton (Fig. 35). The evaluation of the "mean" proton range and the corresponding energy should then be carried out as before, but the result should be increased by

$$\tfrac{1}{2}E_2\chi_0^2. \tag{809a}$$

XVII. Results of Disintegration Experiments

(Closed July 1, 1937)*

§98: Notation

In the following sections (§§99 to 105) the results of disintegration experiments will be discussed in some detail. Attempts to organize the voluminous material in this field have necessitated subdivision of the subject matter. It would be wasteful of space to describe the techniques used by the various experimenters for each process. Accordingly, the experimental techniques most commonly employed are discussed in a separate chapter (Chap. XV), and only brief mention made of these techniques in discussing the results of experiment.

In the discussion of disintegration results general subdivisions into type reactions have been found valuable for the purpose of grouping the results into a form suitable for correlations. A shorthand terminology has been devised for identifying these reactions. The projectiles which have been successfully used for nuclear disintegration are: *alpha-particles*, *protons*, *deuterons*, *neutrons* and *gamma-rays*, denoted, respectively, by the symbols: α, p, d, n, γ. All of these radiations are observed as products of disintegrations as well, and in addition many product nuclei are formed which decay radioactively with emission of electrons (e^-) and positrons (e^+). The reactions produced by one projectile and releasing a common product may be designated by the corresponding symbols, in order. For instance alpha particle disintegrations yielding protons may be symbolized as reactions of the α-p type; neutron bombardment resulting in alpha-particle emission as the n-α type, etc. Under each type reaction the individual processes may be specified by prefixing the chemical symbol of the target, i.e.: B^{10}-α-p, Li^6-n-α, etc. The identity of the resultant products is then obvious from the requisite balance of nuclear charge and mass; in the two instances above they are C^{13} and H^3. A more complete symbolism for such type reactions is to write out the generalized reaction in the usual manner using Z for nuclear charge of the target element and A for its mass number. Using this method of grouping under type reactions we find experimental evidence for the 16 types of primary reactions and 2 radioactive decay processes listed in Table L.

In Table L the number of individual reactions falling under each type reaction is given in the column on the right. There are found to be 385 primary reactions, followed by 220 radioactive decay processes produced through various of the primary ones, giving a total of 605 observed to date (July 1937). Included in the list are those of unusual type which result in more than the usual

* The experimental material presented in this chapter covers publications received prior to July 1, 1937. In addition the Physical Review issues up to Aug. 1 have been included.

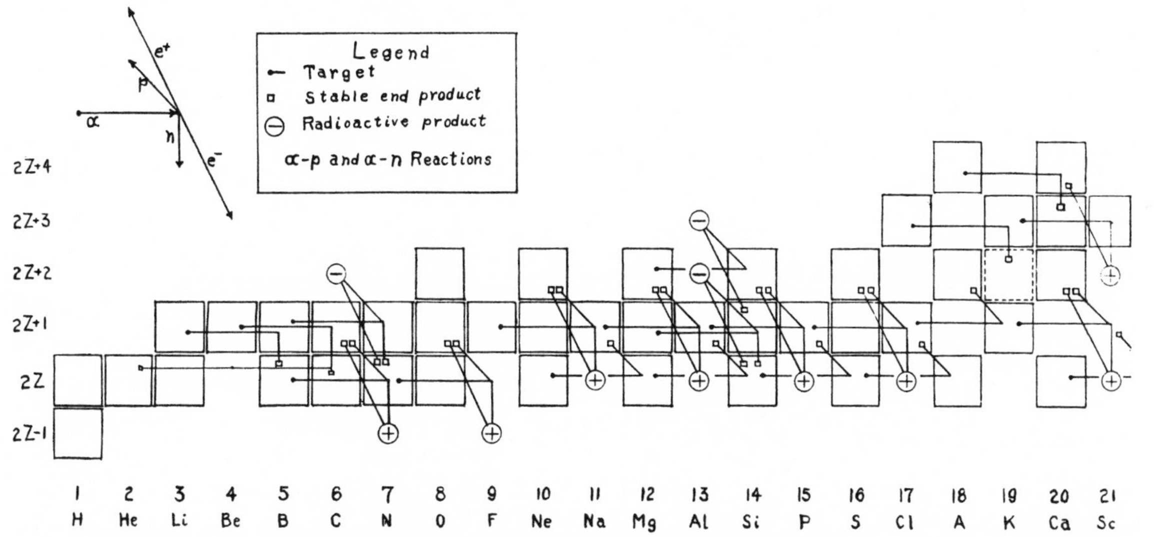

FIG. 39. Isotopic chart showing the reactions produced by α-particles in light elements. Abscissa, nuclear charge; ordinate, isotopic number. The stable isotopes are indicated by heavy squares, the reactions by two lines. The direction and length of the lines give the kind of particle emitted.

two products. These are in general alternative to a process of the ordinary type and are discussed under such reactions in the text. An example: B^{10}-*d*-α results in three alpha-particle products as well as in Be^8 and an alpha. It is understood, of course, that the above list of reactions may be incomplete since inconclusive evidence in many cases makes the absolute identification questionable. Only those reactions are entered in the following discussion for which, in the opinion of the authors, there is satisfactory proof.

Another valuable mechanism for correlating nuclear reactions is to plot them on an isotope diagram. For this purpose we have chosen the method used by Evans and Livingston (E11), in which the "isotopic number" ($A-2Z$) is plotted as a function of Z. This method separates the isotopes of all elements. Reactions between elements are plotted by the use of two lines, one representing the absorbed projectile and the other the emitted particle, while the junction of the two lines denotes the compound nucleus. In Figs. 39, 40, 41, 42 the known nuclear reactions are plotted on such charts. The direction and lengths of the lines representing particle absorption or emission are indicated; emission of a neutron reduces the isotopic number by one unit, absorption of an alpha-particle (isotopic number = 0) increases Z by two units, etc. Known stable isotopes are indicated by the large squares on the diagrams. It is noted that the isotopic number, I, varies from -1 for H^1 to $+54$ for U^{238}. The target element is indicated by the small solid circle at the start of a line representing the bombarding particle; the product is a small square if stable or a large open circle if radioactive, and the radioactive process is indicated not only by the direction of the line leading from the circle but also by the + or − sign in the circle. In many instances when the same bombarding particle is used and

TABLE L. *Type reactions and number of nuclear processes.*

(1) α-*p*:	$Z^A+He^4 \to (Z+2)^{A+4} \to (Z+1)^{A+3}+H^1$	20
(2) α-*n*:	$Z^A+He^4 \to (Z+2)^{A+4} \to (Z+2)^{A+3}+n^1$	21
(3) *p*-α:	$Z^A+H^1 \to (Z+1)^{A+1} \to (Z-1)^{A-3}+He^4$	5
(4) *p*-*d*:	$Z^A+H^1 \to (Z+1)^{A+1} \to Z^{A-1} \quad +H^2$	1
(5) *p*-γ:	$Z^A+H^1 \to (Z+1)^{A+1} \to (Z+1)^{A+1}+h\nu$	10
(6) *p*-*n*:	$Z^A+H^1 \to (Z+1)^{A+1} \to (Z+1)^{A} \quad +n^1$	22
(7) *d*-α:	$Z^A+H^2 \to (Z+1)^{A+2} \to (Z-1)^{A-2}+He^4$	23
(8) *d*-*p*:	$Z^A+H^2 \to (Z+1)^{A+2} \to Z^{A+1} \quad +H^1$	50
(9) *d*-*p*, α:	$Z^A+H^2 \to (Z+1)^{A+2} \to (Z-2)^{A-3}+H^1+He^4$	1
(10) *d*-*n*:	$Z^A+H^2 \to (Z+1)^{A+2} \to (Z+1)^{A+1}+n^1$	26
(11) *d*-*n*, α:	$Z^A+H^2 \to (Z+1)^{A+2} \to (Z-1)^{A-3}+n^1+He^4$	2
(12) *n*-α:	$Z^A+n^1 \to Z^{A+1} \quad \to (Z-2)^{A-3}+He^4$	23
(13) *n*-*p*:	$Z^A+n^1 \to Z^{A+1} \quad \to (Z-1)^{A} \quad +H^1$	22
(14) *n*-γ:	$Z^A+n^1 \to Z^{A+1} \quad \to Z^{A+1} \quad +h\nu$	97
(15) *n*-2*n*:	$Z^A+n^1 \to Z^{A+1} \quad \to Z^{A-1} \quad +2n^1$	32
(16) γ-*n*:	$Z^A+h\nu \to Z^A \quad \to Z^{A-1} \quad +n^1$	19
	(Processes leading to more than two products)	11
		385
(17) e^-:	$Z^A \to (Z+1)^A+e^-$	170
(18) e^+:	$Z^A \to (Z-1)^A+e^+$	50
		220
	Total number of nuclear reactions	605

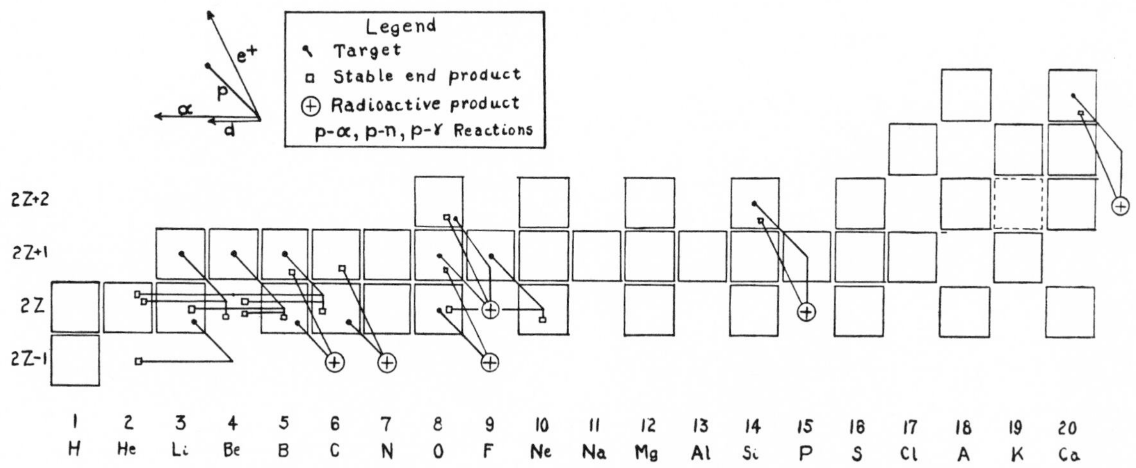

FIG. 40. Isotopic chart showing the reactions produced by protons in light nuclei.

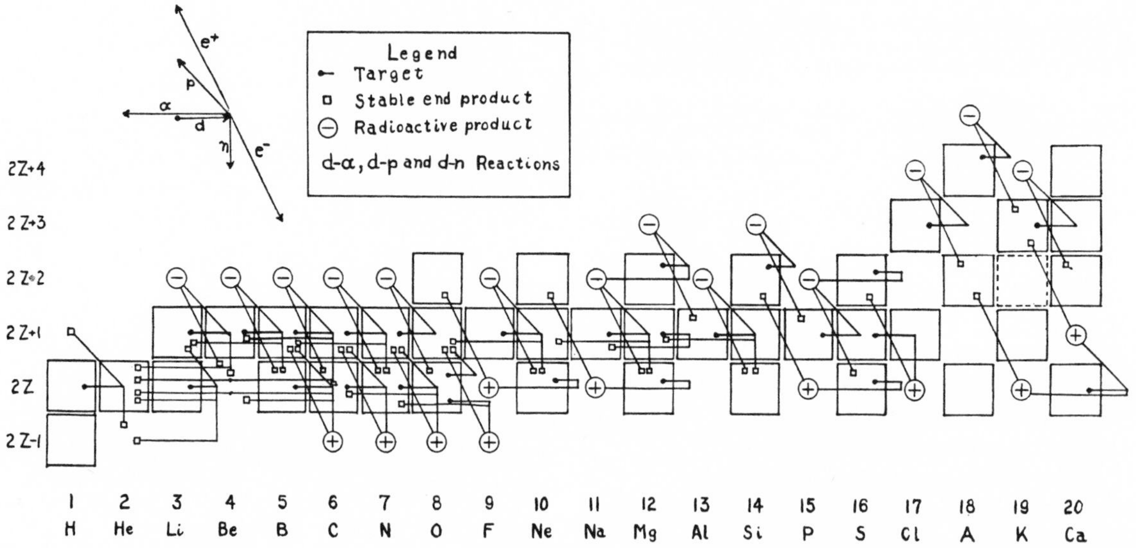

FIG. 41. Isotopic chart of reactions produced by deuterons in light elements.

different products observed this is indicated by lines branching from the common junction representing the compound nucleus. It is interesting to note that all positron emitting radioactive nuclei fall below the band of stable isotopes, while all electron emitters are above it.

Collected at the end of each type reaction are tables of the individual processes in which the more important observational data are listed. These give the observed reaction energy, the resonance levels in the compound nucleus, the excitation levels in the product nucleus and the probability of the reaction. In addition the value of the reaction energy calculated from the masses of the components of the reaction is given in each case, and also for many reactions as yet unobserved.

With this introduction we proceed to the discussion of the experimental evidence.

§99. Disintegration by Alpha-Particles

Rutherford, in 1919, first achieved the artificial disintegration of one of the naturally stable elements. He used the highly energetic alpha-particles from Ra C′ as projectiles and observed protons ejected from a nitrogen gas target. His achievement was not only in visualizing the possible experimental method for obtaining dis-

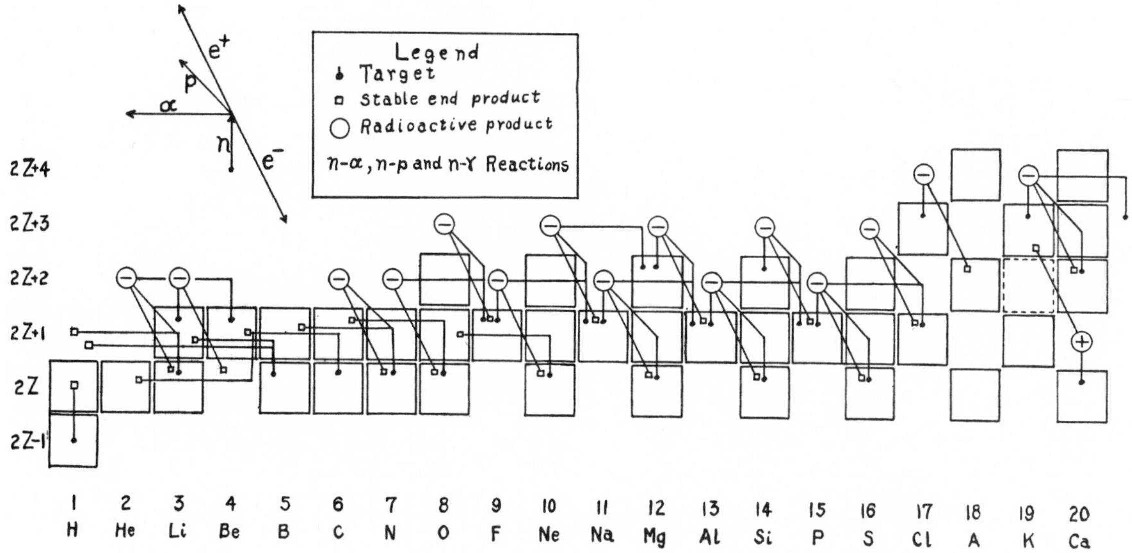

FIG. 42a. Isotopic chart of reactions produced by neutrons in light elements.

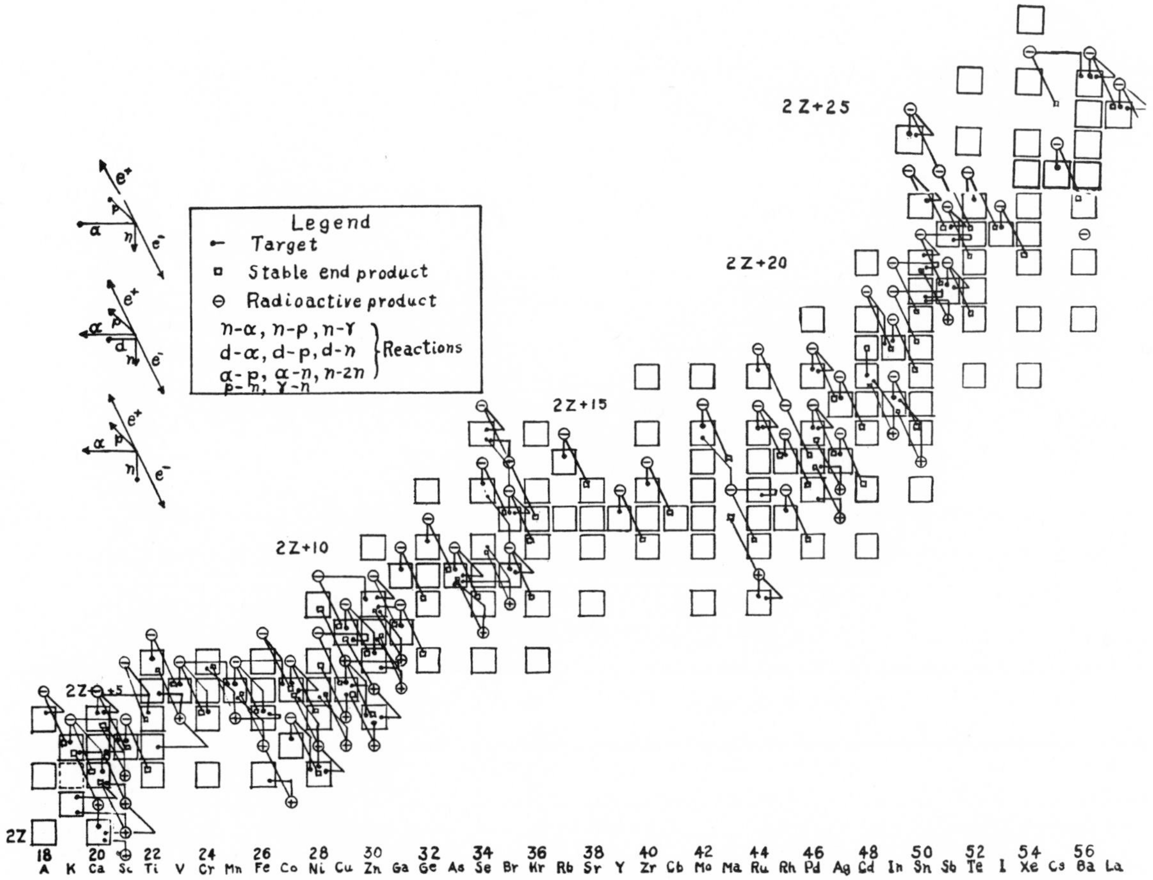

FIG. 42b. Isotopic chart of reactions produced by neutrons, protons, deuterons and gamma-rays in elements of medium atomic weight.

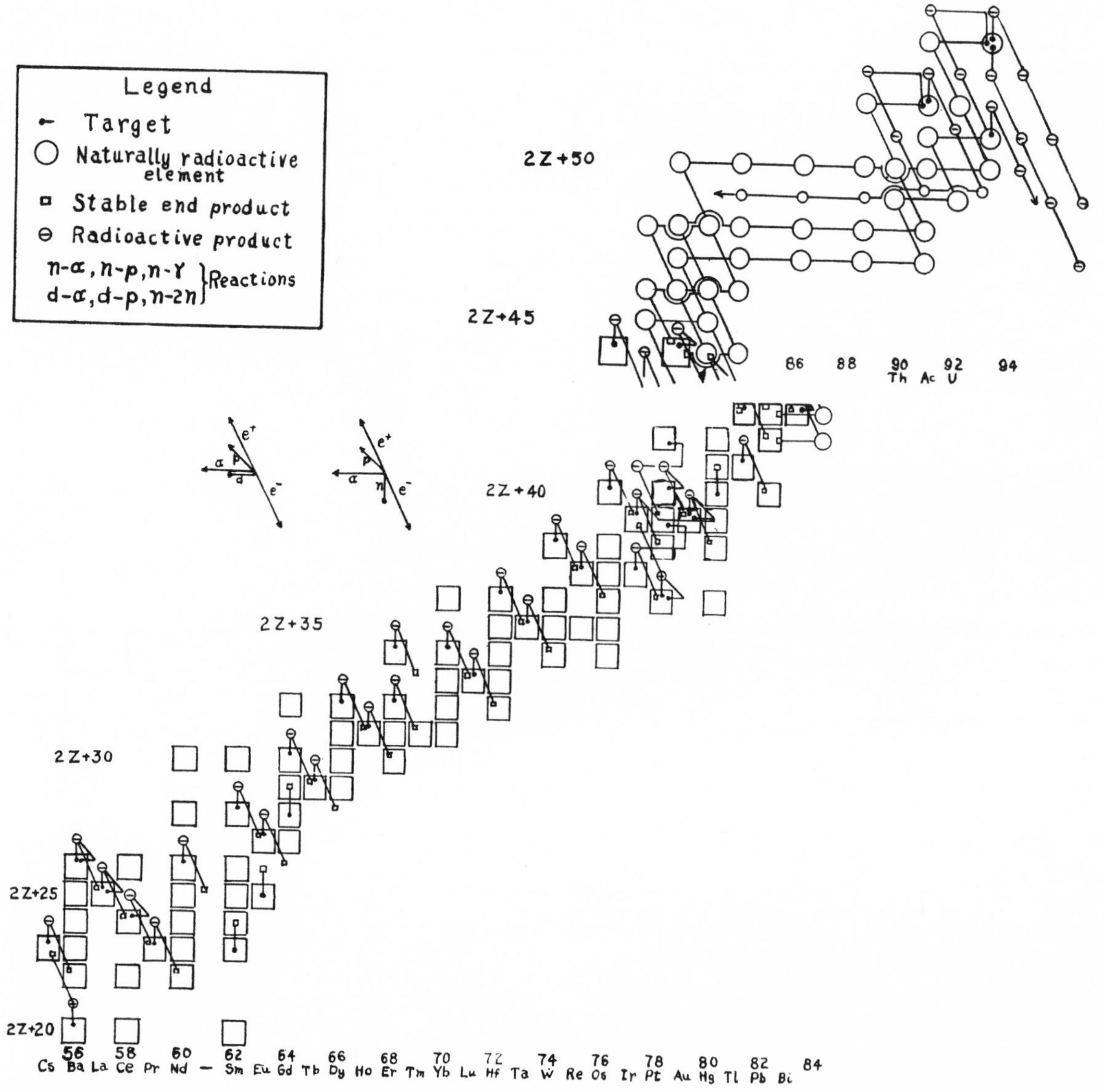

FIG. 42c. Isotopic chart of reactions produced by neutrons and deuterons in heavy elements.

integration, although the sensitivity and simplicity of the device he chose was largely responsible for his success, but also in combating all the good but pessimistic reasons for believing that the experiment could not succeed. These doubts were well grounded. In the first place there was reason to believe from scattering experiments that the lighter nuclei were of less than 10^{-12} cm in diameter, and the alpha-particle still smaller. The possibility of obtaining enough direct hits with the small number of alpha-particles available was small and this feature determined the necessity of observing the individual products of the disintegration. Secondly the energy of the alpha-particle might not have been great enough to cause a disruption of the nucleus. The Coulombian forces of repulsion between the positive charges of the nucleus and the alpha might have prevented sufficient penetration to affect the nucleus, even for a direct hit. Actually the reserve was very small; if the alphas had been moving with speeds of 0.6 their actual value the disintegrations would probably not have been observed. Even so only certain elements with atomic number less than 19 were disintegrated. Thirdly, the products of the disintegration might have been unobservable; they might have consisted solely of neutrons and gamma-rays, or of low energy charged particles which would not have been observed with the instruments used. Again Rutherford's guess was correct, and the fluorescent screen proved to be just the thing to observe the products which appeared, the protons. Finally it could be expected that scattered alphas or recoil nuclei from them might mask the effects, which was indeed the case until better arrangements of apparatus minimized this disturbance.

It is known today that natural alpha-particles will disintegrate all of the lighter elements up to Ca ($Z=20$) and possibly one of much higher charge, Zn, with the exception of H, He, C and O. Many years of experiment and controversy and the discovery of two new nuclear radiations were required to bring the field to this state. The exceptions mentioned are very significant. Neglecting H, which is itself an elementary nucleus, they are composed of 1, 3 and 4 alpha-particle units, resulting in unusually stable nuclei. The importance of this classification fades with heavier elements, for instance Ne^{20}, Mg^{24}, Si^{28}, S^{32} and Ca^{40} have all been reported to be disintegrable.

Using the cyclotron, Lawrence and his collaborators have accelerated [He^{++}] ions to 12 MV energy with intensities thousands of times greater than those available from natural alpha-particle sources. Many induced radioactivities have been observed, of which a few reports are published (W1, W1a, S16a). The Princeton cyclotron has also been used to produce 9 MV alphas (W13a, H24a, H24b, R7b). These results indicate that the field of alpha-particle disintegrations can be expanded indefinitely with such artificial sources.

As products protons, neutrons and gamma-rays are observed, and in some cases the residual nucleus is itself unstable and decays with the emission of electrons or positrons. For general reviews of the subject the reader is referred to Chadwick and Feather's report to the 1934 International Conference on Physics (C10), Darrow's reviews (D5, D4) and the well-known book by Rutherford, Chadwick and Ellis (R21).

Let us now discuss the various type reactions in more detail.

A. Type reaction α-p. (Table LI)

$$Z^A+He^4\rightarrow(Z+1)^{A+3}+H^1+Q:$$

The observation of protons emitted from nitrogen was the first evidence of artificial nuclear disintegration, and was followed by a series of experiments on other elements, chiefly by the Cavendish Laboratory group, which led to the statement in the book mentioned above (R21, p. 285): "In this way Rutherford and Chadwick found evidence of the disintegration of all elements from boron to potassium, with the two exceptions of carbon and oxygen." These processes were all determined by the observation of the ejected protons with a fluorescent screen, and carefully checked to prove that the scintillations were not due to scattered alpha-particles or recoil atoms of hydrogen existing as an impurity. The particles from several of the more intense reactions were found by deflection experiments in electric and magnetic fields to consist of protons. Not all of the disintegrations of this type have been similarly studied but it is assumed, and certainly with safety, that the particles observed on the fluorescent screen from paralleling processes are protons and can be included in the same type reaction. Most of the processes have been verified by more recent experiments,[12] but in the case of argon the only evidence is Rutherford and Chadwick's original observations.

The type reaction we have chosen to start with owes its primary position to the historical arrangement of events, although it is by no means the simplest one. The symbol Q on the right-hand side of the type equation represents the energy

[12] At this place should be mentioned the long-standing controversy between the group at the Cavendish Laboratory and that at Vienna, headed by Kirsch and Petterson. These workers have reported disintegrations of many elements, including a considerable number of heavier atomic weight than potassium, and some lighter than boron. Also they report a much greater yield of protons and deduce a larger probability of disintegration than the Cavendish group. Although the sincerity of these workers cannot be questioned many subsequent experiments have proven their results almost entirely erroneous. The best attempt to explain their anomalous results is given in a paper by Chadwick (C15), indicating that the weak scintillations due to the beta-rays of the radioactive source might be discernible under certain conditions and be confused with the alpha and proton scintillations. In the face of the existing evidence we are forced to eliminate these data from Vienna in the report to follow.

change in the reaction, and can be correlated with the resulting change in mass by means of Einstein's equation: $\Delta E=\Delta mc^2$. This quantity, referred to in certain German papers as the "Tönung," is either positive or negative according to whether the process is "exoergic" or "endoergic," that is, whether energy is emitted or absorbed in the reaction. Nearly all the known reactions of this type are of the endoergic character; the protons which are emitted having in general less energy than the alpha producing the disintegration.

With an "ideal" experimental arrangement, in which a parallel beam of alpha-particles of homogeneous energy impinges upon a thin target (in which the alpha-particle energy absorbed is negligible) the energy transferred to the proton and to the recoil nucleus can be calculated from the usual momentum and energy conservation equations (see §96). If M_1, M_2 and M_3 and E_1, E_2 and E_3 represent the masses and energies, respectively, of the alpha-particle, the proton and the residual nucleus, we find for the special case of observation at 90° to the incident beam (cf. 768, 767a):

$$Q=\left(\frac{M_1}{M_3}-1\right)E_1+\left(\frac{M_2}{M_3}+1\right)E_2. \quad (810)$$

$$E_2=\frac{M_3}{M_2+M_3}Q+\frac{M_3-M_1}{M_2+M_3}E_1, \quad (810a)$$

$$E_3=\frac{M_1}{M_3}E_1+\frac{M_2}{M_3}E_2. \quad (810b)$$

For observation in the direction of the incident beam (0°) these equations become (cf. 769):

$$Q=\left(\frac{M_1}{M_3}-1\right)E_1+\left(\frac{M_2}{M_3}+1\right)E_2 -\frac{2}{M_3}(M_1M_2E_1E_2)^{\frac{1}{2}}, \quad (811)$$

$$(E_2)^{\frac{1}{2}}=\frac{(M_1M_2)^{\frac{1}{2}}}{M_2+M_3}(E_1)^{\frac{1}{2}}+\frac{(M_3)^{\frac{1}{2}}}{M_2+M_3} \times[(M_2+M_3)Q+(M_3+M_2-M_1)E_1]^{\frac{1}{2}}, \quad (811a)$$

$$E_3=[(M_1E_1)^{\frac{1}{2}}-(M_2E_2)^{\frac{1}{2}}]^2/M_3. \quad (811b)$$

These equations show that the energy of the observed proton, E_2, is determined to a considerable extent by the alpha-particle energy. So we would expect to find proton ranges increasing with increasing energy of the alphas. Such is in fact the case for at least one group of protons in each of the disintegration processes of this type for which this phenomenon has been studied. This feature is characteristic of this "normal" type of disintegration and observation of such a variation of proton range is taken as proof of the mechanism of the disintegration.

The magnitude of the proton energy is chiefly dependent upon the value Q. It is possible to conceive of an alternative process resulting in an excited residual nucleus, in which case the Q value would be energetically smaller, and the proton ranges correspondingly less, but the same type of variation of proton range with alpha energy would be expected. The excited nucleus would then revert to the ground state with the emission of gamma-radiation.[13]

Resonance disintegration.—Experimentally we find that not all the protons observed from these processes follow this simple law. Certain groups are observed to have a definite energy which does not vary with the alpha-particle energy. Furthermore, these groups are observed only for a definite energy of the alpha, E_1. This is the phenomenon of resonance first suggested by Gurney (G24) and observed in the bombardment of aluminum by Pose (P12) in 1929. It also explains effects observed as early as 1921 by Rutherford (R24) but not interpreted in this way at that time. It has subsequently been found by these and other observers to exist also in the disintegration of Be, B, N, F, Na and Mg by alphas and in neutron and proton-produced processes.

In order to understand this effect more thoroughly we must consider the actual mechanics of the experiments. It proves to be impossible to perform experiments with the ideal and simple arrangement suggested above. In the first place the collimated beam of alpha-particles

[13] It has been the custom in the past to attempt to visualize this process as due to the entering alpha-particle falling into an excited alpha-particle level in the residual nucleus, or the ejection of a proton from a level lower than that of the most readily removable proton. These visualizations are misleading, and the excited states should be considered as existing in the residual nucleus *as a whole*, rather than for the individual particles (§51).

puts a restriction on intensity which is too severe for the experimental techniques and the quantities of radioactive materials available. The ideally thin target is also impractical since it reduces the number of disintegrations to below the observational limit. However, if a thick target is used in which the alphas are completely absorbed there will always be a lamina in the target in which the alphas do have the prescribed resonance energy, and so the protons produced at this energy are observed. The result is that with increasing alpha-energy, while the groups of normal protons are found to increase in energy correspondingly, the resonance groups will remain essentially fixed in range. See Fig. 43. (In fact, for angles of observation greater than 90°, the increased absorption of the resonance protons in coming out of the thick target results in a decreasing proton range with increasing E_1.) An equivalent technique is to use increasing thicknesses of target, in which case the energy E_1 is reduced by the absorption of the alphas in the target. Range-distribution curves of the protons will show the same "step" structure of the resonance groups, each group appearing as E_1 approaches the resonance value from above.

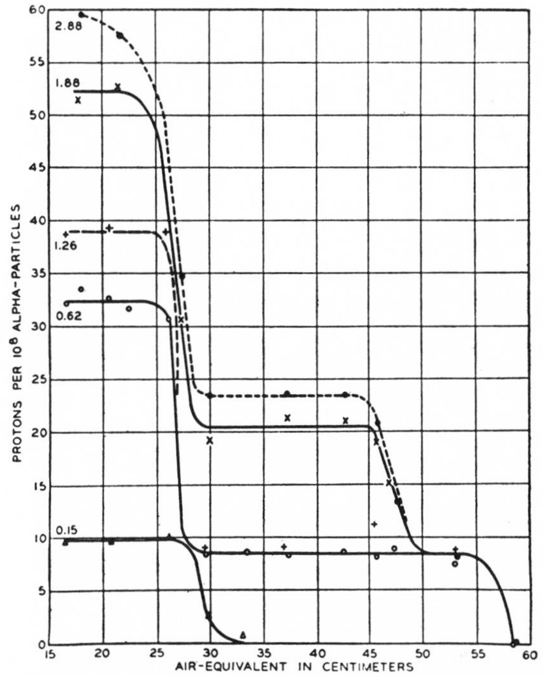

FIG. 43. Distribution-in-range curve of protons from the reaction $Al^{27}+He^4=Si^{30}+H^1$ (Pose) showing group structure dependent on alpha-energy and thus indicating resonance levels in the compound nucleus.

Despite the difficulty of working with thin targets certain observers have been successful in obtaining complementary evidence of resonance by their use. Chadwick and Constable (C4) have measured the disintegration function (number of protons emitted as a function of E_1) for thin foils of aluminum. In this case, as the energy is increased, the total number of protons observed rises to a maximum at the resonance values of E_1.

The Q value for a resonance group should be the same as that calculated for the normal or nonresonance protons. That is, the range of the resonance protons would be the same as the normal protons from alphas of the particular energy required to show the resonance effects. The distinction is in the increased probability of the reaction for that definite energy. So calculations of the Q value from resonance groups require only a knowledge of the proton energy and direction and the value of E_1 for which the probability is the greatest.

The possibility of producing an excited residual nucleus is also present in resonance disintegration, and results in two or more groups of protons of constant range occurring simultaneously for the same energy E_1. The differences between Q values corresponding to these groups would represent the excitation states of the residual nucleus, and gamma-radiation with energies equal to these energy differences should be produced. These differences must agree for the various resonance groups and also for the "normal" groups. Gamma-radiation has been observed experimentally in many of these reactions and certain components might well be associated with such excitation levels. The data are not yet sufficiently precise to allow many accurate correlations, however, and this remains a point for further investigation.

To illustrate the features of resonance disintegration we will cite other experimental observations on aluminum. Chadwick and Constable (C4), using relatively low energy alphas, observed two groups of protons for each of four alpha-particle resonance energies, or eight groups in all. These four resonance levels were at 4.0, 4.44, 4.86 and 5.25 MV and the two proton groups in each case showed Q values of 0.0 and 2.3 MV, respectively. Using higher energy alphas

Haxel (H17) observed nonresonance protons from which the same values of Q were obtained. Duncanson and Miller (D19) have found evidence for two additional higher energy resonance levels, at 5.75 and 6.61 MV. Furthermore, they found four proton groups for each resonance level, corresponding to four Q values, given as 2.07, −0.16, −1.53 and −2.67 MV. The residual nucleus can then be left either in the ground state (Q=2.07 MV) or in one of three excited levels of 2.23, 3.60 or 4.74 MV above the ground state. Duncanson and Miller also found that nonresonance protons were observed for alpha-energies greater than 6.7 MV. These are interpreted as coming from alpha-particles going over the top of the potential barrier (cf. §78) an explanation which is justified by the observed increase of yield up to the limit of 7.7 MV.

The widths of the resonance levels may be stated, as in optical spectra, in terms of the width at half-maximum. Although the poor resolution of the experiments leaves some room for doubt as to the accuracy of the estimates, several methods of interpreting the data lead to a value of about 0.1 to 0.3 MV for this width (D19), for the higher energy levels. Lower energy levels would be expected to be narrower.

The results of these studies of the Al disintegration are summed up graphically in the form of an energy level diagram by Chadwick and Feather (C10) indicating the six resonance levels and the penetration through the barrier. The four Q values are represented as four arbitrarily placed alpha-particle levels in the nucleus. (See reference 13. It should be noted that there are some 22 energetically allowed proton groups from the various combinations of the 6 resonance levels and the 4 excitation levels, in addition to the protons due to normal entry.

Yield.—The values for the absolute yield of protons from a given intensity of alpha-particles on particular targets are subject to considerable error. One factor is the difficulty of estimating the thickness of the target and the solid angle subtended by the recording instrument. If the geometry of the experiment is sufficiently well defined to give such a value, the further question of the angular distribution of the ejected protons arises. An even more important point is that the probability of disintegration is a decided function of the alpha-particle energy, not only for the resonance groups but also for the normal groups. With a given thickness of target (specifying a definite range of energies), a known homogeneous alpha-particle energy and a known angular distribution, it is possible to estimate the probability of disintegration from the numbers of protons counted. This may be expressed in terms of the number of disintegrations per million alpha-particles, or equivalently, as the cross section of the process expressed in sq. cm for the individual reaction. Rutherford gives the rather rough estimate of 20 protons/10^6 alphas in N when alphas of 7 cm range are completely absorbed, and 8/10^6 for Al. Heavier targets, such as Cl, A and K yield smaller numbers of protons under similar conditions, about 1/10^6.

Protons have been observed and measured from alpha-particle bombardment of B^{10}, N^{14}, F^{19}, Ne^{20}, Na^{23}, Mg^{24}, Al^{27}, Si^{28}, P^{31}, S^{32}, Cl^{35}, K^{39} and Ca^{40}. In addition to the above-named processes determined by the observation of the protons, induced radioactivity has been found to occur in $Mg^{25, 26}$, Ca^{40}, Cr^{53} and Ni^{58} (see §105), directly attributable to this type of reaction. We will now discuss the individual reactions in more detail.

B: $B^{10}+He^4\rightarrow C^{13}+H^1$
($B^{11}+He^4\rightarrow C^{14}+H^1$)

Protons were first observed by Rutherford and Chadwick (R24), and were resolved into several groups by later observers. Miller, Duncanson and May (M16) find four groups of protons, each varying in range with the energy of the alpha and the direction of observation and giving Q values of 3.1, 0.4, −0.1 and −1.0 MV. A re-evaluation following the procedure discussed in §97 gives for these groups the values 3.3, 0.5, 0.1 and −0.8 MV. Paton (P3) checks these in principle, giving Q values of 3.1, 0.35, −0.78 and −1.86 MV. His second group (0.35) is probably a superposition of Miller and Duncanson's 0.5 and 0.1 MV groups, and the −1.86 group represents a higher excitation level. A calculation of the expected energy of disintegration into the ground state of the C^{13} residual nucleus from known masses gives a value of 4.15 MV. There is a discrepancy with the highest observed Q value of almost 1 MV, much greater than is found in any of the other α-p reactions. The mass values are good to ±0.2 MV, and the internal consistency of the α-p reactions is also about that good. This discrepancy suggests that the group representing the transition into the ground state of C^{13} had not been observed. Brubaker and Pollard (B66) have recently searched for such a group and have found one of very weak intensity giving a Q value of 4.7 ± 0.5 MV, compatible with the expected value. If the groups discussed above are all from this process this conclusion would suggest five excitation levels; assuming the calculated 4.1 MV to represent the ground state these would be at 0.8, 3.6, 4.0, 4.9 and 6.0 MV. It may be mentioned at this point that the use of the observed 3.3 MV Q value falsified Bethe's (B11) first attempt to evaluate atomic masses from disintegration data.

One or more of the groups mentioned above and heretofore assumed to all come from the B^{10} isotope may be due to the equivalent reaction on B^{11}. The Q value obtained from masses in this case is 0.94 MV, and the large abundance of this isotope should result in an observable number of protons. It is possible that the proton group yielding the 0.5 MV Q value may be due to this reaction.

Miller, Duncanson and May (M16) also observe a broad group of protons indicating resonance for 2.9 MV alphas. The evidence is not completely satisfactory and it may be the proton group due to normal entry observed by Paton which results in the -1.86 MV Q value.

Gamma-rays, observed by Bothe and Becker (B43) to have an energy of 3 MV, are probably attributable to this reaction although the energy available in the B^{10}-α-n and B^{11}-α-n reactions is also sufficient to explain them. A group of gamma-ray lines would be expected for this reaction and the 3 MV may represent the 4.0 to 0.8 MV transition. A gamma-ray of 3.5 MV observed by Crane and Lauritsen (C43) from the C^{12}-d-p reaction resulting in a C^{13} product was at one time thought to be the gamma observed by Bothe and Becker. It is now known that there is insufficient energy available in the deuteron process to produce this excitation and another explanation of the 3.5 MV gamma-ray is required. (§101).

N^{14}: $N^{14}+He^4 \rightarrow O^{17}+H^1$.

This process is historically interesting as the first artificial disintegration ever observed. There is considerable discrepancy in the more recent literature, an indication of the difficulties involved in experiments using the low intensities of natural alpha-particle sources. The best values are obtained by Haxel (H18) who finds a Q value of -1.26 MV, which checks reasonably well with that expected from the masses, of -1.16 MV. Other evidence by Pollard (P7) and Steudel (S23) seems to confirm Haxel's value. Stegmann (S21) finds Q's of -0.41, -0.95 and -1.60 MV; two of the proton groups (longest and shortest) show characteristics indicating resonance for alphas of 3.6 and 4.1 MV. Stetter (S22) and Fischer-Colbrie (F21) report Q's of -1.4 and -2.8 MV, apparently justified by the observation of gamma-radiation of 1.3 to 1.5 MV energy observed by Savel (S6), and two resonance levels.

The lack of agreement is probably chiefly due to the poor geometry of the experiments, necessitated by the low intensities. That this is the case is also indicated by the low Q values obtained for many α-p reactions. Furthermore the calculation of Q values from groups attributed to resonance disintegration may be invalid. It is dangerous to assume resonance unless a full energy proton group is observed corresponding to the same Q value.

In the O^{16}-d-p reaction from which the O^{17} mass was obtained (used in the evaluation of the expected -1.16 MV Q value), an excitation level of 0.8 MV was observed (§101B). There seems to be no evidence for this excitation level in O^{17} from this reaction.

F^{19}: $F^{19}+He^4 \rightarrow Ne^{22}+H^1$

Six proton groups are found by Chadwick and Constable (C4), occurring in pairs and so indicating two Q values. The corrected values are 1.58 and 0.98 MV. Two resonance levels are found, at 4.1 and 3.7 MV (C10). The Q value representing the ground state transition is found to be 1.53 MV from the masses, in good agreement with the results of the observations. May and Vaidyanathan (M11) have observed at least two additional groups of protons, giving Q values of -3.2, -2.1, -0.1 and 1.4 MV, and suggesting other excitation states. Corrected values indicate excitation states at 1.4, 3.4 and 4.5 MV.

Measurements of the gamma-radiation by different observers give values of 1.2 (S5) and 2.0

MV (W4), and are probably influenced by the neutrons also emitted.

An interesting phenomenon involving the competition of nuclear processes is reported by Haxel (H18). He finds that the proton intensity increases with alpha-energy up to a certain point and then remains constant until the alpha-energy is reached at which neutrons are emitted, at which point the proton intensity decreases.

Ne^{20}: $Ne^{20}+He^4 \rightarrow Na^{23}+H^1$

Pollard and Brasefield (P9) have been the only observers to check the original report of Rutherford and Chadwick, and find a single proton group. Recalculation of their data gives a Q value of -2.54 MV. The mass values of the constituents are fairly well established, and give a Q of -1.42 MV. It is probable that they did not observe the longest range proton group, and that their data represent an excitation level at 1.1 MV.

Na^{23}: $Na^{23}+He^4 \rightarrow Mg^{26}+H^1$

The first work of importance following the preliminary reports of Rutherford and Chadwick is by König (K21). She finds 4 proton groups, whose Q values (recalculated) are 1.91, -0.2 MV and two of smaller energy. May and Vaidyanathan (M11) with higher energy alphas did not find the longest range proton group because of low intensity. For the three shorter groups their results are in sufficient agreement with those of König and if recalculated, give -0.4, -2.1, -3.1 MV. The Mg^{26} mass is determined through this reaction, using the 1.91 MV Q value. Gamma-rays occur but measurements are contradictory. The excitation states are at 2.2, 4.0 and 5.0 MV.

Mg: $Mg^{24}+He^4 \rightarrow Al^{27}+H^1$
$Mg^{25}+He^4 \rightarrow Al^{28}+H^1$
$Mg^{26}+He^4 \rightarrow Al^{29}+H^1$

The protons emitted from Mg have been found by Duncanson and Miller (D19) to have three groups, two of strong intensity and short range and one weak but of longer range. The Q values obtained from the data (recalculated) are -1.05, -1.82 and -2.87 MV for the three groups, assuming the Mg^{24} isotope responsible. More recent data by Haxel (H19) check the Q values of Duncanson and Miller.

In attempting to identify these proton groups with the Mg isotopes responsible it was found necessary to consider the Q values predicted by masses (§108). For Mg^{26} a Q of -4.5 MV is expected so that this isotope is definitely excluded. The calculated Q for the Mg^{25} target based on the reactions Al^{27}-d-α and Al^{27}-d-p is -0.6 MV which seems to agree sufficiently with the Q value observed for the low intensity proton group. (-1.05 MV.) The Q for Mg^{24} cannot be calculated with any certainty because the two nuclei involved (Mg^{24} and Al^{27}) are only connected through two mass spectrograph measurements (of Ne^{21} and Si^{28}) together with five nuclear disintegration reactions, so that the over-all error may be very large. However, an estimate of the Mg^{24} reaction energy may perhaps be obtained from analogous nuclei such as Ne^{20} ($Q=-1.4$), Si^{28} ($Q=-1.8$) and S^{32} ($Q=-2.1$). A Q value of -1.8 MV for Mg^{24} would fit well with this sequence whereas the higher Q value would not. Therefore we attribute the weak group of $Q=-1.05$ to Mg^{25} and the two other strong groups to Mg^{24}.

Duncanson and Miller also find resonance protons for alphas of 5.7 and 6.3 MV but they were unable to determine which proton groups these resonances were associated with. A knowledge of this feature would assist in assigning the groups to the respective isotopes. Nonresonance disintegration was observed to start with alphas of 6.5 MV.

Gamma-radiation expected from the excitations of the product nucleus is observed but reports of its energy vary from 0.5 MV (S5) to 5.0 MV (W4).

More positive evidence for the Mg^{25} reaction has been through the observation of an induced electron radioactivity of 2.36 min. half-life (A2, E10) and the identification of the active material as aluminum. Since the Q value of the reaction is not known to any accuracy the value of the Al^{28} disintegration energy is used for the calculation of the Al^{28} mass.

A weaker electron radioactivity of 11 minutes half-life is attributed (E10) to the Al^{29} coming from the Mg^{26} isotope ($Q=-4.5$ MV). This furnishes evidence for the existence of two radioactive isotopes both heavier than the stable isotope.

Al^{27}: $Al^{27}+He^4\rightarrow Si^{30}+H^1$

In the introductory paragraphs the data of Chadwick and Constable (C4), Haxel (H17), Duncanson and Miller (D19), etc., have been discussed. These lead to the Q values (recalculated) of 2.26, −0.02, −1.32 and −2.49 MV and show excitation levels in the Si^{30} nucleus at 2.3, 3.6 and 4.8 MV. Savel (S4) shows that the gamma-radiation is composed of at least two components and gives a value for one of them of about 2.0 MV. The highest Q value (2.26 MV) is used to obtain the Si^{30} mass. Aluminum has six resonance levels, as discussed earlier, for 4.0, 4.44, 4.86, 5.25, 5.75 and 6.61 MV alphas.

Si^{28}: $Si^{28}+He^4\rightarrow P^{31}+H^1$

Proton groups having ranges of 20, 28 and 37 cm at 90° from 7.68 MV alphas observed by Haxel (H17) indicate Q values (recalculated) of −2.23, −3.28 and −3.92 MV. The large abundance of the Si^{28} isotope and the fact that P^{31} is a stable product determine that this isotope is the only one involved. This reaction allows a direct comparison with the masses of Si^{28} and P^{31} determined by Aston, and from which a Q of −1.4 MV is obtained. The adopted mass values (§108) are chosen to split this discrepancy so that the Q obtained is −1.8 MV. This agreement is fairly satisfactory but is at the same time another indication that the low intensities and poor geometry of the α-p reactions lead to low Q values. However, it is possible that the observed group should be attributed to a transition to an excited state rather than to the ground state.

P^{31}: $P^{31}+He^4\rightarrow S^{34}+H^1$

From the data of Paton (P1), May and Vaidyanathan (M11), and Pollard and Brasefield (P8) a set of average Q values are obtained which are 0.31, −1.0, −2.5, and −4.5. No resonance was found. By analogy with other reactions on this type of nucleus, in which ground state Q values of about 2 MV are common we judge that the longest range proton group was not observed. The Q obtained from masses is 1.8 MV.

S^{32}: $S^{32}+He^4\rightarrow Cl^{35}+H^1$

Brasefield and Pollard (B50) and Haxel (H19) find three groups of protons. An average of the recalculated Q values gives the values −2.10, −2.7 and −3.6 MV. The −2.1 MV value is used for the determination of the Cl^{35} mass. There is no evidence for resonance.

$Cl^{35,\,37}$: $Cl^{35,\,37}+He^4\rightarrow A^{38,\,40}+H^1$

Three well-defined proton groups reported by Pollard and Brasefield (P8) show Q values of 0.1, −2.5 and −4.2 MV and have no resonance characteristics. Mass values give Q values of 2.0 and −1.2 MV for the Cl^{35} and Cl^{37} isotopes, suggesting that the long range protons are from Cl^{35}.

A: (not observed)

Rutherford and Chadwick, in their early experiments, reported protons from argon. Recently Pollard and Brasefield (P9) made an extended study of the supposed process and found no evidence for disintegration. This is the only instance in which the early work of Rutherford and Chadwick has not been checked by subsequent experimenters. The protons observed were probably due to nitrogen or neon contamination.

K^{39}: $K^{39}+He^4\rightarrow Ca^{42}+H^1$

Three proton groups measured by Pollard and Brasefield (P9) result in corrected Q values of −0.89, −2.3 and −3.5 MV. From comparisons with the Q values in analogous nuclei it seems probable that this −0.89 MV does not represent the ground state transition, but the first or second excitation level.

Ca^{40}: $Ca^{40}+He^4\rightarrow Sc^{43}+H^1$

Frisch (F32) found a radioactivity in Ca chemically separable as Sc which has only one unit higher charge than Ca and so determines the reaction. Walke (W1) has checked this observation with the 11 MV He^{++} ions from the cyclotron and finds a period of 4.0 hr.

Following Frisch's report Pollard and Brasefield (P9) have observed a single proton group from which we calculate a Q of −4.3 MV.

$Ti^{46,\,47}$: $Ti^{46,\,47}+He^4\rightarrow V^{49,\,50}+H^1$

An activity identifiable as V has been observed by Walke (W1a) to come from Ti on bombardment with artificially accelerated He^{++} ions of 11 MV. This has a period of 35 min. Other activities, of 2.7 hr., 5.5 hr. and 85 days are not yet identified.

TABLE LI. *Summary of α-p type reaction.*

Target Z	Target El	Product El	Q(MV) (calc.)	Q(MV) (obs.)	Reference	Resonance levels (MV)	Excit. levels (MV)	Yield p/α @ (MV)
2	He^4	Li^7	−17.25					
3	Li^6	Be^9	−2.25					
	Li^7	Be^{10}	−2.58					
4	Be^9	B^{12}	<−7.2					
5	B^{10}	C^{13}	4.15	4.7?	B66	2.9?	{0.8 3.6? 4.0 4.9 6.0	2/10⁶ @ 2.9
	B^{11}	C^{14}	0.94	Obs. ?	M16			
6	C^{12}	N^{15}	−4.79					
	C^{13}	N^{16}	−7.4					
7	N^{14}	O^{17}	−1.16	−1.26	H18	{3.6? 4.1?	?	2/10⁵ @ 7.8
	N^{15}	O^{18}	−2.83					
8	O^{16}	F^{19}	−8.14					
9	F^{19}	Ne^{22}	1.53	1.58	C4	{3.7 4.1	{0.6 1.4 3.4 4.5	2/10⁶ @ 7.8
10	Ne^{20}	Na^{23}	−1.42	(−2.54)	P9		1.1	1.2/10⁵ @ 7.8
	Ne^{21}	Na^{24}	−1.8					
11	Na^{23}	Mg^{26}	1.9*	1.91	K21		{2.2 4.0 5.0	2.5/10⁶ @ 4.4
12	Mg^{24}	Al^{27}	−1.6*	−1.82	D19	{5.7 6.3	1.0	1/10⁶ @ 7.8
	Mg^{25}	Al^{28}	−0.6	−1.05	E10			1/10⁷ @ 7.8
	Mg^{26}	Al^{29}	−4.5	Obs.	E10			1/10⁸ @ 7.8
13	Al^{27}	Si^{30}	2.26*	2.26	D19	{4.0 4.44 4.86 5.25 5.75 6.61	{2.3 3.6 4.8	8/10⁶ @ 7.8
14	Si^{28}	P^{31}	−1.8*	−2.23	H19		{1.05 1.7	1/10⁶ @ 7.8
	Si^{29}	P^{32}	−1.6					
15	P^{31}	S^{34}	1.9	(0.31)	P8		{x(1.5) x+1.3 x+2.8 x+4.8	1/10⁶ @ 8.3
16	S^{32}	Cl^{35}	−2.1*	−2.10	H19		{0.6 1.5	1/10⁶ @ 7.8
	S^{34}	Cl^{37}	−4.0					
17	Cl^{35}	A^{38}	2.0	(0.16)	P8		{x(1.9) x+2.6 x+4.3	1/10⁶ @ 8.3
	Cl^{37}	A^{40}	−1.2					
19	K^{39}	Ca^{42}	—	−0.89	P8		{x x+1.4 x+2.6	1/10⁶ @ 7.8
20	Ca^{40}	Sc^{43}	—	−4.3	P9			5/10⁶ @ 8.3
22	Ti	V	—	Obs.	W1a			
24	Cr^{53}	Mn^{56}	—	Obs.	H24b			
28	Ni^{58}	Cu^{61}	—	Obs.	R7b			

* Q (Observed) used to calculate mass values.

Cr^{53}: $Cr^{53}+He^4 \rightarrow Mn^{56}+H^1$

Henderson and Ridenour (H24b) report a 160-min. electron activity produced in Cr by 8 MV alphas. Mn^{56} is known to have such a period from Mn-*n*-γ, Co-*n*-α, and Fe-*n*-*p* reactions (A7).

Ni^{58}: $Ni^{58}+He^4 \rightarrow Cu^{61}+H^1$

A 3.4-hr. positron activity, chemically identified as Cu, has been found by Ridenour and Henderson (R7b, R7c) after bombardment of nickel by 7 MV alphas. Such an activity had been found previously to be produced in the

Ni-*d*-*n* (T10a) and perhaps Ni-*p*-*n* (B4a) reactions.

Correlations in excited states of similar nuclei.—Before leaving the discussion of the α-p type of disintegration we must mention the apparent correlations in the magnitudes and spacings of the excitation states from similar nuclei. By similar we mean those of the same nuclear type, having mass numbers $4n+2$, $4n+3$, etc. The excitation states exist in the product nuclei so the type designation will be applied to the product nucleus in each case. These correlations have been pointed out and discussed by Haxel (H19), Pollard and Brasefield (P8), May and Vaidyanathan (M11), etc., and have been used as arguments for the alpha-particle sub-unit in the nucleus. In order to study such correlations the corrected Q values and their differences, which represent the spacings between excitation levels, have been tabulated. See Table LII. We see that the similarity in the excitation level spacings is not quantitative, but is still apparent. Furthermore the values of the reaction energies themselves are roughly equivalent, or at least vary consistently. The main point is that the spacings of low excitation levels in *odd* product nuclei are smaller than the spacings of levels in *even* product nuclei, a result which would be expected theoretically (see Feenberg and Wigner (F10)). These similarities add weight to the arguments for the existence of closed proton-neutron shells in the nucleus. They do not give much evidence for periodicities with the period 4 (alpha-particle sub-units), however.

TABLE LII. *Correlations in excitation states of similar nuclei in α-p reactions.*

Product nuclei type $4n+2$						Excitation level		
						Spacings:		
Target	Product	Q_0	Q_1	Q_2	Q_3	(1)	(2)	(3)
Na^{23}	Mg^{26}	1.9	−0.3	−2.1	−3.1	2.2	1.8	1.0
Al^{27}	Si^{30}	2.3	0.0	−1.3	−2.5	2.3	1.3	1.2
P^{31}	S^{34}	?	0.3	−1.0	−2.5	?	1.3	1.5
K^{39}	Ca^{42}	?	−0.9	−2.3	−3.5	?	1.4	1.2
Product nuclei type $4n+3$								
Mg^{24}	Al^{27}	−1.8	−2.9			1.1		
Si^{28}	P^{31}	−2.2	−3.3	−3.9		1.1	0.6	
S^{32}	Cl^{35}	−2.1	−2.7	−3.6		0.6	0.9	

B. Type reaction α-*n*. (Table LIII)

$$Z^A + He^4 \rightarrow (Z+2)^{A+3} + n^1 + Q:$$

Certain anomalous properties and effects of the supposed gamma-radiation from Be under alpha-particle bombardment had been under investigation by several experimenters when Chadwick (C6) concluded that these properties were due, not to gamma-radiation, but to a new elementary particle of unit mass and zero charge, the neutron. Other workers immediately started searching for these disintegration products from other targets, and they have been found as well in the alpha-bombardment of Li, B, F, Na, Mg and Al. The Be reaction yields the greatest intensities and also the highest energy neutrons, and constitutes at present the source used by many investigators for studies of the properties of and disintegrations produced by neutrons.

In some reactions the neutrons are observed and measured through the recoil protons they produce in hydrogen. Such protons have been studied with ionization chambers behind paraffin absorbers (D21) and with hydrogen or methane filled cloud chambers (B40). Other reactions of this type have been determined through the observation of positron radioactivity, utilizing chemical analysis and comparison of half-life values. This has resulted in establishing the process for several elements in addition to those from which neutrons have been recognized.

As in the α-p type reaction a few processes have been found to have characteristics suggesting resonance entry of the alpha-particle, of which Be^9 and B^{10} are most certain.

H^2: $H^2+He^4 \rightarrow H^1+n^1+He^4$ (noncapture)

Schultz (S8a) reports that neutrons are produced when deuterium is bombarded by Th C alphas (8.78 MV). This is interpreted as due to a noncapture disintegration of the deuteron for which the energy required is 2.20 MV.

Li^7: $Li^7+He^4 \rightarrow B^{10}+n^1$

Large intensities of neutrons were observed from Li both before (B43) and after (C56) Chadwick gave the new particle a name. The neutron energies have not been measured, but calculations from masses indicate that the Q value is −2.99 MV and that the neutron energies should extend up to a maximum of 1.0 MV for Po alphas.

Gamma-rays of about 0.4–0.6 MV (W4, B47) have been the subject of controversy. Curie and Joliot (C59), Schnetzler (S7) and Savel (S6) have found that gammas were observed for alphas of more than about 3 MV, while they did not detect neutrons until alphas of about 5 MV were used. This seemed to indicate two processes and since Li was not known to yield protons of any intensity a *noncapture* excitation of the Li was suggested (C10) to explain the gamma-rays.

$$Li^7+He^4 \rightarrow {}^*Li^7+He^4; \quad {}^*Li^7 \rightarrow Li^7+h\nu.$$

This conclusion was forced by the impossibility of balancing the mass-energy equation of the α-n type process under discussion if the gamma-ray was the result of an excited B^{10} nucleus. The Q value calculated from the present mass values is -2.99 MV. Since only the relative kinetic energy of the alpha-particle is available for excitation or disintegration, alpha-particles of $11/7 \times 2.99$ MV $=4.7$ MV are required to produce neutrons, in accord with the experimental results of Savel, etc. An even higher energy would be required to leave the B^{10} nucleus in an excited state. Other possible reactions caused by alpha-particles on Li would have similar Q values (e.g., Li^6-α-n, $Q<-4.2$ MV; Li^6-α-p, $Q=-2.3$ MV; Li^7-α-p, $Q=-2.6$ MV), and can therefore not be produced by 3 MV alphas, especially if the residual nucleus is to be left in an excited state. There remain therefore only two processes, namely the "noncapture" excitation of Li^7 and the radiative capture of the alpha-particle forming B^{11} (B53a). According to our present knowledge, the probability of emission of particles from a light compound nucleus is always much larger (by a factor of 10^4 or more) than the probability of emission of gamma-rays (§81), unless the particle emission is forbidden by a rigorous selection rule which is not the case here. Therefore the noncapture excitation seems to be the more probable process, which is also confirmed by the energy of the gamma-rays (B47) which should be 9 MV for the radiative capture process but is actually less than 1 MV.

From the Li^6-d-p reaction (R19) it is known that Li^7 possesses an excited state with 0.44 MV excitation energy (§101B). The γ-ray measurements of Bothe (B47) indicate γ-rays of approximately this energy. Bothe analyzes his curve as showing two γ-rays of 0.4 and 0.6 MV[14] and possibly a third of 0.2 MV. It is hard to understand how such a spectrum could originate from the Li^7 nucleus because theoretically this nucleus should have only two low levels ($^2P_{\frac{3}{2},\frac{1}{2}}$) including the ground state, and the next higher level should have an excitation energy of at least 2 MV.

We expect, therefore, that only one gamma-ray (of 0.44 MV) will be produced by alpha-particles of energies from 0.7 MV up to about 5.6 MV. Above 5.6 MV the B^{10} product nucleus of the α-n reaction might be left in the excited state at 0.6 MV known from the Be^9-d-n reaction.

Be^9: (a) $Be^9+He^4 \rightarrow C^{12}+n^1$
(b) $Be^9+He^4 \rightarrow 3He^4+n^1$

Because of its intensity and prominence as a source of neutrons for studies of induced radioactivity and other processes, this reaction has received considerable attention. The first good distribution in range curve of projected protons was published by Dunning (D21), using a paraffin layer in front of a shallow ionization chamber actuating a linear amplifier. With a radon alpha-source in which there are three groups of alphas, of 5.44, 5.97 and 7.68 MV from Rn, Ra A and Ra C′, respectively, he observed many groups: 13.7, 12.0, 7.6, 6.2, 4.6 MV and large intensities in unresolved groups between 0.5 and 1.5 MV. See Fig. 44. The proton groups expected from the first two alpha-groups are probably unresolved, so two groups should be expected for each Q value of the process. Although the intensity of the longest range proton group is extremely small (from 7.68 MV alphas), it can be used to obtain a Q value (recalculated) of 6.3 MV. The value obtained from the masses is 5.56 MV. A better experimental value is given by Bernardini (B7a) who finds neutrons of 11 MV using Po alphas and obtains a Q of 5.8 MV.

The group structure of the neutrons has been most carefully studied by Bernardini and

[14] Crane and Lauritsen (C43) have observed a gamma-ray of essentially the same energy (0.7 MV) from deuterons on Be, supposedly from a B^{10} isotope. Evidence from neutron groups in the same reaction (B35) also shows a 0.6 MV excited level in the B^{10} nucleus. This was originally correlated (C43) with the Li-α gamma-ray and said to be due to B^{10}. Since this is now shown to be energetically impossible the coincidence in values must be regarded as fortuitous.

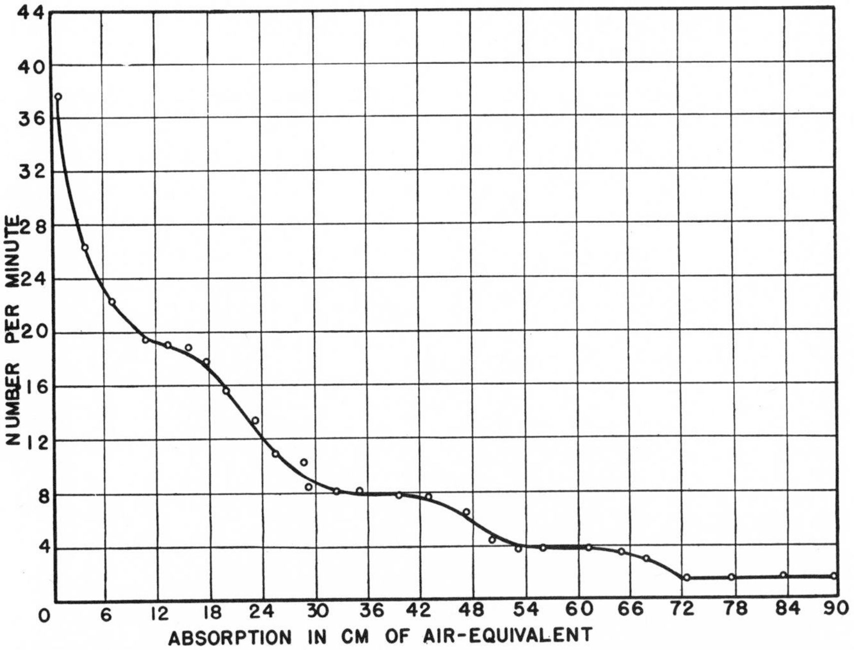

FIG. 44. Distribution-in-range curve of the recoil protons projected by the neutrons from the Be^9-α-n reaction, according to Dunning. The curve shows various energy groups indicating excitation levels in the final nucleus C^{12}, and a very great number of slow neutrons which probably originate from the four-particle disintegration $Be^9+He^4=3He^4+n^1$.

Bocciarelli (B7b). Using Po alphas, which have the advantage of having only one alpha-group (5.3 MV), they find neutron groups representing Q values of 2.8, 1.4 and -0.6 MV. Using Bernardini's value for the highest energy group (5.8 MV Q) this indicates excitation levels in C^{12} at 3.0, 4.4 and 6.4 MV. Blau (B64), using the photographic emulsion technique, reports a total of 10 neutron groups of which three are strong enough to be considered (28, 47 and >100 cm). These lead to Q values of -0.7, 1.1 and >4.1 MV. Bonner and Mott-Smith (B35), using the pressure cloud chamber and the same source, observe a large number of groups (22 in all) very few of which are justified statistically. They were not able to observe the longest range recoils representing the highest Q value.

From the measurements of Bothe (B47) the gamma-radiation from this process is found to have three components, at 6.7, 4.2 and 2.7 MV. Gammas of essentially the same energy (6.7, 4.2 and 2.4 MV) have been found by Crane, Delsasso, Fowler and Lauritsen (C44) in the bombardment of B by deuterons (in addition lines at 5.6 and >10 MV were observed). The first three mentioned may well be interpreted as coming from $^*C^{12}$ in the B^{11}-d-n reaction and be identical with those of Bothe. They also agree fairly well with the excitation states of the C^{12} nucleus as determined from neutron groups. From groups in the B^{11}-d-n reaction yielding the same product nucleus (§102) an excitation level is found at 4.4 MV, and another at 9.5 MV. Rasetti (R2) observes that the excitation curves for the production of neutrons and gamma-rays are identical, indicating that they come from the same process.

In all observations there are found large numbers of slow neutrons, usually not resolved into groups. From the "evaporation model" of nuclear disintegrations (§54) not many neutrons of energies below 2 MV would be expected. It was suggested by Bohr that the slower neutrons are due to a different process, namely:

$$Be^9+He^4\rightarrow{}^*C^{13}\rightarrow{}^*Be^9+He^4;\quad {}^*Be^9\rightarrow Be^8+n^1.$$

The final products would be 3 alpha-particles and a neutron; the same set of products have been observed in other processes where the $^*C^{13}$ compound nucleus is formed, e.g., $C^{12}+n^1$ and $B^{11}+H^2$. The calculated Q value for the complete reaction is -1.59 MV.

Bernardini (B7) has observed definite evidence

for a resonance level for 3.3 MV alphas. Chadwick (C7) reports two levels, at 1.5 and 2.6 MV, but with much less satisfactory data.

The complete evidence indicates four neutron groups, for Q's of 5.8, 2.8, 1.4 and -0.6 MV and one resonance level at 3.3 MV, besides the slow neutrons attributed to an alternate reaction.

B^{10}: $B^{10}+He^4\rightarrow N^{13}+n^1$

Of the large intensities of neutrons observed from B about 1/10 are thought to come from the B^{10} isotope and lead to the radioactive N^{13} isotope (B35). Although this number may be observable as groups superimposed on those originating in the B^{11} disintegration, discussed in the following reaction, no information is at hand to identify them. The calculated Q value of this reaction is 1.11 MV.

The Curie-Joliots found an induced radioactivity with positron emission and half-life of about 14 min. in their first reports of this phenomenon. Ellis and Henderson (E5), with more careful measurements, have shown the period to be 11.0 min. (see §105). The most accurate values of positron energy are obtained from N^{13} produced by the C^{12}-d-n reaction (§102) and show a maximum of 1.25 MV.

The alternative reaction yielding protons is known to result in the same ultimate product, C^{13}. The calculated Q is 4.1 MV, the measured value was 3.3 MV. By combining these twe primary reactions with the energy release of tho radioactive process a value may be obtained for the neutron mass. This method was proposed by Curie and Joliot (C55) before the more accurate methods now employed were known. The assumption made by Curie and Joliot was that all neutrons originated from B^{10} and that their experimentally determined energy could be used in calculating masses. Since only about 1/10 of the neutrons seem to so originate the cycle is invalidated unless the particular group of neutrons can be identified.

B^{11}: $B^{11}+He^4\rightarrow N^{14}+n^1$

The greater number of neutrons from B are from the B^{11} isotope, yielding a stable N^{14} nucleus (B35). Chadwick (C5) first identified this radiation as neutrons and in following work by the Curie-Joliots (C59) and Chadwick (C7) the maximum neutron energy was found to be 3.3 MV for Po alphas. Bonner and Mott-Smith (B35) have made a cloud chamber study of the neutron energy distribution using Po alphas and find many groups, of which the most energetic is 4.1 MV. The statistical error involved in plotting curves through such a small amount of data makes most of these groups uncertain. Assuming the most energetic group to come from B^{11} with alphas of maximum energy and without excitation of the residual nucleus, $Q=-0.7$ MV. The calculated value is $+0.32$ MV.

Gamma-rays from B have been measured by Bothe and Becker (B43) and others to be about 3.0 MV but are attributed to the excited C^{13} product of the α-p reaction. Chadwick (C7) reports a resonance level for alphas of >2.6 MV.

N^{14}: $N^{14}+He^4\rightarrow F^{17}+n^1$

Danysz and Zyw (D3) and Wertenstein (W12) have found a 1.2 min. positron activity in nitrogen due to the F^{17} resulting from this process. Other observers (E5) have found it as a spurious period on other targets bombarded in air. The same activity has more recently been found by Newson (N6) in deuteron bombardment of O^{16} with a half-life of 1.16 min. The positron energy spectrum has been measured by Kurie, Richardson and Paxton (K33) to have a 2.1 MV maximum.

In studies of the α-p reaction on nitrogen Haxel (H18) found a positron radioactivity of 1.2 min. period for alphas of greater than 7 MV energy. The fraction of this energy available for the reaction is 5.45 MV, so this suggests a Q value of -5.45 MV for the primary process. The Q calculated from masses is -4.8 MV.

F^{19}: $F^{19}+He^4\rightarrow Na^{22}+n^1$

The neutrons from F were first observed and studied by Chadwick and the Curie-Joliots. Bonner and Mott-Smith (B35) find 5 groups, the most energetic being 2.12 and 2.54 MV, and resulting in Q values (re-evaluated) of -2.3 and -2.6 MV. Some of the groups can be explained by considering the two resonance levels found in the α-p reaction at 3.7 and 4.1 MV. From the mass of Na^{22} obtained by using the radioactive energy evolution we judge the primary reaction to have a Q value of about -0.7, so the highest energy neutrons were not observed.

Meitner (M14) and Frisch (F32) have found a

long lived positron activity ($T \sim 6$ months) in F attributed to this process. Meitner reported a positron energy maximum of 0.4 MV which is of questionable accuracy and probably low, while Frisch estimated the period and performed chemical separation tests to identify the product as Na.

Na^{23}: $Na^{23}+He^4 \rightarrow Al^{26}+n^1$

Although neutrons have been observed (C59), (S5), no energy measurements are available. The positron radioactivity of 7 sec. half-life has been studied by Frisch (F31) to identify the reaction.

Mg^{24}: $Mg^{24}+He^4 \rightarrow Si^{27}+n^1$

Savel (S5) reports neutrons from Mg, corroborating the reaction discovered by the Curie-Joliots (C54) leading to positron radioactivity. Mass values give a Q of -5.3 MV. The half-life is 6.7 min. (F1) and the maximum positron energy is given as 2.0 MV (A2a). Ellis and Henderson (E10) and Fahlenbrach (F1) suggest resonance levels for alphas of 5.4 and 6.1 MV; these check the resonance levels found in the α-p process at 5.7 and 6.3 MV, and indicate excitation levels in the compound nucleus Si^{28}.

Al^{27}: $Al^{27}+He^4 \rightarrow P^{30}+n^1$

Curie and Joliot (C54) reported radioactivity from Al as one of their three original reactions. The half-life is 2.5 min. (R7a) and the radioactive energy 3.6 MV (A1). The neutrons from the primary reaction have been observed by Savel (S4) and others. Waring and Chang (W3) have studied the resonance levels for neutron production and finds 6 or 7 levels agreeing closely with those found for proton emission. A Q of -3.3 MV is expected.

P^{31}: $P^{31}+He^4 \rightarrow Cl^{34}+n^1$

Frisch (F31) observed a positron activity of 40 min. half-life from P activated by alphas and estimated the positron energy to be 1.8 MV. His comparison with the energy of positrons from Al, however, suggests a much higher value. Although neutrons are not observed, due to the low intensities, chemical tests prove the validity of the reaction. The calculated Q value is -1.6 MV.

Cl^{35}: $Cl^{35}+He^4 \rightarrow K^{38}+n^1$

A 7.7 min. positron activity resulting from the α-particle bombardment of Cl (H24a, H39a) can only be due to K^{38}, since other likely reactions produce stable isotopes.

Pollard, Schultz and Brubaker (P10) report the observation of neutrons from chlorine under alpha-particle bombardment for alpha energies of more than 6.6 MV. The observers assign the neutrons to a Cl^{37}-α-n reaction. It is possible that both occur.

A^{40}: $A^{40}+He^4 \rightarrow Ca^{43}+n^1$

Neutrons have been observed by Pollard, Schultz and Brubaker (P10) for alphas of more than 6.8 MV.

$K^{39,41}$: $K^{39}+He^4 \rightarrow Sc^{42}+n^1$;
$K^{41}+He^4 \rightarrow Sc^{44}+n^1$

Zyw (Z3) first observed a positron activity, of 3 hr. period, in K after alpha-bombardment. Walke (W1), using 11 MV He^{++} ions from the cyclotron, finds periods of 4.1 hr. and 52 hr. The 52 hr. period is also found in the Ca-d-n reaction and so must be Sc^{44}; the 4.1 hr. period is then Sc^{42}.

Cr^{50}: $Cr^{50}+He^4 \rightarrow Fe^{53}+n^1$

The 8.9 min. positron activity chemically identified with Fe by Henderson and Ridenour (H24b) is assigned to Fe^{53} because a 91 min. period found by the Rochester group (B4a) to result from a Mn-p-n reaction must be due to Fe^{55}.

Co^{59}: $Co^{59}+He^4 \rightarrow Cu^{62}+n^1$

Ridenour and Henderson (R7b, H24b) find the 10 min. positron activity characteristic of Cu^{62} (see Tables LXVI and LXVIa) to result from the bombardment of Co by alphas.

Ni^{60}: $Ni^{60}+He^4 \rightarrow Zn^{63}+n^1$

Only the greater abundance of Ni^{60} over Ni^{62} favors this assignment for the 37 min. activity found by Ridenour and Henderson (R7c) and already known from Zn-n-$2n$ and Zn-γ-n reactions (§102, 103).

$Cu^{63,65}$: $Cu^{63}+He^4 \rightarrow Ga^{66}+n^1$
$Cu^{65}+He^4 \rightarrow Ga^{68}+n^1$

Henderson and Ridenour (H24b) find that two positron activities (9.4 hr. and 68 min.), identified chemically with Ga, are produced from Cu under alpha-bombardment. The 68 min. period is probably the same as the 60 min. period found in a Ga-γ-n reaction (§103) and is there-

TABLE LIII. *Summary of α-n type reaction.*

Target		Product	Q(MV)	Q(MV)		Resonance levels	Excit. levels	Yield
Z	El	El	(calc.)	(obs.)	Reference	(MV)	(MV)	n/α(Rn α's)
2	H^2	He^4+H^1	−2.20	obs.	S8a			
3	Li^7	B^{10}	−2.99	−5?	C59			
4	Be^9	C^{12}	5.56	5.8	B7a	3.3	3.0, 4.4, 6.4	2.5/10^4
	Be^9	$3He^4$	−1.59	obs.?				
5	B^{10}	N^{13}	1.11	obs.	C54			1/10^5
	B^{11}	N^{14}	0.32	−0.7?	B35	3?	many	1/10^4
6	C^{12}	O^{15}	−8.3					
	C^{13}	O^{16}	2.3					
7	N^{14}	F^{17}	−4.8	−5.5?	H18			
	N^{15}	F^{18}	−5.4					
8	O^{18}	Ne^{21}	−0.99					
9	F^{19}	Na^{22}	−0.7	−2.3?	B35		1.6, 1.9	3/10^5
10	Ne^{21}	Mg^{24}	2.3					
	Ne^{22}	Mg^{25}	0.0					
11	Na^{23}	Al^{26}	−1.8	obs.	C59			5/10^6
12	Mg^{24}	Si^{27}	−5.3	obs.	C54	5.4, 6.1		
	Mg^{25}	Si^{28}	2.0					
	Mg^{26}	Si^{29}	−1.7					
13	Al^{27}	P^{30}	−3.3	obs.	W3	4.0, 4.5, 5.0, 5.25 (5.55), 6.0, 6.7		1/10^5
14	Si^{29}	S^{32}	−0.7					
15	P^{31}	Cl^{34}	−1.6	obs.	F31			2/10^6
17	Cl^{35}	K^{38}	—	obs.	H24a, P10			2/10^6
18	A^{40}	Ca^{43}	—	obs.	P10			
19	K^{39}	Sc^{42}	—	obs.	Z3			
	K^{41}	Sc^{44}	—	obs.	W1			
24	Cr^{50}	Fe^{53}	—	obs.	H24b			
27	Co^{59}	Cu^{62}	—	obs.	R7b			
28	Ni^{60}	Zn^{63}	—	obs.	R7c			
29	Cu^{63}	Ga^{66}	—	obs.	H24b			
	Cu^{65}	Ga^{68}	—	obs.	H24b			
33	As^{75}	B^{78}	—	obs.	S16a			

fore assigned to Ga^{68}. This leaves Ga^{66} to account for the 9.4-hr. half-life.

As^{75}: $As^{75}+He^4\rightarrow Br^{78}+n^1$

In a report by Snell (S16a) a positron active period of 6.3 min. from As bombarded by artificially accelerated He^{++} ions is found to be chemically Br, so specifying this type reaction.

§100. Disintegration by Protons

A. Type reaction *p*-α. (Table LIV)

$$Z^A+H^1\rightarrow(Z-1)^{A-3}+He^4+Q:$$

It was early in 1932 that Cockcroft and Walton (C21) were first successful in producing the disintegration of lithium by protons. The apparatus used was a condenser-rectifier voltage multiplier to produce the high voltage, applied to a discharge tube through which hydrogen ions were accelerated (§92). These experimenters, as were many others, were striving for a really high voltage source, but upon the introduction of the quantum-mechanical penetration idea, development was halted long enough to allow search for disintegrations. The observations were made on that time-honored instrument, the scintillation screen. An exhaustive series of experiments, using elements throughout the periodic table, showed that only in Li, B and F were any definite results then obtainable (C19). Many elements were found to give alpha-particles, but as was suggested at that time, they were later identified as due to impurities (of boron). With improvements in experimental techniques other

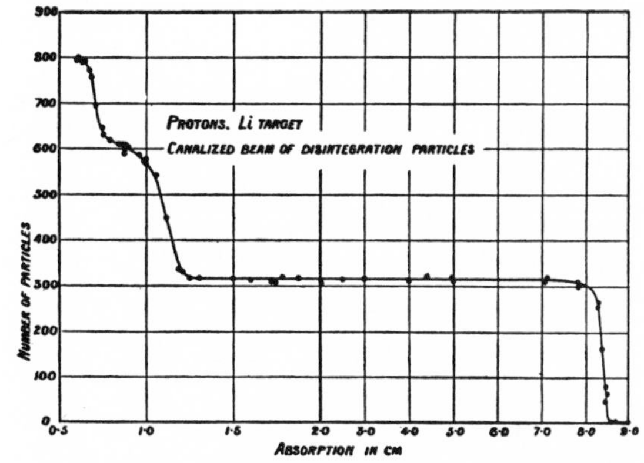

FIG. 45. Distribution in range of the α-particles produced by proton bombardment of lithium, showing the 8 cm alphas from $Li^7+H^1=2He^4$ and the short range particles He^3 (1.2 cm) and He^4 (0.8 cm) from $Li^6+H^1=He^4+He^3$. (Oliphant, Kinsey and Rutherford.)

reactions have been added to the list. It will be well to take up these elements individually.

Li^7: $Li^7+H^1 \rightarrow He^4+He^4$

In the case of Li^7 the residual nucleus is itself an alpha-particle, so the energy evolved is divided equally between the two. These alphas were found to have a range of slightly over 8 cm, and are commonly referred to as the "8 cm group from Li." See Fig. 45. As we shall see later, this range depends to some extent on the bombarding proton energy. Oliphant, Kempton and Rutherford (O8) report that the normal process results in alpha-particles of 8.31 cm mean range (our corrected value, obtained from the 8.40 cm extrapolated range given by the observers) for protons of 0.19 MV if observed at 90° to the proton beam. The momentum-energy considerations for this case (§92) show that $Q=2E_2-\frac{3}{4}E_1$ or twice the energy of the alpha measured at 90° less $\frac{3}{4}$ the bombarding proton energy. Using the range-energy relation of §95 we find $Q=2(8.63)-\frac{3}{4}(.19)=17.13$ MV. This is in good agreement with the value of 17.25 MV obtained from the masses and justifies the masses and range-energy relations used. When higher energy protons are used alphas of longer range are observed, as for instance the 9.1 cm alphas (extrapolated range) observed at 90° by Livingston, Henderson and Lawrence (L28) for 1.4 MV protons. This extrapolated range[15] is equivalent to about 9.0 cm mean range or 9.1 MV and so $Q=17.1$ MV with somewhat less accuracy.

The assumed mechanism of the reaction is justified by the cloud chamber photographs of Dee and Walton (D6) using protons of less than 0.2 MV, in which the two oppositely directed alpha-particles are observed to originate from the same point on the thin target. The alphas will be co-linear only for very low energy protons. The momentum relations show that if one alpha is emitted at 90° the other alpha will make an angle to the direction of the proton beam of 79° for 1.4 MV protons, or 86° for 0.2 MV.

The alpha-particles from Li^7 have been reported for proton energies ranging from 8 kv (Burhop (B68)) for which one-half milliampere of ions was required, to 1.4 MV (L28) at which energy a microampere gave large yields. The excitation function (intensity *vs.* proton energy) has been a subject of considerable investigation, because of its theoretical importance (O15). Studies in the extreme low voltage range are reported by Doolittle (D16) and Burhop (B68). For somewhat higher voltages (up to 0.40 MV) accurate data have been taken for both thick and thin targets by Herb, Parkinson and Kerst (H27). In this voltage range both the thick and thin target curves show the approximately exponential rise with voltage expected from a penetration function. Hafstad and Tuve (H4), with thin targets, find the curve tending toward a maximum at a point above 1.0 MV. Ostrofsky, Breit and Johnson (O15) have calculated the depth and radius of the potential hole required to explain the data of Herb, Parkinson and Kerst.

Li^6: $Li^6+H^1 \rightarrow He^4+He^3$

A search for shorter range particles from lithium led Oliphant, Kinsey and Rutherford (O3) to the discovery of two groups of doubly-charged ions of about 0.65 and 1.15 cm range. (See Fig. 45.) They suggested that the Li^6 isotope was responsible and was yielding an alpha-particle and a He^3 nucleus. This reaction was verified by later work of Oliphant, Shire and Crowther (O1) using separated isotopes of lithium. A more accurate determination by Neuert (N4) in a cloud chamber gives values of 0.82 and 1.19 cm mean range

[15] It should be noted that the mean range, not the extrapolated range usually measured, represents the average energy of the alpha. This correction is discussed and a curve given in §97.

for the two particles, when produced by 0.12 MV protons.

The qualitative agreement of the two range groups with that expected from He^4 and He^3 particles justifies the suggested interpretation. The He^3, being the lighter product, will get 4/7 of the reaction energy, while the He^4 has 3/7. The energy of the He^3 can be obtained from the alpha-particle range-energy relations in a region in which they are fairly satisfactory. This energy is found to be 2.12 MV, from which a reaction energy of 3.72 MV is obtained. The quantitative check of this value with that obtainable from masses (3.76 MV) justifies the use of the shorter He^4 range to determine the range-energy relation for this region (§95).

Be^9: $Be^9+H^1\rightarrow Li^6+He^4$

Several early reports of low intensities of alpha-particles of about 3 cm range have been ascribed to impurities of B or Li. Dee first observed particles of about 8 mm range and judged them to be singly-charged ions, probably deuterons. Döpel (D17) and also Kirchner and Neuert (K16) observed these particles and called them alpha-particles, Kirchner and Neuert being the first to suggest the now accepted reaction resulting in Li^6, and measuring a range of 7.4 mm for the alphas. Oliphant, Kempton and Rutherford (O9) then performed a series of experiments in which they found two groups of particles, one of alphas and one of deuterons (§100B) of almost identically the same range of 7.4 mm at 90°. They based their conclusion on the different specific ionizations observed in the ionization chamber, and found the e/m values to be the same in deflection experiments in electric and magnetic fields. Near the end of their range H^2 and He^4 particles have very nearly the same specific ionization, which explains why the two particles were at first confused. The alphas, of 7.4 mm mean range, have energies of about 1.4 MV. We find that $Q=2.33$ MV, while the calculated value is 2.25 MV. Allen (A4) reports results in agreement with those above and measures the probability cross section to be 5×10^{-29} cm^2 at 0.1 MV. This reaction is important as supporting evidence for the mass of Be^9 and also in checking the validity of the alpha-particle range-energy relation in this low energy region.

B^{11}: (a) $B^{11}+H^1\rightarrow He^4+He^4+He^4$
(b) $B^{11}+H^1\rightarrow Be^8+He^4$

Boron yields large intensities of alpha-particles, of diverse ranges. The first observation, by Cockcroft and Walton (C19), indicated an ill-defined range of about 3.5 cm. Further experiments (C23) showed a distribution of energies with a few alphas of nearly 5 cm range, and they suggested reaction (a), considering a mechanism involving simultaneous emission of the three alphas. Oliphant and Rutherford (O2) measured the maximum range to be 4.7 cm but found that the observed distribution-in-range curve could not be explained by the simplest assumption, that the angular distribution was symmetrical and centered about a most probable value of angle of 120° between the particles. Using this picture the Q value calculated from the masses was not in accord with that calculated from the maximum range.

Dee and Gilbert (D9) have resolved these difficulties in a report of cloud chamber investigations in which the supposed simultaneity and angular distribution of tracks were carefully studied. They find no evidence for simultaneity, or of a preferred 120° angle, but rather, that one alpha is always emitted with a range of approximately 2.4 cm. They conclude that in the process this alpha is first emitted, leaving an excited Be^8 nucleus which subsequently disintegrates into two other alphas. The Be^8 is in a 2.8 MV excited state with a width of 0.77 MV (see §85A). This large width is due to its short life time, and explains the variation in range of the initial 2.4 cm alphas. The life-time of this particle disintegration can be calculated from the width and is found to be 8.5×10^{-22} sec. On this assumption the calculated distribution-in-energy and the angular distribution of tracks in the individual reactions are found to agree with the experimental observations (Fig. 25 in §85). They calculate the reaction energy from pairs of associated tracks, of which many are observed, and obtain 8.7 ± 0.2 MV. There are two other less accurate methods of using the observations to obtain the reaction energy. From the cut-off of the continuous distribution, using the interpretation above, we obtain the value 8.62 MV. From the average energy of all alphas emitted Oliphant and Ruther-

TABLE LIV. *Summary of p-α type reaction.*

Target		Product	Q(MV)	Q(MV)	Reference	Resonance	Excit. levels	Yield
Z	El	El	(calc.)	(obs.)		levels (MV)	(MV)	α/p @ (MV)
3	Li^6	He^3	3.76	3.72	N4			$1/10^7$ @ 0.4
	Li^7	He^4	17.25	17.13	O8			$1/10^7$ @ 0.4
4	Be^9	Li^6	2.25	2.28	K16			$2/10^9$ @ 0.1
5	B^{11}	Be^8	8.60*	8.60	O9	0.18		$1/10^{14}$ @ 0.15
	B^{11}	$2He^4$	8.72	8.7	D9		2.9 (Be^8)	$1/10^{12}$ @ 0.15
6	C^{13}	B^{10}	−4.15					
7	N^{14}	C^{11}	−3.3					
	N^{15}	C^{12}	4.79					
8	O^{16}	N^{13}	−5.40					
	O^{17}	N^{14}	1.3					
	O^{18}	N^{15}	2.82					
9	F^{19}	O^{16}	8.14	8.15	H22			$1/10^7$ @ 1.8
10	Ne^{20}	F^{17}	−4.2					
	Ne^{22}	F^{19}	−1.52					
11	Na^{23}	Ne^{20}	1.42					
12	Mg^{25}	Na^{22}	−2.0					
	Mg^{26}	Na^{23}	−1.9					
13	Al^{27}	Mg^{24}	1.6					
14	Si^{29}	Al^{26}	−1.9					
	Si^{30}	Al^{27}	−2.26					
15	P^{31}	Si^{28}	1.8					
16	S^{34}	P^{31}	−1.9					
17	Cl^{35}	S^{32}	2.2					
18	A^{38}	Cl^{35}	−2.0					

* Q (observed) used to calculate mass values.

ford (O2) get a value of 8.60 MV. These methods are subject to interpretational and experimental errors, so we use the value of 8.7 MV obtained from the data of Dee and Gilbert to compare with that calculated from masses of 8.72 MV.

The existence of an alternate mode of disintegration in which only two products are formed (reaction *b*) is suggested by the observations of Kirchner and Neuert (K15) and Oliphant, Kempton and Rutherford (O9) of a homogeneous group of alphas of 4.4 cm mean range. This is interpreted as due to the formation of a Be^8 nucleus in the ground state rather than the excited state assumed for the three particle process. The Q value obtained from this evidence is 8.60 MV, only slightly lower than that for the alternative reaction; it is used to determine the mass of Be^8, which proves to be unstable (heavier than two He^4 atoms) by 0.13 MV.

The excitation curve for the production of alphas from B exhibits a somewhat different character than that for Li, in that it is lower for energies of less than 100 kv and rises more steeply in the 200–400 kv region (N4). It can be fitted to a Gamow probability function with reasonable assumptions. Williams, *et al.* (W21a) studied the excitation function of the homogeneous alpha group from reaction (b) and found good evidence for a resonance level at 180 kv with a half-width of slightly less than 10 kv. This represents a level in the excited C^{12} compound nucleus at 16.1 MV. Gentner (G26) has observed the same resonance state in the production of gamma-radiation. A higher resonance level at about 0.4 MV proton energy is also indicated.

Neuert (N4a) has studied the angular distribution of the alpha-particles of discrete range from this reaction and finds it nonisotropic, with a strong minimum for an angle of 90° to the proton beam.

F^{19}: $F^{19}+H^1 \rightarrow O^{16}+He^4$

The observation of alpha-particles from fluorine was first indicated by Cockcroft and Walton (C19) and somewhat better measurements were made later by Oliphant and Rutherford (O2). At such low proton energies (O. and R. used 0.20 MV) the yields are small and range measurements are not precise. Henderson, Livingston and Lawrence (H22) have made the most complete survey of this reaction for higher energies, showing alphas of 6.95 cm extrapolated range for 1.69 MV protons. Indications of a shorter range of about 3 cm are ascribed to boron impurities. In this case the particles showed a definite group structure, indicating only two resulting particles, and the assumed reaction is that given above. For this reaction $Q=5/4E_2-15/16E_1$ (§99), from which we find a value of 8.15 MV by taking an average of the results obtained at several bombarding energies (H22). This is in excellent agreement with the value of 8.14 MV obtained from the mass values.

The excitation curve of fluorine is found to be lower for low proton energies than that for B or Li, but once again there is no indication of a sharp threshold. In the curves given for protons of high energy and rather low intensity (H22) there is no observable emission below 0.5 MV, but Oliphant has observed alphas at 0.2 MV with considerably larger proton currents. Observational data (H22) show some deviations from a correct Gamow curve (see §78), which are best interpreted as due to changes in the internal disintegration probability.

B. Type reaction *p-d*: $Z^A+H^1 \rightarrow Z^{(A-1)}+H^2+Q$

This type reaction has been observed for only one element, Be^9, and is based chiefly on the data and conclusions of a single paper by Oliphant, Kempton and Rutherford (O9). An earlier report by Oliphant (O6) of investigations on a 7 mm group of particles from Be observed by Dee concluded that they were singly charged, i.e., hydrogen ions. With scant evidence at that time he suggested the reaction:

$$Be^9+H^1 \rightarrow Be^8+H^2.$$

Döpel (D17), Kirchner and Neuert (K16) also observed these particles, but identified them as alpha-particles, having a double charge. In the paper mentioned above (O9) Oliphant, Kempton and Rutherford reported experiments from which they concluded that both H^2 and He^4 particles were present in about equal numbers, thus establishing the reaction above. In magnetic and electrostatic deflection experiments they found the particles to have the same e/m ratio, and obtained estimates of the energy assuming the character of the ions. Furthermore, both groups disappeared at almost exactly the same thickness of absorber: 7.4 mm. (Further experiments with different bombarding energies might assist in separating the two groups.) When the particles traversed a shallow ionization chamber, however, and the pulses were amplified linearly by a vacuum tube amplifier, they were found to be of two sizes, one corresponding to alpha-particles and the other to singly charged ions. The reactions are highly probable, and both groups occur with about the same intensity at 100 kv. The Q value obtained from the reaction is 0.46 MV, while that obtained from the masses is 0.48 MV. The excitation curve (O6) is of the type expected for an element of such low atomic number.

C. Type reaction *p*-γ. (Table LV)

$$Z^A+H^1 \rightarrow (Z+1)^{(A+1)}+h\nu:$$

This type of simple capture reaction has been the subject of much controversy. It involves the emission of gamma-radiation, which must come from the residual nucleus. It is now apparent that the reaction occurs usually only for protons of discrete energies—the phenomenon of resonance. This resonance characteristic plays an important role in the general scheme of the theories of Breit and Wigner (B51) and of Bohr (B32). (Chapter IX, and §81.)

Following the announcement by the Curie-Joliots of the α-n type reactions yielding neutrons and radioactive elements such as N^{13} and Si^{27}, and their suggestion that the same radioactive products might be expected from deuteron and proton bombardment of other elements, several laboratories started to search for this effect. The C^{12}-d-n process was readily observed to give the N^{13} activity (see §101C). When proton bombardment of carbon was found (in three laboratories (H21, C24, C39) to give the same activity, so strong was the analogy that the process: $C^{13}+H^1 \rightarrow N^{13}+n^1$ was at first suggested. It is now known to be energetically impossible, as indeed are all such p-n reactions, except for very high energy protons.

Hafstad and Tuve (H2), however, failed to observe radioactivity with protons. Deuteron contamination of the proton beam was suggested and eliminated as an explanation of the other observations. With improved mass values it was possible for Cockcroft (C26) to conclude that the reaction was simple capture in the C^{12} isotope, but still without any suggestion of resonance. This feature was discovered experimentally by

Hafstad and Tuve (H3, H4), who found that the excitation curve of the production of N^{13} was not the monotonic increase observed in other nuclear processes and explained by the Gamow penetration theory, but consisted of several sharp maxima. Since their first data were taken with thin (gas cell) targets in which the proton energies were higher than those of the resonance maxima, their original negative results are explained and their data are found to be compatible with those from other laboratories in which thick targets were used.

With this knowledge other processes in which resonance features have been observed have been entered under this type reaction.

Li^7: $Li^7+H^1 \rightarrow Be^8+h\nu$

The gamma-rays observed from Li under proton bombardment have been the subject of much speculation and research, and their complete interpretation is still uncertain. A summary of their history will serve to illustrate the changes and progress in the techniques of gamma-ray measurements.

They were first observed by Traubenberg, Eckardt and Gebauer (T13), and later, independently, by Lauritsen and Crane (L6). The first absorption measurements in Pb indicated a quantum energy of about 1.6 MV, which was changed to 6.3 MV (C42) with the application of the theory of pair production to absorption measurements (§93E). Crane and Lauritsen (C42) then introduced the cloud chamber method of measurement of gamma-energies and showed that there were at least two groups, of 4 and 12 MV. With improvements in technique and more data Crane, Delsasso, Fowler and Lauritsen (C51) found what seemed to be evidence for as many as 11 gamma-ray lines, ranging from 2.9 MV to 16 MV. Statistical factors make many of these uncertain but later evidence indicates that at least a few of the higher energy lines were valid. Development of the method of measuring gamma-rays from the sum of the energies of electron and positron pairs suggested for a time that the radiation was almost entirely a single gamma-ray of 17.5 MV energy (D12, G2). A recent report by Gaertner and Crane (G3) gives evidence for gamma-ray lines at 17.5, 14.5, 11.0 and 8.5 MV from the Compton secondary electrons, but only the 17.5 MV line from the pairs. On the other hand Delsasso, Fowler and Lauritsen (D13) find evidence from studies of pairs and of recoil electrons for a line at 17.1 MV, possibly one or more between 10 and 17 MV, but none of lower energy. They show that the gammas of 2 to 6 MV are due to absorption of the 17.1 MV line, and found no lines of energies between 2 and 10 MV. When a gamma-ray of less then 17.5 MV is emitted the residual $^*Be^8$ breaks up into two alpha-particles of energy 1.5 MV or higher (Lauritsen, private communication) similar to the Be^8 nucleus formed in the β-decay of Li^8 (§101B).

In studying the yield of gammas from Li Crane, Delsasso, Fowler and Lauritsen (C51) found a decided resonance maximum in the excitation curve. This has been more carefully studied by Hafstad and Tuve (H4) who find two very sharp resonances, at 440 and 850 kv. Analysis of the shape of the resonance peaks using very thin targets and calibrating the energy spread of the proton beam shows that the peak at 440 kv has a natural width of not more than 11 kv (H5) (see §81). Herb, Kerst and McKibben (H28a) check the first peak but observe the second to be at 1.0 MV and much broader than Hafstad and Tuve's findings indicated. Bothe and Gentner (B47g) report a resonance level at 0.20 MV. The resonance character of these gamma-rays suggests that they are due to this simple capture type process, arising from an excited Be^8 nucleus. From the alternative process resulting in two alphas and having a Q value of 17.25 MV, the mass difference between Be^8 and two He^4 atoms (0.13 MV) and the fraction of the proton energy absorbed by the Be^8 nucleus ($\frac{7}{8}$ of 0.44 MV or 0.38 MV) we can calculate the resonance level in Be^8 to be at: $17.25-0.13+0.38$, or 17.50 MV, a sufficiently good check of the observed gamma-ray energy. The normal Be^8 nucleus is then thought to break up into two low energy alphas. The only assumption required for this explanation is that the excited nucleus cannot break up into the usual 8 cm alpha-particles According to the selection rules (§81) the protons responsible for the resonance gammas have even orbital momentum while those yielding alphas have odd. The two reactions are therefore quite independent and this resonance process cannot be detected in the excitation curve of the 8 cm alphas. The observed yield at resonance is about 1 gamma per 2 alpha pairs.

Be^9: $Be^9+H^1 \rightarrow B^{10}+h\nu$

The gamma-rays observed from Be on proton bombardment have a maximum energy of about 6 MV (C45, C47). This is too large to be due to the p-α type process, for which the particle energies are measured, and no other processes are known which would explain it. The conclusion is that it must be due to this simple capture reaction resulting in stable B^{10}. The energy available for the gamma-ray, calculated from the masses involved, is 6.39 MV plus the resonance energy of the proton, a satisfactory check of the proposed explanation. Herb, Kerst and McKibben (H28a) find a single broad resonance for protons of 0.99 MV. Other lower energy gamma-ray groups have been reported, but may be largely due to statistical fluctuations of the data.

B^{10}: $B^{10}+H^1 \rightarrow C^{11}+h\nu$

When boron was bombarded by protons a positron activity with the half-life characteristic of C^{11} was observed by Crane and Lauritsen (C39). Since the proton energies used were insufficient to cause the B^{11}-p-n type process in the heavier isotope ($Q=-3$ MV), the conclusion must be that the activity is due to simple capture. The assignment is justified by the observation of a resonance level at 0.18 MV by Shepherd, Haxby and Williams (S11a). The energy evolution in the gamma-ray, calculated from masses, should be 8.55+ MV, and there is some slight evidence for a gamma-ray of approximately this energy from cloud chamber studies of Crane, Delsasso, Fowler and Lauritsen (C50).

B^{11}: $B^{11}+H^1 \rightarrow C^{12}+h\nu$

The evidence for this reaction on the heavier boron isotope is entirely from gamma-ray measurements, since the product C^{12} is stable. Crane, Delsasso, Fowler and Lauritsen (C50) have observed and measured the gamma-rays produced by proton bombardment and find a maximum energy of 14.5 MV. No other reaction yields sufficient energy to account for this gamma-ray, but assuming simple capture the Q value calculated from the masses is 15.89 MV. Herb, Kerst and McKibben (H28a) observe a weak resonance effect for protons of 0.82 MV. This may indicate an excitation level in the C^{12} nucleus, or in C^{11} if the B^{10} isotope is responsible for the γ-rays observed.

C^{12}: $C^{12}+H^1 \rightarrow N^{13}+h\nu$

As discussed in the introductory paragraphs, the observation of the N^{13} radioactivity suggested this process, which has been confirmed by Hafstad and Tuve (H4) in the observation of resonance maxima in the excitation function at 0.40 and 0.48 MV, with an experimental width of the levels of about 10 kv. The Q value obtained from masses is 1.93 MV, so the gamma-ray energy will be 2.30 or 2.37 MV for the two resonances. No direct measurements are available, but absorption estimates by the above authors indicate a value of about 2 MV. The properties of the N^{13} radioactivity are best determined from the more probable deuteron reaction, and are discussed in §101. The N^{13} radioactivity appears as a contamination when targets are bombarded by fast protons (§100C).

O^{16}: $O^{16}+H^1 \rightarrow F^{17}+h\nu$
($O^{17}+H^1 \rightarrow F^{18}+h\nu$)

Radioactivities having the known periods of F^{17} and F^{18} were observed when O was bombarded by fast protons by DuBridge, Barnes and Buck (D28). The F^{18} activity may be produced through this reaction or through the O^{18}-p-n process (§100D). These activities are prominent as a contamination when other targets are bombarded in air, eliminated by activation in an atmosphere of hydrogen.

F^{19}: $F^{19}+H^1 \rightarrow Ne^{20}+h\nu$

The gamma-rays from fluorine were originally detected by McMillan (Mc4) who found a Pb absorption indicating 2 MV which was raised to 5.4 MV by using different absorbers and considering pair production. Assuming this radiative capture process the energy available, as shown by the mass values, is 12.89 MV.

Hafstad and Tuve (H5) found resonance maxima in the excitation curve for gamma-ray production at 0.328, 0.892 and 0.942 MV. (See Fig. 46.) The width of the first resonance level at 0.328 MV proves to be exceedingly narrow, only 4 kv actually observed, which is just the known

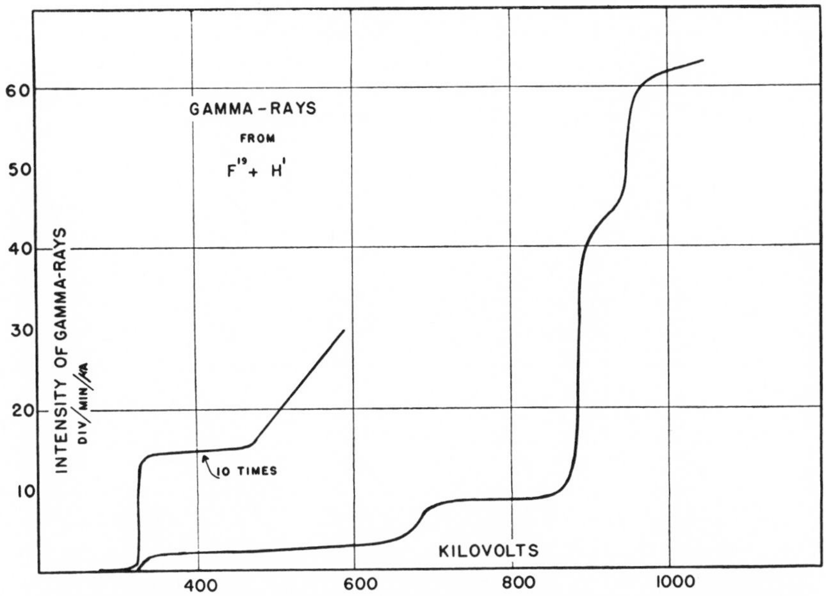

FIG. 46. Excitation curve of the γ-radiation from the radiative capture of protons by F^{19} showing three pronounced resonance levels. (Hafstad, Heydenburg and Tuve.)

inhomogeneity of the proton beam so that the width of the level must be considerably less than 4 kv. Herb, Kerst and McKibben (H28a) have extended these data to higher energies and find a broad resonance at 0.6–0.7 MV, a sharp one at 1.40 MV and indications of one at 1.76 MV.

Delsasso, Fowler and Lauritsen (D14) find a single line at 6.0 MV by analysis of pairs observed in a cloud chamber. They were using protons of 0.75 MV, so only the 0.328 MV resonance level was effective. No evidence for high energy gammas was obtained.

Although there is sufficient energy available in the *p*-α reaction to give the gamma-ray, the resonance features and especially the narrow resonance maxima favor the interpretation as a simple capture reaction.

Na^{23}: $Na^{23}+H^1 \rightarrow Mg^{24}+h\nu$

In studies of the gamma-radiation from proton bombardment, Herb, Kerst and McKibben (H28a) found a broad resonance for protons of 1.15 to 1.32 MV. This requires a capture reaction for its explanation and indicates a level in the Mg^{24} nucleus.

Al^{27}: $Al^{27}+H^1 \rightarrow Si^{28}+h\nu$

Herb, Kerst and McKibben (H28a) find that the γ-ray intensity from Al has resonance maxima for protons of 0.75, 0.99, 1.16, 1.37 (strong), 1.62 and 1.85 MV. The resonance feature specifies this capture reaction.

D. Type reaction *p*-*n*. (Table LVa)

$$Z^A+H^1 \rightarrow (Z+1)^A+n^1+Q:$$

Exploratory work with high energy protons (3.8 MV) carried out at the University of Rochester (B4a, B67, D28) has resulted in the discovery of a great number of radioactivities some of which can be identified with known radioactive substances while the greater part represent new radioactive nuclei. It seems likely that with sufficiently fast protons and sufficiently heavy nuclei the type reaction giving neutrons will be the most probable. This hypothesis is confirmed by chemical analysis and a consideration of the possible radioactive isotopes which may be formed. The *p*-*n* reactions will all be endoergic. If the radioactive substance formed emits positrons and thus returns to the target substance, the energy required is equal to the difference between n^1 and H^1 mass (0.8 MV) plus the mass of two electrons (1.0 MV) plus the upper limit of the positron energy spectrum from the radioactive substance produced. The excitation functions should therefore show a definite threshold of 2 MV or more which should be directly correlated with the positron energy. The type

TABLE LV. *Summary of p-γ type reaction.*

Target Z	Target El	Product El	Q(MV) (calc.)	Q(MV) (obs.)	Reference	Resonance levels (MV)	Excit. levels (MV)	Yield γ/p @ (MV)
1	H^2	He^3	5.39					
3	Li^7	Be^8	17.12	17.1	G3	0.20 0.44 0.85	3–7?	$7/10^{10}$ @ 0.44
4	Be^9	B^{10}	6.39	6	C47	0.99?		
5	B^{10}	C^{11}	8.55	obs.	C50	0.18	?	
	B^{11}	C^{12}	15.89	14.5	C50	0.82?	?	
6	C^{12}	N^{13}	1.93	2	H4	0.40 0.48		$1/10^{10}$ @ 0.52
	C^{13}	N^{14}	7.7					
7	N^{14}	O^{15}	7.3					
	N^{15}	O^{16}	12.12					
8	O^{16}	F^{17}	0.5	obs.	D28			
	O^{17}	F^{18}	6.5	obs.?	D28			
	O^{18}	F^{19}	6.8					
9	F^{19}	Ne^{20}	12.89	obs.	H5 H28a	0.328 0.6–0.7 0.892 0.942 1.40 1.76	7.2	$1.2/10^9$ @ 1.0
10	Ne^{21}	Na^{22}	7.1					
	Ne^{22}	Na^{23}	9.9					
11	Na^{23}	Mg^{24}	11.0	obs.	H28a	1.2		
12	Mg^{25}	Al^{26}	8.3					
	Mg^{26}	Al^{27}	7.5					
13	Al^{27}	Si^{28}	10.6	obs.	H28a	0.75 0.99 1.16 1.37 1.62 1.85		
14	Si^{29}	P^{30}	6.1					
	Si^{30}	P^{31}	6.6					
15	P^{31}	S^{32}	9.4					
16	S^{34}	Cl^{35}	5.4					
17	Cl^{37}	A^{38}	11.3					

reaction should then be particularly suitable for a study of the relation between energy evolution, K-U limit and observational limit in beta-transformations. If the product nucleus is electron active the threshold may be lower (by 1.0 MV). In this case the measurement of the threshold and the electron energy gives a direct comparison of the masses of three isobaric nuclei.

In a few instances among the lighter elements the identification of the activity indicates simple capture of the proton (§100C). This *p*-γ type reaction will have very small probability if it has to compete with the *p*-*n* reaction and can therefore only be observed if the *p*-*n* reaction is energetically impossible. It is to be expected that in some instances a *p*-α reaction might also occur, and might sometimes lead to a radioactive product.

Although at the time of writing this report the Rochester experiments are in a preliminary stage the list as given in Table LVa will give an indecation of the results obtained to date:

§101. Disintegration by Deuterons

The isotope of hydrogen of mass 2, now called deuterium, was discovered by Urey, Brickwedde and Murphy (U1) following the prediction of its existence by Birge and Menzel (B20) from considerations of the discrepancies between the physical and chemical atomic weights. It was first produced in quantity by Lewis and Macdonald (L20). A thorough discussion of the history and properties of deuterium is contained in the book by Farkas and Farkas (F2) and in the review paper of Urey and Teal (U2). The atomic ion, the "deuteron,"[16] was first used for dis-

[16] Other names which have been used for this projectile are: (a) "deuton," introduced by Lawrence, Livingston

TABLE LVa. *Evidence for p-n reactions.*†

Target					
Z	El	Periods*	Threshold En. (MV)	Assumed Product	Evidence
8	O	107 min.	2.6	F^{18}	Ne^{20}-*d*-α, F^{19}-*n*-2*n* (108 min.)
14	Si	1.9 min.	>2	P^{30}	Al^{27}-α-*n*, S^{32}-*d*-α (2.5 min.)
20	Ca	40 hr.	>2	Sc^{44}	Ca-*d*-*n*, K-α-*n*, Sc-*n*-2*n* (52 hr.)
24	Cr	40 min.	>2	Mn^{53}	Cr-*d*-*n* (46 min.)
25	Mn	91 min.	>2	Fe^{55}	
27	Co	2.4 hr.	>2	Ni^{59}	Ni-*n*-γ, Ni-*d*-*p* (3 hrs.)
28	Ni	4.1 hr.	>2	Cu^{61}	Ni-α-*p*, Ni-*d*-*n* (3.4 hr.)
		20 hr.		Cu^{64}	Cu-*n*-γ, Cu-*d*-*p*, Zn-*n*-*p*, Zn-*d*-α (12.8 hr.)
30	Zn	26 min.	>2	Ga^{70}	Ga-*n*-γ, Ga-γ-*n* (20 min.)
		40 hr.		Ga	
33	As	1.3 min.	>2	Se(?)	(Only As^{75} known)
		113 min.			
34	Se	19.5 min., 7 hr.	2.9	Br^{80}	Br-*n*-γ, Br-*d*-*p*, Br-γ-*n* (18 min., 4.2 hr.)
		14 hr.		Br^{82}	Br-*n*-γ, Br-*d*-*p* (36 hr.)
42	Mo	0.5 min.	>2		
		31 min.		Ma	
48	Cd	1.2 min.	>2	In^{112}	In-γ-*n* (1.1 min.)
		6 min.		In	In-*n*-γ (13 sec., 54 min., 3 hr.)
		37 min.			
		128 min.			
49	In	14 min.	>2	Sn(?)	

* An activity of 11–12 min. was found in Si, Mn, Co, Ni, Zn, and As. It is very probable that this is to be ascribed to N^{13} formed from carbon contamination through the *p*-γ reaction.
† Preliminary data (private communication), also (B4a, B67, D28).

integration experiments by Lawrence, Livingston and Lewis, using the large magnetic resonance accelerator at the University of California as the accelerating device. In a rapid series of survey experiments (L21, L12, L22) they observed the production of alpha-particles and protons by these new projectiles from many targets, and later observed neutrons and induced radioactivity. The first valuable summary papers were Oliphant's and Cockcroft's reports to the 1934 International Conference on Physics, (O6) (C26) while others of more recent date are by Kirchner (K14), Fea (F4), and Flügge and Krebs (F26, F26a).

A. Type Reaction *d*-α. (Table LVI)

$$Z^A+H^2\rightarrow(Z-1)^{A-2}+He^4+Q:$$

When deuterons were first used to bombard elements, large numbers of protons were observed from many elements. In addition, however, in certain of the lighter elements, alpha-particles were found, of ranges distinctly different than those known to occur with proton bombardment. In most instances the alphas showed the decided group structure indicating two resultant particles and defining the type reaction given above. With the increasing availability of deuterium more and more laboratories have utilized it for nuclear studies. It has been particularly valuable in the hands of the Cavendish Laboratory group who, with their accurately calibrated low voltage apparatus, have been able to secure more detailed and accurate data on these processes.

With the low deuteron energies available in those laboratories which have specialized in accurate measurement of disintegration reactions the processes observed have been limited to those which are exoergic, and in the lighter elements. This is most readily understood by considering that the ejected alpha must penetrate a potential barrier twice as high as the entering deuterons. The group at the University of California, on the other hand, with deuterons of up to 6 MV (C33), have been able to identify processes yielding alphas from the heavier nuclei, chiefly through the identification of induced radioactivities.

Li^6: $Li^6+H^2\rightarrow He^4+He^4$

One of the first experiments performed by Lewis, Livingston and Lawrence (L21) with

and Lewis and used by American physicists for a time, but now replaced by the more widely accepted term "deuteron"; (b) "diplon," suggested by Lord Rutherford and used until recently by English physicists. These terms apply only to the H^2 nucleus or the high speed ionic projectile, paralleling the usage of "proton" and "alpha-particle" in this sense.

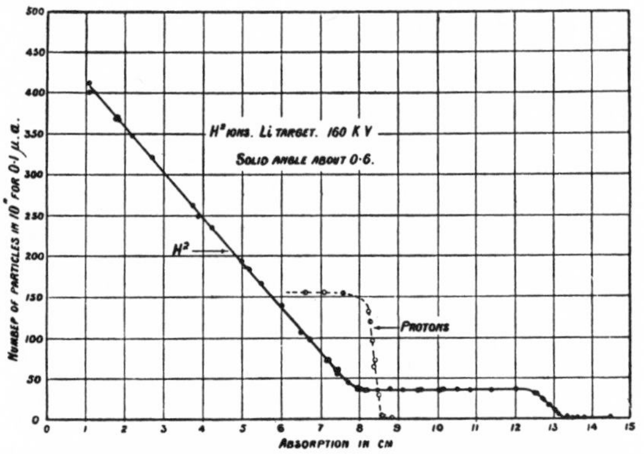

FIG. 47. Distribution-in-range of the α-particles from deuteron bombardment of lithium showing the 13 cm alphas from $Li^6+H^2=2He^4$ (homogeneous group) and the continuous distribution from the three-product reaction $Li^7+H^2=2He^4+n^1$. (Oliphant, Kinsey and Rutherford.)

deuteron projectiles was on Li. A group of alphas were observed of 14.5 cm extrapolated range, or 12.0 MV, the highest energy alphas observed up to that time. The incident deuteron energy was about 1.33 MV and the alphas were observed at 90°. The existence of such a group of alpha-particles of definite range indicated only two products. Furthermore, these particles were present in numbers about 1 : 10 of the 8 cm alphas from an equivalent current of protons (the approximate ratio of abundance of Li^6 and Li^7) so the above reaction was suggested with Li^6 as the disintegrated isotope. Using these data, the energy evolved is: $Q=2(12.0)-\frac{3}{4}(1.3)=23.0$ MV, which is the largest reaction energy measured for any nuclear process.

Oliphant, Kinsey and Rutherford (O3), using deuterons of 0.16 MV, observed these long range alphas to have a mean range of 13.0 cm, while in an accompanying paper by Dee and Walton (D6) the assumed mechanism was justified by cloud chamber photographs of the two oppositely directed alphas of equal range. Using a separated sample of the Li^6 isotope, Oliphant, Shire and Crowther (O1) were able to prove conclusively that this isotope was responsible for the long range alphas (of 13.2 cm). The most careful measurement of their range has been made by Oliphant, Kempton and Rutherford (O8) using a gas cell in which pressure could be varied to give accurate range values. On the basis of this determination of 12.70 cm mean range measured at 90° to the direction of the deuterons of 0.19 MV, they calculate a Q value of 22.06 MV, which is corrected to 22.07 MV by the more recent considerations discussed in §97. This is considered more accurate than mass-spectrograph measurements and is used to determine the mass of Li^6, in conjunction with the Li^6-d-p results. The adopted masses give the value 22.17 MV for this reaction.

An excitation curve of the process at low voltages (O3) shows a much more rapid increase with deuteron energy in the 100–200 kv region than the paralleling process of Li^7-p-α, but can also be fitted by a Gamow probability function with suitable constants. The total probability of disintegration is the product of two factors, the probability of penetration of the barrier and the internal probability of disintegration after the particle has entered the nucleus. The first involves the mass of the projectile, being given by $e^{-(m)^{\frac{1}{2}}\cdots}$. This means that the heavier deuteron has a higher momentum and so a smaller wavelength, resulting in a smaller probability for low energies where this is the determining factor. The second factor, the internal probability, depends on the effect of nuclear spins, (§83). Goldhaber's (G15) analysis indicates that the Li^6 reaction is allowed and the Li^7 reaction forbidden. Indeed the experiments show a higher probability for the Li^6 case at high deuteron energies where the internal probability predominates. From the experimental observations, Goldhaber predicted the spin of Li^6 to be 1, later confirmed experimentally by Fox and Rabi (F29), while the spin of Li^7 is $\frac{3}{2}$.

Li^7: (a) $Li^7+H^2\rightarrow He^4+He^4+n^1$
(b) $Li^7+H^2\rightarrow He^5+He^4$

The first reaction may also be classified under the d-n type process, since neutrons are also products. The classification really depends upon the mechanism of the reaction; if it belongs to the d-α type the mechanism would have to be:

$$Li^7+H^2\rightarrow {}^*Be^9\rightarrow {}^*He^5+He^4\rightarrow He^4+n^1+He^4;$$

if it belongs to the d-n type the mechanism would be:

$$Li^7+H^2\rightarrow {}^*Be^9\rightarrow {}^*Be^8+n^1\rightarrow He^4+He^4+n^1.$$

From studies of the experimental data it seems possible that both mechanisms are involved. The discovery of the process occurred through observations of the alphas and many features of the reaction have been studied through their properties. These features will be discussed at this time and those having to do more specifically with the neutron properties are entered under the *d-n* process.

Using a pure deuteron beam (magnetically separated from proton contamination) Oliphant, Kinsey and Rutherford (O3) first observed this process as a group of alphas having a continuous distribution of ranges from 1 cm up to 7.8 cm, superimposed on the 13 cm alphas from the Li^6 isotope. (See Fig. 47.) Such a distribution indicated a three-particle process and they suggested reaction (**a**) above. They observed the excitation function to be a rapidly rising curve with deuteron energy, of a type similar to that in the Li^6 process previously mentioned. The subsequent work with separated isotope targets (O1) confirmed their suggestion, naming 8 cm as the end of the range for 0.16 MV deuterons. In the more exact range measurements of Oliphant, Kempton and Rutherford (O8) this value is given as 7.6 to 7.8 cm (8.2 to 8.3 MV). The neutron should have the most energy when the two alphas have paralleling paths (a Be^8 nucleus), while one alpha has the greatest energy when the other alpha and the neutron take the same direction. In this case the alpha observed will have maximum energy when it has 5/9 Q, corresponding to the other alpha and the neutron having the same velocity (a temporary He^5 nucleus). This can be shown by considering the motion of the center of gravity of the (He^4+n^1) system and the relative motion of the He^4 and n^1 separately, in which case the maximum energy of the observed alpha corresponds to zero relative motion of the other components. From the momentum considerations and the experimental data the Q value (recalculated) is estimated to be 14.9 MV with considerable allowance for error, and is in sufficient agreement with the mass value of 15.04 MV.

An equally important aspect of this reaction is the neutron produced, discovered independently by Crane, Lauritsen and Soltan (C38), and discussed in detail in the section on *d-n* processes. Evidence from neutron studies by Bonner and Brubaker (B37) shows a continuous distribution of neutron energies and so justifies the assumed three particle process. The excitation curve for neutron production observed by Crane, Lauritsen and Soltan (C41) is for a higher energy range than that observed by Oliphant for alpha-particles, which makes direct comparison difficult, but apparently fits the Gamow probability function. It has been suggested (K2) that the true process involves a two-stage disintegration, going through a phase in which an excited Be^8 nucleus supplies the energy for the recoil nucleus, later breaking up to give the alphas. The expected neutron groups due to the discrete energy levels expected in this $^*Be^8$ nucleus are not observed, however, so this suggestion does not seem valid in principle, and the possibility of an excited He^5 must be considered.

Williams, Shepherd and Haxby (W21c) have recently found evidence for a homogeneous group of α-particles having a most probable range of 7.10 cm. These are attributed to the reaction (**b**) above. Applying the corrections of §97 to the observed range, we find a Q value of 14.3 MV. From this the mass of He^5 is found to be 5.0137, indicating that this isotope is unstable against neutron emission by 0.78 MV. This fits in with the fact that the observed alpha group is not strictly homogeneous but has a natural width. No group representing the He^5 particles was observed.

Although gamma-rays are observed from Li targets by McMillan (Mc4) and others, Lauritsen and Crane (L8) are satisfied from intensity considerations that they arise only from hydrogen impurity of the deuteron beam, indicating that the particle ranges observed correspond to the full energy evolution.

Be^9: $Be^9+H^2 \to Li^7+He^4$

Lewis, Livingston and Lawrence (L21) first observed alphas from Be, of about 3 cm range and in numbers 100 times those from proton bombardment. In Oliphant's report to the International Conference (O6) he reports experiments in which two groups of particles were observed, of 1 and 3 cm, respectively, in nearly equal numbers, indicating that they are the two products of a single reaction. The mass values available at that time gave a Q value which would predict ranges much larger than those observed and Oliphant was forced to suggest the reaction resulting in Li^6 and He^5. With the development of a consistent set of mass values it was found unnecessary to assume the production of the unknown nucleus He^5, and it became possible to make the much more natural assumption that Li^7 and He^4 were the products. Oliphant, Kempton and Rutherford (O9) have measured the particle ranges and with the corrections discussed in §97 the Q value of the process is found to be 7.19 MV, which is used together with the B^{11}-d-α results to determine the mass of Be^9. The adopted masses predict a Q of 7.17 MV.

B^{10}: (a) $B^{10}+H^2 \rightarrow He^4+He^4+He^4$
(b) $B^{10}+H^2 \rightarrow Be^8+He^4$

It might be expected that this process would result in three alpha-particles in a manner similar to the $B^{11}+H^1$ reaction, and so result in a continuous distribution of ranges. Such was observed by Cockcroft and Walton (C25, C27) who measured the maximum range to be 14.3 cm or 12.0 MV with deuterons of 0.5 MV. As in the proton reaction the end of this distribution is obscured by the homogeneous group due to reaction (b), and exact range measurements cannot be made. Cockcroft and Lewis (C28), although they are not able to separate this continuous group from the homogeneous one, have analyzed the energy distribution and find the average energy compatible with that calculated from the assumed three-alpha reaction. The Q value calculated from masses is 18.03 MV. The distribution shows two maxima. It is probable that the reaction passes through an excited Be^8 stage, and is similar to that suggested by Dee and Gilbert (D9) for the proton disintegration. Probably more than one excited level of Be^8 takes part in the process. Wheeler (private communication) has analyzed the alpha-particle distribution function and has found it compatible with the assumption of two excitation levels, namely the 3 MV level known from the B^{11}-p-α process and another level at about 6 MV. Evidence for the latter has also been obtained from the scattering of alphas by alphas (§74) and the existence of such a level seems to be in agreement with theoretical expectations (F10).

The homogeneous group observed by Cockcroft and Lewis (C28) having an extrapolated range of 14.75 cm at 90° to the deuteron beam of 0.55 MV is believed to follow reaction (b), resulting in a Be^8 product. In this case an exact value of the reaction energy can be obtained, given as 17.5 MV, corrected to 17.76 MV. The calculated value is 17.90 MV.

Gamma-rays from B by deuteron bombardment have been observed by Crane, Delsasso, Fowler and Lauritsen (C44). They were not successful in measuring the energy of the line of highest energy (given as >10 MV). It may arise from the reaction under discussion, from that yielding neutrons from B^{11}, or from a simple capture of the deuteron, all of which yield sufficient energy to produce it.

B^{11}: (a) $B^{11}+H^2 \rightarrow Be^9+He^4$
(b) $B^{11}+H^2 \rightarrow 3He^4+n^1$

In their experiments on the deuteron disintegration of B Cockcroft and Lewis (C28) observed a group of alphas of 4.60 cm range which best fit reaction (**a**) above, and a continuous distribution of less than 4.5 cm thought to be due to reaction (**b**). These are similar to the groups attributed to the B^{10} isotope except for their range. The Q value for the two particle disintegration is found to be 8.13 MV; it is used (together with Be^9-d-α) to determine the Be^9 mass from which the calculated value of 8.11 MV is obtained. Although the continuous distribution cannot be analyzed to give an experimental value of Q, the results are compatible with the assumption of the (b) reaction, from which, with the use of the known masses, the reaction energy of 6.53 MV and the maximum energy of 4.5 MV expected for the alphas can be computed. These are the same products found in the C^{12}-n-α and the Be^9-α-n (b) reactions; the mechanism may involve $^*Be^9$ or Be^8 and $^*He^5$.

C^{13}: $C^{13}+H^2 \rightarrow B^{11}+He^4$

Although the Q value for the C^{12}-d-α reaction is negative (-1.39 MV) and the process has not been observed, Cockcroft and Lewis (C29) have found an alpha-group of 2.7 cm range from carbon targets bombarded with 0.55 MV deuterons which fits the expected value for the C^{13} reaction. The recalculated Q value is 5.24 MV while mass values predict 5.14 MV.

N^{14}: $N^{14}+H^2 \rightarrow C^{12}+He^4$
$N^{14}+H^2 \rightarrow 4He^4$

Lewis, Livingston and Lawrence (L21) in preliminary experiments observed alpha-particles from a nitrogen compound bombarded by deuterons of 1.2 MV, which had ranges of about 6.8 cm. Later work by Lawrence, McMillan and Henderson (L16) with the N in air as a target showed two valid groups of alphas, the most energetic of which gives a Q value of 13.4 MV if corrected for forward direction of the alphas. The best measurements are those of Cockcroft and Lewis (C29), using 0.575 MV deuterons, in which two groups of alphas were found, and attributed to this reaction, at 6.20 and 11.37 cm, from which corrected Q values of 9.08 and 13.40 MV are obtained. The calculated Q value is 13.37 MV. The two alpha-groups indicate an excitation state of the C^{12} nucleus at 4.32 MV.

Cockcroft and Lewis also observed a continuous distribution of alphas extending up to 4.0 MV. This suggests a multiple product reaction, probably leading to four alphas through the steps: $N^{14}+H^2 \rightarrow {}^*O^{16} \rightarrow {}^*C^{12}+He^4 \rightarrow Be^8+2He^4 \rightarrow 4He^4$. The initial O^{16} compound nucleus would have an excitation energy of 20.7 MV at zero bombarding energy; the secondary C^{12} compound nucleus would have an excitation energy of more than 8 MV. Such a level which is unstable against alpha-emission is known from the B^{11}-d-α reaction and has an excitation energy of 9.5 to 14.5 MV. The over-all energy evolution in this multiple disintegration is 6.2 MV as calculated from the known mass values.

Gamma-radiation, observed and measured by Crane, Delsasso, Fowler and Lauritsen (C49), shows indications of several lines, the most energetic of which is of nearly 7.0 MV energy. Other evidence for the C^{12} excitation states comes from Bothe's (B47) measurements of the gamma-radiation from the Be^9-α-n reaction. He finds a gamma of 6.7 MV which corresponds to that of Crane, etc. and one of 4.2 MV which fits with the 4.3 MV excitation state from the alpha groups of Cockcroft and Lewis in this reaction, and also with the 4.4 MV level found by Bonner and Brubaker (B40) in the neutron groups from B^{11}-d-n. The same two excitation levels are indicated in the neutron groups from the Be^9-α-n process (§99B).

O^{16}: $O^{16}+H^2 \rightarrow N^{14}+He^4$

Cockcroft's original observation of this reaction (C26) was made definite by further experiments with Lewis (C29) in which alphas of 1.59 cm range were observed. This gives a Q value of 3.13 MV when corrected for geometrical factors. The value obtained from masses is 3.11 MV.

F^{19}(?): $F^{19}+H^2 \rightarrow O^{17}+He^4$

Alphas have been observed of about 3.8 cm range with 1.33 MV deuterons by Lewis, Livingston and Lawrence (L21). If these are not due to an impurity the Q value indicated is 2.1 MV while that obtained from masses is 10.10 MV. Certainly the maximum range group was not observed.

Ne^{20}: $Ne^{20}+H^2 \rightarrow F^{18}+He^4$

A positron activity of 112 min. half-life was found by Snell (S16) from deuteron bombardment of Ne gas. It was identified as F and so is probably F^{18} following this reaction. This is supported by the observations of the same period in O-p-γ and O-p-n processes. The alphas have not been observed.

Na^{23}: $Na^{23}+H^2 \rightarrow Ne^{21}+He^4$

In the course of a series of experiments investigating all the processes produced in Na on deuteron bombardment, Lawrence (L15) found a 6.5 cm alpha-group with 2.15 MV deuterons. A recalculation of the reaction energy gives 6.85 MV, and it is used for the determination of the Na^{23} mass from the mass-spectroscopic value of Ne^{21}. A discrepancy in the mass values obtained from disintegration data as compared to mass spectrographic data in this region is divided among the several reactions used (see §108). This results in a mass of Na^{23} from which a Q value of 6.76 MV is obtained for this reaction.

The probability of the reaction was found to be considerably smaller than for reactions yielding protons or neutrons; since the alpha is energetic enough to escape over the top of the potential barrier this must be due to a small internal disintegration probability.

Mg: $Mg^{24}+H^2 \rightarrow Na^{22}+He^4$
$Mg^{25}+H^2 \rightarrow Na^{23}+He^4$
$Mg^{26}+H^2 \rightarrow Na^{24}+He^4$

The only observation of alphas from Mg is by Lewis, Livingston and Lawrence (L21). These alphas were of about 6 cm range. The values of Q calculated from masses are 2.8, 7.9 and 3.0 MV for the Mg^{24}, Mg^{25} and Mg^{26} isotopes respectively. The Mg^{25} isotope is the only one yielding sufficient energy to result in the observed alphas; interpreted in this way the data show a Q of about 7.2 MV.

The process on the Mg^{24} isotope has been verified through the long period ($T\sim 9$ months) positron activity observed by Laslett (L2), no doubt correctly interpreted as being due to radioactive Na^{22} which is also known to result from the F-α-n reaction.

Another induced radioactivity, of more measurable intensity, has been found by Henderson (H24) to yield electrons and to have the characteristic 15 hr. half-life of the Na^{24} isotope. This identifies the reaction as due to the third isotope, Mg^{26}. The calculated Q value is 3.0 MV, so the alphas should have a range of 3.5 cm in Henderson's experiments; this is too low in energy to be the group observed and which has been ascribed to the Mg^{25} reaction. The excitation function for the production of radioactivity is found by Henderson to be much steeper than that for the production of protons in the d-p process. This seems to be connected with an increase in the number of possible excited states of the product nucleus with increasing deuteron energy (cf. §79).

Al^{27}: $Al^{27}+H^2\rightarrow Mg^{25}+He^4$

Again the preliminary experiments of Lewis, Livingston and Lawrence (L21) showed alpha-particles of about 6 cm range resulting from 1.2 MV deuterons on Al. An improved measurement is given by McMillan and Lawrence (Mc5) showing two groups of 5.7 and 6.5 cm at 90° with 2.2 MV deuterons, corresponding to reaction energies of 5.8 and 6.46 MV (corrected). The higher of these is used for obtaining the mass of Mg^{25}. The yield is about 6 per 10^9 deuterons for each group and the groups indicate an excitation level of 0.7 MV in the resultant Mg^{25} nucleus. Experimental difficulties in observing these alphas in the presence of longer range protons of about 100 times the intensity make these range measurements somewhat uncertain.

A possible variation of this type reaction is indicated by the report of Pool and Cork (P11) who find a 12.8 hr. positron activity from Al under deuteron bombardment which is separable as Mg. If this is Mg^{23} as is suggested it represents the emission of a He^6 particle in place of the more usual alpha. He^6 has been found as a product of the Be^9-n-α reaction and is itself radioactive with a 1 sec. period. Unless the 12.8 hr. activity is due to some spurious effect it may represent the production of two radioactive products in the same reaction, the only known instance.

$S^{32,\,34}$: $S^{32}+H^2\rightarrow P^{30}+He^4$
$S^{34}+H^2\rightarrow P^{32}+He^4$

Sagane (S1), using the high energy deuterons from the Berkeley cyclotron, has observed a positron activity with the 3 min. half-life characteristic of P^{30} and previously observed in the Al-α-n reaction, and also an electron activity of 14 days period known from several other processes to be P^{32}. Chemical separations which show that both activities are phosphorus verify the assignment.

Ca^{40}: $Ca^{40}+H^2\rightarrow K^{38}+He^4$

Hurst and Walke (H39a) found an activity of 7.6 min. period in the K fraction of a chemical separation of the products of deuteron bombardment of Ca. It is attributed to K^{38}, a positron emitter also produced by a Cl-α-n reaction.

Fe: $Fe^{54,\,56}+H^2\rightarrow Mn^{52,\,54}+He^4$

Deuteron bombardment of chemically pure Fe has been found to result in an activity of 21 min. period separable as Mn by Darling, Curtis and Cork (D3a). The experiment was repeated by Livingood, Seaborg and Fairbrother (L27a), who detected in addition, a 5 day positron activity and an electron activity of several months' period. Since only two Fe isotopes can be transformed into radioactive Mn isotopes by means of (d-α) reactions, one of the three periods is not accounted for. It may be due to neutrons known to be released from the Fe target.

Cu^{65}?: $Cu^{65}+H^2\rightarrow Ni^{63}+He^4$

An electron activity of 130 days half-life found by Livingood (L26) in copper is tentatively ascribed to an alpha-emission process. It cannot be due to the known d-p reaction giving Cu^{64} (12.8

hr.) or Cu^{66} (5 min.). Chemical analysis of the active products was inconclusive.

Zn^{66}: $Zn^{66}+H^2 \rightarrow Cu^{64}+He^4$

Livingood (L26) reports an electron activity of 12 hr. half-life produced in Zn by high energy deuterons and chemically separable as copper. The same period is found in the Cu-*n*-γ, Cu-*d*-*p*, Zn-*n*-*p* and, perhaps, Ni-*p*-*n* reactions.

Sn:

Livingood (private communication) refutes the activities ascribed to In in a previous report (L27). The In fraction is found to be inactive.

Sb^{123}: $Sb^{123}+H^2 \rightarrow Sn^{121}+He^4$

An activity of 24 hr. half-life found by Livingood (L26) seems to be the same as a 28 hr. period in the Sn-*d*-*p* reaction so suggesting this type reaction. The only Sb isotope available to give a radioactive Sn by a (*d*-α) process is Sb^{123}, which yields Sn^{121}.

B. Type reaction *d*-*p*. (Tables LVII and LVIII)

$$Z^A+H^2 \rightarrow Z^{A+1}+H^1+Q:$$

The first reports of the use of deuterons as nuclear projectiles (L12, L22) included the observation of large numbers of protons from all targets investigated.[17] Of these protons there were found, in general, one or more groups characteristic of each target element. The reaction follows the usual laws, in some instances resulting in excitation states of the residual nucleus and gamma-radiation. The protons are in all instances in homogeneous groups, indicating only two particle products, and measurements of the range lead to accurate Q values, since the proton range is much more extended than that of alphas of the same energy and so can be more exactly determined.

The binding energy of the deuteron (2.20 MV) is in general much smaller than the average energy of binding of neutrons in nuclei. That is, the energy released by a nucleus in absorbing a neutron is larger than the energy required to separate the proton and neutron of the deuteron. This means that the reaction energy will in general be positive. The potential barrier of a nucleus for a proton is relatively small (half that for alphas). These factors suggest that the reaction should be quite probable and should be observed for most of the light elements, while with the high energy particles from the magnetic accelerator type of apparatus this would be extended to quite heavy elements.

Studies of the excitation curves of processes of this type by Lawrence, McMillan and Thornton (L17) revealed a less rapid increase of yield with increasing deuteron energy than for processes resulting in alpha-emission, and they did not fit a Gamow curve calculated on the usual assumption that the finite extension of nuclear forces could be neglected. Oppenheimer (O10) suggested a mechanism to explain the results and Oppenheimer and Phillips (O11) have calculated probabilities on the basis of the new concept. This suggestion is that the deuteron is essentially "polarized" and split in the nuclear field and that the neutron only is absorbed by the nucleus while the proton is repelled by the field. With this assumption the shapes of the curves could be explained (H24), (L17).

As is indicated in §80 the Oppenheimer-Phillips penetration function can not be used to explain these experimental results in most cases. Since the energies of the bombarding deuterons used are not small compared to the potential barriers of the elements chosen, the simplified Gamow formula (see §78) is not sufficient. When the more exact formula (§80) is used, the experimental excitation curves are found to fit the Gamow probability function sufficiently. The steeper curves obtained from alpha-particle emission processes are readily explained because of the necessary factor involving the probability of penetration of the potential barrier by the outgoing alpha.

The O-P theory can be used, however, for very heavy nuclei ($A > 100$). As indicated in §80 an entry of the deuteron as a whole into the nucleus will in general be followed by emission of a neutron, whereas the *d*-*p* process usually follows the O-P mechanism. The probabilities of the two processes are comparable.

[17] A group of protons common to all targets, of 18 cm range with 1.3 MV deuterons, was at first incorrectly interpreted as being due to disintegration of the deuterons without capture. These are now identified with the $H^2+H^2 \rightarrow H^3+H^1$ reaction and are found in all disintegration experiments with deuterons due to the deposition of deuterium on the target by the beam. Neutrons from the accompanying process $H^2+H^2 \rightarrow He^3+n^1$ are also observed from all targets.

TABLE LVI. *Summary of d-α type reaction.*

Target								
Z	El	Product El	*Q*(MV) (calc.)	*Q*(MV) (obs.)	Reference	Resonance levels (MV)	Excit. levels (MV)	Yield *α/d* @ (MV)
3	Li^6	He^4	22.17*	22.07	O8			$3/10^{11}$ @ 0.16
	Li^7	He^5	14.3*	14.3	W21c			$2/10^{12}$ @ 0.16
	Li^7	He^4+n^1	15.04	14.9	O8			$2/10^{10}$ @ 0.16
4	Be^9	Li^7	7.17*	7.19	O9		?	
5	B^{10}	Be^8	17.90	17.76	C28			$1/10^9$ @ 0.57
	B^{10}	$2He^4$	18.03	obs.	C28		2.9(Be^8)	$2/10^8$ @ 0.57
	B^{11}	Be^9	8.11*	8.13	C28			$2/10^9$ @ 0.57
	B^{11}	$2He^4+n^1$	6.53	obs.	C28			$1/10^9$ @ 0.57
6	C^{12}	B^{10}	−1.39					
	C^{13}	B^{11}	5.14	5.24	C29			
7	N^{14}	C^{12}	13.37	13.40	C29		4.32	$5/10^{10}$ @ 0.57
	N^{15}	C^{13}	7.56					
8	O^{16}	N^{14}	3.11	3.13	C29			$3/10^{10}$ @ 0.57
	O^{17}	N^{15}	10.5					
	O^{18}	N^{16}	3.0					
9	F^{19}	O^{17}	10.10	2.1?	L21			
10	Ne^{20}	F^{18}	3.8	obs.	S16			
	Ne^{21}	F^{19}	5.59					
	Ne^{22}	F^{20}	<4.9					
11	Na^{23}	Ne^{21}	6.76*	6.85	L15			$6/10^8$ @ 2.15
12	Mg^{24}	Na^{22}	2.8	obs.	L2			
	Mg^{25}	Na^{23}	7.9	7.2	L21			
	Mg^{26}	Na^{24}	3.0	obs.	H24			
13	Al^{27}	Mg^{25}	6.46*	6.46	Mc5		0.7	$1/10^8$ @ 2.2
14	Si^{28}	Al^{26}	4.2					
	Si^{29}	Al^{27}	7.3					
	Si^{30}	Al^{28}	3.5					
15	P^{31}	Si^{29}	7.0					
16	S^{32}	P^{30}	4.6	obs.	S1			
	S^{34}	P^{32}	4.3	obs.	S1			
	etc.							

* *Q* (observed) used to calculate mass values.

H^2: $H^2+H^2 \rightarrow H^3+H^1$

Oliphant, Harteck and Rutherford (O7) first recognized a group of 14.3 cm protons as coming from this reaction and also observed the H^3 particles to have 1.6 cm range. Dee (D7) took cloud chamber photographs and found two tracks of these ranges to be co-linear in each case. The best values are reported by Oliphant, Kempton and Rutherford (O8), who find 14.7 cm extrapolated range for the proton when measured at 90° to the beam of 0.20 MV deuterons. With the application of small corrections for the effect on the straggling of the thick target and of the finite angular resolution of the recording instruments the *Q* value is found to be 3.98 ± 0.02 MV. This is the most accurately measured of any nuclear disintegration. The accurate knowledge of the H^3 mass, determined through this reaction, is of great theoretical value for the calculation of nuclear forces.

The calculated energy of the H^3 particle may be used to check the range energy relation for protons in the low energy region, and the observed range of 1.6 cm checks well with the 1.55 cm obtained from the range energy relation of §95.

The reaction is extremely probable, even for low energies, due to the low potential barrier for the deuterons, and has been observed for energies as low as 8 kv (B68). About 1 disintegration/10^6 deuterons is observed at 0.1 MV (O6). Kempton, Browne and Maasdorp (K3) have studied the angular distribution of protons and find it not spherically symmetrical. Neuert (N4a) has studied the asymmetry and finds a minimum near 90°.

Whenever deuterons are used the deposition of deuterium on the targets may produce this group of protons as a contaminant of other reactions.

Li^6: $Li^6+H^2 \rightarrow Li^7+H^1$

Protons of about 40 cm range were first detected by Lawrence, Livingston and Lewis (L12) with deuterons of 1.3 MV. The reaction was identified by Oliphant, Shire and Crowther (O4) by the observation of 30 cm protons from bombardment of a separated Li^6 target with 0.16 MV deuterons. A better value seems to be that of Cockcroft and Walton (C25) of 30.5 cm at 90° to a beam of 0.5 MV deuterons. A recent measurement by Delsasso, Fowler and Lauritsen (D11) gives an extrapolated range of 31.7 cm with 0.7 MV ions. Using this data we obtain a Q value (recalculated) of 5.02 MV. The other values are all reasonably consistent if the different bombarding energies are considered. The Q value predicted by the masses is 4.92 MV.

This reaction is of most value in checking the accuracy of the mass values and reaction energy determinations in the light element region. This can be done most readily by considering the cycle: $(Li^6\text{-}d\text{-}p)+(Li^7\text{-}p\text{-}\alpha)-(Li^6\text{-}d\text{-}\alpha)=5.02+17.13-22.07=0.08$ MV. This cycle should give exactly zero; the difference is well within the limits of error of the measurements.

Studies of the reactions produced by deuterons on separated isotopes of Li^6 and Li^7 by Rumbaugh and Hafstad (R19) show two proton groups from Li^6, at 27.2 and 31.4 cm extrapolated range for 0.54 MV deuterons. The longer range group is that observed by Cockcroft and Walton, while the shorter range group indicates an excitation level in Li^7 of 0.44 MV and predicts a gamma-ray of this energy. The value given by Rumbaugh and Hafstad is 0.40 MV but becomes 0.44 when the recoil of the Li^7 and the new range-energy relations are considered. Williams, Shepherd and Haxby (W21b) have studied the relative yields of the two proton groups as a function of deuteron energy and find distinct differences. They also obtain evidence for the expected 0.44 MV gamma-ray (S11a). A small asymmetry in the angular distribution of the protons has been observed by Neuert (N4a) but not identified as to behavior of the separate groups.

Li^7: $Li^7+H^2\rightarrow Li^8+H^1$

This reaction has been identified by the observation of Crane, Delsasso, Fowler and Lauritsen (C48) of an electron radioactivity of 0.5 sec. period from a Li target. The period was more accurately measured by Lewis, Burcham and Chang (L24) to be 0.88 sec. As would be expected from such a short period activity, the electrons were found to have an exceedingly high maximum energy. The visually extrapolated limit was originally reported to be about 10.0 MV while a K-U extrapolation gave 11.2 MV (F28). A recent report by Bayley and Crane (B5a) gives a visual limit of 12.0 MV (Cf. below).

A proton group of 26 cm extrapolated range with 0.7 MV deuterons was attributed by Delsasso, Fowler and Lauritsen (D11) to this process, and used to calculate the Q value. It was later shown to be due to the Li^6 isotope (see above). Rumbaugh, Roberts and Hafstad (R19, R19a) found no proton groups of range greater than 1.7 cm from the Li^7 isotope using deuterons of 0.86 MV, which gives an upper limit to the reaction energy of 0.28 MV corresponding to a Li^8 mass of >8.0245.

An upper limit for the mass of Li^8 may be obtained from the fact that deuterons of 0.36 MV give an appreciable yield of Li^8 (B5a, R19a). Therefore $Q>-0.3$ MV and the upper limit to the mass is <8.0251. It is quite possible that this value represents the actual mass. The disintegration function shows an abrupt rise above 0.36 MV which cannot be explained satisfactorily by the penetrability of the deuteron but suggests a definite threshold, i.e., an endoergic reaction. Both of the above masses are higher than that calculated from the radioactive beta-energy of 12 MV which is 8.0208. This suggests that the beta-decay results in an excited Be^8 product which transforms to the ground state with the emission of gamma-radiation (3.5–4.0 MV). In fact, Lewis, Burcham and Chang (L24) have shown that the residual Be^8 does break up into alpha-particles. Their distribution curve (private communication) shows a broad group with maximum intensity at an alpha-energy of 2.4 MV and a width at half-maximum of about 0.7 MV indicating that Be^8 is left in a level of 4.7 MV excitation energy and 1.4 MV width. In addition to this group there is a small number of alphas (about 2–3 percent of the total) having energies extending up to 6 MV, corresponding to a Be^8 excitation of 12 MV. The beta-particles of maximum energy should correspond to the lower limit of the main level (about 4 MV). This would agree with a $Q=\pm0.3$ MV in the primary reaction together with a limit of 12.0 MV for the beta-spectrum. The spectrum is expected to be a superposition of simple beta-spectra with their upper limits varying between 10 and 12 MV.

Fowler and Lauritsen (F28a) have also obtained a distribution curve for the alphas. All its features agree with the above except that the maximum occurs at 1.3 MV. Rumbaugh, Roberts and Hafstad (R19a) find no maximum at all, down to energies as low as 1 MV. This means that the Be^8 level in question may lie much lower than 4.7 MV. There is a possibility that it is identical with the level at 2.9 MV with a width of 0.8 MV which is observed in the B^{11}-p-α reaction. Wigner and Breit (W16) have shown that selection rules forbid beta-disintegration from the ground state of Li^8 to the ground state of Be^8 but would allow transitions to an excited state of angular momentum 2, assuming that the theoretical determinations of the angular momenta of the states involved is correct.

Be^9: $Be^9+H^2\rightarrow Be^{10}+H^1$

Oliphant (O6) reported the observation of three or more groups of protons from Be, the most prominent of which were at 8, 14 and 26 cm. He attributed the 8 cm group to oxygen present as a contaminant and the 14 cm group to deuterium in the target, but offered no explanation for the 26 cm ones. Later Oliphant, Kempton and Rutherford (O9) measured the range to be 25.6 cm and suggested this process. Assuming the bombarding energy to be 0.20 MV (not stated) the Q value of this reaction is found to be 4.59 MV, and is used to obtain the mass of Be^{10}. From this value the expected electron energy is calculated to be 0.37 MV. The electron radioactivity has been observed by McMillan (Mc7) to have a half-life of more than 10 years and a very low electron energy (estimated to be 0.3 MV).

In the same paper Oliphant, Kempton and Rutherford found again a broad group of singly charged ions (8 to 10 cm), but rather than referring it to oxygen they prefer to consider it a H^3 particle coming from a new reaction: $Be^9+H^2\rightarrow Be^8+H^3$. Whether this can be justified must be determined by e/m measurements. The calculated reaction energy is 4.47 MV and the H^3 particle would have 3.25 MV and a range of 7.8 cm. An alternative interpretation is that it is a proton corresponding to an excitation state of the Be^{10} nucleus, in which case the energy difference would be 1.4 MV.

B^{10}: $B^{10}+H^2\rightarrow B^{11}+H^1$

Cockcroft and Walton (C25) found three groups of protons from boron under deuteron bombardment which have been more accurately studied and reported in a later paper by Cockcroft and Lewis (C28). The extrapolated ranges are found to be 90.7, 58.5 and 30.7 cm, and the Q values obtained for the three groups are 9.14 7.00, and 4.71 MV, suggesting excitation states of the B^{11} nucleus at 2.14 and 4.43 MV. No protons of ranges of greater than 91 cm have been found so this must represent the disintegration into the ground state for which the calculated Q value is 9.30 MV. We have for comparison the measurements of the gamma-ray lines by Crane, Delsasso, Fowler and Lauritsen (C44). They report five lines, all well resolved although not allowing very accurate measurements. Two of these at 2.4 and 4.2 MV suggest a correspondence with the proton group differences, although they have an alternative interpretation in the B^{11}-d-n reaction. A 6.7 MV line is also attributed to the B^{11}-d-n reaction, while one of >10 MV may be from B^{10}-d-α. The fifth line, of 5.5 MV, is not yet identified.

B^{11}: $B^{11}+H^2\rightarrow B^{12}+H^1$

Lawrence and Thornton first found evidence of high energy radioactive electrons from B, and suggested the B^{11} isotope as responsible. Following this suggestion Crane, Delsasso, Fowler and Lauritsen (C46) arranged a cloud chamber to photograph the electron tracks and found them to have a distributed energy spectrum with a practical maximum limit at 11 MV. The K-U extrapolation has been reported by Fowler, Delsasso and Lauritsen (F28) to be 13 MV. A recent report by Bayley and Crane (B5a) gives an observational limit of 12 MV. By using an automatic timing device to turn off the ion beam a short time before the cloud chamber expansion and counting the number of tracks formed after different time intervals it was possible to estimate the half-life as 0.02 sec. This activity has the highest energy electrons and the shortest half-life yet reported.

If the mass of the B^{12} obtained from the radioactive process into C^{12} with the emission of electrons of 12 MV energy is used to calculate the primary reaction energy we obtain the value of 2.46 MV, predicting protons of over 2.28 MV (9 cm range). (The electrons were found to have 20 times the abundance of the 91 cm proton group from B^{10}.) Cockcroft and Lewis (C28) searched for such a group and found none with ranges greater than 2.7 cm with 0.55 MV deuterons. This indicates a reaction energy of less than 0.90 MV and a mass of the B^{12} formed in the primary reaction of >12.0186, or >1.7 MV higher than that from the radioactive beta-energy. This result is similar to that obtained for the Li^7-*d*-*p* reaction, and may be interpreted as due to a β-ray of less energy than the maximum available and the formation of an excited C^{12} product, and predicts an accompanying gamma-ray of >1.7 MV.

This activity was observed with deuterons of <0.3 MV (B5a) so that certainly Q is >-0.25 MV. This gives an upper limit to the B^{12} mass of 12.0198 and an upper limit to the gamma-ray energy of 2.7 MV.

There is some slight evidence (§99B) for a C^{12} excitation level at 3.0 MV which may represent the level at <2.7 MV discussed above. However, it seems desirable to repeat the search for proton groups with separated B isotopes.

C^{12}: $C^{12}+H^2\rightarrow C^{13}+H^1$

The protons from this reaction were first recognized by Cockcroft and Walton (C25) who measured an extrapolated range of 14.0 cm with 0.5 MV deuterons. The more recent results of Cockcroft and Lewis (C29) show this range to be 13.9 cm from which we get a Q value of 2.71 MV. The calculated value is 2.76 MV.

A gamma-ray from carbon targets bombarded by deuterons played a large part in the early discussions of the disintegration mass scale. This was measured by Lauritsen and Crane (L7), McMillan (Mc4) and others to be of about 3.7 MV, using absorption techniques. With cloud chamber methods Tuve and Hafstad (T15) found it to consist of a strong line of 2.7 MV and a much weaker one extending to 4 MV or higher. It was thought to have the proper intensity to be associated with the protons of this reaction and in the first analyses was added to the proton energy to determine the reaction energy. In calculating masses from disintegration data Bethe (B11) found that a consistent set of values would be obtained only if this gamma-ray were not included in the reaction energy. The recent mass spectroscopic measurements by Bainbridge and Jordan (B2) of C^{12} and C^{13} show definitely that the observed proton group corresponds to the formation of C^{13} in the ground state. Moreover, there is not sufficient energy to produce the more energetic gamma-ray in this reaction and it has been tentatively ascribed to C^{13}-*d*-*n*. The 4 MV gamma-ray is of sufficiently low intensity to be due to the less abundant isotope. Whether the 2.7 MV radiation belongs to the C^{12}-*d*-*p* reaction and is associated with a very slow proton is still an open question. A gamma-ray of similar energy was observed from B^{10}-α-*p*, and there is also some evidence for a level of C^{13} near 3 MV from the proton groups of that reaction.

Carbon is a common contaminant of targets, especially in apparatus using oil diffusion pumps, and this group of protons share with those from imbedded H^2 in producing a contamination group observed in many other experiments.

C^{13}?: $C^{13}+H^2\rightarrow C^{14}+H^1$

McMillan (Mc7) suggests this reaction to explain a 3 months electron activity found on many samples of Mo and brass after H^2 bombardment and supposedly due to some common contaminant. The electrons were estimated to have energies of about 0.3 MV. From the mass of C^{14} obtained from other disintegration data the Q for the primary reaction is found to be 6.08 MV and the energy available for the radioactive decay process is 0.16 MV.

N^{14}: $N^{14}+H^2\rightarrow N^{15}+H^1$

Preliminary investigations at Berkeley showed protons from N, followed by the report of Lawrence, McMillan and Henderson (L16) of two groups, of 24 and 85 cm range. Cockcroft and Lewis (C29) measure these ranges to be 18.3 and 85.1 cm with 0.5 MV deuterons. The corrected Q values corresponding to these groups are 8.55 and 3.11 MV, while that obtained from masses is 8.57 MV. The two experimental Q values must represent an excitation state of N^{15} of 5.4 MV. One of the gamma-ray lines observed by Crane, Delsasso, Fowler and Lauritsen (C49) is reported to have an energy of 5.3 MV and may be tentatively attributed to this excitation.

N^{15}(?): $N^{15}+H^2 \rightarrow N^{16}+H^1$

Some slight evidence for this reaction is in the observation by Fowler, Delsasso and Lauritsen (F28) of a group of radioactive electrons from a sodium nitrite target with an energy maximum of 6 MV. These were of low intensity but separable from the Na^{24} electrons and O^{15} positrons also observed. N^{16} is also produced in the F^{19}-n-α reaction and its half-life measured. The predicted Q value is 0.2 MV.

O^{16}: $O^{16}+H^2 \rightarrow O^{17}+H^1$

Cockcroft and Walton (C25) observed protons from a target of oxidized tungsten. In the more recent work of Cockcroft and Lewis (C29) two groups are found which can be associated with this process, at 9.22 and 4.65 cm. The corrected values for the Q's of these two groups give 1.95 and 1.12 MV, specifying an excitation level in the O^{17} nucleus at 0.83 MV. The higher value is assumed to lead to a normal nucleus and is used to calculate the mass of O^{17}. This mass value is of theoretical interest in determining the shell structure of nuclei, and shows the high mass expected from the assumption of a completed shell in O^{16}. See §33.

F^{19}: $F^{19}+H^2 \rightarrow F^{20}+H^1$

A target of CaF was found by Henderson, Livingston and Lawrence (H21) to yield an electron activity of short half-life, supposedly from this reaction. Crane, Delsasso, Fowler and Lauritsen (C47) found a half-life of 12 sec., and later measured the energy of the radioactive electrons to be 5.0 MV. The production of the same activity by slow neutrons on F and by fast neutrons on Na makes this assignment definite.

Protons of 10.0 cm range are reported by Fowler, Delsasso and Lauritsen (F28) in the bombardment of F by 0.9 MV deuterons. The reaction energy obtained from this datum is 1.75 MV. If this proton group represents the full energy evolution the mass of the F^{20} formed is 20.0092. On the other hand the mass calculated from the radioactive energy and Ne^{20} is 20.0042. This again indicates that the Ne^{20} product of the radioactive process is left in an excited state and a gamma-ray is expected, of 4.6 MV, in this case.

Na^{23}: $Na^{23}+H^2 \rightarrow Na^{24}+H^1$

Lawrence (L10, L15) observed an electron radioactivity from Na having a half-life of 15.5 hr., chemically identifiable as Na^{24}. A better value of the half-life, obtained by Van Voorhis (V6) is 14.8 hr. The same period activity had previously been produced by neutrons in three different ways, and also identified as this Na^{24} isotope. The protons were found to have ranges of 49 and 17 cm for deuteron energies of 2.15 MV. The assumed stopping power of the Al foils used for range measurements is 12 percent low so a correction to the reported ranges is required. This gives a recalculated Q value for the primary reaction of 4.92 MV. The Q for the shorter range group is 1.72 MV, indicating an excited level in Na^{24} of 3.20 MV. The 4.92 MV Q value is used to obtain the Na^{24} mass.

The excitation curve of the intensity of the observed radioactivity (supposedly the same as that for proton emission) was found not to fit the usual Gamow probability function (L17) and the Oppenheimer-Phillips theory was introduced as an explanation. As explained in the introductory paragraphs to this type reaction, these observations are adequately interpreted by a correct Gamow function.

The electron radiation from Na^{24} has been found by Kurie, Richardson and Paxton (K33) to have an energy distribution with maximum energy at 1.7 MV. Gamma-radiation of high intensity accompanies the electrons (the same half-life) and Lawrence (L15) first measured it by its absorption coefficients in several absorbers to be 5.5 MV. Using the cloud chamber method of magnetic deflection of secondary electrons Richardson and Kurie (R7) observed gamma-ray lines at 0.95, 1.93 and 3.08 MV. Later measurements of Richardson and Emo (R6a) based on the photo-disintegration of deuterium by the gamma-ray give the maximum energy the value 2.8 MV. It is plausible to assume that Mg^{24} is left in an excited state of about 2.9 MV and goes over to the ground state with the emission of a quantum of this energy or of two of lesser energy. The total energy evolution in the secondary process would be 4.6 MV, and represents a

TABLE LVII: *Evidence for d-p type reaction in heavy elements.*

Target Z	Target El	Raa Product	T	Chem. Identif.	Also Produced by	Ref.
14	Si^{30}	Si^{31}	160 min.	Yes	P-n-p, Si-n-γ	N5, N7b
15	P^{31}	P^{32}	14.5 da.	Yes	P-n-γ, S-n-p, Cl-n-α, S-d-α	N5, N7b
17	$Cl^{35,37}$	$Cl^{36,38}$	37 min.		Cl-n-γ	V6
18	A^{40}	A^{41}	110 min.	Yes	K-n-p, A-n-γ	S15
19	K^{41}	K^{42}	12.2 hr.		K-n-γ, Sc-n-α	K33
20	Ca^{44}	Ca^{45}	2.4 hr.	Yes	Ca-n-γ, Ti-n-α	W1
22	Ti^{50}	Ti^{51}	2.8 min.	Yes	Ti-n-γ	W1a
25	Mn^{55}?	Mn^{56}	2.5 hr.	Yes	Fe-n-p, Co-n-α, Mn-n-γ, Cr-α-p	D3a
26	Fe^{58}	Fe^{59}	40 da.	Yes	Co-n-p	L27a
27	Co^{59}	Co^{60}	~ yr.		(Co-n-γ)(?)	T10
28	Ni^{58}	Ni^{59}	3.5 hr.		Ni-n-γ, Co-p-n(?)	T10
29	Cu^{63}	Cu^{64}	12.8 hr.	Yes	Cu-n-γ, Zn-d-α, Zn-n-p, Ni-p-n	V7
30	$Zn^{64,68}$	$Zn^{65,69}$	{1 hr. 97 hr.			L26
33	As^{75}	As^{76}	27 hr.		As-n-γ	T10
34	Se^{82}	Se^{83}	10–20 min.	Yes	Se-n-γ?	S16a
35	Br^{79}	Br^{80}	18 min.	Yes	Br-n-γ, Br-γ-n	S16a
	Br^{79}	Br^{80}	4.2 hr.	Yes	Br-n-γ, Br-γ-n	S16a
	Br^{81}	Br^{82}	35 hr.	Yes	Br-n-γ	S16a
44	$Ru^{96,102}$	$Ru^{97,103}$	{39 hr. 11 da.			L26
46	Pd	Pd^{+1}	10 hr.	Yes	Pd-n-γ	K25
48	Cd^{114}	Cd^{115}	{4.3 hr.	Yes	}Cd-n-γ	C36a
	Cd^{116}	Cd^{117}	58 hr.	Yes		C36a
50	Sn^{120}	Sn^{121}	28 hr.	Yes	(Sb-d-α)	L27
51	$Sb^{121,123}$	$Sb^{122,124}$	{2.5 da. 60 da.		Sb-n-γ	L26, L27b
56	Ba^{138}	Ba^{139}	85.6 min.	Yes	Ba-n-γ	P11a
57	La^{139}	La^{140}	31 hr.	Yes	La-n-γ	P11a
58	Ce	Ce(?)	2.4 hr.			P11a
78	$Pt^{192,196}$	$Pt^{193,197}$	{49 min. 14.5 hr.	Yes	Pt-n-γ	C35
79	Au^{197}	Au^{198}	(2.7 da.)	Yes	Au-n-γ	C36
82	Pb	Pb^{+1}	8.6 da.	Yes		T11
83	Bi^{209}	Bi^{210}	5 da.		RaE	L26

case in which the full energy beta-emission is a forbidden transition. This value is used to obtain the Mg^{24} mass from Na^{24}.

Mg^{26}: $Mg^{26}+H^2 \rightarrow Mg^{27}+H^1$

Preliminary experiments (H21) showing an induced radioactivity in Mg have been followed up in other experiments by Henderson (H24) in which he finds two radioactive periods. One of these (15.8 hr.) is identifiable as Na^{24}, while the second, having a half-life of 10.2 min. is doubtless the Mg^{27} isotope observed by Fermi under neutron bombardment of Mg and Al. The radiations were found to be negative electrons. Absorption curves of these electrons in Al give the somewhat inconclusive maximum energy value of 2.05 MV. Cloud chamber determinations of the energy spectrum have not been reported. Gamma-rays are also emitted from the radioactive process and rough determinations of their energy with Pb absorbers indicate about 1.3 MV. The protons from the primary process have not been observed but the calculated Q value is 4.2 MV.

Al^{27}: $Al^{27}+H^2 \rightarrow Al^{28}+H^1$

Al was first found to emit protons under deuteron bombardment by Lawrence and Livingston (L13). These protons have been found by McMillan and Lawrence (Mc5), with 2.2 MV deuterons, to consist of several groups, two strong ones at 10 and 21 cm, two weaker ones at 30 and 53 cm and another extending up to 62 cm range. Assuming this last group to be due to the normal disintegration a reaction energy of 5.79 MV is indicated, corrected for geometry and the stopping power of the Al foils used to measure the range. The successively smaller Q values of 5.11, 3.10, 2.12 and 0.64 MV for the shorter range groups suggest excitation levels of the Al^{28} nucleus. Gamma-radiation has been observed by

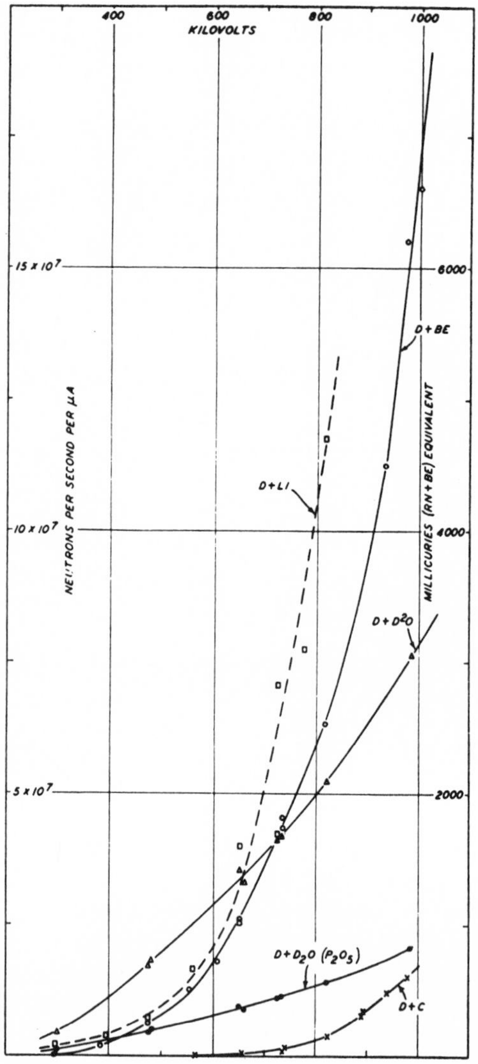

FIG. 48. Neutron yields from various targets bombarded by deuterons. Abscissa: deuteron energy in kilovolts. Ordinate: Neutron yield. Left-hand scale: Number of neutrons per microampere of deuterons. Right-hand scale: Strength of radon-beryllium source in millicuries equivalent to one microampere of deuterons. (Amaldi, Hafstad and Tuve (A11a).)

McMillan (Mc4) but its line structure has not been analyzed. The highest Q value is used to determine the Al^{27} mass.

The radioactive half-life of Al^{28} is found to be 2.6 min. and the maximum of the energy distribution is shown by Cork, Richardson and Kurie (C34) to be 3.3 MV. This is used to connect the mass of Al^{28} with that of Si^{28}. Gamma-radiation of 2.3 MV is also observed in the radioactive process. In this case the gammas are probably alternative to and not consecutive to the beta-emission. The excitation function for the production of the radioactivity (assumed to be the same for the primary reaction) has been found by Lawrence, McMillan and Thornton (L17) and follows a correct Gamow relation.

Targets of higher atomic number.—Using the large magnetic accelerator a group of investigators chiefly at the University of California have studied the radioactivities produced through this type of reaction in many heavy elements. Chemical identification of the radioactive products in many reactions have shown these products to be isotopes of the target element, so specifying this type reaction. In other cases the observation of a half-life period also observed in neutron produced reactions has been sufficient to determine the reaction. The processes in the heavier elements may follow the O-P mechanism (§80). Table LVII indicates the nature and extent of these observations:

A point of interest in the reactions in Table LVII is the observation of positrons. Van Voorhis (V7) has observed positrons and electrons in approximately equal numbers from activated copper. They have similar energy distributions and exactly identical half-life periods, and are interpreted as coming from a branching process in the radioactive C^{64} isotope leading alternately to Ni^{64} and Zn^{64}. A few positrons were observed by Paxton (P4) from irradiated phosphorus, and may possibly have a similar explanation. Furthermore, Cork and Lawrence (C35) have observed positrons from activated platinum, and have identified them as coming from the 49 min. period activity.

C. Type reaction *d-n*. (Table LIX)

$$Z^A+H^2 \rightarrow (Z+1)^{A+1}+n^1+Q:$$

The production of neutrons by deuteron bombardment was discovered by Crane and Lauritsen (C37, C38) in Li and Be targets. The neutrons were found to be emitted in large intensities, increasing with bombarding energy. Before that time neutrons had been produced only by alpha-particle bombardment of certain targets such as Be, and intensities were dependent upon the relatively weak natural sources of such alpha-particles. This new method of production is hundreds of times more intense than the earlier alpha-particle sources, and is limited only by the

TABLE LVIII. *Summary of d-p type reaction.*

Target Z	Target El	Product El	Q(MV) (calc.)	Q(MV) (obs.)	Reference	Resonance levels (MV)	Excit. levels (MV)	Yield p/d @ (MV)
1	H^2	H^3	3.98*	3.98	O8			$1/10^6$ @ 0.1
2	He^4	He^5	−3.0					
3	Li^6	Li^7	4.92*	5.02	D11		0.44	1/2(Li^6-*d*-α @ 0.5)
	Li^7	Li^8	−0.3*	−0.3?	B5a			
4	Be^9	Be^{10}	4.59*	4.59	O9		2.4?	
5	B^{10}	B^{11}	9.30	9.14	C28		2.14, 4.43	$6/10^9$ @ 0.57
	B^{11}	B^{12}	<0.9	obs.	C46			
6	C^{12}	C^{13}	2.76	2.71	C29			$3/10^9$ @ 0.57
	C^{13}	C^{14}	6.08	obs.	Mc7			
7	N^{14}	N^{15}	8.57	8.55	C29		5.42	$2/10^{10}$ @ 0.57
	N^{15}	N^{16}	0.2	obs.	F28			
8	O^{16}	O^{17}	1.95*	1.95	C29		0.82	$1/10^9$ @ 0.57
9	F^{19}	F^{20}	>1.8*	1.75	F28			
10	Ne^{20}	Ne^{21}	5.33					
	Ne^{21}	Ne^{22}	7.11					
11	Na^{23}	Na^{24}	4.94*	4.92	L15		3.20	$6/10^7$ @ 1.7
12	Mg^{24}	Mg^{25}	4.7					
	Mg^{25}	Mg^{26}	9.9					
	Mg^{26}	Mg^{27}	4.2	obs.	H24			$5/10^7$ @ 3.0
13	Al^{27}	Al^{28}	5.79*	5.79	Mc5		0.68, 2.69, 3.67, 5.15	$3/10^7$ @ 1.9
14	Si^{28}	Si^{29}	6.1					
	Si^{29}	Si^{30}	9.3					
	Si^{30}	Si^{31}	3.3	obs.	N5			
15	P^{31}	P^{32}	6.3	obs.	N5			
17	Cl^{37}	Cl^{38}	—	obs.	V6			
18	A^{40}	A^{41}	—	obs.	S15			

* Q (observed) used to calculate mass values.

intensity and voltage of the accelerating apparatus. Using Be targets and the 5 to 6 MV deuterons available in the magnetic accelerator Lawrence estimates that as many as 10^{10} neutrons per second are produced with about 10 microamperes of deuterons. These intensities make this a superior source for studying the properties of and disintegrations produced by neutrons.

The deuteron must itself be absorbed in this process, while the neutron, having no potential barrier to penetrate, would be readily emitted. The probability of disintegration should, therefore, be proportional to the probability of the deuteron penetrating the nuclear barrier, following the simple Gamow theory.

H^2: $H^2+H^2 \rightarrow He^3+n^1$

These neutrons were first recognized by Oliphant, Harteck and Rutherford (O5, O7) in addition to the protons and H^3 particles also emitted from deuterium. The necessary balance of mass and charge in the reaction require He^3 as the residual nucleus. The neutrons were found to be essentially monokinetic, as would be expected from the simple two particle reaction, and should have energies of 2.38 MV plus $\frac{1}{4}$ the deuteron energy for 90° observation. Dee and Gilbert (D8) have observed the He^3 residual nucleus to have a range of 4.3 mm if extrapolated to zero bombarding energy. Bonner and Brubaker (B38) observe a 2.53 MV maximum for the neutron energy at 90° to the incident 0.5 MV deuterons. This indicates a Q of 3.18 MV when corrected and seems more accurate than the previous results. Using the masses of H^2 and H^1 we can calculate the mass of He^3.

The probability of disintegration is quite large for low voltages. The most complete information about absolute neutron yields from various targets was obtained by Amaldi, Hafstad and Tuve (A11a) using the Amaldi and Fermi (A11) method for measuring neutron intensities. The results are given in Fig. 48.[18] At 300 kv the

[18] We are indebted to the Director of the Department of Terrestial Magnetism of the Carnegie Institution for permission to publish these results.

number of neutrons per 10^6 deuterons is about 0.7 from a "heavy" ice target. The increase with voltage is not very rapid, so that above 0.7 MV the neutron yield from Li or Be targets becomes larger than from the ice target. The excitation curve above 0.5 MV is nearly linear with the range of the deuterons, but at low voltages exhibits the Gamow exponential-type increase. At 60 kv Zinn and Seeley (Z2) report a neutron intensity equivalent to 125 millicuries of Rn-Be per milliampere of ion current containing both molecular and atomic ions. They find the neutron yield to double for each 20 kv voltage increase in this voltage range.

The low voltages for which this reaction occurs make it possible to apply standard low voltage techniques to the artificial production of neutrons. Most attempts to date have been handicapped by the lack of a suitable heavy hydrogen target. Heavy water ice has been used, but requires liquid-air cooling and limits the allowable ion beam. Moreover, only a small fraction of the beam as usually used consists of atomic ions, while the molecular ions have only half the equivalent energy and so are very inefficient compared to their heating effect on the target.

A particular advantage of the neutrons from this reaction is their monokinetic character, in which property they are unique. This feature makes them invaluable in studies of neutron scattering and disintegrations by fast neutrons.

Li^6(?): $Li^6+H^2 \rightarrow He^4+He^3+n^1$

In their experiments on separated isotopes of Li Rumbaugh and Hafstad (R19) observed neutrons from the Li^6 target in sufficient intensities to preclude the possibility of contamination of Li^7, H^2, O, C, etc. The excitation function shows significant differences from that of the most reasonable impurity (C), and leads to a value for the yield of $6/10^8$ deuterons at 0.75 MV. A reasonable guess as to the reaction is the formation of He^4 and He^3 in a manner similar to the Li^7 reaction below. The calculated Q value for this reaction is 1.56 MV, of which a maximum of $\frac{7}{8}$ would be available to the neutron. The observations show more energetic neutrons than from C, partially justifying the assumption. An alternative would be the formation of Be^7, in which case the Q value can be estimated as 3.1 MV if the Be^7 mass is chosen 1 MV greater than that of Li^7.

Li^7: (a) $Li^7+H^2 \rightarrow He^4+He^4+n^1$
(b) $Li^7+H^2 \rightarrow Be^8+n^1$

The alpha-particle products of reaction (**a**) have been discussed under the *d*-α type reaction. The accompanying neutrons were first observed by Crane, Lauritsen and Soltan (C38), in intensities nearly as great as from Be, and they independently suggested this reaction as responsible. The excitation function and yield relative to other processes is shown in Fig. 48.

In cloud chamber measurements of the proton recoils from the neutrons in this process, Bonner and Brubaker (B37) observed a continuous distribution in energy compatible with the three particle products, with a maximum of about 13.6 MV from which a Q value of 14.6 MV can be estimated. In addition to the neutrons of continuous energy a monokinetic group was found at 13.5 MV, which indicates the alternative formation of a Be^8 nucleus with a Q of 14.5 MV. The calculated values for the two modes of disintegration are 15.05 and 14.91 MV.

Be^9: $Be^9+H^2 \rightarrow B^{10}+n^1$

Be was the first element in which Crane, Lauritsen and Soltan (C37) observed the production of neutrons by deuteron bombardment, and is the source from which Lawrence and his colleagues obtain their extremely high neutron intensities. The reaction is widely used as a source of neutrons in high voltage apparatus.

Only recently have neutron energy measurements been made which are free from criticism on the grounds of scattering and statistical errors. These are obtained by Bonner and Brubaker (B40) using the cloud chamber technique and measuring only those proton recoils in the forward direction. The highest energy group shows a Q value of 4.20 MV while the Q calculated from masses is 4.18 MV. Other neutron groups having Q values of 3.7, 2.2, and 0.9 MV suggest excitation states of 0.5, 2.0 and 3.3 MV in the B^{10} product nucleus. Gamma-radiation from this (and the other $Be+H^2$ reactions) was first noted by McMillan (Mc4). From cloud chamber studies Crane, Delsasso, Fowler and Lauritsen (C45) and later Kruger and Green (K26b) each report as many as six gamma-ray lines. In both cases there is some evidence for lines which may represent the same excitation states as the neutron groups. Only a small fraction of the neutrons have the maximum energy. The excitation curve relative to other targets is given in Fig. 48.

B^{10}: $B^{10}+H^2 \rightarrow C^{11}+n^1$

Crane and Lauritsen (C40) observed a positron radioactivity of 20 minute half-life from B and also observed neutrons (L5) although these are probably chiefly from the B^{11} isotope. After some confusing reports the half-life has been found to be 20.5 minutes (Y1) and the radioactive isotope chemically identified as *C*. The positron energy spectrum has a maximum of 1.15 MV (F28). Yields are small, about 1 positron being observed for 10^8 deuterons of 0.57 MV at equilibrium.

One group of neutrons from B observed by Bonner and Brubaker (B40) have been identified with this process and have a Q value (recalculated) of 6.08 MV. This compares favorably with the mass value of 6.34 MV. The observed group may be superposed on one from the B^{11} isotope leading to the 7.0 MV excited level of C^{12}, so that the reaction energy deduced from it may not be very accurate. Another group attributed by Bonner and Brubaker to this reaction would give a Q value of 4.0 MV, indicating an excitation state of 2.2 MV in C^{11}. Since this second group is even stronger than the highest energy one it is possible that it belongs rather to the B^{11}-d-n process.

B^{11}: (a) $B^{11}+H^2 \rightarrow C^{12}+n^1$
(b) $B^{11}+H^2 \rightarrow 3He^4+n^1$

The neutrons observed when boron is bombarded by deuterons are largely due to the two reactions above. Bonner and Brubaker (B40) have obtained cloud chamber data of recoil protons and He atoms which they analyze into the various processes. The most probable reaction is (b), giving rise to a continuous group of neutrons with energies below 3 MV. This is in accord with the alpha-particle observations of Cockcroft and Lewis (C11) (§101A) who find a similar continuous distribution of alphas. The shape of the curve cannot be analyzed to give an experimental Q value, but using mass values this is found to be 6.53 MV.

Reaction (a) is thought responsible for the group of highest energy neutrons, from which a corrected Q value of 13.4 MV is obtained; the calculated value is 13.68 MV. Another group gives a Q of 9.0 MV; it can only be ascribed to this reaction since no other has sufficient energy. This predicts an excitation level at 4.4 MV in the C^{12}, for which there is also other evidence (see below). Two additional neutron groups with Q values of 6.0 and 3.9 MV can on energetic grounds be attributed either to B^{10} or B^{11}. If attributed to B^{11} they would indicate two further excitation levels of C^{12} at 7.4 and 9.5 MV. For the first-named level there is again evidence from other reactions and from gamma-rays so that the observed neutron group is probably a superposition of groups from B^{10} and B^{11}. The second group is of very high intensity which may indicate that its source is the more abundant B^{11} isotope.

The gamma-radiation has been studied by Crane, Delsasso, Fowler and Lauritsen (C44), and is found to have components at 2.4, 4.2, 5.6 and 6.7 MV. What may be the same gamma-rays from C^{12} are also found (C49) in the N^{14}-d-α process, with energies of 4.0, 5.3, 7.0 MV and two of lower energy. Bothe (B47) has observed gamma-radiation of 2.7, 4.2 and 6.7 MV from the C^{12} product of the Be^9-α-n reaction. Neutron groups in this reaction also fit the two excitation levels at 4.4 and 6.7 MV (§99B).

The complete evidence for possible excitation states in the C^{12} nucleus shows the following levels:

(1) 3.0 MV(?): Be^9-α-n group.

(2) *4.3* MV: N^{14}-d-α group, B^{11}-d-n group, Be^9-α-n gamma, B^{11}-d-n gamma, N^{14}-d-gamma (compatible with Be^9-α-n groups). *7.0−4.3=2.7* MV: Be^9-α-n gamma, B^{11}-d-n gamma.

(3) *7.0* MV: Be^9-α-n group, Be^9-α-n gamma, B^{11}-d-n (group may be present), B^{11}-d-n gamma, N^{14}-d-α gamma.

7.0−4.3=2.7 MV: Be^9-α-n gamma, B^{11}-d-α gamma.

(4) *9.5* MV: B^{11}-d-n group (may be from B^{10}-d-n, however), relative intensity of group to positrons from C^{11} makes it seem likely that this group is due to C^{12}.

9.5−4.3=5.2: B^{11}-d-n gamma, N^{14}-d-α gamma (may be a N^{15} level as indicated by N^{14}-d-p group).

(5) *14.5* MV: B^{11}-p-γ gamma (no other reasonable interpretation).

(6) *16.16* MV: B^{11}-p-α resonance level.

(7) Another level between *9.5* and *14.5* MV is indicated for the alpha-emission process: $C^{12}\rightarrow Be^{8}+He^{4}$. Observed in B^{11}-d-α as a discrete alpha-group in the continuous distribution, also as groups in B^{11}-d-n and N^{14}-d-α.

C^{12}: $C^{12}+H^{2}\rightarrow N^{13}+n^{1}$

A positron radioactivity chemically separable as nitrogen was observed by Henderson, Livingston and Lawrence (H21) and by Crane and Lauritsen (C40) when C was bombarded by deuterons. The best determination of the half-life of the N^{13} produced is the value of 11.0 minutes observed by Ellis and Henderson (E5) in the B^{10}-α-n reaction. The excitation function for the production of this activity shows the exponential-type rise represented by the Gamow function (Fig. 46). Newson (N7a) has obtained an excitation curve up to 5 MV and it exhibits a flat maximum at about 3.3 MV. Since the potential barrier for deuterons is of the order of 2.2 MV the further increase beyond this value indicates the variation of the internal probability of disintegration.

The neutrons were found by Tuve and Hafstad (T15) to have very low energies. Bonner and Brubaker (B40) obtain a measurement of recoil proton range which gives a Q value of -0.37 MV, a reasonable check with the statement by Cockcroft and Lewis (C29) that neutron emission begins sharply at 0.32 MV deuteron. The latter value is assumed to be the most accurate since it must be the lower limit. Thus $Q=-0.28$ MV (the relative energy), and it is used to obtain the N^{13} mass.

Using the mass values of N^{13} and C^{13} we are able to predict the energy expected in the positron radioactive process (considering the mass of two electrons) and find it to be 1.24 ± 0.07 MV. The limit of the positron spectrum has been measured by Fowler, Delsasso and Lauritsen (F28) to be 1.25 MV, and the K-U extrapolated limit of 1.45 MV was also obtained. In this case it seems quite certain that the visual limit is closer to the correct value than the K-U limit, a conclusion that is borne out with somewhat less certainty by several other such cycles. Alichanian, Alichanow and Dzelepow (A1) find an even higher "visual limit" of 1.45 MV, using a magnetic beta-ray spectrograph. This is hard to reconcile with the evidence given above but is even stronger proof that the K-U limit cannot be valid.

C^{13}: $C^{13}+H^{2}\rightarrow N^{14}+n^{1}$

Bonner and Brubaker (B40) find a neutron group representing about 1 percent of the total number of neutrons from *C* from which a Q value of 5.2 MV is obtained. The calculated value is 5.47 MV. They suggest this reaction as the source of the 4 MV gamma-rays which have been a disturbing factor in the carbon disintegrations (§101B). This would predict a second neutron group at 1 MV which could not be observed in the experiments.

N^{14}: $N^{14}+H^{2}\rightarrow O^{15}+n^{1}$

A positron activity of 2.1 minutes half-life was found by Livingston and McMillan (L30, Mc6) to be produced in N, and chemically identified as an oxygen isotope. Any target bombarded in air or N_2 gas was found to have this radioactive element deposited on its face by recoil. An excitation curve by Newson (N7) shows the Gamow exponential rise and a break at 3.2 MV followed by a slow decrease. The positron energy has been measured by Fowler, Delsasso and Lauritsen (F28) to have a maximum at 1.7 MV. The neutrons have been observed only qualitatively. From the mass of O^{15} determined through the radioactive energy evolution the Q value can be estimated as 5.1 MV, which would give neutrons of 4.8 MV energy for zero energy deuterons.

O^{16}: $O^{16}+H^{2}\rightarrow F^{17}+n^{1}$

Newson (N6) has observed a positron radioactivity identifiable as F, of 1.16 minutes half-life, from an oxygen target. It is found to occur only for deuterons of more than 1.8 MV, indicating a -1.7 MV value of Q which is used to determine the F^{17} mass. Using this mass to calculate the expected positron energy we find nearly sufficient energy (2.0 MV) to explain the observed limit of 2.1 MV (K33). This result suggests again that the K-U limit (2.4 MV) is untenable.

The neutrons from the reaction were not detected above the general contamination background, but a study of the angles of recoil of the F^{17} gave results compatible with neutron emission and a Q of -1.7 MV. With increased energies the excitation curve is found to have a break at 3.7 MV followed by a slower increase (N7a).

F^{19}: $F^{19}+H^2 \rightarrow Ne^{20}+n^1$

Although the neutrons observed from CaF_2 by Lawrence and Livingston (L13) might have been partially due to deuterium in the target, a later check of this reaction by Tuve and Hafstad (T14) reports neutrons from only two targets, Be and CaF_2, indicating that F is undergoing this reaction. Strong gamma-rays from F observed by Hafstad, Tuve and Brown (H1) may come from excited levels in the Ne^{20} nucleus, or from the alternative F^{20} nucleus from the *d-p* reaction. The calculated Q value is 10.68 MV.

Na^{23}: $Na^{23}+H^2 \rightarrow Mg^{24}+n^1$

Neutrons have been observed by Lawrence (L15) in numbers about equal to the protons occurring in the alternate process, that is, 1 per 10^6 deuterons of 1.7 MV. From the known masses this process should have a Q of 8.8 MV, and yield neutrons of 8.5 MV maximum energy with zero energy deuterons.

Al^{27}: $Al^{27}+H^2 \rightarrow Si^{28}+n^1$

McMillan and Lawrence (Mc5) have observed neutrons with yields of 1 per 5×10^8 deuterons of 1.7 MV energy. The calculated Q is 8.4 MV, predicting high energy neutrons.

S^{33}: $S^{33}+H^2 \rightarrow Cl^{34}+n^1$

Positron activity of 33 min. half-life was found by Sagane (S1) to result from deuteron bombardment of sulphur. Chemical separations showed this activity to be Cl. It is probably Cl^{34} which was found by Frisch (F31) in the P-*α*-*n* reaction to have a period of 40 min.

$Ca^{40,\,42,\,43}$: $Ca^{40}+H^2 \rightarrow Sc^{41}+n^1$
$Ca^{42}+H^2 \rightarrow Sc^{43}+n^1$
$Ca^{43}+H^2 \rightarrow Sc^{44}+n^1$

Walke (W1) reports three radioactive periods from Ca under deuteron bombardment which are chemically separable as Sc and yield positrons. A 4.0 hr. period is no doubt the 4.4 hr. period found by Frisch in the Ca-*α*-*p* reaction, and so is identified as Sc^{43}. The other two periods, of 53 min. and 52 hr. half-life represent Sc^{41} and Sc^{44} respectively; this assignment follows from the observation of the 52 hr. period from K-*α*-*n* and Sc-*n*-2*n*.

Ti: $Ti^{48,\,49}+H^2 \rightarrow V^{49,\,50}+n^1$
Ti $+H^2 \rightarrow$ V $+n^1$

Walke (W1a) has reported periods of 33 min. and 16 days from Ti under deuteron bombardment which he ascribes to this type of reaction. The 33 min. period is also observed in the Ti-*α*-*p* reaction suggesting either V^{49} or V^{50} as the active body. If due to V^{49} positrons would be expected, while V^{50} might yield either positrons or electrons. The 16 day period, which is positron active, may be $V^{47,\ 48\text{ or }49}$.

Fe^{54}: $Fe^{54}+H^2 \rightarrow Co^{55}+n^1$

Darling, Curtis and Cork (D3a) find an 18.2 hr. positron period from Fe under deuteron bombardment which is chemically separable as Co. The only Fe isotope which could result in radioactive Co other than the known active and stable isotopes is Fe^{54}. This Co^{55} should decay into Fe^{55} (radioactive, T=91 min.). Besides confirming the above findings, Livingood, Seaborg and Fairbrother (L27a) report other Co activities (100–200 da.; e^+, e^-) as yet unidentified.

Cr^{52}: $Cr^{52}+H^2 \rightarrow Mn^{53}+n^1$

The 46 min. positron period reported by Livingood, Seaborg and Fairbrother (L27a) as due to Mn produced from Cr by deuterons has previously been found to result from a Cr-*p*-*n* reaction. This makes the assignment definite, since Mn^{54} is occupied by one of the Fe-*d*-*α* activities.

Ni^{60}: $Ni^{60}+H^2 \rightarrow Cu^{61}+n^1$

Thornton (T10a) has observed a positron activity of 3.4 hr. half-life, chemically separable as copper, in the deuteron bombardment of Ni. This activity has also been observed in Ni-*α*-*p* and Ni-*p*-*n* reactions.

Se^{82}: $Se^{82}+H^2 \rightarrow Br^{83}+n^1$
($Se^{82}+H^2 \rightarrow Br^{82}+2n^1$?)

Snell (S16a) reports an activity from Se separable as Br with a period of 2.5 hrs. It is attributed to Br^{83} Confirmation of this assignment comes through the observation of a 10–20 min. Se period also produced by deuterons, which decays into the same active Br isotope found in this *d-n* reaction. The only Br isotope that could result from these two reactions is Br^{83}. Snell also reports the observation of the 35 hr. Br activity (known to be Br^{82}) from Se; this does not follow the usual type reactions from any of the known Se isotopes; it may be an instance of a *d-2n* reaction.

$Pd^{104,\,105}$: $Pd^{104}+H^2\rightarrow Ag^{105}+n^1$
$Pd^{105}+H^2\rightarrow Ag^{106}+n^1$

Two activities chemically separable as Ag were found by Kraus and Cork (K25) to result from deuteron bombardment of Pd. The periods of 32 min. and 7.5 days (reported earlier as 2 hr. and 150 hr.) are probably Ag^{106} and Ag^{105}, respectively. The shorter period is probably the 24 min. period known from the Ag-n-$2n$ and Ag-γ-n reactions. Positrons would be expected.

Sn^{119}: $Sn^{119}+H^2\rightarrow Sb^{120}+n^1$
$Sn\ +H^2\rightarrow Sb^{+1}+n^1$

Livingood and Seaborg (L27b) find radioactive Sb isotopes from a Sn target. A 16 min. positron activity is also observed in the Sb-n-$2n$ and Sb-γ-n reactions but not in Sb-n-γ (slow neutrons) so it is certainly Sb^{120}. Livingood (private communication) indicates that the 13 hr. period reported earlier (L27) is due to a Cu contamination, but that other activities are present, with half-lives of 3–5 hr., 1–4 da., (60–80 da.) and (>100 da.).

No *d-n* process has been observed for elements heavier than Sn. This is probably due to an instability of the nucleus produced in the *d-n* reaction against alpha-decay, leading generally to a d-n,α reaction (cf. §101D, below).

D. Type reaction d-n,α.

$$Z^A+H^2\rightarrow(Z-1)^{A-3}+n^1+He^4+Q:$$

For heavy nuclei the particle most frequently emitted in the primary nuclear process is a neutron. According to the Bohr evaporation model (§54) the neutron will in general have low energy ($\sim$1 MV) and the residual nucleus will thus be left in a highly excited state. It will then be capable of emitting an alpha-particle. See §79 for a more complete discussion.

Pt: $Pt^{195}+H^2\rightarrow Ir^{192}+n^1+He^4$
$Pt^{198}+H^2\rightarrow Ir^{195}+n^1+He^4$

The bombardment of Pt by 4 to 5 MV deuterons by Cork and Lawrence (C35) has been found to result in both positron and electron activities. Two of the electron activities, with half-lives of 28 min. and 8.5 hr. are chemically separable as Ir, necessitating an alpha-emission process. The mechanism consists probably in an emission of a neutron of about 1 MV followed by an alpha-particle (§79). From the known stable isotopes of Pt (192, 194, 195, 196, 198) the Ir isotopes 189, 191, 192, 193, 195 can be formed. Of these 191 and 193 are stable and 189 would emit positrons. Thus the two activities should be attributed to Ir^{192} and Ir^{195}. At least two periods are found with slow neutrons (A7, S20) on Ir, one of which seems to be identifiable with the deuteron activities (50 min$\backsimeq$28 min.), and so should be Ir^{192}. This means that the 8.5 hr. period is probably Ir^{195}.

The excitation function shows several broad maxima in the region between 3.5 and 5.0 MV which are interpreted as due to a resonance effect. Present theories (see §§53, 60) indicate that resonance levels in such a heavy nucleus as Pt should be very closely spaced, so it is most satisfactory to interpret the broad maxima as due to fluctuations in the density of such resonance levels. The probability was high for such a heavy element disintegration, having an activation cross section of about 10^{-27} cm^2. This favors the large nuclear radius proposed by Bethe (B14) which would predict correspondingly lower and narrower potential barriers.

E. Type reaction d-p,α.

$$Z^A+H^2\rightarrow(Z-2)^{A-3}+H^1+He^4+Q:$$

As in the *d-n* type reaction, the residual nucleus formed in a *d-p* reaction may also be left with sufficient excitation energy to be unstable against alpha-emission (§80), giving rise to a d-p, α reaction.

Au^{197}: $Au^{197}+H^2\rightarrow Ir^{194}+H^1+He^4$

Cork and Thornton (C36) have observed an activity (half-life not reported) separable as Ir from deuteron bombardment of Au. The chemical identification of Ir necessitates the emission

TABLE LIX. *Summary of d-n type reaction.*

Target Z	Target El	Product El	Q(MV) (calc.)	Q(MV) (obs.)	Reference	Resonance levels (MV)	Excit. levels (MV)	Yield n/d @ (MV)
1	H^2	He^3	3.18*	3.18	B38			$1/10^7$ @ 0.1 $8/10^6$ @ 0.5
3	Li^6	He^4+He^3	1.56	obs.	R19			$1.7/10^5$ @ 0.8
	Li^7	$2He^4$	15.05	14.6	B37			
	Li^7	Be^8	14.91	14.55	B37			$1/10^6$ @ 0.8
4	Be^9	B^{10}	4.18	4.20	B40		0.5 2.0 3.3	$2.7/10^5$ @ 1.0
5	B^{10}	C^{11}	6.34	6.08	B40			$1/10^8$ @ 0.57
	B^{11}	C^{12}	13.68	13.4	B40		4.4 7.4 9.5	
	B^{11}	$3He^4$	6.53	obs.	B40			
6	C^{12}	N^{13}	−0.28*	−0.28	C29			$3/10^6$ @ 1.0
	C^{13}	N^{14}	5.47	5.2	B40		4.0(γ)	
7	N^{14}	O^{15}	5.1	obs.	Mc6			$2/10^{10}$ @ 0.57
	N^{15}	O^{16}	9.92					
8	O^{16}	F^{17}	−1.7*	−1.7	N6			$1/10^6$ @ 2.5
	O^{18}	F^{19}	4.59					
9	F^{19}	Ne^{20}	10.68	obs.	T14		?(γ)	
10	Ne^{21}	Na^{22}	4.9					
	Ne^{22}	Na^{23}	7.73					
11	Na^{23}	Mg^{24}	8.8	obs.	L15			$3/10^7$ @ 2.15
12	Mg^{25}	Al^{26}	6.2					
	Mg^{26}	Al^{27}	5.3					
13	Al^{27}	Si^{28}	8.4	obs.	Mc5			$2/10^7$ @ 2.15
14	Si^{29}	P^{30}	3.9					
	Si^{30}	P^{31}	4.3					
15	P^{31}	S^{32}	7.2					
16	S^{34}	Cl^{35}	3.2					
17	Cl^{37}	A^{38}	9.1					
etc.								

* Q (observed) used to calculate mass values.

of three units of nuclear charge (if the deuteron is absorbed), and thus suggests the type reaction. The high energy of the deuterons used (6 to 7 MV) probably explains the emission of alpha-particles in a secondary process. (Cork and Lawrence used only 4 to 5 MV deuterons in their experiments on Pt in which this type reaction was not observed.) Since Au has only one known stable isotope the product must be Ir^{194}, known to have a half-life of 3 days from neutron capture experiments.

§102. Disintegrations by Neutrons

Disintegrations produced by neutrons are of four types, three yielding particle products (alphas, neutrons and protons), and another resulting in simple capture with gamma-ray emission. The first neutron disintegration reported ($N^{14}+n^1 \rightarrow B^{11}+He^4$) was by Feather (F5) and followed immediately upon Chadwick's report of the identification of this new particle. The cloud chambers used to measure neutron recoils served to show the forks characteristic of fast neutron disintegration processes. The forks showed alpha-particle tracks of considerable length originating from the same point as the more dense, short-ranged track of the residual nucleus. Other experimenters followed and discovered similar reactions in other elements. This method proved to be tedious, as it required thousands of photographs to detect a few of the rare forks. Its peculiar advantage, however, is that a single track is sufficient to identify the reaction under suitably controlled conditions. The chief difficulty is in the momentum determinations. Since the neutron produces no track its direction must be assumed and its momentum must be determined from the momenta of the two observed branches of the fork. Conclusions based on forks caused by scattered neutrons may lead to serious errors of interpretation.

The discovery of radioactivity induced by

alpha-particles was the clue which led Fermi (F14) to search for a similar effect with neutrons and disclose this simple and most satisfying method of studying neutron processes. The neutron, having no charge, experiences no potential barrier, and so the probabilities of penetration and of disintegration are large. Furthermore, the detection and observation of such induced radioactivities, which can be carried on in the absence of the source, has proven to be unique and simple. Chemical identification of the active materials showed that three of the processes named above occur. Rapid development of the field has led to a knowledge of scores of neutron reactions and has materially broadened the field of nuclear physics. Theoretical analysis has kept step with the development and it is safe to say that there is a better knowledge of disintegrations produced by neutrons than of any other type (cf. Chap. X).

The simple capture reactions yield the most direct evidence on nuclear structure because the results are not influenced by potential barriers. Recent discoveries have led to some knowledge of the resonance levels of nuclear systems. The new nuclear model proposed by Bohr (B32) is based on these experiments.

In addition to the three main types of disintegrations produced by neutrons (n-α, n-p and n-γ) there is some evidence for another type involving the emission of two neutrons. This means, in effect, that the incident neutron ejects another neutron from the nucleus and is itself not absorbed. Another prominent process is the inelastic scattering of neutrons by which no actual transmutation of the scattering nucleus is obtained but only an excitation which is followed by emission of gamma-radiation. Both of these are most readily visualized as leading to a compound nucleus which then releases a neutron of low energy; the excitation energy goes to liberate a second neutron in the first instance and a gamma-ray in the inelastic scattering process.

A. Type reaction n-α. (Tables LX and LXI)

$$Z^A+n^1\rightarrow(Z-2)^{A-3}+\mathrm{He}^4+Q:$$

The first type of neutron disintegration to be recognized, that yielding alphas, has been observed in three ways. For those elements available in the gaseous form (C, N, O, F, Ne) the cloud chamber is readily adaptable. Fast neutrons penetrate the chamber walls easily and produce reactions in the gas which are observable as forked tracks. Up to the present the sources have usually been limited to the intensities available in naturally radioactive materials (radon +Be). Photographs are sometimes confused by the many tracks of atoms recoiling from the neutrons. The neutron is assumed to come from the source (preferably small to subtend a small angle at the cloud chamber). If the two observed prongs of the fork are co-planar with the line from the source this assumption is partially justified. Some tracks are discarded by the observers because this condition is obviously not fulfilled, indicating a scattered neutron coming from a direction other than the source. Even for co-planar tracks there is still the possibility of scattering in that plane. Such tracks are useless since the information obtained from the observed tracks is not sufficient to determine both neutron energy and direction uniquely.

In order to analyze forks which are deemed satisfactory the neutron energy is determined from the momenta of the two particles. It is first necessary to assume a reaction mechanism, so naming the masses of the charged particles. From the measured components of alpha-particle momentum and the angle of the heavy track the energies of the neutron and residual nucleus can be estimated. Since the heavy particle carries most of the momentum and the angle of its short track with the assumed neutron direction cannot be measured with any accuracy, the errors in this method have been considerable.

In a large share of tracks measured the sum of observed particle energies is less than the neutron energy calculated from the momenta. It is extremely doubtful how much of the evidence in this respect can be trusted. Often extremely large energies of incident neutrons are concluded, certainly much larger than the energies available determined through other, more recent, evidence on masses. Examples: Kurie (K28), 17 MV instead of 10.7 MV from $\mathrm{Be}^9+\mathrm{He}^4$(Po); Harkins (H12), 15.8 MV instead of 10.7 MV from $\mathrm{Be}^9+\mathrm{He}^4$(Po). However, in the remaining evidence there is still enough to indicate energy losses.

Attempts to explain this energy loss have taken several forms. Feather (F5) and for a time Harkins, Gans and Newson (H12), favored a noncapture reaction, in which varying amounts of energy are absorbed from the neutron. Later Harkins and Gans (H13) disproved the noncapture reaction directly by statistical analysis of neutron directions. There is at present no good evidence for noncapture disintegration but it remains a possibility which should not be discarded. Kurie (K31) hypothecated an intermediate stage in the N^{14} reaction passing through a radioactive N^{15} with subsequent emission of the alpha-particle. This has no justification either theoretically or experimentally, and has subsequently been discarded. All workers have considered the possibility of excitation levels in the residual nucleus as responsible, but the data failed to show the unique values of Q that this would predict. One paper by Jaeckel (J1) on the disintegration of Ne, however, indicates a grouping of the amounts of the energy loss in the individual forks about two distinct values. This suggests that the process is actually producing excited residual nuclei, that the observed deviations from these values are unavoidable experimental errors inherent in the method, and that the processes follow the same rules as apply to the other types of disintegration. In the discussion to follow we consider the processes as conserving mass-energy and resulting in excited states and gamma-radiation.

The second method of observation of neutron disintegrations yielding heavy particles is through the use of electrical recording devices such as ionization chambers to observe the alphas. These have been successful only in those processes produced by slow neutrons. By definition this means exoergic reactions, and the energy of the resulting alpha is limited by the reaction energy. Therefore these reactions occur with appreciable probability only for light elements, whose low potential barriers allow the comparatively slow alphas to escape. Several such processes have extremely large cross sections (Li^6, B^{10}), and have the great advantage of having no neutron momentum to influence the accuracy of the results. Lacking the impediment of indeterminate neutron energies these processes have resulted in accurate observations of reaction energy in spite of difficulties in the technique. Such slow neutron processes have also been observed in cloud chambers and in photographic emulsions, in which case a single straight track is observed, of a length equal to the sum of the ranges of the ejected particle and the recoil nucleus.

The observation of induced radioactivity has been useful as a third method of establishing this type reaction. Either chemical identification of the radioactive element (of Z two less than the target) or the observation of a half-life period characteristic of a known element has been considered sufficient evidence to justify the existence of the process. Where a target is known to have only one isotope the observation of three half-life periods is good evidence for the existence of the three type reactions n-α, n-p, and simple capture. For example Al^{27} is known to result in three distinctive radioactive decay periods, characteristic of Na^{24}, Mg^{27} and Al^{28}.

The potential barrier for the escaping alphas from such processes would lead us to expect a relatively low probability for elements of high atomic number, as discussed by Bethe (B12). The observations show a limit at present at $Z=31$ (Ga). Only for the heaviest of heavy elements (U and Th) is this process again observed, due in these cases to the small binding energy for alphas in this region of the atomic table, as indicated by the natural alpha-radioactivity observed. The probability should be large for slow neutrons if the reaction energy Q is positive (the $1/v$ law) and should increase again for neutrons of several MV energy, for which the increased energy of the alphas would make penetration of the potential barrier more probable. We would expect, therefore, that the observations would be separated rather definitely into those produced by slow and by very fast neutrons.

Li^6: Li^6+n^1 (slow)$\rightarrow H^3+He^4$

In their early experiments with slow neutrons Amaldi, D'Agostino, Fermi, Pontecorvo, Rasetti and Segrè (A7) found a large absorption of slow neutrons in Li, but no radioactivity or gamma-radiation, which led to the suggestion of a process of this type. Chadwick and Goldhaber (C12) searched for heavy particle products with an ionization chamber and linear amplifier and found singly charged ions of about 5.5 cm range and doubly charged ions of less than 1.5 cm. Taylor and Dabholkar (T3) have observed single straight tracks in a photographic emulsion from this reaction and measure a total range (in air-cm equivalent) of 6.64 cm. Rotblat (R16) reports a H^3 range of 5.36 cm and a Q value of 4.5 MV. The best measurement is that of Livingston and Hoffman (L32) using the shallow ionization chamber technique and giving a range of 5.73 cm for the H^3 particles. By comparison of the range of the H^3 with that of a proton of the same velocity, the particle energy is obtained, from which they obtain a Q value for the reaction of 4.67 MV. The calculated value of Q is 4.56 MV.

This process should follow the simple theory of neutron capture (B12) discussed in §64, and should have a probability inversely proportional to the velocity of the neutron. Thermal energy neutrons will be most effective. The most interesting feature of this reaction is its large cross section, measured by Mitchell (M21) to be 70×10^{-24} cm² for the mixed isotopes which means 900×10^{-24} cm² for the reaction on Li^6. This is larger than the geometrical cross section of the Li^6 nucleus by a factor of nearly 1000 and can only be understood in terms of the long associated wave-lengths of the slow neutrons and their corresponding uncertainty of position. The reaction is valuable as a detecting mechanism for slow neutrons; ionization chambers lined with Li are used (D24) to measure their numbers and relative velocities (§93D).

The process also occurs with fast neutrons, but with a much smaller probability. This results in a fast neutron "background" which is present even with Cd shielding and which must be subtracted to obtain the slow neutron effects.

Be^9: $Be^9+n^1\rightarrow He^6+He^4$

An electron radioactivity of about 1 sec. half-life which had the properties of an inert gas was reported by Bjerge (B23) to result from fast neutron bombardment of Be. Nahmias and Walen (N1) have also observed the activity, finding a period of 0.7 sec. Further work by Bjerge and Broström (B24) gives a value for the maximum of the beta-spectrum of 3.7 MV, from which a mass of He^6 is obtained relative to Li^6. The Q value for the primary reaction is then found to be -0.63 MV, requiring fast neutrons, as observed.

B^{10}: (a) $B^{10}+n^1$ (slow)$\rightarrow Li^7+He^4$
(b) $B^{10}+n^1 \rightarrow H^3+He^4+He^4$

Amaldi, D'Agostino, Fermi, Pontecorvo, Rasetti and Segrè (A7) found a strong absorption of slow neutrons in B, but no gamma-radiation, indicating a heavy particle disintegration. The heavy disintegration products were observed by Chadwick and Goldhaber (C12, C16), who found doubly charged ions of up to 0.5 cm range (alphas) and also singly charged ions of longer range, and suggested the three particle reaction given above. Amaldi (A8) suggested the alternative two particle reaction to explain his observations of alphas of 5 to 10 mm range with slow neutrons. Taylor's (T2) photographic emulsion technique has shown that both processes occur, that is both straight tracks and three prong forks are observed. He calculated a small negative Q for the three particle process, indicating that slow neutrons would not produce it, and checking with the observation that the total momentum of the three prongs of the forks was not equal to zero. The present mass values show a Q of 0.41 MV, so low that in order to explain the observed lengths of the tracks it is necessary to assume neutrons of several million volts energy.

The straight tracks observed by Taylor and Dabholkar (T3) were found to have a range in air of 1.14 cm, representing the combined Li^7 and He^4 ranges from reaction (a). Assuming a range-velocity relation for the Li^7 atoms they find a reaction energy of 2.25 MV. Rotblat (R16) measures an alpha-range of 0.82 cm and a Li^7 range of 0.36 cm and calculates a Q of 2.24 MV which becomes 2.43 MV with the range energy relations of §95. These results have been checked by Fünfer (F33a) who finds an alpha-range of 0.86 cm and a reaction energy of 2.52 MV. Haxel (H19a) has studied this reaction using low intensity neutron sources and finds some evidence for two groups of alphas (of 0.94 and 0.64 cm range). He calculates Q values of 2.6 and 1.7 MV, which become 2.8 and 1.8 MV with the new range energy relations, and indicates an excitation state of Li^7 at 0.9 MV. The low intensity and consequent poor collimation possible in the experiment makes his method of extrapolation questionable; a single group of alphas of 8.5 cm range ($Q=2.55$ MV) would be a reasonable alternate interpretation.

The experimental evidence points to a reaction energy of about 2.5 MV; that calculated from mass values, however, is 2.99 MV. This is too large a discrepancy (0.5 MV) to be due to experimental errors. A possible explanation is that the transition to the ground state of Li^7 is forbidden and that the reaction leads to an excitation state in the Li^7 nucleus, with different angular momentum, for which the transition is allowed. Two pieces of experimental evidence support this interpretation. Firstly, the noncapture excitation of Li by alphas (§99A) results in gamma-radiation of about 0.4–0.6 MV, indicating such a level; the excitation energy of the compound B^{11} nucleus formed in each case is equivalent. Secondly, from proton groups in the Li^6-*d*-*p* reaction (§101B) an excitation level in Li^7 at 0.44 MV is indicated. It is probable that these values represent a single level, since only one low lying level would be expected in such a light nucleus.

Kikuchi, Aoki and Husimi (K8) have observed gamma-radiation under slow neutron bombardment with a cross section of 1/20 the total boron absorption cross section. This may represent the excitation state or may indicate a simple capture reaction.

This slow neutron reaction is even more probable than the Li^6 process, having a cross section of 3000×10^{-24} cm^2 (A11), measured by the absorption of slow neutrons in B. It is of correspondingly greater value as a slow neutron detector and has had its most successful application in the form of a large ionization chamber filled with boron trifluoride gas (C30) in which the alpha-particles are detected and counted with a linear amplifier and counter. It is also useful as a neutron absorber, following the $1/v$ dependence of cross section on neutron velocity, and in this respect differs from Cd which absorbs chiefly thermal energy neutrons and has a constant cross section with neutron velocity in that region of energies. This feature has been utilized in the methods in use for the determination of the selective energy regions responsible for induced radioactivity by neutron capture in many elements (see §60).

C^{12}: $C^{12}+n^1 \rightarrow Be^9+He^4$
$C^{12}+n^1 \rightarrow 3He^4+n^1$

Evidence for the fast neutron disintegration of C is based upon photographs of forks produced in cloud chambers filled with CO_2 gas. Feather (F8) has reported three such forks, and Harkins, Gans and Newson (H14) have added one more. The two product reaction is the reverse of the well-known Be^9-α-n process and is endoergic by 5.57 MV, which explains the infrequency of the process.

Another mode of disintegration is indicated by the observation of a "trident" track (three heavy particle tracks originating from the same point) by Chadwick, Feather and Davies (C8). This is interpreted as a break-up of the compound nucleus, $*C^{13}$, into three alphas and a neutron, a process well known from the $B^{11}+H^2$ reaction and also found in the Be^9-α-n process (§101A). It may be written out in detail:

$$C^{12}+n^1 \rightarrow *C^{13} \rightarrow *Be^9+He^4 \rightarrow *Be^8+n^1+He^4 \rightarrow 3He^4+n^1.$$

Since a neutron is a product as well as a projectile it has been referred to as a noncapture disintegration. However, in the light of the present interpretation involving compound nuclei it is better to view this as a normal alternative of the two particle reaction, yielding multiple products. The Q value is found from masses to be -7.16 MV.

N^{14}: $N^{14}+n^1 \rightarrow B^{11}+He^4$

Feather (F5) first reported disintegration forks in an air filled cloud chamber and suggested the reaction. Harkins, Gans and Newson (H12), Kurie (K31) and others have also studied the reaction, which is quite probable, and have observed enough forks upon which to base statistical calculations. The measurements of energy and momentum of the tracks shows a widely varying Q value, reported as a "loss in energy" in the disintegration.

Mass values indicate a Q of -0.32 MV and excitation levels in the B^{11} nucleus have been suggested from proton groups in the B^{10}-*d*-*p* reaction at about 2 and 4 MV above the ground state. These would allow alternate Q's of -2.3 and -4.3 MV. Bonner and Brubaker (B41) have recently studied and tabulated the results from cloud chamber forks, and find that with low energy neutrons no loss in energy is indicated and an average Q of -0.3 MV is obtained. They also find several cases of excitation in which an

TABLE LX. *Evidence for n-α reactions in heavy elements.*

Target						
Z	Element	Raa El	T	Chem. Identif.	Also produced by	Ref.
11	Na^{23}	F^{20}	12 sec.		F-*d*-*p*, F-*n*-γ	N1
12	Mg^{26}	Ne^{23}	33 sec.	Yes	Na-*n*-*p*	A7, B24a
13	Al^{27}	Na^{24}	14.8 hr.	Yes	Na-*n*-γ, Mg-*n*-*p*; Na-*d*-*p*, Mg-*d*-α	A7, K17
15	P^{31}	Al^{28}	2.36 min.		Al-*n*-γ, Si-*n*-*p*; Al-*d*-*p*, Mg-α-*p*	A7
17	Cl^{35}	P^{32}	15 days	Yes	P-*n*-γ, S-*n*-*p*; P-*d*-*p*, S-*d*-α	A7
19	$K^{39,\,41}$	$Cl^{36,\,38}$	37.5 min.	Yes	Cl-*n*-γ, Cl-*d*-*p*	H39a
21	Sc^{45}	K^{42}	12.2 hr.	Yes	K-*n*-γ, K-*d*-*p*	A7, W1
22	Ti^{48}	Ca^{45}	2.3 hr.		Ca-*n*-γ, Ca-*d*-*p*	W1
25	Mn^{55}	V^{52}	3.7 min.	Yes	V-*n*-γ, Cr-*n*-*p*	A7
27	Co^{59}	Mn^{56}	2.5 hr.	Yes	Mn-*n*-γ, Fe-*n*-*p*; Mn-*d*-*p*, Cr-α-*p*	A7
30	Zn^{68}	Ni^{65}	100 min.	Yes	Ni-*n*-γ(?), Cu-*n*-*p*	M1
31	Ga^{69}	Cu^{66}	5 min.		Cu-*n*-γ, Zn-*n*-*p*	C16a
56	Ba	Xe ?	3 min.			A7
90	Th^{232}	Ra^{229}	1 min.	Yes		C60
92	U^{238}	Th^{235}	4 min.	Yes		M15

"energy loss" of about 6 MV is observed.

Chadwick and Goldhaber (C16) and Bonner and Brubaker (B36) reported particles from N on slow neutron bombardment which they at first considered to be alphas. However, energies calculated from the data on this assumption led to a serious discrepancy in the mass-energy balance, and Bonner and Brubaker (B39) have since shown that the particles are protons and so do not belong to this reaction.

O^{16}: $O^{16}+n^1\rightarrow C^{13}+He^4$

Meitner and Phillipp (M12) observed the first disintegration of oxygen by neutrons and later reported several more forks (M13). Feather (F6) has observed and measured a total of 8 disintegrations in the cloud chamber. These have alpha-tracks of the order of 1 cm range and recoil C^{13} tracks of about 2 to 3 mm, depending on the angles and the neutron energy. From the best mass values the process is found to have a Q of -2.36 MV.

F^{19}: $F^{19}+n^1\rightarrow N^{16}+He^4$

Harkins, Gans and Newson (H14) reported on 13 or more disintegration forks observed in a cloud chamber filled with a F gas, and suggested the reaction above, predicting a possible instability of the N^{16} product. The tracks show the characteristic loss in kinetic energy of similar reactions and can not be critically analyzed.

An estimate of the average "energy loss" in Harkins' data is 2 MV. If we choose $Q=-2$ MV to represent the ground state reaction energy we get a mass of N^{16} of 16.0113 MV. This is 4.5 MV heavier than that calculated from the beta-energy and so the full energy beta disintegration must be a forbidden transition; a gamma-ray of about 4.5 MV would be expected.

Fermi, Pontecorvo and Rasetti (F13) first observed a radioactivity of 9 sec. half-life in F by neutron bombardment and ascribed it, rather arbitrarily at that time, to this process. A second activity of 40 sec. period was found by Bjerge and Westcott (B22) with 5 percent of the intensity of the 9 sec. period. A lack of evidence of the "water effect" (increase of intensity with slowing down of the neutrons) made it difficult to determine whether either of these periods was of the *n*-γ type, or whether they belonged to the *n*-α type resulting in N^{16} and the *n*-*p* type yielding O^{19}. The F^{20} product of the *n*-γ reaction is also obtained from the F^{19}-*d*-*p* process, with a half-life of 12 sec. (C48). Nahmias and Walen (N1) found two periods, of 8.4 and 31 sec. half-life, and observed that the intensity of the short period was slightly increased with interposition of paraffin (slow neutrons) and at the same time the period increased to 8.9 sec. This is interpreted as due to two nearly equal periods, one of which is water sensitive and has a slightly longer half-life ($F^{19}+n^1\rightarrow F^{20}$; $T\simeq 12$ sec.). The remaining activity comprising most of the intensity is not water sensitive and has a shorter half-life ($T\simeq 8.4$ sec.). From the cloud chamber evidence which indicates that the *n*-α reaction is more probable than the *n*-*p* process we can ascribe this short period to N^{16}. This leaves the weak 31 sec. period (also measured as 40 sec.) to the otherwise unobserved *n*-*p* reaction resulting in O^{19}. Further confirmation of this explanation comes from Chang, Goldhaber and Sagane (C16a) who observed the 8 sec. N^{16} period in an O-*n*-*p* reaction (producible only by very fast neutrons).

Ne^{20}: $Ne^{20}+n^1\rightarrow O^{17}+He^4$

Cloud chamber studies by Harkins, Gans and Newson (H14) show eleven forks indicating this reaction. An "energy loss" was observed in all instances, roughly correlated with the neutron energy. Jaeckel (J1) has made a careful study of this process with some exceedingly valuable results. Although he observes an energy loss in each case, the losses group closely around two distinct values, -0.7 and -5.3 MV. This is to be expected from the general principles involved in other type reactions, and indicates an excitation level in the O^{17} nucleus at 4.6 MV. It suggests that the other disintegrations of this type in which non-constant energy losses were observed will show similar groupings with sufficient experimental resolution. The value for the disintegration into the ground state is found from masses to be -0.58 MV, in sufficient agreement with the observed value.

Other targets.—The other reactions of this type are all inferred from electron radioactivity produced by neutrons. Fermi's laboratory is chiefly responsible for the list given in Table LX:

B. Type reaction *n-p*. (Tables LXII and LXIII)

$$Z^A+n^1 \rightarrow (Z-1)^A+H^1+Q:$$

In this type of reaction the product formed is an isobar of the target and will transform back into the target element with electron radioactivity. The neutron absorbed in the primary process is 0.78 MV heavier than the combined masses of the ejected proton and the radioactive electron, so slow neutrons can cause this reaction only when the mass of the radioactive product is less than 0.78 MV heavier than the target element, in which case the induced radioactivity would be of long life and low energy. With fast neutrons the energy of the escaping proton and that of the radioactive process must be supplied (less 0.78 MV) by the kinetic energy of the neutron. In heavier elements the escaping proton must have an energy comparable with the height of the potential barrier for the reaction to have any observable probability. Thus we can expect it to occur in general only for very fast neutrons and only in the lighter elements (see Chap. X). If the neutron and proton energies could be accurately determined and good measurements made on the radioactive energy, this would be an ideal reaction for checking the beta-ray theory, since the *p-n* mass difference is well known.

The identification has been accomplished chiefly through the induced radioactivities, with the assistance of chemical analysis or correlating activities from other processes.

TABLE LXI. *Summary of n-α type reaction.*

Target		Product	Q(MV)	Q(MV)		Resonance	Excit. levels	Cross Section in
Z	El	El	(calc.)	(obs.)	Reference	levels (MV)	(MV)	10^{-24} cm²
3	Li^6	H^3	4.56	4.67	L32			900
4	Be^9	He^6	−0.63	obs.	B24			
5	B^{10}	Li^7	2.99	2.43	R16		0.5	3000
	B^{10}	He^4+He^3	0.41	obs.	T2			
	B^{11}	Li^8	−6.6					
6	C^{12}	Be^9	−5.57	obs.	F8			
	C^{12}	$2He^4+n^1$	−7.16	obs.	C8			
	C^{13}	Be^{10}	−3.75					
7	N^{14}	B^{11}	−0.32	−0.3	B41		∼6	
	N^{15}	B^{12}	<−8.0					
8	O^{16}	C^{13}	−2.36	obs.	F6			
9	F^{19}	N^{16}	−2*	−2	H14			2.5
10	Ne^{20}	O^{17}	−0.58	−0.7	J1		4.6	
	Ne^{21}	O^{18}	1.0					
11	Na^{23}	F^{20}	<−2.8	obs.	N1			
12	Mg^{24}	Ne^{21}	−2.3					
	Mg^{25}	Ne^{22}	0.2					
	Mg^{26}	Ne^{23}	—	obs.	A7			
13	Al^{27}	Na^{24}	−2.3	obs.	A7			
14	Si^{28}	Mg^{25}	−2.0					
	Si^{29}	Mg^{26}	1.7					
	Si^{30}	Mg^{27}	−3.6					
15	P^{31}	Al^{28}	−0.8	obs.	A7			
16	S^{32}	Si^{29}	0.7					
17	Cl^{35}	P^{32}	1.3	obs.	A7			
etc.								

* Q (observed) used to calculate mass values.

Li^6?: $Li^6+n^1 \rightarrow He^6+H^1$

Knol and Veldkamp (K20, V8) have reported the observation of an electron radioactivity in Li under *n* bombardment. Using a rotating wheel method they measure the half-life to be 0.8 sec. In their report they assumed the activity to come from the reaction: $Li^7+n^1 \rightarrow Li^8+\gamma$, since Li^8 was known at that time to be produced in the Li^7-*d*-*p* reaction with a period of about 0.5 sec. If the reaction is produced by slow neutrons the reaction is doubtless the one they suggest; however, they apparently did not search for a fast neutron effect. Subsequently Bjerge and Bronström (B24) and Nahmias and Walen (N1) have identified a radioactive He^6 as a product of the Be^9-*n*-α reaction and find a half-life of 0.7 sec. This suggests that the activity observed by Knol and Veldkamp may be due to He^6, produced from Li^6 following this type reaction. The simple capture process is less probable than the *n*-*p* process for such a light element (see §57). This feature, plus the coincidence in half-life periods, makes it seem more satisfactory to attribute the activity to this process. Using the mass of He^6 obtained from the radioactive energy of the Be^9-*n*-α reaction we can predict a reaction energy in this case of -2.9 MV.

N^{14}: $N^{14}+n^1 \rightarrow C^{14}+H^1$

This reaction was first observed by Kurie (K30), in an air-filled cloud chamber bombarded by fast neutrons. He found forked tracks in which one prong was long range and of the low density characteristic of protons.

Using slow neutrons Chadwick and Goldhaber (C16) measured the energy of the disintegration particles by an ionization method, and at first called them alphas. Bonner and Brubaker (B36) observed the ranges of the particles in a cloud chamber (1.06 cm) also using slow neutrons, and found that the energy indicated by the range measurement did not check with Chadwick and Goldhaber's ionization value. This led them to re-study the photographs and assign the tracks to protons (B39), in which case the Q value is found to be 0.58 MV. A recalculated value of 0.62 MV is used to obtain the mass of C^{14}. Burcham and Goldhaber (B63) have verified the mechanism by use of the photographic emulsion technique.

McMillan (Mc7) has observed the radioactivity of C^{14} produced through the C^{13}-*d*-*p* reaction, and finds the long half-life (*ca.* 3 months) and low energy electrons (*ca.* 0.3 MV) expected, as discussed in the introductory paragraphs.

The reaction probability is quite large, as would be expected for a slow neutron process, and in air filled ionization chambers used to record other slow neutron disintegration processes it results in a large "background" effect, removable by substituting a nondisintegrable gas.

Other targets.—Other processes which have been identified through their radioactivities are listed in Table LXII:

TABLE LXII. *Evidence for n-p reactions.*

Z	TARGET	Raa El	T	CHEM. IDENTIF.	ALSO PRODUCED BY	REF.
8	O^{16}	N^{16}	8 sec.		F-*n*-α, N-*d*-*p*	C16a
9	F^{19}	O^{19}	31 sec.			N1
11	Na^{23}	Ne^{23}	33 sec.	Yes	Mg-*n*-α	A7, N 1
12	Mg^{24}	Na^{24}	14.8 hr.	Yes	Al-*n*-α, Na-*n*-γ Na-*d*-*p*, Mg-*d*-α	A7
13	Al^{27}	Mg^{27}	10.2 min.		Mg-*n*-γ, Mg-*d*-*p*	A7
14	Si^{28}	Al^{28}	2.36 min.	Yes	P-*n*-α, Al-*n*-γ Al-*d*-*p*, Mg-α-*p*	A7
15	P^{31}	Si^{31}	2.4 hr.	Yes	Si-*n*-γ, Si-*d*-*p*	A7
16	S^{32}	P^{32}	15.0 da.	Yes	Cl-*n*-α, P-*n*-γ P-*d*-*p*, S-*d*-α	A7
17	Cl^{35}	S^{35}	80 da.			A14
19	K^{39}	A^{39}?	4 min.			P11d
	K^{41}	A^{41}	110 min.		A-*n*-γ, A-*d*-*p*	H39a
20	Ca^{42}	K^{42}	12.5 hr.	Yes	K-*n*-γ, K-*d*-*p*, Sc-*n*-α	H39a
22	Ti	Sc	1.7 hr. 28 hr.			W1a, P11d
24	Cr^{52}	V^{52}	3.75 min.	Yes	Mn-*n*-α, V-*n*-γ	A7
26	Fe^{56}	Mn^{56}	2.5 hr.	Yes	Co-*n*-α, Mn-*n*-γ Mn-*d*-*p*, Cr-α-*p*	A7
27	Co^{59}	Fe^{59}	40 da.	Yes	Fe-*d*-*p*	L27a
28	Ni^{58}	Co^{58}	20 min.		Co-*n*-γ	R15
29	Cu^{65}	Ni^{65}	160 min.	Yes	Ni-*n*-γ, Zn-*n*-α	O½
30	Zn^{64}	Cu^{64}	12.8 hr.	Yes	Cu-*n*-γ, Cu-*d*-*p* Zn-*d*-α, Ni-*p*-*n*	A7
	Zn^{66}	Cu^{66}	5 min.	Yes	Cu-*n*-γ	B21, A7, F16

C. Type reaction *n*-γ. (Tables LXIV and LXV)

$$Z^A+n^1 \rightarrow Z^{A+1}+h\nu:$$

This type reaction, known as "simple capture" of the neutron, was discovered by Fermi, *et al.* (F13) through the radioactivity induced in many substances, and the identification of the radioactive element as an isotope of the target. If the product nucleus of the primary reaction exists the reaction must be exoergic by an amount approximately equal to the binding energy of the neutron (5 to 10 MV), which energy is given off in the form of gamma-radiation. Since there is no emission of charged particles it can occur for heavy elements with even higher probability than for light ones, since heavy nuclei possess many closely spaced resonance levels (see Chap. X).

The product nucleus may be either radioactive

TABLE LXIII. *Summary of n-p type reaction.*

Target		Product	Q(MV)	Q(MV)	Reference	Resonance levels (MV)	Excit. levels (MV)	Cross section in 10^{-24} cm²
Z	El	El	(calc.)	(obs.)				
3	Li^6	He^6	−2.9	obs.?	K20			
5	B^{10}	Be^{10}	0.41					
6	C^{12}	B^{12}	<−12.8					
7	N^{14}	C^{14}	0.62*	0.62	B39			11.3
8	O^{16}	N^{16}	−9.7					
9	F^{19}	O^{19}	—	obs.	N1			
10	Ne^{20}	F^{20}	<−4.2					
11	Na^{23}	Ne^{23}	—	obs.	N1			
12	Mg^{24}	Na^{24}	−3.9	obs.	A7			
13	Al^{27}	Mg^{27}	−1.4	obs.	A7			
14	Si^{28}	Al^{28}	−2.7	obs.	A7			
15	P^{31}	Si^{31}	−1.0	obs.	A7			
16	S^{32}	P^{32}	−0.9	obs.	A7			
17	Cl^{35}	S^{35}	—	obs.	A14			
etc.								

* *Q* (observed) used to calculate mass values.

or stable. In the first case the radioactivity is used as an indicator of the process, in the second the occurrence of the reaction may be inferred from an observed absorption of neutrons (as in Cd). With fast neutrons the probability is small, but observable (§65). When the neutrons are slowed down by the interposition of paraffin or other light materials the intensity of the observed activity is increased. This increase was found by Fermi to be as much as 40 times in some substances. The activities observed with fast neutrons (no water or paraffin) are probably in some part due to the slow neutrons at all times present in the beam, produced by scattering in the target and adjacent materials. Since the reactions with slow neutrons yielding particle products are in general not probable in heavy elements the observation of a "water sensitive" activity is taken to be strong evidence for this type of reaction.

The reaction energy of the primary process is equal to the binding energy of the neutron in the product nucleus, and must appear in the form of gamma-radiation. This may be emitted in a single gamma-ray, or more probably, as several successive rays determined by the arrangement of energy levels in the product nucleus (§90). Certain of these processes have been found by Rasetti (R3), Kikuchi (K6 *et seq.*) and Fleischmann (F25) to result in gamma-radiation, and measurements on the average energy made. This capture radiation may be confused with that due to a noncapture excitation of the nucleus, discussed in a later section of this paper. Where actual observations of the gammas are available, it is so indicated in the table to follow.

The increased activity due to paraffin indicated that the neutrons most effective in this process were of thermal energies. Experiments with absorbers mounted on a rotating wheel (R4) and with neutrons slowed down in scatterers at liquid-air temperatures proved that this was the case. Cd was found to absorb these thermal neutrons although it was not rendered radioactive in the process; it was found to have an absorption coefficient essentially constant throughout the thermal energy region.

The discovery of the selective absorption of neutrons of different energies in various targets by Moon and Tillman (M26) initiated a new field of research and has led to a most promising theoretical interpretation. These experiments were followed by others by Szillard (S29) and by Amaldi and Fermi (A11) who separated the neutron groups of different energies responsible for the residual activity (Cd filtered) in several elements. To give an example of this effect: the activity in Ag (22 sec.) caused by slow neutrons is reduced to about $\frac{1}{3}$ by interposing a Cd absorber; this residual activity may be largely suppressed by Ag absorbers, but to a much smaller degree by other elements such as Rh and Hg; using a Rh target the opposite is true, Rh absorbs the neutrons strongly while Ag and Hg do not, etc. This is interpreted as indicating regions of neutron energy, somewhat above the thermal region, selectively responsible for the activities.

TABLE LXIV. *Evidence for n-γ reactions.*

Target Z	Target El	Raa El	T	Water Sensit.	Res. Levels Chap. X	Chem. Identif.	Also produced by	COMMENTS	REF.
3	Li^{7}	Li^{8}?	0.88 sec.	>1			Li-d-p		K20
9	F^{19}	F^{20}	12 sec.	>1			Na-n-α, F-d-p	gammas	B21, A7, N1
11	Na^{23}	Na^{24}	14.8 hr.	>1			{Al-n-α, Mg-n-p; Na-d-p, Mg-d-α}	gammas	A7, B21
12	Mg^{26}	Mg^{27}	10.2 min.	>1			Mg-d-p, Al-n-p	gammas	A7
13	Al^{27}	Al^{28}	2.36 min.	>1			{P-n-α, Si-n-p; Al-d-p, Mg-α-p}	$\gamma=5.8$ MV	A7, F25
14	Si^{30}	Si^{31}	2.4 hr.	>1			Si-d-p, P-n-p	gammas	A7
15	P^{31}	P^{32}	15 da.				{Cl-n-α, S-n-p; P-d-p, S-d-α}		P15
17	Cl^{37}	Cl^{38}	37 min.	>1			Cl-d-p, K-n-α	$\gamma=6.6$ MV	A3, A7, R3
18	A^{40}	A^{41}	110 min.				A-d-p, K-n-p		S15
19	K^{41}	K^{42}	12.2 hr.	>1			Sc-n-α, K-d-p	gammas	A7, W1
20	Ca^{44}	Ca^{45}	2.4 hr.	>1			Ca-d-p, Ti-n α		H29, W1
21	Sc^{45}	Sc^{46}	(long)						H32
22	Ti^{50}	Ti^{51}	3 min.				Ti-d-p	gammas	A7
23	V^{51}	V^{52}	3.75 min.	40			Mn-n-α, Cr-n-p	gammas	A7
25	Mn^{55}	Mn^{56}	2.5 hr.	23	*	Yes	{Co-n-α, Fe-n-p; Cr-α-p, Mn-d-p}	gammas	A7
27	Co^{57}	Co^{58}	11 min.	>1			Ni-n-p	$\gamma=5.0$ MV	R15, A7, R3
	Co^{59}	Co^{60}	~ 1 year	>1		Yes	Co-d-p		R9a, L27a
28	Ni^{58}	Ni^{59}	3 hr.	>1				gammas	R15, N3
29	Cu^{65}	Cu^{66}	5 min.	15	*		Zn-n-p, Ga-n-α	} $\gamma=7.4$ MV	A7, L29, F25
	Cu^{63}	Cu^{64}	12.8 hr.	>1			{Cu-d-p, Zn-n-p; Zn-d-α, Ni-p-n}		A7, B21
31	Ga^{69}	Ga^{70}	20 min.	3			Zn-p-n, Ga-γ-n	gammas	A7
	Ga^{71}	Ga^{72}	23 hr.	>1			Zn-d-γ(?)	gammas	A7
32	$Ge^{(70), 7}$	$Ge^{(71), 75}$	30 min.						A7, S25
33	As^{75}	As^{76}	26 hr.	6	*	Yes	As-d-p	gammas	A7
34	Se	Se^{+1}	22 min.	4			Se-d-p	$\gamma=5.8$ MV	A7, F25, H33b
35	Br^{79}	Br^{80}	18 min.	10	*	Yes	Br-d-p, Se-p-n, Br-γ-n	} gammas	A7, J2
	Br^{79}	Br^{80}	4.2 hr.	>1		Yes	Br-d-p, Se-p-n, Br-γ-n		K34
	Br^{81}	Br^{82}	36 hr.?			Yes	Br-d-p, Se-p-n, Se-d-$2n$(?)		
37	$Rb^{85, 87}$	$Rb^{86, 88}$	20 min.				γ-n-α		F13
39	Y^{89}	Y^{90}	70 hr.	≫1				{Strong abs'n. $\gamma=4$ MV}	A7, R3, H31
40	Zr^{92}	Zr^{93}	40 hr.	>1					H29
42	Mo	Mo^{+1}	{25 min.; 36 hr.}	>1; >1					A7, Mc1, H33b
44	Ru	Ru^{+1}	{40 sec.; 100 sec.; 11 hr.; 75 hr.}				Ru-d-p (39 hr., 11 da.)		K35
45	Rh	Rh^{+1}	{44 sec.; 3.9 min.}	15; >1	**; *				A7
46	Pd	Pd^{+1}	{15 min.; 3 min.; 12 hr.; 60 hr.}	>1			Pd-d-p (10 hr.)		A7, Mc1, K36
47	Ag^{107}	Ag^{108}	2.3 min.	15	*		Ag-γ-n	$\gamma=3.7$ MV	A7, F25
	Ag^{109}	Ag^{110}	22 sec.	30	**				
48	Cd	Cd^{+1}	{——; 5 hr.; 52 hr.}	≫1; >1; >1	**		} Cd-d-p	{strong abs'n. $\gamma=4.1$ MV}	A7, F25; M20a
49	In^{113}; In^{115}	{In^{114}; In^{116}}	{13 sec.; 54 min.; 3 hr.}	15; >1	**		Cd-p-n, Sn-n-p		A7, S28
50	Sn	Sn^{+1}	8 min.				Sn-d-p		N3
	Sn^{112}	Sn^{113}	18 min.				Sn-d-p, In-p-n		N3
51	$Sb^{121, 123}$	$Sb^{122, 124}$	{2.5 da.; 60 da.}	>1			Sb-d-p	gammas	A7, L27b
52	Te	Te^{+1}	45 min.	>1			Te-n-$2n$, Te-γ-n		A7
53	I^{127}	I^{128}	25 min.	5	**	Yes		gammas	A7
55	Cs^{133}	Cs^{134}	1.5 hr.	>1					A7, L3

* Resonance energy determined.
** Energy and width of resonance level determined.

TABLE LXIV.—*Continued.*

Target Z	Target El	Raa El	T	Water Sensit.	Res. Levels Chap. X	Chem. Identif.	Also produced by	COMMENTS	REF.
56	Ba	Ba^{+1}	80 min.	8			Ba-*d*-*p*		A7, A8
57	La^{139}	La^{140}	1.9 da.	12			La-*d*-*p*		M7, H31
59	Pr^{141}	Pr^{142}	19 hr.				Nd-*n*-*p*		A7, H30
60	Nd	Nd^{+1}	1 hr.						A7, H30
62	Sm	Sm^{+1}	40 min.; 2 da.					Very strong abs'n. $\gamma=3.3$ MV	A7, H30, F25 H31a
63	Eu	Eu^{+1}	9.2 hr.	40				Very strong abs'n.	S25, M7, H31
64	Gd	Gd^{+1}	8 hr.	>1				$\gamma=4.0$ MV *Very strong* abs'n.	A7, H30, F25
65	Tb^{159}	Tb^{160}	3.9 hr.						H30, M7, S25
66	Dy^{164}	Dy^{165}	2.5 hr.	20					H30, M7
67	Ho^{165}	Ho^{166}	35 hr.	15					H31a, M7, Mc3
68	Er	Er^{+1}	5.8 min.; 12 hr.						Mc3, S25, H31a
69	Tm^{169}	Tm^{170}	8 mo.						C61
70	Yb	Yb^{+1}	3.5 hr.						H30, M7, S25
71	Lu^{175}	Lu^{176}	3.6 hr.; 6 da.						H31, M7, Mc3
72	Hf^{180}	Hf^{181}	*ca.* mo.						H29
73	Ta^{181}	Ta^{182}	(200 da.)						F27
74	W	W^{+1}	23 hr.	15				gammas	A7, Mc1
75	Re	Re^{+1}	20 hr.; 85 hr.	>1	*				A7, K36
76	Os	Os^{+1}	40 hr.						K36
77	Ir	Ir^{+1}	19 hr.; 2 mo.; 50 min.; 3 da.	>1	**	Yes	Pt-*d*-α, Pt-*n*-*p*; Au-*d*-α, *p*	$\gamma=3.3$–4.4MV	A7, 3R, S20, F27
78	Pt	Pt^{+1}	50 min.	>1			Pt-*d*-*p*		A7, Mc1
79	Au^{197}	Au^{198}	2.7 da.	>1	**	Yes	Au-*d*-*p*	gammas	A7
80	Hg	Hg^{+1}	40 hr.	>1				absorption $\gamma=4.5$ MV	A7, R3, A13, F25
81	Tl	Tl^{+1}	1.3 hr.; 4 min.						P15, Mc2
90	Th	Th^{+1}	25 min.	>1		Yes		New Raa Series	C60
92	U^{238}	U^{239}	——	>1		Yes		New Raa elements with $Z>92$	A7, M15

The true significance of these selective effects has been realized in the theory of Breit and Wigner (B51) in which they are interpreted as "resonance" effects. Bohr (B32) has formulated a nuclear model which extends the interpretation to other type reactions. On this picture the compound nucleus formed from the original nucleus and the entering neutron may exist in a variety of quasi-discrete excitation states, becoming more closely spaced for heavy nuclei. The states above the dissociation energy will correspond to definite neutron energies, so these neutrons are selectively absorbed. The probability of re-emission of a slow neutron from this excited nucleus (scattering) is extremely small compared with the probability of radiation of energy to form a normal nucleus (capture). This would predict resonance capture of neutrons of certain definite energies in a particular nucleus; these resonance levels would be in general closer spaced and therefore closer to zero energy in heavier nuclei. Bethe and Placzek (B15) have calculated the probability of neutron capture due to these levels and have shown how the widths and spacings of these levels may be determined from experiment.

A method for the measurement of the neutron energies responsible for these selective effects was suggested by Frisch and Placzek (F33), and independently by Weekes, Livingston and Bethe

(W5). In this method the $1/v$ dependence of capture probability in the B^{10}-n-α process was used to determine the neutron velocities responsible, compared to the known velocities of thermal neutrons. Goldsmith and Rasetti (G17) have measured such resonance levels for many elements, and find values ranging from fractions of a volt to 100 volts. These values probably represent the resonance level nearest zero or some average of the low levels (cf. §60). Other methods for measuring resonance energies have been presented, such as Fermi's technique of determining the "age" of the slow neutron through the thickness of water required to give the highest intensity of this residual activity (§59D). The results of all these measurements are presented in Table XXII of §60, with corrections for scattering.

H^1: $H^1+n^1 \rightarrow H^2+\gamma$

Lea (L19) has observed gamma-radiation from many elements under neutron bombardment, explaining it as due to inelastic scattering, i.e., noncapture excitation of the nucleus by fast neutrons. See §65. In paraffin Lea observed such gamma-radiation, from the hydrogen. The elementary nature of the proton makes noncapture excitation seem impossible and it is assumed that the radiation comes from the synthesis of H^2. This is the reverse of the photoelectric disintegration of the deuteron which is known to require 2.20 MV (see §103), so this process should be exoergic by this amount. The gammas have been tentatively measured by Lea as 2 to 4 MV, by Fleischmann (F25) to be 2.26 MV and by Kikuchi, Aoki and Husimi (K6) to be 2.2 MV. The probability of the process has been calculated (§17) and has been shown to be very small for fast neutrons and fairly large for slow ones. With slow neutrons it is the only known method of ultimate absorption in hydrogen, and it can be assumed that only slow neutrons are effective.

Other targets.—In Table LXIV are listed the processes of the n-γ type which have been identified through the radioactivity induced by or the strong absorption for slow neutrons.

In Table LXIV are certain features of particular interest. The observed activities are spread uniformly throughout the atomic table with the exception of the lighter elements. They are found for as many as 64 of the known elements, suggesting that the lack of evidence for the other elements is probably due to experimental difficulties such as extremely long or short periods or low intensities. In many instances two or more activities of different half-lives have been observed, indicating processes from two or more isotopes. In practically all these cases the stable isotopes are known to be spaced two mass units apart. In addition to the activities the list contains 6 elements, Y, Cd, Sm, Eu, Gd and Hg, for which the absorption of slow neutrons is so great that although they show no strong radioactivities the reaction must be assumed to occur leading to stable isotopes. The expected gamma-radiation is observed in most of these cases.

Several reactions are marked with a question, indicating incomplete evidence. In Li^7 the evidence is insufficient to determine whether the activity is due to this type reaction or the n-p reaction (see §101B), for which the half-life value is essentially the same. Other questionable cases are due to insufficient chemical evidence or observations of the water sensitivity. The complex activities found in U and Th will be more fully discussed in §105.

D. Type reaction n-$2n$: $Z^A+n^1 \rightarrow Z^{A-1}+2n^1$ (Table LXVI)

This reaction has been assumed in several instances to explain the observation of more discrete half-life periods due to isotopes of the target element than the known number of stable isotopes. The earliest positive proof was the observation by Heyn (H33, H33a) of positron activity chemically analyzed as the target element in Cu and Zn after neutron bombardment indicating a light isotope. It has also been suggested by Rusinow (R20) to explain an increase in the total number of neutrons (as indicated by total intensity of radioactivity produced) when a neutron source is surrounded by Be. Such reactions require fast neutrons and should not be invoked, as has been done in some instances (J2), to explain an excessive number of slow neutron activities; these require another explanation, in terms of isomeric nuclei (see §105).

This reaction amounts in effect to a noncapture disintegration of the target element, but its mechanism is no doubt analogous to the usual reactions (§85). The n-$2n$ reaction leads to an

TABLE LXV. *Summary of n-γ type reaction.*

Target Z	Target El	Product El	Q(MV) (calc.)	Q(MV) (obs.)	Reference	Resonance levels (MV)	Excit. levels (MV)	Cross section in 10^{-24} cm^2
1	H^1	H^2	2.20	2.26	F25			
	H^2	H^3	6.19					
3	Li^6	Li^7	7.12					
	Li^7	Li^8	>2.0	obs.?	K20			
4	Be^9	Be^{10}	6.80					
5	B^{10}	B^{11}	11.51					
	B^{11}	B^{12}	>1.9					
6	C^{12}	C^{13}	5.0					
	C^{13}	C^{14}	8.3					
7	N^{14}	N^{15}	10.78					
	N^{15}	N^{16}	2.4					
8	O^{16}	O^{17}	4.15					
9	F^{19}	F^{20}	>4.0	obs.	N1			
10	Ne^{20}	Ne^{21}	7.54					
	Ne^{21}	Ne^{22}	9.32					
11	Na^{23}	Na^{24}	7.2	obs.	A7			4.2
12	Mg^{24}	Mg^{25}	7.0					
	Mg^{25}	Mg^{26}	11.9					
	Mg^{26}	Mg^{27}	6.2	obs.	A7			3.5
13	Al^{27}	Al^{28}	8.0	obs.	A7			1.5
14	Si^{28}	Si^{29}	8.4					
	Si^{29}	Si^{30}	11.5					
	Si^{30}	Si^{31}	5.6	obs.	A7			2.5
15	P^{31}	P^{32}	8.5	obs.	P15			14.7
17	Cl^{37}	Cl^{38}	5.6	obs.	A7			39.
18	A^{40}	A^{41}	—	obs.	S15			
19	K^{41}	K^{42}	—	obs.	A7			8.2
etc.								

isotope of the target element of lower mass number and so results in the same product as the *γ-n* reaction. Observation of the same period produced by gamma-rays and by fast neutrons (but not by slow neutrons) indicates this type reaction.

Pool, Cork and Thornton (P11d) have studied the radioactivity produced in some 60 targets by the high energy neutrons from the Li-*d-n* reaction, using the 6.3 MV deuterons from the cyclotron. They observe some 113 radioactive periods and make many assignments to the isotopes responsible. Four of the activities (Al, Cl, Mn and Co targets) can be definitely ascribed to *n-α* reactions, through chemical analysis and recognition of known half-life periods, and are entered in §102A. Twelve or thirteen are *n-p* reactions (§102B). Of the remainder, many are found to have periods identical with known slow neutron (*n-γ*) periods, in some instances from targets known to have only one isotope (i.e., V^{52}, Mn^{56}, Y^{90}, I^{128} etc.). This suggests that the *n-γ* reaction is responsible for a large share of the unidentified activities isotopic with the target element especially since slow neutrons are always present in the bombardment. There are many other activities, however, which can only be ascribed to the *n-2n* reaction, through recognition of a known period, the observation of a new period isotopic with the target and not found with slow neutrons, etc. In some instances the period is also observed in a *γ-n* reaction. Most of these are positron active. In Table LXVI are listed those for which the evidence seems sufficient to justify designation of this type reaction. (Activities reported by other observers are included in the table.) Those activities for which there is not sufficient evidence to include in this table are not listed under a primary reaction but only as activities (usually under the target element) in the tabulation of induced radioactivities (Table LXIX).

The most conclusive evidence for the *n-2n* type reaction is through the technique used by Heyn (H33a) and others of using neutrons of different maximum energy from several reactions (H^2-*d-n*, Be-*d-n*, Li-*d-n*, etc.). An observation of a new period with very fast neutrons (Li-*d-n*) but not with slower neutrons (H^2-*d-n*),

TABLE LXVI. *Evidence for n–2n reactions.*

TARGET						
Z	El	Raa El	*T*	Raa	COMMENTS	REFS.
4	Be^{9}	Be^{8}	—	—	Increase in number of *n*'s	R20
6	C^{12}	C^{11}	20 min.	e^+	Known *T*	P11d
7	N^{14}	N^{13}	10.5 min.	e^+	Known *T*	P11d
8	O^{16}	O^{15}	2.1 min.	e^+	Known *T*	P11d
9	F^{19}	F^{17}	1.2 min.	e^+	Known *T*, (F^{19}-*n*-3*n*)	P11c
	F^{19}	F^{18}	108 min.	e^+	Known *T*, chem.	P11d
14	Si^{28}	Si^{27}	6 min.	e^+	Known *T*	P11d
15	P^{31}	P^{30}	3 min.	e^+	Known *T*	P11d
16	S^{32}	S^{31}	26 min.	e^+	New *T*, best assignment	P11d
17	Cl^{35}	Cl^{34}	33 min.	e^+	Known *T*, chem.	P11c
19	K^{39}	K^{38}	7.5 min.	e^+	Known *T*	P11d
20	Ca^{40}	Ca^{39}	4.5 min.	e^+	New *T*, best assignment	P11d
21	Sc^{45}	Sc^{43}	4 hr.	e^+	Known *T*, (Sc^{45}-*n*-3*n*)	P11c, P11d
	Sc^{45}	Sc^{44}	52 hr.	e^+	Known *T*	P11c, P11d
28	Ni^{58}	Ni^{57}?	2 hr.		Fast *n*'s only	H33a, P11d
29	Cu^{63}	Cu^{61}	3.5 hr.	e^+	Known *T*, (Cu^{63}-*n*-3*n*)	P11c
	Cu^{63}	Cu^{62}	10.5 min.	e^+?	Known *T* (Cu-γ-*n*), fast *n*'s	C16a, H33a, R16a
30	Zn^{64}	Zn^{63}	40 min.	e^+	Known *T* (Zn-γ-*n*), chem.	P11d, H33a, R16a
31	Ga^{69}	Ga^{68}	55 min.	e^+	Known *T* (Ga-γ-*n*?), chem.	C16a, P11d
33	As^{75}	As^{74}?	13 da.		New *T*, best assignment	P11d
34	Se^{80}	Se^{79}?	56 min.	e^-	Fast *n*'s only, chem.	H33b, P11d
35	Br^{79}	Br^{78}	5 min.	e^+	Known *T* (Br-γ-*n*)	C16a, P11d, H33b
37	Rb^{85}	Rb^{84}?	22 hr.	e^-	New *T*, best assignment	P11d
38	Sr^{86}	Sr^{85}?	3 hr.	e^+	New *T*, chem.	P11d
42	Mo^{92}	Mo^{91}?	17 min.	e^-	Known *T* (Mo-γ-*n*)	H33b, P11d
47	Ag^{107}	Ag^{106}	25.5 min.	e^+	Known *T* (Ag-γ-*n*)	{R16a, P11d C16a, H33b
48	Cd^{110}	Cd^{109}?	33 min.	e^+	Fast *n*'s only	H33b, P11d
49	In^{113}	In^{112}	1.1 min.	e^-	Known *T* (In-γ-*n*)	C16a, P11d
51	Sb^{121}	Sb^{120}	15.4 min.	e^+	Known *T* (Sb-γ-*n*, Sn-*d*-*n*)	{C16a, P11d H33b, L27b
52	Te^{128}	Te^{127}?	1.1 hr.	e^-	Known *T* (Te-γ-*n*)	H33b, P11d
57	La^{139}	La^{138}?	2.2 hr.		New *T*, best assignment	P11d
59	Pr^{141}	Pr^{140}?	3 min.	e^+	New *T*, best assignment	P11d, H31a
79	Au^{197}	Au^{196}	17 min.		New *T*, best assignment	P11d
80	Hg^{198}	Hg^{197}	45 min.	e^-	New *T*, fast *n*'s only	H33b, P11d
92	U^{238}	U^{237}?	40 sec.?	e^-	Too many U activities	M15

and chemical analysis of the radioactive isotope, make the assignment definite.

In line with the theoretical suggestions of Bohr that "cascade" disintegrations are probable with sufficiently high excitation energy, Pool, Cork and Thornton (P11c) have found evidence for a reaction of the type *n*-3*n* in Sc, in which the known Sc^{43} period was found to be produced by neutrons of >8 MV bombarding the Sc^{45} target. They suggest the same explanation for the F^{17} and Cu^{61} activities. These are also listed in Table LXVI.

E. Excitation without capture

Experiments by Ehrenberg (E1), Fermi (F17), Lukirsky and Careva (L37) and others on the slowing down of fast neutrons in heavy elements such as Pb indicate an energy absorption in excess of that expected from elastic scattering (§65). Furthermore, gamma-radiation has been observed by Lea (L19), Kikuchi, Aoki and Husimi (K11), and others to come from all elements under fast neutron bombardment, and with intensities much greater than that associated with the observed activities representing capture. Part of the radiation is no doubt due to the simple capture processes, but the larger share must be interpreted as due to noncapture excitation, or inelastic scattering. Bohr's theory is more definite than experiments on this point and predicts more inelastic than elastic collisions for neutrons with heavy nuclei. Many excitation states and a complexity of the radiation are to be expected. Further evidence for the production of excitation radiation is in the observation by Kikuchi, Aoki and Husimi (K12) of high energy electrons from all substances when bombarded by neutrons. These are most readily interpreted as secondary electrons from this radiation, and not from the break-up of the neutron into proton and electron in nuclear fields as was suggested by Kikuchi, *et al.*

§103. Photoelectric Disintegration

Type reaction γ-n: $Z^A + h\nu \rightarrow Z^{A-1} + n^1$ (Tables LXVIa and LXVII)

With two of the light elements (*viz.* H^2 and Be^4) this type of disintegration can be produced by the natural gamma-rays from Th C″. All other nuclei, as far as is known, require more energetic gamma-rays and in all cases except the two mentioned, the gamma-rays of 17.5 MV from the capture of protons by Li^7 have been used. It should occur with gamma-energies equal to or greater than the binding energy of the neutron in the nucleus (see Table XLVII).

Gamma-rays may also produce other type reactions, particularly in heavy nuclei. With comparatively low energy gamma-rays (6–8 MV) there may result the splitting off of an alpha-particle, while with very energetic gamma-rays a cascade disintegration may take place in which a neutron and an alpha-particle or two neutrons may be emitted in succession (§101D).

The cross section for most photoelectric disintegrations seems to be of the order of 10^{-27} cm^2 (B47c) in agreement with theoretical expectations (§91).

H^2: $H^2 + \gamma \rightarrow H^1 + n^1$

Chadwick and Goldhaber (C14) discovered the effect experimentally using Th C″ gamma-rays of 2.62 MV energy, by measuring the ionization produced by the protons released. If the proton energy is doubled (to include the equal neutron energy) and subtracted from 2.62 MV, the value obtained for the Q of the reaction and hence the binding energy of the neutron in H^2 is -2.16 MV. In a report before the British Association, Feather (F9) gives what seems to be a better value of proton range, and obtains a Q of -2.26 MV. When this is corrected for the range energy relations used it becomes -2.23 MV.

Chadwick and Goldhaber and also Banks, Chalmers and Hopwood (B3) observed neutrons using radium gammas on H^2. More careful work by Mitchell, Rasetti, Fink and Pegram (M22) with radium sources shows a neutron intensity so low that it can only be separated from neutrons produced in the glass containers with some difficulty. The neutrons show the very low energy expected. The highest energy gamma-rays from Rn are from Ra C, of 2.198 MV energy (E8), which is lower by 0.03 MV than the Q value obtained from the Th C″ gammas. Although this is only slightly outside the errors of Feather's experiment it is significant in that this binding energy value determines the relative masses of H^1 and n^1. Either the experimental evidence with radium gammas is spurious or Feather's results are incorrect by this amount. We feel it necessary to compromise with these experiments by choosing the value -2.20 MV for the binding energy.

Knowledge of the binding energy of the deuteron is of great theoretical importance. It is used to calculate the neutron mass from the mass spectroscopic values of H^2 and H^1, and the n^1-H^1 mass difference so obtained (0.78 MV) will be of vital importance in theories of the elementary particles. The binding energy is also invaluable in the determination of n-p forces in nuclei and in interpreting the results of scattering experiments. Bethe and Peierls (B10), Hall (H11) and others have calculated the probability cross section of the reaction and find results in accord with the experiments. Breit and Condon (B52) have shown that from studies of the cross section at higher energies the nature of the forces between proton and neutron (exchange *vs.* ordinary) and the range of such forces can be easily inferred.

Richardson and Emo (R6a) have reported on the photoelectric disintegration of H^2 with the radioactive gamma-rays from Na^{24}. They use the observed proton energies (0.30 MV) to determine the gamma-ray energy, which proves to be 2.8 MV. The cross section at this energy is found to be 1×10^{-27} cm^2. This result suggests the use of this method as a method for measurement of gamma-ray energies.

Be^9: $Be^9 + \gamma \rightarrow Be^8 + n^1$

This process was discovered by Szilard and Chalmers (S26) and studied more thoroughly by Chadwick and Goldhaber (C14). They determined the energy of gamma-ray required to produce the disintegration by measuring the maximum energy of He recoils from the neutrons in a cloud chamber. From these observations they find a Q of -1.6 MV. The value calculated from masses is -1.72 MV.

Radium gamma-rays produce neutrons through this process, both the 1.761 and 2.198 MV radiations being effective. The neutrons have lower energy than those from Th C″ gammas, as shown by Mitchell, Rasetti, Fink and Pegram (M22) and others. Studies of the reaction probability with different gamma-energies have been made by Fleischmann and Gentner (F24), etc.

The reaction has also been observed with x-rays of energy less than 2 MV (B48) through the neutron induced radioactivity.

Goloborodko and Rosenkewitsch (G18b) study the angular distribution and find it to be isotropic.

Table LXVIa contains the results obtained with the gamma-rays from the Li-p-α reaction. Most of these data have been obtained by Bothe and Gentner (B47b, B47c, B47d). Some of the activities observed have been confirmed in the Cavendish Laboratory by Chang, Goldhaber and Sagane (C16a) and additional activities found. In some instances the activity is the same as that produced by neutrons, which is valuable in that it allows identification of the active isotope in certain instances (cf. §105).

§104. Disintegrations by Other Particles and Radiations

Disintegrations by Li ions

If Li ions bombard a hydrogen-containing substance the same disintegration should be observed as for protons bombarding Li, with a probability determined by the relative velocity of the two ions: $H^1+Li^7\rightarrow He^4+He^4$. Zeleney, Brasefield, Bock and Pollard (Z1) first observed this expected result with Li ions of 240 kv energy. This is equivalent to proton bombardment by 34 kv protons, which is known to be observable. Kinsey (K13) used Li ions of 1 MV (143 kv protons) and found large intensities of the expected 8 cm alphas, from all targets, due to the hydrogen impurities present, but found no indication of any other type of disintegration using Li and Be targets.

Whitmer and Pool (W14) reported alphas and neutrons from bombardment of Li by Li ions of 120 kv energy. As indicated by the results of Kinsey these observations are spurious.

Coates (C17) reports studies of the effects of bombardment of various targets with 2.8 MV Hg^{++} ions obtained with the "linear accelerator" of Sloan and Coates (S13). He finds characteristic x-rays in low intensities but no evidence for disintegration, as expected.

A search for radioactivity produced by high speed electrons is reported by Livingood and Snell (L25) to give negative results. Thomson and Saxton (T9) find no radioactivity to be produced by positrons.

Disintegrations by cosmic rays

The production of "showers" by certain components of the cosmic radiation has been studied by many investigators using cloud chambers, coincidences between three or more counters, etc. Certain arguments indicating that these showers may be due to nuclear disintegrations have been advanced, but the most probable explanation is that they are due to a multiplication of the complementary processes of pair production and radiative energy loss of the electrons and positrons formed (C1). In addition to showers, however, Anderson and Neddermeyer (A12) have observed many heavy particles, usually ejected from Pb by nonionizing radiation. These may well be interpreted as due to disintegration processes. It is known that high energy gamma-radiation does produce particle disintegration of certain of the lighter elements, and since in the secondary or tertiary products of the degradation of cosmic radiation there are large intensities of high energy gamma-rays, it is to be expected that similar disintegrations will be produced. Neutrons may also accompany cosmic radiation, possibly themselves produced through disintegration processes. It is to be expected that such heavy-particle tracks will be directly attributable to neutron recoils and to disintegrations.

§105. Induced Radioactivity

The phenomenon of induced radioactivity is among the most striking of recent developments in nuclear physics and has done much to clarify concepts regarding the constitution and stability of nuclei. At the time of its discovery in January of 1934 by Curie and Joliot (C54) the field of

TABLE LXVIa. *Evidence of γ-n reactions in heavy elements.*

Target						
Z	El	Prod. El	*T*	Raa particle	Ref.	Other evidence
1	H^2	H^1	—		F9	
4	Be^9	Be^8	—		C14	
8	O^{16}	O^{15}	2.1 min.	e^+	C16a	$T(O^{15})=126$ sec.
15	P^{31}	P^{30}	2–3 min.		B47b	$T(P^{30})=2.5$ min.
29	Cu^{63}	Cu^{62}	10.5 min.	e^+	B47b, C16a	Chem., Cu-*n*-2*n*, Co-α-*n* (*T*=10.5 min.)
30	Zn^{64}	Zn^{63}	38 min.	e^+	B47d, C16a	Zn-*n*-2*n*, Ni-α-*n* (*T*=40 min.)
31	Ga^{69}	Ga^{68}	60 min.	e^+	B47d, C16a	Ga-*n*-2*n*, Ga-*n*-γ (*T*=20 min., 23 hr.)
	Ga^{71}	Ga^{70}	20 min.	e^-		
35	Br^{79}	Br^{78}	5 min.		B47c, C16a, B47e	Chem. { Br-*n*-γ (*T*=18 min., 4.2 hr., 36 hr.); Br-*n*-2*n* (*T*=5 min.) }
	Br^{81}	Br^{80}	{ 18 min.; 4.2 hr. }	e^-		
42	Mo^{92}	Mo^{91}	17 min.		B47d	Mo-*n*-2*n* (*T*=21 min.)
47	Ag^{107}	Ag^{106}	24 min.		B47c, B47c, C16a	{ Ag-*n*-γ (*T*=22 sec., 2.3 min.); Ag-*n*-2*n* (*T*=25.5 min.) }
	Ag^{109}	Ag^{108}	2.3 min.	e^-		
49	In^{113}	In^{112}	1 min.		B47d,C16a	In-*n*-2*n* (*T*=1 min.)
51	Sb^{121}	Sb^{120}	13 min.	e^+	B47c,C16a	Sn-*d*-*n*, Sb-*n*-2*n* (*T*=13 min.)
52	$Te^{128,\ 130}$	$Te^{127,\ 129}$	60 min.		B47d	Te-*n*-γ (*T*=45 min.)
73	Ta^{181}	Ta^{180}	15 min.		B47d	Only one stable Ta isotope

TABLE LXVII. *Summary of γ-n type reaction.*

Target								
Z	El	Product El	*Q*(MV) (calc.)	*Q*(MV) (obs.)	Reference	Resonance levels (MV)	Excit. levels (MV)	Cross section in 10^{-24} cm²
1	H^2	H^1	− 2.20*	−2.20	F9			6.6×10^{-4}@2.62
2	He^4	He^3	−20.61					
3	Li^7	Li^6	− 7.12					
4	Be^9	Be^8	− 1.72	−1.6	C14			$>1\times10^{-4}$@2.62
	Be^9	$2He^4$	− 1.59					
5	B^{11}	B^{10}	−11.51					
6	C^{12}	C^{11}	−18.85					
7	N^{14}	N^{13}	−10.71					
8	O^{16}	O^{15}	−15.61					
9	F^{19}	F^{18}	− 4.0	Obs.	C16a			
10	Ne^{21}	Ne^{20}	− 7.54					
	Ne^{22}	Ne^{21}	− 9.32					
11	Na^{23}	Na^{22}	−12.2					
12	Mg^{25}	Mg^{24}	− 7.0					
	Mg^{26}	Mg^{25}	−11.9					
13	Al^{27}	Al^{26}	−11.1					
14	Si^{28}	Si^{27}	−14.4					
	Si^{29}	Si^{28}	− 8.4					
	Si^{30}	Si^{29}	−11.5					
15	P^{31}	P^{30}	−12.0	Obs.	B47b			

* *Q* (observed) used to calculate mass values.

nuclear physics was so well developed that many laboratories were immediately able to repeat and extend their observations. Sources to produce it and instruments capable of detecting it had been in existence for many years. One of the most amazing features was that such an easily observable phenomenon had remained undiscovered for so long.

Soon after the discovery of the positive electron the Curie-Joliots (C57) reported the observation of these new particles from targets bombarded by alphas. They suggested that in part they might be due to the "materialization" of gamma-radiation produced in the disintegration process, that is, the formation of electron pairs by the rays, which no doubt occurs. They also suggested at first that the positrons might result from the alternative formation of a neutron and positive electron instead of the usual proton, and as such would be a product of the disintegration itself. For a time it was thought that this represented a reaction yielding three particle products, and was reported as such (D4). Subsequent experiments showed, however, that the positrons resulted from a delayed disintegration of the product nucleus in each case, a phenomenon known as "artificial" or "induced" radioactivity.

The Curie-Joliots (C54, C58), in their first papers, reported the observation of such delayed positron emission from B, Mg and Al. The initial intensity of the radioactivity was found to decay in the exponential manner common for the naturally radioactive elements and could be accurately expressed by the same equation: $I=I_0e^{-\lambda t}$, where λ is the decay constant and related to the half-life T by the usual relation: $T=\log 2/\lambda=0.693/\lambda$. Chemical tests determining the radioactive element were performed in each case by the separation of the target element from the product element and the observation of the radioactivity only in the latter. For example, after boron nitride (BN) had been irradiated by alphas for some minutes it was heated with caustic soda, liberating all the nitrogen as gaseous ammonia whereupon the ammonia was found to have all the radioactivity. The half-life was the same as that found in other B targets, and was not found in nitrogen targets. This constitutes chemical proof of the "artificial" transformation of one element to another by disintegration and was the first such definite chemical evidence, although earlier physical evidence was so conclusive that the validity of such processes was unquestioned. The usefulness of having such chemical means of determining the products of nuclear reactions is evident, and they supply the most definite proof of the validity of reactions in doubtful cases. The nature of the ejected particles was determined by their ionization and by deflection in a magnetic field. With this evidence it was possible to write the disintegration equations with confidence.

The announcement of the discovery of this type of induced radioactivity was accompanied by the suggestion that the same radioactive elements might be produced in other ways; e.g., the activity produced in B: $B^{10}+He^4\rightarrow N^{13}+n^1$; $N^{13}\rightarrow C^{13}+e^+$ might also be produced by deuteron or proton bombardment following the reactions: $C^{12}+H^2\rightarrow N^{13}+n^1$ or $C^{13}+H^1\rightarrow N^{13}+n^1$. Three laboratories (H21, C24, C40) equipped for using deuteron projectiles looked for the effect and found the expected radioactivity in C, of the same half-life,[19] almost simultaneously.

A study of a wider range of targets under alpha-bombardment by other observers revealed certain new features. Activities were found in many elements, of the positron type, and supposedly from the same α-n type reaction. The radioactivity has been found to be easier to observe than the neutrons accompanying the reaction, and in many of the processes is the only evidence of their validity. A variation was the observation that Mg emits electrons as well as positrons (A2), identified as coming from a radioactive Al instead of the radioactive Si which yields the positrons. A simple consideration of the mass and charge values of the elements involved showed that this was due to a different primary reaction yielding protons (see §99A). involving a Mg^{25} target and yielding Al^{28}. A weaker electron activity from Mg (A2, E6) of different half-life is interpreted as coming from Mg^{26} and yielding radioactive Al^{29}.

Although the second alternative suggested by the Curie-Joliots for the production of N^{13} activity is now known to be energetically impossible, another method was found to be through the capture of protons in C^{12} without particle emission. It was first observed by Henderson, Livingston and Lawrence (H21) as an activity in C under proton bombardment with an identical half-life as the deuteron produced activity. Many similar reactions are now known to occur.

A new chapter to induced radioactivity was added by Fermi (F13), in the observation of radioactivity from neutron bombardment of many elements. For the three most important processes which are known to cause this effect (n-α, n-p and n-γ type reactions) the radioactivity was found to consist of electrons in all cases investigated. The active isotope was always heavier

[19] An early discrepancy in the half-life values of the N^{13} produced by alpha and by deuteron bombardment led Gamow (G4) to suggest the existence of "isomeric" nuclei and the "negative" proton. This discrepancy has been removed by the work of Ellis and Henderson (E5).

than the stable isotopes of the element, and in a few cases the same activity was produced by each of the three processes. Chemical analysis was used to prove the nature of the active product in many cases, but in others the identification of the half-life itself was sufficient. For example the observation of the same period activity in each of three consecutive elements was proof of the simple capture, proton emission and alpha-emission type reactions, in order, for the three elements.

The climax came with the discovery by the group at Rome (F15) of the "water effect," the increased activity of the simple capture process when the neutrons were slowed down to thermal energies by the interposition of hydrogenous materials. This was found to increase the yields in some cases by as much as 40 times, and to make definitely observable many reactions on the heavy elements. Nearly a hundred activities are found to be produced by this simple capture reaction with slow neutrons.

With the development of the magnetic accelerator to produce deuterons of as much as 5 to 6 MV, Lawrence and his collaborators observed radioactivity yielding electrons and in some instances positrons for elements as heavy as Bi and Pt. The high energies reduce the problem of penetration of the potential barrier. The large yields from the *d-p* type process have resulted in intensities of radioactivities comparable with the naturally radioactive sources, and suggest a possible utilization as chemical indicators and in biological and medical research.

Classifying the reactions resulting in induced radioactivity we find eleven type reactions producing electron activity and nine giving positrons. Those producing electrons are

$$\left.\begin{array}{lllll}
\alpha\text{-}p: & (Z-1)^{A-3} & +\mathrm{He}^4 & \rightarrow Z^A+\mathrm{H}^1 \\
p\text{-}n: & (Z-1)^{A} & +\mathrm{H}^1 & \rightarrow Z^A+n^1 \\
d\text{-}\alpha: & (Z+1)^{A+2} & +\mathrm{H}^2 & \rightarrow Z^A+\mathrm{He}^4 \\
d\text{-}p: & Z^{A-1} & +\mathrm{H}^2 & \rightarrow Z^A+\mathrm{H}^1 \\
d\text{-}p,\ \alpha: & (Z+2)^{A+3} & +\mathrm{H}^2 & \rightarrow Z^A+\mathrm{H}^1+\mathrm{He}^4 \\
d\text{-}n,\ \alpha: & (Z+1)^{A+3} & +\mathrm{H}^2 & \rightarrow Z^A+n^1+\mathrm{He}^4 \\
n\text{-}\alpha: & (Z+2)^{A+3} & +n^1 & \rightarrow Z^A+\mathrm{He}^4 \\
n\text{-}p: & (Z+1)^{A} & +n^1 & \rightarrow Z^A+\mathrm{H}^1 \\
n\text{-}\gamma: & Z^{A-1} & +n^1 & \rightarrow Z^A+h\nu \\
n\text{-}2n: & Z^{A+1} & +n^1 & \rightarrow Z^A+2n^1 \\
\gamma\text{-}n: & Z^{A+1} & +h\nu & \rightarrow Z^A+n^1
\end{array}\right\} Z^A\rightarrow(Z+1)^A+e^-.$$

Those producing positrons are[20]

$$\left.\begin{array}{lllll}
\alpha\text{-}p: & (Z-1)^{A-3} & +\mathrm{He}^4 & \rightarrow Z^A+\mathrm{H}^1 \\
\alpha\text{-}n: & (Z-2)^{A-3} & +\mathrm{He}^4 & \rightarrow Z^A+n^1 \\
p\text{-}\gamma: & (Z-1)^{A-1} & +\mathrm{H}^1 & \rightarrow Z^A+h\nu \\
p\text{-}n: & (Z-1)^{A} & +\mathrm{H}^1 & \rightarrow Z^A+n^1 \\
d\text{-}\alpha: & (Z+1)^{A+2} & +\mathrm{H}^2 & \rightarrow Z^A+\mathrm{He}^4 \\
d\text{-}p: & (Z)^{A-1} & +\mathrm{H}^2 & \rightarrow Z^A+\mathrm{H}^1 \\
d\text{-}n: & (Z-1)^{A-1} & +\mathrm{H}^2 & \rightarrow Z^A+n^1 \\
n\text{-}2n: & Z^{A+1} & +n^1 & \rightarrow Z^A+2n^1 \\
\gamma\text{-}n: & Z^{A+1} & +h\nu & \rightarrow Z^A+n^1
\end{array}\right\} Z^A\rightarrow(Z-1)^A+e^+.$$

The same radioactive isotope may be produced by several different disintegration processes, but in all such cases the properties of the activity such as the disintegration energy, the half-life and the mass of the isotope are identical within the limits of experimental error. Na^{24}, Al^{28} and Cu^{64}, for instance, are each produced by five of the reactions listed above.

The *α-n* type radioactivity (positrons) occurs more often than the *α-p* kind (electrons in general). This is understandable through a consideration of the types and relative abundance of the stable isotopes in the light element region. C^{13} and $\mathrm{O}^{17 \text{ and } 18}$ are rare and so the possible N^{16} and $\mathrm{F}^{20 \text{ and } 21}$ radioactive nuclei from the *α-p* reactions are not observed. Transformations of the same type from the more common isotopes of B^{10}, C^{12}, N^{14} and O^{16}, etc. would result in stable nuclei; N^{15} would give a stable O^{18}, while the C^{14} that results from B^{11} has a very long life and so is hard to observe. For heavier elements the higher potential barriers, both for the entering alpha and the escaping proton would be prohibitive. Mg is the only such element for which there is good proof of the reaction.

The number of excess neutrons (the isotopic number) in the stable nuclei increases more or less regularly with increasing charge, observable in the isotopic charts of Figs. 39 to 42. The unstable, radioactive nuclei, above or below this region of stability, will tend to return to the stable region with the emission of those particles which best accomplish this result. Isotopes heavier than the known stable ones are electron active, resulting in a product nucleus of the same mass but higher charge. Isotopes too light to be stable will emit positrons, thus lowering their charge. This results in either case in the produc-

[20] The type reaction *p-α* should also, in some instances, give e^+ radioactive products, but no instance has yet been observed.

tion of a nucleus having a more stable ratio of charge to mass. In most cases observed the unstable isotope is found to have a mass number differing by one unit from the extremes of the stable isotopes of that element. N^{14} and N^{15}, for example, are flanked by positron-active N^{13} and electron-active N^{16}. In a few instances the known unstable isotope differs by two or more mass units ($F^{17,18}$, $Al^{28,29}$, $Sc^{41,42,43,44}$ etc.). In many instances, a radioactive isotope fills the gap between two known stable ones. Ag, for instance, has two stable isotopes of mass numbers 107 and 109; the two water sensitive periods observed with neutrons (22 sec. and 2.3 min.) can occur only through the n-γ reaction and so represent radioactive Ag^{108} and Ag^{110}.

Electron emission is common for activities produced by the n-γ, n-p n-α, d-p and d-α processes, since these lead generally to an isotope heavier than those in the stable band of isotopes. In a few cases, however, positrons are observed; they come from lighter isotopes which have isobars of lower atomic number (cf. Pt^{193}). In several instances branching processes occur, yielding both electrons and positrons from the same radioactive isotope. These are found in all cases to have stable isobars of both lower and higher Z.

In several instances, the products of radioactive decay of the initial isotope are also radioactive, leading to subsequent decay processes, continuing until a stable product is reached. This involves alpha-radioactivity in the Bi, Th and U processes.

The measurement of the half-life periods of the decaying substances has proven to be relatively simple in most cases, and can be observed with an accuracy dependent on the intensity of the radiation and the sensitivity of the instruments. With thin walled Geiger counters the decreasing number of counts with time is directly observable. Ionization chambers equipped with sensitive electrometers or electroscopes are equally adaptable and somewhat more reliable for measuring long periods. Each radioactive element was found to have a characteristic half-life, and values range from 0.02 sec. (B^{12}) to more than 10 years (Be^{10}). These limits have both been obtained from deuteron induced processes, in which case the high voltages and high intensities have made these extreme cases just observable. The shortest lived element named was observed by counting the number of electron tracks in a cloud chamber arranged to be automatically expanded at short intervals following the instant when the deuteron beam was cut off. The longest lived one was found by studying the activity in a Be target which had been bombarded by deuterons for a period of 6 months as a neutron source. It is to be expected that with suitable arrangements of apparatus still longer and shorter half-lives will be found.

Certain discrepancies have appeared in measurements of the period, in particular that of N^{13} which was said to be 14 minutes by observers using alphas to produce it and to be 10.5 minutes by those using deuterons. These have been shown to be identical by Ellis and Henderson (E5), with a value of 11.0 ± 0.1 minutes. Such errors are in part experimental, as pointed out by Van Voorhis (V6), and the apparatus and potentials for collecting the ionization must be arranged to eliminate effects of ionization space charge. The methods of analysis of decay data taken in various ways are known, and Peierls (P5) gives methods for determining the most accurate values of the half-life.

In cases where contamination is suspected the measurement of the half-life provides an ideal method of distinguishing activities. In cases where an activity of known half-life is being searched for through a different reaction the observation of the expected period is considered satisfactory proof.

It can be seen that the production of radioactivity in elements provides a very sensitive method for the detection of a disintegration of those elements, even though the heavy particle produced in the reaction cannot be observed. Instruments for the detection of this activity can be made compact and extremely sensitive and can be isolated from the source of the particles used for producing the original disintegration. Observation of such an activity is just as complete proof of the validity of the primary reaction as the observation of the heavy particles, and sometimes more definite in interpretation, especially if supplemented by chemical separations.

The electron and positron radiations were

found by absorption and deflection techniques to have maximum energies varying from below 0.3 MV to 13 MV. They have been found, in all cases studied, to have the same general shape of energy distribution curve as the beta-particles from naturally radioactive elements except for the obvious influence of the different nuclear charge. This type of curve has puzzled physicists for years and has resulted in the neutrino hypothesis (see §39). One consequence is that the maximum energy must be used to balance the mass-energy equations.

The chief difficulty with the original Fermi (F12) theory of beta-decay (see Chap. VI), in which the existence of the neutrino was postulated[21] in order to conserve energy and angular momentum, is that it leads to an energy distribution curve essentially symmetrical about the half-energy value and distinctly different from the observed distributions, which show in all cases an optimum beta-energy of much less than this half-value and tailing off to high energies. The Konopinski and Uhlenbeck (K22) modification of this theory usually seems to give a much better fit to the observed distribution curves. The neutrino mass is put equal to zero in the theoretical formulae. Kurie, Richardson and Paxton (K32) have indicated how the K-U theory can be used to obtain a value for the theoretical energy maximum from experimental data, and such a value has been obtained from many of the observed distributions. On the other hand, in those few cases in which it is possible to predict the energy of the beta-decay from data on heavy particle reactions (*viz.* C^{12}-*d*-*n*, O^{16}-*d*-*n*, Al^{27}-*d*-*p*), the visually extrapolated limit has been found to fit the data better than the K-U value. The fact that the shape of the distribution curves is checked by the theory over wide ranges suggests, however, that the K-U limit has some significance, and it is recorded in the tabulations to follow wherever observed. For calculation of atomic masses we

TABLE LXVIII. *Forbidden beta-ray transitions.*

Raa El	Product	Beta En. (MV)	Excit. En. (MV)	Total En. (MV)	Evidence
Li^8	Be^8	12.0	3.5–4.0	15.5–16.0	Li^7-*d*-*p*; $Li^8 \rightarrow e$- (L24)
B^{12}	C^{12}	12.0	1.7–2.7	13.7–14.7	B^{11}-*d*-*p*
N^{16}	O^{16}	6.0	~ 5	~ 11	F^{19}-*n*-α
F^{20}	Ne^{20}	5.0	4.6	9.6	F^{19}-*d*-*p*
Na^{24}	Mg^{24}	1.7	2.9	4.6	Na^{23}-*d*-*p*
Al^{28}	Si^{28}	3.3	$(\gamma = 2.3)$	3.3	Al^{27}-*d*-*p*
P^{32}	S^{32}	1.69	—	1.69	Raa. En.
K^{40}	Ca^{40}	0.7	—	0.7	Nat. Raa.

have decided to use the visually extrapolated limit. This procedure gives satisfactory results only if the experimental data are based on the observation of a sufficiently large number of electrons. Otherwise, the visual limit is apt to be considerably too low.

In several of the radioactive processes gamma-rays are also observed to be emitted with the betas. These are much too strong in intensity to be due to conversion of the betas themselves, and appear to be in general monoenergetic. They must be considered as coincident with the beta-emission and as representing a transition of the residual nucleus from an excited state in which it is left after the beta-emission. In some cases (Li^8, B^{12}, N^{16}, F^{20} and Na^{24}) the beta-emission leads practically always to an excited state of the residual nucleus (Be^8, etc.) (see Table LXVIII) so that the gamma-energy is additive to the beta-maximum. This will result when the beta-transition to the ground state is highly forbidden. This is indicated experimentally by a simple beta-ray distribution and measurements indicating that the number of gamma-quanta equals the number of radioactive electrons. In other cases (Cl^{38}, A^{41}, K^{42}, Mn^{56}, and As^{76}) the beta-decay leads sometimes to the ground state and sometimes to an excited state. This gives rise to a complex beta-spectrum (K33, B62) consisting of several "groups," analogous to the heavy particle groups in ordinary nuclear reactions. The absolute maximum of the beta-spectrum corresponds then to the total energy evolution while the gamma-ray energy should be correlated with the difference between the maximum energies of the beta-ray groups. The observed shape of beta-ray distribution curves does not lend itself to analysis into groups, and in this

[21] It should be mentioned at this point that the neutrino suggested by Pauli and used by Fermi and by Konopinski and Uhlenbeck in their theories has not been observed. The properties attributed to it by the theories are essentially zero mass, zero charge, spin $\frac{1}{2}$ and zero or very small magnetic moment. The only observable property is its momentum. It may be hoped that studies of the recoil of light nuclei in beta disintegrations will give more direct evidence about the neutrino.

instance the use of the Konopinski-Uhlenbeck theory makes it possible to separate them even though it may not predict the correct maximum energies.

Of considerable interest because of their theoretical importance are the data on the energies involved in the beta-ray transitions from nuclei of the "4*n* type." In many of these processes the beta-ray does not represent the full energy, due probably to forbidden transitions. Estimates of the total energy available can be made from analysis of the primary reactions; the difference is then entered as excitation energy. In a few instances this gamma-radiation is observed and measured and found to decay with the half-life of the beta-decay. In other instances it is inferred from other evidence.

The total energy is found to decrease regularly from the light to the heavy nuclei. For Al^{28} and the two heavier elements it seems probable that the gamma-radiation is alternative rather than consecutive to the beta-ray, and is so indicated. The similarity in nuclei of this type and the general trend of the total energy evolution are to be expected theoretically (see Part A, §30). A thorough discussion of the theory is given by Wigner (W18).

A few generalizations can be drawn concerning the methods useful in the analysis of the evidence from the various type reactions to give specific information about the radioactivities involved. These are more readily followed by referring to the stable isotopes and their abundance (Table LXXVII) and to the isotopic charts (Figs. 39 to 42).

Activities showing the "water effect" with neutrons (slow neutrons) give generally isotopes of the target element of one unit higher mass, and are usually electron active. Deuteron bombardment resulting in the same chemical element as the target (the *d-p* reaction) will in general duplicate the slow neutron activities. Whenever positron activities are obtained by either of these two reactions it indicates reactions in light isotopes of the elements. *d-n* reactions lead to an element of one unit higher charge and in general are positron activities; *d-α* reactions give elements of lower charge; *n-p* and *n-α* reactions result in elements of one and two units lower *Z*, respectively. So observation of identical periods produced in these various ways from adjacent chemical elements will usually result in the positive identification of the isotope responsible for the activity. The results of proton and gamma-ray activation are of even greater value. If a target element has only two stable isotopes and two periods are observed with slow neutrons and two periods with gamma-rays, one of which is common, this common activity will belong to the isotope between the two stable isotopes. The other neutron period is one unit heavier than the heavy stable isotope and the other gamma-period one unit lighter than the lighter stable isotope. In this way the 2.3 min. period of Ag is identified as Ag^{108}, the 18 min. period of Br is Br^{80}, and the 20 min. period of Ga is Ga^{70}. In other cases the observation of a period different from any observed neutron period specifies an isotope other than those which can be produced by neutron capture (e.g., In).

In Table LXIX are collected the best data on the half-lives, disintegration energies and other features of all the radioactive elements produced in nuclear disintegrations for which these quantities have been measured. The natural radioactivities are not included. In many cases the evidence is sufficient to lead to a unique interpretation; in others the particular isotope responsible cannot be determined. In order to indicate the certainty each reaction is marked with a letter in the column headed "class" according to the following code:

A—Isotope certain
B—Isotope probable
C—One of few isotopes
D—Element certain
E—Element probable
F—Insufficient evidence

For a large share of the radioactive elements the evidence given in the table (chiefly through the indicated primary reactions) is sufficient to justify classification, and such cases are not discussed in detail. In other instances the interpretation is based upon more varied and uncertain evidence and these cases are discussed individually in the following pages.

The assignments in Table LXIX requiring a more involved justification are discussed individually below.

TABLE LXIX. *Induced radioactivities.*

Raa Element Z	Raa Element A	Class	Raa Particle	Half-life T	Raa Energy Obs. (MV)	Raa Energy K−U (MV)	Raa γ's (MV)	Produced by	Observers
2	He^{6}	A	e^-	0.7 sec.	3.7			Be-n-α, (Li-n-p)	B24, N1, K20
3	Li^{8}	A	e^-	0.88 sec.	12			Li-d-p, (Li-n-γ)	B5a, L24, C48
4	Be^{10}	A	e^-	≥10 yr.	<0.3(?)			Be-d-p	Mc7
5	B^{9}	F	e^+	——	0.3(?)			Li-α-n	M14
	B^{12}	A	e^-	0.02 sec.	12	13.0		B-d-p	C46, F28, B5a
6	C^{11}	A	e^+	20.5 min.	1.15	1.3		B-d-n, B-p-γ, C-n-2n	C39, F28, P11d
	C^{14}	F	e^-	∼3 mo.	0.3(?)			(N-n-p) (C-d-p)(B-α-p)	Mc7, M16, B39
7	N^{13}	A	e^+	11.0 min.	1.25	1.45		B-α-n, N-n-2n C-d-n, C-p-γ	E5, H4, F28, P11d
	N^{16}	A	e^-	8.4 sec.	6.0	6.5		N-d-p, F-n-α, O-n-p	F28, N1, C16a
8	O^{15}	A	e^+	126 sec.	1.7	2.0		N-d-n, O-γ-n, O-n-2n	Mc6, F28, P11d
	O^{19}	A	e^-	31 sec.				F-n-p	N1
9	F^{17}	A	e^+	70 sec.	2.1	2.4		O-d-n, N-α-n, O-p-γ, F-n-3n	N6, W12, K33, D28, P11c
	F^{18}	A	e^+	112 min.				Ne-d-α, (O-p-γ), O-p-n, F-n-2n	S16, D28, P11d
	F^{20}	A	e^-	12 sec.	5.0	5.9		Na-n-α F-d-p, F-n-γ	C48, N1, F28
10	Ne^{23}	A	e^-	33 sec.				Mg-n-α, Na-n-p	A7, N1
11	Na^{22}	A	e^+	≃6 mo.	0.4(?)			Mg-d-α, F-α-n	L2, F32
	Na^{24}	A	e^-	14.8 hr.	1.7	1.95	0.95 1.93 3.08	Na-d-p, Na-n-γ Al-n-α, Mg-n-p, Mg-d-α	A7, H24, V6 K33, R7, L15
12	Mg^{27}	A	e^-	10.2 min.	2.05(?)		1.3	Mg-d-p Al-n-p, Mg-n-γ	A7, H24
13	Al^{26}	A	e^+	7.0 sec.	1.8(?)			Na-α-n	F31
	Al^{28}	A	e^-	2.36 min.	3.3		2.3	Si-n-p, Al-d-p, Al-n-γ Mg-α-p, P-n-α	E10, A7, C34
	Al^{29}	B	e^-	11.0 min.	>3.3(?)			Mg-α-p	E10
14	Si^{27}	A	e^+	6.7 min.	2.0			Mg-α-n, Si-n-2n	E10, F1, P11d
	Si^{31}	A	e^-	170 min.	1.8	2.05		Si-d-p, P-n-p, Si-n-γ	A7, N5, K33 N7b, P11d
	Si	F	e^+	11 min.				(Si-n-3n)	P11d
15	P^{30}	A	e^+	2.5 min.	3.6			Al-α-n, S-d-α, P-n-2n Si-p-n, P-γ-n	C54, A1, R7a B4a, P11d
	P^{32}	A	e^-	14.5 da.	1.69	2.15(?)		P-d-p, S-d-α P-n-γ, S-n-p, Cl-n-α	N5, K33, P11d A7, S12, L36, N7b
16	S^{31}	F	e^+	26 min.				S-n-2n	P11d
	S^{35}	A	e^-	80 da.				Cl-n-p	A14
17	Cl^{34}	A	e^+	33 min.	1.8(?)			P-α-n, S-d-n, Cl-n-2n	F31, S1, P11d
	Cl^{38}	B	e^-	37.0 min.	4.8	1.5, 6.1	2.0, 2.5	Cl-d-p, Cl-n-γ, K-n-α	A3, R6, K33, H39a
18	A^{39}	F	e^-	4 min.				(K-n-p)	P11d
	A^{41}	A	e^-	110 min.	2.7	1.5, ∼5	1.37	A-d-p, K-n-p, A-n-γ	S15, R7, K33, H39a
19	K^{38}	A	e^+	7.7 min.				Cl-α-n, Ca-d-α, K-n-2n	H24a, H39a, P11d
	K^{42}	A	e^-	12.2 hr.	3.5	1.4, 4.4		K-n-γ, Ca-n-p K-d-p, Sc-n-α	W1, A7, K33, H39a
20	Ca^{39}	B	e^+	4.5 min.				Ca-n-2n	P11d
	Ca^{45}	A	e^-	2.4 hr.	1.9(?)		<0.5	Ca-n-γ, Ca-d-p Ti-n-α	H29, W1
21	Sc^{41}	A	e^+	53 min.				Ca-d-n	W1
	Sc^{42}	A	e^+	4.1 hr.	1.5(?)			K-α-n	A7, Z3, W1
	Sc^{43}	A	e^+	4.0 hr.	0.4(?)			Ca-α-p, Ca-d-n, Sc-n-3n	F32, W1, P11c
	Sc^{44}	A	e^+	52 hr.				Ca-d-n, K-α-n, Ca-p-n, Sc-n-2n	W1, B4a, P11d
	Sc^{46}	F	e^-(?)	—long—	low			Sc-n-γ	H32
	Sc	F	e^-	1.7 hr. 28 hr.				Ti-n-p, (V-n-α)	W1a, P11d
22	Ti^{51}	A	e^-	2.8 min.				Ti-n-γ, Ti-d-p	A7, W1a
23	V$^{49, 50}$	C	?	35 min.				Ti-α-p, Ti-d-n	W1a
	V^{52}	A	e^-	3.75 min.				Mn-n-α, V-n-γ, Cr-n-p	A7, P11d
	V	F	e^+	16 da.				Ti-d-n	W1a
	V	F		1.8 da.				V-n-?	P11d
24	Cr	F	e^-	1.7 hr.				Cr-n-?	P11d
25	Mn^{52}	B	e^+	5 da.				Fe-d-α	L27a
	Mn^{53}	A	e^+	40 min.				Cr-p-n, Cr-d-n	B4a, L27a
	Mn^{54}	B	e^-, e^+(?)	21 min.			1.65	Fe-d-α	D3a, M20b
	Mn^{56}	A	e^-	2.5 hr.	2.8	(1.2, 2.9)		Fe-n-p, (Mn-d-p?) Co-n-α, Mn-n-γ, Cr-α-p	A7, G1, B62, H24b, P11d
	Mn	F	e^-	*ca.* mo.				(Fe-d-α)	L27a
26	Fe^{53}	B	e^+	8.9 min.				Cr-α-n	H24b
	Fe^{55}	D	e^+(?)	91 min.				Mn-p-n	B4a
	Fe^{59}	A	e^-	40 da.				Co-n-p, Fe-d-p	A15, L27a
27	Co^{55}	A	e^+	18.2 hr.				Fe-d-n	D3a
	Co^{58}	B	e^-	11 min.				Ni-n-p, Co-n-γ	R15
	Co^{60}	B	e^-	∼ yr.	0.15?		2.0	Co-n-γ, Co-d-p	S2, T10
	Co	D	e^+, e^-(?)	100–200 da.				Fe-d-n	L27a
28	Ni^{57}	D		2 hr.				Ni-n-2n(?)	P11d
	Ni^{59}	B	e^+	3 hr.				Ni-n-γ	N3
	Ni^{63}	F	e^-	130 da.				Cu-d-α	L26
	Ni^{65}	B	e^-	160 min.				Zn-n-α, Cu-n-p	M1, O½
	Ni	F		6 da.				Ni-n-?	P11d
29	Cu^{61}	B	e^+	3.4 hr.	0.9			Ni-p-n, Ni-α-p, Ni-d-n, Cu-n-3n	B4a, T10a, R7b, P11c
	Cu^{62}	B	e^+	10.5 min.				Cu-n-2n, Cu-γ-n, Co-α-n	H33, B47b, R7b
	Cu^{64}	A	e^-, e^+	12.8 hr.	0.7	e^-=0.83 e^+=0.79	No	Zn-n-p, Ni-p-n Cu-d-p, Zn-d-α, Cu-n-γ	V7, L26, B21, B4a
	Cu^{66}	A	e^-	5 min.				Zn-n-p, Cu-n-γ, Ga-n-α	A7, F16, C16a

TABLE LXIX.—*Continued.*

Raa Element Z	Raa Element A	Class	Raa Particle	Half-life T	Raa Energy Obs. (MV)	Raa Energy K−U (MV)	Raa γ's (MV)	Produced by	Observers
30	Zn^{63}	*B*	e^{+}	37 min.				Zn-*n*-2*n*, Ni-α-*n*, Zn-γ-*n*	H33, R7c, P11d, B47d
	Zn^{65}	*B*	e^{+}(?)	1 hr.				Zn-*d*-*p*	L26
	Zn^{69}	*B*	e^{-}	100 hr.				Zn-*d*-*p*	L26
31	Ga^{66}	*A*	e^{+}	9.4 hr.				Cu-α-*n*	H24b
	Ga^{68}	*A*	e^{+}	60 min.				Ga-γ-*n*, Cu-α-*n*, Ga-*n*-2*n*	B47d, H24b, P11d
	Ga^{70}	*B*	e^{-}	20 min.				Ga-*n*-γ, Ga-γ-*n*, Zn-*p*-*n*	A7, B47d, B4a
	Ga^{72}	*B*	e^{-}	23 hr.				Ga-*n*-γ, (Zn-*d*-γ), Zn-*p*-*n*, Ge-*n*-*p*	A7, L26, B4a, P11d
	Ga	*F*	e^{+}	1.7 hr.				Ga-*n*-?	P11d
32	Ge^{75}	*D*	e^{-}	30 min.				Ge-*n*-γ	A7, S25
	Ge	*F*	e^{-}	1.3 hr.				Ge-*n*-?	P11d
33	As^{74}	*E*		13 da.				As-*n*-2*n*	P11d
	As^{76}	*A*	e^{-}	26 hr.	1.5(?)	1.1, 3.4	Yes	As-*n*-γ, As-*d*-*p*	{A7, T10 B62
34	Se	*F*	e^{+}(?)	1.3 min., 113 min.				As-*p*-*n*	B4a
	Se^{79}	*D*	e^{-}	56 min.				Se-*n*-2*n*	H33b
	Se^{83}	*B*	e^{-}	10–20 min.				Se-*d*-*p*, Se-*n*-γ	S16a, A7
35	Br^{78}	*A*	e^{+}	6.3 min.				As-α-*n*, Br-γ-*n*, Br-*n*-2*n*	S16a, B47c, C16a, P11d
	Br^{80}	*A*	e^{-}	18 min.	2.00			{Br-*n*-γ, Se-*p*-*n* Br-*d*-*p*, Br-γ-*n*	A7, A1, B4a, B47b
	Br^{80}	*A*	e^{-}	4.2 hr.	2.05			{Br-*n*-γ, Se-*p*-*n* Br-*d*-*p*, Br-γ-*n*	A7, A1, B4a, B47e
	Br^{82}	*A*	e^{-}	36 hr.	0.85		Yes	{Br-*n*-γ, Se-*p*-*n* Br-*d*-*p*, (Se-*d*-2*n*)	{K34, A1, J2 B4a, S16a
	Br^{83}	*B*	e^{-}?	2.5 hr.				Se-*d*-*n*	S16a
37	Rb^{84}	*E*	e^{-}	22 hr.				Rb-*n*-2*n*	P11d
	Rb^{86}	*B*	e^{-}	11 min.				Rb-*n*-γ, γ-*n*-α	F13, P11d
38	Sr^{85}	*B*	e^{+}	3 hr.				(Sr-*n*-2*n*)	P11d
	Sr	*F*	e^{-}	18 min.				Sr-*n*-?	P11d
39	Y^{90}	*A*	e^{-}	70 hr.				Y-*n*-γ	H31, P11d
	Y	*F*	{e^{-}	1.2 hr. 6.5 hr.				Y-*n*-?	P11d
40	$Zr^{93, 95}$	*D*	e^{-}	40 hr.				Zr-*n*-γ	H29
	Zr	*F*		{10 min. 5 hr.				Zr-*n*-?	P11d
41	Cb	*F*		{7.3 min. 3.8 da.				Cb-*n*-?	P11d
42	Mo^{91}	*D*	e^{-}	17 min.				Mo-γ-*n*, Mo-*n*-2*n*	B47d, H33b
	Mo^{93}	*B*	e^{+}	25 min.				Mo-*n*-γ	A7
	$Mo^{99, +}$	*D*	e^{-}(?)	36 hr.				Mo-*n*-γ	A7, Mc1
	Mo	*F*	e^{-}	5 da				Mo-*n*-?	P11d
43	Ma	*F*	e^{-}	{4 hr. 46 da.				Ru-*d*-α	L26
	Ma	*F*	e^{+}(?)	{0.5 min. 31 min.				Mo-*p*-*n*	B4a
44	$Ru^{97, 103}$	*F*	e^{-}(?)	{11 hr. (39 hr.) 75 hr. (11 da.)				Ru-*n*-γ, Ru-*d*-*p*	K35, L26
	Ru^{105}	*F*	e^{-}	40 sec.→100 sec.				Ru-*n*-γ	K35
	Ru	*F*		{24 min. 3.6 hr.				Ru-*n*-?	P11d
45	Rh^{102}	*B*	e^{-}	3.9 min.				Rh-*n*-γ	A7, S3
	Rh^{104}	*B*	e^{-}	44 sec.	2.5	2.8		Rh-*n*-γ	A7, A1, G1
	Rh^{105}	*F*	e^{-}	100 sec.				Ru-*n*-γ-β	K35
	Rh	*F*	e^{-}	1.1 hr.				Rh-*n*-?	P11d
46	Pd	*D*	e^{-}(?)	{3 min. 15 min. 12 hr. 60 hr.				Pd-*n*-γ, Pd-*d*-*p*	A7, K36, K25
47	Ag^{104}	*B*	e^{-}	7.5 da.				Pd-*d*-*n*, Ag-*n*-2*n*	K25, P11d
	Ag^{106}	*A*	e^{+}	25.5 min.				Pd-*d*-*n*, Ag-γ-*n*, Ag-*n*-2*n*	K25, B47c, P11d
	Ag^{108}	*A*	e^{-}	2.3 min.	2.8?			Ag-*n*-γ, Ag-γ-*n*	A7, G1, B47c
	Ag^{110}	*A*	e^{-}	22 sec.	2 6	2.8		Ag-*n*-γ	A7, G1, N2
48	Cd^{109}	*E*	e^{+}	33 min.				Cd-*n*-2*n*	P11d, H33b
	Cd^{115}	*B*	e^{-}	4.3 hr.				Cd-*d*-*p*, Cd-*n*-γ	C36a, M20a
	Cd^{117}	*B*	e^{-}	58 hr.				Cd-*d*-*p*, Cd-*n*-γ	C36a, M20a
49	In^{111}	*D*	e^{-}	2 mo.				(In-*n*-3*n*)	P11d
	In^{112}	*B*		1 min.				In-γ-*n*, In-*n*-2*n*, Cd-*p*-*n*	B47d, P11d
	{In^{114}	{*C*	e^{-}	13 sec.	3.0	3.2		In-*n*-γ, (Cd-*p*-*n*)	A7, G1, B4a
	In^{116}}	*C*	e^{-}	54 min.	1.1	1.3	1.39	In-*n*-γ, (Cd-*p*-*n*), Sn-*n*-*p*	{A7, G1, B4a, M20b, P11b, P11d
		C	e^{-}	3.5 hr.				(In-*n*-γ)	S28
	In^{117}	*A*	e^{-}	2.3 hr.	1.0			Cd-*d*-*p*, Cd-e^{-}, (Cd-*p*-*n*)	C36a
	In	*D*	e^{-}	3 hr.				In-*n*-γ	B47f
50	Sn^{113}	*B*	e^{+}(?)	14 min.				In-*p*-*n*, Sn-*n*-γ, Sn-*d*-*p*	B4a, N3
	Sn^{121}	*A*	e^{-}	28 hr.				Sn-*d*-*p*, Sb-*d*-α	L27, L26
	Sn^{123}	*D*	e^{-}	∼ mo.				Sn-*d*-*p*	L27
	Sn^{125}	*D*	e^{-}	8 min.→18 min.?				Sn-*n*-γ, Sn-*d*-*p*	N3, L27
51	Sb^{120}	*A*	e^{+}	13 min.				Sn-*d*-*n*, Sb-γ-*n*, Sb-*n*-2*n*	L27, B47c, P11d
	Sb^{122}	*C*	e^{-}	2.5 da.				Sb-*n*-γ, Sb-*d*-*p*	A7, L26
	Sb^{124}	*C*	e^{-}	60 da.				Sb-*d*-*p*, Sb-*n*-γ	L26, P11d
	Sb	*D*	e^{-}	112 da.				Sn-*d*-*n*	L27
52	$Te^{127, 129}$	*C*	e^{-}	60 min.				Te-*n*-γ, Te-γ-*n*, Te-*n*-2*n*	A7, B47d, P11d, H33b
	Te	*F*		30 da.				Te-*n*-?	P11d
53	I^{128}	*A*	e^{-}	25 min.	2.10			I-*n*-γ	A7, A11, A1, P11d
54	Xe	*F*	e^{-}?	3 min.				Ba-*n*-α	A7
55	Cs^{124}	*A*	e^{-}	1.5 hr.				Cs-*n*-γ	A7, L3

TABLE LXIX.—*Continued.*

Raa Element					Raa Energy				
Z	A	Class	Raa Particle	Half-life T	Obs. (MV)	K−U (MV)	Raa γ's (MV)	Produced by	Observers
56	Ba^{139}	A	e^-	85.6 min.				Ba-n-γ, Ba-d-p	A7, A8, P11a, d
57	La^{138}	E		2.2 hr.				La-n-$2n$	P11d
	La^{140}	A	e^-	31 hr.				La-n-γ, La-d-p	M7, H31, P11a
58	Ce	F	e^-?	2.4 hr.				Ce-d-p, Nd-n-α	P11a, P11d
	Ce	F	e^-	40 min.				Ce-n-?	P11d
59	Pr^{140}	E	e^+	3 min.				Pr-n-$2n$	P11d
	Pr^{142}	D	e^-	19 hr.				Pr-n-γ, Nd-n-p	A7, P11d
60	Nd	D	e^-	1 hr.				Nd-n-γ	A7, H30
62	Sm	D	e^-	{40 min. 2 da.				Sm-n-γ	A7, H30, H31a
63	Eu	D	e^-	9.2 hr.	2.6(?)			Eu-n-γ	S25, M7, N2
64	Gd	D	e^-	8 hr.				Gd-n-γ	A7, H30
	Gd	F	e^-	19 hr.				Nd-n-?	P11d
65	Tb^{160}	A	e^-	3.9 hr.				Tb-n-γ	H30, M7
66	Dy^{165}	A	e^-	2.5 hr.	1.2	1.4		Dy-n-γ	H30, M7, G1
67	Ho^{166}	A	e^-	35 hr.				Ho-n-γ	M7, H31, H31a
68	Er	D	e^-	{5.8 min. 12 hr.				Er-n-γ	Mc3, H31a, S25
69	Tm^{170}	A	e^-	8 mo.				Tm-n-γ	C61
70	Yb	D	e^-	3.5 hr.				Yb-n-γ	H30, M7
71	Lu^{176}	D	e^-	{3.6 hr. 6 da.				Lu-n-γ	M7, Mc3, H31
72	Hf^{181}	A	e^-	~ mo.				Hf-n-γ	H29
73	Ta^{180}	A	e^+(?)	15 min.				Ta-γ-n	B47d
	Ta^{182}	A	e^-	200 da.				Ta-n-γ	F27
	Ta	F		9.1 hr.				Ta-n-?	P11d
74	W$^{185, 187}$	D	e^-	23 hr.				W-n-γ	A7, Mc1
75	Re$^{186, 188}$	D	e^-	{20 hr. 85 hr.				Re-n-γ	A7, K36
76	Os$^{191, 193}$	D	e^-	40 hr.				Os-n-γ	K36
77	Ir	C	e^-	28 min.	1.1		0.4	Ir-n-γ; Pt-d-n, α	S20, C35
	Ir	C	e^-	{3 da. 19 hr.? 2 mo.	2.20?			{Ir-n-γ; Au-d-p, α Pt-n-p	S20, C36, A1, F27 P11d
	Ir	C	e^-	8.5 hr.				Pt-d-n, α	C35
78	Pt^{193}	B	e^+	49 min.				Pt-d-p, Pt-n-γ	A7, C35
	Pt^{197}	B	e^-	14.5 hr.				Pt-d-p	C35
	Pt	F	e^-	1.8 hr.				Pt-n-?	P11d
79	Au^{196}	E		17 min.				Au-n-$2n$	P11d
	Au^{198}	A	e^-	2.7 da.	1.15			Au-n-γ, Au-d-p	A7, A1, C36
80	Hg^{197}	E	e^-	45 min.				Hg-n-$2n$	H33b
	Hg^{205}	C	e^-	40 hr.				Hg-n-γ	A13
81	Tl$^{204, 206}$	C	e^-	{4 min. 1.3 hr.				Tl-n-γ	P15
82	Pb^{209}	D	e^-	8.6 da.				Pb-d-p	T11
	Pb	F	e^-	{5 min. 1.5 hr.				Pb-n-?	P11d
83	Bi^{210}	A	$e^- \rightarrow \alpha$	5 da.				Bi-d-p	L26
88	Ra^{229}	D	e^-	1 min.				Th-n-α	C60
89	Ac^{229}	D	e^-	{15 min. 3.5 hr.				Ra^{229}-β	C60
90	Th^{233}	D	e^-	25 min.				Th-n-γ	C60
	Th^{235}	E	e^-	4 min.				U-n-α	M15
	Th	F	e^-	{5 min. 1.4 hr.				Th-n-?	P11d
91	Pa^{235}	E	e^-	very short				Th^{235}-β	M15
	Pa^{233}	D	e^-	2 min.				Th^{233}-β(?)	C60
92	U^{239}	E	e^-	15 sec.				U-n-γ	M15, A7
	U^{237}	E	e^-	40 sec.				U-n-$2n$	M15
	U^{236}	D	e^-	24 min.				U-n-γ	M15
	U	F	e^-	{4 hr. 13 hr.				U-n-?	P11d
93	93$^{227, 239}$	D	e^-	{2.2 min. 16 min.				U-β	M15
94	94$^{237, 239}$	E	e^-	{59 min. 6–12 hr.				93-β	M15
95} 96}		F	e^-	{3 da. 3 hr.				94-β	M15

N^{16}; O^{19}; F^{20}

A discussion of the evidence whereby the activities of these three isotopes were identified was given in §102*A* under the F^{19}-n-α reaction, in order to justify the primary process. The evidence is largely in a report by Nahmias and Walen (N1) who found a water-sensitive period (F^{20}) with a half-life only slightly greater than a nonsensitive period (8.4 sec.) which is identified with N^{16} by considerations of relative probabilities of the n-α and n-p processes in F. The remaining weak activity with a longer period (31 sec. must then be due to O^{19}. The interpretation is verified by the d-p reactions on N and F which give activities of nearly equal half-lives (F28).

(Si)

Pool, Cork and Thornton (P11d) report that fast neutrons produce a 11 min. positron activity in Si besides the 6 min. period characteristic of Si^{27}. A 11 min. electron activity could be understood as due to a Si-*n*-*p* reaction producing Al^{29} or a Si-*n*-α process giving Mg^{27}, but the reported positron period cannot be ascribed to a definite element at present. It may be Si^{26} from an *n*-3*n* process.

Cl^{38}

It seems probable that the Cl^{38} rather than the Cl^{36} isotope should be responsible for the observed activity which has a rather short period and high energy because Cl^{36} should have a very low energy and long period if it is similar to the analogous nuclei P^{32} and K^{40}.

Ca^{45}

The 2.4 hr. period found by Walke (W1) with deuterons on Ca and neutrons on Ti is probably the same as the 4 hr. period reported by Hevesy and Levy (H29) from slow neutron bombardment of Ca. Walke found this period in the Ca fraction of the chemical separation, specifying Ca-*d*-*p* and Ti-*n*-α reactions. The radioactive particles were electrons, thus determining Ca^{45} as the isotope responsible, since the only alternative, Ca^{41}, would be positron active.

$Sc^{41, 42, 43, 44, 46, 48}$

Walke (Wl) has observed three periods chemically separable as Sc from Ca after deuteron bombardment. The periods were 53 min., 4.0 hr. and 52 hr. From the known isotopes of Ca and Sc these must be Sc^{41}, Sc^{43} and Sc^{44} if the reaction is Ca-*d*-*n*. The radioactive particles were found to be positrons in all cases. The 4.0 hr. period is certainly the positron activity of 4.4 hr. period obtained by Frisch (F32) from the Ca-α-*p* reaction and Sc^{43} is the only possibility for this common period. In a note added in proof Walke reports the production of the 52 hr. period from alpha-particle bombardment of K, defining it as due to Ca^{44}. The 53 min. period must then be Sc^{41}. Sc^{44} and Sc^{43} are also produced (P11c) by very fast neutrons on Sc. Zyw (Z3) first found a 3 hr. positron activity from K-α-*n*; Walke, using He^{++} ions accelerated to 11 MV in the Berkeley cyclotron, has repeated the experiment, naming 4.1 hr. for the period. Since Sc^{44} is excluded by the results discussed above this must be due to Sc^{42}. A slow neutron produced activity in Sc reported by Hevesy (H32) to have a long period and low energy can only be Sc^{46} unless due to a contamination. Finally a fast neutron period of 16–28 hr. in Ti (W1a, P11d) may be either a *n*-α or *n*-*p* reaction but seems most probably a *n*-*p* reaction leading to Sc^{48}.

V, Cr

It is as yet impossible to assign to definite elements a 1.8 day activity found in V and a 1.7 hour electron period found in Cr (P11d) with very fast neutrons.

$Co^{58, 60}$

The evidence given by Rotblat (R15) for a 20 min. period (later corrected to 11 min. (L27a)) from Ni under neutron bombardment which was not water sensitive (Ni-*n*-*p*), and the same period from Co which was water sensitive (Co-*n*-γ), is sufficient to identify this period with a radioactive Co even though chemical tests were not made. The intensity was too low to account for the large neutron absorption coefficient. This feature was cleared up by Sampson, Ridenour and Bleakney (S2) who found mass-spectroscopic evidence for a stable Co^{57} isotope (0.2 percent abundance) and a second neutron produced activity of some years period. The long period has a much greater total intensity than the short one and so is ascribed to Co^{60}, while the 11 min. period is due to the rare isotope and is Co^{58}. The long period should also be produced by the Cu^{63}-*n*-α reaction if this interpretation is correct.

$Ni^{59, 63}$

Rotblat (R15) first reported a water sensitive neutron induced activity in Ni, measured by Naidu (N3) to have a 3 hr. half-life. Thornton (T10a) finds an activity with a similar period (3.4 hr.) produced by deuterons on Ni, but chemical separations identify it as Cu (Cu^{61}). The same period (3.6 hr.) is found by Thornton from Co under deuteron bombardment. Unless due to a Ni impurity which is unlikely, or a coincidence of half-life values, it would require an unusual reaction to explain it, possibly; $Co^{59}+H^2 \rightarrow Ni^{59}+2n^1$,

A period of 130 days observed by Livingood (L26) from Cu with deuterons is probably Ni^{63} from a Cu-*d*-*α* reaction. Madsen (M1) finds a 100 min. period from Zn-*n*-*α*, probably Ni^{65}. Two fast neutron induced periods (2 hr. and 6 da.) found in Ni (P11d) cannot at present be assigned to definite elements.

(Cu)$^{61, 62, 64, 66}$

A 12.8 hr. period identified as Cu from the Cu-*d*-*p* reaction has been shown to consist of both electrons and positrons, by Van Voorhis (V7). This isotope should have stable isobars of both Ni and Zn, and so is certainly Cu^{64}. It is also observed in the Zn-*d*-*α* reaction (L26) and the Zn-*n*-*p* and Cu-*n*-*γ* reactions (6 hr.) (B21), the Cu-*n*-2*n* process (P11d) and also probably in a Ni-*p*-*n* reaction (B4a). An electron activity of 5 min. period observed with slow neutrons on Cu and fast neutrons on Zn (F16) is chemically copper, and so must be Cu^{66}. Heyn (H33) finds an activity which yields positrons from Cu under only fast neutron bombardment which has a period of 10.5 min. but separates chemically as Cu. As discussed in §102D this is interpreted as due to a *n*-2*n* reaction, and so is Cu^{62}. It is also produced by the Cu-*γ*-*n* reaction which confirms the interpretation (B47b). Cu^{61} probably has the 3.4 hr. positron period found in Ni-*d*-*n*, Ni-*α*-*p* and Ni-*p*-*n* reactions.

Zn$^{63, 65, 69}$

In the deuteron bombardment of Zn Livingood (L26) found a variety of radioactive periods. Certain of these can be shown by coincidence of half-life values to be isotopes of Cu, others may be Ga. The *d*-*p* reaction would be expected to result in Zn^{65}, Zn^{69} and possibly Zn^{71} although the target isotope is in this last case extremely rare. Of the two periods which cannot readily be explained as due to other elements the 100 hr. period yields electrons and so is probably Zn^{69} while the 1 hr. period may be positron active and so be Zn^{65}. A 37 min. positron activity observed in the Ni-*α*-*n* (R7c), Zn-*γ*-*n* (B47d) and Zn-*n*-2*n* (P11d) reactions is most probably Zn^{63}.

Ga$^{68, 70, 72}$

Amaldi, *et al.* (A7) observed two slow neutron activities of 20 min. and 23 hr. periods yielding electrons, from the known two stable isotopes of Ga. Bothe and Gentner (B47d) find periods of 20 min. and 60 min. with gamma-rays on Ga. This identifies the 20 min. period as Ga^{70}, the 60 min. as Ga^{68} and the 23 hr. as Ga^{72}. An activity of 25 hr. period found by Livingood (L26) in the deuteron bombardment of Zn is probably identical with Ga^{72} from neutrons. If so it represents a *d*-*γ* reaction, otherwise unknown. A 1.7 hr. positron activity produced in Ga by fast neutrons (P11d) cannot as yet be given a definite assignment.

Ge, As

A 1.3 hr. electron activity in Ge and a 13 day activity in As produced by fast neutrons (P11d) cannot be assigned to any definite element at present.

Br$^{78, 80, 82, 83}$

In addition to the two slow neutron produced periods in Br originally observed by Amaldi, *et al.* (A7), Kurtschatov, etc. (K34) have found a third, also produced by slow neutrons. The periods and beta energy distributions have been studied by Alíchanian, Alichanow and Dzelepow (Al). They find the 18 min. period to have an energy maximum of 2.00 MV, the 4.2 hr. period 2.05 MV and the 36 hr. (also measured as 24 hr. (J2)) period 0.85 MV, accompanied by gamma-radiation. Only two stable isotopes of Br are known and a search for another has been unsuccessful (B28). The measurements and chemical identification cannot be questioned so some unusual interpretation is required. The *n*-2*n* process has been suggested (K34), but is quite impossible for slow neutrons. The best interpretation seems to be the formation of an isomeric pair in one instance (either Br^{80} or Br^{82}) which decay with different periods into the same product nucleus (see §87). When Br is bombarded by high energy gamma-rays (B47b, B47c, B47d) three periods are observed (18 min., 4.2 hr. and 3.5 min.) two of which are identical with slow neutron periods. The isomeric nuclei must then be Br^{80}; the 3.5 min. period is Br^{78}. This leaves the 35 hr. period for Br^{82}. A still heavier isotope, Br^{83}, is formed in the Se-*d*-*n* reaction with a period of 2.5 hr. (S16a).

Rb, Y, Zr, Cb, Mo

Activities produced in these elements by fast neutrons (P11d) and which have not so far been attributed to definite radioactive elements have the half-lives: 22 hr. (e^-) in Rb, 1.2 hr. and 6.5 hr. (e^-) in Y, 10 min. and 5 hr. in Zr, 7.3 min. and 3.8 days in Cb, and 5 days (e^-) in Mo.

$Ru^{97, 103, 105}$

Kurtschatow, Nemenow, and Selinow (K35) found four activities in Ru produced by slow neutrons; 40 sec., 100 sec., 11 hr., 76 hr. There are only three vacancies in the known stable isotopes of Ru, at 97, 103 and 105. Intensities of the two short periods are identical and the two-step transition is suggested:

$$Ru^{104}+n^1 \rightarrow Ru^{105}\ (T=40\text{ sec.}) \rightarrow Rh^{105}\ (T=100\text{ sec.}) \rightarrow Pd^{105}.$$

The first period must be the shorter lived one if the second is to be observable. It should be possible to check this feature by analysis of the half-life curves with short activation times. The two longer periods are then Ru^{97} and Ru^{103}. The 97 isotope would probably be positron active, and so would supply a means of identification, but this feature has not been reported. Livingood (L26) observes periods of 4 hr. 39 hr., 11 da. and 46 da. from Ru on deuteron bombardment. It is probable that the 11 hr. (inaccurately measured) and 75 hr. periods of Kurtschatow are identical with the 39 hr. and 11 da. (very weak intensity) periods of Livingood; such errors in measurement are easily possible under the circumstances. The short periods were not observed by Livingood. This leaves the 4 hr. and 46 da. periods to be assigned to other elements. $Rh^{101, 103}$ are stable, $Rh^{102, 104}$ are known from Rh-n reactions and Rh^{105} is excluded by the two-step reaction of Kurtschatow. This makes it improbable that Rh is the source of these activities and they are ascribed to Ma through d-α reactions and for which there are many possibilities.

$Rh^{102, 104, 105}$

Two water sensitive neutron produced activities in Rh (A7) suggested a second isotope, which was found by Sampson and Bleakney (S3). The strong period (44 sec.) is ascribed to the more abundant isotope and so is Rh^{104}; the weaker period (3.9 min.) is from the new isotope and is Rh^{102}. The observation of a 100 sec. activity in Ru by neutrons has required the assumption of another radioactive isotope, Rh^{105} (see preceding paragraph). An electron activity of 1.1 hr. half-life, induced in Rh by fast neutrons (P11d), cannot as yet be assigned to a particular element.

$Pd^{107, 109, 111}$

Four radioactive periods are found in Pd on slow neutron bombardment (A7, K36) which are 3 min., 15 min., 12 hr. and 60 hr. There are four available spaces for radioactive isotopes, at 103, 107, 109 and 111, although the first is from an isotope of very low abundance. The 12 hr. period has also been observed in bombardments by deuterons (K25).

$Ag^{105, 106, 108, 110}$

The two usual Ag periods of 22 sec. and 2.3 min. ($Ag^{108, 110}$) are produced by slow neutrons on Ag (A7). The observation of a 24 min. and a 2.3 min. period from gamma-bombardment (B47b, B47c) and fast neutron bombardment (H33b) through what must be Ag-γ-n and Ag-n-$2n$ reactions indicates that the 2.3 min. period, which is common, is Ag^{108}. The 24 min. must then be Ag^{106} and the 22 sec. Ag^{110}. Kraus and Cork (K25) have observed two periods of 32 min. and 7.5 days from the Pd-d-n reaction. The 32 min. period is probably identical with the 24 min. half-life found in the Ag-γ-n and Ag-n-$2n$ reactions (Ag^{106}). The 7.5 day period has also been produced in Ag by *fast neutrons* (P11d) (the original report of 13 days has been corrected to 7.9 days in a private communication) and was found to emit *electrons*. A tentative explanation might be that Ag^{106} is isomeric, emitting positrons with a period of 24 min. and electrons with the longer half-life.

$In^{112, 114, 116}$

Two periods are observed with slow neutrons (13 sec., 54 min.) and three with fast neutrons (3.5 hr., 1 min., 2 mo.) (A7, S28, C16a, P11d). Gaerttner, Turin and Crane (G1) have measured the beta-spectra of the two slow neutron periods and find 3.0 and 1.1 MV for the limits. A gamma-ray of about 1 MV from the 54 min. period suggests an excitation state of the product nucleus in this case.

The 1 min. period is observed in the In-γ-n (B47d), the In-n-$2n$ (C16a) and the Cd-p-n reactions and seems certainly to be In^{112}. The 2 mo. activity may be In^{111} through an n-$3n$ process. This still leaves three periods for In^{114} and In^{116} and suggests the existence of an isomeric pair.

$Sn^{121,\ 123,\ 125}$

Slow neutron periods of 8 min. and 18 min. are reported by Naidu (N3). Livingood (L27) finds periods of 28 hr. and several months from deuterons on Sn, chemically separable as Sn. The vacancies in the isotopic table for n-γ and d-p reaction products are 113, 121, 123, 125; Sn^{113} would be positron active and such has not been observed. The strong 28 hr. period is also observed in a Sb-d-α reaction and so is Sn^{121}. This leaves the 8 min., 18 min. and several months activities to the two isotopes Sn^{123} and Sn^{125}. It may be that the 8 min. and 18 min. periods are a consecutive process (as in Ru):

$$Sn^{124}+n^{1}\rightarrow Sn^{125}(T=8m)\rightarrow Sb^{125}(T=18m)\rightarrow Te^{125},$$

which is possible from the arrangement of stable isotopes, and if so the period of some months is Sn^{123}. Livingood also finds short periods of 12 min. and 45 min. on short bombardment without chemical analysis; these are probably the same as Naidu observed, the latter one possibly intermixed with the 54 min. period of In, which might be produced through a d-α reaction.

$Sb^{120,\ 122,\ 124,\ 125}$

Periods of 2.5 da. and 60 da. are observed with slow neutrons (A7, L27b). These are checked by Livingood (L26) with Sb-d-p who measures them as 68 hr. and 50 days. These then represent Sb^{122} and Sb^{124}, coming from the known stable isotopes. A 25 hr. period also observed, but not chemically separated, may well be the 28 hr. period of Sn^{121}, produced in this case by Sb^{123}-d-α. A 16 min. period in Sn bombarded with deuterons (L27b) and separable as Sb must be from a Sn-d-n reaction. The same period is observed with γ-rays and fast neutrons and must be Sb^{120}.

Te, La, Ce, Gd, Ta

Fast neutrons produce in Te a 30 day period, in La a 2.2 hr. period, in Ce a 40 min. period (e^-), in Gd a 19 hr. (e^-) period and in Ta a 9.1 hr. period of which none can be definitely assigned to particular elements at present (P11d).

$Ir^{192,\ 194,\ 195}$

A 19 hr. water sensitive period in Ir on neutron bombardment (A7) has not been checked by Sosnowski (S20), who finds two periods of 50 min. and 3 da. Fomin and Houtermans (F27) report still another slow neutron activity of about 2 months half-life. Pool, Cork and Thornton (P11d) find that a 15 hr. electron activity is induced in Ir and a 3 day activity in Pt by fast neutrons. In the disintegration of Pt by deuterons Cork and Lawrence (C35) find electron activities of 28 min. and 8.5 hr., shown by chemical separations to be Ir. These are thought to follow the d-n, α reaction (§102D). The short period is the one for which certain resonance effects are observed in the energy-activation curves. A radioactive Ir isotope is also reported by Cork and Thornton (C36) from Au following the Au-d-p,α reaction discussed in §102E. Since Au has only one isotope this must be Ir^{194}. No assignments are possible with the present complexity of data.

$Pt^{193,\ 197}$

Platinum shows a 50 min. water sensitive activation with neutrons (A7). Cork and Lawrence (C35) find a 49 min. positron activity from Pt-d-p (supposedly the same activity), and also a 14.5 hr. electron activity, both proven chemically. The 49 min. period would then be Pt^{193} and the long period probably Pt^{197}. A 1.8 hr. electron activity induced in Pt by fast neutrons (P11d) can not be assigned to an element.

Bi^{210}

Livingood (L26) gives satisfactory evidence for the synthesis of Ra E (Bi^{210}) from Bi under deuteron bombardment. It shows the characteristic 5 day electron period and produces an alpha active product identical with Po. A report by Sosnowski (S19) of a 1 hr. period from Bi with neutrons is not confirmed by other observers (A13).

Th

The activity produced in Th is complex, yielding at least 4 distinct periods of beta decay and also alpha-particles. Curie, von Halban and Preiswerk (C60) have made a complete separation and identification of all the activities and find them to consist of a radioactive family of mass type $4n+1$, where n is integral. This is a new type of radioactive family, and adds to the U, Th and Ac series to make a complete paralleling set. It can be described as follows:

$$90^{233} \xrightarrow[25\text{ m}]{\beta} 91^{233} \xrightarrow[2.5\text{ m}]{\beta} 92^{233} \xrightarrow{\alpha}$$

$$90^{229} \xrightarrow{\alpha} 88^{225} \xrightarrow{\alpha} 86^{221} \xrightarrow{\alpha} \text{etc.}$$

Slow neutrons are found to lead to a member of this family though the reaction: $Th^{232}+n^1 \rightarrow Th^{233}$; the resultant beta-activity has a period of 25 min., followed by the subsequent activities of the chain. Fast neutrons are found to emit alphas through the reaction: $Th^{232}+n^1 \rightarrow 88^{229} + He^4$; this isotope has a beta-activity of 1 min. half-life, and is isotopic with radium. A subsequent beta disintegration of period either 15 min. or 3.5 hr. leads to 90^{229}, a member of the above chain:

$$Th^{232}+n^1 \longrightarrow He^4+88^{229}$$

$$\xrightarrow[1\text{ m}]{\beta} 89^{229} \xrightarrow[\begin{cases}15\text{ m}\\3.5\text{ hr}\end{cases}]{\beta} 90^{229} \longrightarrow \text{etc.}$$

More extensive analysis will be necessary before periods of 5 min. and 1.4 hr. induced in Th by fast neutrons (P11d) can be placed.

U

A total of 10 radioactive periods have been reported by Fermi and his associates (A7) and by Meitner and Hahn (M15, M15a). At the most three of these can be listed as direct products of neutron disintegrations, the others being radioactive decay products. The experiments are difficult; they must be performed in the presence of the natural UX_1 and UX_2 beta-active decay products of U, and require some of the most difficult radioactive chemistry. Some of the new activities can be chemically identified with Ac, Pa, or U (90, 91 and 92), but others are not separable with any of the known heavy elements and are certainly due to elements of atomic number 93, 94, 95, etc. Meitner and Hahn have attempted to analyze the results into the radioactive series which must result. Several improbable assumptions are required, but in the present stage it seems the best that can be done with the data. They suggest:

$$U^{238}+n\begin{pmatrix}\text{slow}\\\text{fast}\end{pmatrix} \longrightarrow U^{239} \xrightarrow[10\text{ s}]{\beta} 93^{239} \xrightarrow[2.2\text{ m}]{\beta}$$

$$94^{239} \xrightarrow[59\text{ m}]{\beta} 95^{239} \xrightarrow[66\text{ hr}]{\beta} 96^{239} \xrightarrow[2.5\text{ hr}]{\beta} ?$$

$$U^{238}+n(\text{fast}) \longrightarrow 2n^1+U^{237} \xrightarrow[40\text{ s}]{\beta}$$

$$93^{237} \xrightarrow[16\text{ m}]{\beta} 94^{237} \xrightarrow[6\text{ hr}]{\beta} 95^{237} \longrightarrow ?$$

$$U^{238}+n(\text{slow}) \longrightarrow He^4+90^{235} \xrightarrow[4\text{ m}]{\beta}$$

$$91^{235} \xrightarrow[\text{very short}]{\beta} 92^{235} \rightarrow ?$$

It is also possible that U^{235} (AcU), which is present to about 5 percent, may be involved to give some of the activities:

$$U^{235}+n(\text{slow}) \longrightarrow U^{236} \xrightarrow[23\text{ m}]{\beta} 93^{236} \longrightarrow ?$$

XVIII. Nuclear Masses

§106. The Mass Spectrograph

The term "mass spectrograph" is applied to instruments designed to study the mass and the relative abundance of atomic species. By the use of electric and magnetic fields these instruments separate ions of different e/m values; ions having a common e/m but differing in direction and energy within certain limits are brought to a common focus. Masses are determined by comparison of the deflections obtained to that of O^{16} ions or suitable sub-standards.

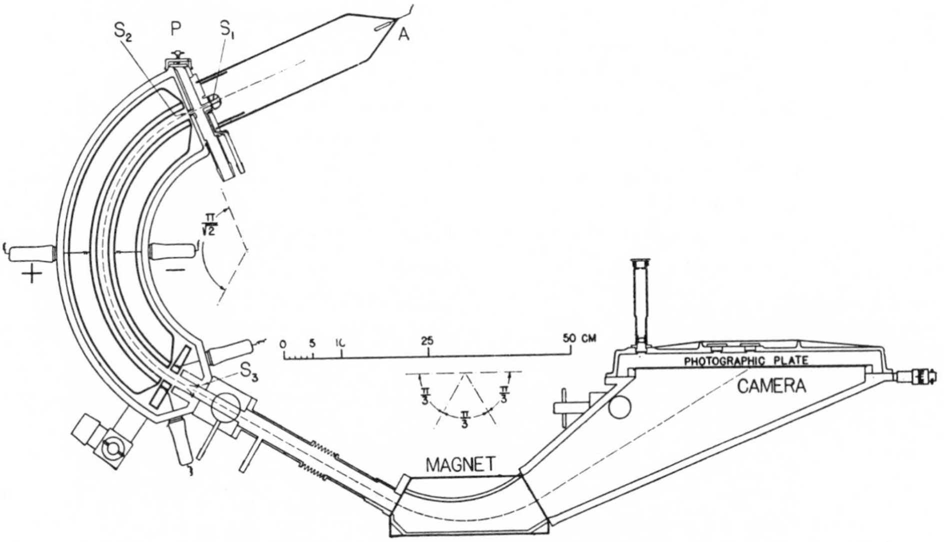

FIG. 49. Scale diagram of mass spectrograph. (Bainbridge and Jordan.)

A variety of focusing schemes have been utilized, some resulting in velocity focusing, some in direction focusing (of noncollimated ion beams), some aimed more directly at achieving a linear mass scale. The instrument was originally developed by Aston (A18), using consecutive electrostatic and magnetic focusing of a well collimated ion beam. If θ is the average angle of deflection of the electric field and ϕ that of the magnetic field the focusing condition requires that $d\theta/\theta = 2d\phi/\phi$. This results in a focus of the different ions along a flat photographic plate; the mass scale is not linear and must be determined from calculations of the geometry of the focusing conditions. The resolution is determined by the accuracy of collimation of the slits defining the beam of ions from the source.

A method perfected by Dempster (D27) uses the focusing of ions of different primary directions by a 180° magnetic field. The ions must have homogeneous energy, which limits the resolution in this case. The e/m values of ions traversing the fixed slit systems are inversely proportional to the square of the magnetic field required to bring the ions to the exit slit.

Many subsequent improvements and modifications, involving several different types of focusing, have resulted in instruments adaptable to a variety of purposes. Excellent reviews of the subject and discussions of the focusing principles are contained in the publications of Brüche and Scherzer (B65) and of Mattauch (M10a).

The instrument which has been most successful in obtaining accurate values of atomic masses is that developed by Bainbridge and Jordan (B2a). The focusing principle involves the radial electric field velocity analyzer of Hughes and Rojansky and the magnetic focusing of ions which differ in velocity. A radial electrostatic field of $\pi/(2)^{\frac{1}{2}}$ radians extent and a magnetic field of $\pi/3$ radians are used. The fields are so placed that the dispersion of ions of different velocities produced by the electric field is just compensated by the dispersion of the magnetic field. A scale drawing of the instrument is given in Fig. 49. Ions entering the radial electric field having a spread in velocities of 3 percent are brought to focus on a flat photographic plate. Furthermore the method allows the use of ion beams diverging in direction by a considerably greater amount than possible in other types of apparatus, thus resulting in higher intensities.

The ions are produced in a low pressure discharge tube of standard design. Ions of gaseous elements are obtained by introducing the gas directly into the discharge tube. Metallic elements are used as cathodes or introduced into other cathodes to provide ions of other elements; neon gas is used to stabilize the discharge in such cases. The ions have energies of 15,000 to 20,000 volts, and are deflected through the electrostatic analyzer (radius 25 cm) by fields of the order of 1200 volts/cm. The magnet provides a suitable field across a 0.32 cm. gap, and the edges of the poles are machined to correct for the effect of stray fields. This focusing principle results in a linear mass scale along the photographic plate; over a region 14 cm long the divergences from linearity are of the order of 1 part in 5000. The resolving power of the instrument can be calculated from the geometry and shows a maximum of $M/\Delta M = 5000$. This instrument has the special advantages of a linear mass scale, a high resolving power and a simultaneous focusing of velocity and direction of the ions.

In the analysis of the results of mass-spectrographic observations the direct comparison of the photographic traces of the ions with e/m values differing from that of O^{16} is handicapped by any nonlinearity of the mass scale and by the many sources of experimental error. This has resulted in the past in mass values now known to be in error by amounts greatly exceeding those computed from the internal consistency of the data. The method of bracketing ion "doublets" of essentially the same e/m values has eliminated most of these difficulties. This involves the simultaneous analysis of ions differing in nature but having the same integral values of e/m. In this way two closely-spaced photographic traces can be obtained whose spacing represents the mass difference of the two ions, and is relatively free from errors in the absolute calibration of the mass scale. For instance $(O^{16})^+$ and $(S^{32})^{++}$ may be compared directly since they have the same e/m values (the electronic masses must be added to obtain the atomic masses usually quoted).

The method of using ion doublets has been most satisfactorily employed by Bainbridge and Jordan (B2 etc.). They use doublets matched in intensity so that the spacings can be measured accurately. A variety of molecular ion doublets have been studied, such as: $(O^{16}H^1)^+$ and $(N^{14}H^1{}_3)^+$, $(O^{16})^+$ and $(C^{12}H^1{}_4)^+$, $(N^{14})^+$ and $(C^{12}H^1{}_2)^+$, etc. Triplets such as $(C^{12})^{++}$, $(He^4H^2)^+$ and $(H^2{}_3)^+$ are also obtained. By grouping and comparing mass differences from such doublets accurate values of the atomic masses of many of the light atoms have been obtained. The errors involved are in general smaller than those of mass values calculated from disintegration data.

An interesting application of this method is in the possibility of obtaining the mass difference representing the reaction energy of a nuclear reaction, enabling a direct comparison with disintegration data. For example, Bainbridge and Jordan (B1) have obtained the Q value of the C^{12}-d-p reaction from the two doublets $(C^{12}H^1)^+ - (C^{13})^+$ and $(H^2)^+ - (H^1{}_2)^+$ and find it to be 0.00297 mass units, or 2.76 MV, while that obtained by an evaluation of the disintegration data of Cockcroft and Lewis (C29) is 2.71 MV.

§107. Masses from Mass Spectrograph

The most extensive and accurate mass determinations are those made during the last year by Bainbridge and Jordan (B1, B2, J5, J6, and private communication[22]) using the doublet method. Care was always taken that each "line" was symmetrical and that the lines of a doublet were of equal intensity. The region of "linear dispersion" of the mass spectrograph was used. Many internal cross checks are provided by comparing connected doublets, such as $C^{12}H_4 - O^{16}$, $C^{12}H_2 - N^{14}$, $N^{14}H_2 - O^{16}$ and $N_2 - C^{12}O^{16}$, and all of them check within the given "experimental error." The agreement with the best disintegration data (§108) is also gratifying, including the few cases (e.g., $C^{12}H_4 - O^{16}$) in which Bainbridge and Jordan's data do not agree with Aston's. With very few exceptions (see below), we have used Bainbridge and Jordan's measurements as a basis of our mass table (Table LXX).

Aston's measurements (A17, A18, A19, A20, A21) agree, in general, with those of Bainbridge and Jordan within the experimental errors given. The most notable exceptions are the doublets $C^{12}H_4 - O^{16}$ and $C^{12}H_2 - N^{14}$. Our decision in favor of Bainbridge and Jordan's values (cf. also Mattauch, M10b) was made on the basis of the cross checks mentioned above which confirm Bainbridge and Jordan's values. In some other cases, the combined evidence of all mass spectrograph and disintegration data agrees slightly better with Aston's determination than with Bainbridge and Jordan's. This is true of the doublet $B^{10} - \frac{1}{2}Ne^{20}$ (Table LXXI) for which the

[22] We are very much indebted to Professor Bainbridge for communicating his most recent (March 24, 1937) results to us which made it possible to bring this section up to date.

value derived from the "adopted masses" (Table LXXIII) differs by 0.16 units from Bainbridge-Jordan, but only 0.07 units from Aston. Also Aston's value of the He^4 mass seems to give better agreement with the numerous disintegration data from α-particle reactions (§108) therefore we have used the average between his determination and that of Bainbridge-Jordan for the mass of He^4.

Aston's measurements extend to higher atomic weights, including F^{19} Si^{28} Si^{29} P^{31} S^{32} Cl^{35} Cl^{37}. These data were used in Table LXX, due corrections being made for the atomic weights of the "standards": E.g., Si^{28} was measured against $C^{12}O^{16}$; the value for the doublet separation CO-Si was taken from Aston's data but the mass of C^{12} inserted from Bainbridge's determination.

Mattauch (M10) measured the masses of the rare isotopes N^{15} and O^{18} against $N^{14}H$ and $O^{16}H_2$, respectively. The former was also measured by Bainbridge and Jordan, the agreement being sufficient.

The key measurements for the determination of all the masses lighter than oxygen are the three doublets measured by Bainbridge and Jordan (rechecked values, private communication):

$$H_2-D=1.53\pm0.04, \qquad (a)$$

$$D_3-\tfrac{1}{2}C^{12}=42.19\pm0.05, \qquad (b)$$

$$C^{12}H_4-O^{16}=36.49\pm0.08. \qquad (c)$$

All values here, and in Table LXXI, are given in thousandths of a mass unit. The errors are those given by Bainbridge and Jordan in cases (*a*) and (*b*) while we have given a smaller error for CH_4 $-O$. This doublet is based on 24 independent measurements, from 4 different plates, none of the measurements deviating more than 0.3 unit, so that 0.08 is more than twice the mean square error. The value given in (*c*) agrees, within the error given, with the sum of the two doublets

$$C^{12}H_2-N^{14}=12.74\pm0.11, \qquad (d)$$

$$N^{14}H_2-O^{16}=23.69\pm0.15, \qquad (e)$$

viz. 36.43 ± 0.19. These doublets have also been measured many times. Another check was provided by

$$N_2-CO=11.17\pm0.2, \qquad (f)$$

based on not quite so many measurements. The values derived are

$$2(d)+(f)=36.65\pm0.3,$$
$$2(e)-(f)=36.21\pm0.4.$$

The larger deviation from the direct determination (*c*) does not seem serious, in view of the fact that the mean error in $2(d)+(f)$ and $2(e)-(f)$ is enhanced by the necessity of multiplying the result for one of the doublets by two, and by the smaller intrinsic accuracy of (*f*).

From the doublets (*a*) (*b*) (*c*) we find

$$H=\tfrac{1}{16}O^{16}+\tfrac{3}{8}(a)+\tfrac{1}{8}(b)+\tfrac{1}{16}(c),$$

$$D=\tfrac{1}{8}O^{16}-\tfrac{1}{4}(a)+\tfrac{1}{4}(b)+\tfrac{1}{8}(c),$$

$$C^{12}=\tfrac{3}{4}O^{16}-\tfrac{3}{2}(a)-\tfrac{1}{2}(b)+\tfrac{3}{4}(c),$$

i.e., for the atomic weights

$$H=1.008\,13\pm0.000\,017,$$
$$D=2.014\,73\pm0.000\,019,$$
$$C^{12}=12.003\,98\pm0.000\,09.$$

From these "secondary standards," the determination of other masses is straightforward. N^{14} is found by averaging the results obtained from *d*, *e*, and *f*. (The result differs by 0.04 unit from that obtained by B. and J. from *d* and *e* only.) Li^7 is connected to N^{14} by the doublet $Li^7-\tfrac{1}{2}N^{14}$ (cf. Table LXXI). He^4 is obtained from the doublet D_2-He, using the average of the determinations of Aston and Bainbridge-Jordan. Ne^{20} comes directly from $O^{16}D_2-Ne^{20}$, Ne^{21} from $Ne^{20}H-Ne^{21}$.

The boron isotopes are the only ones for which the agreement between various determinations is not quite perfect. B^{10} is compared to nuclei already mentioned in two ways, *viz.* by means of the doublets

$$B^{10}H_2-C^{12}=28.74\pm0.14, \qquad (g)$$

$$B^{10}-\tfrac{1}{2}Ne^{20}=16.75\pm0.15. \qquad (h)$$

From these doublets and the masses of C^{12} and Ne^{20} as previously determined, we find

$$B^{10}=10.016\,46\pm0.000\,18 \text{ (from } C^{12} \text{ and g)},$$
$$B^{10}=10.016\,16\pm0.000\,16 \text{ (from } Ne^{20} \text{ and } h).$$

The average is

$$B^{10}=10.016\,31,$$

TABLE LXX. *Masses from mass spectrograph.*

NUCLEUS	MASS	AUTHOR	ASTON MASS
H^1	1.008 13±0.000 017	B	1.008 12±0.000 04
H^2	2.014 73±0.000 02	B	2.014 71±0.000 07
He^4	4.003 89±0.000 07	B+A	4.003 91±0.000 16
Li^7	7.018 18±0.000 12	B	
Be^9	9.015 16±0.000 2	B	
B^{10}	10.016 31±0.000 20	B	10.016 1 ±0.000 3
B^{11}	11.012 92±0.000 16	B	
C^{12}	12.003 98±0.000 09	B	12.003 55±0.000 15
C^{13}	13.007 61±0.000 15	B	
N^{14}	14.007 50±0.000 08	B	14.007 3 ±0.000 5
N^{15}	15.004 89±0.000 2	B	
O^{16}	16.000 000		
O^{18}	18.003 69±0.000 20	M	18.005 7 ±0.000 2*
F^{19}	19.004 52±0.000 3	A	19.004 5 ±0.000 6
Ne^{20}	19.998 81±0.000 11	B	19.998 6 ±0.000 6
Ne^{21}	20.999 68±0.000 23	B	
Ne^{22}	21.998 64±0.000 36	B	
Si^{28}	27.986 8 ±0.000 6	A	27.986 3 ±0.000 7
Si^{29}	28.986 6 ±0.000 7	A	28.986 4 ±0.000 8
P^{31}	30.984 1 ±0.000 5	A	30.983 6 ±0.000 6
S^{32}	31.982 3 ±0.000 3	A	31.982 3 ±0.000 3
Cl^{35}	34.981 3 ±0.000 7	A	34.980 0 ±0.000 8
Cl^{37}	36.978 8 ±0.000 8	A	36.977 5 ±0.000 8
A^{36}			35.978 0 ±0.001 0*
A^{40}	39.975 04±0.000 26	B	39.975 4 ±0.001 4

A = Aston
B = Bainbridge and Jordan
M = Mattauch
* Added in proof. (A21a)

which is just within the experimental error of the two determinations mentioned. Aston's determination of the $B^{10}-\frac{1}{2}Ne^{20}$ doublet (16.84) gives better agreement with the average. For B^{11}, we have then again two determinations, one in terms of C^{12} through the doublet $B^{11}H-C^{12}$ and one in terms of B^{10} through $B^{10}H-B^{11}$. The results for the mass are 11.012 99 and 11.012 84, respectively; the average was taken in Table LXX. Finally, there are again two determinations for Be^9, from the doublets Be^9H-B^{10} and $Be^9H-\frac{1}{2}Ne^{20}$: They agree almost perfectly, giving 9.015 18 and 9.015 14, respectively, for the mass of Be^9.

A^{40} was found by averaging the "direct" determination $OD_2-\frac{1}{2}A^{40}$ with the "indirect" OD_2-Ne^{20} plus $Ne^{20}-\frac{1}{2}A^{40}$; Ne^{22} from $B^{11}-\frac{1}{2}Ne^{22}$. The other nuclei ($C^{13}$ N^{15} O^{18} F^{19} Si^{28} Si^{29} P^{31} S^{32} Cl^{35} Cl^{37}) appear in just one "doublet" each (cf. Table LXXI), so that their determination is straightforward, inserting for the "standards" to which these nuclei are compared ($C^{12}H$, $N^{14}H$, etc.) the values obtained from the foregoing analysis. The masses determined in this way are listed in Table LXX, together with those given by Aston.

The masses given in Table LXX were adopted for the final mass table (LXXIII) and used as a basis for obtaining additional masses from disintegration data (§108), with the exception of Be^9, Si^{28}, P^{31}, Cl^{35} and Cl^{37}. For Be^9, the disintegration data are very good and give consistently lower results for the Be^9 mass, *viz.* 9.015 04 instead of 9.015 16. Si^{28} and P^{31} are connected by the well-investigated reaction $Si^{28}+He^4=P^{31}+H^1$ whose observed reaction energy (−2.23 MV) does not agree completely with that calculated from spectrograph masses (−1.45 MV); the discrepancy was distributed on the masses of Si^{28} (0.2 MV) and of P^{31} (0.2 MV) and on the reaction energy (0.4 MV). For Cl^{35}

TABLE LXXI. *Measured mass-spectrograph "doublets." (In thousandths of a mass unit.)*

DOUBLET	AUTHOR	OBSERVED	CALCUL.	DIFF.
H_2-D	B	1.53±0.04	1.53	*
	A	1.52±0.04	1.53	0.01
D_2-He	B	25.61±0.04	25.56	0.05Δ
	A	25.51±0.08	25.56	0.05Δ
Be^9H-B^{10}	B	6.96±0.20	6.86	0.10
$B^{10}H-B^{11}$	B	11.60±0.10	11.52	0.08Δ
$D_3-\frac{1}{2}C^{12}$	B	42.19±0.05	42.20	*
	A	42.36±0.18	42.20	0.16
$B^{10}H_2-C^{12}$	B	28.75±0.20	28.59	0.16Δ
$B^{11}H-C^{12}$	B	17.14±0.10	17.07	0.07Δ
$C^{12}H-C^{13}$	B	4.5 ±0.1	4.51	*
$Li^7-\frac{1}{2}N^{14}$	B	14.43±0.1	14.43	*
$C^{12}H_2-N^{14}$	B	12.74±0.08	12.74	0.00Δ
	A	12.45±0.07	12.74	0.29
$C^{12}H_3-N^{15}$	M	23.82±0.08	23.48	0.34
$N^{14}H-N^{15}$	B	10.74±0.2	10.74	*
$C^{12}H_4-O^{16}$	B	36.49±0.08	36.50	*
	A	36.01±0.24	36.50	0.49
$N^{14}H_2-O^{16}$	B	23.69±0.15	23.76	0.07Δ
$N_2^{14}-C^{12}O^{16}$	B	11.17±0.20	11.02	0.15Δ
$O^{16}H_2-O^{18}$	M	12.57±0.18	12.57	*
$O^{16}DH-F^{19}$	A	18.33±0.26	18.33	*
$Be^9H-\frac{1}{2}Ne^{20}$	B	23.91±0.20	23.77	0.14
$B^{10}-\frac{1}{2}Ne^{20}$	B	16.75±0.15	16.91	0.16Δ
	A	16.84±0.15	16.91	0.07
$O^{16}D_2-Ne^{20}$	B	30.65±0.10	30.65	*
	A	30.82±0.40	30.65	0.17
$Ne^{20}H-Ne^{21}$	B	7.26±0.20	7.26	*
$B^{10}H-\frac{1}{2}Ne^{22}$	B	25.1 ±0.5	25.12	0.02
$B^{11}-\frac{1}{2}Ne^{22}$	B	13.60±0.15	13.60	*
$C^{12}O^{16}-Si^{28}$	A	17.2 ±0.6	17.4	0.2Δ
$B^{10}F^{19}-Si^{29}$	A	34.2 ±0.6	34.2	*
$C^{12}F^{19}-P^{31}$	A	24.4 ±0.5	24.2	0.2Δ
$O_2^{16}-S^{32}$	A	17.7 ±0.3	17.7	*
$C_3^{12}-Cl^{35}H$	A	22.5 ±0.7	23.4	0.9
$C_3^{12}H-Cl^{37}$	A	41.2 ±0.7	42.1	0.9Δ
$O^{16}D_2-\frac{1}{2}A^{40}$	B	41.89±0.20	41.94	0.04Δ
$N_e^{20}-\frac{1}{2}A^{40}$	B	11.30±0.20	11.29	0.01Δ
	A	10.88±0.3	11.29	0.41

A = Aston
B = Bainbridge and Jordan
M = Mattauch
* used as standard to calculate one of the masses involved.
Δ used in combination with other data to calculate masses.

TABLE LXXII. *Energy evolution in important nuclear disintegrations.*
A. Reactions produced by deuterons and protons and yielding charged particles.

Reaction	Reference	Energy Inc. Part. MV	Obs. Range cm	Straggling and Thick Target Corrections: Straggling % Range	Straggling % Angle	β	Corr. cm	Mica Corr. cm	Obliquity Corr. cm	Mean Range cm	Energy Prod. Part. MV	Q MV	Q Calc. MV
$H^2+H^2=H^3+H^1$	O8	0.20	14.70±0.05	2.07	2.58	2.7	0.36	0.00_5	0.02_5	14.37 ±0.10	3.04 $\pm 0.01_2$	3.98±0.02	3.98*
$Li^6+H^1=He^4+He^3$	N4	0.12	1.19 M	for He^3†			+0.03†			1.22 ±0.04	$2.17_5\pm0.04$	3.72±0.08	3.76
			0.82 M	for He^4†			+0.03†			0.85 ±0.04	1.62 ±0.08	3.72±0.18	‡‡
$Li^6+H^2=2He^4$	O8	0.19	12.70±0.05	1.05	1.36	4.32	0.18	0.00_5	0.02	12.50 ±0.07	$11.05_5\pm0.03_5$	22.07±0.07	22.17*
$Li^6+H^2=Li^7+H^1$	C25	0.500	30.5±0 6	1.96	1.31	0.82	0.35	0.34		30.5 ±0.8	4.71 ±0.10	5.02±0.12	4.92*
$Li^7+H^1=2He^4$	O8	0.20	8.40±0.03	1.10	1.13	1.82	0.09			8.31 ±0.04	$8.63_5\pm0.03$	17.13±0.06	17.25
$Be^9+H^1=Li^6+He^4$	K16	0.12	0.74 M†				+0.02†			0.76 ±0.04	1.43 ±0.08	2.28±0.13	2.25
$Be^9+H^1=Be^8+H^2$	O9	0.20	0.74 M	2.2	1.45	0.26	+0.01			0.75 ±0.06	0.51 ±0.06	0.46±0.08	0.49
$Be^9+H^2=Li^7+He^4$	O9	0.20	‡	1.22	1.31	1.46	0.00					7.19±0.12	7.17*
$Be^9+H^2=Be^{10}+H^1$	O9	0.20	25.6		large			0.2		25.8 ±1.0	4.32 ±0.10	4.59±0.11	4.59*
$B^{10}+H^2=Be^8+He^4$	C28	0.55	14.75±0.07	1.03_5	1.89	2.26	0.22	0.08		14.61 ±0.10	$12.11_5\pm0.05$	17.76±0.08	17.90
$B^{10}+H^2=B^{11}+H^1$	C28	0.55	90.7 ±0.6	1.81	1.20	1.66	1.25	1.9_5		91.4 ±1.0	8.79 ±0.05	9.14±0.06	9.30
			58.5 ±0.7	1.86	1.34	1.36	0.8	1.0		58.7 ±1.0	6.83 ±0.07	7.00±0.08	—
			30.7 ±0.3	1.96	0.83	0.88	0.32	0.35		30.7 ±0.5	$4.72_5\pm0.04_5$	4.71±0.05	—
$B^{11}+H^1=Be^8+He^4$	O9	0.20	4.40±0.06 M	1.15	0.79	1.28	$+0.03_5$		−0.02	$4.41_5\pm0.08$	5.83 ±0.07	8.60±0.11	8.60*
$B^{11}+H^2=Be^9+He^4$	C28	0.55	4.60±0.07	1.15	2.30	1.07	0.06	−0.01		4.53 ±0.09	5.92 ±0.08	8.13±0.12	8.11*
$C^{12}+H^2=C^{13}+H^1$	C29	0.563	13.85±0.12	2.08	1.73	0.72	0.17	0.08		13.76 ±0.12	$2.95_5\pm0.017$	2.71±0.05§	2.76
$C^{13}+H^2=B^{11}+He^4$	C29	0.55	2.70±0.05	1.24	2.23	0.76	0.03	−0.01		2.66 ±0.05	4.17 ±0.05	5.24±0.11	5.14
$N^{14}+H^2=C^{12}+He^4$	C29	0.575	11.37±0.10	1.06	1.54	1.17	0.12	0.04		11.29 ±0.18	10.41 ±0.10	13.40±0.15	13.37
			6.20±0.10	1.11_5	1.75	1.28	0.07	0.00		6.13 ±0.10	7.17 ±0.065	9.08±0.09	—
$N^{14}+H^2=N^{15}+H^1$	C29	0.575	85.1 ±1.0	1.81	0.91	1.74	1.1	1.75		85.7_5 ±1.3	8.48 ±0.07	8.55±0.08	8.57
			18.24±0.3	2.04	1.41	0.82	0.20	0.13		18.17 ±0.3	$3.48\pm0.03_3$	3.11±0.04	—
$O^{16}+H^2=N^{14}+He^4$	C29	0.575	1.59±0.05	1.25	2.30	0.26	0.01	−0.01		1.57 ±0.07	$2.81_5\pm0.09_5$	3.13±0.13	3.11
$O^{16}+H^2=O^{17}+H^1$	C29	0.575	9.22±0.07	2.15	1.45	0.62	0.10	0.02		9.14 ±0.07	2.32 ±0.01	1.95±0.06	1.95*
			4.65±0.12	2.30	1.72	0.40	0.04_5	−0.01		$4.59_5\pm0.15$	1.54 ±0.03	1.12±0.07	—
$F^{19}+H^1=O^{16}+He^4$	H22	1.69‖	6.95±0.15	1.11	2.06	0.92	0.02			6.93 ±0.15	7.72 ±0.10	8.15±0.12¶	8.14
$Na^{23}+H^2=Ne^{21}+He^4$	L15	2.15	6.5±0.2**	1.1	2.75	0.15	0.13			6.44 ±0.2	7.39 ±0.14	6.85±0.20	6.76*
$Na^{23}+H^2=Na^{24}+H^1$	L15	2.15	53.1±4††	1.89	1.6	0.29	0.4	1.5		55.4 ±4	6.61 ±0.27	4.92±0.30	4.94*
$Al^{27}+H^2=Mg^{25}+He^4$	Mc5	2.20	6.37±0.2**				0.05			6.32 ±0.2	7.31 ±0.14	6.46±0.14	6.46*
$Al^{27}+H^2=Al^{28}+H^1$	Mc5	2.20	67.2 ±4				0.5	2.2		68.9 ±4	7.50 ±0.3	5.79±0.3	5.79*

M = Mean range. The straggling and thick target corrections must be *added* instead of subtracted.
* Used for calculation of masses.
† Mean range measured in cloud chamber. The average depth in the target at which the reaction occurs was estimated (0.03 and 0.02 cm for Li^6 $p\alpha$ and Be^9 $p\alpha$, respectively) from the Gamow disintegration function, and added to the observed mean range.
‡ Range not given by authors, but only reaction energy (7.21 MV). The straggling correction applied by Oliphant, Kempton and Rutherford (range straggling only) is exactly equal to that calculated from our scheme. The only correction to be made is thus the change in the range-energy relation.
§ Average of five determinations at different deuteron energies.
‖ The energy of the protons was deduced by the authors from the range and given as 1.63 MV. The change in the range-energy relation increases this figure to 1.69 MV.
¶ Average of six determinations at different proton energies.
** The ranges were given by the authors for air at 20°C. The value given here is corrected to the standard temperature of 15°C.
†† The stopping power of Al was assumed to be 4.4 cm air of 15°C per mil of Al rather than 4.0 cm air of 20°C as assumed by the authors.
‡‡ Used for determining the range-energy relation.

the mass was determined from the reaction $S^{32}+He^4=Cl^{35}+H^1$. Cl^{37} was obtained by assuming that it differs from the value given by Aston by the same amount (+0.4 unit) as Cl^{35}.

The masses obtained in this way are given in Table LXXIII, §108, together with estimated errors. The latter are based on the estimates given by the respective authors and are believed to be rather larger (about by a factor two) than the mean errors. Table LXXI gives, for each measured doublet, besides the observed value also the one calculated from the adopted masses in Table LXXIII. The agreement is gratifying.

§108. Masses from Disintegration Data

As is well known, disintegration data gave the first indication of an error in the then accepted nuclear mass values (B11, O9). The extent of the error has since been found to be greater than had been supposed at first (mass of He^4 according to old scale 4.002 16; suggested by Bethe in 1935, 4.003 36; new value, 4.003 89). Subsequently, improved disintegration masses were given by Cockcroft and Lewis (C29) and by Bonner and Brubaker (B40), both of them in substantial agreement with the now accepted masses. With the improvement of mass spectrograph technique, it seems that the mass spectrograph determinations are now superior to the disintegration values, especially for comparing atoms of widely different mass such as D and C. The chief value of the disintegration data consists now in four points:

1. To provide a check of the mass spectrograph data.

2. To deduce the masses of radioactive nuclei and of nuclei too rare to be measured in the mass spectrograph.

TABLE LXXII*B*. *Reactions of the α-p type.*

Reaction	Ref.	α-part. Energy MV	Angle between α and Part.	Proton Range cm	Mica Correction	Proton Energy MV	Recoil Energy MV	Q MV	Q Calc. MV
$(B^{10}+He^4=C^{13}+H^1$	M16	5.30	0°	82	1.8	8.37	0.22	3.29 }	4.15)
		5.30	90°	61	1.1	7.06	1.54	3.30 }	
$N^{14}+He^4=O^{17}+H^1$	P9	8.36	45°	48	0.7	6.15	1.05	−1.16	−1.16‡
$F^{19}+He^4=Ne^{22}+H^1$	C4	5.30	0°	56	0.8	6.71	0.17	1.58	1.53
$(Ne^{20}+He^4=Na^{23}+H^1$	P9	8.36	45°	32.5	0.4	4.92	0.90	−2.54	−1.42)
$Na^{23}+He^4=Mg^{26}+H^1$	K21	4.44	0°	49	0.7	6.22	0.13	1.91	1.9
$Mg^{24}+He^4=Al^{27}+H^1$†	D19	7.68	0°	39.5	0.5	5.49	0.37	−1.82	−1.6
$Mg^{25}+He^4=Al^{28}+H^1$†	D19	7.68	0°	50	0.8	6.30	0.33	−1.05	−0.6
$Al^{27}+He^4=Si^{30}+H^1$	D19	7.68	0°	107	2.5	9.74	0.20	2.26	2.26*
$Si^{28}+He^4=P^{31}+H^1$	H19	8.78	75°	38.0	0.5	5.48	1.07	−2.23	−1.8*
$(P^{31}+He^4=S^{34}+H^1$	M11	7.68	75°	62	1.1	7.13	0.88	0.31	1.9)
$S^{32}+He^4=Cl^{35}+H^1$	H19	8.78	75°	42.5	0.6	5.73	0.95	−2.10	−2.1*
$(Cl^{35,\,37}+He^4=A^{38,\,40}+H^1$	P8	8.36	0°	88	1.8	8.71	0.23	0.16	?)
$K^{39}+He^4=Ca^{42}+H^1$	P8	8.36	0°	70	1.6	7.65	0.24	−0.89	?
$Ca^{40}+He^4=Sc^{43}+H^1$	P9	8.36	45°	19	—	3.59	0.60	−4.27	?

( ) Parentheses indicate that the longest proton group has probably not yet been observed.
* Used for calculating masses.
† Concerning the assignment of the proton groups to the nuclei Mg^{24} Mg^{25}, cf. §99A.
‡ Standard reaction for evaluating effective angle in experiments of Pollard and Brasefield.

TABLE LXXII*C*. *Reactions producing neutrons.*

Reaction	Ref.	Energy of Incid. Part.	Neutron Energy Given (MV)	Stopping Power of Gas	Corrections to Energy (MV)			Corrected Neutron Energy	Q MV	Q Calc. MV
					Stopp. Power	Range-En. Rel.	Straggl., Obliquity etc.			
$H^2+H^2=He^3+n^1$	B38	0.50	2.53	0.960‡	+0.02	−0.03	−0.01	2.51	3.18	*
$Li^7+H^2=Be^8+n^1$	B37	0.85	13.5	0.906‡	0†	−0.01	0.03	13.5	14.55	14.91
$Be^9+H^2=B^{10}+n^1$	B40	0.9	4.52	0.934‡	−0.05	−0.02	+0.02	4.47	4.20	4.18
$B^{10}+H^2=C^{11}+n^1$	B40	0.9	6.35	0.918‡	−0.13	−0.02	+0.05	6.25	6.08	6.34
$B^{11}+H^2=C^{12}+n^1$	B40	0.9	13.2	0.896‡	−0.17†	−0.01	+0.05	13.07	13.4	13.68
$C^{12}+H^2=N^{13}+n^1$	C29	0.32		Threshold	of	reaction		0	−0.28	*
$O^{16}+H^2=F^{17}+n^1$	N6	1.95		"	"	"		0	−1.7	*
$O^{18}+H^1=F^{18}+n^1$	D28	2.7		"	"	"		0	-2.5_5	*

* Used as standard for computing masses.
† Part of the path is in mica, part in the gas.
‡ Assumed = 0.950 by Bonner and Brubaker.

3. To find excited levels in nuclei (§109) and, most of all,

4. To establish and confirm the interpretation of nuclear reactions.

A. Computation of energy evolution

The method for deriving accurate values of the energy evolved in a nuclear reaction was described in Chapter XVI. The results for all reactions in which the ranges of the produced particles were measured, were given in the tables at the end of §§99–102. In Table LXXII, we have listed the most important nuclear reactions and given an account of how the energy evolution Q was calculated. The first column gives the reaction, the second the reference to the paper in which the observations were reported, the third the energy of the incident particles. In all reactions listed in part *A* of Table LXXII, the observation of the range recorded in the fourth column was at right angles to the incident beam and the geometry was "good" in the sense of §§96, 97. The following columns in Table LXXII*A* give the ordinary (range) straggling (§97*A*) and the angle straggling (§96*b*) in percent, the quantity β determining the thick target correction (§97*C*), the difference between extrapolated and mean range deduced from straggling and thick target correction in cm, the mica correction (§95*E*) where necessary, and the obliquity correction (§96*b*) which was only computed for the most accurate experiments. With the three last named corrections applied to the observed range, the corrected mean range is obtained; from this the corresponding energy of the produced particles is calculated with the help of the range-energy relations of Figs. 29, 30. This energy corresponds to emission exactly at right angles, so that the energy evolution Q follows from (768). This "observed" energy evolution and the energy evolution calculated from the masses of Table LXXIII are listed in the last two columns.

In Table LXXII*B*, we have listed some of the more important α-p-reactions. In these cases, the geometry is necessarily poor because of the low intensity. It was customary in experimental papers to assume that the particles of maximum range observed are emitted at the most forward angle permissible by the experimental arrangement. This is certainly not the case (§§96, 97) because so few particles will have this maximum possible energy that they cannot be observed. We have therefore estimated a probable angle which we believe to correspond approximately to the fastest disintegration products observed. This angle was estimated from the given geometrical arrangement in all cases except for the experiments of Pollard and Brasefield. For these, the estimate was based on the reaction $N^{14}+He^4=O^{17}+H^1$ whose reaction energy is very exactly known from mass-spectrograph data combined with the reaction energy of $O^{16}+H^2=O^{17}+H^1$. The N^{14}-α-p reaction was measured by Pollard and Brasefield with the same apparatus as were the reactions with heavier targets. Therefore we computed the "probable angle" so that P. and B.'s measurements gave the correct answer for the N^{14} reaction, and used this value throughout. In all cases, only the data for the longest proton group were given.

In Table LXXII*C*, we have given the most reliable data on reactions in which neutrons are produced. All these data are based on cloud chamber measurements of the ranges of recoil protons emitted within a small angle about the direction of the neutrons. In column 4 we have listed the neutron energy given by the authors (Bonner and Brubaker). In the next column the stopping power of the gas used in the cloud chamber is given; this was a mixture of 85.1 percent CH_4, 13.5 percent C_2H_6 and 1.4 percent N_2. Bonner and Brubaker assumed a stopping power of 0.950 for this mixture; the correction caused by the difference between this figure and the actual stopping power is given in column 6. The range-energy relation used by Bonner and Brubaker was that given by Mano (M6); the necessary correction is listed in column 7. The following column gives all other corrections, for straggling, for thick target and for the fact that

TABLE LXXIII. *Atomic masses.*

NUCLEUS	MASS	ERROR IN 10^{-5} M.U.	SOURCE
n^1	1.008 97	6	$H^2+\gamma=H^1+n^1$
H^1	1.008 13	2	B
H^2	2.014 73	2	B
H^3	3.017 05	7	$H^2+H^2=H^3+H^1$
He^3	3.017 07	12	$H^2+H^2=He^3+n^1$
He^4	4.003 89	7	B and A
He^5	5.013 7	40	$Li^7+H^2=He^5+He^4$
He^6	6.020 8	50	$He^6=Li^6+\epsilon^-$
Li^6	6.016 86	20	$\{Li^6+H^2=2He^4$ $\{Li^6+H^2=Li^7+H^1$
Li^7	7.018 18	18	B
Li^8	~8.025 1	100	$Li^7+H^2=Li^8+H^1$
Be^8	8.007 92	28	$B^{11}+H^1=Be^8+He^4$
Be^9	9.015 04	25	$\{Be^9+H^2=Li^7+He^4$ $\{B^{11}+H^2=Be^9+He^4$
Be^{10}	10.016 71	30	$Be^9+H^2=Be^{10}+H^1$
B^{10}	10.016 31	25	B
B^{11}	11.012 92	17	B
B^{12}	$\{<$12.019 9 $\{>$12.018 6		$B^{11}+H^2=B^{12}+H^1$
C^{11}	11.015 26	35	$C^{11}=B^{11}+\epsilon^+$
C^{12}	12.003 98	10	B
C^{13}	13.007 61	15	B
C^{14}	14.007 67	12	$N^{14}+n^1=C^{14}+H^1$
N^{13}	13.010 04	13	$C^{12}+H^2=N^{13}+n^1$
N^{14}	14.007 50	8	B
N^{15}	15.004 89	20	B
N^{16}	~16.011	200	$F^{19}+n^1=N^{16}+He^4$
O^{15}	15.007 8	40	$O^{15}=N^{15}+\epsilon^+$
O^{16}	16.000 000		Standard
O^{17}	17.004 50	7	$O^{16}+H^2=O^{17}+H^1$
O^{18}	18.003 69	20	M
F^{17}	17.007 6	30	$O^{16}+H^2=F^{17}+n^1$
F^{18}	18.005 6	40	$O^{18}+H^1=F^{18}+n^1$
F^{19}	19.004 52	17	A
F^{20}	$\{<$20.009 2 $\{>$20.004 2		$F^{19}+H^2=F^{20}+H^1$ $F^{20}=Ne^{20}+\epsilon^-$
Ne^{20}	19.998 81	11	B
Ne^{21}	20.999 68	23	B
Ne^{22}	21.998 64	36	B
Na^{22}	22.000 2	50	$Na^{22}=Ne^{22}+\epsilon^+$
Na^{23}	22.996 1	35	$Na^{23}+H^2=Ne^{21}+He^4$
Na^{24}	23.997 4	45	$Na^{23}+H^2=Na^{24}+H^1$
Mg^{24}	23.992 4	60	$Na^{24}=Mg^{24}+\epsilon^-$
Mg^{25}	24.993 8	90	$Al^{27}+H^2=Mg^{25}+He^4$
Mg^{26}	25.989 8	50	$Na^{23}+He^4=Mg^{26}+H^1$
Mg^{27}	26.992 1	150	$Mg^{27}=Al^{27}+\epsilon^-$
Al^{26}	25.992 9	200	$Al^{26}=Mg^{26}+\epsilon^+$
Al^{27}	26.989 9	80	$Al^{27}+H^2=Al^{28}+H^1$
Al^{28}	27.990 3	70	$Al^{28}=Si^{28}+\epsilon^-$
Al^{29}	28.990 4	200	$Al^{29}=Si^{29}+\epsilon^-$
Si^{27}	26.993 1	150	$Si^{27}=Al^{27}+\epsilon^+$
Si^{28}	27.986 6	60	A, and $Si^{28}+He^4=P^{31}+H^1$
Si^{29}	28.986 6	60	A
Si^{30}	29.983 2	90	$Al^{27}+He^4=Si^{30}+H^1$
Si^{31}	30.986 2	60	$Si^{31}=P^{31}+\epsilon^-$
P^{30}	29.988 2	150	$P^{30}=Si^{30}+\epsilon^+$
P^{31}	30.984 3	50	A, and $Si^{28}+He^4=P^{31}+H^1$
P^{32}	31.984 1	50	$P^{32}=S^{32}+\epsilon^-$
S^{32}	31.982 3	30	A
S^{34}	33.978	200	$P^{31}+He^4=S^{34}+H^1$ (estimate)
Cl^{34}	33.981	300	$Cl^{34}=S^{34}+\epsilon^+$
Cl^{35}	34.980 3	60	$S^{32}+He^4=Cl^{35}+H^1$
Cl^{37}	36.977 9	120	A (comparison with Cl^{35})
Cl^{38}	37.981	300	$Cl^{38}=A^{38}+\epsilon^-$ (Raa $+\gamma$)
A^{38}	37.974	250	$Cl^{35}+He^4=A^{38}+H^1$ (estimate)
A^{40}	39.975 04	30	B

A = Aston
B = Bainbridge
M = Mattauch

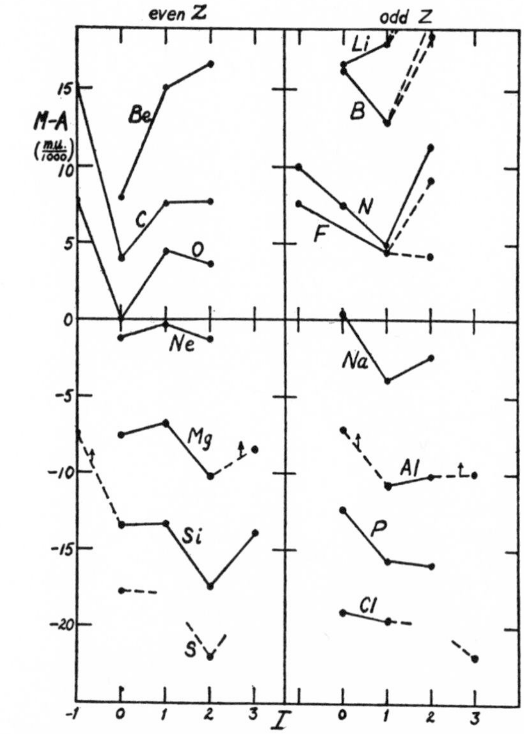

FIG. 50. The mass excess of the lighter atoms. Abscissa: isotopic number. Ordinate: Difference between exact atomic mass and mass number in thousandths of a mass unit. The lines join isotopes of the same element. Left-hand part, even nuclear charge; right-hand side, odd charge. The diagram shows clearly the minima for even number of neutrons. The general trend of the curves shows a minimum at or near zero isotopic number for small nuclear charge which shifts to larger I for the heavier nuclei.

the recoil proton does not move exactly in the same direction as the neutron and therefore does not have quite the full energy. The neutron energy thus corrected is given next, and the energy evolution is calculated (second last column) assuming the neutron to be emitted exactly at right angles to the incident deuteron. The last column gives, as in parts *A* and *B*, the energy evolution calculated from the nuclear masses of Table LXXIII.

B. Agreement of disintegration data with mass spectrograph and with each other

The agreement with the mass spectrograph values is good to excellent for all reactions for which exact determinations are available (cf. Table LXXII). Among the less favorable agreements would be the reactions involving Be^9, either as target or as product, if we took the spectrograph mass of that atom which is 0.11 MV higher than the adopted disintegration mass. With the latter, the agreement is almost perfect for all beryllium reactions.—The agreement for the two reactions involving Li^7, *viz.* Li^7-*p*-α and Li^6-*d*-*p*, is also not quite as good as might be expected in view of the excellent disintegration data and might be improved by adopting a slightly lower mass for Li^7. For most other reactions, the agreement is within the limits of error of mass spectrograph and disintegration measurements.

Of special interest are the reactions connecting the nuclei C^{12}, N^{14} and O^{16} with respect to which the mass spectrograph results of Bainbridge and Jordan and those of Aston disagree seriously. The disintegration data are decidedly in favor of Bainbridge and Jordan, *viz.*

$N^{14}+H^2=C^{12}+He^4+13.40$ MV observed,
13.41 MV using all masses from Bainbridge and Jordan,
13.37 MV using the adopted masses,
13.59 MV using Aston's masses throughout,
13.64 MV using Aston's doublet CH_2-N combined with adopted masses for H, D and He^4.

$O^{16}+H^2=N^{14}+He^4+3.13$ MV observed,
3.14 MV Bainbridge-Jordan,
3.11 MV adopted masses,
3.26 MV Aston.

Of importance is also the very good agreement in the case of C^{12}-*d*-*p* because it showed that in this reaction the C^{13} nucleus is formed in the ground state (B1), a matter which had been in question before (C25, B11, C29, H4).

The internal agreement is best shown by "cycles" involving three or more reactions. The oldest example is

I $\quad Li^7+H^1=2He^4+17.13$ MV, (a)
$\quad Li^6+H^2=2He^4+22.07$, (b)
$\quad Li^6+H^2=Li^7+H^1+5.02$. (c)

The agreement between (*a*)+(*c*) and (*b*) is satisfactory (difference 0.08 MV). Other examples are

II $\quad Be^9+H^1=Li^6+He^4+2.28$ MV,
$\quad Be^9+H^2=Li^7+He^4+7.19$,
$\quad Li^6+H^2=Li^7+H^1+5.02$.

Difference = 0.11

II $Be^9+H^1=Be^8+H^2+0.46$ MV,
$B^{11}+H^1=Be^8+He^4+8.70$,
$B^{11}+H^2=Be^9+He^4+8.13$.

Difference: 0.11

III $N^{14}+He^4=O^{17}+H^1-1.26$ MV,
$O^{16}+H^2=N^{14}+H^4+3.13$,
$O^{16}+H^2=O^{17}+H^1+1.95$.

Difference: 0.08

The most notable disagreements among light nuclei are the reactions

$B^{10}+He^4=C^{13}+H^1$,
$B^{10}+n^1=Li^7+He^4$ (slow neutrons, cf. Table LXI).

In both cases, the energy evolution is much less (by 0.85 and 0.55 MV, respectively) than is expected from masses. This discrepancy cannot be due to an error in the B^{10} mass, not only because this mass is well established by many cross checks with the mass spectrograph (§107) but also by the reactions

$B^{10}+H^2=B^{11}+H^1$,
$B^{10}+H^2=Be^8+He^4$,

which both check with the masses.

C. Determination of additional masses

Table LXXIII gives all the atomic weights derived from both mass-spectrograph (§107) and disintegration data. After each element, we have indicated whether its mass was derived from the mass spectrograph or from disintegrations, and, in the latter case, which nuclear reaction was used. For many nuclei, one single reaction is sufficient to express the mass of the nucleus in terms of masses already determined. (Examples: n^1, H^3, Be^8, Be^{10} etc.) When the masses of these nuclei are known, it is possible to express others in terms of them, such as He^3 which is produced in the reaction $H^2+H^2=He^3+n^1$ and is therefore determinable from the energy evolution in this reaction and the neutron mass.

A longer chain of reactions is needed to calculate the masses of heavier nuclei. There is no reliable mass spectrograph measurement between neon and silicon; therefore, all the intermediate nuclei must be obtained from disintegration data. The chain of reactions connecting Ne^{21} to Si^{28} is the following:

$Na^{23}+H^2=Ne^{21}+He^4$, I (L15)

$Na^{23}+H^2=Na^{24}+H^1$, II (L15)

$Na^{24}=Mg^{24}+\epsilon^-$, III (K33)

$Mg^{24}+He^4=Al^{27}+H^1$, IV (D19)

$Al^{27}+H^2=Al^{28}+H^1$, V (Mc5)

$Al^{28}=Si^{28}+\epsilon^-$. VI (C34)

Of these six reactions, the most accurately measured ones are I, II, III. In all these cases the measurement of the range of the produced particle is straightforward and the geometry satisfactory. The only correction necessary is for the stopping power of the Al used for measuring the proton range in II and V: In accord with the considerations in §95*E*, we assumed one mil of Al to be equivalent to 4.4 cm air of 15°C rather than 4.0 cm of 20°C as assumed by Lawrence and McMillan (L15, Mc5).

The β-transformations III and VI involve the well-known uncertainty regarding the validity of the Konopinski-Uhlenbeck or the inspection limit. Moreover, in both cases the β-decay is accompanied by γ-rays: Na^{24} emits β-particles of 1.9 MV (inspection limit) and γ-rays of 2.9, 0.95 and 1.93 MV. It seems reasonable to assume that the β-transformation leads always to a state of 2.9 MV excitation, which may go over into the ground state either by emission of one quantum of 2.9 MV or two of 0.95 and 1.93 MV. This would mean a total energy evolution of 4.6 MV which fits in fairly well with similar nuclei (N^{16}, F^{20}, Al^{28} etc., cf. Table LII). Al^{28} emits β-rays of 3.3 MV and γ-rays of 2.3 MV: Here it is possible to assume that the β-spectrum is complex, leading either to the ground state (emission of 3.3 MV betas) or to a state with 2.3 MV excitation energy (1.0 MV betas). Although such a complexity is not apparent in the β-spectra, it may very well be present since it is quite possible that only a fairly small fraction of the β-processes lead to the excited states. The main argument is again the comparison of the energy evolution of the analogous nuclei F^{20}, Na^{24}, Al^{28}, P^{32}, K^{40}: The energy evolution in this series decreases in a regular way with increasing mass, as should be

TABLE LXXIV. *Nuclear excitation levels.*

Nucleus	Level: No.	Energy MV	Width kv	Nuclear Mass	Spectr. Symbol	Class	Source	γ-rays
Li^7	1	0.44	—	7.018 65	$^2P_{1/2}\,u$	*A*	Li^6-*d-pP*	~0.4 Li^7-α-α
Be^8	1	2.9	780	8.011 1		*A*	B^{11}-*p*-α*P*, B^{10}-*d*-α*P*	
	2	~4.8	~1400	8.013 1	$^1D_2\,g$	*B*	Li^8-ϵ^--α*P*	17.5 MV 4→0
	3	6–12	Large	8.014–20		*C*	B^{10}-*d*-α*P*, Li^8-ϵ^--α*P*	10–14 MV 4→1, 2
	4	17.50	9	8.026 72	1 *u*	*A*	Li^7-*p*-γ*R*	(from Li^7-*p*-γ)
	5	17.86	Large	8.027 11		*B*	Li^7-*p*-γ*R*	(from Li^7-*p*-γ)
Be^{10}	1	2.4	Small	10.019 3	$^1D\,g$?	*C*	Be^9-*d-pP* ?	
B^{10}	1	0.5	"	10.016 9	*S g* ?	*B*	Be^9-*d-nP*	
	2	2.0	"	10.018 5	*D g* ?	*B*	"	
	3	3.3	"	10.019 8	*D g* ?	*B*	"	
	4	7.28	Large	10.024 13		*B*	Be^9-*p*-γ*R*	(from Be^9-*p*-γ)
B^{11}	1	2.14	Small	11.015 22	*D u* ?	*A*	B^{10}-*d-pP*	
	2	4.43	"	11.017 68	F *u* ?	*B*	"	
C^{12}	1	4.3	"	12.008 6	$^1D_2\,g$	*B*	N^{14}-*d*-α*P*, B^{11}-*d-nP* Be^9-α-*nP* ?	2.7 MV 2→1 (Be^9-α-*n*, B^{11}-α-*n*)
	2	7.0	Small	12.011 5	1 *u* ?	*B*	Be^9-α-*nP*, B^{11}-*d-nP* ?	4.3 MV 1→0 (Be^9-*d-n*, B^{11}-*d-n*, N^{14}-*d*-α)
	3	9.5	?	12.014 2		*D*	B^{11}-*d-nP* ?	7.0 MV 2→0(Be^9-α-*n*, B^{11}-*d-n*, N^{14}-*d*-α)
	4	16.07	~10	12.021 25	1 *u* ?	*A*	B^{11}-*p*-α*R*, B^{11}-*p*-γ*R*	15 MV 4→0 (B^{11}-*p*-γ)
C^{13}	1	0.8	Small	13.008 5	$^1P_{3/2}$?	*B*	B^{10}-α-*pP*	
	2	3.6	"	13.011 5	$^2F_{3/2}$?	*B*	"	
	3	4.0	"	13.011 9	$^2F_{5/2}$?	*C*	"	
	4	4.9	"	13.012 9		*B*	"	
	5	6.0	?	13.014 1		*B*	"	
	6	12.9	300	13.021 5		*B*	Be^9-α-*nR*	
N^{13}	1	2.25	~20	13.012 46		*A*	C^{12}-*p*-γ*R*	2 MV 1,2→0 (C^{12}-*p*-γ)
	2	2.32	~20	13.012 56		*A*	C^{12}-*p*-γ*R*	
N^{14}	1	0–4	Small	several levels		*D*	B^{11}-α-*nP*	4.0 MV 2→0 (C^{13}-*d-n*)
	2	4.0	"	14.011 8		*C*	C^{13}-*d-n*γ	
	3	14.8	500	14.023 4		*B*	B^{10}-α-*pR*	
N^{15}	1	5.42	Small	15.010 72		*A*	N^{14}-*d-pP*	
	2	13.4	~400	15.019 2		*D*	B^{11}-α-*nR*	
O^{16}	1	10.7	700	16.011 5		*B*	C^{12}-α scatt. *R*	
O^{17}	1	0.82	Small	17.005 38		*A*	O^{16}-*d-pP*	
	2	4.6	"	17.009 4		*D*	Ne^{20}-*n*-α*P*	
F^{18}	1	8.2	~150	18.014 4		*D*	N^{14}-α-*pR*	
Ne^{20}	1	7.2	?	20.006 5		*A*	F^{19}-*p*-γ*P*	6.0 MV 2→1 (F^{19}-*p*-γ)
	2	13.20	<4	20.012 98		*A*	" *R*	
	3	13.5	Large	20.013 3		*B*	" *R*	
	4	13.73	<12	20.013 56		*A*	" *R*	
	5	13.78	<15	20.013 59		*A*	" *R*	
	6	14.22	Small	20.014 08		*B*	" *R*	
	7	14.85	?	20.014 44		*D*	" *R*	
Ne^{22}	1	0.6	Small	21.999 3		*B*	F^{19}-α-*pP*	
	2	1.5	"	22.000 3		*B*	"	
	3	3.5	"	22.002 4		*B*	"	
	4	4.6	"	22.003 6		*B*	"	
Na^{22}	1	1.3	"	22.001 6		*C*	F^{19}-α-*nP*	
	2	1.6	"	22.001 9		*C*	"	
Na^{23}	1	14.5	100	23.011 7		*B*	F^{19}-α-*pR*	
	2	14.8	130	23.012 0		*B*	"	
Na^{24}	1	3.20	Small	24.000 8		*A*	Na^{23}-*d-pP*	
Mg^{24}	1	1.0 or 2.0	Small	23.993 5 or 9945		*B*	Na^{24}-ϵ^--γ	0.95 MV 1→0 or 2→1
	2	2.9	"	23.995 4		*B*	"	1.93 MV 2→1 or 1→0
	3	12.2	Large	24.005 5		*B*	Na^{23}-*p*-γ*R*	2.9 MV 2→0 (Na^{24}-ϵ^-)
Mg^{25}	1	0.7	Small	24.994 6		*A*	Al^{27}-*d*-α*P*	
Mg^{26}	1	2.3	"	25.992 3		*B*	Na^{23}-α-*pP*	
	2	4.0	"	25.994 1		*B*	"	
	3	5.0	"	25.995 2		*B*	"	
Al^{27}	1	0.98	"	26.991 0		*B*	Mg^{24}-α-*pP*	1.3 MV by absorp. meth. (Mg^{27}-ϵ^-)
Al^{28}	1	0.68	"	27.991 0		*A*	Al^{27}-*d-pP*	
	2	2.69	"	27.993 2		*A*	"	
	3	3.67	"	27.994 2		*A*	"	
	4	5.15	"	27.995 8		*A*	"	

TABLE LXXIV.—*Continued.*

Nucleus	Level						Source	γ-rays
	No.	Energy MV	Width kv	Nuclear Mass	Spectr. Symbol	Class		
Si^{28}	1	2.3	Small	27.989 1		*B*	Al^{28}-ϵ^--γ	2.3 MV 1→0 (Al^{28}-ϵ^-)
	2	12.0	"	27.999 5		*B*	Al^{27}-*p*-γ*R*	
	3	13.7	100	28.001 3		*C*	Mg^{24}-α-*pR*	
	4	14.2	130	28.001 9		*C*	"	
Si^{30}	1	2.3	Small	29.985 7		*B*	Al^{27}-α-*pP*	
	2	3.6	"	29.987 1		*B*	"	
	3	4.8	"	29.988 4		*B*	"	
P^{31}	1	1.05	"	30.985 4		*B*	Si^{28}-α-*pP*	
	2	1.7	"	30.986 1		*B*	"	
	3	12.0	~100	30.997 2		*B*	Al^{27}-α-*pR*	
	4	12.4	~100	30.997 6		*B*	Al^{27}-α-*pR*	
	5	12.7_5	80	30.998 0		*B*	"	
	6	13.1	130	30.998 4		*B*	"	
	7	13.5_5	~120	30.998 9		*B*	"	
	8	14.3	~120	30.999 7		*B*	"	
S^{34}	1	x	Small	33.980 2		*C*	P^{31}-α-*pP*	
	2	$x+1.3$	"	33.981 6		*C*	"	
	3	$x+2.8$	"	33.983 2		*C*	"	
	4	$x+4.8$	"	33.985 3		*C*	"	
Cl^{35}	1	0.6	Small	34.981 0		*B*	S^{32}-α-*pP*	
	2	1.5	"	34.982 0		*B*	"	
A^{38}	1	2.1	"	37.976 1		*C*	Cl^{38}-ϵ^--γ, Cl^{35}-α-*pP*	2.0 MV 1→0
	2	4.5	"	37.978 7		*C*	Cl^{38}-ϵ^--*P*, Cl^{35}-α-*pP*	2.5 MV 2→1
	3	6.2	"	37.980 5		*C*	Cl^{35}-α-*pP*	β-groups with en. diff. of 4.6 MV

expected theoretically (W18), therefore it seems highly improbable that it should increase between Na^{24} and Al^{28} from 5.0 to 5.6 MV. This forces us to the conclusion that the total energy evolution in VI is equal to the β-energy, i.e. 3.3 MV. The agreement finally obtained (see below) confirms this decision.

The remaining reaction IV is complicated by the fact that magnesium represents a mixed element and it is not known to which isotope a given proton group should be attributed. The observed proton groups from Mg-α-*p* correspond to energy evolutions $Q=-1.04$, -1.82 and -2.80 MV. The longest group ($Q=-1.04$ MV) is weak compared to the two others. The isotopes of Mg which may be responsible for these groups, are Mg^{24} and Mg^{25} while Mg^{26} is excluded because the energy evolution would in this case be -4.4 MV. The calculated energy evolution for Mg^{25} is -0.6 MV which agrees sufficiently with the observed Q for the longest group. This group might still be a superposition of protons from Mg^{24} and Mg^{25} although this is unlikely in view of the small intensity of the group. The main reason against assigning the longest group to Mg^{24} is again the analogy to other nuclei of the 4*n*-type: For Ne^{20}-α-*p*, the Q calculated from masses is -1.4 MV, for Si^{28}, the observed energy evolution is -2.2, for S^{32} also -2.2. A Q value of -1.8 for the Mg^{24}-α-*p* reaction would fit in very well with this sequence, while -1.0_5 would definitely not. Therefore we decided to accept -1.82 MV for the energy evolution in reaction IV.

With the energy evolutions in I to VI thus determined, the mass difference between Si^{28} and Ne^{21} comes out about 0.5 MV too great which must be considered as a very good agreement in view of the many uncertain data involved. For the determination of the masses, we "distributed" the error among the reactions I to VI, assuming the observed reaction energy to be 0.1 MV too high in I, 0.1 MV too low in III and VI, and 0.2 MV too low in IV, the general experience being that α-*p* reactions give generally too low energy evolutions owing to small intensity and poor geometry.

With the main chain thus established, it is easy to link additional masses to it such as Mg^{25} by the Al^{27}-*d*-α reaction etc. For the elements heavier than silicon, reaction energies have been used in some cases to correct the mass spectroscopic values which have a rather large probable error: Thus, e.g. Si^{28} and P^{31} have both been measured

TABLE LXXV. *Binding energies of neutrons, protons and α-particles, in thousandths of a mass unit.*

A. NEUTRONS

Z	$I=0$ N	odd	even	$I=1$	odd	even	$I=2$	odd	even	$I=3$	odd	even
1	H^2	2.37		H^3		6.65						
2	He^4		22.15	He^5	<0.8		He^6		1.9			
3				Li^7		7.65	Li^8	~2				
4				Be^9	1.85		Be^{10}		7.30			
5				B^{11}		12.36	B^{12}	~3				
6	C^{12}		20.25	C^{13}	5.34		C^{14}		8.91			
7	N^{14}	11.51		N^{15}		11.58	N^{16}	~3				
8	O^{16}		16.8	O^{17}	4.47		O^{18}		9.8			
9	F^{18}	11.0		F^{19}		10.1	F^{20}	~6				
10				Ne^{21}	8.10		Ne^{22}		10.0			
11				Na^{23}		13.1	Na^{24}	7.7				
12				Mg^{25}	7.6		Mg^{26}		13.0	Mg^{27}	6.7	
13				Al^{27}		12.0	Al^{28}	8.6		Al^{29}		8.9
14	Si^{28}		15.5	Si^{29}	9.0		Si^{30}		12.4	Si^{31}	6.0	
15				P^{31}		12.9	P^{32}	9.2				

B. PROTONS

N	$I=2$ Z	odd	even	$I=1$	odd	even	$I=0$	odd	even	$I=-1$	odd	even
1							H^2	2.37				
2							He^4		21.29	He^3		5.79
3				Li^7	10.7		Li^6	4.9				
4				Be^9		~18	Be^8		18.39			
5				B^{11}	11.92		B^{10}	6.86				
6				C^{13}		~20	C^{12}		17.07	C^{11}		9.18
7				N^{15}	10.91		N^{14}	8.24		N^{13}	2.07	
8				O^{17}		15	O^{16}		13.02	O^{15}		7.8
9				F^{19}	7.30		F^{18}	7.0		F^{17}	0.5	
10				Ne^{21}		16	Ne^{20}		13.84			
11				Na^{23}	10.7		Na^{22}	7.6				
12				Mg^{25}		11.7	Mg^{24}		11.8			
13	Al^{28}	9.9		Al^{27}	8.0		Al^{26}	9.0				
14	Si^{30}		15.3	Si^{29}		11.8	Si^{28}		11.4			
15				P^{31}	7.0		P^{30}	6.5				
16							S^{32}		10.1			

C. α-PARTICLES

Z	$I=-1$		0		1		2		3	
3			Li^6	1.76	Li^7	2.76				
4			Be^8	−0.14	Be^9	2.5	Be^{10}	8.0		
5			B^{10}	4.44	B^{11}	9.15	B^{12}	~10		
6			C^{12}	7.83	C^{13}	11.32	C^{14}	12.93		
7			N^{14}	12.70	N^{15}	11.92	N^{16}	~12		
8	O^{15}	11.3	O^{16}	7.87	O^{17}	7.00	O^{18}	7.87		
9	F^{17}	6.4	F^{18}	~5.8	F^{19}	4.26	F^{20}	~8		
10			Ne^{20}	5.08	Ne^{21}	8.71	Ne^{22}	8.94		
11			Na^{22}	9.3	Na^{23}	12.3	Na^{24}	~14		
12			Mg^{24}	10.3	Mg^{25}	9.8	Mg^{26}	12.8		
13			Al^{26}	11.2	Al^{27}	10.1	Al^{28}	11.0		
14			Si^{28}	9.7	Si^{29}	11.1	Si^{30}	10.5	Si^{31}	9.8
15			P^{30}	8.6	P^{31}	9.5	P^{32}	10.1		
16			S^{32}	8.2			S^{34}	~9		
17			Cl^{34}	~11	Cl^{35}	7.9				
18							A^{38}	~8		

TABLE LXXVI. *Energy difference between isobars (weight of nucleus with lower charge minus that with higher charge in thousandths of a mass unit).*

$I=+1/-1$		$I=2/0$, Z and N odd		$I=2/0$, Z and N even		$I=1/3$	
n^1-H^1	$+0.84$	He^6-Li^6	3.9	Li^8-Be^8	17	$Mg^{27}-Al^{27}$	2.2
H^3-He^3	-0.02	$Be^{10}-B^{10}$	0.40	$B^{12}-C^{12}$	~ 15	$Al^{29}-Si^{29}$	3.8
$B^{11}-C^{11}$	-2.34	$C^{14}-N^{14}$	0.17	$N^{16}-O^{16}$	~ 11	$Si^{31}-P^{31}$	1.9
$C^{13}-N^{13}$	-2.34	$O^{18}-F^{18}$	-1.9	$F^{20}-Ne^{20}$	~ 8		
$N^{15}-O^{15}$	-2.9	$Ne^{22}-Na^{22}$	-1.6	$Na^{24}-Mg^{24}$	5.0		
$O^{17}-F^{17}$	-3.1	$Mg^{26}-Al^{26}$	-3.1	$Al^{28}-Si^{28}$	3.7		
		$Si^{30}-P^{30}$	-5.0	$P^{32}-S^{32}$	1.8		
		$S^{34}-Cl^{34}$	$-3??$	$K^{40}-Ca^{40}$	0.8		

in the mass spectrograph and are also connected by the reaction $Si^{28}+He^4=P^{31}+H^1$; the observed reaction energy is 0.8 MV less than that obtained from the masses; it was therefore assumed that the spectrograph masses of Si^{28} and P^{31} are 0.2 units too high and too low, respectively, and the actual energy evolution in the reaction is 0.4 MV higher than the observed one. Cl^{35} has been based entirely on the reaction $S^{32}+He^4=Cl^{35}+H^1$ which reduces the probable error considerably since the mass of S^{32} is very accurately known.

§109. Excited States of Nuclei

The evidence for excited states of nuclei is obtained from three sources, *viz.*

(1) groups of outgoing particles from nuclear reactions,
(2) γ-rays emitted by the residual nucleus,
(3) resonance effects.

From (3), only states above the dissociation energy may be obtained, from (2), primarily states below the dissociation energy, because the higher states will ordinarily disintegrate with emission of particles rather than γ-rays. The particle groups (1) give evidence of high as well as low states.

The evidence is reported and discussed together with the various reactions in chapter XVII, that on resonance levels also in Tables XXXIX and XXXX (§§81, 82). In Table LXXIV, we give a summary of the excited levels for which there is experimental evidence. In the first column, the nucleus is given, in the second, the number of the excitation level, in the third, the excitation energy in MV above the ground state; then follows the width of the level, and the mass of the nucleus in that particular state (which may be useful for comparing the levels of isobars and other similar nuclei). The following column gives the probable spectroscopic character of the level according to the theory of Feenberg and Wigner. Where it is reasonably certain the full spectroscopic character is given, e.g. $^2P_{\frac{1}{2}}$, sometimes only the orbital momentum (S, P, D level etc.) or the total angular momentum (e.g., 1). The parity of the term is indicated by g (even) or u (odd). The "class" A, B, C, D denotes the reliability of the evidence about the level, A denoting the most reliable. The next column gives the reaction from which the evidence is obtained; the letter P means that the evidence comes from a group of outgoing particles, γ that it comes from γ-rays, and R that the level is a resonance level. The last column indicates the γ-ray transitions observed, giving first the quantum energy of the γ-ray, then the transition to which it is ascribed (using the numbers of the levels introduced in the second column) and finally the nuclear reaction in which the γ-rays are produced.

It need hardly be pointed out that practically each of the nuclei listed will have many more levels than are known at present.

§110. Discussion

The masses given in Table LXXIII show, with greater precision than earlier mass tables, the often-discussed features of nuclear masses: For very light nuclei, we have the well-known periodicity with the period 4, with the "multiples of α-particles" having an especially low mass. For higher atomic weight, this periodicity becomes

Table LXXVII. *Stable isotopes of the elements.**

Element	Z	A	Relative Abundance
H	1	1	99.98
		2	0.02
		3	$\sim 10^{-8}$
He	2	4	—
Li	3	6	7.9
		7	92.1
Be	4	9	—
B	5	10	20.6
		11	79.4
C	6	12	99
		13	~1
N	7	14	99.7
		15	0.3
O	8	16	99.76
		17	0.04
		18	0.20
F	9	19	—
Ne	10	20	90.00
		21	0.27
		22	9.73
Na	11	23	—
Mg	12	24	77.4
		25	11.5
		26	11.1
Al	13	27	—
Si	14	28	89.6
		29	6.2
		30	4.2
P	15	31	—
S	16	32	97.0
		33	0.8
		34	2.2
Cl	17	35	76
		37	24
A	18	36	0.31
		38	0.06
		40	99.63
K	19	39	93.2
		40	0.012
		41	6.8
Ca	20	40	96.76
		42	0.77
		43	0.17
		44	2.30
Sc	21	45	—
Ti	22	46	8.5
		47	7.8
		48	71.3
		49	5.5
		50	6.9
V	23	51	—
Cr	24	50	4.9
		52	81.6
		53	10.4
		54	3.1
Mn	25	55	—
Fe	26	54	6.5
		56	90.2
		57	2.8
		58	0.5
Co	27	57	0.2
		59	99.8
Ni	28	58	67.2
		60	27.0
		61	~1
		62	3.8
		64	~1
Cu	29	63	68
		65	32

Element	Z	A	Relative Abundance
Zn	30	64	50.9
		66	27.3
		67	3.9
		68	17.4
		70	0.5
Ga	31	69	61.2
		71	38.8
Ge	32	70	21.2
		72	27.3
		73	7.9
		74	37.1
		76	6.5
As	33	75	—
Se	34	74	0.9
		76	9.5
		77	8.3
		78	24.0
		80	48.0
		82	9.3
Br	35	79	50.7
		81	49.3
Kr	36	78	0.42
		80	2.45
		82	11.79
		83	11.79
		84	56.85
		86	16.70
Rb	37	85	72.7
		87	27.3
Sr	38	84	0.5
		86	9.6
		87	7.5
		88	82.4
Y	39	89	—
Zr	40	90	48
		91	11.5
		92	22
		94	17
		96	1.5
Cb	41	93	—
Mo	42	92	14.2
		94	10.0
		95	15.5
		96	17.8
		97	9.6
		98	23.0
		100	9.8
Ma	43	—	—
Ru	44	96	5
		98	—
		99	12
		100	14
		101	22
		102	30
		104	17
Rh	45	101	0.1
		103	99.9
Pd	46	102	0.8
		104	9.3
		105	22.6
		106	27.2
		108	26.8
		110	13.5
Ag	47	107	52.5
		109	47.5
Cd	48	106	1.4
		108	1.0
		110	12.8

Element	Z	A	Relative Abundance
Cd	48	111	13.0
		112	24.2
		113	12.3
		114	28.0
		116	7.3
In	49	113	4.5
		115	95.5
Sn	50	112	1.1
		114	0.8
		115	0.4
		116	15.5
		117	9.1
		118	22.5
		119	9.8
		120	28.5
		122	5.5
		124	6.8
Sb	51	121	56
		123	44
Te	52	120	<0.1
		122	2.9
		123	1.6
		124	4.5
		125	6.0
		126	19.0
		128	32.8
		130	33.1
I	53	127	—
Xe	54	124	0.08
		126	0.08
		128	2.30
		129	27.13
		130	4.18
		131	20.67
		132	26.45
		134	10.31
		136	8.79
Cs	55	133	—
Ba	56	130	0.16
		132	0.015
		134	1.72
		135	5.7
		136	8.5
		137	10.8
		138	73.1
La	57	139	—
Ce	58	136	<1
		138	<1
		140	89
		142	11
Pr	59	141	—
Nd	60	142	36
		143	11
		144	30
		145	5
		146	18
		148	<1
		150	<1
—	61	—	—
Sm	62	144	3
		147	17
		148	14
		149	15
		150	5
		152	26
		154	20
Eu	63	151	50.6
		153	49.4

Element	Z	A	Relative Abundance
Gd	64	155	21
		156	23
		157	17
		158	23
		160	16
Tb	65	159	—
Dy	66	161	22
		162	25
		163	25
		164	28
Ho	67	165	—
Er	68	166	36
		167	24
		168	30
		170	10
Tm	69	169	—
Yb	70	171	9
		172	24
		173	17
		174	38
		176	12
Lu	71	175	—
Hf	72	176	5
		177	19
		178	28
		179	18
		180	30
Ta	73	181	—
W	74	182	22.6
		183	17.3
		184	30.2
		186	29.9
Re	75	185	38.2
		187	61.8
Os	76	186	1.0
		187	0.6
		188	13.4
		189	17.4
		190	25.1
		192	42.5
Ir	77	191	38.5
		193	61.5
Pt	78	192	0.8
		194	30.2
		195	35.3
		196	26.6
		198	7.2
Au	79	197	—
Hg	80	196	0.15
		198	10.1
		199	17.0
		200	23.3
		201	13.2
		202	29.6
		204	6.7
Tl	81	203	30.5
		205	69.5
Pb	82	204	1.50
		206	28.3
		207	20.1
		208	50.1
Bi	83	209	—
Th	90	232	—
U	92	235	<1
		238	>99

* Main reference: Rasetti, *Nuclear Physics*, pp. 157–160 (R3a, M10a). Other references: *A*, Zn, Cd (N8); Co (S2); Ni (D27a); Ga, Ba (S3a); Br (S28); Rh, Pd, Ir, Pt (S3); Hg, Pb (N9).

less pronounced, and the minimum mass shifts gradually to nuclei containing more neutrons than protons.

Figure 50 shows the mass excess of the nuclei contained in Table LXXIII. Each curve corresponds to a certain value of Z, and gives the mass excess as a function of the isotopic number I. The general tendency of the mass excess to decrease with increasing mass number is apparent. The curves in the left-hand half of Fig. 50 refer to even, those in the right-hand half to odd Z. All curves, but especially those for even Z, have minima for even number of neutrons, maxima for odd number of neutrons. These maxima and minima are most pronounced for very light nuclei. Apart from this period of two, the curves have a minimum for equal number of neutrons and protons for light nuclei, and a general trend to fall off towards the right for heavier nuclei. The latter effect is, of course, due to the Coulomb forces.

Another useful representation of nuclear masses was given by Bothe and Maier-Leibnitz (B47a). They plot the mass excess of nuclei of the same isotopic number $I=-1$, 0, 1 and 2 against the atomic weight. For odd isotopic number, smooth curves are obtained, while for even isotopic number the nuclei with even Z lie much lower than those with odd Z, each group forming again a smooth curve.

Table LXXV gives the binding energies of neutrons, protons and α-particles in nuclei. The nuclei are listed according to their isotopic number ($I=-1$ to $+3$), and, in the case of neutrons (protons), the binding energies are listed in separate columns for even and odd number of neutrons (protons). It is evident from these tables that the binding energy of "even" neutrons and protons is much greater than of "odd" ones, the difference being, in the average, of the order of 5 units. This difference is an effect of the Pauli principle (cf. §10, and F10, W17). The binding energies of neutrons (protons) decrease from the left to the right of the tables, i.e. with increasing number of neutrons (protons) already present in the nucleus, which is also primarily an effect of the Pauli principle (§6, W8, W17). In each column, the binding energy of protons decreases generally with increasing size of the nucleus (due to Coulomb forces, §8). The binding of neutrons decreases steadily for isotopic number 0 while for $I=1$ the binding energy increases at first up to a maximum near Mg and decreases afterwards. For $I=2$ the maximum binding energy occurs approximately at the end of the table. This behavior is again a consequence of the competition between nuclear and electric forces: There is a general tendency towards a higher binding energy of neutrons for higher atomic weight, but when the nuclei become less stable because of the Coulomb forces, their density will become smaller and therefore the binding energy of neutrons will decrease slowly.

The binding energies of α-particles do not show such marked trends as those of neutrons and protons but stay in general fairly constant except for an increase in the beginning and, perhaps, a very slight decrease at the end. This absence of trends makes it easier to recognize fluctuations such as the minimum in the binding energies at the fluorine isotopes: This minimum is evidence for the completion of a neutron-proton shell at O^{16} (cf. §33).[22a]

Table LXXVI contains the mass differences between some pairs of isobars. The weight of the isobar containing more protons has been subtracted from that containing more neutrons. In accordance with general rules (Coulomb forces!), the figures decrease with increasing atomic number in practically all instances. Some irregularities are apparent, but unfortunately many experimental data are still very uncertain.

[22a] The effect of the completion of the shell, though noticeable, is much less marked than would be expected on the grounds of the elementary individual particle-picture. According to that picture, the energy difference between subsequent levels of an individual particle, is about 5×10^{-3} mass units near O^{16}. This would make the difference in the binding energy of an α-particle before and after completion of the shell 20 units instead of, at the best, the 8 units observed.

References

The following list of references is not intended to be a complete list. It does contain all papers referred to explicitly in the text, both in Part B and in Part C. The references are arranged alphabetically and are denoted by a letter indicating the initial of the primary author and a number representing the position in the alphabetical list. The paragraphs in which the references appear are indicated.

A1. Alichanian, Alichanow and Dzelepow, Physik. Zeits. Sowjetunion **10**, 78 (1936). (§87, 99, 101 105)

A2. Alichanow, Alichanian and Dzelepow, Nature **133**, 871 (1934). (§99, 105)

A2a. Alichanow, Alichanian and Dzelepow, Nature **133**, 950 (1934). (§99)

A3. Alichanow, Alichanian and Dzelepow, Nature **135**, 393 (1935). (§102, 105)

A4. Allen, Phys. Rev. **51**, 182 (1937). (§100)

A5. Allison, Proc. Camb. Phil. Soc. **32**, 179 (1936). (§81)

A6. Alvarez, McMillan and Snell, Phys. Rev. **51**, 148 (1937). (§92)

A7. Amaldi, D'Agostino, Fermi, Pontecorvo, Rasetti and Segrè, Proc. Roy. Soc. **149**, 522 (1935). (§57, 59, 65, 101, 102, 105)

A8. Amaldi, Nuovo Cimento **12**, 223 (1935). (§57, 102, 105)

A9. Amaldi and Fermi, Ricerca Scient. **1**, 56 (1936). (§57)

A10. Amaldi and Fermi, Ricerca Scient. **1**, 310 (1936). (§57)

A11. Amaldi and Fermi, Phys. Rev. **50**, 899 (1936). (§58, 59, 60, 61, 62, 92, 101, 102, 105)

A11a. Amaldi, Hafstad and Tuve, Phys. Rev. **51**, 896 (1937). (§102)

A12. Anderson and Neddermeyer, Phys. Rev. **50**, 263 (1936). (§104)

A13. Andersen, Nature **137**, 457 (1936). (§102, 105)

A14. Andersen, Zeits. f. physik. Chemie **32**, 237 (1936). (§102, 105)

A15. Andersen, Nature **138**, 76 (1936). (§102, 105)

A16. Arsenjewa-Heil, Heil and Westcott, Nature **138**, 462 (1936). (§60)

A17. Aston, Nature **135**, 541 (1935). (§107)

A18. Aston, *Mass Spectra and Isotopes* (E. Arnold, 1933). (§107)

A19. Aston, Nature **137**, 357 (1936). (§107)

A20. Aston, Nature **137**, 613 (1936). (§107)

A21. Aston, Nature **138**, 1094 (1936). (§107)

A21a. Aston, Nature **139**, 922 (1937). (§107)

A22. Aoki, Nature **139**, 372 (1937). (§65)

B1. Bainbridge and Jordan, Phys. Rev. **49**, 883 (1936). (§107)

B2. Bainbridge and Jordan, Phys. Rev. **51**, 384 (1937). (§101, 107)

B2a. Bainbridge and Jordan, Phys. Rev. **50**, 282 (1936). (§107)

B3. Banks, Chalmers and Hopwood, Nature **135**, 99 (1935). (§103)

B4. Bardeen, Phys. Rev. **51**, 799 (1937). (§53)

B4a. Barnes, DuBridge, Wiig, Buck, and Strain, Phys. Rev. **51**, 775 (1937). (§100, 105)

B5. Bayley, Curtis, Gaerttner and Goudsmit, Phys. Rev. **50**, 461 (1936). (§63)

B5a. Bayley and Crane, Phys. Rev. **51**, 1012 (1937). (§101, 105)

B6. Beams and Trotter, Phys. Rev. **45**, 849 (1934). (§92)

B6a. Bennet, Proc. Roy. Soc. **155**, 419 (1936). (§95)

B7. Bernardini, Zeits. f. Physik **85**, 555 (1933). (§82, 99)

B7a. Bernardini, Ricerca Scient. **8**, 33 (1937). (§99)

B7b. Bernardini and Bocciarelli, Accad. Lincei, Atti. **24**, 132 (1936). (§99)

B8. Bethe, Ann. d. Physik **5**, 325 (1930). (§95)

B9. Bethe and Heitler, Proc. Roy. Soc. **146**, 83 (1934). (§93)

B10. Bethe and Peierls, Int. Conf. Phys., London (1934). (§103)

B11. Bethe, Phys. Rev. **47**, 633 (1935). (§99, 101, 107)

B12. Bethe, Phys. Rev. **47**, 747 (1935). (§57, 65, 102)

B13. Bethe, Phys. Rev. **50**, 332 (1936). (§52)

B14. Bethe, Phys. Rev. **50**, 977 (1936). (§101)

B15. Bethe and Placzek, Phys. Rev. **51**, 450 (1937). (§52, 54, 61, 71, 91, 102)

B16. Bethe, *Handbuch der Physik*, Vol. 24, 273 (1933). (§51, 95)

B17. Bethe, Phys. Rev. to be published shortly. (§80)

B18. Bethe, Phys. Rev. **51**, 1004 (1937). (§79, 80)

B19. Bethe, Zeits. f. Physik **76**, 293 (1932). (§95)

B20. Birge, Phys. Rev. **37**, 1669 (1931). (§101)

B21. Bjerge and Westcott, Nature **134**, 286 (1934). (§102, 105)

B22. Bjerge and Westcott, Nature **134**, 177 (1934). (§102)

B23. Bjerge, Nature **137**, 865 (1936). (§102)

B24. Bjerge and Broström, Nature **138**, 400 (1936). (§102, 105)

B24a. Bjerge, Nature **139**, 757 (1937). (§102)

B25. Blackett and Lees, Proc. Roy. Soc. **134**, 658 (1932). (§95)

B26. Blackett, Proc. Roy. Soc. **135**, 132 (1932). (§95)

B27. Blackett and Champion, Proc. Roy. Soc. **130**, 380 (1931). (§74)

B28. Blewett, Phys. Rev. **49**, 900 (1936). (§87, 105)

B29. Bloch, Ann. d. Physik **16**, 285 (1933). (§95)

B30. Bloch and Gamow, Phys. Rev. **50**, 260 (1936). (§87)

B31. Bloch, Zeits. f. Physik **81**, 363 (1933). (§95)

B32. Bohr, Nature **137**, 344 (1936). (§51, 57, 100, 102)

B33. Bohr and Kalckar, Kgl. Dansk Acad. (1939). (§51–56, 77, 79, 81–90)

B33a. Bohr, Phil. Mag. **25**, 10 (1913). (§95)

B34. Bohr, Phil. Mag. **30**, 581 (1915). (§97)

B35. Bonner and Mott-Smith, Phys. Rev. **46**, 258 (1934). (§99)

B36. Bonner and Brubaker, Phys. Rev. **48**, 469 (1935). (§102)

B37. Bonner and Brubaker, Phys. Rev. **48**, 742 (1935). (§85, 97, 101)

B38. Bonner and Brubaker, Phys. Rev. **49**, 19 (1936). (§97, 101)

B39. Bonner and Brubaker, Phys. Rev. **49**, 778 (1936). (§102, 105)

B40. Bonner and Brubaker, Phys. Rev. **50**, 308 (1936). (§74, 85, 97, 99, 101, 108)

B41. Bonner and Brubaker, Phys. Rev. **50**, 781 (1936). (§65, 95, 102)

B42. Born, Zeits. f. Physik **58**, 306 (1929). (§66)

B43. Bothe and Becker, Zeits. f. Physik **66**, 289 (1930). (§99)

B44. Bothe and Frãnz, Zeits. f. Physik **49**, 1 (1928). (§95)

B45. Bothe and Baeyer, Gött. Nachr. **1**, 195 (1935). (§94)

B46. Bothe, Physik. Zeits. **36**, 776 (1935). (§99)

B47. Bothe, Zeits. f. Physik **100**, 273 (1936). (§99, 101)

B47a. Bothe and Maier-Leibnitz, Niturwiss. **25**, 25 (1937). (§110)

B47b. Bothe and Gentner, Naturwiss. **25**, 90 (1937). (§91, 103)

B47c. Bothe and Gentner, Naturwiss. **25**, 126 (1937). (§91, 103)

B47d. Bothe and Gentner, Naturwiss. **25**, 191 (1937). (§91, 103)

B47e. Bothe and Gentner, Naturwiss. **25**, 284 (1937). (§103)

B47f. Bothe and Gentner, Zeits. f. Physik **106**, 236 (1937). (§102)

B47g. Bothe and Gentner, Zeits. f. Physik **104**, 685 (1937). (§100)

B48. Brasch, Naturwiss. **21**, 82 (1933). (§92)

B49. Brasch, Lange, Waly, Banks, Chalmers, Szilard and Hopwood, Nature **134**, 880 (1934). (§92)

B50. Brasefield and Pollard, Phys. Rev. **50**, 296 (1936). (§99)

B51. Breit and Wigner, Phys. Rev. **49**, 519 (1936). (§52, 57, 86, 100, 102)

B52. Breit and Condon, Phys. Rev. **49**, 904 (1936). (§103)

B53. Breit, Condon and Present, Phys. Rev. **50**, 825 (1936). (§72)

B53a. Breit, R. S. I. **8**, 95 (1937). (§99)

B54. Breit, Phys. Rev. **51**, 248, 778 (1937). (§84)

B54a. Breit and Wigner, Phys. Rev. **51**, 593 (1937). (§105)

B55. Briggs, Proc. Roy. Soc. **114**, 313 (1927). (§97)

B56. Briggs, Proc. Roy. Soc. **114**, 341 (1927). (§96)

B57. Briggs, Proc. Roy. Soc. **118**, 549 (1928). (§95)

B58. Briggs, Proc. Roy. Soc. **139**, 638 (1933). (§95)

B59. Briggs, Proc. Roy. Soc. **143**, 604 (1934). (§95)

B60. Briggs, Proc. Roy. Soc. **157**, 183 (1936). (§92)

B61. Brillouin, Comptes rendus **183**, 24 (1926). (§66)

B62. Brown and Mitchell, Phys. Rev. **50**, 593 (1936). (§105)

B63. Burcham and Goldhaber, Proc. Camb. Phil. Soc. **32**, 632 (1936). (§102)

B64. Blau, J. de phys. et rad. **61** (1934). (§99)

B65. Bruche and Scherzer, *Geometrische Elektronenoptik* (J. Springer, 1934). (§106)

B66. Brubaker and Pollard, Phys. Rev. **51**, 1013 (1937). (§99)

B67. Buck, Strain, and Valley, Phys. Rev. **51**, 1012 (1937). (§100, 105)

B68. Burhop, Proc. Camb. Phil. Soc. **32**, 643–7 (1936). (§92, 100, 101)

C1. Carlson and Oppenheimer, Phys. Rev. **51**, 220 (1937). (§104)

C2. Chadwick and Bieler, Phil. Mag. **42**, 923 (1921). (§76)

C3. Chadwick, Proc. Roy. Soc. **128**, 114 (1930). (§74)

C4. Chadwick and Constable, Proc. Roy. Soc. **135**, 48 (1931). (§82, 99)

C5. Chadwick, Proc. Roy. Soc. **136**, 692 (1932). (§99)

C6. Chadwick, Nature **129**, 312 (1932). (§99)

C7. Chadwick, Proc. Roy. Soc. **142**, 1 (1933). (§82, 99)

C8. Chadwick, Feather and Davies, Proc. Camb. Phil. Soc. **30**, 357 (1934). (§85, 112)

C9. Chadwick and Goldhaber, Nature **134**, 237 (1934). (§91, 92)

C10. Chadwick and Feather, Int. Conf. Phys., London (1934). (§82, 99)

C11. Chadwick, Phil. Mag. **40**, 734 (1920). (§70)

C12. Chadwick and Goldhaber, Nature **135**, 65 (1935). (§102)

C13. Chadwick, Constable and Pollard, Proc. Roy. Soc. **130**, 463 (1931). (§82)

C14. Chadwick and Goldhaber, Proc. Roy. Soc. **151**, 479 (1935). (§93, 103)

C15. Chadwick, Phil. Mag. **2**, 1056 (1926). (§99)

C16. Chadwick and Goldhaber, Proc. Camb. Phil. Soc. **31**, 612 (1935). (§57, 102)

C16a. Chang, Goldhaber and Sagane, Nature **139**, 962 (1937). (§102, 103)

C17. Coates, Phys. Rev. **46**, 542 (1934). (§104)

C18. Cockcroft and Walton, Proc. Roy. Soc. **129**, 477 (1930). (§92)

C19. Cockcroft and Walton, Proc. Roy. Soc. **137**, 229 (1932). (§92, 100)

C20. Cockcroft and Walton, Nature **129**, 242 (1932). (§95)

C21. Cockcroft and Walton, Nature **129**, 649 (1932). (§100)

C22. Cockcroft and Walton, Proc. Roy. Soc. **136**, 619 (1932). (§92)

C23. Cockcroft and Walton, Nature **131**, 23 (1933). (§100)

C24. Cockcroft, Gilbert and Walton, Nature **133**, 328 (1934). (§100, 105)

C25. Cockcroft and Walton, Proc. Roy. Soc. **144**, 704 (1934). (§92, 96, 101)

C26. Cockcroft, Int. Conf. Phys., London (1934). (§96, 100, 101)

C27. Cockcroft, Gilbert and Walton, Proc. Roy. Soc. **148**, 225 (1935). (§92, 101)

C28. Cockcroft and Lewis, Proc. Roy. Soc. **154**, 246 (1936). (§85, 96, 101)

C29. Cockcroft and Lewis, Proc. Roy. Soc. **154**, 261 (1936). (§85, 96, 101, 107)
C30. Collie, Nature **137**, 614 (1936). (§93, 102)
C31. Condon and Breit, Phys. Rev. **49**, 229 (1936). (§59)
C32. Condon and Gurney, Phys. Rev. **33**, 127 (1929). (§66, 92)
C33. Cooksey and Lawrence, Phys. Rev. **49**, 866 (1936). (§101)
C34. Cork, Richardson and Kurie, Phys. Rev. **49**, 208 (1936). (§101, 105)
C35. Cork and Lawrence, Phys. Rev. **49**, 788 (1936). (§79, 84, 85, 101, 105)
C36. Cork and Thornton, Phys. Rev. **51**, 59 (1937). (§79, 85, 92, 101, 105)
C36a. Cork and Thornton, Phys. Rev. **51**, 608 (1937). (§101, 105)
C37. Crane, Lauritsen and Soltan, Phys. Rev. **44**, 514 (1933). (§101)
C38. Crane, Lauritsen and Soltan, Phys. Rev. **44**, 692 (1933). (§78, 101)
C39. Crane and Lauritsen, Phys. Rev. **45**, 497 (1934). (§81, 100, 105)
C40. Crane and Lauritsen, Phys. Rev. **45**, 430 (1934). (§101, 105)
C41. Crane, Lauritsen and Soltan, Phys. Rev. **45**, 507 (1934). (§101)
C42. Crane, Delsasso, Fowler and Lauritsen, Phys. Rev. **46**, 531 (1934). (§100)
C43. Crane and Lauritsen, Int. Conf. Phys., London (1934). (§99)
C44. Crane, Delsasso, Fowler and Lauritsen, Phys. Rev. **46**, 1109 (1934). (§99, 101)
C45. Crane, Delsasso, Fowler and Lauritsen, Phys. Rev. **47**, 782 (1935). (§100, 101)
C46. Crane, Delsasso, Fowler and Lauritsen, Phys. Rev. **47**, 887 (1935). (§101, 105)
C47. Crane, Delsasso, Fowler and Lauritsen, Phys. Rev. **48**, 484 (1935). (§100, 101)
C48. Crane, Delsasso, Fowler and Lauritsen, Phys. Rev. **47**, 971 (1935). (§101, 105)
C49. Crane, Delsasso, Fowler and Lauritsen, Phys. Rev. **48**, 100 (1935). (§101)
C50. Crane, Delsasso, Fowler and Lauritsen, Phys. Rev. **48**, 102 (1935). (§81, 90, 100)
C51. Crane, Delsasso, Fowler and Lauritsen, Phys. Rev. **48**, 125 (1935). (§90, 94, 100)
C52. Crane, Phys. Rev. **52**, 11 (1937). (§92)
C53. Curie and Joliot, J. de phys. et rad. **5**, 153 (1934). (§94)
C54. Curie and Joliot, Comptes rendus **198**, 254 (1934). (§99, 105)
C55. Curie and Joliot, Comptes rendus **197**, 237 (1933). (§99)
C56. Curie and Joliot, J. de phys. et rad. **4**, 21 (1933). (§99)
C57. Curie and Joliot, J. de phys. et rad. **4**, 494 (1933). (§105)
C58. Curie and Joliot, Comptes rendus **198**, 559 (1934). (§105)
C59. Curie and Joliot, J. de phys. et rad. **4**, 278 (1933). (§99)
C60. Curie, Halban and Preiswerk, J. de phys. et rad. **7**, 361 (1935). (§102, 105)
C61. Curie and Preiswerk, Comptes rendus **203**, 787 (1936). (§102)
D1. Dahl, Hafstad and Tuve, R. S. I. **4**, 373 (1933). (§94)
D2. Danysz, Rotblat, Wertenstein and Zyw, Nature **134**, 970 (1934). (§65)
D3. Danysz and Zyw, Acta Phys. Polonica **3**, 485 (1934). (§99)
D3a. Darling, Curtis and Cork, Phys. Rev. **51**, 1010 (1937). (§101, 105)
D4. Darrow, Bell S. Tech. J. **10**, 628 (1931). (§99, 105)
D5. Darrow, R. S. I. **5**, 66 (1934). (§99)
D6. Dee and Walton, Proc. Roy. Soc. **141**, 733 (1933). (§100, 101)
D7. Dee, Nature **133**, 564 (1934). (§101)
D8. Dee and Gilbert, Proc. Roy. Soc. **149**, 200 (1935). (§101)
D9. Dee and Gilbert, Proc. Roy. Soc. **154**, 279 (1936). (§85, 100, 101)
D10. Delbruck and Gamow, Zeits. f. Physik **72**. 492 (1931). (§87)
D11. Delsasso, Fowler and Lauritsen, Phys. Rev. **48**, 848 (1935). (§101)
D12. Delsasso, Fowler and Lauritsen, Phys. Rev. **50**, 389 (1936). (§100)
D13. Delsasso, Fowler and Lauritsen, Phys. Rev. **51**, 391 (1937). (§81, 90, 100)
D14. Delsasso, Fowler and Lauritsen, Phys. Rev. **51**, 527 (1937). (§90, 100)
D15. Dolch, Zeits. f. Physik **100**, 401 (1936). (§84)
D16. Doolittle, Phys. Rev. **49**, 779 (1936). (§84, 100)
D17. Döpel, Zeits. f. Physik **91**, 796 (1934). (§100)
D18. Duncanson, Proc. Camb. Phil. Soc. **30**, 102 (1934). (§95)
D19. Duncanson and Miller, Proc. Roy. Soc. **146**, 396 (1934). (§82, 99)
D20. Dunning and Pegram, Phys. Rev. **45**, 295 (1934). (§92)
D21. Dunning, Phys. Rev. **45**, 586 (1934). (§59, 85, 99)
D22. Dunning, R. S. I. **5**, 387 (1934). (§94)
D23. Dunning, Pegram, Fink and Mitchell, Phys. Rev. **48**, 265 (1935). (§57, 58, 60, 62, 63, 74)
D24. Dunning, Pegram, Fink and Mitchell, Phys. Rev. **47**, 970 (1935). (§102)
D25. Dunning, Pegram, Fink and Mitchell, Phys. Rev. **48**, 704 (1935). (§60)
D26. Dunning, Fink, Pegram and Segrè, Phys. Rev. **49**, 199 (1936). (§60)
D27. Dempster, Phys. Rev. **11**, 316 (1918). (§106)
D27a. Dempster, Phys. Rev. **50**, 98 (1936). (§107)
D28. DuBridge, Barnes and Buck, Phys. Rev. **51**, 995 (1937). (§100, 105)
D29. Dyer, *The Long Death* (Chas. Scribner's Sons, 1937). (§92)
E1. Ehrenberg, Nature **136**, 870 (1935). (§65, 102)
E2. Ellis, Int. Conf. Phys., London (1934). (§69)

E3. Ellis and Mott, Proc. Roy. Soc. **139**, 369 (1933). (§69)
E4. Ellis, Proc. Roy. Soc. **136**, 396 (1932). (§69)
E5. Ellis and Henderson, Nature **135**, 429 (1935). (§99, 101, 105)
E6. Ellis and Henderson, Nature **136**, 755 (1935). (§105)
E7. Ellis, Proc. Roy. Soc. **138**, 318 (1932). (§88)
E8. Ellis, Proc. Roy. Soc. **143**, 350 (1934). (§69, 103)
E9. Ellis and Aston, Proc. Roy. Soc. **129**, 180 (1930). (§88)
E10. Ellis and Henderson, Proc. Roy. Soc. **156**, 358 (1936). (§99, 105)
E11. Evans and Livingston, Rev. Mod. Phys. **7**, 229 (1935). (§98)
F1. Fahlenbrach, Zeits. f. Physik **96**, 503 (1935). (§99, 105)
F2. Farkas and Farkas, *Ortho-hydrogen, Para-hydrogen and Heavy hydrogen*. (§101)
F3. Fay, Phys. Rev. **50**, 560 (1936). (§65)
F4. Fea, Nuovo Cimento **12**, 368 (1935). (§101)
F5. Feather, Proc. Roy. Soc. **136**, 709 (1932). (§102)
F6. Feather, Nature **130**, 257 (1932). (§102)
F7. Feather, Proc. Roy. Soc. **141**, 194 (1933). (§95)
F8. Feather, Proc. Roy. Soc. **142**, 689 (1933). (§93, 94, 102)
F9. Feather, Nature **136**, 468 (1935). (§103)
F9a. Feenberg, Phys. Rev. **49**, 328 (1936). (§84)
F10. Feenberg and Wigner, Phys. Rev. **51**, 95 (1937). (§81, 83, 89, 99, 101, 110)
F11. Feenberg and Phillips, Phys. Rev. **51**, 597 (1937). (§110)
F12. Fermi, Ric. Sci. **2**, 12 (1933). (§102, 105)
F13. Fermi, Amaldi, D'Agostino, Rasetti and Segrè, Proc. Roy. Soc. **146**, 483 (1934). (§57, 65, 94, 102, 105)
F14. Fermi, Nature **133**, 757 (1934). (§102)
F15. Fermi, Pontecorvo and Rasetti, Ricerca Scient. **2** (1934). (§105)
F16. Fermi and Rasetti, Nuovo Cimento **12**, 201 (1935). (§105)
F17. Fermi, Ric. Sci. **7**, 13 (1936). (§59, 102)
F18. Fermi, *Zeeman Jubilee* (1935), p. 128. (§59)
F19. Fink, Dunning, Pegram and Mitchell, Phys. Rev. **49**, 103 (1936). (§60, 61)
F20. Fink, Phys. Rev. **50**, 738 (1936). (§59)
F21. Fischer-Colbrie, Ak. W. Wien, **145**, 283 (1936). (§99)
F22. Fisk, Schiff and Shockley, Phys. Rev. **50**, 1090 (1936). (§59)
F23. Fisk and Morse, Phys. Rev. **51**, 54 (1937). (§59)
F24. Fleischmann and Gentner, Zeits. f. Physik **100**, 440 (1936). (§102, 103)
F25. Fleischmann, Zeits. f. Physik **103**, 113 (1936). (§57, 90, 102)
F26. Flügge, and Krebs, Physik. Zeits. **36**, 466 (1935). (§101)
F26a. Flügge and Krebs, Physik. Zeits. **38**, 13 (1937). (§101)
F27. Fomin and Houtermans, Physik. Zeits. Sowjetunion **9**, 273 (1936). (§102, 105)
F28. Fowler, Delsasso and Lauritsen, Phys. Rev. **49**, 561 (1936). (§101, 105)
F28a. Fowler and Lauritsen, Phys. Rev. **51**, 1103 (1937). (§101)
F29. Fox and Rabi, Phys. Rev. **48**, 746 (1935). (§101)
F30. Frenkel, Physik. Zeits. Sowjetunion **9**, 533 (1936). (§53)
F31. Frisch, Nature **133**, 721 (1934). (§99, 101, 105)
F32. Frisch, Nature **136**, 220 (1935). (§99, 105)
F33. Frisch and Placzek, Nature **137**, 357 (1936). (§58, 61, 102)
F33a. Fünfer, Ann. d. Physik **29**, 1 (1937). (§102)
F34. Furry, Phys. Rev. **51**, 592 (1937). (§59)
G1. Gaerttner, Turin and Crane, Phys. Rev. **49**, 793 (1936). (§105)
G2. Gaerttner and Crane, Phys. Rev. **51**, 58 (1937). (§100)
G3. Gaerttner and Crane, Phys. Rev. **51**, 49 (1937). (§100)
G4. Gamow, Nature **133**, 833 (1934). (§87, 105)
G5. Gamow, Phys. Rev. **49**, 946 (1936). (§65)
G6. Gamow, Physik. Zeits. **32**, 651 (1931). (§69)
G7. Gamow, Zeits. f. Physik **52**, 510 (1929). (§66, 92)
G8. Gamow, Nature **122**, 805 (1929). (§66)
G9. Gamow, *Atomic Nuclei and Radioactivity* (Cambridge, 1937). (§68)
G10. Gamow and Houtermans, Zeits. f. Physik **52**, 496 (1929). (§66, 68)
G11. Geiger, Proc. Roy. Soc. **83**, 492 and 505 (1910). (§95)
G12. Geiger, *Handbuch der Physik*, Vol. 24 (1927), 137. (§95)
G13. Geiger and Nuttall, Phil. Mag. **22**, 613 (1911). (§68)
G13a. Gentner, Naturwiss. **25**, 12 (1937)
G14. Gerthsen and Reusse, Physik. Zeits. **34**, 478 (1933). (§95)
G14a. Giarratana and Brennecke, Phys. Rev. **49**, 35 (1936). (§84)
G15. Goldhaber, Proc. Camb. Phil. Soc. **30**, 560 (1934). (§83, 101)
G16. Goldsmith and Cohen, Phys. Rev. **45**, 850 (1934). (§102)
G17. Goldsmith and Rasetti, Phys. Rev. **50**, 328 (1936). (§58, 60, 61)
G18. Goldsmith and Manley, Phys. Rev. **51**, 382 (1937). (§58, 60)
G18a. Goldsmith and Manley, Phys. Rev. **51**, 1022 (1937). (§60)
G18b. Goloborodko and Rosenkewitsch, Physik. Zeits. Sowjetunion **11**, 78 (1937). (§103)
G19. Goudsmit, Phys. Rev. **49**, 406 (1936). (§59)
G20. Goudsmit, Phys. Rev. **51**, 64 (1937). (§53)
G21. Gray, Proc. Camb. Phil. Soc. **27**, 103 (1931). (§93)
G22. Greinacher, Zeits. f. Physik **36**, 364 (1926). (§94)
G23. Griffiths and Szilard, Nature **139**, 323 (1937). (§90)
G24. Gurney, Nature **123**, 565 (1929). (§99)
G25. Güttinger and Pauli, Zeits. f. Physik **67**, 743 (1931). (§83)
G26. Gentner, Naturwiss. **25**, 12 (1937). (§100)

G27. Grahame, Seaborg and Gibson, Phys. Rev. **51**, 590 (1937). (§65)

G28. Gibson, Seaborg and Grahame, Phys. Rev. **51**, 370 (1937). (§65)

H1. Hafstad, Tuve and Brown, Phys. Rev. **45**, 746 (1934). (§101)

H2. Hafstad and Tuve, Phys. Rev. **47**, 506 (1935). (§92, 100)

H3. Hafstad and Tuve, Phys. Rev. **47**, 507 (1935). (§100)

H4. Hafstad and Tuve, Phys. Rev. **48**, 306 (1935). (§81, 92, 100, 105)

H5. Hafstad, Heydenburg and Tuve, Phys. Rev. **49**, 866 (1936). (§92, 100)

H6. Hafstad, Heydenburg and Tuve, Phys. Rev. **50**, 504 (1936). (§56, 81, 84)

H7. Hahn and Meitner, Naturwiss. **23**, 320 (1935). (§65, 85)

H8. von Halban and Preiswerk, Nature **137**, 905 (1936). (§58, 60)

H9. von Halban and Preiswerk, Helv. Phys. Acta **9**, 318 (1936). (§58, 60)

H10. von Halban and Preiswerk, J. de phys. et rad. **8**, 29 (1937). (§60)

H11. Hall, Phys. Rev. **49**, 401 (1936). (§103)

H11a. Halpern, Lueneburg and Clark, Phys. Rev., in press. (§59)

H12. Harkins, Gans, Newson, Phys. Rev. **44**, 529 (1933). (§102)

H13. Harkins and Gans, Phys. Rev. **46**, 397 (1934). (§102)

H14. Harkins, Gans and Newson, Phys. Rev. **47**, 52 (1935). (§102)

H15. Harkins, Kamen, Newson and Gans, Phys. Rev. **50**, 980 (1936). (§59)

H16. Hartree, Proc. Roy. Soc. **139**, 311 (1933). (§95)

H17. Haxel, Zeits. f. Physik **83**, 323 (1933). (§99)

H18. Haxel, Zeits. f. Physik **93**, 400 (1935). (§99)

H19. Haxel, Physik. Zeits. **36**, 804 (1935). (§99)

H19a. Haxel, Zeits. f. Physik **104**, 540 (1937). (§102)

H20. Henderson, M. C., Phys. Rev. **43**, 98 (1935). (§84)

H21. Henderson, Livingston and Lawrence, Phys. Rev. **45**, 428 (1934). (§100, 101, 105)

H22. Henderson, Livingston and Lawrence, Phys. Rev. **46**, 38 (1934). (§78, 100)

H23. Henderson, Phys. Rev. **48**, 480 (1935). (§78, 80)

H24. Henderson, Phys. Rev. **48**, 855 (1935). (§101, 105)

H24a. Henderson, W. J., Ridenour, White and M. C. Henderson, Phys. Rev. **51**, 1107 (1937). (§99)

H24b. Henderson and Ridenour, Phys. Rev. **52**, 40 (1937). (§99)

H25. Henneberg, Zeits. f. Physik **83**, 555 (1933). (§51)

H26. Henneberg, Zeits. f. Physik **86**, 592 (1933). (§95)

H27. Herb, Parkinson and Kerst, Phys. Rev. **48**, 118 (1935). (§84, 100)

H28. Herb, Parkinson and Kerst, Phys. Rev. **51**, 75 (1937). (§92)

H28a. Herb, Kerst and McKibben, Phys. Rev. **51**, 691 (1937). (§100)

H29. Hevesy and Levi, Nature **135**, 580 (1935). (§102, 105)

H30. Hevesy and Levi, Nature **136**, 103 (1935). (§102, 105)

H31. Hevesy and Levi, Nature **137**, 185 (1936). (§60, 62, 102, 105)

H31a. Hevesy and Levi, Kgl. Danske Acad. (1936). (§102)

H32. Hevesy, Nature **135**, 1051 (1935). (§102, 105)

H33. Heyn, Nature **138**, 723 (1936). (§85, 102, 105)

H33a. Heyn, Physica **4**, 160 (1937). (§65, 102)

H33b. Heyn, Nature **139**, 842 (1937). (§102)

H34. Hoffman and Bethe, Phys. Rev. **51**, 1021 (1937). (§61)

H35. Hönl, Zeits. f. Physik **84**, 1 (1933). (§95)

H36. Hopwood and Chalmers, Nature **135**, 341 (1935). (§93)

H37. Horsley, Phys. Rev. **48**, 1 (1935). (§74)

H38. Horvay, Phys. Rev. **50**, 897 (1936). (§59)

H38a. Hulme, McDougall, Buckingham and Fowler, Proc. Roy. Soc. **149**, 131 (1935). (§93)

H39. Hulme, Mott, Oppenheimer and Taylor, Proc. Roy. Soc. **155**, 315 (1936). (§88)

H39a. Hurst and Walke, Phys. Rev. **51**, 1033 (1937). (§99, 101, 102)

I1. Inglis, Phys. Rev. **50**, 783 (1936). (§83)

J1. Jaeckel, Zeits. f. Physik **96**, 151 (1935). (§102)

J2. Johnson and Hamblin, Nature **138**, 504 (1936). (§102, 105)

J3. Johnson and Johnson, Phys. Rev. **50**, 170 (1936). (§94)

J4. Jordan and Bainbridge, Phys. Rev. **49**, 883 (1936). (§107)

J5. Jordan and Bainbridge, Phys. Rev. **50**, 98 (1936). (§107)

J6. Jordan and Bainbridge, Phys. Rev. **51**, 385 (1937). (§107)

K1. Kapitza, Proc. Roy. Soc. **106**, 602 (1924). (§95)

K2. Kempton, Browne and Maasdorp, Proc. Roy. Soc. **157**, 372 (1936). (§83, 85, 101)

K3. Kempton, Browne and Maasdorp, Proc. Roy. Soc. **157**, 386 (1936). (§84, 85, 101)

K4. Kikuchi, Aoki and Husimi, Proc. Phys. Math. Soc. Japan **17**, 369 (1935). (§57)

K5. Kikuchi, Husimi and Aoki, Proc. Phys. Math. Soc. Japan **18**, 35 (1936). (§57, 89)

K6. Kikuchi, Aoki and Husimi, Nature **137**, 186 (1936). (§102)

K7. Kikuchi, Aoki and Husimi, Nature **137**, 398 (1936). (§65)

K8. Kikuchi, Aoki and Husimi, Nature **137**, 745 (1936). (§102)

K9. Kikuchi, Aoki and Husimi, Proc. Phys. Math. Soc. Japan **18**, 115 (1936). (§57, 65)

K10. Kikuchi, Husimi and Aoki, Nature **137**, 992 (1936). (§57)

K11. Kikuchi, Husimi and Aoki, Proc. Phys. Math. Soc. Japan **18**, 188 (1936). (§57, 102)

K12. Kikuchi, Aoki and Husimi, Nature **138**, 84 (1936). (§102)

K12a. Kikuchi, Aoki and Husimi, Proc. Phys. Math. Soc. Japan **18**, 727 (1936). (§65)

K12b. Kikuchi, Husimi and Aoki, Zeits. f. Physik **105**, 265 (1937). (§102)

K13. Kinsey, Phys. Rev. **50**, 386 (1936). (§92, 104)
K14. Kirchner, Ergeb. d. exact. Naturwiss. **13**, 57 (1934). (§101)
K15. Kirchner and Neuert, Physik. Zeits. **35**, 292 (1934). (§85, 100)
K16. Kirchner and Neuert, Physik. Zeits. **36**, 54 (1935). (§95, 100)
K17. Klarmann, Zeits. f. Physik **95**, 221 (1935). (§102)
K18. Klarmann, Zeits. f. Physik **87**, 411 (1933). (§94)
K19. Klein and Nishina, Zeits. f. Physik **52**, 853 (1928). (§93)
K20. Knol and Veldkamp, Physica **3**, 145 (1936). (§102, 105)
K21. Konig, Zeits. f. Physik **90**, 197 (1934). (§99)
K22. Konopinski and Uhlenbeck, Phys. Rev. **48**, 7 (1935). (§105)
K23. Konopinski and Bethe, Phys. Rev., to be published shortly. (§78)
K23a. Konopinski and Bethe, Phys. Rev. **51**, 1004 (1937). (§54).
K24. Kramers, Zeits. f. Physik **39**, 828 (1926). (§66)
K25. Kraus and Cork, Phys. Rev. **51**, 383 (1937). (§101, 105)
K26. Kruger and Green, Phys. Rev. **51**, 699 (1937). (§92)
K26a. Kruger, Shoupp and Stallmann, Phys. Rev. **51**, 1021 (1937). (§59)
K26b. Kruger and Green, Phys. Rev. **52**, 247 (1937). (§101)
K27. Kurie, R. S. I. **3**, 655 (1932). (§94)
K28. Kurie, Phys. Rev. **43**, 771 (1933). (§102)
K29. Kurie, Phys. Rev. **43**, 1056 (1933). (§59)
K30. Kurie, Phys. Rev. **45**, 904 (1934). (§102)
K31. Kurie, Phys. Rev. **47**, 97 (1935). (§65, 102)
K32. Kurie, Richardson and Paxton, Phys. Rev. **48**, 167 (1935). (§105)
K33. Kurie, Richardson and Paxton, Phys. Rev. **49**, 368 (1936). (§99, 101, 105)
K34. Kurtchatov, Kurtchatov, Myssowsky and Roussinow, Comptes rendus **200**, 1201 (1935). (§87, 102, 105)
K35. Kurtschatow, Nemenow and Selinow, Comptes rendus **200**, 2162 (1935). (§102, 105)
K36. Kurtschatow, Latyschew, Nemenow and Selinow, Physik. Zeits. Sowjetunion **8**, 589 (1935). (§102, 105)
K37. Kronig, Physica **4**, 171 (1937). (§85)
L1. Ladenburg, Roberts and Sampson, Phys. Rev. **48**, 467 (1935). (§92)
L1a. Ladenburg and Kanner, Phys. Rev. **51**, 1022 (1937). (§105)
L1b. Laporte, Phys. Rev., in press (§61)
L2. Laslett, Phys. Rev. **50**, 388 (1936). (§101, 105)
L3. Latimer, Hull and Libby, J. Am. Chem. Soc. **57**, 781 (1935). (§102, 105)
L4. Laue, Zeits. f. Physik **52**, 726 (1929). (§66)
L5. Lauritsen and Crane, Phys. Rev. **45**, 493 (1934). (§85, 101)
L6. Lauritsen and Crane, Phys. Rev. **45**, 63 (1934). (§100)
L7. Lauritsen and Crane, Phys. Rev. **45**, 345 (1934). (§94, 101)
L8. Lauritsen and Crane, Phys. Rev. **45**, 550 (1934). (§101)
L9. Lauritsen and Crane, Phys. Rev. **32**, 850 (1928). (§92)
L10. Lawrence, Phys. Rev. **46**, 746 (1934). (§101)
L11. Lawrence, Livingston and White, Phys. Rev. **42**, 150 (1932). (§92)
L12. Lawrence, Livingston and Lewis, Phys. Rev. **44**, 56 (1933). (§101)
L13. Lawrence and Livingston, Phys. Rev. **45**, 220 (1934). (§101)
L14. Lawrence and Livingston, Phys. Rev. **45**, 608 (1934). (§92)
L15. Lawrence, Phys. Rev. **47**, 17 (1935). (§101, 105)
L16. Lawrence, McMillan and Henderson, Phys. Rev. **47**, 273 (1935). (§96, 101)
L17. Lawrence, McMillan and Thornton, Phys. Rev. **48**, 493 (1935). (§80, 101)
L18. Lawrence and Cooksey, Phys. Rev. **50**, 1131 (1936). (§92)
L19. Lea, Proc. Roy. Soc. **150**, 637 (1935). (§65, 83, 102)
L20. Lewis, G. N., and Macdonald, J. Chem. Phys. **1**, 341 (1933). (§101)
L21. Lewis, G. N., Livingston and Lawrence, Phys. Rev. **44**, 55 (1933). (§101)
L22. Lewis, G. N., Livingston and Lawrence, Phys. Rev. **44**, 317 (1933). (§101)
L23. Lewis, W. B. and Wynn-Williams, Proc. Roy. Soc. **136**, 349 (1932). (§69, 92)
L24. Lewis, W. B., Burcham and Chang, Nature **139**, 24 (1937). (§101, 105)
L25. Livingood and Snell, Phys. Rev. **48**, 851 (1935). (§92, 104)
L26. Livingood, Phys. Rev. **50**, 425 (1936). (§101, 105)
L27. Livingood and Seaborg, Phys. Rev. **50**, 435 (1936). (§101, 105)
L27a. Livingood, Seaborg and Fairbrother, Phys. Rev. **52**, 135 (1937). (§101, 102, 105)
L27b. Livingood and Seaborg, Phys. Rev. **52**, 135 (1937). (§101, 102, 105)
L28. Livingston, Henderson and Lawrence, Phys. Rev. **44**, 316 (1933). (§100)
L29. Livingston, Henderson and Lawrence, Phys. Rev. **46**, 325 (1934). (§102)
L30. Livingston and McMillan, Phys. Rev. **46**, 437 (1934). (§101)
L31. Livingston, R. S. I. **7**, 55 (1936). (§92)
L32. Livingston and Hoffman, Phys. Rev. **50**, 401 (1936). (§102)
L33. Livingston and Hoffman, Phys. Rev. **51**, 1021 (1937). (§61)
L34. Livingston, Genevese and Konopinski, Phys. Rev. **51**, 835 (1937). (§95)
L35. Lyford and Bearden, Phys. Rev. **45**, 743 (1934). (§68)
L36. Lyman, Phys. Rev. **51**, 1 (1937). (§105)
L37. Lukirsky and Careva, Comptes rendus (U.S.S.R.) **3**, 411 (1936). (§102)
L38. Li, Proc. Roy. Soc. **158**, 571 (1937). (§88)
M1. Madsen, Nature **138**, 722 (1936). (§102, 105)

M2. Manley and Millman, Phys. Rev. **51**, 19 (1937). (§83)

M2a. Manley, Goldsmith and Schwinger, Phys. Rev. **51**, 1022 (1937). (§61)

M3. Mano, Comptes rendus **197**, 47 (1933). (§95)

M4. Mano, Ann. d. Physik **1**, 407 (1934). (§95)

M5. Mano, Comptes rendus **194**, 1235 (1932). (§95)

M6. Mano, J. de phys. et rad. **5**, 628 (1934). (§95)

M7. Marsh and Sugden, Nature **136**, 102 (1935). (§102, 105)

M8. Massey and Mohr, Proc. Roy. Soc. **148**, 206 (1935). (§74)

M9. Massey and Burhop, Phys. Rev. **48**, 468 (1935). (§51)

M10. Mattauch, Phys. Rev. **50**, 617, 1089 (1936). (§107)

M10a. Mattauch, Physik. Zeits. **35**, 567 (1934). (§107)

M10b. Mattauch, Naturwiss. **25**, 156 (1937). (§107)

M10c. Mattauch, Naturwiss. **25**, 170 (1937). (§107)

M11. May and Vaidyanathan, Proc. Roy. Soc. **155**, 519 (1936). (§99)

M12. Meitner and Philipp, Naturwiss. **20**, 929 (1932). (§102)

M13. Meitner and Philipp, Zeits. f. Physik **87**, 484 (1934). (§102)

M14. Meitner, Naturwiss. **22**, 420 (1934). (§99, 105)

M15. Meitner and Hahn, Naturwiss. **24**, 158 (1936). (§65, 85, 105)

M15a. Meitner, Ann. d. Physik **29**, 246 (1937). (§105)

M16. Miller, Duncanson and May, Proc. Camb. Phil. Soc. **30**, 549 (1934). (§82, 99, 105)

M17. Mitchell, A. C. G., and Murphy, Phys. Rev. **47**, 881 (1935). (§58, 63)

M18. Mitchell, A. C. G., and Murphy, Phys. Rev. **48**, 653 (1935). (§58, 63)

M19. Mitchell, A. C. G., Murphy and Langer, Phys. Rev. **49**, 400 (1936). (§58, 63)

M20. Mitchell, A. C. G., Murphy and Whitaker, Phys. Rev. **50**, 133 (1936). (§58, 63)

M20a. Mitchell, A. C. G., Phys. Rev. **51**, 995 (1937). (§102, 105)

M20b. Mitchell, A. C. G. and Langer, Phys. Rev. **52**, 137 (1937). (§105)

M21. Mitchell, D. P., Phys. Rev. **49**, 453 (1936). (§93, 102)

M22. Mitchell, D. P., Rasetti, Fink and Pegram, Phys. Rev. **50**, 189 (1936). (§91, 103)

M23. Mitchell, D. P., and Powers, Phys. Rev. **50**, 486 (1936). (§60)

M24. Mohr and Pringle, Nature **137**, 865 (1936). (§74)

M25. Møller, Ann. d. Physik **14**, 531 (1932). (§95)

M26. Moon and Tillman, Proc. Roy. Soc. **153**, 476 (1936). (§57, 102)

M28. Morse, Fisk and Schiff, Phys. Rev. **50**, 748 (1936). (§59)

M29. Morse, Fisk and Schiff, Phys. Rev. **51**, 706 (1937). (§59)

M30. Marsden and Taylor, Proc. Roy. Soc. **88**, 443 (1913). (§95)

M31. Mott, Proc. Roy. Soc. **126**, 259 (1929). (§74)

M32. Mott and Massey, *Theory of Atomic Collisions* (Oxford, 1933). (§51, 59, 70)

M33. Mott, Proc. Camb. Phil. Soc. **27**, 553 (1931). (§95)

M34. Mott-Smith and Bonner, Phys. Rev. **45**, 554 (1934). (§92)

Mc1. McLennan, Grimmett and Read, Nature **135**, 147 (1935). (§102, 105)

Mc2. McLennan, Grimmett and Read, Nature **135**, 505 (1935). (§102)

Mc3. McLennan and Rann, Nature **136**, 831 (1935). (§102, 105)

Mc4. McMillan, Phys. Rev. **46**, 868 (1934). (§93, 100, 101)

Mc5. McMillan and Lawrence, Phys. Rev. **47**, 343 (1935). (§101)

Mc6. McMillan and Livingston, Phys. Rev. **47**, 452 (1935). (§94, 101, 105)

Mc7. McMillan, Phys. Rev. **49**, 875 (1936). (§101, 102, 105)

N1. Nahmias and Walen, Comptes rendus **203**, 71 (1936). (§102, 105)

N2. Naidu and Siday, Proc. Phys. Soc. London **48**, 330 (1936). (§105)

N3. Naidu, Nature **137**, 578 (1936). (§102, 105)

N4. Neuert, Physik. Zeits. **36**, 629 (1935). (§95, 100)

N4a. Neuert, Physik. Zeits. **38**, 122 (1937). (§100, 101)

N5. Newson, Phys. Rev. **48**, 482 (1935). (§101, 105)

N6. Newson, Phys. Rev. **48**, 790 (1935). (§96, 99, 105)

N7. Newson, Phys. Rev. **49**, 208 (1936). (§101)

N7a. Newson, Phys. Rev. **51**, 620 (1937). (§101)

N7b. Newson, Phys. Rev. **51**, 624 (1937). (§101, 105)

N8. Nier, Phys. Rev. **50**, 1041 (1936). (§107)

N9. Nier, Phys. Rev. **51**, 1007 (1937). (§107)

O½. Oeser and Tuck, Nature **139**, 1110 (1937). (§102)

O1. Oliphant, Shire and Crowther, Proc. Roy. Soc. **146**, 922 (1934). (§100, 101)

O2. Oliphant and Rutherford, Proc. Roy. Soc. **141**, 259 (1933). (§85, 92, 100)

O3. Oliphant, Kinsey and Rutherford, Proc. Roy. Soc **141**, 722 (1933). (§78, 85, 100, 101)

O4. Oliphant, Shire and Crowther, Nature **133**, 377 (1934). (§101)

O5. Oliphant, Harteck and Rutherford, Nature **133**, 413 (1934). (§101)

O6. Oliphant, Int. Conf. Phys., London (1934). (§84, 100, 101)

O7. Oliphant, Harteck and Rutherford, Proc. Roy. Soc. **144**, 692 (1934). (§92, 101)

O8. Oliphant, Kempton and Rutherford, Proc. Roy. Soc. **149**, 406 (1935). (§96, 100, 101)

O9. Oliphant, Kempton and Rutherford, Proc. Roy. Soc. **150**, 241 (1935). (§85, 96, 100, 101)

O10. Oppenheimer, Phys. Rev. **43**, 380 (1933). (§101)

O11. Oppenheimer and Phillips, Phys. Rev. **48**, 500 (1935). (§80, 101)

O12. Oppenheimer and Serber, Phys. Rev. **50**, 391 (1936). (§53)

O13. Ortner and Stetter, Zeits. f. Physik **54**, 449 (1929). (§94)

O14. Ostrofsky, Breit and Johnson, Phys. Rev. **49**, 196 (1936). (§84)

O15. Ostrofsky, Breit and Johnson, Phys. Rev. **49**, 22 (1936). (§84, 100)

O16. Ostrofsky, Bleick and Breit, Phys. Rev. **49**, 352 (1936). (§84)
P1. Paton, Phys. Rev. **46**, 229 (1934). (§99)
P1a. Paneth and Lollett, Nature **136**, 950 (1935). (§99)
P2. Paton, Zeits. f. Physik **90**, 586 (1934). (§82)
P3. Paton, Phys. Rev. **47**, 197 (1935). (§99)
P4. Paxton, Phys. Rev. **49**, 206 (1936). (§101)
P5. Peierls, Proc. Roy. Soc. **149**, 467 (1935). (§105)
P6. Perrin and Elsasser, J. de phys. et rad. **6**, 194 (1935). (§57)
P6a. Polanyi and Wigner, Zeits. f. physik. Chemie **139**, 439 (1928). (§51)
P7. Pollard, Proc. Leeds Phil. Lit. Soc. **2**, 324 (1932). (§99)
P8. Pollard and Brasefield, Phys. Rev. **50**, 890 (1936). (§99)
P9. Pollard and Brasefield, Phys. Rev. **51**, 8 (1937). (§99)
P10. Pollard, Schultz and Brubaker, Phys. Rev. **51**, 140 (1937). (§99)
P11. Pool and Cork, Phys. Rev. **51**, 383 (1937). (§101, 105)
P11a. Pool and Cork, Phys. Rev. **51**, 1010 (1937). (§101, 105)
P11b. Pool, Cork and Thornton, Phys. Rev. **51**, 890 (1937). (§102, 105)
P11c. Pool, Cork and Thornton, Phys. Rev. **52**, 41 (1937). (§102, 105)
P11d. Pool, Cork and Thornton, Phys. Rev. **52**, 239 (1937). (§102, 105)
P12. Pose, Physik. Zeits. **30**, 780 (1929). (§99)
P13. Pose and Diebner, Zeits. f. Physik **90**, 773 (1934). (§74)
P14. Powers, Fink and Pegram, Phys. Rev. **49**, 650 (1936). (§61, 62)
P15. Preiswerk and Halban, Comptes rendus **201**, 722 (1935). (§102, 105)
R1. Rabi, Phys. Rev. **43**, 838 (1933). (§65)
R2. Rasetti, Zeits. f. Physik **78**, 165 (1932). (§99)
R3. Rasetti, Zeits. f. Physik **97**, 64 (1935). (§57, 90, 93, 94, 102)
R4. Rasetti, Segrè, Fink, Dunning and Pegram, Phys. Rev. **49**, 104 (1936). (§58, 61, 102)
R5. Rasetti, Mitchell, Fink and Pegram, Phys. Rev. **49**, 777 (1936). (§58, 60)
R5a. Rasetti, *Elements of Nuclear Physics* (Prentice Hall, 1936). (§69, 88)
R6. Richardson, Phys. Rev. **49**, 203 (1936). (§105)
R6a. Richardson and Emo, Phys. Rev. **51**, 1014 (1937). (§103)
R7. Richardson and Kurie, Phys. Rev. **50**, 999 (1936). (§101, 105)
R7a. Ridenour, Henderson, Henderson and White, Phys. Rev. **51**, 1013 (1937). (§99)
R7b. Ridenour and Henderson, Phys. Rev. **51**, 1102 (1937). (§99)
R7c. Ridenour and Henderson, Phys. Rev. **52**, 139 (1937). (§99)
R8. Riezler, Proc. Roy. Soc. **134**, 154 (1932). (§75)
R9. Riezler, Ann. d. Physik **23**, 198 (1935). (§75)
R9a. Risser, Phys. Rev. **51**, 1013 (1937). (§102, 105)
R10. Rose and Bethe, Phys. Rev. **51**, 205 (1937). (§83)
R10a. Rose and Bethe (Errata), Phys. Rev., in press. (§91)
R11. Rose, Phys. Rev. **51**, 1024 (1937). (§90)
R12. Rosenblum and Chamie, Comptes rendus **194**, 1154 (1932). (§69)
R13. Rosenblum, Ann. d. Physik **10**, 408 (1928). (§95)
R14. Rosenblum, Comptes rendus **202**, 1274 (1936). (§69)
R14a. Rosenblum, Guillot and Perey, Comptes rendus **204**, 175 (1937). (§69)
R15. Rotblat, Nature **136**, 515 (1935). (§102, 105)
R16. Rotblat, Nature **138**, 202 (1936). (§102)
R16a. Rotblat, Nature **139**, 1110 (1937). (§102).
R17. Ruark and Fussler, Phys. Rev. **48**, 151 (1935). (§68)
R17a. Ruben and Libby Phys. Rev. **51**, 776 (1937). (§60)
R18. Rüchardt, Ann. d. Physik **71**, 377 (1923). (§95)
R19. Rumbaugh and Hafstad, Phys. Rev. **50**, 681 (1936). (§101)
R19a. Rumbaugh, Roberts and Hafstad, Phys. Rev. **51**, 1106 (1937). (§101)
R20. Rusinow, Physik. Zeits. Sowjetunion **10**, 219 (1936). (§65, 85, 102)
R21. Rutherford, Chadwick and Ellis, *Radiations from Radioactive Substances* (1930). (§94, 99)
R21a. Rutherford, Phil. Mag. **47**, 277 (1924). (§95)
R22. Rutherford, Lewis, Bowden, Proc. Roy. Soc. **142**, 347 (1933). (§69, 95)
R23. Rutherford, Wynn-Williams, Lewis and Bowden, Proc. Roy. Soc. **139**, 617 (1933). (§69, 92, 95)
R24. Rutherford and Chadwick, Phil. Mag. **42**, 809 (1921). (§99)
R25. Rutherford and Chadwick, Phil. Mag. **4**, 605 (1927). (§74)
R26. Rutherford, Phil. Mag. **21**, 669 (1911). (§70)
S1. Sagane, Phys. Rev. **50**, 1141 (1936). (§101, 105)
S2. Sampson, Ridenour and Bleakney, Phys. Rev. **50**, 382 (1936). (§105)
S3. Sampson and Bleakney, Phys. Rev. **50**, 732 (1936). (§105)
S3a. Sampson and Bleakney, Phys. Rev. **50**, 456 (1936). (§107)
S4. Savel, Comptes rendus **198**, 368 (1934). (§99)
S5. Savel, Ann. d. Physik **4**, 88 (1935). (§99)
S6. Savel, Comptes rendus **198**, 1404 (1934). (§99)
S7. Schnetzler, Zeits. f. Physik **95**, 302 (1935). (§99)
S8. Schüler, Zeits. f. Physik **66**, 431 (1930). (§83)
S8a. Schultz, Phys. Rev. **51**, 1023 (1937). (§99)
S9. Segrè, Nuovo Cimento **12**, 232 (1935). (§94)
S10. Sexl, Zeits. f. Physik **81**, 163 (1933). (§66)
S11. Sexl, Zeits. f. Physik **56**, 62 and 72: **59**, 579 (1929). (§66)
S11a. Shepherd, Haxby and Williams, Phys. Rev. **52**, 247 (1937). (§100, 101)
S12. Sizoo and Koene, Physica **3**, 1053 (1936). (§105)
S13. Sloan and Coates, Phys. Rev. **46**, 539 (1934). (§92, 104)
S14. Sloan, Phys. Rev. **47**, 62 (1935). (§92)
S15. Snell, Phys. Rev. **49**, 555 (1936). (§101, 105)
S16. Snell, Phys. Rev. **51**, 142 (1937). (§101, 105)
S16a. Snell, Phys. Rev. **51**, 1011 (1937). (§99, 101, 105)

S17. Soden, Ann. d. Physik **19**, 409 (1934). (§51)
S18. Sosnowski, Comptes rendus **200**, 391 (1935). (§101)
S19. Sosnowski, Comptes rendus **200**, 1027 (1935). (§105)
S20. Sosnowski, Comptes rendus **200**, 922 (1935). (§102, 105)
S21. Stegmann, Zeits. f. Physik **95**, 72 (1935). (§99)
S22. Stetter, Zeits. f. Physik **100**, 652 (1936). (§99)
S23. Steudel, Zeits. f. Physik **77**, 139 (1932). (§99)
S24. Stone, Livingston, Sloan and Chaffee, Radiology **24**, 153 (1935). (§92)
S25. Sugden, Nature **135**, 469 (1935). (§102, 105)
S26. Szilard and Chalmers, Nature **134**, 494 (1934). (§92, 103)
S27. Szilard and Chalmers, Nature **134**, 462 (1934). (§94)
S28. Szilard and Chalmers, Nature **135**, 98 (1935). (§87, 105)
S29. Szilard, Nature **136**, 950 (1935). (§57, 102)
T1. Taylor and Goldhaber, Nature **135**, 341 (1935). (§94)
T2. Taylor, Proc. Phys. Soc. London **47**, 873 (1935). (§102)
T3. Taylor and Dabholkar, Proc. Phys. Soc. London **48**, 285 (1936). (§102)
T4. Taylor and Mott, Proc. Roy. Soc. **142**, 215 (1933). (§88)
T5. Taylor and Mott, Proc. Roy. Soc. **138**, 665 (1932). (§88)
T6. Taylor, Proc. Roy. Soc. **136**, 605 (1932). (§73, 74)
T7. Taylor, Proc. Roy. Soc. **134**, 103 (1931). (§74)
T8. Teller, Phys. Rev. to be published shortly? (§78)
T9. Thomson and Saxton, Phil. Mag. **23**, 241 (1937). (§104)
T10. Thornton, Phys. Rev. **49**, 207 (1936). (§101, 105)
T10a. Thornton, Phys. Rev. **51**, 893 (1937). (§101)
T11. Thornton and Cork, Phys. Rev. **51**, 383 (1937). (§101, 105)
T12. Traubenberg, Eckardt and Gebauer, Naturwiss. **21**, 26 (1933). (§92, 100)
T13. Traubenberg, Eckardt and Gebauer, Zeits. f. Physik **80**, 557 (1933). (§100)
T14. Tuve and Hafstad, Phys. Rev. **45**, 651 (1934). (§101)
T15. Tuve and Hafstad, Phys. Rev. **48**, 106 (1935). (§101)
T16. Tuve, Hafstad and Dahl, Phys. Rev. **48**, 315 (1935). (§92)
U1. Urey, Brickwedde and Murphy, Phys. Rev. **40**, 1 (1932). (§101)
U2. Urey and Teal, Rev. Mod. Phys. **7**, 34 (1935). (§101)
V1. Van Atta, Northrup, Van Atta and Van de Graaff, Phys. Rev. **49**, 761 (1936). (§92)
V2. Van Atta, Van de Graaff and Barton, Phys. Rev. **43**, 158 (1933). (§92)
V3. Van de Graaff, Compton and Van Atta, Phys. Rev. **43**, 149 (1933). (§92)
V4. Van de Graaff, Phys. Rev. **38**, 1919 (1931). (§92)
V5. Van Vleck, Phys. Rev. **48**, 367 (1935). (§57)
V6. Van Voorhis, Phys. Rev. **49**, 889 (1936). (§101, 105)
V7. Van Voorhis, Phys. Rev. **50**, 895 (1936). (§101, 105)
V8. Veldkamp and Knol, Physica **4**, 166 (1937). (§102, 105)
W1. Walke, Phys. Rev. **51**, 439 (1937). (§99, 101, 102, 105)
W1a. Walke, Phys. Rev. **51**, 1011 (1937). (§99, 101, 102, 105)
W2. Ward, Wynn-Williams and Cave, Proc. Roy. Soc. **125**, 715 (1929). (§94)
W3. Waring and Chang, Proc. Roy. Soc. **157**, 652 (1936). (§99)
W4. Webster, Proc. Roy. Soc. **136**, 428 (1932). (§99)
W5. Weekes, Livingston and Bethe, Phys. Rev. **49**, 471 (1936). (§58, 102)
W6. Weisskopf and Wigner, Zeits. f. Physik **63**, 54 (1930). (§52)
W7. Weisskopf, Phys. Rev., in press. (§54, 56, 65, 79)
W8. Weizsäcker, Zeits. f. Physik **96**, 431 (1935). (§110)
W9. Weizsäcker, Naturwiss. **24**, 813 (1936). (§87)
W10. Wentzel, Zeits. f. Physik **38**, 518 (1926). (§66)
W11. Wenzel, Zeits. f. physik. Chemie **90**, 754 (1934). (§75)
W12. Wertenstein, Nature **133**, 564 (1934). (§99, 105)
W13. Wheeler, Phys. Rev., to be published shortly, (§73)
W13a. White, Henderson, Henderson and Ridenour, Phys. Rev. **51**, 1012 (1937). (§99)
W14. Whitmer and Pool, Phys. Rev. **47**, 795 (1935). (§104)
W15. Wick, Phys. Rev. **49**, 192 (1936). (§59)
W16. Wigner and Breit, Phys. Rev. **50**, 1191 (1936). (§101)
W17. Wigner, Phys. Rev. **51**, 106 (1937). (§87, 110)
W18. Wigner, Phys. Rev. **51**, 947 (1937). (§105, 110)
W19. Wilson, Phil. Trans. Roy. Soc. **193**, 289 (1913). (§94)
W20. Williams, E. J., Proc. Roy. Soc. **135**, 108 (1932). (§51, 97)
W21. Williams, J. H., and Wells, Phys. Rev. **50**, 187 (1936). (§100)
W21a. Williams, J. H., Wells, Tate and Hill, Phys. Rev. **51**, 434 (1937). (§100)
W21b. Williams, J. H., Shepherd and Haxby, Phys. Rev. **51**, 1011 (1937). (§101)
W21c. Williams, J. H., Shepherd and Haxby, Phys. Rev. **51**, 888 (1937). (§101)
W22. Wright, Proc. Roy. Soc. **137**, 677 (1932). (§72)
W23. Wynn-Williams, Proc. Roy. Soc. **132**, 295 (1931). (§94)
Y1. Yost, D. M., Ridenour and Shinohara, J. Chem. Phys. **3**, 133 (1935). (§101)
Y2. Yost, F. L., Wheeler and Breit, Phys. Rev. **49**, 174 (1936). (§72)
Z1. Zeleny, Brasefield, Bock and Pollard, Phys. Rev. **46**, 318 (1934). (§92, 104)
Z1a. Zahn, Phys. Rev., in press. (§61)
Z2. Zinn and Seeley, Phys. Rev. **50**, 1101 (1936). (§92, 101)
Z3. Zyw, Nature **134**, 64 (1934). (§99, 105)